Highlights in Theoretical Chemistry

Vol. 10
Series Editors: Ch.J. Cramer • D.G. Truhlar

More information about this series at http://www.springer.com/series/11166

Angela K. Wilson • Kirk A. Peterson • David E. Woon
Volume Editors

Thom H. Dunning, Jr.

A Festschrift from Theoretical Chemistry Accounts

With contributions from

Rodney Bartlett • Charles W. Bauschlicher Jr. • Jeffery S. Boschen
Joel M. Bowman • Benoît Braïda • David H. Bross • Bun Chan
David L. Cooper • T. Daniel Crawford • Wibe A. de Jong • Dong Die
Thom H. Dunning Jr. • David Feller • Marek Freindorf • G. Frenking
Niranjan Govind • Ireneusz Grabowski • Robin Haunschild
Lawrence B. Harding • Philippe C. Hiberty • Leo F. Holroyd
Kimihiko Hirao • So Hirata • Ida-Marie Høyvik • Karl K. Irikura
P. Jerabek • Kenneth D. Jordan • Poul Jørgensen • Apostolos Kalemos
Peter B. Karadakov • Manoj K. Kesharwani • R. Bruce King
Thomas Kjærgaard • Stephen J. Klippenstein • Wim Klopper • Elfi Kraka
Kasper Kristensen • Jeff Leiding • William A. Lester Jr. • Beth A. Lindquist
Hans Lischka • Taylor J. Mach • Jan M. L. Martin • Aristides Mavridis
Evangelos Miliordos • Kirk A. Peterson • Ajith Perera • Rita Prosmiti
Chen Qu • Leo Radom • Alessandra Ricca • Varun Rishi • Klaus Ruedenberg
Branko Ruscic • Henry F. Schaefer III • Thomas Sexton • Sason Shaik
Ron Shepard • Jack Simons • Jong-Won Song • Matthew K. Sprague
Daniel Theis • Tanja van Mourik • Athanassios A. Vassilakis
Konstantinos D. Vogiatzis • Vamsee K. Voora • Hongyan Wang • Hui Wang
Zhiping Wang • Angela K. Wilson • Theresa L. Windus • David E. Woon
Wei Wu • Sotiris S. Xantheas • Yaoming Xie • Dmitry Yu Zubarev
Huaiyu Zhang • Wenli Zou

Springer

Volume Editors
Angela K. Wilson
Department of Chemistry
University of North Texas
Denton, TX, USA

Kirk A. Peterson
Department of Chemistry
Washington State University
Pullman, WA, USA

David E. Woon
Department of Chemistry
University of Illinois
Urbana, IL, USA

Originally published in Theor Chem Acc, Volume 132 (2013), Volume 133 (2014), Volume 134 (2015)

ISSN 2194-8666 ISSN 2194-8674 (electronic)
Highlights in Theoretical Chemistry
ISBN 978-3-662-47050-3 ISBN 978-3-662-47051-0 (eBook)
DOI 10.1007/978-3-662-47051-0

Library of Congress Control Number: 2015939351

Springer Heidelberg New York Dordrecht London

Printed on acid-free paper

Springer-Verlag GmbH Berlin Heidelberg is part of Springer Science+Business Media (www.springer.com)

Contents

Theor Chem Acc (2015) 134:1
DOI 10.1007/s00214-014-1584-4

INTRODUCTION

Thom H. Dunning, Jr.: Contributions to chemical theory and computing

Angela K. Wilson · Kirk A. Peterson · David E. Woon

Published online: 27 December 2014

Thom H. Dunning, Jr., provides an outstanding example of success in science, computing, and scientific leadership. From his scientific contributions to knowledge to his developments and accomplishments that have enabled discovery, Thom's influence upon science, computing, and the scientific community has been significant.

Published as part of the special collection of articles celebrating the career of Professor Thom Dunning upon his retirement.

A. K. Wilson (✉)
Department of Chemistry, University of North Texas, 1155 Union Circle, #305070, Denton, TX 76203-5017, USA
e-mail: akwilson@unt.edu

K. A. Peterson
Department of Chemistry, Washington State University, Pullman, WA 99164-4630, USA
e-mail: kipeters@wsu.edu

D. E. Woon
Department of Chemistry, University of Illinois Urbana-Champaign, Urbana, IL 61801, USA
e-mail: davidewoon@gmail.com

As a chemistry undergraduate at the Missouri University of Science and Technology, which was then called the University of Missouri at Rolla, Thom was fascinated by the ability of organic chemists to rationalize the course of a broad range of chemical reactions by shuttling electrons around in the molecules. However, he was also somewhat skeptical—this seemed too good to be true. The following year he took a course in quantum mechanics in the Physics Department and realized that quantum mechanics was the means for fact-checking the explanations in organic chemistry. Thus began his passion for chemical theory. In the quantum mechanics course, he encountered the differential equation for the states and wavefunctions of the harmonic oscillator. Since he had just had a course in numerical analysis and the university was touting its new, "very powerful" computer that it had just installed (an IBM 1620!), Thom decided to solve the harmonic oscillator equation on the computer. After teaching himself the new programming language developed by IBM—FORTRAN—and coding the problem, he was amazed to find that his numerical solution agreed perfectly with that obtained analytically. Thus began his passion for computing.

1 Contributions to chemical theory

Thom's research career has focused on topics in chemical theory (e.g., open-shell Hartree–Fock theory, generalized valence bond theory), numerical techniques for solving the Schrödinger equation (e.g., basis sets for both orbital and correlated calculations), and applications of computation to important chemistry problems (molecular physics of lasers, species and reactions involved in combustion, structure and energetics of aqueous clusters, and, more recently, the unusual chemistry of the late p-block elements in the second row and beyond). Below is a brief summary of Thom's contributions to chemical theory and computation.

1.1 Generalized valence bond theory

Thom began his graduate career at Caltech in 1965, just as Dr. William A. Goddard III accepted an appointment as an Arthur Amos Noyes Fellow in the Chemistry Division. Although Bill could not accept students at that point, Thom was impressed by the insights provided by the valence bond-based theory that he was developing (later called generalized valence bond, GVB, theory). The GVB wavefunction corrects the main deficiencies of the Hartree–Fock/molecular orbital wavefunction while providing a conceptually appealing, chemistry-oriented description of the electronic structure of molecules. Although Thom's thesis research focused on the excited states of the ethylene molecule with Professor B. Vincent McKoy, as a postdoctoral fellow, he contributed to the development of the perfect pairing, strong orthogonality version of GVB theory. Imposition of these constraints made it possible to extend GVB calculations to a broad range of molecules, often with only modest sacrifices in accuracy. A paper summarizing early work in this area was published in *Accounts of Chemical Research* (Goddard et al., *Acc. Chem. Res.* **1973**, *6*, 368–376).

More recently, Dr. Dunning, Dr. David E. Woon, and the rest of his research group at the University of Illinois at Urbana-Champaign used the GVB method, as well as more sophisticated coupled cluster and multireference configuration interaction methods, to study the electronic structure of hypervalent molecules, e.g., PF_5, SF_4/SF_6, and ClF_3/ClF_5. It was found that a new type of bond, the *recoupled pair* bond, provides the basis for the formation of hypervalent molecules. In fact, the ability of the second row elements to form this new type of bond explains many of the other anomalies associated with the chemistry of the late p-block elements in the second row and beyond: new bound low-lying excited states, edge versus vertex inversion in heavily halogenated species, and unusually rapid chemical reactions to name just a few of these anomalies. A paper summarizing early work in this area was published in *Accounts of Chemical Research* (Dunning et al., *Acc. Chem. Res.* **2013**, *46*, 359–368). These full GVB calculations were made possible by the CASVB program, which was developed by T. Thorsteinsson and D. Cooper and is included in both the MOLPRO (http://www.molpro.net/) and MOLCAS (http://www.molcas.org/) electronic structure packages.

1.2 Basis sets for molecular computations

Dr. Dunning is perhaps best known for the development of basis sets for use in molecular calculations. In graduate school, Thom found the existing Gaussian basis sets for Hartree–Fock calculations on molecules of interest to him—ethylene, formaldehyde, and other organic molecules—to be lacking. Thom systematically explored the construction of new basis sets, laying out a set of rules that enabled the construction of high-quality basis sets for Hartree–Fock calculations. In the late 1980s, at Argonne National Laboratory, he became frustrated at the inability to converge configuration interaction calculations with regard to the basis set. Then Jan Almlöf and Peter Taylor reported that the occupations of the natural orbitals from atomic calculations grouped naturally into sets with similar occupation numbers. Thom had no explanation for this finding and set out on a systematic exploration of correlation in atoms. The result of this study was the development of the correlation-consistent basis sets for the first row atoms (H, He, B–Ne), a family of basis sets that provide systematic, increasingly accurate solutions of the electronic Schrödinger equation for atoms and molecules. After his move to Pacific Northwest National Laboratory, this work was extended to other elements of the Periodic Table and to the use of these basis sets to quantify the intrinsic error in electronic structure methods (see following paragraph). These basis sets are used in almost all computational chemistry codes. Six scientific publications in this series are "Citation Classics," i.e., they have received more than 1000 citations (*J. Chem. Phys.* **1970**, *53*, 2823–2833; *ibid.* **1971**, *55*, 716–723; *ibid.* **1989**, *90*, 1007–1023; *ibid.* **1992**, *96*, 6796–6806; *ibid.* **1993**, *98*, 1358–1371; *ibid.* **1995**, *103*, 4572–4585). This work was summarized in a chapter in the *Encyclopedia of Computational Chemistry* (Dunning et al., *Encyclopedia of Computational Chemistry*, Ed. P. v. R. Schleyer, John Wiley & Sons Ltd, 1997).

1.3 Assessment of electronic structure methods

The development of the correlation-consistent basis sets, which systematically approach the complete basis set limit, made it possible to separate the basis set error from the error due to the electronic structure method. This made it possible to determine the *intrinsic* error associated with a given electronic structure method (configuration interaction, perturbation theory, coupled cluster, etc.). Dr. Dunning and his co-workers used this approach to rigorously assess the accuracy of the quantum chemical methodologies used to describe the electronic structure of molecules. This work yielded a number of major surprises, *e.g.*, the poor/non-convergence of the perturbation expansion for a broad range of molecules, and established the intrinsic accuracy of the coupled cluster method. A "Feature Article" summarizing some of this work was published in the *Journal of Physical Chemistry A* (T. H. Dunning, Jr., *J. Phys. Chem. A* **2000**, *104*, 9062–9080).

1.4 Characterization of molecular lasers

At Los Alamos National Laboratory, Dr. Dunning and Dr. P. Jeffrey Hay were involved in characterizing the electronic states responsible for laser action in a number of exotic molecular systems (rare has halides, rare gas oxides, etc.), which were being developed for laser fusion and laser isotope separation applications. These molecules have unstable ground states and, thus, are very difficult to characterize experimentally. The computational studies provided information on these molecules (e.g., excited state lifetimes) that were otherwise unavailable and guided subsequent experimental studies of these species. As a result of the computational studies, a new laser transition in the rare gas halides was predicted and later observed. A review article describing this work was published in the *Annual Reviews of Physical Chemistry* (P. J. Hay, T. H. Dunning, Jr. and W. R. Wadt, *Annu. Rev. Phys. Chem.* **1979**, *30*, 311–346).

1.5 Chemical species and reactions in combustion

When Dr. Dunning moved from Los Alamos National Laboratory to Argonne National Laboratory, he led a research group focused on characterizing the molecular species and reactions involved in the combustion of hydrogen and hydrocarbon fuels. These studies provide new insights into the chemical reactions involved in flames, answering some questions that had puzzled chemical kineticists and combustion modelers for decades. This work also provided information on the complex pathways involved in the reactions of hydrocarbon species, including a wealth of heretofore unavailable information on reaction products and branching ratios. Using the information on quantum chemical methods/basis sets referred to above developed the protocol to compute the heats of many chemical reactions to an accuracy that is comparable to (and often better than) that obtained from experiment.

2 Contribution to computing

Thom's leadership abilities were recognized early on when he was appointed Associate Group Leader for Molecular Theory in the Laser Theory Group at Los Alamos National Laboratory. With the untimely passing of Chris Wahl, Thom was recruited to Argonne National Laboratory, where he was appointed the Group Leader of the Theoretical and Computational Chemistry Group. During Thom's tenure at Argonne, the group became an international focal point for theoretical and computational studies of the species and reactions involved in combustion. The success of this effort led to Thom's recruitment to the Pacific Northwest National Laboratory, which was planning the development of a new molecular science research center—an activity that eventually matured into the Environmental Molecular Sciences Laboratory (EMSL), a $230 million project funded by the Office of Biological and Environmental Research (BER) in the US Department of Energy (DOE). Although the EMSL project was funded by BER, it also had strong support and guidance from Dr. Robert S. Marianelli in the Chemical Sciences Division in the Office of Basic Energy Sciences.

The EMSL Project began just as computing technologies were transitioning from the fast vector supercomputers built by Cray Research to "massively" parallel computers based on microprocessors. The trouble was that to take full advantage of these new computers, the codes used to model molecular systems had to be rewritten to distribute the compute load over the processors. This was a major undertaking. Fortunately, the EMSL project provided the funding to do this, and Thom recruited a top team of computational chemists, including Robert Harrison, Jeff Nichols, and Ricky Kendall as well as computer scientists and applied mathematicians to tackle this problem. This effort led to the development of NWChem, an application that has maintained its position at the leading edge of parallel chemistry software for more than 15 years and has scaled from the 100's of processors available in the mid-1990s to the 100,000's of cores available today.

The success of the NWChem project led the US DOE to ask Thom to help them plan a new DOE-wide initiative in scientific simulation. Although they stated this would only be a temporary assignment, Thom was, in fact, the Assistant Director for Scientific Simulation in DOE's Office of Science for more than 2 years. The culmination of this effort was DOE's "Scientific Discovery through Advanced Computing" (SciDAC) program, which brought together disciplinary computational scientists, computer scientists, and applied mathematicians to enable a broad range of scientific computational disciplines—from biology and chemistry through materials science to fusion energy, nuclear physics and particle physics—to take full advantage of parallel computers, which at the end of the 1990s were capable of executing one trillion arithmetic operations per second (a teraflop) to a quadrillion operations per second (a petaflop) by the end of the 2000s.

After Thom left his assignment at the Office of Science, he briefly held the position of Vice President for High Performance Computing & Communications at MCNC and was responsible for high performance computing and networking for the University of North Carolina System as well as Duke University and Wake Forest University. He was then appointed the founding director of the Joint Institute for Computational Sciences, Distinguished Professor of Chemistry and Chemical Engineering at the University of Tennessee, and Distinguished Scientist at Oak Ridge National Laboratory. In 2004, he accepted the position of director of the National Center for Supercomputing Applications and Distinguished Chair for Research Excellence in the Department of Chemistry at the University of Illinois at Urbana-Champaign. In October 2006, Dr. Dunning was also appointed the founding director of a new Institute for Advanced Computing Applications and Technology at UIUC. This Institute has a charter similar to that of UIUC's Beckman Institute, namely, to foster multi- and interdisciplinary research, in this case, at the interface between disciplinary computational science, computer science, and applied mathematics.

As NCSA director, Dr. Dunning led the effort to obtain funding for and then the deployment of the largest supercomputer system ever to be sited at a university—the *Blue Waters Sustained Petascale Computer*. This $208 million project for the National Science Foundation was visionary in that it clearly defined a computing system that would enable the computational science and engineering community to make a major advance in their research. The project not only developed the specifications for the computer system, it also established a technical user assistance program to ensure that the research teams granted access to Blue Waters could take full advantage of its capabilities. Also during Thom's tenure, NCSA was selected to lead NSF's national cyberinfrastructure project—XSEDE, the Extreme Science & Engineering Discovery Environment. NCSA is now widely recognized as one of the world's preeminent supercomputing centers and the only one of its kind on a university campus anywhere in the world.

At the end of 2013, Thom retired from the University of Illinois at Urbana-Champaign. But, retirement does not mean that he has lost interest in chemistry and computing. As soon as he had retired from UIUC, Thom accepted a part-time position with the Pacific Northwest National Laboratory to help PNNL and the University of Washington establish the new Northwest Institute for Advanced Computing. He was also appointed a Distinguished Faculty Fellow in UW's Department of Chemistry. We look forward to Thom's continuing contributions to the field.

Thom's recognitions include the American Chemical Society Award for Computers in Chemical and Pharmaceutical Research (2011), Distinguished Associate Award (2001) and E. O. Lawrence Award in Chemistry (1996) from the U.S. DOE, Award for Excellence in Technology Transfer (2000) from the Federal Laboratory Consortium for Technology Transfer, Fellow of the American Chemical Society (2011), Fellow of the American Physical Society (1992), and Fellow of the American Association for the Advancement of Science (1992).

Angela K. Wilson, Kirk A. Peterson, and David E. Woon

Theor Chem Acc (2014) 133:1443
DOI 10.1007/s00214-013-1443-8

REGULAR ARTICLE

The nature of the SO bond of chlorinated sulfur–oxygen compounds

Beth A. Lindquist · Thom H. Dunning Jr.

Received: 17 October 2013 / Accepted: 16 December 2013 / Published online: 8 January 2014

Abstract Sulfur–oxygen chemistry encompasses a rich variety of chemical species and reactions. Sulfur–oxygen bonds can be quite short and strong, and historically, there has been disagreement as to the nature of the SO bond in sulfinyl groups. Early work invoked participation by the 3*d* orbitals of sulfur to explain the apparent double-bond character of sulfinyl bonds, but modern calculations have clearly established that sulfur 3*d* atomic orbitals do not participate as valence orbitals in hypervalent sulfur compounds. In prior work, we used generalized valence bond (GVB) theory to explain the features of the SO bond in the HSO/SOH structural isomers, and we extend that work here to the chlorinated analogs (ClSO/SOCl). We also use GVB theory to elucidate the nature of the bonding in Cl_2SO and its higher energy structural isomer ClSOCl. We find that recoupled pair bonding, which we first introduced in our study of sulfur fluorides, is integral to describing the SO bond in all of these species. We also connect our analysis to the use of hyperconjugation to explain the back-bonding in the π system in the sulfinyl halides.

Keywords Recoupled pair bonding · Thionyl chloride · Hyperconjugation · Sulfinyl group · Generalized valence bond (GVB) theory

Published as part of the special collection of articles celebrating the career of Professor Thom Dunning upon his retirement.

B. A. Lindquist · T. H. Dunning Jr. (✉)
Department of Chemistry, University of Illinois at Urbana–Champaign, 600 S. Mathews Avenue, Urbana, IL 61801, USA
e-mail: thdjr@illinois.edu

B. A. Lindquist
e-mail: blindqu2@illinois.edu

1 Introduction

Sulfur and oxygen atoms interact to form a rich variety of molecules with varying bond lengths and strengths that contribute to the diversity of sulfur–oxygen chemistry. Of interest here are molecules containing a single sulfinyl (SO) group with two additional ligands, which have the general formula XYSO. The SO bonds in these species are typically shorter and stronger than a standard single SO covalent bond [1], and their bond dissociation energies can vary significantly depending on the identity of the substituents (X, Y) [2]. Both the electronegativity and aromaticity of X and Y affect the SO bond dissociation energy. In general, electronegative substituents tend to correlate with especially strong SO bonds [3]. In fact, the SO bond in such compounds is often drawn as a double bond in recognition of its shortness and strength, but the presence of multiple bond character is controversial, and the origin of this multiple bond character has been debated [4]. For example, some texts depict XYSO species with a dative, or hypercoordinate covalent SO bond between the $S3p^2$ orbital and the empty $O2p^2$ arising from the $O(^1D)$/$O(^1S)$ states with back-bonding being invoked to explain the shortness and strength of the bond.

As is typical of sulfur-containing species, the atomic 3*d* orbitals of the sulfur atom were employed as a possible explanation for the variation in SO bond lengths and strengths. It was suggested that the doubly occupied O2*p* lone pair orbitals back-bond to the S3*d* orbitals of the central atom [5, 6]. However, detailed calculations of various hypervalent molecules show that S3*d* functions only provide polarization and correlation effects [7–9]; these functions do not participate as 3d valence orbitals in the bonding in the sulfinyls. In light of this finding, other bonding schemes have been proposed. It has been suggested that back-bonding in the SO bonds (and related bonds involving late *p*-block elements and oxygen) is a consequence of hyperconjugation [10]. So-called anionic hyperconjugation occurs when a bond pair and lone pair interact, and the lone pair orbital energy is stabilized by interaction with the empty anti-bonding orbital of the bond [11]. Other studies have described these types of bonds as composed of a polar covalent σ bond and a nearly ionic π bond, especially when the (X, Y) substituent is a very electronegative element (e.g., a halide), without necessarily addressing the origin of the multiple bond character [12–14]. Yet other studies concluded that there in fact is no multiple bond character, and the SO bond is strengthened by purely electrostatic interactions [4].

In this work, we focus on thionyl chloride (Cl_2SO) and its structural analog, ClSOCl, as well as their parent triatomic molecules: ClSO and SOCl. Cl_2SO is interesting from a theoretical perspective but is also an important reagent in a wide variety of chlorination reactions [15] and in electrochemistry as a component in lithium/sulfinyl chloride batteries [16]. We use generalized valence bond (GVB) theory to provide insights into the nature of the bonding in these molecules. A prior study of sulfur–oxygen compounds in our group showed that recoupled pair bonding involving the electrons in the π orbitals accounts for the strength of the SO bond in the ground $X^3\Sigma^-$ state of diatomic SO, as well as for the large differences in geometry and SO bond strength of the $\tilde{X}^2A''$ states of HSO and SOH [17]. We found that a recoupled pair π bond is formed by the interaction of the electrons in the S3$p\pi^2$ lone pair and the O2$p\pi^1$ orbital in SO($X^3\Sigma^-$). In the S(^{3}P) atom, the two electrons in the S3p_y^2 (or S3p_{y-}–3p_{y+}) lone pair are singlet-coupled, but in SO($X^3\Sigma^-$), one of the electrons in the S3$p\pi$-like orbital (3p_{y+}) is singlet-coupled to the O2$p\pi^1$-like orbital to form a recoupled pair π bond. The remaining S3$p\pi$-like orbital (3p_{y-}) is mostly centered on the sulfur atom. As a result, the two singly occupied orbitals of SO($X^3\Sigma^-$) are largely localized on the S atom and the recoupled pair π bond is maintained in HSO but must be broken to form SOH. This is shown in GVB diagrams of the $\tilde{X}^2A''$ states of HSO and SOH in Fig. 1. Consequently, the SO bond in HSO is 0.15 Å shorter and 21.5 kcal/mol stronger than that in SOH.

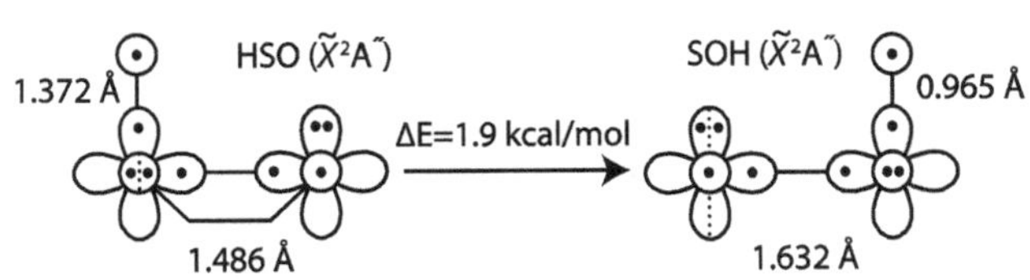

Fig. 1 GVB diagrams for the ground ($\tilde{X}^2A''$) states of HSO and SOH. Valence *p* orbitals for the O and S atoms and the 1*s* orbital of hydrogen atom are shown. We indicate the 3p^2 (3p_-, 3p_+) lone pair on sulfur with a *dashed line*; this pair can be recoupled to form an SO π bond

Despite this difference, the ground states of HSO and SOH are very close in energy ($\Delta E = 1.9$ kcal/mol), which is due to a near cancellation of two effects: (1) OH bonds are stronger than SH bonds (favoring SOH), and (2) the recoupled pair bond is maintained in HSO but not SOH (favoring HSO). We expect that similar effects are present in ClSO/SOCl and Cl_2SO/ClSOCl and may provide an explanation for the preference of ClSO over SOCl and of Cl_2SO over ClSOCl as well as the shortness and strength of the SO bond in the former species.

The remainder of this work is organized as follows. In Sect. 2, we describe the computational methodology, including a brief overview of GVB theory. In Sect. 3, we compare the structure, energetics, and GVB orbitals of the ClSO and SOCl structural isomers. We compare the corresponding dichlorinated species (Cl_2SO and ClSOCl) in Sect. 4. Finally, in Sect. 5, we conclude.

2 Computational methods

Prior studies in our group have utilized GVB theory to gain insights into the nature of the bonding in a variety of molecules of the second-row elements. Unlike Hartree–Fock (or molecular orbital, MO) theory, GVB theory has the advantage of being inherently multi-reference and therefore able to describe bond dissociation and formation. A subset of the GVB orbitals (n_a) are singly occupied, instead of doubly occupied, and the GVB wave function can be written as follows [18–21].

$$\Psi_{GVB} = \hat{A}\phi_{d1}\phi_{d1}\cdots\phi_{dn_d}\phi_{dn_d}\varphi_{a1}\varphi_{a2}\cdots\varphi_{an_a}\alpha\beta\cdots\alpha\beta\Theta^{n_a}_{SM} \quad (1)$$

In Eq. (1), the $\{\phi_{di}\}$ are the doubly occupied core and valence orbitals and the $\{\varphi_{ai}\}$ are the active GVB orbitals. The $\{\phi_{di}\}$ orbitals can be considered as orbital pairs that are singlet-coupled and overlap perfectly with one another. The active orbitals, by contrast, are singly occupied with overlaps less than unity and can be spin-coupled in various ways as described by the spin-coupling coefficients, the sum of which is $\Theta^{n_a}_{SM} = \sum_k c_{sk}\Theta^{n_a}_{SM;k}$. Because the active spatial orbitals, $\{\varphi_{ai}\}$, as well as the spin-coupling coefficients, $\{c_{Sk}\}$, are optimized at each nuclear configuration, the GVB wave function provides an accurate, yet compact description of the changes in the electronic structure of the molecule with changes in the nuclear configuration. This is in contrast to traditional VB calculations that typically require several covalent and ionic structures to be included for the calculation to be sufficiently accurate. Characterizing the evolution of the spatial orbitals and the associated spin-coupling patterns as a function of internuclear distance provides a clear bridge between the electronic structure of the molecule and that of its constituent fragments.

The spin-coupling coefficients, $\{c_{Sk}\}$, are related to the relative weight, w_k, of a particular spin function ($\Theta^{n_a}_{SM;k}$) in the wave function. There are various choices for the spin basis functions used in the GVB calculation. The Kotani spin basis is a popular choice because the basis is orthonormal, and the squares of the resulting coefficients $\{c_{Sk}\}$ yield a direct measure of the contribution of the spin-coupling patterns to the total GVB wave function, $w_k = c^2_{Sk}$ [22]. Any weights reported in this work will be computed in the Kotani spin basis. A related spin basis is the Rumer spin basis [23, 24]. The Rumer spin functions have the advantage of being interpretable in terms of singlet pairs. The Kotani spin functions (in reverse order) can be obtained by Gram–Schmidt orthogonalization of the Rumer spin functions [25]. The disadvantage of using Rumer spin functions is that, because they are not orthogonal, the contribution of each spin function to the total wave function cannot be uniquely defined (though various definitions are available [26–28]). Nonetheless, the magnitudes of the coefficients provide valuable insights into the relative importance of the various Rumer spin-coupling patterns. For species in which the bonding pattern changes as a function of geometry, the Rumer spin basis is often viewed as a natural choice to describe this transition.

Generalized valence bond (GVB) calculations are inherently more accurate than Hartree–Fock calculations because they contain the major nondynamical correlation effects in a valence CASSCF wave function, e.g., those associated with the *s*–*p* near-degeneracy in the atoms and those associated with the incorrect dissociation of the Hartree–Fock wave function in the molecule. However, GVB calculations do not include dynamic correlation and are thus not as accurate as MO-based multi-reference configuration interaction (MRCI) or coupled cluster (CC) methods. Therefore, in this work, we will use a hybrid approach: (1) we use very accurate MO-based methods, such as the MRCI and CC methods, to optimize the geometries and compute the energetics of the species of interest here, and (2) we combine these results with calculations of the GVB wave function to examine the electronic structure of the molecule.

We will make use of GVB orbital diagrams to schematically represent the GVB wave function as shown in Fig. 1 in Sect. 1. In this work, the valence *p* orbitals of the atoms will be represented as follows: two lobes in the plane of the paper for each of the p_x and p_z orbitals, and a small circle to represent the p_y orbital, which is the *p* orbital that has a node in the plane of the paper. The SO bonding axis is defined to be the *z*-axis. The dots in the lobes represent the electronic occupation of the orbitals, and singlet coupling between two orbitals centered on different atoms (a chemical bond) is shown as a line connecting the orbitals.

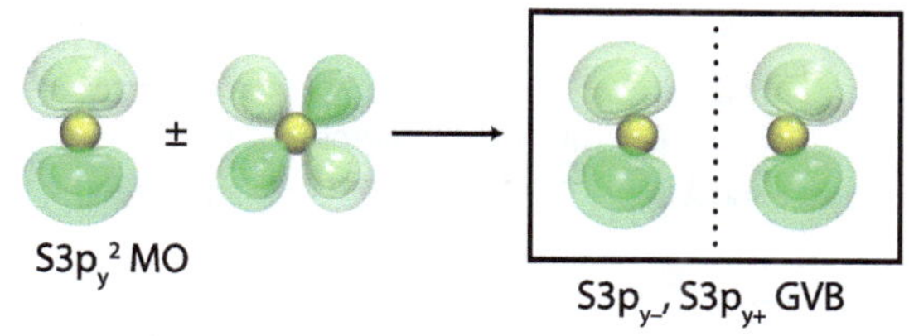

Fig. 2 MO and GVB representations of the S3p^2 pair

We indicate the possibility of recoupling the S3p^2 pair with a dotted line drawn through this doubly occupied orbital. This S3p^2 pair is shown in Fig. 2 in both the MO and GVB representations. In MO theory, the 3*p* orbital is doubly occupied; in GVB theory, this lone pair is represented by two lobe orbitals. The 3*p* lobe orbitals result from the inclusion of an S3*d* orbital in the GVB wave function, which provides the GVB orbitals with angular correlation such that they have some spatial separation, though their overlap is still high [29].

All calculations were performed with the Molpro suite of quantum chemical programs [30]. The GVB calculations were performed using the CASVB program of Thorsteinsson et al. [31]. We generally computed the full GVB wave function (also called the spin-coupled VB (SCVB) wave function by Gerratt et al. [18, 19]). To compute the GVB wave function, our general strategy was to localize the Hartree–Fock orbitals in terms of atomic orbitals [32] and then perform a CASSCF calculation with a small active space if needed to further refine these orbitals, which were then used to generate a starting guess for the CASVB program. The main constraint on the wave function was orthogonality between orbitals of different symmetry.

All geometries were optimized, and bond energies were calculated with explicitly correlated coupled cluster theory including a perturbative triples correction [CCSD(T)-F12 and RCCSD(T)-F12 with the "*a*" approximation] [33–36]. For geometry optimizations, the augmented correlation consistent double-zeta basis set with tight *d*-functions on sulfur and chlorine [AV($D + d$)Z] was used [37–39]. Because the explicitly correlated coupled cluster methodology contains terms that depend explicitly on interelectronic distances, convergence with respect to basis size is accelerated relative to traditional approaches. A calculation including explicitly correlated terms is typically at least as accurate as a calculation using a basis set that is one zeta higher without the explicitly correlated terms [40]. So we expect that the CCSD(T)-F12/AV($D + d$)Z calculations performed here are comparable in accuracy to standard CCSD(*T*)/AV($T + d$)Z calculations. For calculating molecular energies, we used the AV($T + d$)Z basis set, which, using the explicitly correlated methodology, should have an accuracy between quadruple or quintuple zeta for

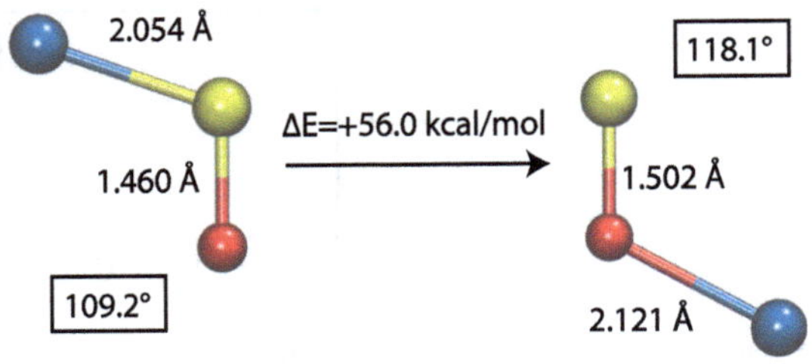

Fig. 3 Optimized geometries and relative energetics of the ClSO and SOCl isomer. Throughout this work, the chlorine atom is *blue*, the sulfur atom is *yellow*, and the oxygen atom is *red*

Table 1 Comparison of bond lengths (in Å) and dissociation energies (in kcal/mol) for the SCl and OCl bonds in ClSO/SOCl versus the diatomic species

	R_e	D_e
Cl–SO($\tilde{X}^2$A″)	2.054	56.1
S–Cl($X^2\Pi$)	1.985	67.6
SO–Cl($\tilde{X}^2$A″)	2.121	0.1
O–Cl($X^2\Pi$)	1.569	63.8

calculations performed without the explicitly correlated terms. For potential energy scans over a range where a single-reference method was not appropriate, a CASSCF calculation [41, 42] was performed or MRCI with the Davidson correction was employed (MRCI + Q) [43, 44]. For these calculations, the AV($T + d$)Z basis set was used and a full valence active space was used unless otherwise stated.

3 The ClSO/SOCl isomers

3.1 The structures and energetics of ClSO/SOCl

The geometries and relative energetics of the ground ($\tilde{X}^2$A″) states of the ClSO and SOCl isomers are shown in Fig. 3. The structure for ClSO($\tilde{X}^2$A″) is in good agreement with a prior study [45]. The ClSO($\tilde{X}^2$A″) isomer is substantially more stable than the SOCl($\tilde{X}^2$A″) isomer, $\Delta E = 56.0$ kcal/mol. This is in contrast to the nearly energetically degenerate HSO/SOH structural isomers shown in Fig. 1. We cannot attribute the increased stability of the ClSO isomer to any inherent difference in the strengths of the SCl and OCl bonds; the dissociation energies of these bonds in the diatomic species (Table 1) are very similar, with the SCl($X^2\Pi$) state being only 3.8 kcal/mol more strongly bound than the OCl($X^2\Pi$) state. We also compare the bond lengths and energies of the diatomic molecules to those of the ClSO/SOCl molecules in Table 1. Clearly, additional factors are at play in the ClSO/SOCl isomers compared to the HSO/SOH isomers.

The SCl bond energy in the ClSO($\tilde{X}^2$A″) isomer, 56.1 kcal/mol, is slightly weakened relative to that in the SCl($X^2\Pi$) state, 67.6 kcal/mol, but similar in magnitude. Similarly, the SCl bond is only slightly longer in the ClSO($\tilde{X}^2$A″) isomer, 2.054 Å, than that in the SCl($X^2\Pi$) state, 1.985 Å. Therefore, we consider this bond to be a typical SCl covalent bond, and attribute its lengthening and weakening relative to SCl($X^2\Pi$) to the additional electronic repulsion among the SCl bond pair, the SO bond pair, and the O2p^2 pair in the ClSO($\tilde{X}^2$A″) isomer. By contrast, in the SOCl($\tilde{X}^2$A″) isomer, the OCl bond is weak almost to the point of nonexistence, D_e(SO–Cl) $\approx$ 0.1 kcal/mol. The OCl bond length is correspondingly much longer in the SOCl($\tilde{X}^2$A″) isomer than that in OCl($X^2\Pi$): R_e(SO–Cl) = 2.121 Å versus R_e(O–Cl) = 1.569 Å. Clearly, the interaction between the O and Cl atoms cannot be described as a covalent bond. We will discuss the nature of bonding in this species in Sect. 3.3.

The SO bonds in the triatomic molecules are quite similar in terms of length, with the SO bond length only increased by 0.04 Å in the SOCl isomer relative to the ClSO isomer. This is in distinct contrast to the hydrogen-substituted case where the SO bond in HSO was 0.15 Å shorter than in SOH. However, both the HS bond length in the HSO($\tilde{X}^2$A″) isomer and the OH bond length in the SOH($\tilde{X}^2$A″) isomer were very close to that of the $X^2\Pi$ states of the corresponding diatomic molecules, XH (X = O, S). Thus, the shortness of the SO bond and the extraordinary length of the OCl bond in the SOCl($\tilde{X}^2$A″) isomer are peculiar to chlorine substitution.

3.2 The GVB orbitals of ClSO

In our prior study of HSO/SOH, we found that the recoupled pair π bond in SO was maintained in HSO and was broken upon formation of SOH. In contrast, for ClSO and SOCl, based on the bond lengths alone, it seems that the recoupled pair π bond in SO is maintained in both isomers. The GVB orbitals for the $\tilde{X}^2$A″ states of the ClSO and SOCl isomers support this conclusion. The five GVB valence orbitals of a'' symmetry in ClSO, as well as the associated GVB diagram, are shown in Fig. 4. In the dominant spin-coupling pattern (95.6 %), the electrons of the Cl3p^2 pair (φ_1, φ_2) are singlet-coupled, which is denoted by the dotted line between the orbitals. The other singlet-coupled pair (φ_3, φ_4) is a recoupled pair π bond between the O2p^1-like orbital and one of the S3p-like lobe orbitals, which has delocalized onto the O atom and represents the highly polarized ($S^{\delta+}O^{\delta-}$) nature of the bond. The remaining S3p-like lobe orbital has α spin, yielding an overall doublet state. This bonding scheme is depicted in the GVB orbital diagram in Fig. 4. The bonding scheme for

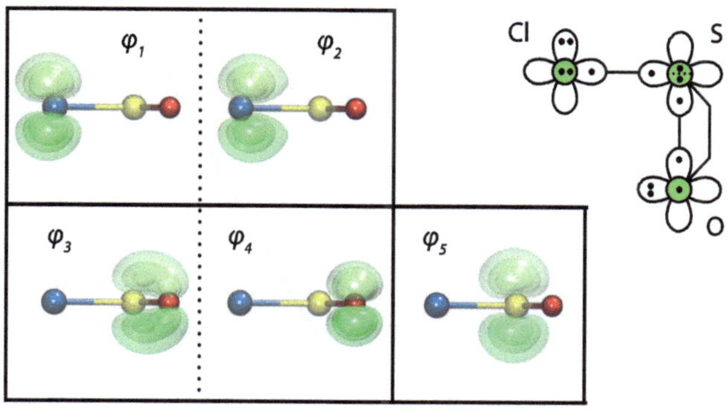

Fig. 4 GVB orbitals for those orbitals shaded green in the accompanying GVB diagram for ClSO($\tilde{X}^2A''$). A *dotted line* separating two orbitals indicates that they are singlet-coupled in the dominant spin-coupling pattern of the wave function

the S and the O atoms in the ClSO($\tilde{X}^2A''$) isomer is essentially the same as that found in HSO previously.

Note that the Cl3p^2 lone pair is an example of a GVB description of a 3p^2 lone pair that is not recoupled. The lone pair consists of two lobe orbitals, ($\varphi_1 \approx$ Cl3p_{y-}, $\varphi_2 \approx$ Cl3p_{y+}), with high overlap ($S = 0.88$) that are singlet-coupled to each other.

3.3 The GVB orbitals of SOCl

Unlike HSO and ClSO, the features of SOCl are qualitatively different than those of SOH, indicating a different bonding pattern in the two species. In SOH, the SO bond was significantly lengthened relative to that in the HSO isomer. By contrast, in SOCl, the SO bond is still fairly short, 1.502 Å in SOCl versus 1.460 Å in ClSO. However, the OCl bond is very long and weak. The GVB orbitals for the SOCl($\tilde{X}^2A''$) isomer are shown in Fig. 5a. In the dominant spin-coupling pattern (87.3 %), the O2p^1-like GVB orbital (φ_3) is singlet-coupled to one of the S3p lobes (φ_4) and not to the Cl3p GVB orbital, resulting in the same type of recoupled pair π bond that we observed in the ClSO($\tilde{X}^2A''$) isomer. The other singlet pair consists of the Cl3p-like GVB orbital (φ_2) and the remaining S3p lobe orbital (φ_5). There is a slight tail of these orbitals on the O atom, but φ_2 is largely centered on the chlorine atom and φ_5 is largely centered on the S atom. Therefore, these two orbitals do not have a large spatial overlap ($S = 0.24$) with one another and do not form a conventional chemical bond. In essence, the recoupled pair bond is maintained at the expense of forming a covalent OCl bond. The π bond delocalizes slightly onto the Cl atom, which could be the genesis of the slightly longer SO bond in SOCl relative to that in the ClSO isomer.

Ongoing work in our group has observed a similar bonding pattern in the $\tilde{A}^2A'$ state of NOF (Takeshita and Dunning, to be published). We will refer to this weakly

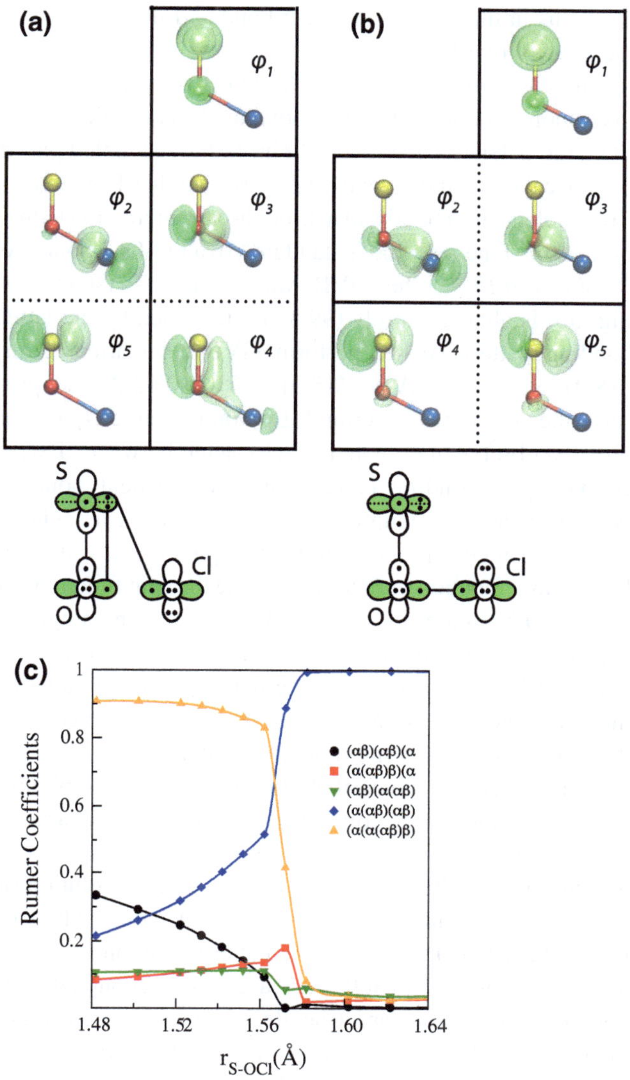

Fig. 5 **a** GVB orbitals for those orbitals shaded green in the accompanying GVB diagram for optimized geometry of SOCl($\tilde{X}^2A''$). **b** The same GVB orbitals for $R(\text{SO}) = R_e(\text{SO}) + 0.12$ Å with optimization of the remaining degrees of freedom. Note that the dotted line indicating singlet coupling between orbitals has switched from vertical to horizontal. **c** Magnitudes of the coefficients of the Rumer spin functions for the GVB wave function of SOCl($\tilde{X}^2A''$) as a function of R(SO) for the orbital ordering indicated in (**a**) and (**b**)

overlapping singlet-coupled pair as a *through-pair* interaction. Despite the small overlap of this singlet pair, the $\tilde{X}^2A''$ state of SOCl is 10.7 kcal/mol lower in energy than the corresponding quartet state at the equilibrium geometry for the doublet state. (The O–Cl bond dissociates on the quartet surface). A more comprehensive study of the through-pair interaction will be the subject of a future article (Takeshita and Dunning, to be published); however, we speculate that the singlet coupling is energetically favorable not because a traditional covalent bond is formed, but rather because the overlap of the singlet-coupled pair reduces the electronic repulsion between the electrons in the recoupled pair π bond and the electrons in the weakly overlapping pair.

The electronic structure of the SOCl isomer is actually even more complex than the above analysis implies. If the SO bond is lengthened (even slightly), the recoupled pair π bond weakens and the bonding pattern switches from that of a recoupled pair π bond and a through-pair interaction to that of an S3p^2 lone pair and a standard covalent OCl bond. At R_e(SO) + 0.12 Å, the GVB wave function (shown in Fig. 5b) is almost entirely (99.8 %) described by the latter bonding pattern. If we perform GVB calculations from R(SO) = R_e – 0.02 Å to R(SO) = R_e + 0.12 Å and preserve the ordering of the orbitals by atomic character, we can directly observe the switch in character between the recoupled pair bonding pattern and the covalent bonding pattern in the spin-coupling coefficients, $\{c_{Sk}\}$, associated with the Rumer spin functions. We plot these values in Fig. 5c as a function of R(SO). Because the spin functions in the Rumer basis are nonorthogonal, there is no unambiguous way to relate the spin-coupling coefficient to its contribution to the GVB wave function. Nevertheless, the magnitudes of the spin function coefficients are certainly correlated with the importance of the spin-coupling patterns.

At intermediate values of R(SO), there is some ambiguity in the assignment of the orbitals as they exchange character through both the spin and spatial degrees of freedom, and an orbital ordering that resulted in the smoothest evolution of the Rumer spin functions was chosen. At larger values of R(SO), the covalent OCl bonding pattern (diamonds) totally dominates, but as R(SO) decreases, the recoupled pair bonding pattern (upward triangles) becomes increasingly important, and becomes dominant at R(SO) values below 1.56 Å. A third spin-coupling pattern (circles) also becomes important as R(SO) decreases. This spin function corresponds to coupling φ_3 and φ_4 into a recoupled pair π bond, and then singlet coupling the out-of-plane S3p GVB orbital (φ_1) to the Cl3p in-plane orbital (φ_2) and high spin-coupling the other in-plane S3p GVB orbital (φ_5). This spin function does not describe an OCl bond. This spin-coupling coefficient increases because the triplet-coupling of the orthogonal valence S3p-like orbitals (φ_1 and φ_5 in Fig. 5a) on the SO($X^3\Sigma^-$) fragment becomes important as the Cl atom becomes increasingly weakly bound. (In the Rumer spin basis, there is no spin function that corresponds directly to triplet-coupling these two orbitals). The increasing contribution of this spin function as R(SO) decreases is not surprising given the weakness of the OCl bond.

The presence of two distinct bonding patterns has consequences for the potential energy curve as a function of SO bond length for SOCl. For ClSO, the analogous potential energy curve acts like a Morse oscillator around the minima as the SO bond is stretched; see Fig. 6, where the energies of the respective minima have been set to zero.

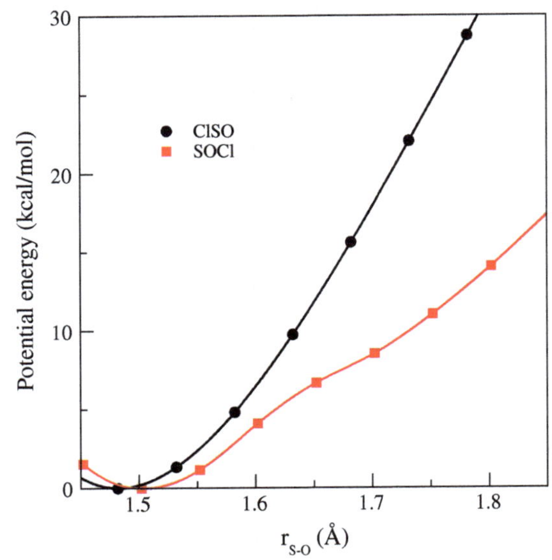

Fig. 6 Potential energy curves as a function of R(SO) for the $\tilde{X}^2A''$ states of ClSO (*circles*) and SOCl (*squares*) with all other degrees of freedom optimized at the MRCI + Q/AV(T + d)Z level of theory. The energy at the minima is set equal to zero for both curves

However, for SOCl, while there is only one minimum, there is an obvious change in the character of the potential energy curve around 1.65 Å. We attribute this feature to the competition of the two bonding motifs, which possess two distinct equilibrium bond lengths and strengths, depending on whether the recoupled pair π bond is present or absent.

4 The Cl_2SO/ClSOCl isomers

Similarly to ClSO/SOCl, we can understand the energetic differences between the Cl_2SO and ClSOCl isomers in terms of recoupled pair π bonding. Consider first ClSO($\tilde{X}^2A''$) + Cl(^{2}P). In order to form an OCl bond to yield ClSOCl, the Cl atom must bond via a through-pair interaction or it has to break the π bond just as we observed in SOCl. However, in Cl_2SO, the second SCl bond forms with the orbital left over from the formation of the recoupled pair π bond, φ_5 in Fig. 4, and the recoupled pair π bond is maintained. These pathways are depicted in Fig. 7. By this logic, we expect the energy difference between the Cl_2SO and ClSOCl isomers to be similar in magnitude to that between the triatomic ClSO and SOCl molecules, and it is: the Cl_2SO isomer is 47.0 kcal/mol lower in energy than the ClSOCl isomer.

We can also understand the variation in SO bond strength as a function of X and Y in the XYSO molecules in terms of recoupled pair π bonding. We saw in our prior studies of recoupled pair bonding that the orbital left over from the formation of the recoupled pair bond has large

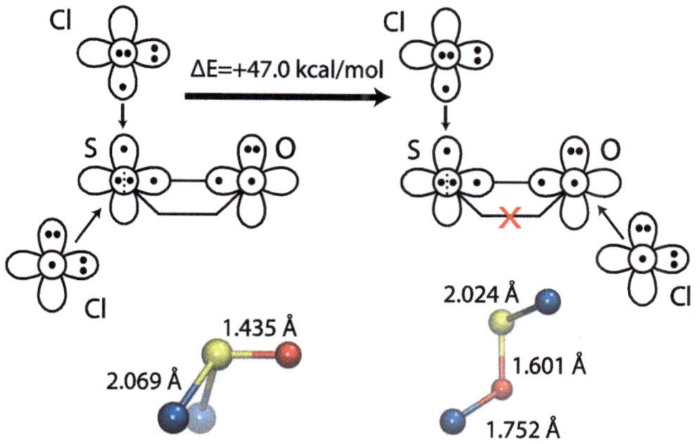

Fig. 7 GVB orbital diagrams for the addition of two chlorine atoms to $SO(X^3\Sigma^-)$ to form either Cl_2SO or ClSOCl, and optimized geometries and relative energetics for these isomers

unfavorable overlaps with the bond pair [17, 46]. For ClSO, we find that φ_5 in Fig. 4 has a large energetically unfavorable overlap with the π bond pair (φ_3 and φ_4). Therefore, bond formation with this orbital will be more favorable if the incoming ligand is strongly electronegative and can polarize this orbital away from the SO bond pair, but would be less favorable or even unfavorable if the ligand is weakly electronegative and thus augments the electronic repulsion in the π system. A more in-depth discussion of Cl_2SO will be deferred until Sect. 4.2.

4.1 The ClSOCl isomer

Given the similarity in the GVB diagrams of SOCl and ClSOCl, it should come as no surprise that the potential energy curve associated with stretching the ClS–OCl bond also has contributions from two bonding patterns: recoupled pair π bond/through-pair interaction and $S3p^2$ lone pair/covalent OCl bond. However, for ClSOCl, there is increased electron repulsion between the SCl bond and the OCl bond and π orbitals. Because lengthening the SO bond reduces this repulsion, the more stable bonding pattern is shifted from the recoupled pair π bond/through-pair interaction to the $S3p^2$ lone pair/covalent OCl bond motif. As a result, ClSOCl has a longer SO bond length ($\Delta R_e = 0.10$ Å) but much shorter OCl bond length ($\Delta R_e = -0.37$ Å) than in SOCl. In addition, the potential energy curve for stretching the ClS–OCl bond is very flat, increasing by only 0.3 kcal/mol at $R(\text{SO}) = R_e(\text{SO}) - 0.1$ Å, compared to 8.5 kcal/mol in ClSO over the same range. In ClSOCl, the presence of the two bonding patterns is especially obvious because the balance between the two bonding motifs yields two distinct minima at the CCSD(T)-F12/AV($D + d$)Z level of theory (confirmed by frequency calculations); see Fig. 8a.

The two minima for ClSOCl are shown in Fig. 8b. As anticipated, the major geometric differences between these two isomers are the SO and OCl bond lengths; the SCl bond length and the angles are effectively the same. This pair of isomers can thus be described as bond stretch isomers, although the barrier separating the two isomers is extremely small. The OCl bond length is highly dependent on the SO bond length for this entire region of the potential energy curve: as the SO bond shortens, the recoupled pair π bond strengthens at the expense of the OCl bond. Figure 8c shows the strong inverse correlation between R(SO) and R(OCl). In contrast, the R(SCl) bond length is effectively independent of R(SO).

While the miniscule barrier between these two isomers makes distinguishing between them experimentally impossible, the unusual potential energy curve shown in Fig. 8a will lead to a distinct, if complicated, infrared spectrum that could, in principle, be observed. The atypical features of this potential energy curve are similar to what has been observed previously for the H_2PO radical, where, at lower levels of theory, there was a double well in the potential energy as a function of R(PO) [47]. In that case, increasing the amount of dynamic correlation in the calculation eliminated the barrier completely yielding a very flat potential energy curve. There was also a change in the character of the singly occupied orbital as the PO bond length decreased, from that consistent with a single PO σ bond (singly occupied orbital localized on oxygen) to that suggestive of a σ bond plus a recoupled pair π bond (singly occupied orbital on phosphorus).

4.2 The Cl_2SO isomer

At the start of this section, we considered the formation of Cl_2SO from ClSO and Cl, which suggested a covalent σ and a recoupled pair π bond for the SO bond and, through resonance, two predominantly covalent ClS bonds. This pathway is consistent with prior studies that showed a polar covalent σ SO bond and a nearly ionic π SO bond in related molecules [9, 12–14] (recoupled pair bonds tend to be quite polarized toward the recoupling ligand). This bonding scheme would predict Cl_2SO to have two ClS bonds similar in length to that in ClSO (2.054 Å) with a ClSCl angle near 90°. It would also predict an SO bond length close to that in ClSO (1.460 Å) with ∠ClSO like that in ClSO (109.2°). This compares well with the calculated structure of Cl_2SO: $R_e(\text{ClS}) = 2.069$ Å, ∠ClSCl = 95.5°, $R_e(\text{SO}) = 1.435$ Å, and ∠Cl_2S–SO = 115.9° (the latter is the angle between the bisector of the Cl_2S plane and the SO bond); see Table 2. These geometric parameters compare well with prior computational and experimental studies of Cl_2SO [48–51]. Clearly, this approach provides a compelling description of the bonding in the Cl_2SO molecule. But, there are alternative descriptions of the SO bond in the sulfinyls.

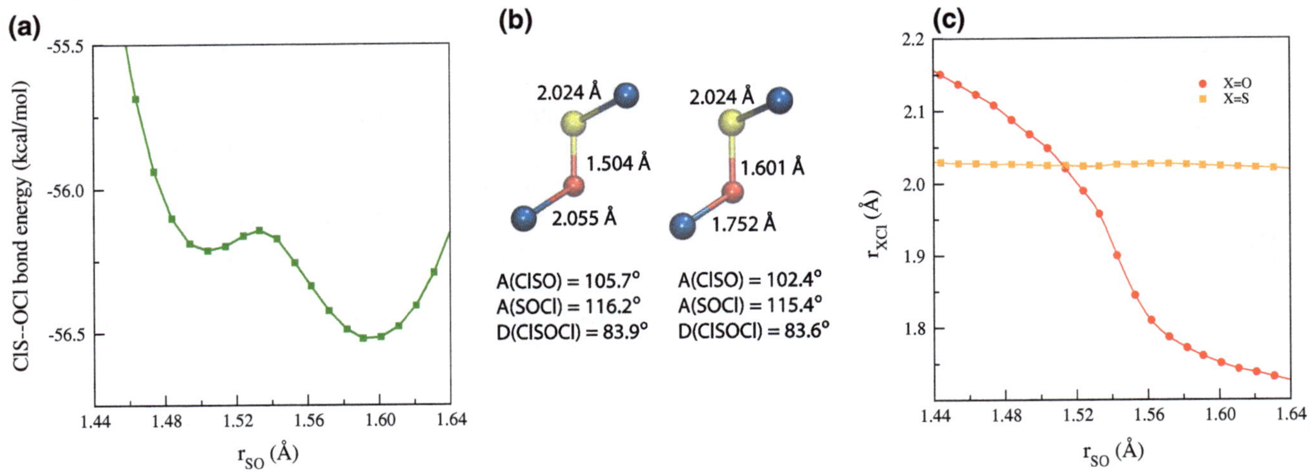

Fig. 8 **a** Potential energy for ClSOCl as a function of *R*(SO) from CCSD(*T*)-F12/AV(*T* + *d*)*Z* calculations with all other geometric parameters optimized at the CCSD(*T*)-F12/AV(*D* + *d*)*Z* level of theory. b Structures and geometric parameters of the two minima at the same level of theory as (**a**). Optimized SCl (*squares*) and OCl (*circles*) bond lengths as a function of *R*(SO) in ClSOCl from the geometries in (a)

Table 2 Optimized geometric parameters (in Å and degrees) of the $\tilde{X}^1A_1$, $\tilde{a}^3B_1$, and $\tilde{b}^3A_2$ states of Cl_2S and Cl_2SO at the minimum and the planar transition state for inversion

	R(S–Cl)	*A*(Cl–S–Cl)	*R*(S–O)	*A*(Cl_2S–S–O)[a]
$Cl_2S(\tilde{X}^1A_1)$	2.017	100.3	–	–
$Cl_2S(\tilde{a}^3B_1)$	2.117	147.8	–	–
$Cl_2S(\tilde{b}^3A_2)$	2.146	87.6	–	–
Cl_2SO(Opt.)	2.069	95.5	1.435	115.9
Cl_2SO(T.S.)	2.189	161.9	1.433	180.0

[a] The angle between the O atom, S atom, and the bisector of the Cl–S–Cl angle

We can, for example, consider the formation of Cl_2SO from Cl_2S and O instead of ClSO and Cl. However, the ground-state fragment, $Cl_2S(\tilde{X}^1A_1)$ and $O(^3P)$, cannot yield the ground singlet state of Cl_2SO, the $\tilde{X}^1A'$ state. As a result, bonding in XYSO species is often assumed to originate from the $XYS(\tilde{X}^1A) + O(^1D)$ asymptote. This suggests that the SO bond is a dative bond, where the S3p^2 pair of Cl_2S is donated to the empty O2p_z orbital that results from mixing the $^1A'$ states arising from the $XYS(\tilde{X}^1A) + O(^1D)$ and $XYS(\tilde{X}^1A) + O(^1S)$ asymptotes, i.e., the $O(2p_x^2 2p_y^2)$ configuration. The optimized structure of $Cl_2SO(\tilde{X}^1A')$ is in reasonable agreement with expectations based on this bonding motif: the geometric parameters of the Cl_2S group are reasonably close to those in the $Cl_2S(\tilde{X}^1A_1)$ state (see Table 2). Further, the angle between the Cl_2S plane and the oxygen atom 115.9° is consistent with what would be anticipated from bonding with one of the lone pairs of Cl_2S, a mix of S3s^2 and S3p^2 orbitals.

However, the $Cl_2S(\tilde{X}^1A_1) + O(^1D)$ limit is actually not the lowest energy asymptote for the dissociation of $Cl_2SO(\tilde{X}^1A')$. The lowest energy asymptote that can form a singlet state is $Cl_2S(\tilde{a}^3B_1) + O(^3P)$. The energy of $Cl_2S(\tilde{a}^3B_1)$ at its optimum geometry is only 36.2 kcal/mol higher in energy than that of the $Cl_2S(\tilde{X}^1A')$ ground state, compared to 49.1 kcal/mol for the $O(^3P)$ to $O(^1D)$ excitation energy [RCCSDT-F12/AV(*T* + *d*)*Z* calculations]. We have described bonding in $Cl_2S(\tilde{a}^3B_1)$ elsewhere [52]. In short, the two SCl bonds comprise a σ recoupled pair bond dyad, where both Cl atoms form bonds to the 3p^2 pair of the $S(^3P)$ atom. As we have shown, recoupled pair bond dyads are very stable, consistent with the small excitation energy observed in Cl_2S. The angle between the two ClS bonds in the $Cl_2S(\tilde{a}^3B_1)$ state is 147.8° and R_e(ClS) = 2.117 Å, 0.10 Å longer than in the $Cl_2S(\tilde{X}^1A_1)$ ground state. The two singly occupied orbitals, one in the plane of the molecule and one out of the molecular plane, resemble two S3*p* orbitals, with the in-plane orbital having significant S3*s* admixture that polarizes it toward the side of the sulfur atom containing the Cl atoms; see Fig. 9a. The next lowest energy electronic state of Cl_2S, the $\tilde{b}^3A_2$ state, is bound by one recoupled pair bond and one covalent bond, yielding a strongly bent geometry, 87.6°, with R_e(ClS) = 2.146 Å. The $Cl_2S(\tilde{b}^3A_2)$ state lies 57.0 kcal/mol above the ground state.

The geometric parameters listed above for the optimized Cl_2SO molecule deviate significantly from those optimal for $Cl_2S(\tilde{a}^3B_1)$; see the comparison in Table 2. However, the planar transition state for the inversion of Cl_2SO possesses structural features similar to the $\tilde{a}^3B_1$ state of Cl_2S, namely R_e(ClS) = 2.189 Å and ∠ClSCl = 161.9°. Figure 9b shows

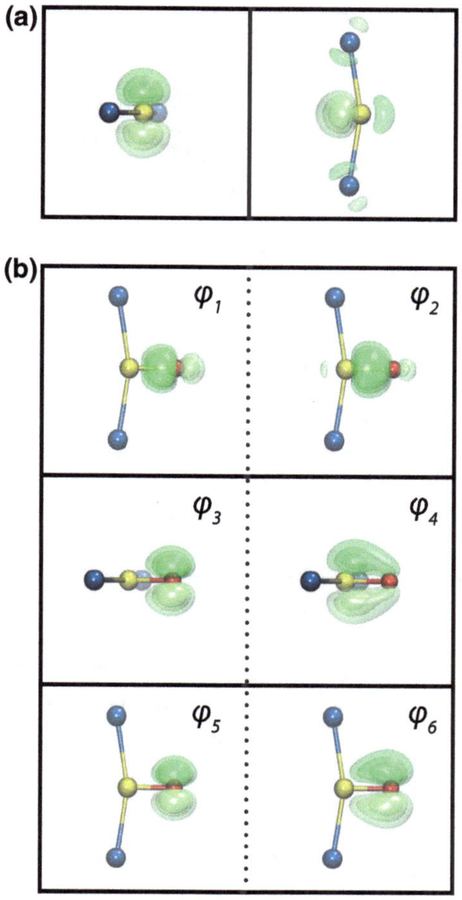

Fig. 9 **a** Singly occupied orbitals of $Cl_2S(\tilde{a}^3B_1)$ at its optimized geometry. **b** Select GVB orbitals for the transition state to inversion of Cl_2SO

the relevant orbitals of the planar Cl_2SO transition state. The in-plane S3*p*-like orbital from Fig. 9a forms an extremely polar σ bond with the in-plane $O2p^1$ orbital, and the out-of-plane S3*p* orbital forms a polar π bond (though less polar than the σ bond). The remaining $O2p^2$ pair is largely localized on the oxygen atom. While both the σ and π bonds are polarized toward the more electronegative oxygen atom, the σ bond is much more polar because polarizing toward oxygen reduces the anti-bonding character present in the S3*p*-like in-plane orbital. Despite the polar nature of these bonds, it seems reasonable that bonding in this state should be interpreted as an SO double bond, and the strength of this bond relative to the $Cl_2S(\tilde{a}^3B_1) + O(^3P)$ asymptote (100.8 kcal/mol) is consistent with this interpretation. For reference, the $a^1\Delta$ state of SO, which contains a σ and a π bond, is bound by 103.2 kcal/mol. The transition state for inversion is 40.1 kcal/mol higher in energy than the minimum—similar to many other XYSO molecules where X and Y are not bonded together via rings [2, 53–57].

The equilibrium geometry of Cl_2SO leads to two questions. (1) Why do the geometries for the minimum and transition state appear to be associated with two different asymptotes? Further, (2) why is the calculated dissociation energy for $Cl_2SO(\tilde{X}^1A') \rightarrow Cl_2S(\tilde{X}^1A_1) + O(^1D)$ so large? It is 153.8 kcal/mol—substantially larger even than that for the strong bond in the ground, $X^3\Sigma^-$, state of SO. See Table 3 for the RCCSD(T)-F12 bond dissociation energies for the various asymptotes relative to the optimized $Cl_2SO(\tilde{X}^1A')$ energy. The answers to both of these questions are related.

Table 3 Energy (kcal/mol) of the molecular fragments (Cl_2S and O) that are capable of forming $Cl_2SO(\tilde{X}^1A')$ relative to the optimized Cl_2SO geometry

	Fixed Cl_2S	Optimized Cl_2S
$Cl_2S(\tilde{X}^1A_1) + O(^1D)$	155.8	153.8
$Cl_2S(\tilde{a}^3B_1) + O(^3P)$	170.1	141.0
$Cl_2S(\tilde{b}^3A_2) + O(^3P)$	165.4	161.8

The Cl_2S molecule is either at the same geometry as in the optimized Cl_2SO structure (first column) or at optimized for the given electronic state of Cl_2S (second column)

We can understand the variation in Cl_2S angle between the optimized geometry and the planar transition state for inversion of Cl_2SO from the orbitals of the different electronic states of Cl_2S. When the Cl_2S group is constrained to be planar with the O atom, an a_1 orbital must be present and available in Cl_2S to form a σ bond with oxygen. Figure 9a shows that $Cl_2S(\tilde{a}^3B_1) + O(^3P)$ clearly meets this criterion. For the $Cl_2S(\tilde{X}^1A_1) + O(^1D)$ limit, the situation is not nearly as favorable. From Fig. 10, which shows the relevant orbitals of the Cl_2S group at the optimized Cl_2SO geometry, we can see that, while there is an a_1-symmetric lone pair (a slightly polarized $S3s^2$-like orbital) associated with the $\tilde{X}^1A_1$ state of Cl_2S, bonding between this orbital and $O(^1D)$ is likely to be disfavored. This is a result of the electronic repulsion among the $O2p^2$ pairs and the $S3p^2$ lone pair, coupled with the short SO bond length that would be required for the empty O2*p* orbital to have a large overlap with the mostly spherical 3*s*-like lone pair. The third lowest-lying state of Cl_2S, the $\tilde{b}^3A_2$ state of Cl_2S, does not have any a_1-symmetric singly occupied orbitals that can bond with $O(^3P)$. The dominance of the $Cl_2S(a^3B_1) + O(^3P)$ asymptote is reflected in the large (161.9°) Cl–S–Cl angle that is near optimal for the $\tilde{a}^3B_1$ state, but not the $\tilde{X}^1A_1$ or $\tilde{b}^3A_2$ states, of Cl_2S.

However, if the oxygen atom approaches from an approximately 90° angle relative to the plane of the Cl_2S group, then an SO σ bond can be formed with the highest-lying b_1 orbital of Cl_2S. For $Cl_2S(\tilde{X}^1A_1) + O(^1D) \rightarrow Cl_2SO(\tilde{X}^1A')$, because the symmetry is C_S instead of C_{2V}, the S3*s* and S3*p* orbitals shown in Fig. 10 can mix to minimize electronic repulsion, so dative bonding from this asymptote will be more stable than in the

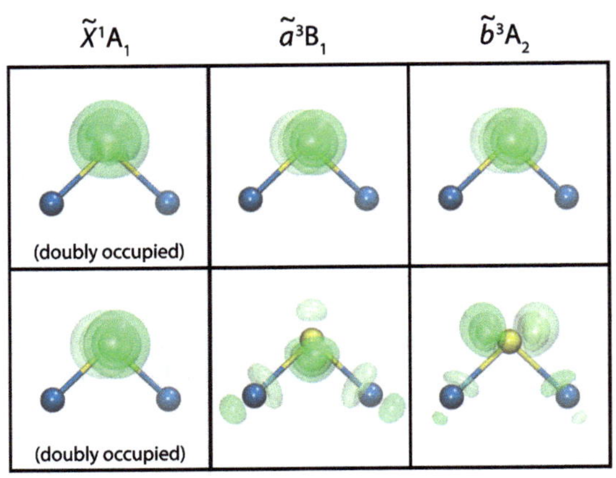

Fig. 10 Orbitals of Cl_2S relevant to SO bond formation to yield Cl_2SO for the $\tilde{X}^1A_1$, $\tilde{a}^3B_1$, and $\tilde{b}^3A_2$ states. The geometry is that of the Cl_2S group in Cl_2SO

planar configuration. Moreover, the singly occupied orbitals of the $Cl_2S(\tilde{a}^3B_1)$ and $Cl_2S(\tilde{b}^3A_2)$ states are well positioned to form double bonds with $O(^3P)$, as is apparent from Fig. 10. At the optimum Cl–S–Cl angle in $Cl_2SO(\tilde{X}^1A')$, all of these asymptotes are similar in energy, see the left column in Table 3. As we saw for the transition state, bonds originating from an orbital possessing antibonding character are likely to be quite polar to minimize this unfavorable character. The upshot of this situation is that all three asymptotes have comparable energies and have the same symmetry and basic orbital structure, and thus they all have the potential to mix to stabilize the ground state—lowering its energy and leaving the excited states with weaker bonds.

We have investigated the impact of the higher-lying asymptotes in Fig. 11 with a 6-state CASSCF/AV($T + d$)Z potential energy scan as a function of R(SO), where all of the other geometric parameters are fixed at the optimized Cl_2SO geometry. We include all valence orbitals except the O2s^2 and Cl3s^2 pairs in the active space and include two a' and one a'' virtual orbitals. At large R(SO), the first three states collapse to the $Cl_2S(\tilde{X}^1A_1) + O(^1D)$ asymptote, the fourth and fifth asymptotes correspond to $Cl_2S(\tilde{b}^3A_2) + O(^3P)$, and the sixth state is the $Cl_2S(\tilde{a}^3B_1) + O(^3P)$ asymptote. (The next asymptote is 50.9 kcal/mol higher in energy). Figure 11a shows the results of the CAS calculation, where, for clarity, several data points from the four highest-lying states at small values of R(SO) have not been plotted due to their interactions with even higher-lying electronic states. The ground state is bound by 119.8 kcal/mol, recovering about 77 % of the CCSD(T)-F12/AV($T + d$)Z energy reported in the upper left cell of Table 3.

Clearly, these states are interacting and curve crossings are present that complicate the interpretation of these results. However, if we assume that all states corresponding to aligning the O2p^2 pair with an occupied S3p-like orbital are

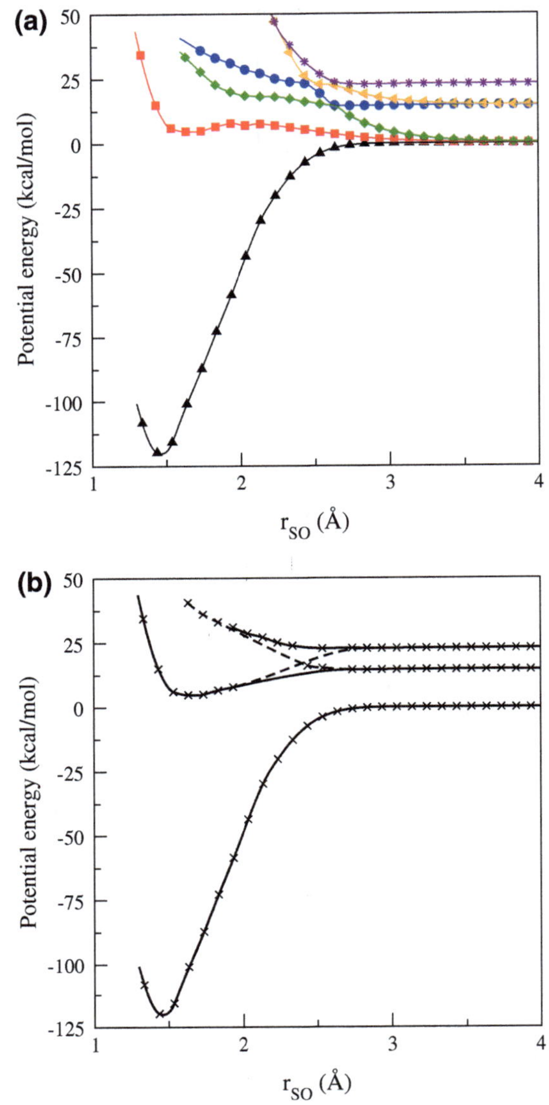

Fig. 11 **a** Potential energy (CASSCF/AV($T + d$)Z) for the first six electronic states of Cl_2SO as a function of R(SO) with all other geometric parameters fixed. For clarity, some of the data points for the four highest-lying states at small values of R(SO) that interact with even higher-lying electronic states have been removed. **b** Fits to the parts of the potential energy curves corresponding to "bonding" asymptotes; data points from (**a**) shown as "x"s

purely repulsive and smooth, we can isolate the parts of the calculation related to the asymptotes that we expect to bond. We show these data (denoted by "x"s) with corresponding fits with splines in Fig. 11b. It is clear in both Fig. 11a, b that there is a dearth of favorable bonding interactions associated with the triplet asymptotes, two of which (those in Fig. 11b) are aligned to form double bonds. The remaining favorable character is mixed into the second lowest electronic state at small R(SO)—squares in Fig. 11a. This state is bound by 9.9 and 18.2 kcal/mol relative to the asymptotes involving $Cl_2S(\tilde{b}^3A_2)$ and $Cl_2S(\tilde{a}^3B_1)$, respectively. For reference, with the same active space, the similarly double-bonded transition state is bound by 82.8 kcal/mol. Therefore, the majority of the bonding character associated with the triplet asymptotes is not found in the higher-lying electronic states.

The stabilizing effects of these triplet asymptotes can form an anomalously strong SO bond in the ground state, with the ground state effectively robbing these excited states of their favorable character because the orbitals between the three asymptotes are so similar. The double bonds arising from the triplet asymptotes are similar to those of the transition state, only the σ and π orbitals have switched. In this case, it is the π bonds that are extremely polar because of the anti-bonding character present in the a_1- and b_2-symmetry Cl_2S orbitals of the $\tilde{a}^3B_1$ and $\tilde{b}^3A_2$ states, respectively. This is direct evidence that multi-bond character is present in the ground state, significantly lowering its energy. Based on these results, we suggest that the multi-bond character owing to the triplet asymptotes dominates SO bonding in Cl_2SO, with dative bonding being of secondary importance.

Before concluding this discussion, we will consider one other point. The polarization/delocalization of the orbitals on the oxygen atom in sulfinyl groups has been previously noted in both MO and GVB calculations [9, 10, 12–14] and had been attributed to π–3d back-bonding in early work. However, consistent with more modern calculations, we find that this description does not explain the polarization of these orbitals for $Cl_2SO(\tilde{X}^1A')$. Figure 12a, b both show the π GVB orbitals of Cl_2SO but those in Fig. 12b are computed with only s and p basis functions on the sulfur atom. While the polarization toward sulfur is somewhat reduced in the latter orbitals, one of the GVB lobes in each direction remains highly delocalized. This comparison is consistent with the idea that the S3d functions provide additional polarization and correlation corrections but do not to act as valence orbitals for sulfur. As stated above, we attribute this polarization (and the energetic stabilization that it imparts) to the participation of the triplet asymptotes in the ground state of Cl_2SO; furthermore, we can actually see that some of the features of the low-lying triplet states of Cl_2S are present in the orbitals of Cl_2SO. The asymmetric electron density of the a_1-symmetric orbital of $Cl_2S(\tilde{a}^3B_1)$ in Fig. 10 with respect to reflection through the xz plane (out of the plane of the paper) explains the similar asymmetric nature of φ_1 and φ_2 in Fig. 12. This asymmetric π bond may also be connected to the bent-bond description of SO bonds observed by Cooper et al. [14] for some short sulfur–oxygen bonds.

5 Conclusion

In this work, we have explained the difference in stability between two pairs of structural isomers: ClSO/SOCl and Cl_2SO/ClSOCl. While the bond strengths of the SCl and OCl diatomic molecules are comparable, bonding the Cl atoms to sulfur allows the recoupled pair π bond present in $SO(X^3\Sigma^-)$ to be maintained, whereas forming a covalent bond with oxygen causes the recoupled pair π bond to break. (This is what occurred in our prior work on SOH). However, we saw that in SOCl, instead of breaking the recoupled pair π bond, the orbital on the Cl atom is singlet-coupled with the singly occupied orbital largely localized on sulfur. We refer to this bonding motif here as a through-pair interaction and the associated bond energy is very small (essentially 0 kcal/mol in SOCl, but see the work on NOF by Takeshita and Dunning, to be published).

In ClSOCl, competition between the two modes of bonding present in SOCl yields a very flat potential energy surface when the SO bond is stretched and contains a pair of bond stretch isomers separated by a very small barrier.

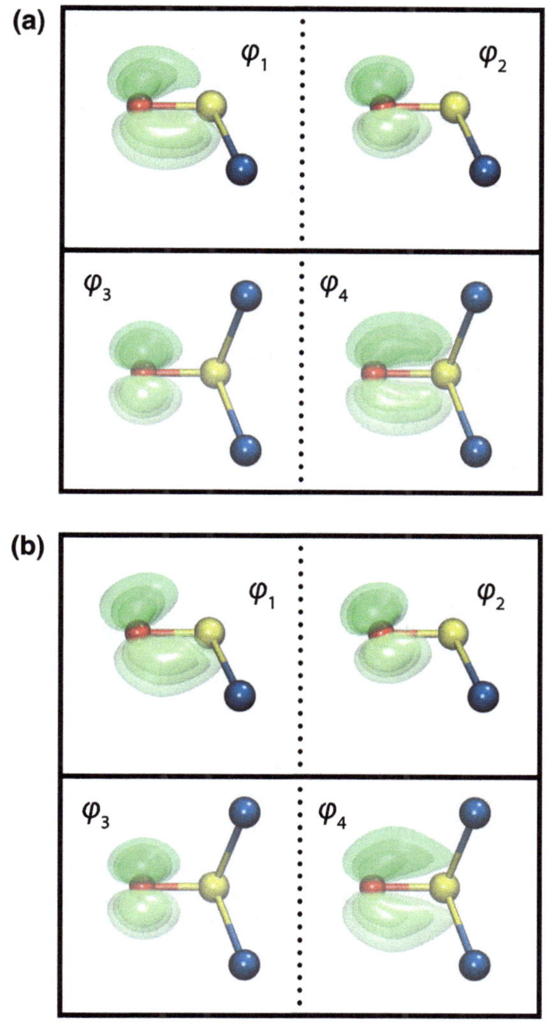

Fig. 12 GVB orbitals in the π space of the SO bond for Cl_2SO **a** with and **b** without d basis functions on the sulfur atom

The ClSOCl molecule has many commonalities with the H_2PO radical, and we suspect that there may be other instances of this behavior in compounds involving oxygen bonded to an element in the second-row late p-block where a competition between a recoupled pair bonded scheme and traditional covalent bonding yields quite interesting and unexpected features in the potential energy surface of these compounds.

For Cl_2SO, we rationalized the short SO bond length and very strong SO dissociation energy relative to the $Cl_2S(\tilde{X}^1A_1) + O(^1D)$ asymptote by showing that other (slightly) higher-energy asymptotes also contribute to the ground-state Cl_2SO molecule. The polarization of the SO π orbitals can be attributed to the importance of two triplet asymptotes in the electronic structure of Cl_2SO: $Cl_2S(\tilde{a}^3B_1) + O(^3P)$ and $Cl_2S(\tilde{b}^3A_2) + O(^3P)$. This is a useful application of our previous work using recoupled pair bonding to describe the energetics and bonding in the a^3B_1 and b^3A_2 states of Cl_2S. Understanding the nature of the bonding in Cl_2SO provides an important motivation for fully understanding the electronic structure of not only the ground state, but also the lower-lying excited states, of the fragments that compose a molecule.

The nature of the bonding in Cl_2SO is consistent with donation from the O2p orbitals to the SCl anti-bonding orbitals—or "anionic" hyperconjugation—in the MO framework, which has been suggested for similar hypervalent molecules (F_2SO and H_3PO for instance) [10]. The low-lying triplet states of Cl_2S can be generated by excitation of one of the S3p electrons into one of the SCl anti-bonding orbitals of either a_1 or b_2 symmetry to yield the $\tilde{a}^3B_1$ and $\tilde{b}^3A_2$ states of Cl_2S, respectively. The MO description of these bonds is sometimes referred to as single bonds plus hyperconjugation, but from the VB perspective, it is clear that the SO bond should be considered double bonds since they arise from the triplet asymptotes, which form true double bonds. Recoupled pair bonding explains why these triplet states are so low in energy, or said in the MO framework, why some molecules have lower energy σ^* orbitals. These results unify the MO and VB perspectives (single bond and hyperconjugation versus polar σ bond and nearly ionic π bond) reported in previous work by demonstrating that hyperconjugation in this case arises from higher-lying asymptotes that are capable of forming double bonds.

The Cl_2S orbitals that contribute to the π bonding have significant anti-bonding character, which causes these bonds to be extremely polarized toward oxygen, which may make them look like lone pairs from an electron density perspective as observed in the prior studies of similar compounds [4, 58]. But the weakness of the SO bonds in the states evolving from the higher-lying triplet asymptotes as a function of SO internuclear distance provides direct evidence that multiple bond character is a critical feature of SO bonding in sulfinyl halides. As the electronegativity of the substituents (X and Y) decreases, the energy of triplet XYS states will increase, as will the polarizability of the S3p-like orbital of XYS. This will both decrease the importance of multi-bond character in the SO bond and increase the strength of the dative SO bond. So for compounds such as H_2SO and organic sulfoxides, the single bond plus electrostatic stabilization view may very well be appropriate. However, for the sulfinyl halides, multiple bond character is an essential feature in the description of the SO bond.

Acknowledgments This work was supported by the Distinguished Chair for Research Excellence in Chemistry and the National Center for Supercomputing Applications at the University of Illinois at Urbana–Champaign.

References

1. Kucsman AK, Kapovits I (1985) Organic sulphur chemistry. Elsevier, Amsterdam
2. Jenks WS, Matsunaga N, Gordon M (1996) J Org Chem 61(4):1275–1283
3. Benson SW (1978) Chem Rev 78(1):23–35
4. Dobado JA, Martinez-Garcia H, Molina JM, Sundberg MR (1999) J Am Chem Soc 121(13):3156–3164
5. Kwart H, King K (1977) d-Orbitals in the chemistry of silicon, phosphorus and sulfur. Springer, Berlin
6. Mayer I, Revesz M (1983) Inorg Chim Acta 77(6):L205–L206
7. Magnusson E (1990) J Am Chem Soc 112(22):7940–7951
8. Magnusson E (1993) J Am Chem Soc 115(3):1051–1061
9. Cooper DL, Cunningham TP, Gerratt J, Karadakov PB, Raimondi M (1994) J Am Chem Soc 116(10):4414–4426
10. Reed AE, Schleyer PV (1990) J Am Chem Soc 112(4):1434–1445
11. Schleyer PV, Kos AJ (1983) Tetrahedron 39(7):1141–1150
12. Cioslowski J, Mixon ST (1993) Inorg Chem 32(15):3209–3216
13. Cioslowski J, Surjan PR (1992) J Mol Struct Theochem 87:9–33
14. Cunningham TP, Cooper DL, Gerratt J, Karadakov PB, Raimondi M (1997) J Chem Soc Faraday Trans 93(13):2247–2254
15. Page EM (2007) In: Zuckerman JJ, Hagen AP (eds) Inorganic reactions and methods. Wiley, Hoboken, NJ
16. Gangadharan R, Namboodiri PNN, Prasad KV, Viswanathan R (1979) J Power Sources 4(1):1–9
17. Lindquist BA, Takeshita TY, Woon DE, Dunning TH (2013) J Chem Theory Comput 9(10):4444–4452
18. Cooper DL, Karadakov PB (2009) Int Rev Phys Chem 28(2):169–206
19. Gerratt J, Cooper DL, Karadakov PB, Raimondi M (1997) Chem Soc Rev 26(2):87–100
20. Goddard WA, Dunning TH, Hunt WJ, Hay PJ (1973) Acc Chem Res 6(11):368–376
21. Hiberty PC, Shaik S (2007) J Comput Chem 28(1):137–151

22. Cooper DL, Thorsteinsson T, Gerratt J (1997) Int J Quantum Chem 65(5):439–451
23. Rumer G (1932) Gottinger Nachr 3:337
24. Simonetta M, Gianinetta E, Vandoni I (1968) J Chem Phys 48(4):1579
25. Karadakov PB, Gerratt J, Cooper DL, Raimondi M (1995) Theor Chim Acta 90(1):51–73
26. Chirgwin BH, Coulson CA (1950) Proc R Soc A 201(1065): 196–209
27. Gallup GA, Norbeck JM (1973) Chem Phys Lett 21(3):495–500
28. Thorsteinsson T, Cooper DL (1998) J Math Chem 23(1–2): 105–126
29. Dunning TH, Woon DE, Leiding J, Chen L (2013) Acc Chem Res 46(2):359–368
30. MOLPRO, version 2010.1, a package of ab initio programs, Werner HJ, Knowles PJ, Knizia G, Manby FR, Schutz M, Celani P, Korona T, Lindh R, Mitrushenkov A, Rauhut G, Shamasundar KR, Adler TB, Amos RD, Bernhardsson A, Berning A, Cooper DL, Deegan MJO, Dobbyn AJ, Eckert F, Goll E, Hampel C, Hesselmann A, Hetzer G, Hrenar T, Jansen G, Koppl C, Liu Y, Lloyd AW, Mata RA, May AJ, McNicholas SJ, Meyer W, Mura ME, Nicklass A, O'Neill DP, Palmieri P, Pfluger K, Pitzer R, Reiher M, Shiozaki T, Stoll H, Stone AJ, Tarroni R, Thorsteinsson T, Wang M, Wolf A, see http://www.molpro.net
31. Thorsteinsson T, Cooper DL, Gerratt J, Karadakov PB, Raimondi M (1996) Theor Chim Acta 93(6):343–366
32. Pipek J, Mezey PG (1989) J Chem Phys 90(9):4916–4926
33. Knizia G, Adler TB, Werner HJ (2009) J Chem Phys 130(5):054104
34. Knizia G, Werner HJ (2008) J Chem Phys 128(15):154103
35. Knowles PJ, Hampel C, Werner HJ (1993) J Chem Phys 99(7):5219–5227
36. Manby FR (2003) J Chem Phys 119(9):4607–4613
37. Dunning TH (1989) J Chem Phys 90(2):1007–1023
38. Dunning TH, Peterson KA, Wilson AK (2001) J Chem Phys 114(21):9244–9253
39. Woon DE, Dunning TH (1993) J Chem Phys 98(2):1358–1371
40. Adler TB, Knizia G, Werner HJ (2007) J Chem Phys 127(22):221106
41. Knowles PJ, Werner HJ (1985) Chem Phys Lett 115(3):259–267
42. Werner HJ, Knowles PJ (1985) J Chem Phys 82(11):5053–5063
43. Knowles PJ, Werner HJ (1988) Chem Phys Lett 145(6):514–522
44. Werner HJ, Knowles PJ (1988) J Chem Phys 89(9):5803–5814
45. Chu LK, Lee YP, Jiang EY (2004) J Chem Phys 120(7): 3179–3184
46. Lindquist BA, Dunning TH (2013) J Phys Chem Lett 4(18):3139–3143
47. Wesolowski SS, Johnson EM, Leininger ML, Crawford TD, Schaefer HF (1998) J Chem Phys 109(7):2694–2699
48. Mata F, Carballo N (1983) J Mol Struct 101(3–4):233–238
49. Mayer PM, Baer T (1996) Chem Phys Lett 261(1–2):155–159
50. Meloni S, Pieretti A, Bencivenni L, Albertini VR, Sadun C, Caminiti R (2001) Comput Mater Sci 20(3–4):407–415
51. Mootz D, Merschenz-Quack A (1988) Acta Crystallogr. Sect C: Cryst Struct Commun 44:926–927
52. Leiding J, Woon DE, Dunning TH (2011) J Phys Chem A 115(18):4757–4764
53. Cubbage JW, Jenks WS (2001) J Phys Chem A 105(46): 10588–10595
54. Fueno H, Ikuta S, Matsuyama H, Kamigata N (1992) J Chem Soc. Perkin Trans 2(11):1925–1928
55. Miller EG, Rayner DR, Thomas HT, Mislow K (1968) J Am Chem Soc 90(18):4861
56. Rayner DR, Gordon AJ, Mislow K (1968) J Am Chem Soc 90(18):4854
57. Rayner DR, Miller EG, Bickart P, Gordon AJ, Mislow K (1966) J Am Chem Soc 88(13):3138
58. Chesnut DB, Savin A (1999) J Am Chem Soc 121(10): 2335–2336

Theor Chem Acc (2014) 133:1434
DOI 10.1007/s00214-013-1434-9

REGULAR ARTICLE

Correlation consistent, Douglas–Kroll–Hess relativistic basis sets for the 5*p* and 6*p* elements

David H. Bross · Kirk A. Peterson

Received: 11 October 2013 / Accepted: 2 December 2013 / Published online: 17 December 2013

Abstract New sets of all-electron correlation consistent triple- and quadruple-zeta basis sets have been developed for the 5*p* and 6*p* elements (In–Xe, Tl–Rn). For the 5*p* elements, the spin-free Douglas–Kroll–Hess (DKH) Hamiltonian truncated at second order was used, while for the 6*p* row, DKH3 was employed. The resulting cc-pV*m*Z-DK sets ($m = T, Q$) are designed to correlate the valence *ns* and *np* electrons, but both core–valence sets (cc-pwCV*m*Z-DK) for $(n-1)spd$ correlation and diffuse-augmented sets (aug-cc-pV*m*Z-DK) for weak interactions have also been included. Benchmark DKH CCSD(T) calculations were carried out on the atoms for their first ionization potentials and electron affinities. Coupled cluster calculations of the near-equilibrium potential energy functions of 18 selected diatomic molecules were also carried out to determine their spectroscopic and thermodynamic properties. These results are extensively compared to those obtained using the analogous aug-cc-p(wC)V*m*Z-PP basis sets with their associated small-core pseudopotentials. For the quadruple-zeta quality basis sets, the mean unsigned differences were found to be just 1.4 mÅ for r_e, 0.7 cm^{-1} for ω_e, and 0.2 kcal/mol for D_e with corresponding maximum differences of 4.8 mÅ, 4.3 cm^{-1}, and 0.7 kcal/mol, respectively. Using all-electron DKH calculations with the present basis sets as corrections to the pseudopotential approximation appears to be most accurate when $(n-1)d$ correlation is considered in both cases using aug-cc-pwCVQZ quality basis sets. The new DK basis sets exhibit similar basis set convergence toward the complete basis set (CBS) limit as the PP-based sets and hence should find utility in all-electron [*T*, *Q*] basis set extrapolations.

Keywords Basis sets · Douglas–Kroll · Correlation consistent · Pseudopotentials

1 Introduction

Theoretical chemistry has been developed over the last 30 years or so into a quantitative instrument capable of making predictions that can sometimes rival experiment. The most important development has probably been the massive increases in computational resources, but theoretical breakthroughs were of course necessary to make use of these. Two of the key developments behind quantitative ab initio theory were an improved understanding of basis set convergence and the implementation of efficient, highly correlated methods such as those based on coupled cluster theory [1–3]. Basis set convergence is regular and relatively well understood when the correlation consistent basis sets, pioneered by Dunning [4], are used, and this allows the complete basis set (CBS) limit of the chosen method to be estimated [5, 6]. This eliminates one of the major sources of errors in ab initio calculations. Coupled cluster theory is size extensive and systematically convergent toward full configuration interaction (FCI) as higher excitations are included [7]. By combining the systematic characteristics of basis set convergence and electron correlation treatment, composite schemes [8–14] allow various approximations to be accounted for in

Dedicated to Professor Thom Dunning and published as part of the special collection of articles celebrating his career upon his retirement.

D. H. Bross · K. A. Peterson (✉)
Department of Chemistry, Washington State University, Pullman, WA 99164, USA
e-mail: kipeters@wsu.edu

separate, tractable calculations, and these have been shown to be a reliable strategy for obtaining high accuracy in applications such as ab initio thermochemistry or spectroscopy.

One of the approximations often invoked in ab initio calculations, which is particularly relevant for heavy systems, is the use of a non-relativistic Hamiltonian. Spin–orbit and scalar relativistic effects have been shown to be important in obtaining chemical accuracy (1 kcal/mol) in thermochemistry applications, as well as obtaining accurate spectroscopic properties [1, 15]. Even in a system as light as water, scalar relativistic effects can account for about 3 cm^{-1} in the stretching harmonic frequencies and nearly 0.3 kcal/mol in its atomization energy [16–19]. Fortunately, there are ways to introduce relativistic effects without needing to use the computationally expensive four-component Dirac–Coulomb Hamiltonian. Scalar relativistic effects can be treated using methods such as the X2C or Douglas–Kroll Hamiltonians [20–23], or even a posteriori with the Cowan–Griffin or mass–velocity–Darwin (MVD) method [24, 25], which involves a calculation of the expectation values of the mass velocity and one-electron Darwin operators. Relativistic pseudopotentials (PPs) or effective core potentials (ECPs) can account for both scalar and spin–orbit relativistic effects with high accuracy. In particular, the energy-consistent small-core pseudopotentials of the Stuttgart–Köln variety, which have been adjusted to atomic multi-configurational Dirac–Hartree–Fock (MCDHF) results using the Dirac–Coulomb–Breit Hamiltonian, have been shown to reliably reproduce a variety of all-electron data [26–30]. Recent reviews of pseudopotentials and their uses in quantum chemistry calculations are available [31, 32]. Also, correlation consistent basis sets have been developed in conjunction with Stuttgart–Köln pseudopotentials [27–30, 33], which conceptually makes investigating systems involving atoms as heavy as Rn essentially the same as those containing Kr.

There are properties, however, which depend on the electron density near the nucleus, e.g., chemical shielding constants for NMR or the even multipole moments [34], where PP treatments are typically not amenable (although certainly in some cases they are, see, e.g., Ref. [35] and references therein). In other cases, explicit treatment of some of the electrons subsumed by the pseudopotential can be important, e.g., correlation of the $4f$ electrons in $5d$ transition metals [30, 36]. Of course, the pseudopotential approximation itself introduces some error and the ability to calibrate this for any given system is important when performing high accuracy work. Hence, it is important to have all-electron relativistic treatments available. A variationally stable a priori method, which reliably recovers relativistic effects, can be used in one- or two-component calculations, and also allows control over the level of accuracy in treating these effects, is the Douglas–Kroll–Hess Hamiltonian (DKH) [21, 22, 37]. Basis sets for use within this framework are the primary focus of the present paper. For many of the published correlation consistent basis sets employing pseudopotentials, an equivalent triple-zeta all-electron DKH basis set was also included, primarily for the purposes of benchmarking the pseudopotential results. Comparing DKH and PP results does introduce some ambiguity, since, for example, the DKH Hamiltonian does not include certain contributions such as the Breit interaction. On the other hand, the current generation of Stuttgart–Köln pseudopotentials was adjusted to four-component MCDHF reference data using a Dirac–Coulomb Hamiltonian and perturbatively including the Breit correction to the two-electron terms [29, 38]. But even so, attempts at such a calibration can be useful. Comparison of the PP-based results to all-electron DKH calculations for $4d$ transition metals showed differences in electron affinities generally <0.5 kcal/mol, but for excitation energies, these reached about 1–2 kcal/mol for valence correlation in the late metals Tc-Pd, and just over 2 kcal/mol when outer-core $4s4p$ electron correlation was considered [29]. In $5d$ atomic calculations, the difference between ionization potentials (IPs) and EAs reached about 1 kcal/mol [30].

A few basis sets developed for use in DKH calculations on molecules involving the $5p$ and $6p$ elements have already been made available in the literature. Specifically, these include the atomic natural orbital (ANO) sets of Roos [39], the double-zeta HF DK3 sets of Hirao and co-workers [40], the Sapporo DKH sets of Koga and co-workers [41, 42], and the segmented DKH sets of Jorge and co-workers for the $5p$ elements [43–45]. The relativistic correlation consistent-style basis sets developed by Dyall [46] for four-component calculations should also be mentioned in this context. In the present work, all-electron DKH correlation consistent basis sets have been developed at both the triple- and quadruple-zeta levels for the $5p$ and $6p$ elements, i.e., In–Xe and Tl–Rn, respectively. The outermost composition of these sets was designed to be as similar as possible to their corresponding PP counterparts, i.e., cc-pVmZ-PP, cc-pwCVmZ-PP, and aug-cc-p(wC)VmZ-PP, so that they could be reliably used to benchmark these PP results and also be of utility in all-electron calculations estimating CBS limits. Sets designed for valence-only correlation (nsp correlation, cc-pVmZ-DK) and outer-core correlation [valence + $(n-1)spd$, cc-pwCVmZ-DK] are included, as well as their diffuse-augmented versions (aug-cc-p(wC)VmZ-DK).

2 Methodology

2.1 Computational details

The MOLPRO ab initio suite of programs was used throughout this work [47]. All orbitals were symmetry equivalenced, and pure spherical harmonics were used throughout, although at the correlated level symmetry equivalencing was not enforced. A one-electron, spin-free DKH implementation was employed at either second or third order (DKH2 or DKH3) [23]. All of the benchmark atomic and molecular correlated calculations used the CCSD(T) method [48, 49]. For the open-shell systems, the R/UCCSD(T) method was used, which utilizes restricted open-shell Hartree–Fock (ROHF) orbitals but with relaxed spin restrictions in the CCSD [50–53].

The basis sets used in pseudopotential calculations corresponded to aug-cc-pV*m*Z-PP [27, 33, 54] and aug-cc-pwCV*m*Z-PP [55], including the revised iodine pseudopotential and basis sets [54]. The basis sets for the light atoms were aug-cc-pV*m*Z (H,N,O) [3, 4] and aug-cc-pV($m + d$)Z (Si and Cl) [2, 56]. While the difference between all-electron DKH- and PP-based calculations in a comparable basis set can yield estimates of the errors due to the pseudopotential approximation, it also contains the scalar relativistic effects of any light atoms. To separate these two contributions, the scalar relativistic effects of the light atoms were estimated using the MVD method. These utilized the triple-zeta basis sets described above [aug-cc-pVTZ(-PP) or aug-cc-pwCVTZ(-PP)], but completely uncontracted, in CISD calculations. Of course, this assumes the contribution to the MVD result from the atom with the PP is negligible, but this has been validated previously by Feller et al. [57].

To assess the new all-electron DK basis sets (see below), the spectroscopic properties of the diatomic molecules originally treated in Refs. [27, 33] using PP-based correlation consistent sets were determined with the new basis sets in all-electron DKH2 and DKH3 calculations. Specifically, the near-equilibrium potential energy functions of InCl, TlCl, In_2, Tl_2, SnO, PbO, SbN, BiN, Sb_2, Bi_2, SiTe, SiPo, HI, HAt, I_2, At_2, XeH^+, and RnH^+ were obtained by fitting 7 energies distributed around their equilibrium values ($r - r_e = -0.3, -0.2, -0.1, 0.0, +0.1, +0.3, +0.5$ a.u.) to a fifth- or sixth-order polynomial in internal displacement coordinates. Spectroscopic constants were obtained from the usual Dunham analysis [58]. All of these molecules involved closed-shell singlet ground states, except for In_2 and Tl_2 where $^3\Pi_u$ states were utilized. Unless otherwise noted, the frozen-core approximation was used throughout, i.e., only the valence *s* and *p* electrons were correlated, leaving the $(n-1)d$ electrons of the post-*d* elements in the frozen core. This led to a few problems with molecules containing early post-*d* elements, i.e., the d_σ orbital of the heavy element tended to mix with the valence *s* orbital of the lighter atom. Therefore, in order to obtain accurate frozen-core calculations for InCl, SnO, SbN, TlCl, PbO, and BiN, these d_σ and *s* orbitals were localized using the Pipek–Mezey method [59] in order to unambiguously leave the $(n-1)d_\sigma$ electrons in the core. This procedure was also carried out previously for these molecules with the PP basis sets [33].

2.2 Basis set construction

The basis set development closely followed this laboratory's previous work [28–30] for the 4*d*, 5*d*, and coinage metal -DK sets, and thus will only be briefly described here. Either a standard BFGS method was employed with convergence set to 10^{-6} a.u. on the exponent gradients or a simplex algorithm was used with a convergence of 10^{-7} E_h on the energy. Overall, the sets were developed to be of the same quality as the existing triple- and quadruple-zeta pseudopotential-based sets for the 5*p* and 6*p* elements [27, 33]. The Hartree–Fock (HF) primitives (*spd* for 5*p* and *spdf* for 6*p*) were optimized using a sixth-order Legendre polynomial scheme as proposed by Petersson [60], although the outermost 5 functions of *s*, *p*, and *d* symmetry were fully optimized. Because of the well-known weak singularity of the Dirac and DKH Hamiltonians for a point nucleus model, which can lead to very large *s*-type exponents (see, e.g., Ref. [61]), the present optimizations were carried out whereby the maximum exponent was constrained to be less than 9×10^{12} in order to avoid numerical problems. The exponents of all the 5*p* elements were optimized at the more commonly used DKH2 level, while for the heavier 6*p* elements DKH3 was utilized. The former was chosen based on the work of Tsuchiya et al. [40] but has also been validated in the present work. Initially, the sizes of the triple-zeta (TZ) primitive sets were chosen to match the coinage metal cc-pVTZ-DK basis sets [28] and then were adjusted during the optimization to achieve a triple-zeta valence HF description of the orbitals, i.e., the outermost part of the valence *s* and *p* atomic orbitals were described by three exponents. The quadruple-zeta basis sets were initially chosen to have two additional functions in each occupied angular momenta, but this was latter revised based on the observed energetic lowerings at the HF level. The HF/TZ optimizations for the 5*p* elements proceeded by bootstrapping each primitive set using the TZ set from the preceding element in the row. Each angular momentum was optimized separately but was iterated until self-consistent. For the 6*p* elements, each angular momentum was also separately optimized, but for computational expediency, the contracted DK3 primitive sets of Tsuchiya et al. [40] were used for the angular momenta not

Table 1 Final DK basis set compositions

	HF [primitives]/ (contracted)	Valence *nsp* correlation (cc-pV*m*Z-DK)	Diffuse augmenting (aug-)	Valence + (*n* − 1)*spd* correlation (cc-pwCV*m*Z-DK)
TZ				
5*p*	[28*s*21*p*14*d*]/(5*s*4*p*2*d*)	+2*s*2*p*2*d*1*f*	+1*s*1*p*1*d*1*f*	+2*s*2*p*2*d*2*f*1*g*
6*p*	[34*s*28*p*19*d*11*f*]/(6*s*5*p*3*d*1*f*)			
QZ				
5*p*	[30*s*23*p*17*d*]/(5*s*4*p*2*d*)	+3*s*3*p*3*d*2*f*1*g*	+1*s*1*p*1*d*1*f*1*g*	+3*s*3*p*3*d*3*f*2*g*1*h*
6*p*	[36*s*30*p*21*d*12*f*]/(6*s*5*p*3*d*1*f*)			

Table 2 Atomic electron affinities (kcal/mol) calculated using DK and the difference between these and analogous PP results (ΔPP)

	Basis set	Hartree–Fock				ΔCorrelation [CCSD(T)-HF]			
		aug-cc-pV*m*Z		aug-cc-pwCV*m*Z		Valence[a]		Valence + (*n* − 1)*d*[b]	
		DK	ΔPP	DK	ΔPP	DK	ΔPP	DK	ΔPP
In	TZ	1.88	−0.02	1.89	0.00	7.78	0.01	7.27	−0.03
	QZ	1.90	0.02	1.90	0.02	8.02	0.00	7.61	−0.05
Sn	TZ	23.53	−0.11	23.54	−0.09	8.56	0.03	8.51	−0.03
	QZ	23.53	−0.08	23.53	−0.08	9.00	0.02	9.05	−0.05
Sb	TZ	−4.73	−0.17	−4.73	−0.08	22.78	0.00	22.90	−0.13
	QZ	−4.75	−0.05	−4.74	−0.05	24.56	0.03	24.65	−0.02
Te	TZ	25.22	−0.11	25.22	−0.06	19.27	0.02	19.59	−0.06
	QZ	25.20	−0.02	25.20	−0.02	21.48	0.03	21.81	−0.03
I	TZ	57.32	−0.17	57.31	−0.12	15.69	−0.02	15.97	−0.09
	QZ	57.27	−0.07	57.27	−0.07	18.38	−0.01	18.64	−0.09
Tl	TZ	−0.41	0.08	−0.40	0.11	7.39	−0.01	7.41	−0.09
	QZ	−0.37	0.14	−0.37	0.14	7.62	−0.03	7.82	−0.09
Pb	TZ	20.51	−0.03	20.52	0.00	8.62	0.00	8.97	−0.10
	QZ	20.53	0.01	20.53	0.01	9.02	−0.02	9.56	−0.09
Bi	TZ	−6.60	−0.20	−6.60	−0.08	22.49	−0.03	22.88	−0.11
	QZ	−6.58	−0.04	−6.57	−0.04	24.13	0.02	24.66	−0.02
Po	TZ	22.30	−0.15	22.30	−0.08	19.18	0.02	19.79	−0.08
	QZ	22.32	−0.05	22.32	−0.05	21.21	0.00	21.88	−0.03
At	TZ	52.82	−0.09	52.82	−0.04	15.84	−0.02	16.33	−0.10
	QZ	52.84	−0.03	52.84	−0.03	18.30	−0.02	18.81	−0.06

[a] aug-cc-pV*m*Z(−DK/−PP)

[b] aug-cc-pwCV*m*Z(−DK/−PP)

being optimized. The chosen primitive HF sets are given in Table 1. These were generally contracted to [5*s*4*p*2*d*] (5*p* elements) and [6*s*5*p*3*d*1*f*] (6*p* elements) based on the HF atomic orbital coefficients of the electronic ground states of the atoms (DKH2 for 5*p* and DKH3 for 6*p*). Note that if these sets are used in calculations employing a different order of the DKH Hamiltonian, it is recommended that they be appropriately recontracted.

The correlating functions for the cc-pV*m*Z-DK basis sets were optimized at the singles and doubles configuration interaction (CISD) level of theory for valence *sp* correlation only and included 2*s*2*p*2*d*1*f* and 3*s*3*p*3*d*2*f*1*g* functions for triple- and quadruple-zeta, respectively. In the corresponding PP sets, the need for an additional *s* correlating function was due to the pseudopotential approximation [33, 62, 63], and thus, just a

Table 3 Atomic ionization potentials (kcal/mol) calculated using DK and the difference between these and analogous PP results (ΔPP)

	Basis set	Hartree–Fock				ΔCorrelation [CCSD(T)-HF]			
		aug-cc-pV*m*Z		aug-cc-pwCV*m*Z		Valence[a]		Valence + $(n-1)d$[b]	
		DK	ΔPP	DK	ΔPP	DK	ΔPP	DK	ΔPP
In	TZ	118.28	−0.03	118.27	−0.03	7.81	0.07	9.54	−0.09
	QZ	118.28	−0.03	118.27	−0.02	8.14	0.05	10.32	−0.12
Sn	TZ	157.76	0.00	157.75	0.00	8.36	0.07	10.13	−0.06
	QZ	157.75	−0.01	157.75	−0.01	8.82	0.08	10.99	−0.07
Sb	TZ	199.36	0.10	199.35	0.10	8.22	0.02	10.04	−0.23
	QZ	199.34	0.08	199.34	0.08	8.90	0.03	10.97	−0.14
Te	TZ	174.84	0.05	174.84	−0.02	21.85	0.06	23.72	−0.07
	QZ	174.82	−0.04	174.81	−0.04	24.39	0.10	26.41	−0.03
I	TZ	220.92	−0.11	220.91	0.00	18.04	0.01	19.81	−0.26
	QZ	220.85	−0.17	220.84	0.00	21.04	0.01	22.94	−0.25
Xe	TZ	268.80	−0.10	268.78	0.00	13.85	0.06	15.23	−0.25
	QZ	268.67	−0.15	268.67	0.00	17.32	0.08	18.80	−0.22
Tl	TZ	113.29	0.46	113.27	0.46	6.76	−0.01	9.59	−0.29
	QZ	113.28	0.45	113.28	0.45	7.13	−0.07	10.65	−0.26
Pb	TZ	150.77	0.00	150.90	−0.16	7.79	−0.12	10.20	0.16
	QZ	150.72	0.00	150.89	−0.18	8.29	−0.17	11.21	−0.14
Bi	TZ	190.03	−0.09	190.03	−0.13	8.03	0.02	10.24	−0.18
	QZ	190.01	−0.13	190.00	−0.13	8.68	0.01	11.34	−0.13
Po	TZ	165.38	−0.07	165.37	−0.24	21.03	0.03	23.37	−0.10
	QZ	165.35	−0.25	165.35	−0.25	23.46	0.04	26.01	0.01
At	TZ	208.14	0.01	208.14	−0.11	17.51	0.06	19.61	−0.09
	QZ	208.11	−0.12	208.11	−0.12	20.32	0.03	22.60	0.01
Rn	TZ	252.16	0.08	252.16	−0.03	13.61	0.07	15.19	−0.09
	QZ	252.12	−0.04	252.12	−0.04	16.84	0.04	18.56	0.00

[a] aug-cc-pV*m*Z(−DK/−PP)

[b] aug-cc-pwCV*m*Z(−DK/−PP)

standard correlation consistent prescription was followed in the present all-electron sets. Care was taken to avoid any linear dependencies between correlating functions and underlying primitives by either substituting an optimal correlating function by a HF primitive or, as in the case of the *d* functions for the TZ sets of the 6*p* elements, substituting the 2 optimal *d* correlating functions for existing HF primitives and reoptimizing the remaining HF primitives in the presence of these fixed correlating functions (the latter step resulted in minimal changes to the HF energy). This was necessary to ensure stability of the DKH integral evaluation, which for MOLPRO involves a resolution of the identity using the completely uncontracted orbital basis sets. The resulting triple-zeta basis sets, denoted cc-pVTZ-DK, consisted of (28*s*21*p*14*d*1*f*)/[7*s*6*p*4*d*1*f*] for In–Xe and (34*s*30*p*19*d*12*f*)/[8*s*7*p*5*d*2*f*] for Tl–Rn. The cc-pVQZ-DK basis sets corresponded to (30*s*23*p*17*d*2*f*1*g*)/[8*s*7*p*5*d*2*f*1*g*] for In–Xe and (36*s*30*p*24*d*14*f*1*g*)/[9*s*8*p*6*d*3*f*1*g*] for Tl–Rn. These are also summarized in Table 1. Diffuse augmenting functions (one for each angular momentum in the basis) were optimized for the total energy of the atomic anions, yielding the aug-cc-pV*m*Z-DK ($m = T, Q$) sets. Specifically, the *s* and *p* exponents were optimized for the HF energy of the anions, while the *d, f, g* functions were optimized for the total CCSD(T) energy.

DKH basis sets appropriate for correlating the outer-core electrons, $(n-1)spd$, were also generated by including additional functions with the valence basis sets described above. The same number of core-correlating functions was added to the cc-pV*m*Z-DK sets as was previously used to augment the cc-pV*m*Z-PP sets [55], i.e., 2*s*2*p*2*d*2*f*1*g* for triple-zeta and 3*s*3*p*3*d*3*f*2*g*1*h* for quadruple-zeta. In general, these exponents were optimized in a weighted core–valence scheme [64] for correlating the valence (*nsp*) and outer-core electrons, $(n-1)spd$, whereby the core–valence pair correlation energy was strongly weighted over the core–core pair correlation energy (0.99:0.01). Linear dependencies among the lower

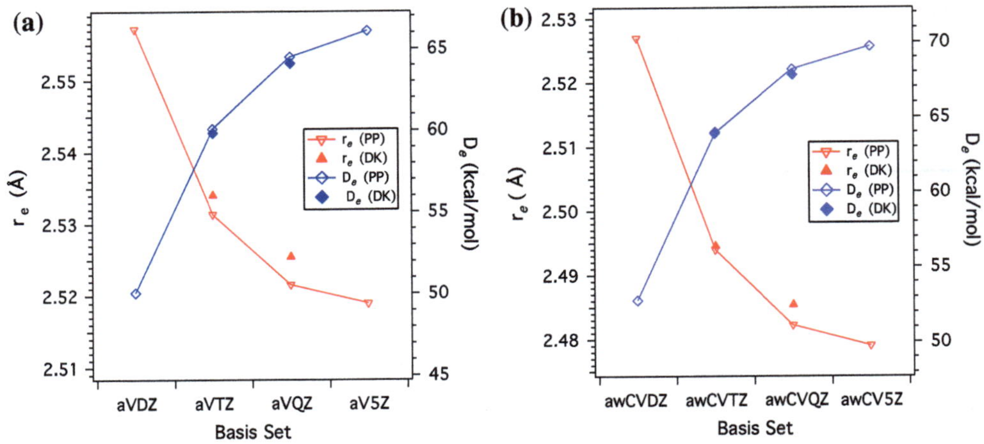

Fig. 1 Convergence of the calculated CCSD(T) equilibrium bond length(r_e) and dissociation energy (D_e) of Sb_2 as a function of aug-cc-pV*m*Z-PP/DK or aug-cc-pwCV*m*Z-PP/DK basis sets for valence and valence + (n − 1) d correlated calculations, respectively. **a** Valence correlated, **b** valence + (n − 1)d correlated

Table 4 Dissociation energies, D_e (kcal/mol), for selected homonuclear diatomics calculated using DK and differences between these and analogous PP calculations (ΔPP)

	Basis set	Hartree Fock[a]		CCSD(T), valence[a]		CCSD(T), valence + (n − 1)d[b]	
		DK	ΔPP	DK	ΔPP	DK	ΔPP
In_2	TZ	15.15	0.06	25.28	0.04	26.89	0.04
	QZ	15.28	0.09	26.01	0.07	27.64	0.05
Sb_2	TZ	−5.48	0.39	59.88	0.23	64.02	−0.09
	QZ	−4.95	0.56	64.11	0.39	67.83	0.37
I_2	TZ	22.07	−0.10	42.19	−0.13	43.86	−0.13
	QZ	22.82	−0.06	45.19	−0.16	46.48	−0.03
Tl_2	TZ	12.92	0.09	22.23	0.14	24.67	0.05
	QZ	13.09	0.13	22.89	0.11	25.45	0.11
Bi_2	TZ	−17.31	−0.06	50.17	−0.08	54.78	−0.12
	QZ	−16.64	0.10	54.50	−0.11	58.51	0.08
At_2	TZ	17.78	0.08	38.11	0.03	40.16	0.02
	QZ	18.64	0.14	41.08	−0.09	42.53	0.15

[a] aug-cc-pV*m*Z(−DK/−PP)

[b] aug-cc-pwCV*m*Z(−DK/−PP)

Table 5 Equilibrium bond lengths, r_e (Å), for selected homonuclear diatomics calculated using DK and the differences between these and analogous PP calculations (ΔPP in mÅ)

	Basis set	Hartree Fock[a]		CCSD(T), valence[a]		CCSD(T), valence + (n − 1)d[b]	
		DK	ΔPP	DK	ΔPP	DK	ΔPP
In_2	TZ	3.1167	0.39	3.0903	4.34	3.0081	−0.26
	QZ	3.1151	−1.52	3.0902	−2.09	2.9956	−0.78
Sb_2	TZ	2.4377	−2.22	2.5341	−2.64	2.4947	−0.67
	QZ	2.4350	−2.93	2.5255	−3.88	2.4854	−3.14
I_2	TZ	2.6674	1.45	2.7024	2.46	2.6718	2.23
	QZ	2.6629	1.02	2.6860	1.01	2.6575	0.60
Tl_2	TZ	3.2088	−5.93	3.2052	−5.11	3.0896	−0.71
	QZ	3.2066	−8.26	3.1988	−6.44	3.0743	−4.75
Bi_2	TZ	2.5759	2.21	2.6885	2.06	2.6412	3.36
	QZ	2.5718	1.45	2.6790	1.18	2.6302	1.06
At_2	TZ	2.8472	−1.28	2.8867	−0.44	2.8483	−0.38
	QZ	2.8411	−1.11	2.8683	−0.89	2.8336	−2.00

[a] aug-cc-pV*m*Z(−DK/−PP)

[b] aug-cc-pwCV*m*Z(−DK/−PP)

angular momenta were avoided by simply uncontracting appropriate *spd* functions from the underlying HF primitive sets instead of adding extra functions with optimal exponents. This yielded the cc-pwCV*m*Z-DK (m = T, Q) basis sets. Of course, the diffuse augmenting functions described above can also be added to these CV sets to yield aug-cc-pwCV*m*Z-DK basis sets.

3 Results

3.1 Atomic electron affinities and ionization potentials

In order to provide an initial assessment of the quality of the basis sets produced in this work, as well as to provide detailed comparisons to the PP-based sets previously developed, atomic electron affinities (EAs) and ionization potentials (IPs) were calculated at the CCSD(T) level of theory. The calculated EAs are shown in Table 2, both for HF and the CCSD(T) correlation contributions for the cases of valence electron correlation (*nsp*) and valence + (n − 1)d electron correlation. It should be noted that the PP results are identical to those of Refs. [27] and [33] except for iodine. At the HF level, there is very little change in the EAs between aug-cc-pVTZ-DK and aug-cc-pVQZ-DK with a maximum difference of just 0.04 kcal/mol. Surprisingly, the PP-based results show generally larger variations as indicated by the dependence of the values of ΔPP, EA(DKH) − EA(PP), with basis set at the HF level. These range to over 0.1 kcal/mol in several cases. This was

Table 6 Harmonic frequencies, ω_e (cm^{-1}), for selected homonuclear diatomics calculated using DK and the differences between these and analogous PP calculations (ΔPP)

	Basis set	Hartree Fock[a]		CCSD(T), valence[a]		CCSD(T), valence + $(n-1)d$[b]	
		DK	ΔPP	DK	ΔPP	DK	ΔPP
In_2	TZ	109.0	0.1	113.8	−1.0	118.6	−0.1
	QZ	109.0	0.0	113.6	0.0	119.6	−0.3
Sb_2	TZ	327.9	0.5	266.5	−0.1	275.1	−0.2
	QZ	327.6	0.5	269.2	0.6	278.8	0.2
I_2	TZ	236.2	−0.4	215.1	−2.2	217.5	−0.6
	QZ	236.6	−0.3	219.2	−0.5	223.1	0.0
Tl_2	TZ	75.6	0.7	75.8	0.9	83.0	0.0
	QZ	75.6	0.5	76.9	0.3	84.1	0.4
Bi_2	TZ	229.9	−0.2	182.1	0.1	191.1	−0.6
	QZ	229.5	−0.3	184.7	−0.3	194.3	−0.3
At_2	TZ	168.8	−0.1	152.5	−0.2	156.6	−0.1
	QZ	169.1	0.0	156.8	−0.2	160.5	0.1

[a] aug-cc-pV*m*Z(−DK/−PP)

[b] aug-cc-pwCV*m*Z(−DK/−PP)

unexpected since the two families of basis sets are very similar in their compositions, especially in the valence regions. Use of the aug-cc-pwCV*m*Z-DK basis sets at the all-electron HF level yields EAs that are essentially unchanged from the aug-cc-pV*m*Z-DK results. Small improvements, however, are observed for the PP values for m = T. In most cases, this removes the differences in HF convergence behavior between DK and PP basis sets for this property. Focusing on the CCSD(T) correlation contributions to the EAs, the DK and PP results are nearly identical when only the valence electrons are correlated, i.e., the ΔPP values lie between 0.00 and 0.03 kcal/mol. The latter do increase slightly once the 4*d* or 5*d* electrons are correlated, but even then, the maximum values are only slightly above 0.1 kcal/mol. Taken as a whole, i.e., HF + Corr., the new TZ and QZ DK basis sets show very similar basis set convergence compared with the PP sets, and in addition, the value of ΔPP has a maximum value of only about 0.2 kcal/mol for the EAs. In general, the DK and PP results also agree better with the QZ quality basis sets. On the whole, this confirms that the new DK basis sets are very comparable in quality to the original PP basis sets, but also that the underlying PPs used in the aug-cc-p(wC)V*m*Z-PP sets are very accurate, at least as compared to scalar DKH calculations.

Table 3 shows the calculated IPs for the 4*p* and 5*p* atoms. Overall, the trends are similar to the EA cases in Table 2 except that the magnitude of ΔPP is larger, reaching just under 0.5 kcal/mol at the HF level (Tl) and 0.29 kcal/mol for the correlation contribution when the outer-core *d* electrons are included. In the majority of cases, however, the differences between DK and PP treatments are between 0.0 and 0.2 kcal/mol. While the results shown in Table 3 utilized the diffuse-augmented sets, calculations of the IPs were also carried out with the un-augmented sets. While not shown here, it was found that while the DK results were very insensitive to this choice for the IPs, the PP values at the HF/TZ level were not. Specifically, the HF IPs using either the cc-pVTZ-PP or cc-pwCVTZ-PP basis sets exhibited basis set errors as large as 1 kcal/mol above those shown in Table 3 when the aug- sets were used. Further investigation indicated that a single diffuse *p* function added to the cc-pVTZ-PP (or wCV) sets eliminated this problem. This seems to be specific to the PP approximation itself and not the basis sets since the outer exponents of the DK and PP basis sets were very similar to each other. Hence, the diffuse-augmented sets have been used throughout the remainder of this work to yield more accurate ΔPP values.

3.2 Spectroscopic properties of homo- and heteronuclear diatomics

3.2.1 Valence and outer-core correlation, DKH versus PP

The pseudopotential-based correlation consistent sets (cc-pV*m*Z-PP) for the post-*d* elements have been previously shown to converge monotonically toward the CBS limit of the CCSD(T) method [27, 33]. These results also usually compared well to experiment, particularly when considering outer-core electron correlation [54, 55]. It is known, however, that in many cases, at least tight *f*-type functions may be important for molecules containing 5*p* and 6*p* elements [65]. This deficiency affects the valence sets, but of course not the core–valence basis sets since the latter contain functions with sufficiently large exponents. It is expected that the current DK sets suffer some of the same problems as the PP sets due to the similarity of their construction. In terms of the spectroscopic properties of diatomic molecules, the overall differences between the aug-cc-pV*m*Z-PP and aug-cc-pV*m*Z-DK results are much smaller than the differences between TZ and QZ, as shown in Fig. 1 for Sb_2.

Comparisons between D_e, r_e, and ω_e calculated with the new DK basis sets and those obtained in PP calculations [27, 33, 55] are shown in Tables 4, 5, and 6, respectively, for six homonuclear diatomics. For the dissociation energies and harmonic frequencies, the DK and PP results are in excellent agreement, differing by at most a few tenths of a kcal/mol or cm^{-1} for D_e and ω_e, respectively. In many cases, the differences between DK and PP (ΔPP) are well reflected at the HF level, but in a few cases, the correlation

Table 7 Dissociation energies, D_e (kcal/mol), for selected diatomics calculated using DK, the difference between these and analogous PP results (ΔPP), and the difference between these and PP results with MVD corrections included (ΔPP + MVD)

	Basis set	Hartree Fock[a]		CCSD(T), valence[a]			CCSD(T), valence + $(n-1)d$[b]		
		DK	ΔPP	DK	ΔPP	ΔPP + MVD	DK	ΔPP	ΔPP + MVD
InCl	TZ	84.65	0.41	103.06	0.28	0.07	104.02	0.42	0.22
	QZ	84.98	0.53	105.59	0.39	0.19	106.44	0.50	0.29
SnO	TZ	46.83	0.76	126.58	0.51	0.36	129.92	0.69	0.51
	QZ	48.39	1.06	130.02	0.75	0.61	132.54	0.84	0.66
SbN	TZ	−34.73	0.65	76.85	0.38	0.33	80.71	0.37	0.29
	QZ	−33.50	0.93	80.86	0.63	0.58	84.09	0.68	0.60
SiTe	TZ	62.24	0.40	103.91	0.31	0.28	106.15	0.24	0.19
	QZ	62.57	0.45	107.41	0.39	0.36	109.24	0.36	0.31
HI	TZ	56.32	−0.11	78.11	−0.16	−0.16	79.42	−0.16	−0.16
	QZ	56.72	−0.07	79.05	−0.11	−0.11	80.10	−0.06	−0.06
XeH^+	TZ	−231.89	−0.01	89.43	−0.02	−0.02	89.55	−0.04	−0.04
	QZ	−231.48	0.06	87.61	−0.03	−0.03	87.31	0.09	0.09
TlCl	TZ	79.62	0.03	98.42	−0.06	−0.28	99.93	0.08	−0.14
	QZ	79.75	0.32	100.81	0.04	−0.18	101.98	0.39	0.18
PbO	TZ	27.71	−0.05	112.17	−0.12	−0.27	116.24	0.14	−0.03
	QZ	29.57	0.43	125.81	0.10	−0.09	128.64	0.38	0.21
BiN	TZ	−54.08	−0.04	63.36	−0.06	−0.11	67.76	0.09	0.02
	QZ	−52.48	0.33	67.19	0.18	0.12	70.74	0.34	0.27
SiPo	TZ	51.78	0.05	95.02	0.11	0.09	97.88	−0.03	−0.06
	QZ	52.20	0.09	98.30	0.02	0.01	101.01	0.13	0.10
AtH	TZ	48.39	0.08	70.73	0.11	0.11	72.44	−0.01	−0.02
	QZ	48.91	0.13	71.78	−0.04	−0.04	72.94	0.07	0.07
RnH^+	TZ	−258.58	0.10	63.73	0.11	0.11	63.92	0.13	0.13
	QZ	−258.07	0.20	62.16	−0.04	−0.05	61.70	0.15	0.14

[a] aug-cc-pV*m*Z(−DK/−PP)

[b] aug-cc-pwCV*m*Z(−DK/−PP)

of the outer-core *d* electrons made a non-negligible impact. In the latter case, this tended to decrease the differences between DK and PP. The results for the equilibrium bond lengths (Table 5) somewhat contrasted these results by showing a bit more sensitivity and larger ΔPP values. The largest difference between the DK and PP r_e's is still only −4.75 mÅ, but correlation of the $(n-1)d$ electrons has a generally larger effect on this quantity. The changes in ΔPP between the TZ and QZ basis sets (reflecting different convergence rates of DK and PP calculations) are also observed to be non-negligible, but generally less than 1 mÅ at the HF level but reaching nearly 4 mÅ for Tl_2 with $(n-1)d$ correlation. In the latter case, however, the total basis set effect (QZ-TZ) on r_e is over 15 mÅ with DK.

Tables 7, 8, and 9 show the analogous results for the series of heteronuclear diatomic molecules. As discussed above, since in these cases only one of the atoms includes scalar relativistic effects via a PP, a MVD correction with uncontracted TZ basis sets has been applied to account for scalar relativistic effects on the lighter atoms in the PP calculations. The impact of the latter correction can be seen from the difference between ΔPP and ΔPP + MVD in the tables, and in general, they are nearly completely negligible for these molecules. Errors due to the PP approximation can be approximately judged by the ΔPP + MVD values. Overall, the trends are very similar to those discussed for the homonuclear diatomics. For the dissociation energies (Table 7), the DK and PP results are in close agreement, differing generally by only a few tenths of a kcal/mol, and the HF results are generally excellent indicators of the magnitude of ΔPP. Electron correlation generally tends to decrease the magnitude of the differences between the DK and PP dissociation energies. In regard to the convergence of the bond lengths (Table 8), larger differences between DK and PP are actually seen at the HF level, although these are still only 1–3 mÅ. Inclusion of electron correlation, particularly with $(n-1)d$ electron correlation, produces ΔPP + MVD values predominately under 1 mÅ. The

Table 8 Equilibrium bond lengths, r_e (kcal/mol), for selected diatomics calculated using DK, the difference between these and analogous PP results (ΔPP in mÅ), and the difference between these and PP results with MVD corrections included (ΔPP + MVD in mÅ)

	Basis set	Hartree Fock[a]		CCSD(T), valence[a]			CCSD(T), valence + $(n-1)d$[b]		
		DK	ΔPP	DK(Å)	ΔPP	ΔPP + MVD[c]	DK	ΔPP	ΔPP + MVD
InCl	TZ	2.4400	−0.42	2.4552	−0.92	−1.78	2.4171	−1.07	−1.95
	QZ	2.4359	−1.76	2.4505	−1.95	−2.81	2.4094	−1.57	−2.45
SnO	TZ	1.7955	−2.55	1.8649	−3.04	−3.04	1.8372	−3.61	−3.65
	QZ	1.7918	−3.42	1.8583	−3.93	−3.93	1.8330	−4.09	−4.13
SbN	TZ	1.7744	−1.51	1.8608	−1.71	−1.67	1.8379	−1.64	−1.65
	QZ	1.7710	−2.20	1.8537	−2.99	−2.95	1.8328	−2.52	−2.53
SiTe	TZ	2.2567	−1.58	2.3015	−1.71	−1.54	2.2866	−1.54	−1.39
	QZ	2.2542	−1.87	2.2934	−2.33	−2.16	2.2780	−1.94	−1.79
HI	TZ	1.6020	1.03	1.6185	1.25	1.19	1.6042	0.52	0.44
	QZ	1.6000	0.52	1.6173	0.58	0.52	1.6047	0.39	0.31
XeH^+	TZ	1.5939	−0.16	1.6067	−0.21	−0.28	1.5953	−0.11	−0.19
	QZ	1.5921	−0.51	1.6148	−0.47	−0.54	1.5973	−0.43	−0.51
TlCl	TZ	2.5451	0.53	2.5681	1.30	0.40	2.5129	0.23	−0.74
	QZ	2.5415	−1.88	2.5656	−0.32	−1.22	2.5034	0.64	−0.33
PbO	TZ	1.8745	2.32	1.9612	2.39	2.48	1.9256	0.90	0.88
	QZ	1.8693	0.44	1.9529	0.53	0.50	1.9214	0.21	0.19
BiN	TZ	1.8506	1.31	1.9547	1.64	1.68	1.9272	0.53	0.54
	QZ	1.8459	0.02	1.9459	−0.27	−0.23	1.9219	−0.24	−0.23
SiPo	TZ	2.3345	0.93	2.3869	0.87	1.03	2.3683	1.42	1.57
	QZ	2.3311	0.54	2.3874	0.38	0.54	2.3602	0.32	0.47
AtH	TZ	1.6886	0.11	1.7108	0.48	0.41	1.6920	−0.19	−0.28
	QZ	1.6855	−0.50	1.7082	−0.31	−0.38	1.6928	−0.74	−0.83
RnH^+	TZ	1.6795	−1.36	1.6977	−1.38	−1.46	1.6826	−0.97	−1.05
	QZ	1.6764	−1.42	1.6965	−0.94	−1.02	1.6848	−1.45	−1.54

[a] aug-cc-pVmZ(−DK/−PP)

[b] aug-cc-pwCVmZ(−DK/−PP)

results for the harmonic frequencies (Table 9) show that for most of the molecules, there is a close correspondence between the DK and PP results. Some of the larger differences occur for the hydrides, where ΔPP + MVD for ω_e change by 2–4 cm^{-1} when increasing the basis set from TZ to QZ in core-correlation calculations. In these cases, though the majority of the differences between DK and PP appear to reside in the TZ sets since at the QZ level, the values of ΔPP + MVD are only a few tenths of a cm^{-1} (with d electrons correlated).

3.2.2 *Third- versus second-order DKH*

Tsuchiya et al. [40] have suggested that for elements heavier than $Z = 81$, at least third-order DKH should be used. To test this assertion in the present molecular benchmarks, all calculations were performed using both DKH2 and DKH3. While it is recommended to recontract a primitive HF basis set when using it in a different order DKH calculation (compared to the order with which the basis set was originally contracted for), the effects of not doing so were also briefly tested. Using the DKH2 contractions in a DKH3 CCSD(T) calculation for the spectroscopic properties of At_2 led to a difference of 150 kcal/mol in the total energies, but only 0.05 kcal/mol in the dissociation energy. Other properties were also relatively insensitive, with differences of only 0.15 mÅ in r_e and 0.1 cm^{-1} in ω_e. Nevertheless, it is still recommended that contractions specific to the DKH order employed should be used (as was done throughout this work).

As seen in Table 10, for the 5p elements there is very little difference in the dissociation energies between using DKH2 or DKH3. Among the molecules containing 5p elements, the difference between the DKH2 and DKH3 D_e values was largest for XeH^+, but this only amounted to 0.07 kcal/mol. The spectroscopic properties of the 5p

Table 9 Harmonic frequencies, ω_e (cm^{-1}), for selected diatomics calculated using DK, the difference between these and analogous PP results (ΔPP), and the difference between these and PP results with MVD corrections included (ΔPP + MVD)

	Basis set	Hartree Fock[a]		CCSD(T), valence[a]			CCSD(T), valence + $(n-1)d$[b]		
		DK	ΔPP	DK	ΔPP	ΔPP + MVD	DK	ΔPP	ΔPP + MVD
InCl	TZ	306.1	−0.1	308.0	0.1	0.3	311.9	0.4	0.5
	QZ	306.4	0.3	309.4	0.3	0.5	315.3	0.2	0.3
SnO	TZ	957.8	1.4	800.2	1.7	1.3	819.7	1.6	1.4
	QZ	964.3	2.2	811.4	2.5	2.2	828.3	1.0	0.7
SbN	TZ	1,097.3	1.6	855.2	1.8	1.5	871.9	1.2	0.5
	QZ	1,100.3	2.0	866.7	2.7	2.4	883.8	1.3	0.6
SiTe	TZ	524.9	1.3	474.9	1.3	0.8	479.0	0.7	0.2
	QZ	525.3	1.1	479.2	1.4	0.8	482.9	0.9	0.3
HI	TZ	2,446.0	−0.3	2,312.8	−0.6	−0.6	2,332.7	−2.9	−2.8
	QZ	2,450.3	−0.9	2,318.5	−2.8	−2.8	2,336.4	−0.2	−0.1
XeH^+	TZ	2,396.5	−0.5	2,297.2	−0.6	−0.5	2,309.6	−3.9	−3.8
	QZ	2,399.8	−0.9	2,290.4	−2.1	−2.0	2,300.7	0.0	0.1
TlCl	TZ	265.3	0.2	271.8	0.0	0.2	277.7	0.0	0.3
	QZ	265.5	0.4	272.6	0.4	0.6	280.7	0.0	0.2
PbO	TZ	868.3	−1.9	713.1	−1.8	−2.3	733.5	−1.4	−1.7
	QZ	874.9	−0.3	723.9	−0.4	−0.3	738.3	0.5	0.2
BiN	TZ	1,002.0	−0.7	757.4	−1.1	−1.4	778.1	−1.0	−1.3
	QZ	1,005.3	−0.1	769.4	0.1	−0.2	787.5	0.6	0.3
SiPo	TZ	478.9	0.4	430.7	0.8	0.4	435.4	−0.2	−0.7
	QZ	479.1	0.2	434.4	0.4	0.0	439.4	0.6	0.1
AtH	TZ	2,278.0	3.9	2,140.4	3.5	3.5	2,164.9	−4.4	−4.3
	QZ	2,282.8	1.3	2,148.9	−2.0	−2.0	2,166.6	0.0	0.1
RnH^+	TZ	2,273.6	4.3	2,158.4	5.4	5.5	2,172.1	−1.7	−1.6
	QZ	2,277.5	2.6	2,159.3	−3.0	−3.0	2,168.6	0.1	0.1

[a] aug-cc-pVmZ(−DK/−PP)

[b] aug-cc-pwCVmZ(−DK/−PP)

Table 10 Statistics of unsigned differences between calculated CCSD(T) DKH2 and DKH3 values of D_e (kcal/mol), r_e (mÅ), and ω_e (cm^{-1})

Molecules	Valence correlated[a]						Valence + $(n-1)d$ correlated[b]					
	D_e		r_e		ω_e		D_e		r_e		ω_e	
	Max	Avg	Max	Avg	Max	Avg	Max	Avg	Max	Avg	Max	Avg
5*p*	0.06	0.02	0.31	0.16	1.12	0.19	0.07	0.02	0.40	0.14	1.47	0.35
6*p*	0.12	0.06	1.68	0.64	2.58	0.73	0.10	0.05	1.19	0.61	2.02	0.71

[a] CCSD(T) with aug-cc-pVTZ-DK for the heteronuclear diatomics and cc-pVTZ-DK for the homonuclear diatomics

[b] CCSD(T) with aug-cc-pwCVTZ-DK with the valence and $(n-1)d$ electrons on 5*p* and 6*p* elements correlated

elements were also relatively insensitive, with the mean unsigned differences (MUD) being 0.1 mÅ for r_e and 0.4 cm^{-1} for ω_e. The maximum difference was 0.4 mÅ for In_2 and 1.5 cm^{-1} for XeH^+. The 6*p* elements showed larger differences for all the properties, almost twice as large for frequencies and energetics, while the changes in equilibrium bond lengths are more notable, being 0.6 mÅ on average and a maximum of 1.2 mÅ for Tl_2. This appears to confirm that DKH2 is probably sufficiently accurate for 5*p* element-containing molecules, while for molecules containing atoms as heavy as 6*p*, DKH3 should be employed.

4 Conclusions

New all-electron correlation consistent basis sets of triple- and quadruple-zeta quality have been optimized for In–Xe and Tl–Rn using the DKH scalar relativistic Hamiltonian. The new sets (cc-pV*m*Z-DK, aug-cc-pV*m*Z-DK, and cc-pwCV*m*Z-DK with $m = $ T and Q) appear to have very similar basis set incompleteness errors as the previously developed PP-based correlation consistent sets. The new sets can be used in general, all-electron scalar relativistic calculations or in PP-based composite thermochemistry approaches to account for scalar relativistic effects on light atoms and provide corrections for the pseudopotential approximation. To perform DKH calculations for 6*p* elements, third-order DKH is recommended, and the present basis sets for these elements have been optimized for this in mind. For routine work, calibration by DK calculations of results obtained with the cc-pV*m*Z-PP-style basis sets does not seem warranted since overall the differences between DK- and PP-based calculations are found to be generally very small in the present work for these elements. If such a PP correction is desired, correlation of the $(n-1)d$ electrons should be considered with cc-pwCVQZ quality basis sets since this appears to give the best converged results for this quantity.

Acknowledgments The authors gratefully acknowledge the support of the US National Science Foundation (Grant No. CHE-0723997).

References

1. Helgaker T, Klopper W, Tew DP (2008) Mol Phys 106:2107–2143
2. Woon DE, Dunning TH Jr (1993) J Chem Phys 98:1358–1371
3. Kendall RA, Dunning TH Jr, Harrison RJ (1992) J Chem Phys 96:6796–6806
4. Dunning TH Jr (1989) J Chem Phys 90:1007–1023
5. Woon DE, Dunning TH Jr (1994) J Chem Phys 101:8877–8893
6. Feller D, Peterson KA, Hill JG (2011) J Chem Phys 135:044102
7. Bartlett RJ, Musiał M (2007) Rev Mod Phys 79:291–352
8. Jiang W, DeYonker NJ, Determan JJ, Wilson AK (2012) J Phys Chem A 116:870–885
9. Mayhall NJ, Raghavachari K, Redfern PC, Curtiss LA (2009) J Phys Chem A 113:5170–5175
10. Feller D, Peterson KA, Dixon DA (2008) J Chem Phys 129:204105
11. Li S, Hennigan JM, Dixon DA, Peterson KA (2009) J Phys Chem A 113:7861–7877
12. Bross DH, Hill JG, Werner H-J, Peterson KA (2013) J Chem Phys 139:094302
13. Tajti A, Szalay PG, Császár AG, Kállay M, Gauss J, Valeev EF, Flowers BA, Vázquez J, Stanton JF (2004) J Chem Phys 121:11599–11613
14. Karton A, Taylor PR, Martin JML (2007) J Chem Phys 127:064104
15. Peterson KA, Feller D, Dixon DA (2012) Theor Chem Acc 131:1–20
16. Császár AG, Kain JS, Polyansky OL, Zobov NF, Tennyson J (1998) Chem Phys Lett 293:317–323
17. Császár AG, Kain JS, Polyansky OL, Zobov NF, Tennyson J (1999) Chem Phys Lett 312:613–616
18. Feller D, Peterson KA (2009) J Chem Phys 131:154306
19. Ruscic B, Wagner AF, Harding LB, Asher RL, Feller D, Dixon DA, Peterson KA, Song Y, Qian X, Ng C-Y, Liu J, Chen W, Schwenke DW (2002) J Phys Chem A 106:2727
20. Iliaš M, Saue T (2007) J Chem Phys 126:4102
21. Douglas M, Kroll NM (1974) Ann Phys 82:89–155
22. Hess BA (1986) Phys Rev A 33:3742
23. Wolf A, Reiher M, Hess BA (2002) J Chem Phys 117:9215–9226
24. Cowan RD, Griffin M (1976) J Opt Soc Am 66:1010
25. Martin RL (1983) J Phys Chem 87:750
26. Metz B, Stoll H, Dolg M (2000) J Chem Phys 113:2563–2569
27. Peterson KA, Figgen D, Goll E, Stoll H, Dolg M (2003) J Chem Phys 119:11113–11123
28. Peterson KA, Puzzarini C (2005) Theor Chem Acc 114:283–296
29. Peterson KA, Figgen D, Dolg M, Stoll H (2007) J Chem Phys 126:124101
30. Figgen D, Peterson KA, Dolg M, Stoll H (2009) J Chem Phys 130:164108
31. Dolg M, Cao X (2012) Chem Rev 112:403–480
32. Schwerdtfeger P (2011) Chem Phys Chem 12:3143–3155
33. Peterson KA (2003) J Chem Phys 119:11099–11112
34. Schwerdtfeger P, Assadollahzadeh B, Rohrmann U, Schäfer R, Cheeseman JR (2011) J Chem Phys 134:204102
35. Mosyagin NS, Titov AV (2005) J Chem Phys 122:234106
36. Spohn B, Goll E, Stoll H, Figgen D, Peterson KA (2009) J Phys Chem A 113:12478
37. Jansen G, Hess BA (1989) Phys Rev A 39:6016
38. Metz B, Schweizer M, Stoll H, Dolg M, Liu WJ (2000) Theor Chem Acc 104:22–28
39. Roos BO, Lindh R, Malmqvist PA, Veryazov V, Widmark P-O (2004) J Phys Chem A 108:2851–2858
40. Tsuchiya T, Abe M, Nakajima T, Hirao K (2001) J Chem Phys 115:4463–4472
41. Noro T, Sekiya M, Koga T (2013) Theor Chem Acc 132:1363
42. Noro T, Sekiya M, Koga T (2012) Theor Chem Acc 131:1124
43. Ceolin GA, Berrêdo RC, Jorge FE (2013) Theor Chem Acc 132:1339
44. Campos CT, Jorge FE (2013) Mol Phys 111:167–173
45. Barros CL, de Oliveira PJP, Jorge FE, Neto AC, Campos M (2010) Mol Phys 108:1965–1972
46. Dyall KG (2006) Theor Chem Acc 115:441
47. Werner HJ, Knowles PJ, Lindh R, Manby FR, Schütz M et al. MOLPRO, version 2012.1, a package of ab initio programs. http://www.molpro.net
48. Raghavachari K, Trucks GW, Pople JA, Head-Gordon M (1989) Chem Phys Lett 157:479–483
49. Hampel C, Peterson KA, Werner H-J (1992) Chem Phys Lett 190:1–12
50. Scuseria GE (1991) Chem Phys Lett 176:27–35
51. Knowles PJ, Hampel C, Werner H-J (1993) J Chem Phys 99:5219–5227
52. Watts JD, Gauss J, Bartlett RJ (1993) J Chem Phys 98:8718–8733
53. Knowles PJ, Hampel C, Werner H-J (2000) J Chem Phys 112:3106–3107
54. Peterson KA, Shepler BC, Figgen D, Stoll H (2006) J Phys Chem A 110:13877–13883
55. Peterson KA, Yousaf KE (2010) J Chem Phys 133:174116
56. Dunning TH Jr, Peterson KA, Wilson AK (2001) J Chem Phys 114:9244–9253
57. Feller D, Peterson KA, de Jong WA, Dixon DA (2003) J Chem Phys 118:3510
58. Dunham JL (1932) Phys Rev 41:721

59. Pipek J, Mezey PG (1989) J Chem Phys 90:4916–4926
60. Petersson GA, Zhong SJ, Montgomery JA, Frisch MJ (2003) J Chem Phys 118:1101–1109
61. Nakajima T, Hirao K (2002) J Chem Phys 116:8270
62. Blaudeau J-P, Brozell SR, Matsika S, Zhang Z, Pitzer RM (2000) Int J Quantum Chem 77:516–520
63. Christiansen PA (2000) J Chem Phys 112:10070–10074
64. Peterson KA, Dunning TH Jr (2002) J Chem Phys 117:10548–10560
65. Weigend F, Ahlrichs R (2005) Phys Chem Chem Phys 7:3297

Theor Chem Acc (2014) 133:1407
DOI 10.1007/s00214-013-1407-z

REGULAR ARTICLE

Improved accuracy benchmarks of small molecules using correlation consistent basis sets

David Feller · Kirk A. Peterson · Branko Ruscic

Received: 3 September 2013 / Accepted: 8 October 2013 / Published online: 22 November 2013

Abstract Improved accuracy benchmark atomization energies, equilibrium structures, and harmonic frequencies were obtained from the composite Feller–Peterson–Dixon procedure applied at the highest possible level permitted by our current hardware and software. Convergence of the 1-particle expansion was achieved through use of correlation consistent basis sets as large as aug-cc-pV8Z and aug-cc-pV9Z, followed by the application of a simple extrapolation formula in order to more closely approximate the basis set limit. Convergence of the *n*-particle expansion was addressed with a systematic sequence of coupled cluster methods up through CCSDTQ5. In 10 cases, coupled cluster theory was augmented with full configuration interaction. Each of the multiple sources of error was carefully monitored in order to minimize the overall uncertainty to the extent possible. Comparison with high-quality experimental values, many of them obtained from the active thermochemical tables, reveals overall close agreement with theory.

Keywords Ab initio · Basis sets · Coupled cluster · Composite methods · Benchmarks

Dedicated to Professor Thom Dunning and published as part of the special collection of articles celebrating his career upon his retirement.

Electronic supplementary material The online version of this article (doi:10.1007/s00214-013-1407-z) contains supplementary material, which is available to authorized users.

D. Feller (✉) · K. A. Peterson
Department of Chemistry, Washington State University, Pullman, WA 99164-4630, USA
e-mail: dfeller@owt.com

B. Ruscic
Chemical Sciences and Engineering Division, Argonne National Laboratory, Argonne, IL 60439, USA

B. Ruscic
Computation Institute, University of Chicago, Chicago, IL 60637, USA

1 Introduction

A "benchmark" is, by definition, a standard for facilitating the comparison of related quantities. In the field of quantum chemistry, benchmarks typically serve to either establish a point of reference whereby less rigorous approaches can be calibrated or illustrate the capabilities of a newly developed technique. Due to the rapid introduction of improvements in hardware and software, time tends to quickly erode the practical value of computational benchmarks. Therefore, it is important that they be updated on a regular basis in order to accurately reflect the evolving state of the art. Quantum chemistry has a long tradition of benchmark studies stretching all the way back to its inception in the 1950s. Their number is too large to attempt a comprehensive listing here, but their study provides an interesting historical perspective on how this branch of science has progressed. The goal of this work is to create a compilation of benchmark data, involving atomization energies (heats of formation), equilibrium structures, and vibrational frequencies for a collection of small molecules. The current approach follows the general strategy outlined in the Feller–Peterson–Dixon (FPD) procedure [1–3]. FPD utilizes multiple levels of coupled cluster theory combined with large correlation consistent basis sets. A second, companion article is devoted to a detailed description of corresponding dissociation energies based on the active thermochemical tables (ATcT) [4].

Of the numerous computational techniques provided by popular electronic structure codes, such as Gaussian [5], GAMESS [6], or MOLPRO [7], relatively few are capable of achieving consistent accuracy across major portions of the Periodic Table for a wide range of properties. A distinction is drawn here between (1) computationally inexpensive approaches which rely on a significant, fortuitous cancelation of errors in order to achieve agreement with experiment (often for a limited collections of molecules) and (2) more expensive approaches that may benefit from some cancelation of error, but do not depend upon it for attaining accurate results. Among the latter are configuration interaction (CI), coupled cluster theory, and quantum Monte Carlo. While each has its own strengths and weaknesses, coupled cluster theory is currently the most widely used method in high-accuracy studies of "well-behaved" chemical systems, i.e., those not suffering from excessive multi-configurational character in their wave functions. The single-reference coupled cluster technique which involves single and double excitations, combined with a noniterative, quasiperturbative estimate of the effect of triple excitations, CCSD(T) [8–11], has been shown to be powerful enough to recover a large fraction of the electron correlation energy [12]. It can even be applied to molecules whose wave functions exhibit moderate multi-configurational character, such as C_2, O_3, and FOO [13, 14]. Unfortunately, the method's computational expense, which scales as $\sim n^2N^4N_{it}$ (with a single n^3N^4 step), is daunting and it requires large 1-particle basis sets to achieve good results. Here, n, N, and N_{it} are the number of occupied and unoccupied molecular orbitals and the number of CCSD iterations, respectively. In practice, CCSD(T) calculations can take hours, days or even weeks, on high-speed computers. A recently developed, explicitly correlated variant, known as CCSD(T)-F12b [15–17], provides more rapid convergence with respect to the basis set, but the current implementation may not be suitable for studies seeking the very highest level of accuracy [18]. Higher levels of coupled cluster theory, which are required for very highest accuracy studies, incur even greater computational expense. These scale as $n^3N^5N_{it}$, $n^4N^6N_{it}$ or even steeper.

2 Approach

2.1 CCSD(T) basis set limit

The FPD method consists of a sequence of (up to) 13 steps, with the dominant contribution to a property arising from the frozen core (FC) CCSD(T) result. An extrapolation is performed in order to better estimate the complete basis set (CBS) limit for the energy. The remaining components are considerably smaller and it is assumed that their contributions are additive. Because they are of varying sign, some cancelation of error is likely. Careful monitoring of the errors arising from each component can help reduce the overall uncertainty in the final answer to a level sufficient to meet the needs of a wide range of studies with different target accuracies. The FPD approach further assumes that when CCSD(T) is carried out with extended basis sets, it is capable of describing most properties of interest sufficiently well to serve as an effective foundation for even higher-order methods. Nevertheless, as a single-reference method, CCSD(T) has its limitations. In terms of the additivity assumption, the FPD method is similar to many of its multi-component contemporaries, such as Gaussian-n [19–21], Wn [22, 23], HEATx [24, 25], ccCA [26–28], Focal Point Analysis [29–31], and Petersson-style Complete Basis Set [32–34] approaches. Statistical analyses performed with the Computational Results Database (CRDB) [35] show that the FPD method is capable of a root mean squared (RMS) deviation of slightly under 0.2 kcal/mol with respect to high-quality experimentally derived atomization energies. The mean absolute deviation (MAD) is roughly half that size. The CRDB currently holds more than 122,000 entries on 434 chemical systems.

Resorting to composite approaches is a practical necessity due to the factorial growth in the size of full CI (FCI) calculations, which would need to be carried out with large basis sets, a relativistic Hamiltonian and involve the correlation of all electrons. FCI represents the exact solution of the electronic Schrödinger equation for any given basis set, but its use remains severely restricted despite tremendous advances in hardware. That is likely to remain true for the foreseeable future.

To illustrate the magnitude of the FCI scaling problem, consider a sequence of calculations on first-row homonuclear diatomics with the small cc-pVDZ basis set. In the following discussion, we will assume the use of the FC approximation in which the $1s^2$ inner shell electrons were excluded from the correlation treatment. For C_2 ($^1\Sigma_g^+$) and N_2 ($^1\Sigma_g^+$), a FCI calculation requires 2.8×10^7 and 5.4×10^8 determinants, respectively. Leininger et al. [36] reported FCI spectroscopic constants at this level as early as 2000. At the time, these calculations represented an imposing computational challenge. Today it is possible to carry out the same calculation on N_2 in a little over 2 h per point on a personal computer. Nonetheless, to the best of our knowledge, after more than a decade, there has been no report in the literature of a similar explicit FCI/cc-pVDZ calculation on O_2 ($^3\Sigma_g^-$) with 5.4×10^{10} determinants, much less F_2 with over 10^{11} determinants. Methods for estimating the FCI energy will be discussed in a subsequent section.

It can be helpful to think about the challenge of obtaining high-accuracy results in terms of the 1-particle and *n*-particle expansions. The former is associated with the quality of the (typically Gaussian) basis set and the latter with the degree of correlation recovery. Demonstrated convergence along each axis is necessary to credibly claim high accuracy. Our goal is more ambitious than either FCI with small basis sets or low-level perturbation theory at the basis set limit. Of the two expansions, deficiencies in the 1-particle basis set normally generate the majority of the error (relative to FCI/CBS) with CCSD(T) calculations, as has been shown in numerous studies [3, 12]. Prior to the introduction of the correlation consistent basis set family of Dunning, Peterson and coworkers [18, 37–56], there were very few efficient tools for systematically addressing the high-accuracy requirements for most elements. Correlation consistent basis sets are available for most of the elements from H through Rn. The quality of the basis sets ranges from 2ζ to as high as 9ζ and they support valence, core/valence (CV), pseudopotential and Douglas-Kroll-Hess [57, 58] scalar relativistic calculations. For neon, a 10ζ basis set has been reported [51], and correlation consistent basis sets have even been developed for pseudopotential-based quantum Monte Carlo [59].

The diffuse function enhanced aug-cc-pV*n*Z, (n = D, T, Q, 5,..., 9) basis sets, and their second row aug-cc-pV(n + d)Z counterparts will be used for all valence CCSD(T) calculations. For the sake of brevity, we will denote these basis sets as aV*n*Z. The aV9Z basis set corresponds to a (23 *s*, 17*p*, 9*d*, 8*f*, 7*g*, 6*h*, 5*i*, 4*k*, 3*l*, 2*m*) → [11*s*, 10*p*, 9*d*, 8*f*, 7*g*, 6*h*, 5*i*, 4*k*, 3*l*, 2*m*] contraction. The new sets are available in Supporting Information. Although some molecular properties converge rapidly with respect to the basis set, such as first-order electrical properties (e.g. dipole and quadrupole moments), others require much greater effort, such as thermochemical properties. Recent studies on more than 180 small-to-medium size molecules demonstrated that basis sets of aV7Z quality or better are required if "chemical accuracy" (±1 kcal/mol) is to be *uniformly* achieved in the FC part of the problem using only the "raw" energies. The term "raw" implies that no secondary extrapolation procedure has been applied.

An unintentional benefit accrued from the correlation consistent basis sets' uniform convergence to the CBS limit in the energy is the practical possibility of accelerating that convergence through the use of simple extrapolation schemes. Through extensive testing, it was possible to measure the performance of six extrapolation techniques chosen from the literature. A seventh approach was also examined, although not as extensively. Even the poorest performing extrapolation provided an improvement of one basis set index or better for problematic atomization energies, i.e. CBS(aV*n*Z) ≈ raw [aV(n + 1)Z] [52, 60]. By using the collection of reference atomization energies as a training set, it proved possible to achieve an increment of 3 in the basis set index, i.e. CBS(aV*n*Z) ≈ raw [aV(n + 3)Z]. This accomplishment represents an enormous savings in computer time [52]. However, the formula that produced the best results is not available for the present study because of the very large sets (aV8Z and aV9Z) that will be employed. Instead, the CBS limit will be estimated by an expression proposed by Martin [61] that involves the maximum angular momentum present in the basis set (ℓ_{max}):

$$E(\ell_{max}) = E_{CBS} + A/(\ell_{max} + 1/2)^4 \qquad (1)$$

This formula built upon earlier work by Schwartz [62, 63], Carroll et al. [64], Hill [65], and Kutzelnigg and Morgan [66]. Among the older extrapolation formulas, Eq. (1) and a multi-formula average proved to be statistically superior to other formulas with RMS deviations of ~0.30 kcal/mol compared to values ranging between 0.36 and 0.74 kcal/mol (592 comparisons) [52]. Equation (1) leads to slightly larger atomization energies than the multi-formula average, which we have tended to use in previous studies. The spread among the various formulas will be adopted as a crude estimate of the uncertainty in our CBS extrapolations. We applied Eq. (1) to the total energy. With basis sets as large as the ones used in this study, the Hartree–Fock component is already converged.

The primary goal of the FPD method is accurate molecular properties. Secondarily we want to accomplish the primary goal in a reasonably efficient manner so that the method can be applied to chemical systems with a range of sizes. The secondary goal has practical consequences in terms of the composition of the multi-step, additive procedure. Ideally, the investment in computer resources should roughly match the target accuracy without undue wasted effort. In particular, it is inefficient to reduce the uncertainty in any component to a level that is much smaller than the least accurate components. In order to accomplish this in practice, it is necessary to exploit the different convergence rates of different components. As a result, we have traditionally separated the slowly convergent valence correlation problem from the more rapidly convergent core/valence (CV) and scalar relativistic (SR) *corrections*. For first-row compounds, the largest of the numerous corrections to the CCSD(T)(FC) energy typically arises from CV effects. This observation has led some researchers to insist that separate treatments of core and valence correlation should be abandoned [25]. A very recent paper on the dissociation of F_2 by one of the original developers of the HEAT model chemistry reported calculations with CV basis sets of 7ζ and 8ζ quality [67]. The largest of these included 574 contracted functions per atom. While it might be possible to perform CV8Z calculations on a diatomic molecule, similar calculations on even slightly larger systems would quickly become prohibitive without a commensurate improvement in overall accuracy.

Our previous examination of this issue indicated that the error resulting from the FC + CV additivity assumption was small compared to other sources of error [14]. In the case of second- and third-row compounds, the SR correction can easily surpass the CV correction in magnitude. For example, in the case of SiF_4, the two corrections for atomization energies are as follows: $\Delta CV = 0.9$ and $\Delta SR = -1.9$ kcal/mol. In the case of AsH, $\Delta CV = 0.8$ and $\Delta SR = -1.1$ kcal/mol. Thus, if magnitude alone was the determining factor for what corrections cannot safely be treated in separate calculations, it appears that neither the CV or SR corrections would qualify. To further gauge the impact of the additivity assumption, we performed CCSD(T) calculations on a collection of first-row molecules that simultaneously included both CV and SR contributions. The CV contribution was obtained with wCV5Z basis sets and the scalar relativistic contribution was modeled with the second-order Douglass–Kroll–Hess (DKH) method [57, 58]. Results for just the homonuclear diatomic spectroscopic properties are shown in Table 1, where multi-component values of D_e, r_e, and ω_e are compared to combined CCSD(T)(CV)-DK/aug-cc-pwCV5Z-DK calculations. It is evident that the magnitudes of the errors are small in an absolute sense. They are also small relative to the residual basis set truncation error, as indicated by the "Best FC + CV + SR" CCSD(T) composite entries. Mean absolute deviations (MADs) for the additivity approximation across all molecules considered in this study are as follows: $D_e = 0.07$ kcal/mol, $r_e = 0.0002$ Å, and $\omega_e = 0.4$ cm^{-1}. For comparison purpose, the MADs of the CCSD(T)(CV)-DK/aug-cc-pwCV5Z-DK results with respect to the Best FC + CV + SR are: $D_e = 0.93$ kcal/mol, $r_e = 0.0006$ Å and $\omega_e = 1.6$ cm^{-1}. On the basis of these and our previous findings, we conclude that in most systems, the error arising from a multi-component (FC + CV + SR) decomposition of the CCSD(T) contribution to a property is acceptably small until the desired uncertainty in the final answers approach an accuracy of ±0.1 kcal/mol, ±0.0003 Å and 0.5 cm^{-1}. In this study, we will include additivity corrections in our best estimates. Very similar numerical findings to those shown in Table 1 can be obtained with the smaller aug-cc-pwCVQZ-DK basis set.

Most of the calculations performed for this study were carried out at the optimal geometries associated with every level of theory/basis set combination. The only exceptions were calculations with the aV9Z basis sets, where bond lengths were estimated from an exponential extrapolation of aV6Z, aV7Z, and aV8Z bond lengths, or levels of theory where a single energy evaluation required multiple days. The latter situation was encountered with certain large basis set, higher-level coupled-cluster calculations, such as CCSDTQ. All CCSD(T) calculations using basis sets up through 6ζ were performed with MOLPRO 2012.1 [7].

Table 1 Comparison of CCSD(T) spectroscopic constants from multi-component and FC + CV + SR combined CCSD(T) calculations

Component	D_e	r_e	ω_e
B_2 ($^3\Sigma_g^-$)			
aV5Z	64.90	1.59276	1048.87
ΔCV(wCV5Z)	0.75	−0.00638	8.59
ΔSR(V5Z-DK)	−0.06	−0.00010	−0.58
Component total	65.59	1.58628	1,056.88
awCV5Z-DK	65.61	1.58604	1,057.81
Error	−0.02	0.00024	−0.93
Best FC + CV + SR[b]	65.9	1.5857	1,057.9
C_2 ($^1\Sigma_g^+$)			
aV5Z	144.45	1.24482	1,858.43
ΔCV(wCV5Z)	0.98	−0.00325	10.63
ΔSR(V5Z-DK)	−0.17	−0.00015	−0.49
Component total	145.26	1.24142	1,868.57
awCV5Z-DK	145.31	1.24123	1,869.22
Error	−0.05	0.00019	−0.65
Best FC + CV + SR[b]	146.2	1.2407	1,870.6
N_2 ($^1\Sigma_g^+$)			
aV5Z	225.59	1.09953	2,359.32
ΔCV(wCV5Z)	0.78	−0.00212	9.89
ΔSR(V5Z-DK)	−0.13	−0.00018	−0.86
Component total	226.24	1.09723	2,368.35
awCV5Z-DK	226.40	1.09706	2,368.96
Error	−0.16	0.00017	−0.61
Best FC + CV + SR[b]	227.9	1.0967	2,370.6
O_2 ($^3\Sigma_g^-$)			
aV5Z	119.07	1.20706	1,601.16
ΔCV(wCV5Z)	0.23	−0.00200	5.82
ΔSR(V5Z-DK)	−0.18	−0.00022	−2.24
Component total	119.12	1.20484	1,604.74
awCV5Z-DK	119.14	1.20511	1604.92
Error	−0.02	−0.00027	−0.18
Best FC + CV + SR[b]	120.0	1.2045	1,607.6
F_2 ($^1\Sigma_g^+$)			
aV5Z	38.08	1.41104	926.95
ΔCV(wCV5Z)	−0.10	−0.00142	1.52
ΔSR(V5Z-DK)	−0.03	−0.00034	−0.44
Component total	37.95	1.40928	928.03
awCV5Z-DK	37.93	1.40989	928.03
Error	0.02	−0.00061	0.00
Best FC + CV + SR[b]	38.4	1.4089	930.2

[a] Dissociation energies in kcal/mol, bond lengths in Å and harmonic frequencies in cm^{-1}. All corrections were computed using optimized bond lengths for every level of theory

[b] Best CCSD(T) component estimates based on: D_e = CCSD(T)(FC)/CBS(aV789Z) + CCSD(T)(CV)/CBS(wCVTQ5)Z + CCSD(T)(FC)-DKH/cc-pV5Z-DK

Calculations involving $k-(\ell_{max}=7)$, $l-(\ell_{max}=8)$, and m-functions ($\ell_{max}=9$) required either Dalton 2.0 [68] for closed shell systems or Gaussian 09 Rev. D.01 [5]. The integral accuracy in Gaussian 09 was increased to 10^{-13} with the Int = (Acc2E = 13) command. Two other commands were used with Gaussian. The first (nobasistransform) instructed the program to disable the automatic transformation of generally contracted basis sets into segmented contractions. The second command tightened the threshold for the elimination of functions associated with small eigenvalues of the overlap matrix. The default in G09 (10^{-6}) was reset to 10^{-7} with IOp(3/59 = 7).

Open-shell calculations were based on the R/UCCSD(T) method, which begins with restricted open-shell Hartree–Fock (ROHF) orbitals, but allows a small amount of spin contamination in the solution of the CCSD equations [69, 70]. Atomic symmetry in the orbitals was imposed in calculations on the atomic asymptotes when computing atomization energies. Because Gaussian only supports UCCSD(T) and what it labels ROHF-CCSD(T), it was necessary to estimate the R/U energies using the scheme described in earlier work [52]. The approach assumes that the difference between UCCSD(T) and R/UCCSD(T) energies for basis sets beyond 6ζ can be determined to an accuracy of $\sim 10^{-6}$ E_h using results up through 6ζ. Spectroscopic properties and multi-component corrections for all of the diatomic species were obtained from a sixth degree Dunham fit of the potential energy curve [71]. Due to differences in the way various codes generate semi-canonical orbitals, the total energies differ by small amounts.

The CCSD(T)(CV) correction was determined with cc-pwCVnZ (n = Q, 5) basis sets and included a CBS extrapolation step. For C_2, the raw cc-pwCV5Z spectroscopic properties were within 0.02 kcal/mol (D_e), 4×10^{-5} Å (r_e) and 0.06 cm^{-1} of the estimated CBS limit. Consequently, there appears to be little reason to push the CV calculations to basis sets beyond 5ζ or 6ζ.

The scalar relativistic correction was obtained from second-order CCSD(T)(FC)-DKH calculations with the cc-pV5Z-DK basis sets. For the systems studied here, a VTZ-DK basis set would have sufficed for most purposes, but the V5Z-DK was adopted for the sake of exhausting the 1-particle contribution to this property. For C_2, the smaller VTZ-DK basis set produces corrections that are within 0.01 kcal/mol (D_e), 1×10^{-5} Å (r_e), and 0.05 cm^{-1} (ω_e) of the much larger cc-pV5Z-DK set.

Atomic spin–orbit (SO) coupling effects shift the energies of the isolated atoms relative to the spin multiplet average values obtained from standard CCSD(T). It is necessary to include the atomic SO contribution for accurate atomization energies. For this purpose, we have chosen to use the tabulated values of Moore [72]. If molecular SO effects are present, they are obtained either from experimental values (if available) or from configuration interaction calculations [73–75]. The only case of a molecular SO correction in the current set of molecules is NO ($^2\Pi$) where a value of 0.17 kcal/mol was used [76]. It is possible to compute the atomic SO correction directly, but there seems little reason to do so purely from an accuracy perspective. As an illustration, in the case of F_2, older Dirac-Coulomb and Gaunt correction calculations by Visscher and Dyall [77] found a Hartree–Fock value of −0.75 kcal/mol, in close agreement with the very recent correlated work of Csontos et al. [67] at −0.77 kcal/mol. The latter value included an estimate of the quantum electrodynamics (QED) contribution. Both of these values are in very close agreement with the value reported in C. Moore tables (−0.77 kcal/mol).

The commonly employed Born–Oppenheimer approximation assumes separation of electronic and nuclear motions. It can contribute a potentially significant error in the case of molecular properties for systems composed of light elements, especially hydrogen. In the present work, a first-order, adiabatic correction, known as the diagonal Born–Oppenheimer correction (DBOC), will be obtained from CCSD(FC)/aVTZ calculations performed with the CFOUR program [78] using CCSD(T)(FC)/aVTZ geometries. The open-shell systems are based on unrestricted Hartree–Fock wave functions and UCCSD.

Vibrational zero point energies (ZPEs) are required for computing heats of formation at 0 and 298 K, $\Delta H_{f,0K}$ and $\Delta H_{f,298K}$, respectively. The Dunham analysis provides both harmonic frequencies ω_e and anharmonic constants $\omega_e x_e$ for diatomics. Anharmonic vibrational zero point energies for polyatomic molecules are based on the expression:

$$\mathrm{ZPE} = 0.5(\mathrm{ZPE_H} + \mathrm{ZPE_F}) + \chi_0 - 0.25(\Sigma\chi_{ii}) \quad (2)$$

where $\mathrm{ZPE_H} = 0.5(\Sigma\omega_i)$, $\mathrm{ZPE_F} = 0.5(\Sigma\nu_i)$, ω_i and ν_i are the harmonic frequencies and the fundamentals and χ_0 and χ_{ii} are given by Barone [79]. The anharmonic corrections are insensitive to the level of theory and can be accurately modeled with frozen core, second-order Møller-Plesset perturbation theory, MP2(FC). In Gaussian, they are based on finite difference evaluations of third and semidiagonal fourth derivatives. The harmonic frequencies for the triatomic molecules were obtained from numerical differentiation of CCSD(T) energies because MOLPRO does not support analytical derivatives for that level of theory. For closed shell systems, such as H_2O, we compared harmonic frequencies produced analytically and via finite differences. The average difference was $\sim$0.3 cm^{-1} with the aVTZ basis set. The difference in the zero point energies was <0.01 kcal/mol. Additionally, we know of no codes that are capable of analytical second derivatives for R/UCCSD(T).

Molecular heats of formation also require knowledge of accurate experimental (or in some cases theoretically derived) atomic heats of formation. The values used here are taken from the following sources: $H = 51.63337 \pm 0.00002$ ATcT 2013; [4] $B = 135.1 \pm 0.2$ kcal/mol, Karton and Martin [80]; $C = 170.027 \pm 0.013$ kcal/mol, ATcT 20 [4, 13] $N = 112.470 \pm 0.005$ kcal/mol, ATcT 20 [4, 13] $O = 58.9971 \pm 0.0005$ kcal/mol, ATcT 20 [4, 13], $F = 18.472 \pm 0.013$ ATcT 20 [4, 13], $P = 75.42 \pm 0.24$ NIST-JANAF [81], and $S = 66.26 \pm 0.07$ Nagy et al. [82] kcal/mol. The procedure for obtaining $\Delta H_f(0\ K)$ and $\Delta H_f(298\ K)$ from the atomization energy has been discussed by Curtiss et al. [83].

2.2 Higher-order contributions

As mentioned above, CCSD(T) has proven to be remarkably accurate. Some early users of the method likened it to FCI [19, 61, 84–86]. However, experience gained over years of use has shown that CCSD(T) alone is incapable of reproducing FCI results to chemical accuracy. Higher-order (HO) correlation recovery can easily contribute more than 1 kcal/mol to atomization energies even in small, first-row molecules. In the present study, HO corrections will be obtained from coupled cluster theory beyond the level of CCSD(T), including CCSDT, CCSDTQ, CCSDTQ5, as well as explicit and estimated FCI. The coupled cluster calculations were performed with the MRCC program of Kállay and co-workers [87, 88] interfaced to MOLPRO. In the limited number of cases where it proved possible, FCI calculations were run with the FCI code in MOLPRO. Atomic symmetry was not imposed in the higher-order coupled-cluster and FCI calculations.

The computational cost of these methods will impose a practical limit on the size of the basis sets that can be used. For diatomics, the largest basis sets will typically be V5Z (CCSDT), VQZ (CCSDTQ), and VDZ (CCSDTQ5 or FCI), unless otherwise noted. Fortunately, each increase in excitation level is accompanied by a sharp, corresponding decrease in the basis set index, n, required to reach a fixed level of convergence [12]. In many cases, the CCSD(T) → CCSDT and CCSDT → CCSDTQ corrections are of opposite sign and of roughly equal magnitude, as seen in Fig. 1 where the relevant data are shown for C_2. Experience has shown that combining CCSDT/V($n + 1$)Z with CCSDTQ/VnZ corrections leads to reasonably balanced ΔCCSDT + ΔCCSDTQ totals.

In order to estimate the remaining difference between CCSDTQ (or CCSDTQ5) and the FCI limit in cases where explicit FCI was impossible, we have previously used a continued fraction (cf) approximant originally suggested by Goodson with Hartree–Fock (HF), CCSD, and CCSD(T) energies [89]. Testing against FCI atomization energies revealed that this formulation lacked sufficient accuracy for our purposes [90]. In many

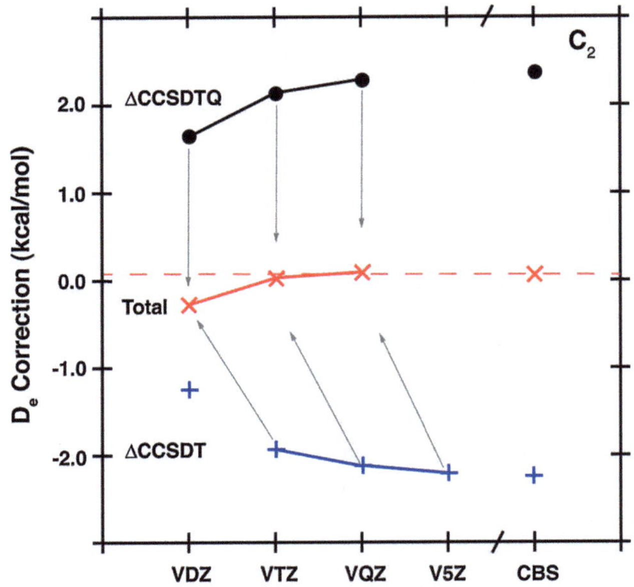

Fig. 1 CCSDT(FC) and CCSDTQ(FC) higher-order corrections (kcal/mol) to the dissociation energy of C_2 ($^1\Sigma_g^+$). The "Total" data points are based on ΔCCSDT/V($n + 1$)Z + ΔCCSDTQ/VnZ

instances, it failed to improve upon the raw CCSD(T) values. However, when the original 3-energy sequence was replaced by CCSD, CCSDT, and CCSDTQ, much better results were found. In 57 out of 61 cases, use of the approximant improved the CCSDTQ value. The RMS deviation was 0.06 kcal/mol. Here we will employ the CCS(DTQ) or CCSD(TQ5) sequences for the continued fraction approximant to FCI. While the method is not perfect, with careful use it can provide a useful adjunct to CCSDTQ calculations.

Bytautas and Rudenberg have proposed a method they call correlation energy extrapolation by intrinsic scaling (CEEIS) to approximate the FCI energy while requiring only a fraction of the number of determinants [91–93]. The authors reported results for first-row homonuclear diatomics with VDZ, VTZ, and VQZ basis sets. An uncertainty of ~0.3 mE_h (0.19 kcal/mol) is quoted in the Conclusion section of their 2005 paper, but it is not clear whether that value just refers to the VQZ results or to the smaller basis set results, as well [93]. The same value is quoted in a follow-up paper focused on the potential energy curve of F_2 [94]. If both the molecule and the isolated atoms were in error by 0.3 mE_h, the overall error might be too large to render CEEIS helpful for high-accuracy studies. Comparisons with coupled cluster methods were presented in the 2007 paper [94], but they were limited to CCSDT or lower levels of CC theory. The same was true of a 2010 paper on the potential energy of O_2 [95].

In order to further compare CEEIS and higher levels of coupled cluster theory for dissociation energies, we carried

Table 2 Comparison of N_2, O_2 and F_2 dissociation energies (kcal/mol) from coupled cluster, continued fraction estimated FCI and CEEIS calculations using the cc-pVDZ basis set

Method	Energy	D_e	# Det.'s
N_2 ($^1\Sigma_g^+$)			
CCSDT	−109.275147	199.630	60,842
CCSDTQ	−109.276751	200.584	969,718
CCSDTQ5	−109.276958	200.714	8,746,130
cf DTQ est. FCI	−109.276982	200.728	540,924,024
cf TQ5 est. FCI	−109.276979	200.726	
FCI	−109.276979	200.726	
CEEIS est. FCI[a]	−109.27698	200.539	
O_2 ($^3\Sigma_g^-$)			
CCSDT(FC)	−149.985713	104.007	90,744
CCSDTQ(FC)	−149.987581	105.081	1,721,238
CCSDTQ5(FC)	−149.987772	105.201	18,907,900
CCSDTQ56(FC)	−149.987794	105.211	127,907,385
cf DTQ est. FCI	−149.987968	105.314	5,408,656,512
cf TQ5 est. FCI	−149.987793	105.214	
cf Q56 est. FCI	−149.987797	105.213	
CEEIS est. FCI[a]	−149.98787	105.265	
F_2 ($^1\Sigma_g^+$)			
CCSDT(FC)	−199.097753	26.686	131,274
CCSDTQ(FC)	−199.099291	27.506	2,905,969
CCSDTQ5(FC)	−199.099367	27.553	37,927,843
cf DTQ est. FCI	−199.099597	27.679	54,087,833,504
cf TQ5 est. FCI	−199.099371	27.556	
CEEIS est. FCI[a]	−199.09935	27.642	
CEEIS est. FCI[b]	−199.09920	27.454	

Calculations were performed at the experimental bond lengths of Huber and Herzberg: $r_e(N_2) = 1.097685$Å, $r_e(O_2) = 1.20752$ Å, $r_e(F_2) = 1.41193$ Å [76]

[a] Energy from Table IX of 2005 paper by Bytautas and Rudenberg [93] Atomic energies are: −54.47870 (*N*), −74.91006 (*O*), −99.52765(*F*)

[b] Energy from Table X of the 2007 paper by Bytautas et al. [94]. The atomic asymptotes were computed at $r = 8.0$ Å

out: (1) explicit FCI on N_2, (2) up through CCSDTQ56 on O_2, and (3) up through CCSDTQ5 on F_2 with the small VDZ basis set. This basis was chosen because it was assumed that the uncertainty in CEEIS would increase as the basis set grew in size. Thus, VDZ is expected to provide the highest level of accuracy. The results are shown in Table 2. For N_2, the CEEIS molecular energy is in exact agreement with the explicit FCI value, but the dissociation energies differ by 0.2 kcal/mol because the nitrogen atom energy is in error: −54.47870 (CEEIS) versus −54.478551 (FCI). For O_2, the coupled cluster results and CEEIS results differ by 0.05 kcal/mol. The cf TQ5 est. FCI D_e falls within 0.001 kcal/mol of the Q56 value. A single point CCSDTQ5/VDZ calculation on O_2 requires 1.4 h on a 2.5 GHz Intel i7 processor. Run times have not been reported for the CEEIS calculations, making performance comparisons impossible. Finally, for F_2, there are two CEEIS values derived from different zeroth-order reference functions which differ by 0.15 mE_h. These lead to D_e values which bracket the cf TQ5 est. FCI value, one being 0.11 kcal/mol smaller and the other 0.09 larger. On the basis of this admittedly limited data, it does not appear that the CEEIS method offers any clear advantage in terms of accuracy or computational efficiency over a continued fraction approximant using higher levels of coupled cluster theory.

Correlation recovery beyond CCSD(T) will also contribute to the CV correction, although the magnitude of the effect for first-row molecules of the size considered here is expected to be small. The only exception would be in cases where there is substantial multi-configurational character to the wave function. For example, the HO(CV) correction to the atomization energy of CO_2 is ~0.07 kcal/mol. For second-row compounds like S_2, the HO(CV) correction increases to 0.23 kcal/mol. For this purpose, we use a combination of CCSDT(CV)/wCVTZ + CCSDTQ(CV)/wCVDZ. Unlike the case with the valence HO correlation correction, where CCSDT and CCSDTQ components are often of opposite signs, here the effects of triple and quadruple excitations are normally of the same sign. Consequently, convergence in the magnitude of the correction is easier to achieve. The HO(CV) correction can quickly become extremely time consuming, especially with second- or third-row elements.

3 Results and discussion

3.1 Homonuclear diatomics

The best theoretical dissociation energies, equilibrium bond lengths, and harmonic frequencies for first-row homonuclear diatomics produced in this study are listed in Tables 3 and 4, along with a breakdown of the components and estimated uncertainties (when available). Also listed are selected experimental values taken from the chemical literature [23, 76, 81, 96–101]. Note that the theoretical and experimental uncertainties are often not directly comparable. Experimentalists prefer to quote uncertainties corresponding to a 95 % confidence interval (roughly two standard deviations), whereas the theoretical uncertainties are an attempt to conservatively estimate the residual error in each component due to basis set truncation error. The total uncertainties are based on the sum of the individual uncertainties, i.e., cancelation of error is not assumed, although due to the alternating sign of the component corrections some is likely to occur. We have discussed alternate approaches for attaching meaningful error bars to theoretical results in a previous work [14].

As expected, the minor corrections to the CCSD(T)/CBS values are all small and of varying sign. For first-row

Table 3 Dissociation energies components for first-row homonuclear diatomics (kcal/mol)

Component	D_e/D_0/correction
B_2 ($^3\Sigma_g^-$)	
D_e R/UCCSD(T)(FC)/CBS(aV78)[a]	65.20 ± 0.03
ΔCCSD(T)(CV)/CBS(wCVTQ5)	0.77 ± 0.01
ΔR/UCCSD(T)-DKH/cc-pV5Z-DK	−0.06 ± 0.00
ΔCCSDT(FC)/CBS(TQ)	0.08 ± 0.04
ΔCCSDTQ(FC)/CBS(DT)	1.26 ± 0.03
ΔFCI(FC)/VTZ	0.09
ΔCCSDT(CV) + CCSDT(Q)(CV)[b]	0.11 ± 0.01
ΔUCCSD(FC)-DBOC/aVTZ	0.01
CCSD(T) additivity correction[c]	0.02
ΔAtomic S.O.	−0.06
Best composite D_e estimate	67.42 ± 0.12
Anharm. ZPE[d]	−1.50 ± 0.00
Best composite D_0 estimate	65.92 ± 0.12
Expt. D_0 (NIST-JANAF)[e]	73.4 ± 8
C_2 ($^1\Sigma_g^+$)	
D_e R/UCCSD(T)(FC)/CBS(aV89)[a]	145.37 ± 0.10
ΔCCSD(T)(CV)/CBS(wCVTQ5)	1.00 ± 0.01
ΔR/UCCSD(T)-DKH/cc-pV5Z-DK	−0.17 ± 0.00
ΔCCSDT(FC)/CBS(Q5)	−2.26 ± 0.03
ΔCCSDTQ(FC)/CBS(TQ)	2.29 ± 0.03
ΔFCI(FC)/VTZ + diff.(s,p)	0.41
ΔCCSDT(CV) + CCSDT(Q)(CV)[b]	0.29 ± 0.01
ΔCCSD(FC)-DBOC/aVTZ	0.03
CCSD(T) additivity correction[c]	0.05
ΔAtomic S.O.	−0.17
Best composite D_e estimate	146.84 ± 0.18
Anharm. ZPE[d]	−2.65 ± 0.01
Best composite D_o estimate	144.19 ± 0.19
Expt. D_0 (ATcT 2013)[f]	144.007 ± 0.066
Expt. D_0 (ATcT 2006)[g]	144.7 ± 0.3
N_2 ($^1\Sigma_g^+$)	
D_e R/UCCSD(T)(FC)/CBS(aV89)[a]	227.25 ± 0.14
ΔCCSD(T)(CV)/CBS(wCVTQ5)	0.77 ± 0.01
ΔR/UCCSD(T)-DKH/cc-pV5Z-DK	−0.06 ± 0.00
ΔCCSDT(FC)/CBS(Q5)	−0.77 ± 0.02
ΔCCSDTQ(FC)/CBS(TQ)	1.01 ± 0.01
ΔFCI(FC)/VDZ	0.13
ΔCCSDT(CV) + CCSDT(Q)(CV)[b]	0.06 ± 0.01
ΔCCSD(FC)-DBOC/aVTZ	0.01
CCSD(T) additivity correction[c]	0.02
ΔAtomic S.O.	0.00
Best composite D_e estimate	228.36 ± 0.19
Anharm. ZPE[d]	−3.37 ± 0.01
Best composite D_0 estimate	224.99 ± 0.20
Expt. D_0 (ATcT 2013)[f]	224.939 ± 0.011
Expt. D_0 (ATcT 2008)[h]	224.938 ± 0.007
Expt. D_0 (NIST-JANAF)[e]	224.94 ± 0.01
O_2 ($^3\Sigma_g^-$)	
D_e R/UCCSD(T)(FC)/CBS(aV89)	119.95 ± 0.08
ΔCCSD(T)(CV)/CBS(wCVTQ5)	0.24 ± 0.00
ΔR/UCCSD(T)-DKH/cc-pV5Z-DK	−0.18 ± 0.00
ΔCCSDT(FC)/CBS(TQ)	−0.66 ± 0.02
ΔCCSDTQ(FC)/CBS(DT)	1.09 ± 0.01

Table 3 continued

Component	D_e/D_0/correction
ΔEst. FCI(FC)/VDZ	0.15
ΔCCSDT(CV) + CCSDT(Q)(CV)[a]	0.00 ± 0.01
ΔUCCSD(FC)-DBOC/aVTZ	0.01
CCSD(T) Additivity Correction[b]	0.02
ΔAtomic S.O.	−0.45
Best composite D_e estimate	120.17 ± 0.12
Anharm. ZPE[c]	−2.29 ± 0.02
Best composite D_o estimate	117.88 ± 0.14
Expt. D_0 (ATcT 2013)[f]	117.994 ± 0.001
Expt. D_0 (ATcT 2008)[h]	117.99 ± 0.001
Expt. D_0 (HH)[i]	117.97 ± 0.002
F_2 ($^1\Sigma_g^+$)	
D_e R/UCCSD(T)(FC)/CBS(aV89)[a]	38.49 ± 0.01
ΔCCSD(T)(CV)/CBS(wCVTQ5)	−0.10 ± 0.00
ΔR/UCCSD(T)-DKH/cc-pV5Z-DK	−0.03 ± 0.00
ΔCCSDT(FC)/CBS(TQ)	−0.30 ± 0.01
ΔCCSDTQ(FC)/CBS(DT)	0.77 ± 0.02
ΔEst. FCI(FC)/VDZ[j]	0.11
ΔCCSDT(CV) + CCSDT(Q)(CV)[b]	0.06 ± 0.01
ΔCCSD(FC)-DBOC/aVTZ	0.00
CCSD(T) Additivity Correction[c]	−0.02
ΔAtomic S.O.	−0.77
Best composite D_e estimate	38.21 ± 0.05
Anharm. ZPE[d]	−1.30 ± 0.00
Best composite D_o estimate	36.91 ± 0.05
Expt. D_0 (ATcT 2013)[f]	36.944 ± 0.026
Expt. D_0 (ATcT 2009)[k]	36.95 ± 0.05
Expt. D_0 (ATcT 2008)[h]	36.93 ± 0.03
Expt. D_0 (ATcT 2006)[l]	36.91 ± 0.07
Expt. D_0 (ATcT 2004)[m]	36.90 ± 0.06
Expt. D_0 (Yang et al.)[n]	37.03 ± 0.02
Expt. D_0 (Colburn et al.)[o]	36.94 ± 0.14
Expt. D_0 (NIST-JANAF/CODATA)[e]	36.94 ± 0.14

[a] Raw CCSD(T)(FC) D_e values (kcal/mol) obtained with the largest basis sets are: $B_2 = 65.14$; $C_2 = 145.19$; $N_2 = 227.01$, $O_2 = 119.82$, $F_2 = 38.45$

[b] Higher-order CV correction based on CCSDT(CV)/wCVTZ + CCSDT(Q)(CV)/wCVDZ

[c] Additivity correction based on CCSD(T)(CV)-DKH/aug-cc-pwCV5Z-DK calculations

[d] ZPE based on CCSD(T)(FC)/aV8Z + ΔCCSD(T)(CV)/wCV5Z + ΔCCSD(T)(FC)-DKH/V5Z-DK + ΔHO FC + Additivity correction. The experimental anharmonic ZPE corrections (kcal/mol), based on the ω_e and $x_e\omega_e$ values from Huber and Herzberg, are: $B_2 = -1.50$, $C_2 = -2.64$, $N_2 = -3.36$, $O_2 = 2.25$, $F_2 = 1.30$ [76]

[e] NIST-JANAF [81] and CODATA [96]

[f] Ruscic et al. [4]

[g] Ruscic [124]

[h] Harding et al. [25]

[i] Huber and Herzberg [76]

[j] Continued fraction CCSDTQ5 estimated full CI

[k] Klopper et al. [97]

[l] Karton et al. [23]

[m] Ruscic [98]

[n] Yang et al. [99, 100]

[o] Colbourn et al. [101]

molecules, this has been well established in our previous work, as well as the work of other groups [1, 12–14, 23, 25, 73, 90, 102–111]. The various corrections to D_e in Table 3 differ qualitatively from the r_e and ω_e corrections due to the local nature (with respect to the potential energy surface) of the latter two. The corrections to the atomization energy tend to grow with the size of the molecule. For example, while the CV/CBS correction to the atomization energy of C_2 is 0.8 kcal/mol, we estimate the value for C_8H_{18}, octane, as ~9.6 kcal/mol. Even ΔDBOC for a molecule the size of octane would approach 0.2 kcal/mol. Thus, while it might be tempting to avoid the burdensome computational expense associated with the various smaller corrections, experience has shown that they are necessary if consistent high quality is desired without relying on fortuitous cancelation of error.

In Table 4, the largest corrections to r_e and ω_e arise from the core/valence and higher-order correlation corrections which are of opposite signs and roughly equal magnitudes. To the extent this observation holds true for polyatomic molecules, it implies that whenever higher-order corrections are unavailable, frozen core CCSD(T) likely provides a better estimate of the full CI limit than a value corrected solely for core/valence effects. The maximum higher-order corrections to r_e and ω_e are 0.004 Å and -22 cm^{-1}, respectively. Average values are roughly half as large.

Most model chemistries focus exclusively on atomization energies and heats of formation. All such calculations are carried out with reference molecular structures obtained from a lower level of theory than the highest level used for the actual thermochemistry. The same is true for the vibrational frequencies. An exception is the work of Karton and Martin who applied W4 theory to the prediction of spectroscopic properties [112]. In this regard, the FPD approach is different in that it tries to use geometries optimized for every method and basis set whenever possible. As a result, accurate molecular structures, frequencies, and zero point vibrational energies are produced as a matter of course. The impact of using optimized structures for atomization energies is not expected to be large, but the precise numerical consequence depends upon the system and the quality of the reference geometry. As an example, consider the C_8H_{18} (n-octane) system which was studied in 2005 with CCSD(T) and the aVDTQZ sequence of basis sets at the optimal second-order Møller-Plesset perturbation theory (MP2) [113] geometry [114]. At that time, the 1,468 basis function CCSD(T)/aVQZ calculation was extremely challenging. It required days of time on a $24 million massively parallel computer using between 600 and 1,400 processors. Today, the identical level of calculation can be performed on a single processor of a personal

Table 4 Equilibrium distances (Å) and harmonic frequencies (cm^{-1}) for first-row homonuclear diatomics

Component	r_e/correction	ω_e/correction
B_2 ($^3\Sigma_g^-$)		
CCSD(T)(FC)/CBS	1.59205	1,049.7
ΔCCSD(T)(CV)/CBS	−0.00638	8.6
ΔR/UCCSD(T)-DKH/V5Z-DK	−0.00010	−0.4
ΔHO(FC)	0.00409	−9.3
ΔHO(CV)	−0.00019	0.0
CCSD(T) additivity correction	−0.00024	0.9
CCSD-DBOC/aVTZ	0.00002	0.01
Best composite estimate	1.5893	1,049.6
Expt.[a]	1.5900	1,051.3
C_2 ($^1\Sigma_g^+$)		
CCSD(T)(FC)/CBS	1.24409	1,860.4
ΔCCSD(T)(CV)/CBS	−0.00329	10.6
ΔR/UCCSD(T)-DKH/cc-pV5Z-DK	−0.00014	−0.5
ΔHO(FC)	0.00195	−15.3
ΔHO(CV)	−0.00004	−0.9
CCSD(T) additivity correction	−0.00019	0.7
Best composite estimate	1.2424	1,855.0
Expt.[a]	1.2425	1,854.7
N_2 ($^1\Sigma_g^+$)		
CCSD(T)(FC)/CBS	1.09900	2,361.5
ΔCCSD(T)(CV)/CBS	−0.00215	10.0
ΔR/UCCSD(T)-DKH/cc-pV5Z-DK	−0.00018	−0.9
ΔHO(FC)	0.00097	−13.3
ΔHO(CV)	0.00030	−0.7
CCSD(T) additivity correction	−0.00017	0.6
Best composite estimate	1.0978	2,357.2
Expt.[a]	1.0977	2,358.6
O_2 ($^3\Sigma_g^-$)		
CCSD(T)(FC)/CBS	1.20629	1,604.0
ΔCCSD(T)(CV)/CBS	−0.00203	5.8
ΔR/UCCSD(T)-DKH/V5Z-DK	0.00022	−2.2
ΔHO(FC)	0.00298	−21.5
ΔHO(CV)	0.00008	−0.5
CCSD(T) additivity correction	0.00027	0.2
Best composite estimate	1.2078	1,585.8
Expt.[a]	1.2075	1,580.2
F_2 ($^1\Sigma_g^+$)		
CCSD(T)(FC)/CBS	1.40939	929.1
ΔCCSD(T)(CV)/CBS	−0.00144	1.5
ΔR/UCCSD(T)-DKH/cc-pV5Z-DK	0.00033	−0.4
ΔHO(FC)	0.00356	−10.8
ΔHO(CV)	0.00030	−2.2
CCSD(T) additivity correction	0.00061	0.0
Best composite estimate	1.4127	917.2
Expt.[a]	1.4119	916.6

[a] Huber and Herzberg [76]

computer, including the effects of geometry optimization. The aVDZ and aVTZ atomization energies increase by 2.4 and 0.6 kcal/mol, respectively. But the final CBS value based on optimized geometries was only 0.3 kcal/mol larger than the 2005 value using the same extrapolation formula. Compared to other sources of uncertainty in the atomization energy, this is a relatively minor contribution.

From a pragmatic point of view, it is relatively easy to compute higher-order corrections for harmonic frequencies (or anharmonicity constants) for diatomic molecules because all of the necessary calculations preserve $C_{\infty v}$ or $D_{\infty h}$ symmetry. Unfortunately, in the general polyatomic case, methods such as CCSDTQ are currently not implemented with either analytical first or second derivatives in any codes to our knowledge. This necessitates the use of two point numerical differentiation using displaced structures with C_1 symmetry. As a concrete example, a CCSDTQ/VDZ calculation on CO_2 at its equilibrium geometry requires 38×10^6 determinants. By comparison, a calculation in C_1 symmetry involves 306×10^6 determinants and a factor of 12 increase in computer time. Consequently, CCSDTQ corrections for polyatomic harmonic frequencies are likely to remain very computationally challenging for some time to come. Note that the CFOUR [78] code does support analytical CCSDT second derivatives, but this level of theory doesn't pick up the majority of the higher-order correction.

The level of agreement between theory and experiment for all three properties is good, with the exception of $D_e(B_2)$ where the NIST-JANAF [81] value is not known

Table 5 Selected theoretical values for the first-row homonuclear diatomics taken from the literature

System	Method	Year	D_0	r_e	ω_e	References
B_2 ($^3\Sigma_g^-$)	FPD	2008	66.15 ± 0.2	1.5899	NA	Feller et al. [14]
C_2 ($^1\Sigma_g^+$)	MRCI/V5Z	1995	140.9	1.2465	1848.2	Peterson [125]
	FPD	2008	143.89 ± 0.3	1.2426	NA	Feller et al. [14]
	W4.4	2007	144.08	NA	NA	Karton et al. [126]
N_2 ($^1\Sigma_g^+$)	W1	1999	227.89	NA	NA	Martin [22]
	W2	1999	228.06	NA	NA	Martin [22]
	W3	2004	225.12	NA	NA	Boese et al. [127]
	W4.4	2007	225.02[a]	NA	NA	Karton et al. [126]
	HEAT	2004	224.92	NA	NA	Tajti et al. [24]
	HEAT-456QP	2008	224.89	NA	NA	Harding et al. [25]
	CC Theory	2006	NA	1.0975	NA	Heckert et al. [128]
	CC + PT[a]	2009	224.86	NA	NA	Klopper et al. [97]
	FPD	2008	224.88 ± 0.3	1.0976	NA	Feller et al. [14]
O_2 ($^3\Sigma_g^-$)	W1	1999	119.89	NA	NA	Martin [22]
	W2	1999	119.58	NA	NA	Martin [22]
	W3	2004	117.99	NA	NA	Boese et al. [127]
	W4.4	2007	117.98[b]	NA	NA	Karton et al. [126]
	HEAT	2004	117.92	NA	NA	Tajti et al. [24]
	HEAT-456QP	2008	118.00	NA	NA	Harding et al. [25]
	FPD	2008	117.92 ± 0.2	1.2077	NA	Feller et al. [14]
F_2 ($^1\Sigma_g^+$)	W1	1999	37.72	NA	NA	Martin [22]
	W2	1999	37.64	NA	NA	Martin [22]
	W3	2004	36.94	NA	NA	Boese et al. [127]
	W4.4	2007	36.95	NA	NA	Karton et al. [126]
	HEAT	2004	36.91	NA	NA	Tajti et al. [24]
	HEAT-456QP	2008	36.92	NA	NA	Harding et al. [25]
	CC Theory	2006	NA	1.4108	NA	Heckert et al. [128]
	CC + PT[a]	2009	36.76	NA	NA	Klopper et al. [97]
	FPD	2008	36.90 ± 0.2	1.4122	NA	Feller et al. [14]
	(HEAT)	2013	37.03 ± 0.11	NA	NA	Csontos et al. [67]

Dissociation energies in kcal/mol. Bond lengths in Å. Harmonic frequencies in cm^{-1}. *NA* not available

[a] Coupled cluster theory augmented with explicitly correlated perturbation theory

[b] Using the partial-wave CBS extrapolation

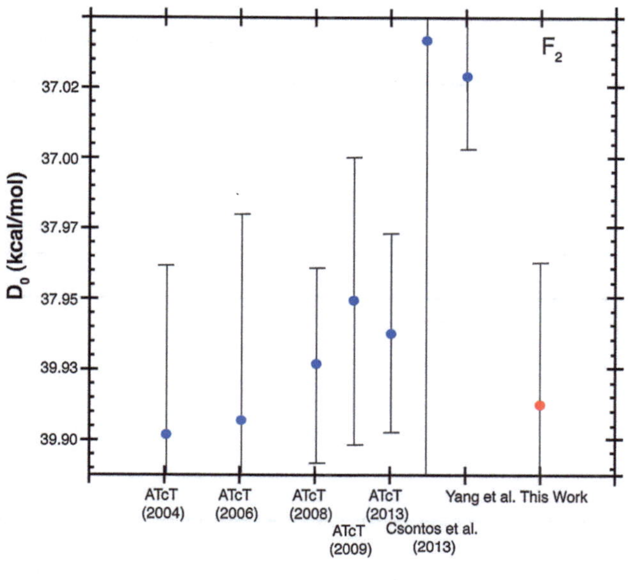

Fig. 2 Experimental and theoretical $D_0(F_2)$ values (kcal/mol)

with high precision. In the case of C_2, which has significant multi-configurational character in its wave function (the $2\sigma_u^2 \rightarrow 3\sigma_g^2$ excitation enters with a CI coefficient of −0.32), the theoretical and experimental uncertainties in D_0 nonetheless overlap each other.

The five first-row homonuclear diatomics are often used in calibration studies undertaken whenever new methods are introduced. Consequently, there are a very large number of theoretical publications reporting spectroscopic properties for these systems. It is beyond the scope of this work to provide a comprehensive survey of literature values. This is especially true of dissociation energies (atomization energies), where there has been a proliferation of thermochemical "model chemistries" in recent years, each with its own predictions for D_0. For example, we are aware of 37 different variants of the Weizmann-n models, alone. Table 5 offers a cursory list of previous high-level results for the first-row homonuclear diatomics. We are aware of relatively few reports dealing with B_2 and C_2, but the other three molecules are well represented. Overall, the spread in values among the approaches shrinks over time as would be expected due to their use of similar computational techniques. The variation among the dissociation energies is well within 1 kcal/mol if we restrict ourselves to data published after 2004.

The recent work on the dissociation energy of F_2 by Csontos et al. [67] attempted to resolve a small discrepancy between the experimental value due to Yang et al. [99, 100] $D_0(F_2) = 37.03 \pm 0.02$ kcal/mol and the 2008 ATcT value of 36.93 ± 0.03 kcal/mol. An older, high-resolution absorption spectra value of $D_0(F_2) = 36.94 \pm 0.14$ kcal/mol had also been reported by Colburn et al. [101]. Csontos et al. used an approach that is similar to HEAT, which involved CV basis sets up through 8ζ, combined with higher-order corrections from coupled cluster theory through CCSDTQ and CCSDTQ(P)$_\Lambda$. Their final $D_0(F_2)$ value was 37.03 ± 0.11 kcal/mol. Overlapping error bars precluded an unequivocal choice between the two experimental values, but fortuitously exact agreement was found with the value of Yang et al. A selection of experimental and theoretical $D_0(F_2)$ values are shown in Fig. 2. Our best theoretical value is in close agreement with the latest ATcT value.

Although results for H_2 are not included in Tables 3 and 4, highly accurate values are possible with FCI/CBS(aV89Z) [18]. We find $D_e = 109.493 \pm 0.003$ kcal/mol, $D_0 = 103.264 \pm 0.003$ kcal/mol using the same $1/(\ell_{max} + ½)^4$ extrapolation formula applied throughout the rest of this work. The $1/\ell_{max}^3$ formula gives a slightly larger value of 103.268 ± 0.003 kcal/mol, compared to the most recent ATcT value of 103.266730 ± 0.000005 kcal/mol. The diagonal Born–Oppenheimer correction for H_2 was computed at the FCI/aVQZ level of theory. The best theoretical bond distance is 0.7416 Å and the best harmonic frequency is 4,403.0 cm^{-1} (atomic mass = 1.007825)' which compare to 0.74144 Å and 4,403.2 cm^{-1} given by Huber and Herzberg [76]. The latter value was given in footnote "i" of the H_2 table in Huber and Herzberg and includes Dunham corrections. The current r_e and ω_e values also agree very well with the explicitly correlated results of Wolniewicz, $r_e = 0.74162$ Å and $\omega_e = 4{,}403.14$ cm^{-1}, using a tenth degree fit to the Born–Oppenheimer + relativistic + adiabatic data points [115].

3.2 Other systems

Several additional molecules were treated with the same procedure applied to the first-row homonuclear diatomics. For P_2 and S_2, it was not possible to compute the higher-order CV corrections for r_e and ω_e due to computational limitation. The atomization energies, equilibrium structures, and harmonic frequency results, shown in Table 6, were found to be similar in terms of their level of agreement with experiment [25, 76, 97, 116–122]. The symbol ΣD_0 denotes the zero point inclusive atomization energy. The optimized structure and harmonic frequencies for H_2O were taken from the recent high-level study of Feller and Peterson [123]. For the 16 molecules examined in this benchmark study, bond lengths were found to have an RMS deviation of 0.0001 Å ($\varepsilon_{MAD} = 0.00004$ Å, $\varepsilon_{Max} = 0.0005$ Å). If the comparison set is expanded to include all molecules in the CRDB for which accurate experimental values are available, $\varepsilon_{RMS} = 0.0027$ Å (124 comparisons). These include many polyatomic species for which semi-experimental structures have been reported. Heats of formation at 0 and 298.15 K for all molecules are listed in Table 7. The wave functions for most of the systems are strongly dominated by the Hartree–Fock configuration. CN ($^2\Sigma$) is an exception, with a leading

Table 6 Results for other molecules

CO ($^1\Sigma^+$)	D_0	r_e	ω_e
Best composite estimate[a]	256.26 ± 0.16	1.1282	2,168.8
Expt. (ATcT 2013)[b]	256.224 ± 0.013	1.12832	2,169.8
Expt. (ATcT 2009)[c]	256.24 ± 0.02		
Expt. (ATcT 2008)[d]	256.25 ± 0.02		
Expt. (NIST-JANAF)[e]	256.16 ± 0.04		

CO_2 ($^1\Sigma_g^+$)	ΣD_0	r_{CO}	$\omega_{\pi u}$	$\omega_{\sigma g}$	$\omega_{\sigma u}$
Best Composite Estimate[a,f]	381.97 ± 0.27	1.1598	669.7	1,350.3	2,390.8
Expt. (ATcT 2,013)[g]	381.977 ± 0.006	1.1601	674.7	1,416.0	2,396.2
Expt. (ATcT 2009)[c]	381.98 ± 0.02				
Expt. (ATcT 2008)[d]	382.00 ± 0.02				
Expt. (NIST-JANAF)[e]	381.91 ± 0.01				

NO ($^2\Pi$)	D_0	r_e	ω_e
Best Composite Estimate[a]	149.78 ± 0.16	1.1507	1,906.6
Expt. (ATcT 2013)[h]	149.816 ± 0.014	1.15077	1,904.2
Expt. (ATcT 2010)[i]	149.82 ± 0.01		
Expt. (NIST-JANAF)[f]	150.06 ± 0.04		

CN ($^2\Sigma^+$)	D_0	r_e	ω_e
Best Composite Estimate[j]	178.20 ± 0.20	1.1715	2,066.8
Expt. (ATcT 2013)[k]	178.120 ± 0.034	1.17182	2,068.6
Expt. (ATcT 2008)[d]	178.18 ± 0.06		
Expt. (NIST-JANAF)[f]	178.1 ± 2.4		
Expt. (Huang)[l]	178.5 ± 0.5		
Expt. (Engleman)[m]	177.2 ± 1.1		
Expt. (Costes)[n]	179.2 ± 1.2		

CH ($^2\Pi$)	D_0	r_e	ω_e
Best Composite Estimate[l]	80.00 ± 0.04	1.1186	2,861.6
Expt. (ATcT 2008)[m]	79.99 ± 0.06	1.1190	2,858.5

HF ($^1\Sigma^+$)	D_0	r_e	ω_e
Best Composite Estimate[n]	135.26 ± 0.04	0.9168	4,138.5
Expt. (ATcT)[o]	135.28 ± 0.00	0.9168	4,138.3

P_2 ($^1\Sigma_g^+$)	D_0	r_e	ω_e
Best Composite Estimate[n]	116.16 ± 0.18	1.8933	779.4
Expt. (NIST-JANAF)[f]	116.07 ± 0.50	1.8934	780.8
Expt. (CODATA)[p]	115.99 ± 0.48		

S_2 ($^3\Sigma_g^-$)	D_0	r_e	ω_e
Best Composite Estimate[a]	102.02 ± 0.17	1.8883	727.1
Expt. (Frederix et al.)[q]	101.89 ± 0.01	1.8892	725.7
Expt. (CODATA)[p]	100.66 ± 0.07		

H_2O ($^1A_1$)	ΣD_0	r_{OH}	ω_{a1}	ω_{a1}	ω_{b2}
Best Composite Estimate[r]	219.35 ± 0.09	0.9577	1,648.6	3,834.8	3,946.3
Expt. (ATcT 2013)[s]	219.369 ± 0.0005	0.9572	1,648.5	3,832.2	3,942.5
Expt. (ATcT 2009)[c]	219.36 ± 0.02				
Expt. (ATcT 2008)[d]	219.37 ± 0.01				
Expt. (NIST-JANAF)[e]	219.36 ± 0.03				

Table 6 continued

CH_2 ($^3B_1$)	ΣD_0	r_{CH}	ω_{a1}	ω_{a1}	ω_{b2}
Best Composite Estimate[r]	179.76 ± 0.07	1.0,754	1,089.8	3,145.9	3,375.9
Expt. (ATcT 2013)[t]	179.85 ± 0.03	1.0748	1,080	3,090	3,220
Expt. (ATcT 2008)[d]	179.90 ± 0.06				
CH_2 ($^1A_1$)	ΣD_0	r_{CH}	ω_{a1}	ω_{a1}	ω_{b2}
Best Composite Estimate[u]	170.98 ± 0.11	1.1061	1,393.3	2,930.3	3,007.0
Expt. (ATcT 2013)[v]	170.85 ± 0.04	1.107			
Expt. (ATcT 2009)[c]	170.87 ± 0.05				

Dissociation energies in kcal/mol. Bond lengths in Å. Harmonic frequencies in cm^{-1}

[a] Based on CCSD(T)(FC)/CBS(aV678Z) + CCSD(T)(CV)/CBS(wCVTQ5Z) + CCSD(T)-DK(FC)/cc-pV5Z-DK + HO(FC)/CBS(VTQZ) + HO(CV)/wCVTDZ) + cf est. FCI/VDZ + CCSD(FC)- DBOC/aVTZ + additivity correction. ZPE(CO) correction = – 3.09 kcal/mol. ZPE(NO) correction = −2.72 kcal/mol

[b] CO experimental values: D_0 Ruscic et al. ATcT 2013;[4] r_e and ω_e Huber and Herzberg [76]

[c] Klopper et al. ATcT 2009 [97]

[d] Harding et al. ATcT 2008 [25]

[e] Chase NIST-JANAF [81]

[f] CO_2 anharmonic ZPE correction (−7.22 kcal/mol) is based on CCSD(T)(FC)/aV6Z + ΔCCSD(T)(CV)/wCV5Z + ΔCCSD(T)(FC)-DKH/V5Z-DK + Additivity correction. with an anharmonic MP2 correction from Pfeiffer et al. [129]

[g] CO_2 experimental values: ΣD_0 Ruscic et al. ATcT 20[4, 13] r_{CO} Pawlowski et al. [116] harmonic frequencies from Herzberg [130]

[h] NO experimental values: D_0 Ruscic et al. ATcT 20[4], [13]r_e and ω_e Huber and Herzberg [76]

[i] Stevens et al. [131]

[j] Based on CCSD(T)(FC)/CBS(aV678Z) + CCSD(T)(CV)/CBS(wCVTQ5Z) + CCSD(T)-DK(FC)/cc-pV5Z-DK + HO(FC)/CBS(VTQZ) + HO(CV)/wCVTDZ) + FCI/VDZ + CCSD(FC)-DBOC/aVTZ + additivity correction. ZPE(CN) correction = −2.95 kcal/mol

[k] CN experimental values: D_0 Ruscic et al. ATcT 2013;[4] r_e and ω_e Huber and Herzberg [76]

[l] Haung et al. [117]

[m] Engleman and Rouse [118]

[n] Costes et al. [119]

[l] Based on CCSD(T)(FC)/CBS(aV678Z) + CCSD(T)(CV)/CBS(wCVTQ5Z) + CCSD(T)-DK(FC)/cc-pV5Z-DK + FCI(F)/VQZ + HO(CV)/wCVTDZ) + CCSD(FC)-DBOC/aVTZ + additivity correction. ZPE(CH) correction = −4.05 kcal/mol

[m] CH experimental values: D_0 ATcT 2008;[25] r_e and ω_e Huber and Herzberg [76]

[n] Based on CCSD(T)(FC)/CBS(aV678Z) + CCSD(T)(CV)/CBS(wCVTQ5Z) + CCSD(T)-DK(FC)/cc-pV5Z-DK + FCI(F)/VTZ + HO(CV)/wCVTDZ) + CCSD(FC)- DBOC/aVTZ + additivity correction. ZPE(HF) correction = −5.87 kcal/mol

[o] Klopper et al. [97]

[p] P_2 experimental values: D_0 CODATA;[96] r_e and ω_e Huber and Herzberg [76]

[q] S_2 experimental values: D_0 Frederix et al.;[132] r_e and ω_e Huber and Herzberg [76]

[r] Based on CCSD(T)(FC)/CBS(aV678Z) + CCSD(T)(CV)/CBS(wCVTQ5Z) + CCSD(T)-DK(FC)/cc-pV5Z-DK + FCI(FC)/VTZ(O),VDZ(H) + HO(CV)/wCVTDZ) + CCSD(FC)-DBOC/aVTZ + additivity correction. The optimal HOH bond angle = 104.5°. ZPE(HF) correction = −13.26 kcal/mol [123]

[s] H_2O experimental values: ΣD_0 Ruscic et al. ATcT 200[4, 9] structure and harmonic frequencies from Benedict et al. [120]

[t] CH_2 ($^3B_1$) experimental values: ΣD_0 Ruscic et al. ATcT 2009,[4] structure and harmonic frequencies from Bunker and Jensen [121]

[u] Based on CCSD(T)(FC)/CBS(aV678Z) + CCSD(T)(CV)/CBS(wCVTQ5Z) + CCSD(T)-DK(FC)/cc-pV5Z-DK + FCI(FC)/VDZ + HO(CV)/wCVTDZ) + CCSD(FC)- DBOC/aVTZ + additivity correction. The optimal HOH bond angle = 102.4°. ZPE(HF) correction = −13.26 kcal/mol

[v] CH_2 ($^1A_1$) experimental values: ΣD_0 Ruscic et al. ATcT 2009,[4] structure from Petek et al. [133]. The experimental HCH angle = 102.4°. Harmonic frequencies are not available. The experimental fundamentals are 1,352.6, 2,805.9, and 2,864.5 cm^{-1} from Jensen and Bunker [122]

FCI coefficient of only 0.89. It was possible to carry out FCI calculations on CN with the VDZ basis set and CH with the VQZ basis set.

If we compare all available experimental atomization energies in the CRDB whose reported uncertainties are ≤0.2 kcal/mol with the corresponding best available theoretical results obtained from the FPD method we find $\varepsilon_{RMS} = 0.16$ (0.67), $\varepsilon_{MAD} = 0.10$ (0.42) and $\varepsilon_{MSD} = 0.02$ (0.08) kcal/mol (kJ/mol in parentheses), where MSD = mean signed deviation. A total of 79 comparisons were included in generating these statistics. Most of the theoretical results were not carried out to quite the same high level as the current findings. For example, they did not include corrections for the additivity assumption or basis sets of aV8Z or aV9Z quality. They did, however, include all of the other major categories of corrections.

4 Conclusion

New benchmark quality spectroscopic properties for a collection of small molecules were obtained from the

Table 7 Summary theoretical heats of formation at 0 and 298.15 K (kcal/mol)

System	$\Delta H_f(0\ K)$	$\Delta H_f(298\ K)$	System	$\Delta H_f(0\ K)$	$\Delta H_f(298\ K)$
B_2 ($^3\Sigma_g^-$)	204.28 ± 0.12	205.81 ± 0.12	CO ($^1\Sigma_g^+$)	−27.24 ± 0.16	−26.46 ± 0.16
C_2 ($^1\Sigma_g^+$)	195.76 ± 0.20	197.77 ± 0.20	CO_2	−93.96 ± 0.27	−94.04 ± 0.27
N_2 ($^1\Sigma_g^+$)	−0.05 ± 0.20	−0.05 ± 0.20	NO ($^2\Pi$)	21.69 ± 0.16	21.81 ± 0.16
O_2 ($^3\Sigma_g^-$)	0.11 ± 0.14	0.11 ± 0.14	CN ($^2\Sigma^+$)	104.29 ± 0.16	103.88 ± 0.16
F_2 ($^1\Sigma_g^+$)	−0.03 ± 0.05	−0.03 ± 0.05	CH ($^2\Pi$)	141.64 ± 0.04	142.46 ± 0.04
HF ($^1\Sigma^+$)	−65.18 ± 0.04	−65.19 ± 0.04	P_2 ($^1\Sigma_g^+$)	34.77 ± 0.16	34.33 ± 0.16
S_2 ($^3\Sigma_g^-$)	30.50 ± 0.17	30.64 ± 0.17	H_2O ($^1A_1$)	−57.09 ± 0.09	−57.79 ± 0.09
CH_2 ($^3B_1$)	93.53 ± 0.07	93.65 ± 0.07	CH_2 ($^1A_1$)	102.31 ± 0.11	102.41 ± 0.11

composite Feller–Peterson–Dixon method applied at the highest currently practical level based on our available hardware and software. The frozen core CCSD(T) contribution to the properties relied on correlation consistent basis sets as large as aV8Z or in some cases aV9Z. The complementary n-particle expansion was modeled with a progression of coupled cluster theory calculations through CCSDTQ5. In ten cases (B_2, C_2, N_2, CN, CH, HF, P_2, H_2O, CH_2 ($^3B_1$) and CH_2 ($^1A_1$)) these results were further augmented with full configuration interaction. Good agreement was found with the best available experimental data. Chief among these are values derived from the active thermochemical tables, whose accuracy continually challenges theoreticians to push the state of the art. By paying careful attention to the various sources of error in the atomization calculations it was possible to assign conservative estimates of the residual uncertainties to our final values. While the same general approach can be applied to larger systems, e.g. octane, practical limitations force us to use smaller basis sets with resulting larger uncertainties. Recent developments in explicitly correlated methods and improved extrapolation techniques should be helpful with the 1-particle expansion.

Acknowledgments One of us (BR) acknowledges the support of the U.S. Department of Energy, Office of Science, Office of Basic Energy Sciences, Division of Chemical Sciences, Geosciences and Biosciences under Contract No. DE-AC02-06CH11357 for work done at Argonne National Laboratory.

References

1. Peterson KA, Feller D, Dixon DA (2012) Theor Chem Acc 131:1079
2. Feller D, Peterson KA, Dixon DA (2012) Mol Phys 110:2381
3. Dixon DA, Feller D, Peterson KA. In: Wheeler RA (eds) Ann Report Comp Chem, (Elsevier, 2012), vol 8
4. Ruscic B, Feller D, Peterson KA (2014) Theor Chem Acc 133:1415
5. Gaussian 09, Revision D.01, Frisch, MJ, Trucks GW, Schlegel HB, Scuseria GE, Cheeseman JR, Robb MA, Scalmani G, Barone V, Mennucci B, Petersson GA, Nakatsuji H, Caricato M, Li X, Hratchian HP, Izmaylov AF, Bloino J, Zheng G, Sonnenberg JL, Hada M, Ehara M, Toyota K, Fukuda R, Hasegawa J, Ishida M, Nakajima T, Honda Y, Kitao O, Nakai H, Vreven T, Montgomery Jr JA, Peralta JE, Ogliaro F, Bearpark M, Heyd JJ, Brothers E, Kudin KN, Staroverov VN, Kobayashi R, Normand J, Raghavachari K, Rendell A, Burant JC, Iyengar SS, Tomasi J, Cossi M, Rega N, Millam JM, Klene M, Knox JE, Cross JB, Bakken V, Adamo C, Jaramillo J, Gomperts R, Stratmann RE, Yazyev O, Austin AJ, Cammi R, Pomelli C, Ochterski JW, Martin RL, Morokuma K, Zakrzewski VG, Voth GA, Salvador P, Dannenberg JJ, Dapprich S, Daniels AD, Farkas Ö, Foresman JB, Ortiz JV, Cioslowski J, Fox DJ (Gaussian, Inc., Wallingford CT, 2009)
6. GAMESS, Schmidt MW, Baldridge KK, Boatz JA, Elbert ST, Gordon MS, Jensen JH, Koseki S, Matsunaga N, Nguyen KA, Su S, Windus TL, Dupuis M, Montgomery JA Jr (1993) J Comp Chem 14:1347
7. MOLPRO H-J, Werner PJ, Knowles G, Knizia FR, Manby M, Schütz P, Celani T, Korona R, Lindh A, Mitrushenkov G, Rauhut KR, Shamasundar TB, Alder RD, Amos A, Bernhardsson A, Berning DL, Cooper MJO, Deegan AJ, Dobbyn F, Eckert E, Goll C, Hampel A, Hesselmann G, Hetzer T, Hrenar G, Jansen C, Koppl Y, Liu AW, Lloyd RA, Mata AJ, May SJ, McNicholas W, Meyer ME, Mura A Nicklass DP, O'Neill P, Palmieri K, Pflüger R, Pitzer M, Reiher T, Shiozaki H, Stoll AJ, Stone R, Tarroni T, Thorsteinsson M, Wang A, Wolf A http://www.molpro.net, (Universität Stüttgart, Stüttgart, Germany, Cardiff University, Cardiff, United Kingdom, 2012)
8. Purvis GD III, Bartlett RJ (1982) J Chem Phys 76:1910–1918
9. Raghavachari K, Trucks GW, Pople JA, Head-Gordon M (1989) Chem Phys Lett 157:479
10. Watts JD, Gauss J, Bartlett RJ (1993) J Chem Phys 98:8718–8733
11. Bartlett RJ, Musiał M (2007) Rev Mod Phys 79:291–352
12. Feller D, Peterson KA, Crawford TD (2006) J Chem Phys 124:054107
13. Feller D, Peterson KA (2007) J Chem Phys 126:114105
14. Feller D, Peterson KA, Dixon DA (2008) J Chem Phys 129:204105–204136
15. Adler TB, Knizia G, Werner H-J (2007) J Chem Phys 127:221106
16. Knizia G, Adler TB, Werner H-J (2009) J Chem Phys 130:054104
17. Rauhut G, Knizia G, Werner H-J (2009) J Chem Phys 130:054105
18. Feller D, Peterson KA (2013) J Chem Phys 139:084110
19. Pople JA, Head-Gordon M, Fox DJ, Raghavachari K, Curtiss LA (1989) J Chem Phys 90:5622

20. Curtiss LA, Raghavachari K, Trucks GW, Pople JA (1991) J Chem Phys 94:7221
21. Curtiss LA, Redfern PC, Raghavachari K (2007) J Chem Phys 126:084108
22. Martin JML, Oliveira GD (1999) J Chem Phys 111:1843
23. Karton A, Rabinovich E, Martin JML, Ruscic B (2006) J Chem Phys 125:144108
24. Tajti A, Szalay PG, Császár AG, Kállay M, Gauss J, Valeev EF, Flowers BA, Vázquez J, Stanton JF (2004) J Chem Phys 121:11599
25. Harding ME, Vázquez J, Ruscic B, Wilson AK, Gauss J, Stanton JF (2008) J Chem Phys 128:114111
26. DeYonker NJ, Cundari TR, Wilson AK (2006) J Chem Phys 124:114104
27. DeYonker NJ, Grimes T, Yokel S, Dinescu A, Mintz B, Cundari TR, Wilson AK (2006) J Chem Phys 125:104111
28. DeYonker NJ, Wilson BR, Pierpont AW, Cundari TR, Wilson AK (2009) Mol Phys 107:1107
29. East ALL, Allen WD (1993) J Chem Phys 99:4638
30. Allen WD, East ALL, Császár AG (1993) In: Laane J, Veken BVD, Overhammer H (eds) Structure and conformations of non-rigid molecules. Kluwer, Dordrecht, pp 343–373
31. Császár AG, Allen WD, Schaefer HF III (1998) J Chem Phys 108:9751
32. Petersson GA, Bennett A, Tensfeldt TG, Al-Laham MA, Shirley WA, Mantzaris J (1988) J Chem Phys 89:2193
33. Petersson GA, Tensfeldt TG, Montgomery JA Jr (1991) J Chem Phys 94:6091
34. Ochterski JW, Petersson GA, Montgomery JA Jr (1996) J Chem Phys 104:2598
35. Feller D (1996) J Comp Chem 17:1571–1586
36. Leininger ML, Allen WD, Schaefer HF III, Sherrill CD (2000) J Chem Phys 112:9213
37. Dunning TH Jr (1989) J Chem Phys 90:1007–1023
38. Kendall RA, Dunning TH Jr, Harrison RJ (1992) J Chem Phys 96:6796–6806
39. Woon DE, Dunning TH Jr (1994) J Chem Phys 100:2975–2988
40. Woon DE, Dunning TH Jr (1995) J Chem Phys 103:4572–4585
41. Peterson KA, Dunning TH Jr (2002) J Chem Phys 117:10548–10560
42. Koput J, Peterson KA (2002) J Phys Chem A 106:9595–9599
43. Woon DE, Dunning TH Jr (1993) J Chem Phys 98:1358–1371
44. Wilson AK, Woon DE, Peterson KA, Dunning TH Jr (1999) J Chem Phys 110:7667–7676
45. Balabanov NB, Peterson KA (2005) J Chem Phys 123:064107–064122
46. Wilson AK, van Mourik T, Dunning TH Jr (1996) J Mol Struct (Theochem) 388:339
47. Feller D, Peterson KA (1999) J Chem Phys 110:8384
48. Peterson KA, Figgen D, Goll E, Stoll H, Dolg M (2003) J Chem Phys 119:11113
49. Peterson KA (2003) J Chem Phys 119:11099
50. Peterson KA, Puzzarini C (2005) Theor Chem Acc 114:283
51. Barnes EC, Petersson GA, Feller D, Peterson KA (2008) J Chem Phys 129:194115
52. Feller D (2013) J Chem Phys 138:074103
53. Peterson KA, Figgen D, Dolg M, Stoll H (2007) J Chem Phys 126:124101
54. Figgen D, Peterson KA, Dolg M, Stoll H (2009) J Chem Phys 130:164108
55. Li H, Feng H, Sun W, Zhang Y, Fan Q, Peterson KA, Xie Y, Schaefer HF, III. Mol. Phys. doi:10.1080/00268976.2013.802818 (2013)
56. Bruss DH, Peterson KA Theor Chem Acc (2013)
57. Douglas M, Kroll NM (1974) Ann Phys (New York) 82:89
58. Jansen G, Hess BA (1989) Phys Rev A 39:6016–6017
59. Xu J, Deible MJ, Peterson KA, Jordan KD (2013) J Chem Theory Comput 9:2170
60. Feller D, Peterson KA, Hill JG (2011) J Chem Phys 135:044102
61. Martin JML (1996) Chem Phys Lett 259:669
62. Schwartz C (1962) Phys Rev 126:1015
63. Schwartz C (1963) In: Alder BJ, Fernbach S, Rotenberg M (eds) Methods computational physics, vol 2. Academic Press, New York, pp 262–265
64. Carroll DP, Silverstone HJ, Metzger RM (1979) J Chem Phys 71:4142
65. Hill RN (1985) J Chem Phys 83:1173
66. Kutzelnigg W, Morgan JD (1992) J Chem Phys 96:4484
67. Csontos B, Nagy B, Csontos J, Kállay M (2013) J Phys Chem A 117:5518
68. Dalton 2011, Angeli C, Bak KL, Bakken V, Christiansen O, Cimiraglia R, Coriani S, Dahle P, Dalskov EK, Enevoldsen T, Fernandez B, Hättig C, Hald K, Halkier A, Heiberg H, Helgaker T, Hettema H, Jensen HJA, Jonsson D, Jørgensen P, Kirpekar S, Klopper W, Kobayashi R, Koch H, Ligabue A, Lutnaes OB, Mikkelsen KV, Norman P, Olsen J, Packer MJ, Pedersen TB, Rinkevicius ERZ, Ruden TA, Ruud K, Sałek P, Meras ASd, Saue T, Sauer SPA, Schimmelpfennig B, Sylvester-Hvid KO, Taylor PR, Vahtras O, Wilson DJ, Ågren H (2005)
69. Knowles PJ, Hampel C, Werner HJ (1993) J Chem Phys 99:5219
70. Deegan MJO, Knowles PJ (1994) Chem Phys Lett 227:321
71. Dunham JL (1932) Phys Rev 41:713
72. Moore CE, (Washington, D.C., 1971), Vol. Natl. Stand. Ref. Data Ser., National Bureau of Standards (U.S.) 35
73. Feller D, Peterson KA, deJong WA, Dixon DA (2003) J Chem Phys 118:3510
74. Shepler BC, Peterson KA (2006) J Phys Chem A 110:12321
75. Peterson KA, Shepler BC, Figgen D, Stoll H (2006) J Phys Chem A 110:13877
76. Huber KP, Herzberg G (1979) Molecular spectra and molecular structure IV: Constants of diatomic molecules. Van Nostrand Reinhold Co. Inc., New York
77. Visscher L, Dyall KG (1996) J Chem Phys 104:9040
78. CFour, a quantum chemical program package written by J. F. Stanton, J. Gauss, M. E. Harding, P. G. Szalay, w. c. from, A. A. Auer, R. J. Bartlett, U. Benedikt, C. Berger, D. E. Bernholdt, Y. J. Bomble, L. Cheng, O. Christiansen, M. Heckert, O. Heun, C. Huber, T.-C. Jagau, D. Jonsson, J. Jusélius, K. Klein, W. J. Lauderdale, D. A. Matthews, T. Metzroth, D. P. O'Neill, D. R. Price, E. Prochnow, K. Ruud, F. Schiffmann, W. Schwalbach, S. Stopkowicz, A. Tajti, J. Vázquez, F. Wang, J. D. Watts and P. P. R. T. and the integral packages MOLECULE (J. Almlöf and P.R. Taylor), ABACUS (T. Helgaker, H.J. Aa. Jensen, P. Jørgensen, and J. Olsen), and ECP routines by A. V. Mitin and C. van Wüllen, http://www.cfour.de/, (2010)
79. Barone V (2004) J Chem Phys 120:3059
80. Karton A, Martin JML (2007) J Phys Chem A 111:5936
81. Chase Jr. MW, NIST-JANAF Thermochemical tables (4th Ed) J Phys Chem Ref Data Mono. 9 (Suppl 1) (1998)
82. Nagy B, Szakács P, Csontos J, Rolik Z, Tasi G, Kállay M (2011) J Phys Chem A 115:7823
83. Curtiss LA, Raghavachari K, Redfern PC, Pople JA (1997) J Chem Phys 106:1063
84. Pople JA, Head-Gordon M, Raghavachari K (1987) J Chem Phys 87:5968
85. Martin JML (1992) J Chem Phys 97:5012
86. Martin JML (1997) J Mol Struct (Theochem) 398–399:135
87. Kállay M (2001) (Budapest University of Technology and Economics)
88. Kállay M, Surján PR (2001) J Chem Phys 115:2945–2954
89. Goodson DZ (2002) J Chem Phys 116:6948

90. Feller D, Dixon DA (2003) J Phys Chem A 107:9641
91. Bytautas L, Ruedenberg K (2004) J Chem Phys 121:10905
92. Bytautas L, Ruedenberg K (2004) J Chem Phys 121:10919
93. Bytautus L, Ruedenberg K (2005) J Chem Phys 122:154110
94. Bytautus L, Nagata T, Gordon MS, Ruedenberg K (2007) J Chem Phys 127:164317
95. Bytautus L, Ruedenberg K (2010) J Chem Phys 132:074109
96. Cox JD, Wagman DD, Medvedev VA (1989) CODATA Key values for thermodynamics. Hemisphere, New York
97. Klopper W, Ruscic B, Tew DP, Bischoff FA, Wolfsegger S (2009) Chem Phys 356:14
98. Ruscic B, Pinzon RE, Morton ML, Laszewski Gv, Bittner S, Nijsure SG, Amin KA, Minkoff M, Wagner AF (2004) J Phys Chem A 108:9979
99. Yang J, Hao Y, Li J, Mo Y (2005) J Chem Phys 122:134308
100. Yang J, Hao Y, Li J, Mo Y (2007) J Chem Phys 127:209901
101. Colbourn AE, Dagenais M, Douglas AE, Raymonda JW (1976) Can J Phys 54:1343
102. Ruscic B, Feller D, Dixon DA, Peterson KA, Harding LB, Asher RL, Wagner AF (2001) J Phys Chem A 105:1
103. Feller D, Dixon DA (2001) J Chem Phys 115:3484
104. Ruscic B, Wagner AF, Harding LB, Asher RL, Feller D, Dixon DA, Peterson KA, Song Y, Qian X, Ng C, Liu J, Chen W, Schwenke DW (2002) J Phys Chem A 106:2727
105. Dixon DA, deJong WA, Peterson KA, Francisco JS (2002) J Phys Chem A 106:4725
106. Feller D, Dixon DA, Francisco JS (2003) J Phys Chem A 107:1604
107. Feller D, Dixon DA (2003) J Phys Chem A 107:10419
108. Balabanov NB, Peterson KA (2003) J Phys Chem A 107:7465
109. Dixon DA, Feller D, Christe KO, Wilson WW, Vij A, Vij V, Jenkins HDB, Olson R, Gordon MS (2004) J Am Chem Soc 126:834
110. Feller D, Peterson KA, Dixon DA (2010) J Phys Chem A 114:613
111. Feller D, Peterson KA, Dixon DA (2011) J Phys Chem A 115:1440
112. Karton A, Martin JML (2010) J Chem Phys 133:144102
113. Moller C, Plesset MS (1934) Phys Rev 46:618
114. Pollack L, Windus TL, deJong WA, Dixon DA (2005) J Phys Chem A 109:6934
115. Wolniewicz L (1993) J Chem Phys 99:1851
116. Pawlowski F, Jørgensen P, Olsen J, Hegelund F, Helgaker T, Gauss J, Bak KL, Stanton JF (2002) J Chem Phys 116:6482–6496
117. Huang Y, Barts SA, Halpern JB (1992) J Phys Chem 96:425
118. Engleman R Jr, Rouse PE (1975) J Quant Spectrosc Radiat Transf 15:831
119. Costes M, Naulin C, Dorthe G (1990) Astron Astrophys 232:425
120. Benedict WS, Gailar N, Plyler EK (1956) J Chem Phys 24:1139
121. Bunker PR, Jensen P (1983) J Phys Chem 79:1224
122. Jensen P, Bunker PR (1988) J Chem Phys 89:1327
123. Feller D, Peterson KA (2009) J Chem Phys 131:154306
124. Ruscic B, (esrlier (interim) Active Thermochemical Tables, ATcT ver. 1.25 and the Core (Argonne) Thermochemical Network ver. 1.056 2006)
125. Peterson KA (1995) J Chem Phys 102:262
126. Karton A, Taylor PR, Martin JML (2007) J Chem Phys 127:064104
127. Boese AD, Oren M, Atasoylu O, Martin JML, Kallay M, Gauss J (2004) J Chem Phys 120:4129
128. Heckert M, Kállay M, Tew DP, Klopper W, Gauss J (2006) J Chem Phys 125:044108
129. Pfeiffer F, Rauhut G, Feller D, Peterson KA (2013) J Chem Phys 138:044311
130. Herzberg G (1991) Molecular spectra and molecular structure III. Electronic spectra and electronic structure of polyatomic molecules. Krieger Publishing Co., Inc., Malabar
131. Stevens WR, Ruscic B, Baer T (2010) J Phys Chem A 114:13134
132. Frederix PWJM, Yang C-H, Groenenboom GC, Parker DH, Alnama K, Western CM, Orr-Ewing AJ (2009) J Phys Chem A 113:14995
133. Petek H, Nesbitt DJ, Darwin DC, Ogilby PR, Moore CB, Ramsay DA (1989) J Chem Phys 91:6566

Theor Chem Acc (2014) 133:1425
DOI 10.1007/s00214-013-1425-x

REGULAR ARTICLE

Accurate ab initio potential energy curves and spectroscopic properties of the four lowest singlet states of C_2

Jeffery S. Boschen · Daniel Theis · Klaus Ruedenberg · Theresa L. Windus

Received: 15 October 2013 / Accepted: 12 November 2013 / Published online: 7 December 2013

Abstract The diatomic carbon molecule has a complex electronic structure with a large number of low-lying electronic excited states. In this work, the potential energy curves (PECs) of the four lowest lying singlet states ($X^1\Sigma_g^+$, $A^1\Pi_u$, $B^1\Delta_g$, and $B'^1\Sigma_g^+$) were obtained by high-level ab initio calculations. Valence electron correlation was accounted for by the correlation energy extrapolation by intrinsic scaling (CEEIS) method. Additional corrections to the PECs included core–valence correlation and relativistic effects. Spin–orbit corrections were found to be insignificant. The impact of using dynamically weighted reference wave functions in conjunction with CEEIS was examined and found to give indistinguishable results from the even weighted method. The PECs showed multiple curve crossings due to the $B^1\Delta_g$ state as well as an avoided crossing between the two $^1\Sigma_g^+$ states. Vibrational energy levels were computed for each of the four electronic states, as well as rotational constants and spectroscopic parameters. Comparison between the theoretical and experimental results showed excellent agreement overall. Equilibrium bond distances are reproduced to within 0.05 %. The dissociation energies of the states agree with experiment to within $\sim$0.5 kcal/mol, achieving "chemical accuracy." Vibrational energy levels show average deviations of $\sim$20 cm^{-1} or less. The $B^1\Delta_g$ state shows the best agreement with a mean absolute deviation of 2.41 cm^{-1}. Calculated rotational constants exhibit very good agreement with experiment, as do the spectroscopic constants.

Keywords Diatomic carbon · Ab initio electronic structure · Dissociation · Configuration interaction · Spectroscopic properties · Multi-configurational wave functions

Dedicated to Professor Thom Dunning and published as part of the special collection of articles celebrating his career upon his retirement.

Electronic supplementary material The online version of this article (doi:10.1007/s00214-013-1425-x) contains supplementary material, which is available to authorized users.

J. S. Boschen · D. Theis · K. Ruedenberg · T. L. Windus (✉)
Department of Chemistry and Ames Laboratory (USDOE), Iowa State University, Ames, IA 50011, USA
e-mail: twindus@iastate.edu

1 Introduction

Diatomic carbon has been studied spectroscopically in sources such as stars [1], comets [2], the interstellar medium [3], and hydrocarbon combustion reactions [4]. The molecule has a large number of low-lying excited states, which have been probed in numerous studies [5–19]. One-band system of note is the Swan system, which involves the $d^3\Pi_g - a^3\Pi_u$ transitions. The high-intensity transitions of this system led to the early inference that the $a^3\Pi_u$ state was the ground state [5]. However, the $X^1\Sigma_g^+$ state was later identified as the ground state with only 700 cm^{-1} separating the two states [6]. C_2 also has several low-lying singlet states, among which the Phillips system ($A^1\Pi_u - X^1\Sigma_g^+$) is well studied. On the basis of these data, the quality of theoretical potential energy curves (PECs) can be assessed by comparing the theoretical rotational–vibrational levels with the experimental values. At the present state of the art, theoretical PECs that reproduce the rotational–vibrational levels to spectroscopic accuracy ($\sim$1 cm^{-1}) or near spectroscopic accuracy ($\sim$10 cm^{-1}) are considered highly accurate.

From a theoretical perspective, the complex electronic structure of C_2 offers a challenge to ab initio methods [20–41]. The large number of low-lying excited states leads to several avoided crossings. Even for the ground state PEC, a reasonable description of C_2 must account for the strong multi-reference character. Indeed, the fundamental nature of the bonding in C_2 is still an active area of discussion in the literature [36, 40]. By virtue of these attributes, C_2 represents a good system for testing new ab initio methods.

Recently, ground state PECs have been calculated using the full configuration interaction quantum Monte Carlo (FCIQMC) method [37], and the explicitly correlated internally contracted multi-reference coupled-cluster method (ic-MRCCSD(F12*)) [41]. The multi-reference correlation consistent composite approach (MR-ccCA) [38] has been applied to the $X^1\Sigma_g^+$, $B^1\Delta_g$, and $B'^1\Sigma_g^+$ states. A comparison of internal contraction schemes in multi-reference configuration interaction (ic-MRCI) [39] was performed on the $X^1\Sigma_g^+$ and $B'^1\Sigma_g^+$ states. Shi et al. [34, 35] recently performed MRCI calculations to obtain the PECs for several ($X^1\Sigma_g^+$, $A^1\Pi_u$, $B^1\Delta_g$, $B'^1\Sigma_g^+$, $C^1\Pi_g$, $D^1\Sigma_u^+$, $E^1\Sigma_g^+$, and $1^1\Delta_u$) electronic states of C_2.

An accurate description of this molecule must account for both static correlation at the reference level and dynamic correlation using a highly correlated method. To account for static correlation within a group of states (ground state and excited states), state-averaged multi-configurational self-consistent field (SA-MCSCF) calculations are commonly used. Often, the states are weighted evenly. However, to ensure smoothly varying reference PECs, dynamically weighted (DW-MCSCF) [42] procedures are also used. The impact of using dynamical weighting versus even weighting on the dynamic correlation will be examined in this work.

In order to recover dynamic correlation, the present study uses the method of correlation energy extrapolation by intrinsic scaling (CEEIS) [43–45]. This approach has been used to obtain highly accurate ground state PECs for the first-row diatomics B_2, O_2, F_2, from which rotational–vibrational energy levels with near spectroscopic accuracy were obtained [46–48]. In contrast to the other ab initio approaches used recently for C_2, the present approach focuses on recovering the correlation energy of higher excitation levels by CEEIS extrapolation for smaller basis sets and then extrapolating to the complete basis set limit. Using a generalized form of this method that extrapolates correlation energies for multiple electronic states, the PECs of the four lowest energy singlet states of C_2 ($X^1\Sigma_g^+$, $A^1\Pi_u$, $B^1\Delta_g$, $B'^1\Sigma_g^+$) are obtained in this study. The spectroscopic constants and rotational–vibrational levels corresponding to these curves are compared to the experimental values.

2 Method

The PEC calculations for the dissociation of C_2 in this work follow the general method developed by Bytautas et al. in earlier studies on the diatomics F_2, O_2, and B_2 [46–51]. In addition to the $X^1\Sigma_g^+$ ground state, PECs are calculated for the three lowest lying excited singlet states: $A^1\Pi_u$, $B^1\Delta_g$, and $B'^1\Sigma_g^+$. The calculations rely on the CEEIS method to obtain the valence correlation with near full configuration interaction (FCI) accuracy. Additional corrections due to core–valence correlation and relativistic effects are added to achieve near spectroscopic accuracy. Where feasible, complete basis set (CBS) extrapolations are performed. All ab initio electronic structure calculations were completed using the GAMESS program suite [52].

2.1 Zeroth-order wave function

An accurate description of the C_2 dissociation requires a multi-reference wave function to capture the static correlation of the system. Of particular note is the strong multi-configurational character of the ground state even at the experimental equilibrium distance of 1.242 Å where the primary determinant is $\left|(\text{core})2\sigma_g^2 2\sigma_u^2 1\pi_{x_u}^2 1\pi_{y_u}^2\right\rangle$. However, a doubly excited configuration $\left|(\text{core})2\sigma_g^2 1\pi_{x_u}^2 1\pi_{y_u}^2 3\sigma_g^2\right\rangle$ has a coefficient of 0.37, which is unusually large for a ground state molecule at equilibrium. The excited states also show multi-configurational character (see the Supporting Information for more details). Therefore, the reference energies and orbitals for all subsequent calculations were obtained from MCSCF calculations in the full valence configuration space. In C_2, the valence space consists of the full optimized reaction space (FORS) [53–56] of eight electrons in eight orbitals (CAS(8,8) in the complete active space notation [57]). The core consists of four electrons in two core orbitals. In the D_{2h} symmetry group used by GAMESS for calculations on linear molecules, three of the investigated states ($X^1\Sigma_g^+$, $B^1\Delta_g$, and $B'^1\Sigma_g^+$) belong to the fully symmetric A_g irreducible representation (irrep). The reference functions of these states were obtained by state-averaged MCSCF (SA-MCSCF) calculations. The $A^1\Pi_u$ state (B_{2u} irrep in D_{2h}) was calculated separately by state-specific MCSCF.

In addition to the evenly weighted SA-MCSCF calculations over the three states (referred to as SA-MCSCF for the rest of the paper), dynamically weighted MCSCF (DW-MCSCF) functions were also optimized at various bond distances. The DW-MCSCF method of Deskevich et al. [42] minimizes a weighted average of the state energies. The weight assigned to state i is given by the formula:

$$w_i = \operatorname{sech}^2[-\beta(E_i - E_0)] = 4/(2 + e^{2\beta(E_i-E_0)} + e^{-2\beta(E_i-E_0)})$$

where β is an adjustable parameter, E_i is the energy of state i, and E_0 is the ground state energy. The reported results used $\beta^{-1} = 2.0$ eV (chosen from the recommended range of values from Deskevich et al.), and other values were tested and showed similar results. DW-MCSCF has been shown to reduce discontinuities and smooth out potential energy surfaces. The impact of using dynamically weighted reference orbitals with the highly correlated CEEIS method will be examined below.

The reference calculations were performed using Dunning's series of correlation consistent basis sets, cc-pVXZ [58, 59] where $X = 4$, 5, and 6. The reference energies were extrapolated to the complete basis set limit using the three-point formula [60, 61]:

$$E(X) = E(\mathrm{CBS}) + a\,e^{-\alpha X}$$

where the three unknowns [E(CBS), a, and α] are determined by calculating $E(X)$ for each of the three different basis sets.

2.2 Valence correlation

2.2.1 *CEEIS method for a single state*

The dynamic valence correlation was determined using the CEEIS method, developed by Ruedenberg and Bytautas [43–45] to approximate the FCI energy. Only a brief description of the method is included here for context. The FCI energy is expressed as a sum of contributions from higher and higher levels of configurational excitation, until all possible configurations have been included. However, since the rapid increase in the number of configurations with increasing system size makes full CI calculations prohibitively expensive for most applications, excitation levels beyond the doubles are typically ignored in favor of the truncated CISD method. The CEEIS method estimates the higher excitation energy contributions from these truncated calculations by means of an extrapolation technique.

The CEEIS procedure can be carried through with respect to single determinant Hartree–Fock reference functions or multi-determinant MCSCF reference functions. A set of correlating virtual orbitals is needed *that is ordered according to decreasing importance*. To this end, the pseudo-natural orbitals [62], i.e., the natural orbitals from a multi-reference CISD calculation (MR-CISD), ordered by their occupation numbers have been found to provide an effective set for the CEEIS procedure. The pseudo-natural orbitals are produced by diagonalizing only the virtual block of the one-particle density matrix. For the rest of this paper, these orbitals will be referred to as natural orbitals even though the full density matrix has not been diagonalized. The number of correlating orbitals, M, is equal to the total number of virtual orbitals.

In the CEEIS procedure, the values of the contributions from the double and triple excitations are used to estimate the energy changes due to the higher excitation levels $x = 4$, 5, 6, etc., i.e., quadruple, quintuple, and sextuple excitations. Note that x is used to denote the excitation level, whereas X was used above to denote the basis set size. Let $E(x)$ be the CI energy when all configurations up to and including excitation level x are taken into account. The *incremental* energy contributions due to the inclusion of excitation level x are then defined as

$$\Delta E(x) = E(x) - E(x-1) \quad \text{for} \quad x > 2,$$

e.g., $\Delta E(3)$ would be the difference between the CISDT and CISD energies.

Bytautas and Ruedenberg [43] found that the energy change $\Delta E(x)$ can be related to the energy change from excitations two levels lower, i.e., $\Delta E(x-2)$. For example, the energy change due to quadruple excitations can be estimated from that due to double excitations. The relation is established by considering correlation energy increments that are analogous to $\Delta E(x)$, but are instead obtained from CI calculations with excitations into smaller subsets of m ($<M$) virtual orbitals. If these increments are denoted as

$$\Delta E(x|m) = E(x|m) - E(x-1|m),$$

then $\Delta E(x|m)$ manifestly becomes $\Delta E(x)$ when m → M. Bytautas and Ruedenberg showed that, in all systems that were examined, the following linear relationship holds with respect to the variation of m:

$$\Delta E(x|m) = a_x \Delta E(x-2|m) + c_x.$$

The linear relationship is not observed at very small values of m, but is seen to hold for m running from some threshold m_0 to M. Consequently, the following extrapolation is found to be effective. Values of $\Delta E(x|m)$ and $\Delta E(x-2|m)$ are calculated *for a range of m values considerably smaller than M*. From these data, the coefficients a_x and c_x in the linear relation are determined by a least-mean-squares fit. The known value of $\Delta E(x-2)$ is then inserted into the linear equation for $\Delta E(x-2|m=M)$, and, thereby, an extrapolated value is produced for the unknown value of $\Delta E(x) = \Delta E(x|m=M)$. The extrapolations must be performed with calculations involving at least m_0 virtual orbitals. Careful selection of the range of m values is necessary to ensure an extrapolation of high accuracy. Detailed information on the choice of effective ranges can be found in past work describing the CEEIS method [43–47].

The contributions from the singles and double excitations $[\Delta E(x = 2) = E(x = 2) - E(x = 0)]$ are computed exactly. If practical, this is also done for the contributions from the triple excitations. If, however, a prohibitive effort is required for the latter, then the value of $\Delta E(x = 3)$ is also obtained by extrapolation from the single and double excitations by an analogous linear extrapolation. The range of m for this extrapolation typically extends to higher values.

2.2.2 *CEEIS method for a set of several states*

Bytautas et al. [48] also showed that the CEEIS method can be applied simultaneously to several states of the same symmetry. Building upon that observation, the analogous multiple state CEEIS approach has been developed further in considerable detail and has been incorporated into GAMESS [63]. Appropriate reference functions are obtained from a SA-MCSCF calculation that includes all the states of interest in the state averaging. Correlating virtual orbitals are obtained from a preliminary MRCISD calculation. In contrast to the single-state case, the one-particle density matrices of the MRCISD wave functions are averaged over the states of interest and the virtual block is diagonalized to give state-averaged natural orbitals. These averaged natural orbitals form the set of M correlating virtual orbitals used in the multistate CEEIS procedure. Multiple-root CI calculations with higher excitations (CISDT, CISDTQ, …) are then computed for a prespecified range of m values as in the single-state case. The eigenvalues of these calculations provide the values $\Delta E_k(x|m)$ for each of the states $|k\rangle$ under investigation, which are then used to extrapolate the total contribution that excitation x will make to the full CI energy for each state. The computation and extrapolation of the multiple states have been automated within GAMESS. However, the user still needs to be careful to correctly identify the CI states when the states are very close in energy. The energy order of near degenerate states may change as calculations are performed throughout the range of correlating orbitals, m. Therefore, one must ensure that all of the energy differences used in the extrapolations are associated with the same reference wave function (*i.e.*, have the same dominant electronic configurations).

2.2.3 *CEEIS procedure for* C_2

For the CEEIS calculations on C_2, the full valence CASSCF(8,8) wave functions described above are used as a reference. The correlating virtual orbitals used are the natural orbitals from MR-CISD calculations with respect to the CASSCF(8,8) references. For the reference functions of the three states $X^1\Sigma_g^+$, $B^1\Delta_g$, and $B'^1\Sigma_g^+$, which are obtained from a SA-MCSCF calculation in the A_g irrep, the correlating orbitals are obtained from the virtual block of the state-averaged density matrix. The number of virtual orbitals is $M = 50$ and $M = 100$ for the cc-pVTZ and cc-pVQZ basis sets, respectively.

The double and triple excitation contributions were calculated exactly in all cases, except for the triple excitations in the cc-pVQZ basis, which was too expensive. These were obtained by extrapolation from the double excitations, as discussed above. The range of virtual orbitals used for the extrapolation of the triples contribution was m:{18–25, 30, 35}.

The CEEIS procedure was performed up to sextuple excitations. For the extrapolation of the quadruple contribution, the range of correlating virtual orbitals used was m:{18–25}. For the quintuple and sextuple excitations, the ranges were m:{13–17} and {10–14}, respectively. In addition, it is important to note that m is chosen so that degenerate orbitals remain paired.

CBS extrapolation of the correlation energy was performed using the cc-pVXZ bases and the two-point formula [64, 65]:

$$\Delta E_{\mathrm{CORR}}(X) = \Delta E_{\mathrm{CORR}}(\mathrm{CBS}) + a_c X^{-3}$$

where ΔE_{CORR} is the difference between the reference and the estimated FCI energy. Using $X = 3$ and 4, the CBS limit for the correlation energy was determined. Addition of the CBS correlation energy to the CBS reference energy yielded the valence-correlated approximate FCI PECs. The uncertainty in the CEEIS energies was estimated to be ~0.1 millihartree.

2.3 Corrections to the PECs

Past studies [50, 51] have shown that reproducing the rotational–vibrational energy levels to near spectroscopic accuracy requires additional corrections to the valence-correlated PECs. The first contribution is the inclusion of core electron correlation. As in previous studies, these effects were captured using MRCISD calculations including the Davidson correction (+Q) [66]. The Dunning triple-zeta basis set modified for core–valence effects, cc-pCVTZ [58], was used. The reference orbitals for the MRCISD+Q calculations were taken from a full valence CASSCF(8,8) calculation. In the valence only MRCISD+Q, excitations are only allowed from the original CAS(8,8) space. The valence plus core MRCISD+Q calculation allowed single and double excitations from a CAS(12,10) space that included the core orbitals. The core–valence correlation correction is obtained from the difference between these two energies.

Relativistic effects are accounted for by two corrections. The first is the scalar relativistic contribution, which was obtained using the one-electron Douglas–Kroll (DK) method [67]. The transformation to third-order (DK3) [68, 69] approach was applied at the CASSCF(8,8) level. A modified DK-contracted basis, cc-pCVQZ [70] was used. The second relativistic correction is due to the spin–orbit (SO) coupling. SO coupling effects were computed using the full one- and two-electron Breit–Pauli operator [71]. Equivalent orbitals were obtained through SA-MCSCF calculations over the 18 states with $M_S = 0$ that dissociate to the ground state ^{3}P atomic terms. The active space for these calculations was a reduced valence space including only the molecular orbitals that arise from the 2p atomic orbitals (CAS(4,6)). These CAS-CI states form the basis for the SO calculation. Addition of the Breit–Pauli operator introduces off-diagonal terms into the Hamiltonian matrix, which generate the spin–orbit couplings between states. Diagonalizing the Hamiltonian produces spin-mixed states. The energy lowering of these states relative to the CAS-CI states provides the SO correction to the PECs. The cc-pVQZ basis was used for these calculations. A more detailed description of the procedure has been given for the molecules F_2 and O_2 [49, 51]. Adding all corrections to the energies calculated with the CEEIS method yields the final potential energy curves.

2.4 Fitting continuous functions to PECs

In order to determine spectroscopic constants and rotational–vibrational levels for comparison with experiment, the potential energy curves must be fitted to a continuous function. Bytautas and Ruedenberg have used even-tempered Gaussian functions to fit PECs of the diatomics O_2, F_2, and B_2 [48, 50, 51]. The terms of the expansion are found using linear least-squares regression. This approach yielded analytic curves that show high-quality fits with two of the states of interest: $A^1\Pi_u$ and $B^1\Delta_g$ (mean absolute deviations of 0.044 and 0.037 millihartree).

However, the even-tempered Gaussian functions fit to the two Σ_g^+ states showed mean absolute deviations (MADs) an order of magnitude greater (0.219 and 0.296 millhartree). Increasing the number of Gaussians in the expansion did not change the quality of the fit to the ab initio data. The even-tempered Gaussian expansions proved incapable of capturing the irregular shape of the Σ_g^+ curves due to an avoided crossing which occurs near 1.70 Å on the PECs. Presumably, these local distortions are difficult to represent in terms of reasonably simple analytic (i.e., everywhere infinitely differentiable) functions. As an alternative, cubic splines were fitted to the ab initio data using the module VIBROT within MOLCAS [72]. The PECs were constructed from 44 ab initio calculations along the dissociation path. The points range from 0.9 to 6.0 Å with an additional point at 20.0 Å to determine the dissociated values. The points were chosen to adequately describe both the minimum and the avoided crossing regions of the potential and to provide a sufficient density of points to fit the cubic spline.

2.5 Calculation of rotational–vibrational energy levels

The rotational–vibrational energy levels are found by solving the nuclear Schrödinger equation for the analytical representations of the PECs. When the analytical function was an even-tempered Gaussian expansion, the discrete variable representation (DVR) method [73] was used to solve for the rotational–vibrational levels. For the functions using cubic splines, the Schrödinger equation was solved using Numerov's method [74] (in the program VIBROT) [72]. The two different approaches yielded similar results for the $^1\Pi_u$ and $B^1\Delta_g$ states (deviations ~ 1 cm^{-1} and ~ 10 cm^{-1}, respectively). However, the energy levels obtained from an even-tempered Gaussian fit and a cubic spline for the Σ_g^+ states show large disagreement (~ 100 cm^{-1}). This supports the inference that the even-tempered Gaussian expansion cannot describe the avoided crossing exhibited by the Σ_g^+ states. The spectroscopic constants were determined by least-mean-squares fitting to the rotational–vibrational levels [9].

3 Results

3.1 Potential energy curves

The ab initio PECs curves are presented in Fig. 1. These energies include the CBS extrapolated reference energies, the CBS extrapolated CEEIS valence correlation energies, the core–valence correlation, and the relativistic corrections. These curves reveal the complexity of the electronic structure of C_2. At distances longer than about 1.6 Å, all four states are close in energy. The $B^1\Delta_g$ state is seen to drop below both the $X^1\Sigma_g^+$ and $A^1\Pi_u$ states near 1.6 Å. Another aspect of interest is the avoided crossing between the two Σ_g^+ states which occurs in the region of $R = 1.6$ Å. This avoided crossing is accompanied by a change in the dominant configurations of the two states as well as a distortion of the shapes of the curves. This distortion is the likely explanation for the failure of the even-tempered Gaussian expansion to accurately fit the Σ_g^+ states. As the

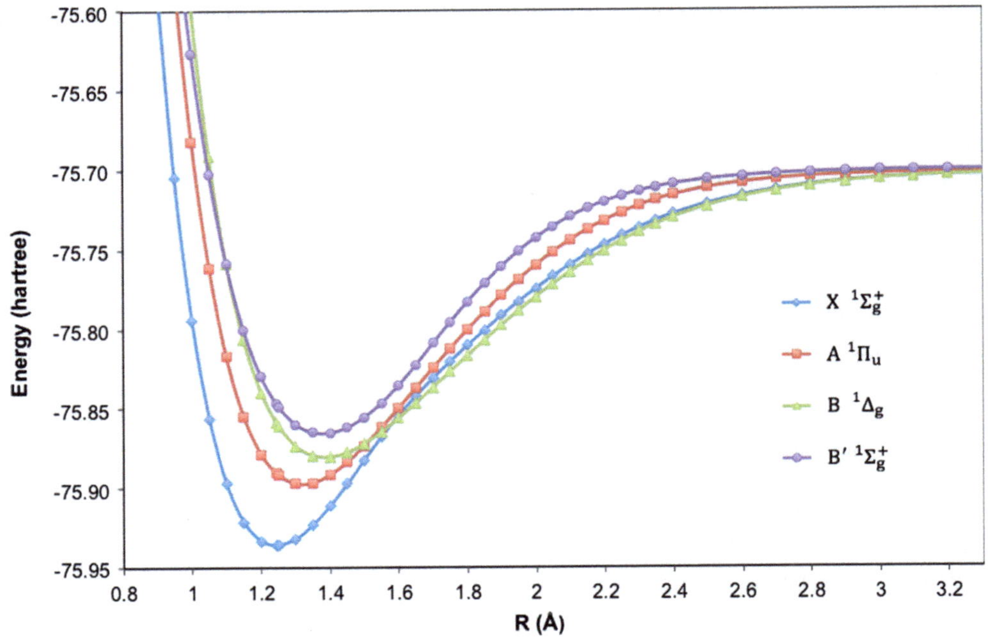

Fig. 1 Ab initio PECs of the lowest energy singlet states of C_2, energies obtained using CBS extrapolated reference, and CEEIS valence correlation energy with core–valence correlation and relativistic corrections

Table 1 Energy differences (millihartree) between SA and DW at the MCSCF reference and CEEIS level

R (Å)	MCSCF			CEEIS		
	$X^1\Sigma_g^+$	$B^1\Delta_g$	$B'^1\Sigma_g^+$	$X^1\Sigma_g^+$	$B^1\Delta_g$	$B'^1\Sigma_g^+$
1.0	13.720	−50.437	−33.531	−0.016	−0.213	−0.079
1.2	12.151	−28.189	−23.413	0.024	−0.199	−0.166
1.25	9.534	−15.148	−12.203	0.028	−0.147	−0.118
1.4	2.507	−1.935	−1.368	0.017	−0.034	−0.026
1.6	0.092	−0.057	−0.037	0.016	−0.018	0.001
1.8	0.023	0.074	−0.101	0.003	0.003	−0.001
2.0	0.042	0.071	−0.118	0.003	0.003	−0.001
2.4	0.007	0.009	−0.016	0.000	0.000	0.000
2.8	0.001	0.001	−0.002	−0.004	0.004	0.000
3.2	0.000	0.000	0.000	−0.005	0.005	0.000

Reported here as the $E(\mathrm{SA}) - E(\mathrm{DW})$

two atoms separate, all four of the states become degenerate, dissociating to the two carbon atoms in their ^3P ground states.

3.2 Dynamic versus even weighting in averaging the reference states

The impact of using DW-MCSCF versus SA-MCSCF reference orbitals with the CEEIS method was examined. CEEIS energies were calculated using the cc-pVTZ basis and DW reference orbitals at 10 points along the dissociation curve. The energy differences between the SA and DW calculations are reported in Table 1. While the reference energies at the MCSCF level are very sensitive to the weighting scheme, the CEEIS energies including dynamic correlation are not significantly changed. Furthermore, the DW weights quickly converge to a near even weighted SA due to the decrease in energy differences between states as the separation distance increases. The maximum impact on the CEEIS energies ($\sim$0.2 millihartree) is of the same order as the uncertainty in the CEEIS extrapolations. For most of the PEC, the change due to using a DW-MCSCF reference is well below the uncertainty in the CEEIS method. Since the FCI energy is independent of the reference wave function, the observed insensitivity to the weighting of states in determining the reference orbitals confirms the soundness of the CEEIS method.

These results are in agreement with Zeng et al. [75] who found little difference between DW- and SA-MCSCF orbitals when used in conjunction with multi-configurational quasidegenerate perturbation theory (MCQDPT). Their work on the Sn_2^+ dissociation showed minimal change in the spectroscopic constants and vibrational energy levels. Based on this previous study and the aforementioned results, the C_2 PECs used in the present study were obtained using SA-MCSCF orbitals.

3.3 Contributions to the PECs

As described above, the PECs include three additional corrections beyond the valence-correlated CEEIS method. Tables with complete information for each contribution at each point on the PECs are provided in the Supporting Information. In the following, all energies are given relative to the value at dissociation.

The most significant correction is the inclusion of the core–valence correlation. Each of the four states exhibits the same qualitative trend in and similar quantitative contributions from the core–valence correlation. The following discussion will therefore focus on the ground state. The core–valence correction grows as the atoms move closer together, and the largest value in these calculations was −6.870 millihartree at 0.9 Å. At equilibrium (1.24244 Å), the correction was −2.906 millihartree.

Of the two relativistic corrections, the scalar Douglas-Kroll method has a greater impact on the shape of the PECs. At most bond distances, the DK3 correction is positive relative to the value at dissociation. For the ground state, this repulsive effect is strongest at 1.4 Å where the value is 0.335 millihartree. As the bond length increases, the correction smoothly decreases. As the bond length decreases from 1.4 Å, the DK3 correction decreases and eventually drops below the dissociated value as the repulsive wall of the PEC is approached. This trend is observed in all the calculated states, although the energy changes are somewhat smaller for the excited states.

The second relativistic contribution, the spin–orbit coupling, causes no significant change in the shape of the PECs. For every state, the contribution is effectively zero inside the potential wells. The SO effects do become nonzero at bond lengths longer than 2.3 Å. As the carbon atoms become well separated, the SO effect lowers the energy of the system. However, even at the dissociation limit, the energy difference is quite small. For the ground state, the difference in energy due to SO coupling between equilibrium and dissociation is only 0.246 millihartree. The impact is slightly smaller for the other three states but on the same order of magnitude.

3.4 Comparison of theoretical and experimental results

3.4.1 *Equilibrium bond distance and electronic energies*

Table 2 reports the experimental and theoretical values for the equilibrium bond distance (R_e), for the dissociation energy from the potential curve minimum (D_e) and for the adiabatic electronic excitation energy from the ground state (T_e). Excellent agreement between experiment and theory is seen for the equilibrium bond distances (percent differences of 0.05 % or less). The experimental value of D_e for the excited states is obtained by adding the experimental values for T_e to the experimentally determined ground state D_e. The theoretical dissociation energies deviate from experiment by <0.6 kcal/mol, demonstrating "chemical accuracy." The dissociation energies in this work show better agreement with the most recent experimental values [76] than other recent high-level ab initio studies [34, 35,

Table 2 Experimental and theoretical spectroscopic constants of C_2 singlet states

	Experiment	Theory	Difference
$X^1\Sigma_g^+$			
R_e (Å)	1.24244	1.2431	0.00066
D_e (kcal/mol)[a]	147.8	147.42	−0.38
T_e (cm^{-1})	–	–	–
ω_e (cm^{-1})	1,855.01	1,850.91	−4.10
$\omega_e x_e$ (cm^{-1})	13.5547	10.5857	−2.969
B_e (cm^{-1})	1.82010	1.82008	−0.00002
α_e (cm^{-1})	0.018012	0.0177588	−0.0002532
D_{rot} (10^{-6} cm^{-1})	6.9640	6.8950	−0.0690
β_e (10^{-8} cm^{-1})	6.41	9.59	3.18
$A^1\Pi_u$			
R_e (Å)	1.318311	1.3176	−0.000711
D_e (kcal/mol)[b]	123.8	123.41	−0.39
T_e (cm^{-1})	8,391.4085	8,413.9989	22.5904
ω_e (cm^{-1})	1,608.20	1,616.52	8.32
$\omega_e x_e$ (cm^{-1})	12.0597	13.1386	1.0789
B_e (cm^{-1})	1.61663	1.61693	0.00030
α_e (cm^{-1})	0.0169691	0.0168091	−0.00016
D_{rot} (10^{-6} cm^{-1})	6.5086	6.4856	−0.0230
β_e (10^{-8} cm^{-1})	2.53	3.12	0.59
$B^1\Delta_g$			
R_e (Å)	1.38548	1.3851	−0.00038
D_e (kcal/mol)[b]	113.3	112.73	−0.57
T_e (cm^{-1})	12,082.3360	12,162.6256	80.2896
ω_e (cm^{-1})	1,407.47	1,403.23	−4.24
$\omega_e x_e$ (cm^{-1})	11.47937	9.7608	−1.71857
B_e (cm^{-1})	1.46369	1.46342	−0.00027
α_e (cm^{-1})	0.0168161	0.0156176	−0.0011985
D_{rot} (10^{-6} cm^{-1})	6.3188	6.3475	0.0287
β_e (10^{-8} cm^{-1})	1.492	0.559	−0.933
$B'^1\Sigma_g^+$			
R_e (Å)	1.37735	1.3771	−0.00025
D_e (kcal/mol)[b]	103.8	103.43	−0.37
T_e (cm^{-1})	15,409.139	15,425.9939	16.8549
ω_e (cm^{-1})	1,424.12	1,413.15	−10.97
$\omega_e x_e$ (cm^{-1})	2.57113	1.62016	−0.95097
B_e (cm^{-1})	1.481006	1.48053	−0.00048
α_e (cm^{-1})	0.011752	0.012027	0.000275
D_{rot} (10^{-6} cm^{-1})	6.8596	6.4618	−0.3978
β_e (10^{-8} cm^{-1})	−15.81	−19.55	−3.74

Experimental values from references [11, 12] unless otherwise noted

[a] Experimental values from Ref. [76]

[b] Experimental value obtained by adding ground state D_e plus excited state T_e

Table 3 Comparison of theoretical and experimental vibrational energy levels of C_2 singlet states (Energies in cm^{-1})

$G(v)$				$G(v) - G(v-1)$		
v	Experiment	Theory	Difference	Experiment	Theory	Difference
$X^1\Sigma_g^+$						
0	923.98	919.49	−4.49	–	–	–
1	2,751.47	2,749.85	−1.62	1,827.48	1,830.36	2.88
2	4,550.67	4,555.98	5.31	1,799.20	1,806.13	6.93
3	6,320.57	6,330.44	9.87	1,769.91	1,774.46	4.55
4	8,060.33	8,071.98	11.65	1,739.76	1,741.54	1.78
5	9,768.11	9,782.08	13.97	1,707.77	1,710.10	2.33
6	11,441.95	11,458.92	16.97	1,673.84	1,676.84	3.00
MAD			9.12			3.58
$A^1\Pi_u$						
0	801.10	806.64	5.54	–	–	–
1	2,385.15	2,394.98	9.83	1,584.05	1,588.34	4.29
2	3,944.97	3,959.64	14.67	1,559.83	1,564.66	4.83
3	5,480.53	5,499.90	19.37	1,535.56	1,540.26	4.70
4	6,991.74	7,014.93	23.19	1,511.21	1,515.03	3.82
5	8,478.54	8,504.38	25.84	1,486.79	1,489.45	2.66
MAD			16.41			4.06
$B^1\Delta_g$						
0	700.95	699.57	−1.38	–	–	–
1	2,085.49	2,081.44	−4.05	1,384.54	1,381.87	−2.67
2	3,447.16	3,445.98	−1.18	1,361.67	1,364.54	2.87
3	4,786.03	4,786.06	0.03	1,338.87	1,340.08	1.21
4	6,102.15	6,104.61	2.46	1,316.12	1,318.55	2.43
5	7,395.60	7,400.96	5.36	1,293.45	1,296.35	2.90
MAD			2.41			2.42
$B'^1\Sigma_g^+$						
0	712.74	706.83	−5.91	–	–	–
1	2,133.23	2,116.28	−16.95	1,420.48	1,409.45	−11.03
2	3,552.75	3,527.01	−25.74	1,419.52	1,410.73	−8.79
3	4,974.08	4,936.74	−37.34	1,421.34	1,409.73	−11.61
MAD			21.48			10.48

Experimental values from references [11, 12]

38]. The theoretical adiabatic excitation energies are in close agreement with experiment, showing percent differences of 0.27, 0.66, and 0.11 % for the $A^1\Pi_u$, $B^1\Delta_g$, and $B'^1\Sigma_g^+$ states, respectively.

3.4.2 Vibrational spectrum

Table 3 shows a comparison between the experimental vibrational levels [11, 12] and the theoretical results obtained in this work. The energy levels reported are given relative to the minimum of the PEC of each respective state. The MADs of the four states are on the order of 10–20 cm^{-1}, demonstrating the near spectroscopic accuracy of the ab initio PECs. However, there is a clear trend of decreasing accuracy with increasing v. Particularly, good agreement (MAD = 2.41 cm^{-1}) is obtained for the $B^1\Delta_g$ state, while the worst agreement is seen for the $B'^1\Sigma_g^+$ state. Also included in Table 3 are the energy differences between the vibrational levels ($G(v) - G(v-1)$). These vibrational spacings show smaller deviations between theory and experiment; however, they cumulatively lead to the larger deviations observed in the absolute vibrational levels. In Table 4, additional ab initio vibrational levels for each state (up to $v = 24$) are reported. The higher vibrational levels currently lack experimental values for comparison. All bound vibrational levels for each PEC are listed in the Supporting Information: 57 levels for the $X^1\Sigma_g^+$, 54 for

Table 4 Theoretical vibrational levels of C_2 singlet states

v	$X^1\Sigma_g^+$	$A^1\Pi_u$	$B^1\Delta_g$	$B'^1\Sigma_g^+$
0	919.49	806.64	699.57	706.83
1	2,749.85	2,394.98	2,081.44	2,116.28
2	4,555.98	3,959.64	3,445.98	3,527.01
3	6,330.44	5,499.90	4,786.06	4,936.74
4	8,071.98	7,014.93	6,104.61	6,344.88
5	9,782.08	8,504.38	7,400.96	7,749.06
6	11,458.92	9,968.70	8,674.54	9,146.88
7	13,100.36	11,408.50	9,924.90	10,535.60
8	14,702.82	12,823.87	11,152.23	11,911.95
9	16,261.81	14,214.39	12,356.91	13,271.97
10	17,773.27	15,579.26	13,539.17	14,611.66
11	19,234.77	16,918.24	14,698.87	15,927.89
12	20,646.17	18,231.53	15,835.77	17,219.13
13	22,009.10	19,519.03	16,949.92	18,484.20
14	23,326.85	20,780.33	18,041.63	19,721.42
15	24,603.22	22,015.14	19,110.63	20,929.34
16	25,841.91	23,223.37	20,156.30	22,106.73
17	27,046.53	24,404.87	21,179.08	23,252.29
18	28,219.42	25,559.47	22,180.00	24,364.26
19	29,362.03	26,687.03	23,158.49	25,440.71
20	30,475.54	27,787.42	24,113.15	26,480.12
21	31,561.58	28,860.20	25,044.74	27,481.17
22	32,620.69	29,904.57	25,954.80	28,442.21
23	33,652.70	30,919.72	26,842.17	29,361.71
24	34,658.45	31,905.30	27,704.66	30,238.18

(Energies in cm^{-1}, relative to the potential curve minimum of each state)

the $A^1\Pi_u$ states, 49 for the $B^1\Delta_g$ state, and 36 for the $B'^1\Sigma_g^+$ state.

The theoretical vibrational levels of this work for the $X^1\Sigma_g^+$ and $A^1\Pi_u$ states show reduced deviations compared to work by Zhang et al. [34] (MADs of 13.1 and 27.60 cm^{-1} for the $X^1\Sigma_g^+$ and $A^1\Pi_u$ states, respectively, for the Zhang work). However, the same group [35] reports vibrational levels for the $B^1\Delta_g$ and $B'^1\Sigma_g^+$ states which show better agreement with experiment (MADs of 1.37 and 5.15 cm^{-1}, respectively) than this work. Kokkin et al. [30] have obtained vibrational levels for the $X^1\Sigma_g^+$ state which are of similar accuracy (MAD 8.43 cm^{-1}) to this work, while their results for the $A^1\Pi_u$ state show better accuracy (MAD 0.77 cm^{-1}) than this work. All of these other studies employed internally contracted MRCISD calculations using the aug-cc-pV6Z basis and included core–valence correlation and scalar relativistic corrections.

It is worth noting the contribution that each additional correction (core–valence correlation and relativistic effects) makes toward achieving accurate vibrational levels. Including core–valence correlation leads to a ~10–20 cm^{-1} reduction in the MAD for each of the states. The scalar relativistic correction (DK3) is less important, with a ~1 cm^{-1} improvement in the MAD of the vibrational levels. Finally, the SO contribution was seen to make no difference in the vibrational levels of the C_2 singlet states. This is not surprising given the magnitude of the SO correction, and the distances at which the contribution becomes significant. The low levels of the vibrational manifold are not strongly impacted by the SO effects that arise near dissociation. The unimportance of SO effects for determining the vibrational levels of C_2 was also seen by Kokkin et al. [30] in calculations on the $X^1\Sigma_g^+$ and $A^1\Pi_u$ states.

3.4.3 Rotational constants

Table 5 reports the rotational constants (B_v and D_v) from both experiment [11, 12] and this work. The rotational energy can be expanded [9] as

$$F_v(J) = B_v[J(J+1)] - D_v[J(J+1)]^2 + \dots$$

A least-squares fit to the lowest 10 rotational levels at each vibrational level was used to determine the theoretical values of B_v and D_v. The overall agreement between the rotational constants of this work and experiment is quite good. For B_v, the largest deviation of the theoretical values from the experimental ones is ~0.1–0.2 %, while most of the deviations are smaller than this. The D_v values are significantly smaller than B_v (reported in 10^{-6} cm^{-1}) and are therefore expected to be less accurate on a percent basis. The agreement between theory and experiment is still very good, with the exception of the highest energy $B'^1\Sigma_g^+$ state (with a 16 % deviation). For the other states, the percent difference between the values is at worst ~1 % and for most values is significantly lower.

3.4.4 Spectroscopic constants

The relationship between the rotational–vibrational levels and the spectroscopic constants is given by the Dunham expansion in terms of $(v + 1/2)$ [9]:

$$G(v) = \omega_e\left(v+\frac{1}{2}\right) - \omega_e x_e\left(v+\frac{1}{2}\right)^2 + \cdots$$

$$B_v = B_e - \alpha_e\left(v+\frac{1}{2}\right) + \cdots$$

$$D_v = D_{\text{rot}} + \beta_e\left(v+\frac{1}{2}\right) + \cdots$$

Using this expansion, the equilibrium spectroscopic constants given in Table 2 were obtained by a least-mean-

Table 5 Comparison of theoretical and experimental rotational constants of the C_2 singlet states (Energies of B_v in cm^{-1}, D_v in 10^{-6} cm^{-1})

B_v				D_v		
v	Experiment	Theory	Difference	Experiment	Theory	Difference
$X^1\Sigma_g^+$						
0	1.81107	1.81106	−0.00001	7.00315	7.00540	0.00225
1	1.79288	1.79329	0.00041	7.05398	6.98626	−0.06772
2	1.77434	1.77397	−0.00037	7.0944	7.12137	0.02697
3	1.75540	1.75534	−0.00006	7.2066	7.25091	0.04431
4	1.73590	1.73590	0.00000	7.2941	7.27346	−0.02064
5	1.71570	1.71601	0.00031	7.499	7.40893	−0.09007
6	1.69381	1.69553	0.00172			
$A^1\Pi_u$						
0	1.60813	1.60871	0.00058	6.52569	6.51101	−0.01468
1	1.59109	1.59113	0.00004	6.53614	6.52156	−0.01458
2	1.57397	1.57475	0.00078	6.5731	6.55940	−0.01370
3	1.55676	1.55747	0.00071	6.6026	6.59281	−0.00979
4	1.53945	1.54010	0.00065	6.6289	6.63691	0.00801
5	1.52205	1.52265	0.00060	6.657	6.65359	−0.00341
$B^1\Delta_g$						
0	1.45527	1.45548	0.00021	6.3259	6.37096	0.04506
1	1.43843	1.43964	0.00121	6.34196	6.30755	−0.03441
2	1.42155	1.42346	0.00191	6.3575	6.41764	0.06014
3	1.40464	1.40646	0.00182	6.3671	6.33773	−0.02937
4	1.38772	1.38928	0.00156	6.4035	6.35229	−0.05121
5	1.37074	1.37223	0.00149	6.3883	6.39922	0.01092
$B'^1\Sigma_g^+$						
0	1.47531	1.47472	−0.00059	6.781	6.39399	−0.38701
1	1.46482	1.46337	−0.00145	6.6208	6.12939	−0.49141
2	1.45614	1.45375	−0.00239	6.744	5.96153	−0.78247
3	1.44786	1.44437	−0.00349	6.881	5.79821	−1.08279

Experimental values from references [11, 12]

squares fitting to the energy levels in Table 3 and the rotational constants in Table 5. The fittings were calculated using VIBROT [72].

The theoretical values for the first terms in each of the expansions, ω_e, B_e, and D_{rot}, exhibit very small deviations from experiment (average percent differences of 0.45, 0.02, and 1.9 %, respectively). The B_e results in particular show excellent agreement. The average percent difference for D_{rot} is 0.6 % if the $B'^1\Sigma_g^+$ state is excluded. The second terms in the expansions are less accurate since the values themselves are quite small to begin with. However, the qualitative trends are correct, and the agreement for the α_e values is still quite good (average difference 3.0 %), while the differences for $\omega_e x_e$ and β_e exhibit average differences of 21 and 40 %, respectively.

The accuracy of the spectroscopic constants in Table 2 is comparable to other recent high-level ab initio studies. Some of these studies (using ic-MRCISD and MR-ccCA) [30, 34, 38] have determined the ground state harmonic frequency (ω_e) with absolute deviations from experiment of 1.5 cm^{-1} or less, while others using ic-MRCISD and ic-MRCCSD(F12*) methods [32, 41] have shown deviations of $\sim$6–8 cm^{-1}. The deviation in the present work is 4.1 cm^{-1}. The first anharmonicity ($\omega_e x_e$) of the $X^1\Sigma_g^+$ state in this work shows slightly worse agreement with experiment (3 cm^{-1} difference) than the previous [30, 34, 39, 41] theoretical results ($\sim$1 cm^{-1} or better). In the $A^1\Pi_u$ state, the 8 cm^{-1} deviation in the current results for ω_e matches the deviation in reference 34, while two previous studies [30, 38] have achieved $\sim$1 cm^{-1} deviations from experiment. The $\omega_e x_e$ result for the $A^1\Pi_u$ state in this work (Table 2) is of a similar quality to past work [30, 34]. The current results for the two highest energy states studied ($B^1\Delta_g$ and $B'^1\Sigma_g^+$) also compare well with the previous work [35, 38]. All theoretical harmonic frequencies exhibit small deviations, especially in terms of percent differences. The first anharmonicities for the $B^1\Delta_g$ and $B'^1\Sigma_g^+$ states from this work are somewhat less accurate than the values found in references [34, 35].

The constants derived from the rotational energy levels (B_e, α_e, and D_{rot}) were only reported in three recent studies [30, 34, 35]. These theoretical studies, as well as this work, reproduce the experimental B_e values to a high level of accuracy for all states. Good agreement is also seen in the α_e results with the exception of the $B^1\Delta_g$ state where the results of this work differ from experiment and the results of reference 35 by 7 %. D_{rot} values from this work and references [34, 35] differ from the experiment by $\sim$1 % or less except for the $B'^1\Sigma_g^+$ state, where all theoretical results deviate by $\sim$5 %.

It is worth noting that all of the previous studies with the highest accuracies employed MRCISD calculations with basis sets up to aug-cc-pV6Z. In contrast, the present work uses only basis sets up to and including cc-pVQZ, but recovers the full correlation up to sextuple excitations by CEEIS and complete basis set extrapolations. Core–valence correction and scalar relativistic effects were included in the previous and the present studies. In general, the present theoretical work and past studies agree very well for the values of the spectroscopic constants. The deviations are on the order of a few cm^{-1}, with percent differences being notably small in most cases.

4 Conclusions

The previous CEEIS methodology for calculating diatomic ground state PECs has been extended to the simultaneous determination of ground and excited states. As was the case for the calculations on ground states, CBS extrapolated CEEIS energies, with the addition of the corrections due to core–valence correlation and relativistic effects have been shown to yield highly accurate PECs for the excited electronic states.

With this method, ab initio PECs of the four lowest lying singlet states ($X^1\Sigma_g^+$, $A^1\Pi_u$, $B^1\Delta_g$, and $B'^1\Sigma_g^+$) have been calculated. In contrast to previous studies that have used MRCISD with aug-cc-pV6Z basis sets, the present work has obtained the extrapolated full correlation energy up to and including sextuple excitations using up to cc-pVQZ basis sets, followed by complete basis set extrapolation. The effect of using dynamically weighted versus evenly weighted MCSCF reference functions was found to be negligible once the dynamical correlation was accounted for by the CEEIS procedure. The calculated PECs exhibit the interesting, complex structure of the low-lying singlet states of C_2, involving multiple curve crossings and the avoided crossing between the two $^1\Sigma_g^+$ states. This avoided crossing causes distortions in the shapes of the $^1\Sigma_g^+$ PECs that make it difficult to fit these curves by analytic functions.

The theoretical PECs show very good agreement with the experimental results. For all four states, equilibrium bond distances, dissociation energies, excitation energies, and spectroscopic constants are obtained with high accuracy. The calculated dissociation energy of the ground state deviates from experiment by only −0.38 kcal/mol, exhibiting "chemical accuracy." After fitting the ab initio energies to an analytical form or to cubic splines, the nuclear Schrödinger equation was solved to obtain rotational–vibrational levels and spectroscopic parameters. The previously unreported full spectra of all bound vibrational states are reported in the Supporting Information for all four electronic states. Comparisons with the available experimental data on the vibrational manifolds (which are limited to 4–7 levels) show MADs of $\sim$10–20 cm^{-1} for the vibrational energy levels, or "near spectroscopic accuracy." Spectroscopic constants were obtained by expanding the vibrational and rotational energy levels in terms of powers of (v + ½) and [J(J + 1)]. The calculated rotational constants (B_v and D_v) show excellent agreement with experiment. The lower-order constants of the vibrational expansions also show excellent agreement with the experimental results, while the higher-order terms are less accurate.

Acknowledgments The authors thank Dr. Laimutis Bytautas and Dr. Luke Roskop for helpful discussions related to this work. This research is supported by the US Department of Energy, Office of Basic Energy Sciences, Division of Chemical Sciences, Geosciences, and Biosciences through the Ames Laboratory. The Ames Laboratory is operated for the US Department of Energy by Iowa State University under Contract No. DE-AC02-07CH11358.

References

1. Vardya MS (1970) Ann Rev Astron Astr 8:87–114
2. Mayer P, O'Dell CR (1968) Astrophys J 153:951–962
3. Souza SP, Lutz BL (1977) Astrophys J 216(1):L49–L51
4. Bleekrode R, Nieuwpoort WC (1965) J Chem Phys 43(10):3680–3687
5. Fox JG, Herzberg G (1937) Phys Rev 52(6):0638–0643
6. Ballik EA, Ramsay DA (1959) J Chem Phys 31(4):1128
7. Ballik EA, Ramsay DA (1963) Astrophys J 137(1):84–101
8. Marenin IR, Johnson HR (1970) J Quant Spectrosc Radiat Transfer 10(4):305–309
9. Huber KP, Herzberg G (1979) Molecular spectra and molecular structure IV. Constants of diatomic molecules. Van Nostrand Reinhold, New York
10. Amiot C (1983) Astrophys J Suppl S 52(3):329–340

11. Douay M, Nietmann R, Bernath PF (1988) J Mol Spectrosc 131(2):250–260
12. Douay M, Nietmann R, Bernath PF (1988) J Mol Spectrosc 131(2):261–271
13. Davis SP, Abrams MC, Phillips JG, Rao MLP (1988) J Opt Soc Am B Opt Phys 5(10):2280–2285
14. Martin M (1992) J Photochem Photobiol A 66(3):263–289
15. Gong MX, Bao YH, Urdahl RS, Jackson WM (1994) Chem Phys Lett 217(3):210–215
16. Wakabayashi T, Ong AL, Kratschmer W (2002) J Chem Phys 116(14):5996–6001
17. Chan MC, Yeung SH, Wong YY, Li YF, Chan WM, Yim KH (2004) Chem Phys Lett 390(4–6):340–346
18. Tanabashi A, Hirao T, Amano T, Bernath PF (2007) Astrophys J Suppl S 169(2):472–484
19. Brooke JSA, Bernath PF, Schmidt TW, Bacskay GB (2013) J Quant Spectrosc Radiat Transfer 124:11–20
20. Fougere PF, Nesbet RK (1966) J Chem Phys 44(1):285–298
21. Langhoff SR, Sink ML, Pritchard RH, Kern CW, Strickler SJ, Boyd MJ (1977) J Chem Phys 67(3):1051–1060
22. Chabalowski CF, Peyerimhoff SD, Buenker RJ (1983) Chem Phys 81(1–2):57–72
23. Kraemer WP, Roos BO (1987) Chem Phys 118(3):345–355
24. Bruna PJ, Wright JS (1991) Chem Phys 157(1–2):111–121
25. Peterson KA (1995) J Chem Phys 102(1):262–277
26. Boggio-Pasqua M, Voronin AI, Halvick P, Rayez JC (2000) J Mol Struct (Theochem) 531:159–167
27. Muller T, Dallos M, Lischka H, Dubrovay Z, Szalay PG (2001) Theor Chem Acc 105(3):227–243
28. Abrams ML, Sherrill CD (2004) J Chem Phys 121(19): 9211–9219
29. Sherrill CD, Piecuch P (2005) J Chem Phys 122(12):124104
30. Kokkin DL, Bacskay GB, Schmidt TW (2007) J Chem Phys 126(8):084302
31. Mahapatra US, Chattopadhyay S, Chaudhuri RK (2008) J Chem Phys 129(2):24–108
32. Varandas AJC (2008) J Chem Phys 129(23):234103
33. Purwanto W, Zhang SW, Krakauer H (2009) J Chem Phys 130(9):094107
34. Zhang X-N, Shi D-H, Sun J-F, Zhu Z-L (2011) Chin Phys B 20 (4)
35. Shi D, Zhang X, Sun J, Zhu Z (2011) Mol Phys 109(11): 1453–1465
36. Su PF, Wu JF, Gu JJ, Wu W, Shaik S, Hiberty PC (2011) J Chem Theory Comput 7(1):121–130
37. Booth GH, Cleland D, Thom AJW, Alavi A (2011) J Chem Phys 135(8):084104
38. Jiang W, Wilson AK (2011) J Chem Phys 134 (3)
39. Angeli C, Cimiraglia R, Pastore M (2012) Mol Phys 110(23):2963–2968
40. Shaik S, Danovich D, Wu W, Su P, Rzepa HS, Hiberty PC (2012) Nat Chem 4(3):195–200
41. Liu W, Hanauer M, Koehn A (2013) Chem Phys Lett 565:122–127
42. Deskevich MP, Nesbitt DJ, Werner HJ (2004) J Chem Phys 120(16): 7281–7289
43. Bytautas L, Ruedenberg K (2004) J Chem Phys 121(22): 10905–10918
44. Bytautas L, Ruedenberg K (2004) J Chem Phys 121(22): 10919–10934
45. Bytautas L, Ruedenberg K (2005) J Chem Phys 122(15):154110
46. Bytautas L, Nagata T, Gordon MS, Ruedenberg K (2007) J Chem Phys 127(16):164317
47. Bytautas L, Ruedenberg K (2010) J Chem Phys 132(7):4307
48. Bytautas L, Matsunaga N, Scuseria GE, Ruedenbereg K (2012) J Phys Chem A 116(7):1717–1729
49. Bytautas L, Matsunaga N, Nagata T, Gordon MS, Ruedenberg K (2007) J Chem Phys 127(20):204301
50. Bytautas L, Matsunaga N, Nagata T, Gordon MS, Ruedenberg K (2007) J Chem Phys 127(20):204313
51. Bytautas L, Matsunaga N, Ruedenberg K (2010) J Chem Phys 132(7):074307
52. Schmidt MW, Baldridge KK, Boatz JA, Elbert ST, Gordon MS, Jensen JH, Koseki S, Matsunaga N, Nguyen KA, Su SJ, Windus TL, Dupuis M, Montgomery JA (1993) J Comput Chem 14(11):1347–1363
53. Cheung LM, Sundberg KR, Ruedenberg K (1979) Int J Quantum Chem 16(5):1103–1139
54. Ruedenberg K, Schmidt MW, Gilbert MM, Elbert ST (1982) Chem Phys 71(1):41–49
55. Ruedenberg K, Schmidt MW, Gilbert MM (1982) Chem Phys 71(1):51–64
56. Ruedenberg K, Schmidt MW, Gilbert MM, Elbert ST (1982) Chem Phys 71(1):65–78
57. Roos BO, Taylor PR, Siegbahn PEM (1980) Chem Phys 48(2):157–173
58. Dunning TH (1989) J Chem Phys 90(2):1007–1023
59. Wilson AK, van Mourik T, Dunning TH (1996) J Mol Struct (Theochem) 388:339–349
60. Klopper W, Kutzelnigg W (1986) J Mol Struct (Theochem) 28:339–356
61. Kutzelnigg W (1994) Int J Quantum Chem 51(6):447–463
62. Lowdin PO (1955) Phys Rev 97(6):1474–1489
63. The multistate CEEIS method is included in all versions of GAMESS that occur after May 1, 2013
64. Helgaker T, Klopper W, Koch H, Noga J (1997) J Chem Phys 106(23):9639–9646
65. Halkier A, Helgaker T, Jorgensen P, Klopper W, Koch H, Olsen J, Wilson AK (1998) Chem Phys Lett 286(3–4):243–252
66. Langhoff SR, Davidson ER (1974) Int J Quantum Chem 8(1):61–72
67. Douglas M, Kroll NM (1974) Ann Phys 82(1):89–155
68. Fedorov DG, Nakajima T, Hirao K (2001) Chem Phys Lett 335(3–4):183–187
69. Nakajima T, Hirao K (2005) Monatsh Chem 136(6):965–986
70. de Jong WA, Harrison RJ, Dixon DA (2001) J Chem Phys 114(1):48–53
71. Fedorov DG, Gordon MS (2000) J Chem Phys 112(13):5611–5623
72. Aquilante F, De Vico L, Ferre N, Ghigo G, Malmqvist P-A, Neogrady P, Pedersen TB, Pitonak M, Reiher M, Roos BO, Serrano-Andres L, Urban M, Veryazov V, Lindh R (2010) J Comput Chem 31(1):224–247
73. Colbert DT, Miller WH (1992) J Chem Phys 96(3):1982–1991
74. Blatt JM (1967) J Comput Phys 1(3):382–396
75. Zeng T, Fedorov DG, Klobukowski M (2011) J Chem Phys 134 (2)
76. Urdahl RS, Bao YH, Jackson WM (1991) Chem Phys Lett 178(4):425–428

Theor Chem Acc (2014) 133:1447
DOI 10.1007/s00214-014-1447-z

REGULAR ARTICLE

Comparative bonding analysis of N_2 and P_2 versus tetrahedral N_4 and P_4

P. Jerabek · G. Frenking

Received: 15 October 2013 / Accepted: 7 January 2014 / Published online: 23 January 2014

Abstract The nature of the chemical bonds in E_2 and tetrahedral E_4 (E = N, P) has been analysed with the help of an energy decomposition method. The π bonds in P_2 are not particularly weak. On the contrary, the contribution of P–P π bonding to the chemical bond in P_2 is even higher than the contribution of N–N π bonding to the chemical bond in N_2. The higher stability of P_4 (T_d) and the much lower stability of N_4 (T_d) with regard to the diatomic species come from the substantially larger weakening of the N–N σ-bonds compared with the P–P σ-bonds in the tetrahedral species. The sum of the six P–P σ-bond energies in P_4 is higher than the sum of two σ- and four π-bonds in two P_2, but the six N–N σ-bonds in N_4 are weaker than two σ- and four π-bonds in two N_2. The crucial factor that leads to the weak N–N bonds in N_4, is the rather long N–N bonds which are 32.8 % longer than in N_2. In contrast, the P–P bonds in P_4 are only 16.2 % longer than in P_2. Since the equilibrium distances in E_2 and E_4 are determined by Pauli repulsion, it can be concluded that the origin for the different stabilities of N_4 and P_4 relative to the diatomic molecules is the exchange repulsion. The nitrogen atoms encounter stronger Pauli repulsion, because the 2*s* and 2*p* valence orbitals have a similar radius while the 3*s* and 3*p* radii are more different.

Dedicated to Professor Thom Dunning and published as part of the special collection of articles celebrating his career upon his retirement.

P. Jerabek · G. Frenking (✉)
Fachbereich Chemie, Philipps-Universität Marburg, Hans-Meerwein-Strasse, 35032 Marburg, Germany
e-mail: frenking@staff.uni-marburg.de; frenking@chemie.uni-marburg.de

Keywords Multiple bonding · π-Bonding of heavy main-group atoms · Energy decomposition analysis · Dipnictogens · Tetrapnictogens

1 Introduction

It is well known that molecules with multiple bonds between heavier main-group atoms are much more difficult to synthesize than their counterparts of the first octal row and that their equilibrium geometries are often very different from the structures of the lighter homologues [1, 2]. For example, the ditetrylenes E_2H_4 possess a trans-bent (E = Si, Ge) or a doubly hydrogen-bridged geometry (E = Sn, Pb) [3–5], while ethene C_2H_4 has a planar C_{2v} structure. Even more exotic are the equilibrium geometries of the heavier ditetrylynes HEEH, which exhibit several unusual structures, none of them being linear like acetylene HCCH. Quantum chemical calculations suggested that the energetically lowest lying form of HEEH (E = Si–Pb) possesses a non-planar doubly hydrogen-bridged butterfly structure E(μ-H)$_2$E and that the next higher lying isomer has a planar singly bridged E(μ-H)EH geometry [6–24]. The theoretical predictions have been verified by low-temperature matrix isolation studies [25–30]. Recently, ditetrylynes E_2R_2, which carry bulky substituents R, could become isolated. X-ray analysis showed that the molecules have a trans-bent geometry where the bending angle is between 95° and 140° [31–35].[1]

The unusual geometries of the heavier ditetrylenes E_2H_4 and ditetrylynes HEEH can be explained in terms of donor–

[1] The bonding situation in the trans-bent structures REER (E = Si–Pb) has been studied. The authors do not discuss the preference of the parents system HEEH for hydrogen-bridged geometries.

acceptor interactions between the respective fragments EH_2 and EH [3*s*]. Chemical bonding in ethylene takes place between the CH_2 fragments in the ($^3B_1$) triplet ground state, while the heavier tetrylenes EH_2 bind through their $^1A_1$ ground state. Similarly, the bonding in acetylene occurs between CH in the $^4\Sigma_g$ state[2], while the bonding in the heavier homologues takes place through the $^2\Pi$ ground state.

Such explanation is not valid when one addresses the question why diatomic N_2 is clearly the most stable isomer of the lightest pnictogen whereas the most stable allotrope of phosphorus is tetra-atomic P_4, which has a tetrahedral (T_d) structure. Unlike N_2, diatomic P_2 is a highly reactive species that can only be observed in the gas phase [36]. Tetra-atomic N_4 (T_d) is an energetically high-lying energy minimum on the potential energy surface that has not been observed so far [37].[3] Chemical bonding in dipnictogens E_2 and tetrapnictogens E_4 (E = N, P) takes place for both elements through the ^4S ground state of atom E yielding a E≡E triple bond in E_2 and E–E single bonds in (T_d) E_4. The higher stabilities of tetrahedral P_4 for phosphorus and of diatomic N_2 for nitrogen are thus an intrinsic property of the atoms in the electronic ground state, which is not related to an excitation energy. What is the driving force for the different equilibrium structures of nitrogen and phosphorus?

Old textbooks suggested that π bonding between heavier main-group atoms is weaker than π bonding between atoms of the first octal row, because the longer bond in the former system would lead to smaller overlap. This intuitive argument does not agree with calculations of the overlap between the $p(\pi)$ AOs of heavier main-group atoms. The more diffuse $p(\pi)$ AOs of the latter atoms lead to overlaps, which are not much smaller than those of the lighter homologues. The chemical bonding of heavier main-group atoms and the contrast to the lighter homologues has been analysed in a seminal paper by Kutzelnigg 30 years ago where the author discusses quantum theoretical findings about the nature of the chemical bond [38, 39] A crucial difference between atoms of the first and higher octal rows concerns the spatial distribution of the s and p valence orbitals, which are engaged in covalent bonding. The 2*p* orbitals of the first octal-row atoms penetrate rather deeply into the core, because there are no p functions in the core. Therefore, the radii of 2*p* and 2*s* orbitals are very similar, which leads to effective sp hybridization. In contrast, the occupied *np* AOs ($n \geq 3$) of the heavier atoms encounter Pauli repulsion with electrons in the p core orbitals, which causes larger radii for the *np* than for the respective *ns* AOs. The spatial regions of *np* and *ns* AOs ($n \geq 3$) are more separate than those of 2*p* and 2*s* AOs, which leads to less effective hybridization of the heavier main-group atoms. Kutzelnigg came to the conclusion that the difference between the spatial distribution of the s and p valence orbitals implies that single bonds between first row elements are weak and multiple bonds are strong, whereas for the second or higher row elements, single bonds are strong and multiple bonds weak [38, 39].

The conclusion by Kutzelnigg offers an explanation for the finding that multiple-bonded N_2 is more stable than N_4, which has N–N single bonds, whereas P_4 is clearly more stable than P_2. However, there are still open questions. Why are single bonds between first octal-row elements weaker than between heavier main-group atoms when the stronger hybridization of the 2*s* and 2*p* AOs means that there is a larger contribution from the energetically lower-lying 2*s* AOs? What is the difference between the pnictogen bonds in diatomic E_2 and tetrahedral E_4 in terms of energy contributions to the chemical bonding? Is there a particular type of interatomic interaction, which can be given as the cause for the different stabilities of E_2 and E_4? In order to address these questions, we analysed the chemical bonds in the diatomic and tetrahedral pnictogens with an energy decomposition scheme, which has been proven to give comprehensive insight into the structures and chemical bonding of a wide variety of main-group compounds [24, 40–72] and transition metal complexes [73–125].

2 Methods

The bond lengths of the diatomic and tetrahedral molecules E_2 and E_4 have been optimized with the generalized gradient approximation (GGA) to density functional theory (DFT) using the exchange functional of Becke [126] in conjunction with the correlation functional of Perdew [127] (BP86). Uncontracted Slater-type orbitals (STOs) were employed as basis functions for the SCF calculations [128]. The basis sets have triple-ζ quality augmented by two sets of polarization functions, i.e. 2*p* and 3*d* functions on hydrogen and 3*d* and 4*f* functions on the other atoms. Core electrons (i.e. 1*s* for nitrogen and 1*s*2*s*2*p* for phosphorus) were treated by the frozen-core approximation. This level of theory is denoted BP86/TZ2P. An auxiliary set of *s*, *p*, *d*, *f* and *g* STOs was used to fit the molecular densities and to represent the Coulomb and exchange potentials accurately in each SCF cycle [129]. The calculations were carried out with the program package ADF [130].

[2] The $a^4\Sigma_g$ state of CH is not the electronic ground state, but the rather small excitation energy from the $X^2\Pi$ ground state is compensated by the stabilization of the π bonds in HC≡CH. The excitation energy $X^2\Pi \rightarrow a^4\Sigma_g$ of the heavier homologues EH (E = Si–Pb) is much higher than for carbon and the stabilization through the π bonds is not as high as in acetylene. For a detailed discussion see reference [24].

[3] A planar isomer of N_4 which has two N_2 moieties that are connected by a long N–N bond has been observed as short-lived isomer in gas-phase experiments.

The interatomic interactions have been analysed by means of an energy decomposition scheme that was developed independently by Morokuma [131] and by Ziegler and Rauk [132, 133]. The focus of the bonding analysis is the instantaneous interaction energy ΔE_{int} between the chosen fragments in the particular electronic reference state and in the frozen geometry of the molecule. In the present case of E_2 and (T_d) E_4, ΔE_{int} is the energy difference between E_2 or E_4 and the atoms E in the ^4S electronic ground state. The interaction energy is divided into three main components:

$$\Delta E_{\text{int}} = \Delta E_{\text{elstat}} + \Delta E_{\text{Pauli}} + \Delta E_{\text{orb}} \qquad (1)$$

ΔE_{elstat} gives the electrostatic interaction energy between the fragments, which are calculated using the frozen electron density of the interacting fragments. The second term in Eq. (1), ΔE_{Pauli}, refers to the repulsive interactions between the fragments, which are caused by the fact that two electrons with the same spin cannot occupy the same region in space. ΔE_{Pauli} is calculated by enforcing the Kohn–Sham determinant on the superimposed fragments to obey the Pauli principle by antisymmetrization and renormalization. The stabilizing orbital interaction term, ΔE_{orb} is calculated in the final step of the energy partitioning analysis when the Kohn–Sham orbitals relax to their optimal form. This term can be further partitioned into contributions by the orbitals belonging to different irreducible representations of the point group of the interacting system.

For molecules where the fragments do not electronically or geometrically relax after bond breaking, the interaction energy ΔE_{int} gives directly (by definition with opposite sign) the bond dissociation energy (BDE) D_e. If the two fragments of the chemical bond are in an electronically excited state or if they have more than one atom which means that there is a relaxation of the fragments during bond rupture into the equilibrium geometry, the preparation energy ΔE_{prep} must be added to ΔE_{int} in order to obtain the bond dissociation energy:

$$\Delta E\,(= -D_e) = \Delta E_{\text{int}} + \Delta E_{\text{prep}} \qquad (2)$$

Because the atomic fragments, which were used in our calculations, are in the electronic ground state it follows that for the E_2 and E_4 molecules $\Delta E_{\text{prep}} = 0$ and $\Delta E_{\text{int}} = -D_e$. Further details about the EDA can be found in the literature [134–136].

3 Results and discussion

Figure 1 shows the results of the geometry optimization and energy calculation of E_2 and (T_d) E_4 (E = N, P). The data provide the framework for the following bonding analysis.

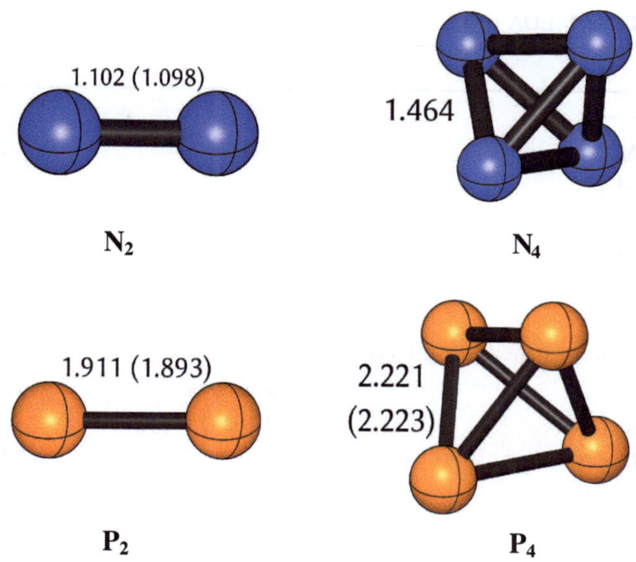

2 N_2	→	$N_4(T_d)$	+158.9 kcal/mol	(1)
2 P_2	→	$P_4(T_d)$	-59.7 kcal/mol	(2)

Fig. 1 Calculated bond lengths (Å) of E_2 and (T_d) E_4 (E = N, P) and dimerization energies D_e of E_2 at BP86/TZ2P. Experimental bond lengths are given in parentheses

The calculated bond lengths at BP86/TZ2P for N_2, P_2 and P_4 are in very good agreement with experimental data [137, 138]. The theoretical values for the reaction energies for formation of the tetrahedral species are striking evidence for the drastic difference between the nitrogen and phosphorous systems. The calculated data for reactions 1 and 2 give only the electronic energies, because we are interested in the variation in the electronic structures, which lead to the different stabilities. Experimental values are available for the dimerization reaction 2 of the phosphorus species. The experimental heats of formation [139][4] of P_2 ($\Delta H_f^0 = 34.8$ kcal/mol) and P_4 ($\Delta H_f^0 = 16.0$ kcal/mol) give a reaction energy of $\Delta E_R^0 = 53.6$ kcal/mol which agrees quite well with the ZPE corrected calculated value of $\Delta E_R^0 = 58.0$ kcal/mol.

Table 1 shows the results of the EDA calculations of E_2 and (T_d) E_4 (E = N, P) where the atoms E in the electronic ground state $(2s^2 2p_x^1 2p_y^1 2p_z^1)$ ^4S are used as interacting fragments. We first discuss the data for N_2 and P_2 [140].[5] The calculated interaction energies ΔE_{int} give after ZPE corrections theoretical bond dissociation energies of $D_o = 236.9$ kcal/mol for N_2 and $D_o = 117.4$ kcal/mol for P_2. This is in reasonable agreement with the experimental values of $D_o = 225.0$ kcal/mol for N_2 and $D_o = 116.1$ kcal/mol for

[4] Calculated from the differences between the heats of formation.

[5] Chemical bonding in diatomic molecules E_2 of atoms of the first and second octal rows.

Table 1 EDA results in kcal/mol for E_2 and (T_d) E_4 (E = N, P) (BP86/TZ2P+)

	E_2	E_4	1/4 E_2	1/2 E_4
$E = N$				
ΔE_{int}	−240.2	−322.6	−120.1	−80.7
ΔE_{Pauli}	802.2	1,636.6	401.1	409.2
ΔE^{a}_{elstat}	−312.8 (30.0 %)	−705.4 (36.0 %)	−156.4 (30.0 %)	−176.4 (36.0 %)
ΔE^{a}_{orb}	−729.6 (70.0 %)	−1,253.7 (64.0 %)	−364.8 (70.0 %)	−313.4 (64.0 %)
ΔE^{b}_{σ}	−478.7 (65.6 %)	−1,253.7 (100.0 %)	−239.4 (65.6 %)	−313.4 (100.0 %)
ΔE^{b}_{π}	−250.9 (34.4 %)	–	−125.5 (34.4 %)	–
Hybridization[c] of the σ bond	*s*: 37.5 %	*s*: 7.0 %	*s*: 37.5 %	*s*: 7.0 %
	p: 61.9 %	*p*: 92.5 %	*p*: 61.9 %	*p*: 92.5 %
R(N–N)	1.102 (1.098)[e]	1.464	1.102	1.464
D^{d}_{o}	236.9 (225.0)[e]	330.9		
$E = P$				
ΔE_{int}	−118.5	−293.1	−59.3	−73.3
ΔE_{Pauli}	317.6	972.3	158.8	243.1
ΔE^{a}_{elstat}	−186.3 (42.7 %)	−545.7 (43.1 %)	−93.2 (42.7 %)	−136.4 (43.1 %)
ΔE^{a}_{orb}	−249.8 (57.3 %)	−719.8 (56.9 %)	−124.9 (57.3 %)	−180.0 (56.9 %)
ΔE^{b}_{σ}	−148.7 (59.5 %)	−719.8 (100.0 %)	−74.4 (59.5 %)	−180.0 (100.0 %)
ΔE^{b}_{π}	−101.1 (40.5 %)	–	−50.6 (40.5 %)	–
Hybridization[c] of the σ bond	*s*: 20.8 %	*s*: 5.2 %	*s*: 20.8 %	*s*: 5.2 %
	p: 77.8 %	*p*: 93.2 %	*p*: 77.8 %	*p*: 93.2 %
R(P–P)	1.911 (1.893)[e]	2.221 (2.223)[f]	1.911	2.221
D^{d}_{o}	117.4 (116.1)[e]	289.2 (285.6)[g]		

The interacting fragment atoms E in the ^4S ground state ($s^2p_x^1p_y^1p_z^1$). Bond lengths R(E–E) are given in Å

[a] The values in parentheses give the percentage contribution to the total attractive interactions $\Delta E_{elstat} + \Delta E_{orb}$

[b] The values in parentheses give the percentage contribution to the total orbital interactions ΔE_{orb}

[c] There is a small contribution from the d polarization functions

[d] Bond dissociation energy into the atoms corrected by ZPE contributions

[e] Experimental value, Ref. [137]

[f] Experimental value, Ref. [138]

[g] Experimental value, Ref. [139]

P_2 [137]. Inspection of the EDA results indicates that the attractive N–N interactions come mainly (70 %) from orbital (covalent) interactions ΔE_{orb}, while 30 % are contributed from electrostatic (Coulombic) attraction ΔE_{elstat}. The P–P bond in P_2 has according to the EDA results a slightly smaller contribution from the orbital term (57.3 %) and a higher electrostatic character (42.7 %). The breakdown of the orbital term into contributions of σ and π interactions suggests that σ bonding provides about two-thirds (65.6 %) to ΔE_{orb} in N_2, while ΔE_{π} contributes about one-third (34.4 %) to covalent bonding. Table 1 shows that the contribution of π bonding to the orbital term in P_2 (40.5 %) is even *higher* than in N_2! This clearly shows that the preference of phosphorus for tetrahedral P_4 does not come from the alleged weak π bonds in P_2. The calculations also show that the overlap integral of the *p*(π) AOs in P_2 (0.63) is slightly smaller than the *p*(π) overlap in N_2 (0.74), but also the overlap of the σ AOs in P_2 (1.49) is a bit smaller than in N_2 (1.59). Note that the hybridization of the N–N σ-bond has a higher s contribution (37.5 %) than the P–P bond (20.8 %), which agrees with the suggestion that atoms of the first octal row exhibit larger s/p hybridization than the heavier homologues [38, 39].

The EDA results in Table 1 disprove the suggestion that π bonding in heavier main-group compounds is intrinsically weak, but this does not provide an explanation for the higher stability of P_4 over P_2. Before we discuss the results for the tetrahedral species, we show in Fig. 2 the EDA results for E_2 at different interatomic distances which are relevant for the further discussion. The variation of the three energy terms as a function of the E–E distance shows that the attractive terms ΔE_{elstat} and ΔE_{orb} become larger when the bond length gets shorter than the equilibrium distance. This holds for a bond shortening up to ~0.3 Å of

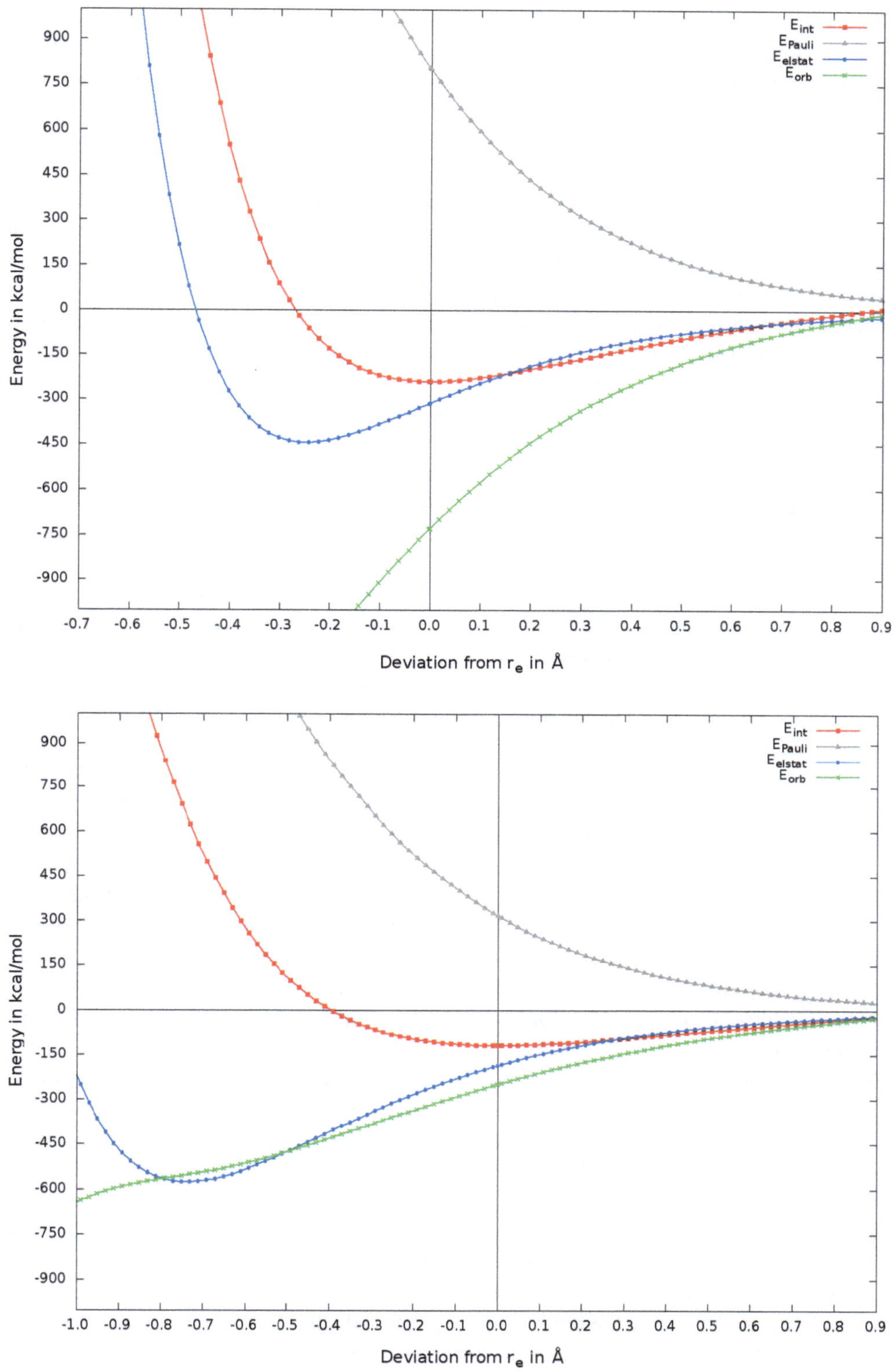

Fig. 2 Calculated energy terms of the EDA at different bond length of N_2 and P_2

N_2 and up to ~0.8 Å in P_2. At shorter interatomic distances the electrostatic interaction decreases because the nuclear–nuclear repulsion becomes very strong. Note that the orbital (covalent) interactions continuously increase up to very short bond lengths. This contradicts the statement in many textbooks that the equilibrium bond lengths of covalent bonds are determined by the size of the overlap of the σ orbitals. It has been shown before that the overlap of the σ orbitals in N_2 and P_2 further increases when the interatomic distance becomes shorter than the equilibrium value [140]. The energy term which prevents further shortening of the E–E bond in E_2 is the Pauli repulsion.

Figure 2 shows that the curve for ΔE_{Pauli} exhibits a sharp rise at shorter distances than the equilibrium value which compensates the increase of the attractive terms. The equilibrium distance of N_2 and P_2 is determined by the Pauli repulsion. This conclusion holds for most covalent bonds [140].

The stage has been set now for discussing the EDA results for tetra-atomic E_4. Table 1 shows also the EDA results for N_4 and P_4. The nature of the bonding in the tetra-atomic molecules which possess E–E single bonds does not change dramatically compared to the triply bonded diatomic species. The electrostatic character increases a bit in E_4 compared with E_2 while the covalent character decreases. The largest change occurs in the hybridization of the E–E σ bonds. The % s contribution to the N–N bond is significantly reduced from 37.5 % in N_2 to 7.0 % in N_4 and it has decreased from 20.8 % in P_2 to 5.2 % in P_4. The change in the hybridization can be explained with the longer bonds in the tetrahedral molecules. The N–N bond of N_2 (1.102 Å) becomes 32.8 % longer in N_4 (1.464 Å), while the P–P bond in P_2 (1.911 Å) is stretched by 16.2 % in P_4 (2.221 Å). The calculated interaction energy ΔE_{int} for P_4 (−293.1 kcal/mol) gives after ZPE correction an atomization energy for fragmentation into four P atoms of $D_o = 289.2$ kcal/mol + ΔZPE which is in good agreement with the experimental value of $D_o = 285.6$ kcal/mol [139]. The calculated ΔE_{int} values for (T_d) E_4 which refer to six E–E bonds suggest that each single bond has an interaction energy of 53.8 kcal/mol for the N–N bond and 48.9 kcal/mol for the P–P bond.

In order to compare the EDA results of E_4 with E_2, we also show in Table 1 the calculated data per atom for the diatomic and tetra-atomic species 1/2 E_2 and 1/4 E_4. The intrinsic interaction energy ΔE_{int} for N_2 (−240.2 kcal/mol) is much higher than for P_2 (−118.5 kcal/mol) while the ΔE_{int} value for N_4 (−322.6 kcal/mol) is only slightly larger than for P_4 (−293.1 kcal/mol). The ΔE_{int} values per atom for the two systems reveal that the interaction energy for a single nitrogen atom in N_2 (−120.1 kcal/mol) is much stronger than in N_4 (−80.7 kcal/mol), while the ΔE_{int} value for a phosphorous atom in P_2 (−59.3 kcal/mol) is weaker than in tetrahedral P_4 (−73.3 kcal/mol). But this is a trivial result, because the differences between the ΔE_{int} values for 4N and 4P are identical to the dimerization energies which are given in Fig. 1. The crucial question concerns the EDA terms which are responsible for the opposite trends in the stabilisation energies of N_2/N_4 and P_2/P_4.

The data in Table 1 show that the ΔE_{Pauli} values per atom slightly increase from N_2 (401.1 kcal/mol) to N_4 (409.2 kcal/mol), but they increase even more from P_2 (158.8 kcal/mol) to P_4 (243.1 kcal/mol). Thus, the differences between the Pauli repulsion do not correlate with the strength of the E–E bond from dimer to tetramer. The electrostatic attraction per atom in P_2 (−93.2 kcal/mol) becomes stronger in P_4 (−136.4 kcal/mol) but the ΔE_{elstat} values also increases from N_2 (−156.4 kcal/mol) to N_4 (−176.4 kcal/mol). Although the increase of ΔE_{elstat} in the phosphorous species is stronger than in the nitrogen molecules, the unidirectional trend of the Coulomb attraction does not agree with the opposite trend of the total interaction energy. A reverse trend is found, however, for the calculated values of the orbital interactions. The calculated ΔE_{orb} values per atom clearly decrease from N_2 (−364.8 kcal/mol) to N_4 (−313.4 kcal/mol), but they significantly increase from P_2 (−124.9 kcal/mol) to P_4 (−180.0 kcal/mol). It follows that the opposite trend in the bond strength between the nitrogen and phosphorous systems N_2/N_4 and P_2/P_4 comes mainly from the change in the covalent (orbital) bonding in the molecules.

It is instructive to analyse the EDA results for the orbital interactions given in Table 1 in more detail. The calculated value of $\Delta E_{orb} = -719.8$ kcal/mol in P_4 which has six σ-bonds means that the ΔE_{σ} value for one P–P bond in the tetramer is −120.0 kcal/mol. This is less than the ΔE_{σ} value in P_2 (−148.7 kcal/mol), but the strength of three P–P σ-bonds (−360.0 kcal/mol) overcompensates the value for the triple bond in P_2 (−249.8 kcal/mol) which consists of a σ and two π-bonds. In N_4, the value of $\Delta E_{orb} = -1{,}253.7$ kcal/mol suggests that the N–N σ-bond strength is −208.9 kcal/mol, which is much weaker than the σ-bond in N_2 (−478.8 kcal/mol). In contrast to P_4, the strength of three N–N σ-bonds in N_4 ($3 \times -208.9 = -626.7$ kcal/mol) does not compensate for the σ- and two π-bonds in N_2 (−729.6 kcal/mol). Thus, the crucial difference between the nitrogen and phosphorous systems comes from the substantial weakening of the N–N σ-bond in N_4, which makes N_2 more stable than the tetra-atomic species.[6]

But what is the driving force for the weaker N–N σ-bond in N_4? The answer is directly related to the question for the factor, which determines the equilibrium bond lengths in covalent bonds, which was discussed above. The bond length in N_2 and P_2 comes from the increase of the Pauli repulsion, which comes from the overlap of the orbitals between the two atoms, which are occupied by electrons that posses the same spin. In N_4, each nitrogen atom encounters Pauli repulsion by three other nitrogen atoms, which makes ΔE_{Pauli} at a given N–N distance much larger than in N_2. This leads to significantly longer N–N

[6] The finding that the σ bond in N_2 is quite strong appears to contradict the suggestion of Kutzelnigg [38, 39] that single bonds between first row elements are weak and multiple bonds are strong, whereas for the second or higher row elements, single bonds are strong and multiple bonds weak. The statement by Kutzelnigg refers to the total interactions ΔE_{int} while we identify σ bonding with the orbital interactions ΔE_{σ}.

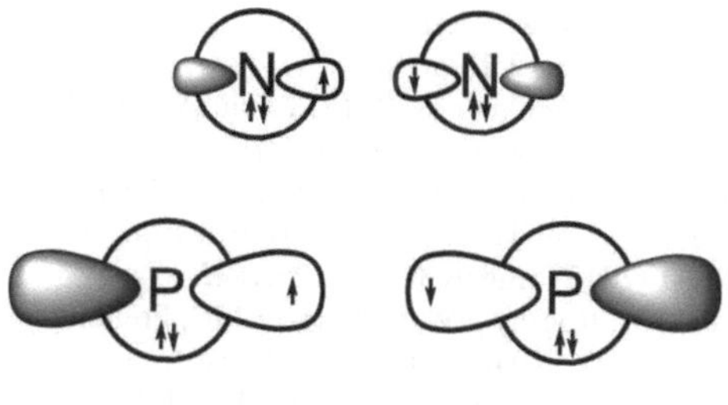

Fig. 3 Schematic view of the *s* and $p(\sigma)$ valence orbitals of nitrogen and phosphorus, which show the different ratios of the radii for the *s* and *p* AOs that cause stronger Pauli repulsion between the phosphorus atoms

bonds in N_4 than in N_2. The N–N bond in N_4 is 32.8 % longer than in N_2. The same effect is operative in P_4, but the bond lengthening is smaller (16.2 %) than in the nitrogen system.

The resume of the EDA results is thus as follows: The transition from 2 E_2 to (T_d) E_4 changes two σ and four π bonds into six σ bonds. The σ bonds in E_4 are weaker than in E_2, because the E–E bonds in the tetra-atomic species are longer than in the diatomic molecules, which is caused by stronger Pauli repulsion. The six longer σ bonds in P_4 are stronger than the two σ and four π bonds of 2 P_2, but the σ bonds of N_4 are weaker than the σ and π bonds of 2 N_2. This is, because the N–N bonds in N_4 are 32.8 % longer than in N_2, while P–P bonds in P_4 are only 16.2 % longer than in P_2. Since the equilibrium distances in E_2 and E_4 are determined by the increase in ΔE_{Pauli}, it can be concluded that the higher stability of P_4 over P_2 and lower stability of N_4 over N_2 is caused by Pauli repulsion.

The latter conclusion is not yet fully satisfactory, because it does not answer the question *why* the Pauli repulsion stretches the N–N bond in N_2 much more than it stretches the P–P bond in P_2. Figure 2 shows that the curve for ΔE_{Pauli} at the equilibrium distance of E_2 exhibits a steeper slope for N_2 than for P_2. This can be explained with the hybridization at nitrogen in N_2, which has a much higher %s character than the phosphorus atom in P_2. The nitrogen atoms encounter overlap of the doubly occupied 2*s* AOs in the bonding region to a higher degree than the phosphorus atoms, which leads to stronger Pauli repulsion in N_2 at the equilibrium distance. This is schematically shown in Fig. 3 that depicts in a nutshell the difference of covalent bonding between first octal-row atoms and heavier homologues. As mentioned in the introduction, the better hybridization of 2*s*/2*p* valence orbitals compared to 3*s*/3*p* comes from the absence of p core AOs in atoms of the first octal row [38, 39]. The higher stability of P_4 over P_2 and lower stability of N_4 over N_2 can thus be traced down to the ratio of the 2*s*/2*p* radii of nitrogen atom versus 3*s*/3*p* of phosphorus atom, which causes stronger Pauli repulsion yielding longer and weaker N–N bonds in N_4.

4 Conclusion

The results of this work can be summarized as follows. The π bonds in P_2 are not particularly weak. On the contrary, the contribution of P–P π bonding to the chemical bond in P_2 is even higher than the contribution of N–N π bonding to the chemical bond in N_2. The higher stability of P_4 (T_d) and the much lower stability of N_4 (T_d) with regard to the diatomic species, comes from the substantially larger weakening of the N–N σ-bonds compared with the P–P σ-bonds in the tetrahedral species. The sum of the six P–P σ-bond energies in P_4 is higher than the sum of two σ- and four π-bonds in two P_2, but the six N–N σ-bonds in N_4 are weaker than two σ- and four π-bonds in two N_2. The crucial factor, which leads to the weak N–N bonds in N_4, are the rather long N–N bonds that are 32.8 % longer than in N_2. In contrast, the P–P bonds in P_4 are only 16.2 % longer than in P_2. Since the equilibrium distances in E_2 and E_4 are determined by Pauli repulsion, it can be concluded that the origin for the different stabilities of N_4 and P_4 relative to the diatomic molecules is the exchange repulsion. The nitrogen atoms encounter stronger Pauli repulsion, because the 2*s* and 2*p* valence orbitals have a similar radius while the 3*s* and 3*p* radii are more different.

Acknowledgments This work was supported by the Deutsche Forschungsgemeinschaft.

References

1. Weidenbruch M (1999) Eur J Inorg Chem 373
2. Power PP (1998) J Chem Soc Dalton Trans 2939
3. Grützmacher H, Fässler T (2000) Chem Eur J 6:2317
4. Jacobsen H, Ziegler T (1994) J Am Chem Soc 116:3667
5. Trinquier G (1991) J Am Chem Soc 113:144
6. Lischka H, Köhler H (1983) J Am Chem Soc 105:6646
7. Binkley JS (1984) J Am Chem Soc 106:603
8. Kalcher J, Sax A, Olbrich G (1984) Int J Quantum Chem 25:543
9. Köhler H-J, Lischka H (1984) Chem Phys Lett 112:33
10. Clabo DA, Schaefer HF (1986) J Chem Phys 84:1664
11. Thies BS, Grev RS, Schaefer HF (1987) Chem Phys Lett 140:355
12. Koseki S, Gordon MS (1988) J Phys Chem 92:364
13. Koseki S, Gordon MS (1989) J Phys Chem 93:118
14. Colegrove BT, Schaefer HF (1990) J Phys Chem 94:5593
15. Grev RS, De Leeuw BJ, Schaefer HF (1990) Chem Phys Lett 165:257
16. Grev RS (1991) Adv Organomet Chem 33:125
17. Colegrove BT, Schaefer HF (1991) J Am Chem Soc 113:1557
18. Grev RS, Schaefer HF (1992) J Chem Phys 97:7990
19. Palagyi Z, Schaefer HF, Kapuy E (1993) J Am Chem Soc 115:6901

20. Han Y-K, Bae C, Lee YS, Lee SY (1998) J Comp Chem 19:1526
21. Nagase S, Kobayashi K, Takagi N (2000) J Organomet Chem 611:264
22. Chen Y, Hartmann M, Diedenhofen M, Frenking G (2001) Angew Chem Int Ed 40:2052
23. Li Q-S, Lü R-H, Xie Y, Schaefer HF (2002) J Comput Chem 23:1642
24. Lein M, Krapp A, Frenking G (2005) J Am Chem Soc 127:6290
25. Bogey M, Bolvin H, Demuyneck C, Destombes J-L (1991) Phys Rev Lett 66:413
26. Cordonnier M, Bogey M, Demuynck C, Destombes J-L (1992) J Chem Phys 97:7984
27. Wang X, Andrews L, Kushto G (2002) J Phys Chem A 106:5809
28. Wang X, Andrews L, Chertihin GV, Souer PF (2002) J Phys Chem A 106:6302
29. Andrews L, Wang X (2002) J Phys Chem A 106:7697
30. Wang X, Andrews L (2003) J Am Chem Soc 125:6581
31. Sekiguchi A, Kinjo R, Ichinohe M (2004) Science 305:1755 (E = Si)
32. Stender M, Phillips AD, Wright RJ, Power PP (2002) Angew Chem Int Ed 41:1785 (E = Ge)
33. Phillips AD, Wright RJ, Olmstead MM, Power PP (2002) J Am Chem Soc 124:5930 (E = Sn)
34. Pu L, Twamley B, Power PP (2000) J Am Chem Soc 122:3524 (E = Pb)
35. Landis CR, Weinhold F (2006) J Am Chem Soc 128:7335
36. Tofan D, Cummins CC (2011) Angew Chem Int Ed 49:7516
37. Cacace F, de Petris G, Troiani A (2002) Science 295:480
38. Kutzelnigg W (1984) Angew Chem 96:262
39. Kutzelnigg W (1984) Angew Chem Int Ed Engl 23:272
40. Esterhuysen C, Frenking G (2004) Theoret Chem Acc 113:294
41. Lein M, Frenking G (2004) Austr J Chem 57:1191
42. Kovács A, Esterhuysen C, Frenking G (2005) Chem Eur J 11:1813
43. Cappel D, Tüllmann S, Krapp A, Frenking G (2005) Angew Chem 117:3683
44. Cappel D, Tüllmann S, Krapp A, Frenking G (2005) Angew Chem Int Ed Engl 44:3617
45. Szabó A, Kovács A, Frenking G (2005) Z Allg Anorg Chem 631:1803
46. Fernández I, Frenking G (2006) Chem Eur J 12:3617
47. Erhardt S, Frenking G (2006) Chem Eur J 12:4620
48. Fernández I, Frenking G (2006) Chem Commun 5030
49. Fernández I, Frenking G (2007) Faraday Discuss 135:403
50. Calhorda MJ, Krapp A, Frenking G (2007) J Phys Chem A 111:2859
51. Fernández I, Frenking G (2007) J Phys Chem A 111:8028
52. Fernández I, Frenking G (2007) J Org Chem 72:7367
53. Fernández I, Uggerud E, Frenking G (2007) Chem Eur J 13:8620
54. Fernández I, Cerpa E, Merino G, Frenking G (2008) Organometallics 27:1106
55. Tonner R, Frenking G (2008) Chem Eur J 14:3260
56. C Elschenbroich, J Six, K Harms, G Frenking, G Heydenrych (2008) Eur J Inorg Chem 3303
57. Ducati LC, Takagi N, Frenking G (2009) J Phys Chem A 113:11693
58. Fernández I, Frenking G, Uggerud E (2010) J Org Chem 75:2971
59. Gonzalez-Gallardo S, Prabusankar G, Cadenbach T, Gemel C, von Hopffgarten M, Frenking G, Fischer RA (2010) Structure Bonding 136:147
60. Fernández I, Duvall M, I-C Wu J, Schleyer P v R, Frenking G (2011) Chem Eur J 17:2215
61. Shimizu T, Frenking G (2011) Theoret Chem Acc 130:269
62. Celik MA, Sure R, Klein S, Kinjo R, Bertrand G, Frenking G (2012) Chem Eur J 18:5676
63. Devarajan D, Frenking G (2012) Chem Asian J 7:1296
64. Goedecke C, Sitt R, Frenking G (2012) Inorg Chem 51:11259
65. Mondal KC, Roesky HW, Schwarzer MC, Frenking G, Neudeck S, Tkach I, Wolf H, Kratzert D, Herbst-Irmer R, Niepötter B, Stalke D (2013) Angew Chem 125:1845
66. Mondal KC, Roesky HW, Schwarzer MC, Frenking G, Neudeck S, Tkach I, Wolf H, Kratzert D, Herbst-Irmer R, Niepötter B, Stalke D (2013) Angew Chem Int Ed 52:1801
67. Holzmann N, Dange D, Jones C, Frenking G (2013) Angew Chem 125:3078
68. Holzmann N, Dange D, Jones C, Frenking G (2013) Angew Chem Int Ed 52:3004
69. Holzmann N, Stasch A, Jones C, Frenking G (2013) Chem Eur J 19:6467
70. Couchman SA, Holzmann N, Frenking G, Wilson DJD, Dutton JL (2013) J Chem Soc, Dalton Trans 42:11375
71. Esterhuysen C, Frenking G (2013) J Chem Soc, Dalton Trans 42:13349
72. Celik MA, Frenking G, Neumüller B, Petz W (2013) ChemPlusChem 78:1024
73. Diefenbach A, Bickelhaupt FM, Frenking G (2000) J Am Chem Soc 122:6449
74. Frenking G, Wichmann K, Fröhlich N, Grobe J, Golla W, Le Van D, Krebs B, Läge M (2002) Organometallics 21:2921
75. Cases M, Frenking G, Duran M, Solà M (2002) Organometallics 21:4182
76. Frenking G, Solà M, Vyboishchikov SF (2005) J Organomet Chem 690:6178
77. Uddin J, Frenking G (2001) J Am Chem Soc 123:1683
78. Dörr M, Frenking G, Allg Z (2002) Anorg Chem 628:843
79. Massera C, Frenking G (2003) Organometallics 22:2758
80. Nechaev MS, Rayón VM, Frenking G (2004) J Phys Chem A 108:3134
81. Lein M, Frunzke J, Timoshkin A, Frenking G (2001) Chem Eur J 7:4155
82. Rayón VM, Frenking G (2003) Organometallics 22:3304
83. Frunzke J, Lein M, Frenking G (2002) Organometallics 21:3351
84. Lein M, Frunzke J, Frenking G (2003) Inorg Chem 42:2504
85. Nemcsok DS, Kovács A, Rayón VM, Frenking G (2002) Organometallics 21:5803
86. Chen Y, Hartmann M, Frenking G (2001) Z Anorg Allg Chem 627:985
87. Kovács A, Frenking G (2001) Organometallics 20:2510
88. Pandey KK, Lein M, Frenking G (2003) J Am Chem Soc 125:1660
89. Spuhler P, Lein M, Frenking G (2003) Z Allg Anorg Chem 629:803
90. Esterhuysen C, Frenking G (2003) Chem Eur J 9:3518
91. Dietz O, Rayón VM, Frenking G (2003) Inorg Chem 42:4977
92. Loschen C, Frenking G (2004) Inorg Chem 43:778
93. Nemcsok D, Wichmann K, Frenking G (2004) Organometallics 23:3640
94. Pandey KK, Frenking G (2004) Eur J Inorg Chem 4388
95. Bessac F, Frenking G (2006) Inorg Chem 45:6956
96. Krapp A, Pandey KK, Frenking G (2007) J Am Chem Soc 129:7596
97. Caramori GF, Frenking G (2007) Organometallics 26:5815
98. Caramori GF, Frenking G (2008) Theor Chem Acc 120:351
99. Erhardt S, Frenking G (2009) J Organomet Chem 694:1091
100. Prabusankar G, Gemel C, Parameswaran P, Flener C, Frenking G, Fischer RA (2009) Angew Chem Int Ed 48:5526
101. Parameswaran P, Frenking G (2009) Chem Eur J 15:8817
102. Gamez JA, Tonner R, Frenking G (2010) Organometallics 29:5676

103. Parameswaran P, Frenking G (2010) J Phys Chem A 114:8529
104. Krapp A, Frenking G (2010) Theor Chem Acc 127:141
105. Takagi N, Krapp A, Frenking G (2010) Can J Chem 88:1079
106. Stander-Grobler E, Schuster O, Heydenrych G, Cronje S, Tosh E, Albrecht M, Frenking G, Raubenheimer HG (2010) Organometallics 29:5821
107. Bollermann T, Cadenbach T, Gemel C, von Hopffgarten M, Frenking G, Fischer RA (2010) Chem Eur J 16:13372
108. Takagi N, Krapp A, Frenking G (2011) Inorg Chem 50:819
109. Senthilnathan D, Vaideeswaran S, Venuvanalingam P, Frenking G (2011) J Mol Model 17:465
110. Rasika Dias HV, Dash C, Yousufuddin M, Celik MA, Frenking G (2011) Inorg Chem 50:4253
111. Petz W, Neumüller B, Klein S, Frenking G (2011) Organometallics 30:3330
112. Bayat M, Hopffgarten Mv, Salehzadeh S, Frenking G (2011) J Organomet Chem 696:2976
113. Esterhuysen C, Frenking G (2011) Chem Eur J 17:9944
114. Bollermann T, Freitag K, Gemel C, Molon M, Seidel RW, von Hopffgarten M, Jerabek P, Frenking G, Fischer RA (2011) Inorg Chem 50:10486
115. von Hopffgarten M, Frenking G (2011) J Phys Chem A 115:12758
116. Bayat M, Salehzadeh S, Frenking G (2012) J Organomet Chem 697:74
117. Das A, Dash C, Yousufuddin M, Celik MA, Frenking G, Rasika Dias HV (2012) Angew Chem 124:4006
118. Das A, Dash C, Yousufuddin M, Celik MA, Frenking G, Rasika Dias HV (2012) Angew Chem Int Ed 51:3940
119. Flener-Lovitt C, Frenking G, Girolami GS (2012) Organometallics 31:4122
120. Nguyen TAN, Frenking G (2012) Chem Eur J 18:12733
121. Celik MA, Dash C, Adiraju VAK, Das A, Yousufuddin M, Frenking G, Dias HVR (2013) Inorg Chem 52:729
122. Mousavi M, Frenking G (2013) Organometallics 32:1743
123. Das A, Dash C, Celik MA, Yousufuddin M, Frenking G, Rasika Dias HV (2013) Organometallics 32:3135
124. Molon M, Gemel C, Seidel RW, Jerabek P, Frenking G, Fischer RA (2013) Inorg Chem 52:7152
125. Nguyen TAN, Frenking G (2013) Mol Phys 111:2640
126. Becke AD (1988) Phys Rev A 38:3098
127. Perdew JP (1986) Phys Rev B 33:8822
128. Snijders JG, Baerends EJ, Vernooijs P (1982) At Nucl Data Tables 26:483
129. Krijn J, Baerends BJ (1984) Fit functions in the HFS-method, internal report. Vrije Universiteit, Amsterdam (in Dutch)
130. ADF Version 2012.01
131. Morokuma K (1971) J Chem Phys 55:1236
132. Ziegler T, Rauk A (1979) Inorg Chem 18:1558
133. Ziegler T, Rauk A (1979) Inorg Chem 18:1755
134. Bickelhaupt FM, Baerends EJ (2000) Rev Comput Chem In: Lipkowitz KB, Boyd DB (eds), vol 15. Wiley-VCH, New York, p 1
135. te Velde G, Bickelhaupt FM, Baerends EJ, van Gisbergen SJA, Fonseca Guerra C, Snijders JG, Ziegler T (2001) J Comput Chem 22:931
136. von Hopffgarten M, Frenking G (2012) WIREs Comput Mol Sci 2:43
137. Huber KP, Herzberg G (1979) Constants of diatomic molecules. Van Nostrand-Reinhold, New York
138. Brassington NJ, Edwards HGM, Long DA (1981) J Raman Spectrosc 11:346
139. Lias SG, Bartmess JE, Liebman JF, Holmes JL, Levin RD, Mallard WG (1988) J Phys Chem Ref Data 17(Suppl 1)
140. Krapp A, Bickelhaupt FM, Frenking G (2006) Chem Eur J 12:9196

Theor Chem Acc (2014) 133:1436
DOI 10.1007/s00214-013-1436-7

REGULAR ARTICLE

Accurate first principles calculations on chlorine fluoride ClF and its ions $ClF^{\pm}$

Athanassios A. Vassilakis · Apostolos Kalemos · Aristides Mavridis

Received: 12 September 2013 / Accepted: 5 December 2013 / Published online: 7 January 2014

Abstract The present work focuses on the first (lightest) of the six diatomic interhalogens, namely ClF and its ions ClF^+ and ClF^-, in an effort to better understand these interesting species. Toward this end, we have performed highly correlated all electron ab initio calculations of multireference (MRCI) and single-reference coupled-cluster calculations, employing quintuple and sextuple correlation consistent basis sets. Within the $\Lambda - S$ ansatz, we have examined all 12 states of ClF correlating adiabatically with the first energy channel, all 23 states of ClF^+ correlating with the first three channels, and three states out of four of ClF^- correlating with the first two channels $Cl^- + F$ and $Cl + F^-$. Full potential energy curves at the MRCI/quintuple zeta level have been constructed for 12, 21, and 3 states of ClF, ClF^+, and ClF^-, respectively. After correcting for core–subvalence and scalar relativistic effects, albeit small as expected, and spin–orbit interactions, most of our results are in excellent agreement with available experimental data. Some lingering questions have been definitely settled. Our final recommended binding energies (D_0 in kcal/mol) and equilibrium bond distances (r_e in Å) for ClF ($X\ ^1\Sigma^+$), ClF^+ ($X\ ^2\Pi$), and ClF^- ($X\ ^2\Sigma^+$) are [60.35, 1.6284], [67.40, 1.5357], and [29.80, 2.151], respectively. The adiabatic electron affinity of ClF, ClF ($X\ ^1\Sigma^+$) + $e^- \rightarrow ClF^-$ ($X\ ^2\Sigma^+$), is $EA_{ad} = 2.25 \pm 0.01$ eV about 0.6 eV smaller than the suggested experimental value which is certainly wrong.

Keywords Interhalogen · ClF · Ab initio

Dedicated to Professor Thom Dunning and published as part of the special collection of articles celebrating his career upon his retirement.

A. A. Vassilakis · A. Kalemos · A. Mavridis (✉)
Laboratory of Physical Chemistry, Department of Chemistry, National and Kapodistrian University of Athens, Panepistimiopolis, Athens 157 71, Greece
e-mail: mavridis@chem.uoa.gr

A. Kalemos
e-mail: kalemos@chem.uoa.gr

1 Introduction

One of the most interesting series of atomic elements are the four stable halogens (X), namely ^{19}F ($Z = 9$), $^{35,37}Cl$ (17), $^{79,81}Br$ (35), and ^{127}I (53) [1]. All possible combinations between the four halogens result to four homonuclear dihalides (X_2) and six unique diatomic interhalogens (XX′), viz. ClF, BrF, IF, BrCl, ICl, and IBr, all of which are known and relatively stable, the most robust being the ClF [2]. The latter is a colorless gas at room temperature with a melting point of −155.6 °C [2] and a dissociation energy (D_0^0) of ∼60 kcal/mol [3]. For some physical properties of the XX′ compounds, see Table 17.12 of Ref. [2].

The present ab initio work focuses on ClF and its ions ClF^+ and ClF^-. We have investigated all 12 $^{2S+1}\Lambda$ molecular states of the neutral species emanating from the ground-state fragments Cl(^{2}P) and F(^{2}P), all 23 (=12 + 9+2) states of ClF^+ related to the first three adiabatic channels of Cl^+(^{3}P, ^{1}D, ^{1}S) + F(^{2}P), and three states of ClF^-, employing variational multireference (MRCI) and single-reference coupled cluster methods combined with large correlation consistent basis sets. In what follows we give a rather complete account of the previous experimental and theoretical work on ClF and $ClF^{\pm}$ related to the present work.

To the best of our knowledge, the first experimental work on ClF was published by Wahrhaftig [4] in 1942 who recorded its absorption spectrum. He cites spectroscopic parameters and dissociation energies, D_e and D_0, for both the ground ($X\,^1\Sigma^+$) and first excited ($a\,^3\Pi$), the only "true" bound valence states correlating with ground-state atoms. The accuracy of Wahrhaftig's results for work published more than seventy years ago is indeed remarkable. Table 1 lists, practically, all the experimental structural information for the $X\,^1\Sigma^+$ and $a\,^3\Pi$ states of ClF concerning the present work (Refs. [4–17]). The work of Schumacher and coworkers published in 1947 and 1950 is not included in Table 1 because of our lack of access to the appropriate journals [18, 19] (see, however, Refs. [6, 7]). It should be mentioned at this point that henceforth the first excited (bound) state, $a\,^3\Pi$, will be tagged either A′ ($^3\Pi_2$), or A ($^3\Pi_1$), or B ($^3\Pi_{0^\pm}$), depending on the $\Omega = \Lambda + \Sigma$ value; see footnote (l) of Table 1. If the Ω coupling is not taken into account, the first excited state of ClF will be simply called $a\,^3\Pi$ according to standard practice [15].

The first theoretical work on the electronic structure and properties of the X state of ClF was published in 1974 by Staub and McLean [20]. These authors reported Hartree–Fock HF/(DZ + P)—Slater results on all diatomic molecules XX′, X_2, and HX (X = Cl, F, Br, and I). Since then, a significant number of ab initio studies on ClF have been appeared in the literature, the most relevant to the present work given in Refs. [21–27, 31–34][1]. In 1985, Scharf and Ahlrichs [22] performed coupled pair functional (CPF) calculations for the X state of ClF (and ClF_3) employing a $[6s6p/_{Cl}\ 6s4p/_F]$ + P(olarizaion) basis sets. They report r_e, $D_e + Q$ ($+Q$ = the Davidson correction), and μ_e (equilibrium electric dipole moment). In 1988, Peyerimhoff and coworkers [23] using the multireference single and double excitation MRDCI method (see below) and a Gaussian $[6s5p2d/_{Cl}\ 4s4p2d/_F]$ + 3s2p bond functions basis set, calculated potential energy curves (PEC) for the first six singlets ($X\,^1\Sigma^+$, $1^1\Pi$, $2^1\Pi$, $1^1\Delta$, $1^1\Sigma^-$, $2^1\Sigma^+$), all repulsive except the X state, correlating with Cl(^{2}P) + F(^{2}P). A number of Rydberg and charge transfer states have also been studied. In 1990, Peterson and Woods [24] using the coupled electron pair approximation (CEPA), singles and doubles electron configuration (CISD), and Møller–Plesset fourth-order perturbation theory with single, double, and quadruple excitations (MP4SDQ) combined with a $[12s8p3d1f/_{Cl}\ 10s3p2d1f/_F]$ Gaussian basis sets, calculated among other things the spectroscopic constants r_e, D_e, ω_e, $\omega_e x_e$, α_e, $\bar{D}_e$, and μ_e of the $X\,^1\Sigma^+$ state of the isoelectronic species ClF, ClO^-, SF^-, and ArF^+ around equilibrium. Their results are in good agreement with corresponding experimental numbers. Perera and Bartlett [25] examined the dependence of the electric dipole moments on scalar relativistic effects of the ground states of all six interhalogens XX′ at the coupled cluster level (CCSD) and effective core potentials.

In 1998, de Jong et al. [26] studied the ground states of the six interhalogens XX′ using augmented correlation consistent aug–cc–pVnZ (n = 2, 3) valence basis sets at the MP2, CCSD, and CCSD(T) fully relativistic level of theory. In particular, correlation effects of dipole moments and polarizabilities were obtained through the CISD relativistic approach.

The most systematic and complete theoretical work so far on ClF published in 2000 is that by Alekseyev, Lieberman, and Buenker [27]. These workers performed multireference MRDCI calculations (see Refs. [28–30]) employing relativistic effective core potentials (RECP), treating explicitly the seven valence electrons ($3s^2 3p^5/2s^2 2p^5$) of each halogen atom through a $[7s7p3d1f/_{Cl,F}]$ basis set. They report full PECs for all 12 $\Lambda - S$ states correlating with Cl(^{2}P) + F(^{2}P) and five ion pair states, which correlate with the ionic fragments $Cl^+(^3P) + F^-(^1S)$. Spin–orbit (SO) effects have been taken into account employing the $^{2S+1}\Lambda$ eigenfunctions as basis for the SO calculations. Their results will be compared to ours in due course. Recall, however, that MRDCI calculations involve a specific number of approximations in addition to the RECPs in the present case, details of which are described in Refs. [27–30].

In a mixture of density functional theory DFT(B3LYP) and coupled cluster CCSD(T) calculations combined with aug–cc–pVnZ (n = T, Q, 5) basis sets, Ricca [31] calculated certain ground-state properties of ClF_x (x = 1, 2, 3) around equilibrium. At the highest level of theory including tight d functions and correcting for scalar relativistic and core effects, her ab initio results r_e, D_e, and IE (ionization energy) are in good agreement with experiment. Three years later, Horný et al. [32] reported a series of coupled cluster calculations around equilibrium on the X states of ClF and ClF^- with the purpose of obtaining an accurate value of the electron affinity (EA) of ClF, a controversial quantity indeed (vide infra), using a series of valence cc-pVnZ (n = D, T, Q, 5) basis sets with and without diffuse functions and a variety of coupled cluster (CC) variants (CCSD, CCSD(T), CCSDT, EOM-CC). Besides EA, they predict r_e and ω_e for the neutral species. Their best values at the CCSD(T)/aug–cc–pV5Z level for the $X\,^1\Sigma^+$ state of ClF are $r_e = 1.6305$ Å and $\omega_e = 783$ cm^{-1} in agreement with experiment.

Very recently Chen et al. [33] in an effort to understand the bonding in the series of molecules ClF_x (x = 1–7), they performed high-level valence MRCI and CCSD(T) calculations in conjunction with aug–cc–pVnZ (n = T, Q, 5)

[1] This just published work by Dunning and coworkers again on ClF_n^+ (n = 1–6) refines the "recoupled pair bonding" model (*rpd*), introduced in order to explain the phenomenon of hypervalency. For another point of view on hypervalency, perhaps more economic and without introducing new models like the rpd, see Refs [35, 36].

Table 1 Experimental results on $^{35}Cl^{19}F$. Bond distances r_e (Å), dissociation energies D_e and D_0 (cm^{-1}), harmonic frequencies and anharmonic corrections ω_e, $\omega_e x_e$, $\omega_e y_e$ (cm^{-1}), rotational vibrational coupling constants α_e (cm^{-1}), electric dipole moments μ_e (D), and separation energies T_e (cm^{-1}) of the $X\,^1\Sigma^+$ and $a\,^3\Pi$ states

Footnotes	$X\,^1\Sigma^+$								
	r_e	D_e^a	D_0^a	ω_e	$\omega_e x_e$	$\omega_e y_e$	$\alpha_e \times 10^3$	μ_e	T_e
b	1.625	21,495 ± 3	21,101 ± 2	793.2	9.9		6 ± 2		0.0
c	1.628 (1)		20,633 ± 2	784.1 ± 1	5.3 ± 0.5		4.5 ± 0.4		
d	1.62831						4.359	0.881 ± 0.02	
e			20,406 ± 323						
f								0.8881 (2)	
g			21,108						
h				785.2 ± 1.3	5.3 ± 0.1	−0.03 ± 0.01			
i	1.628341 (4)			783.553 (30)	5.045 (87)		4.325		
j		21,500	21,126 ± 6						
k			21,110 ± 2						
States	$a\,^3\Pi$ [l]								
$^3\Pi_{0+}$ [b]	1.92	2,946 ± 50	2,790 ± 50	313.484	2.217	−0.400	1.4		18,549 ± 50
$^3\Pi_{0+}$ [d]	2.038 (1)			362.4 ± 1	8.8 ± 0.5	−0.14 ± 0.10	3.8 (1)		
$B^3\Pi_{0+}$ [h]				362.5 ± 1.6	8.5 ± 0.4	−0.13 ± 0.03			18,827 ± 4
$B^3\Pi_{0+}$ [j]	2.031 [m]	3,079.1		363.1 [m]	8.64 [m]	−0.124 [m]	4.7 [m]		18,825.3
$A^3\Pi_1$ [k]				353 ± 1	9.7 ± 1	0.17 ± 0.01			18,841 ± 5
$A'^3\Pi_2$ [n]	2.0245 (16)	3,243 ± 5		363.53 (2.58)	8.3 (67)		6.31 (28)		18,257 ± 5
$A^3\Pi_1$ [o]	2.0247	2,988.4 [p]		361.23	7.74		6.28		18,511.6
$B^3\Pi_{0+}$ [o]	2.0221	3,078.7 [q]		362.578	8.227		7.36		

[a] With respect to the ground-state atoms including spin–orbit interaction, $Cl(^2P_{3/2}) + F(^2P_{3/2})$

[b] Ref. [4]; ClF absorption spectroscopy; most probable dissociation products of the $^3\Pi_{0+}$ state are $Cl(^2P_{3/2}) + F(^2P_{3/2})$

[c] Ref. [6]; visible absorption spectrum; D_0 obtained by assuming that the end products of the $^3\Pi$ state are $Cl(^2P_{1/2}) + F(^2P_{3/2})$

[d] Ref. [5]; microwave spectroscopy; r_e calculated from B_e; see also Ref. [3]

[e] Ref. [7]; photoionization study; most probable end products of the $^3\Pi$ state are $Cl(^2P_{1/2}) + F(^2P_{3/2})$. Ionization energy IE[ClF($X\,^1\Sigma^+$, $v = 0$) → ClF^+ ($X\,^2\Pi_{3/2}$, $v = 0$)] = 12.65 ± 0.01 eV

[f] Ref. [8]; radio frequency spectroscopy

[g] Ref. [9]; chemiluminescence spectroscopy; end products of the $^3\Pi$ state taken as $Cl(^2P_{3/2}) + F(^2P_{1/2})$

[h] Ref. [10]; emission spectroscopy

[i] Ref. [11]; millimeter rotational spectroscopy

[j] Refs. [12, 13]; laser excitation spectrum; upper limit of D_0 with dissociation products of the B state ($^3\Pi_{0+}$) taken as $Cl(^2P_{3/2}) + F(^2P_{1/2})$

[k] Ref. [14]; optical–optical double resonance (OODR) spectroscopy; dissociation products of the B state taken as $Cl(^2P_{3/2}) + F(^2P_{1/2})$, $D_0 = 20{,}633$ cm^{-1} if end products $Cl(^2P_{1/2}) + F(^2P_{3/2})$

[l] According to standard convention(s) (Ref. [15]; see also Ref. [3]), the first higher state with different spin multiplicity from the ground state of a diatomic molecule (here the first excited state $^3\Pi$) should be named $a\,^3\Pi$. However, sanctioned by custom the historical naming is followed, that is, A′ ($^3\Pi_2$), A($^3\Pi_1$), and B($^3\Pi_{0+}$); no letter name has been attached to $^3\Pi_{0-}$. Unfortunately, this practice is still followed for some historical molecules like N_2, X_2, and XX′

[m] Recalculated in Ref. [3] according to data of Ref. [6]

[n] Ref. [16]; OODR + fluorescence spectroscopy

[o] Ref. [17]; OODR spectroscopy

[p] Ref. [17]; dissociation to two ground-state atoms 21,500 (2) above the X state

[q] Ref. [17]; dissociation to $Cl(^2P_{3/2}) + F(^2P_{1/2})$

basis sets and a tight d function on Cl. For the $X\,^1\Sigma^+$ and $a\,^3\Pi$ states of ClF, their equilibrium structural parameters r_e, D_e (D_0), ω_e, and IE will be contrasted to ours later on.

We turn now to the ClF^+ cation. The limited experimental data we are aware of are given in the Huber–Herzberg collection [3], all drawn from Refs. [7, 37–39] published some 40 years ago. The best D_0^0 experimental value of ClF^+ ($X\,^2\Pi_{3/2}$) is indirectly obtained through the energy conservation relation $D_0^0(ClF^+; X^2\Pi_{3/2}) = D_0^0(ClF; X^1\Sigma^+) + IE(Cl; ^2P_{3/2}) - IE(ClF; X^1\Sigma^+) = 2.617 + 12.96763$ (Ref. [40]) -12.66 (Ref. [3]) eV = 2.924 eV (= 67.43 kcal/mol). The Huber–Herzberg value is D_0^0 = 2.93 eV [3].

There are only two relevant ab initio calculations on ClF^+. In 1981, Ewig et al. [21] obtained the vertical ionization energy of ClF within a limited MCSCF-CI/ [$5s4p1d/_{Cl}$ $4s3p1d/_F$] approach. Almost 30 years later, Dunning and coworkers [33] reported D_0, r_e, ω_e, of the $X\,^2\Pi$ and $^4\Sigma^-$ states of ClF^+ at the valence MRCI (+Q)/ aug–cc–pVnZ (n = T, Q, 5) level. Their best results (MRCI + Q/n = 5) are D_0 = 66.14 (13.08) kcal/mol, r_e = 1.536 (1.1745) Å, and ω_e = 919.9 (363.2) cm^{-1} for the $X\,^2\Pi$ ($^4\Sigma^-$) state. At the MRCI + Q/n = 5 level, the ionization energy is predicted to be IE [$ClF(X\,^1\Sigma^+) \rightarrow ClF^+(X\,^2\Pi)$] = 12.56 eV, just 0.1 eV less than the experimental value [3].

Hardly anything is known experimentally for the anion ClF^-. The four experimental publications that refer to the electron affinity EA, the only (indirectly) measured quantity, seems to be off by more than 0.5 eV (see below). Harland and Thynne [41] are the first to give an EA value of ClF (or the IE of ClF^-) after bombarding the pentafluorosulfur chloride with an electron beam; they report EA = 1.5 ± 0.4 eV. Their results were confirmed 10 years later by Dispert and Lacmann [42] who obtained EA = 1.5 ± 0.3 eV. Almost the same time Illenberger et al. [43] through low-energy electron impact experiments on CF_2Cl_2, CF_3Cl, and $CFCl_3$ concluded that EA (ClF) = 1.79 eV, within the (large) error bars of the previous workers. According to Ref. [32], however, Dudlin et al. [44] "in their electron impact investigation of ClF_3 and its dissociation products estimated the lower limit of $EA_{ad(iabatic)}$ to be 2.37 ± 0.21 eV". After further analysis of the dissociation energies, the authors suggest that EA_{ad} = 2.86 ± 0.2 eV (see also Ref. [45]). Obviously, an experimental reinvestigation of the electron affinity of ClF is in order.

We are aware of two ab initio studies on the EA of ClF. Through G3 [46] and G3X [47] calculations, Law et al. [48] report EA = 2.31 and 2.07 eV, respectively. The most recent work on the EA of ClF is that by Horný et al. (vide supra) [32]. Their best results at the CCSD(T)/aug–cc–pV5Z (+zero point energy = ZPE) are $EA_{ad}(ClF)$ = 2.22 (2.25) eV, r_e (ClF^-) = 2.1531 Å, and ω_e (ClF^-) = 378 cm^{-1}, in harmony with the present work (vide infra).

We believe that the above exposition on the past findings on ClF and $ClF^{\pm}$ shows the need for a systematic theoretical investigation on these species, confirming or not certain results and reporting new ones. Hence, we have performed highly correlated all electron ab initio calculations via the MRCI, RCCSD(T), and RCCSDT methods combined with large correlation consistent basis sets, including scalar relativistic and core correlation effects on Cl as well as spin–orbit (SO) couplings. The construction of full potential energy curves for a large number of states allows for a better understanding of bonding interactions and the extraction of accurate spectroscopic parameters and energetics. Our work is structured as follows: In Sect. 2, we outline basis sets and methods, Sect. 3 and subsections 3.1, 3.2, and 3.3 refer to results and discussion on ClF, ClF^+, and ClF^-, respectively, while Sect. 4 epitomizes our general approach and findings.

2 Basis sets and methods

The augmented correlation consistent basis sets of Dunning and coworkers [49–52] for the F and Cl atoms, aug–cc–pVnZ and aug–cc–pV(n + d)Z (n = 5, 6) including a set of tight d functions for Cl were employed in all calculations. Both sets were generally contracted to [$7s6p5d4f3g2h/_F$ $8s7p6d4f3g2h/_{Cl}$] and [$8s7p6d5f4g3h2i/_F$ $9s8p7d5f4g3h2i/_{Cl}$] for n = 5 (quintuple) and n = 6 (sextuple), renamed for brevity A5ζ and A6ζ and consisting of 263 and 387 spherical Gaussians, respectively. To estimate the core effects on ClF for the Cl atom only ($2s^22p^6$), the A5ζ set was augmented by a series of weighted core functions [53], resulting to [$7s6p5d4f3g2h/_F$ $12s11p9d7f5g3h/_{Cl}$] ≡ CA5ζ of order 347. Notice that the CA5ζ basis set does not include the tight d function on Cl. Scalar relativistic valence effects were calculated at the quintuple cardinality level after recontracting the A5ζ basis set accordingly (=A5ζ − rel) [50, 53].

Two methods of correlated calculations are followed, the complete active space self-consistent field + single + double replacements (CASSCF + 1+2 = MRCI), and the (single-reference) restricted coupled cluster + singles + doubles + quasi-perturbative connected triples (RCCSD(T)) [54–57]. The CASSCF reference wavefunctions are defined by allotting 14 (ClF), 13 (ClF^+), and 15 (ClF^-) electrons to 12 orbitals related to ($3s + 3p/_{Cl}$ $2s + 2p + 3s + 3p/_F$) atomic orbitals for ClF, ClF^+, and ClF^- giving rise to 978, 1,196, and 354 reference configuration functions (CF), respectively. Internally contracted (*ic*) [58, 59], valence MRCI wavefunctions are calculated by single + double excitations out of the CASSCF wavefunctions; of course, all of our reported multireference results are of *ic*MRCI quality. RCCSD(T) valence calculations were performed only for the $X\,^1\Sigma^+$ state of ClF, the $X\,^2\Pi$ and $a\,^4\Sigma^-$ states of ClF^+, and the $X^2\Sigma^+$ state of ClF^-. In addition, RCCSDT/A5ζ calculations were performed for the X states of ClF, ClF^+, and ClF^-. Core correlation effects of the $2s^22p^6/_{Cl}$ electrons are taken into account by including them in the CI and CC procedures, tagged C-MRCI and C-RCCSD(T), respectively. The inclusion of the $2s^22p^6/_{Cl}$ subvalence electrons in the CI process increases considerably the number of CFs. For instance, the MRCI (*ic*MRCI)/A5ζ expansion of the $X\,^1\Sigma^+$ wavefunction of ClF contains 3.7×10^8 (8.2×10^6) CFs as contrasted to 1.6×10^9 (3.7×10^7) CFs of the C-MRCI(*ic*C-MRCI)/CA5ζ wavefunctions. Corresponding numbers of the (valence) MRCI (*ic*MRCI)/A6ζ are 8.34×10^8 (9.8×10^6) CFs.

Valence scalar relativistic effects for the $X\,^1\Sigma^+$, $a\,^3\Pi$ (ClF), $X\,^2\Pi$, $a\,^4\Sigma^-$ (ClF^+), and $X^2\Sigma^+$ (ClF^-) were estimated through the Douglas–Kroll–Hess Hamiltonian up to second order (DKH2) [60, 61] using the A5ζ − rel basis set. Spin–orbit effects have been obtained by diagonalizing the $\hat{H}_{el} + \hat{H}_{SO}$ Hamiltonian employing the $\hat{H}_{el}$ MRCI/A5ζ eigenvectors, where $\hat{H}_{SO}$ is the full Breit–Pauli operator. Basis set superposition errors (BSSE) estimated around equilibrium by the usual counterpoise method [62, 63] are small enough not to be taken into account, for example, for the $X\,^1\Sigma^+$ state of ClF BSSE [MRCI/A5ζ (A6ζ)] = 101 (47) cm^{-1}, with similar results at the CC level. Spectroscopic parameters (r_e, ω_e, ω_ex_e, ω_ey_e, α_e, $\bar{D}_e$) have been determined via the Dunham approach [64, 65]. The number of vibrational levels for certain states was determined by solving numerically the one-dimensional Schrödinger equation of the two nuclei. The size non-extensivity error (SNE) is estimated by subtracting the sum of the energies of the atoms from the energy of the supermolecule at an internuclear distance of 20–30 bohr. We find that for the $X\,^1\Sigma^+$ state of ClF SNE = 8.6 (1.7), 8.7 (1.7) kcal/mol at the MRCI (+Q)/A5ζ or A6ζ level, respectively, where +Q refers to the Davidson correction [66, 67]; see also page 3 of Ref. [68]. Finally, C_{2v} constraints have been imposed through all computations.

The RCCSDT/A5ζ calculations were performed by the CFOUR program [69]; all other calculations were carried out through the MOLPRO 2010 package [70].

3 Results and discussion

3.1 ClF

For reasons of convenience, Table 2 lists the most reliable, according to the present authors, experimental values of the $X\,^1\Sigma^+$ and $B\,^3\Pi_{0^+}$ states of $^{35}Cl^{19}F$; see also Table 1. The interaction of the ground-state atoms Cl(^{2}P) + F(^{2}P) gives rise to 12 molecular $^{2S+1}\Lambda$ states, singlets and triplets, that is, $^{1,3}(\Sigma^+[2], \Sigma^-, \Pi[2], \Delta)$. Two of them $X\,^1\Sigma^+$ and $a\,^3\Pi$ are bound, one is a *van der Waals* state ($1^1\Pi$), whereas the rest of the nine states are strongly repulsive. For all 12 states, full MRCI + Q/A5ζ PECs are displayed in Fig. 1. The second adiabatic channel Cl ($3s^23p^44s^1$; ^{4}P) + F(^{2}P) located 8.922 eV higher gives rise to states of Rydberg character; the third and fourth channels are of charge transfer nature, $Cl^+(^3P) + F^-(^1S)$ and $Cl^+(^1D) + F^-(^1S)$, located 9.566 and 11.011 eV above the X state, respectively [71]. From the third and fourth channels, five ion pair states (Cl^+F^-) of $^3\Sigma^-$, $^3\Pi$ and $^1\Sigma^+$, $^1\Pi$, and $^1\Delta$ symmetry emerge. We note that the F^- anion does not have any excited states (vide infra). As was already mentioned, the present study deals exclusively with the first 12 valence Λ − S states related to the first Cl(^{2}P) + F(^{2}P) channel.

Table 2 "Best" experimental properties of the $X\,^1\Sigma^+$ and $B^3\Pi_{0^+}$ states of $^{35}Cl^{19}F$ according to Table 1. Apart from r_e (Å) and μ_e (Debye) all units are in cm^{-1}

Properties	$X\,^1\Sigma^+$	$B^3\Pi_{0^+}$
r_e	1.628341 (4)	2.0221
D_e^0	21,500[a]	3078.7[c]
D_0^0	21,110[b]	
ω_e	783.353	362.6
ω_ex_e	5.045	8.3
ω_ey_e	−0.003 (1)	
$\alpha_e \times 10^3$	4.325	7.4
$\bar{D}_e \times 10^7$	8.98[d]; 8.77[e]	11.0[f]
μ_e	0.8881 (2)	
T_e	0.0	18,825.3

[a] 2.666 eV = 61.47 kcal/mol

[b] 2.616 eV = 60.36 kcal/mol

[c] 0.3817 eV = 8.802 kcal/mol

[d] Refs. [12, 13]

[e] Calculated through $\bar{D}_e = 4B_e^3/\omega_e^2$ where B_e and ω_e are taken from Refs. [4, 5], respectively; see also Ref. [3]

[f] Obtained by us using the B_e and ω_e given in Refs. [12, 13]

3.1.1 $X\,^1\Sigma^+$

Table 3 lists our numerical results on the $X\,^1\Sigma^+$ state of ClF, whereas Fig. 1 displays all 12 MRCI + Q/A5ζ $^{2S+1}\Lambda$ PECs correlating with the ground-state atoms. The bonding of the $X\,^1\Sigma^+$ state is described succinctly by the valence-bond-Lewis (vbL) diagram shown below along with the leading equilibrium MRCI/A5ζ CFs (only valence e^- are counted).

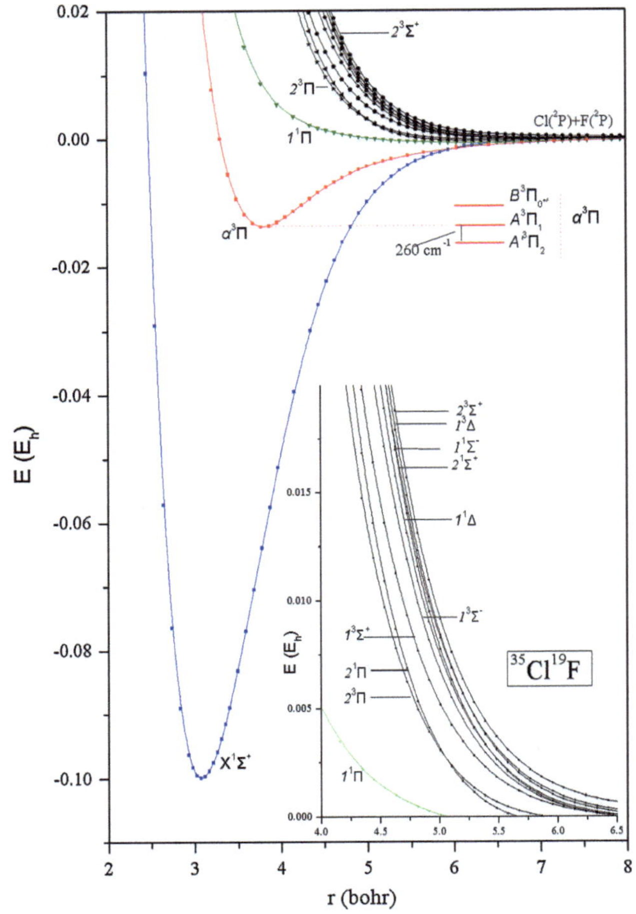

Fig. 1 Relative MRCI + Q/A5ζ PECs of twelve $^{2S+1}\Lambda$ valence states of ClF correlating with Cl(^{2}P) + F(^{2}P). Two states are bound ($X\ ^1\Sigma^+$, $a\ ^3\Pi$), one is of vdW nature ($1^1\Pi$) and the rest strongly repulsive. The first *inset* magnifies the nine repulsive PECs, and the second shows the Ω splittings of the $a\ ^3\Pi$ state

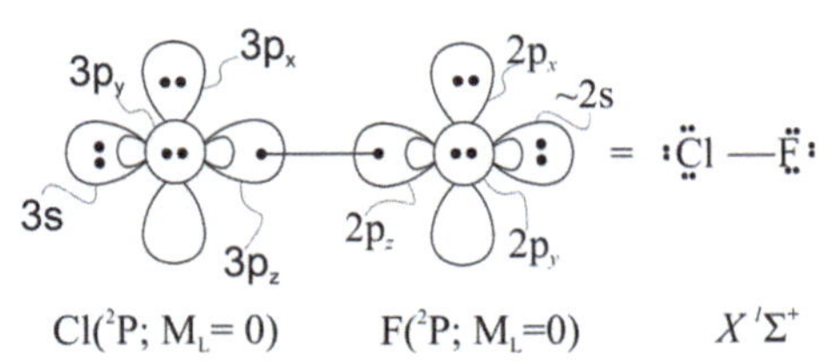

$$|X^1\Sigma^+\rangle \approx \left|1\sigma^2 2\sigma^2[(0.94)3\sigma^2 - (0.11)4\sigma^2]1\pi_x^2 1\pi_y^2 2\pi_x^2 2\pi_y^2\right\rangle.$$

According to the Mulliken population analysis, a total charge of 0.25 e$^-$ is transferred from Cl to F, resulting to a Cl$^+$F$^-$ charge polarity. All our results listed in Table 3, with the exception of the second anharmonicity $\omega_e y_e$, are in very good agreement with experiment. The MRCI + Q/A6ζ and RCCSD(T)/A6ζ properties corrected for core ($2s^2 2p^6/_{Cl}$) and scalar relativistic effects, employing the CA5ζ and A5ζ − rel basis sets at the corresponding methods, are referred as "best" values; Table 3. For instance, at the CC level r_e ("best") ≡ r_e* = r_e [RCCSD(T)/A6ζ] + δr_{core} + δr_{rel} where, δr_{core} = r_e [C-RCCSD(T)/CA5ζ] − r_e [RCCSD(T)/A5ζ] and δr_{rel} = r_e[RCCSD(T) + DK/A5ζ − rel] − r_e [RCCSD(T)/A5ζ]. Therefore, r_e* = 1.6289 + (−0.0022) + 0.0017 = 1.6284 Å in perfect agreement with experiment; Table 3. Exactly the same r_e* (=1.6284 Å) is obtained at the MRCI + Q level. Dissociation D_e* values are 62.78 (MRCI) and 62.21 (RCCSD(T)) kcal/mol with respect to Cl(^{2}P) + F(^{2}P). Taking into account, however, the calculated (experimental [71]) SO splitting $\Delta E(^2P_{3/2} - {}^2P_{1/2})$ = 799.1 (881) and 398.1 (404) cm^{-1} for Cl and F, respectively, the dissociation energy D_e with respect to the lowest Cl($^2P_{3/2}$) + F($^2P_{3/2}$) fragments and disregarding the SO effects of the $X\ ^1\Sigma^+$ state as negligible (~5 cm^{-1}), is D_e^0(MRCI + Q) = D_e* − 1/3 × $\Delta E_{Cl}(^2P_{3/2} - {}^2P_{1/2})$ − 1/3 × $\Delta E_F(^2P_{3/2} - {}^2P_{1/2})$ = 62.78 kcal/mol − 1/3 × 799.1 cm^{-1} − 1/3 × 398.1 cm^{-1} = 61.64 kcal/mol. Also D_0(MRCI + Q) = D_e^0(MRCI + Q) − $\omega_e/2$ + $\omega_e x_e/4$ = 61.64 kcal/mol − 787.7/2 cm^{-1} + 4.70/4 cm^{-1} = 60.35 kcal/mol. Analogously, the CC method gives D_e^0[RCCSD(T)] = 62.21 kcal/mol − 1/3 × 799.1 cm^{-1} − 1/3 × 398.1 cm^{-1} = 61.07 kcal/mol. The MRCI + Q D_e (D_0) = 61.64 (60.35) kcal/mol can be considered as in perfect agreement with the experimental value of 61.47 (60.35) kcal/mol (see Table 3), while the CC value is by 0.40 kcal/mol smaller. Corresponding numbers from Ref. [27] are r_e = 1.641 Å and D_e^0 = 58.62 kcal/mol. Notice that at the CC/A5ζ level triple excitations either perturbatively (RCCSD(T)) or self-consistently (RCCSDT) give identical results; see Table 3. Spectroscopic constants ω_e*, $\omega_e x_e$*, α_e, and $\bar{D}_e$ are in very good agreement with the experimental results as well.

We now turn to the permanent electric dipole moment of the $X\ ^1\Sigma^+$ state of ClF. All calculated values listed in Table 3 either MRCI + Q or CC have been obtained through the finite field approach; see also Ref. [72]. The μ_e*[RCCSD(T)] = 0.891 D is in complete agreement with the experimental value (μ_e = 0.8881(2) D), while μ_e*[MRCI + Q] = 0.865 D is smaller by 0.023 D, a relative error of less than 3 %. It is interesting to note that in 1973 Flygare et al. [73, 74] reported an experimental value μ_e = 2.1 ± 1.4 D and an opposite charge polarity, Cl–F$^+$. The latter was challenged very soon by Green [75] whose CI calculations indicated the reverse polarity and confirmed "experimentally" to be indeed Cl$^+$F$^-$ by Janda et al. [76] and Fabricant and Muenter [77]. For a similar misunderstanding concerning the polarity of boron monofluoride (BF), see Ref. [78] and references therein.

Table 3 Total energies E (E_h), bond distances r_e (Å), dissociation energies D_e (kcal/mol), harmonic frequencies ω_e (cm^{-1}), anharmonicities $\omega_e x_e$ and $\omega_e y_e$ (cm^{-1}), rotational vibrational coupling constants α_e (cm^{-1}), centrifugal distortions $\bar{D}_e$(cm^{-1}), and dipole moments μ_e (D) of the $X\ ^1\Sigma^+$ state of $^{35}Cl^{19}F$ at the MRCI + Q (MRCI), RCCSD(T) and RCCSDT methods. Previous theoretical and experimental results are given for easy comparison

Method	$-E$	r_e	D_e^a	ω_e	$\omega_e x_e$	$\omega_e y_e$	$\alpha_e \times 10^3$	$\bar{D}_e \times 10^7$	μ_e^b
MRCI + Q/A5ζ	559.45806	1.6279	62.79	789.4	5.18	−0.109	4.32	8.86	0.879
MRCI + Q/A6ζ	559.46375	1.6263	62.99	789.5	5.63	0.068	4.64	8.90	0.870
(MRCI/A6ζ)[c]	(559.43428)	(1.6234)	(62.83)	(797.8)	(6.24)	(0.04)	(4.64)	(8.83)	(0.897)
C-MRCI + Q/CA5ζ	559.79451	1.6281	63.01	788.4	4.85	−0.029	4.24	8.89	0.865
MRCI + DK + Q/A5ζ − rel	560.95509	1.6298	62.36	787.0	5.62	0.008	4.50	8.86	0.892
"Best" values[d]		1.6284	62.78	786.1	5.74				0.871
RCCSD(T)/A5ζ	559.46165	1.6299	62.22	791.8	4.97	−0.026	4.22	8.79	0.875
RCCSDT/A5ζ	559.46200	1.6306	61.72	788.2					
RCCSD(T)/A6ζ	559.46760	1.6289	62.49	790.0	4.85	−0.026	4.21	8.82	0.876
C-RCCSD(T)/CA5ζ	559.80680	1.6277	62.25	791.3	4.77	−0.099	4.04	8.83	0.872
RCCSD(T) + DK/A5ζ − rel	560.95866	1.6316	61.91	790.0	5.02	−0.012	4.24	8.78	0.893
"Best" values[d]		1.6284	62.21	787.7	4.70				0.891
MRD − CI[e]		1.641	59.70	785					
MRCI + Q/A5ζ[f]	559.45004	1.627	60.18	797.8					
CCSD(T)[f]	559.46165	1.629	62.21	788.2					
Expt.[g]		1.628341 (4)	61.47	783.353	5.045	−0.003	4.325	8.77	0.8881

[a] With respect to the ground-state atoms $Cl(^2P) + F(^2P)$

[b] Calculated by the finite field approach; fixed strength 5×10^{-5} a.u.

[c] MRCI/A6ζ values (no +Q correction) given for comparison

[d] Corrected MRCI + Q/A6ζ and RCCSD(T)/A6ζ values for core and scalar relativistic effects at the CA5ζ and A5ζ − rel levels of theory; see text

[e] Ref. [27]

[f] Ref. [33]

[g] See Table 2

3.1.2 $a\ ^3\Pi$

Figure 1 displays the MRCI + Q/A5ζ PEC of the $a\ ^3\Pi$ state, while Table 4 lists numerical results corrected for scalar relativistic effects ("best" values). To avoid large relative errors that could possibly be induced by the +Q Davidson correction when treating 14 (valence) + 8 (core) = 22 electrons by the MRCI method (see below) for a state of less than 10 kcal/mol binding energy like the $a\ ^3\Pi$ one, core corrections are not included. Within the $^{2S+1}\Lambda$ ansatz, the $a\ ^3\Pi$ state is the first and only excited bound state of ClF correlating with $Cl(^2P) + F(^2P)$. Its bond strength and distance, however, are much weaker and much longer, respectively, than those of the $X\ ^1\Sigma^+$ state. Clearly, two end combinations $Cl(^2P;\ M = 0) + F(^2P;\ M = \pm 1)$ or $Cl(^2P;\ M = \pm 1) + F(^2P;\ M = 0)$ result to two states of $^3\Pi$ symmetry. According to the Mulliken population analysis, the $a\ ^3\Pi$ state correlates with the first one ($M = 0, \pm 1$), whereas the repulsive $2^3\Pi$ state (see Fig. 1) correlates with the second ($M = \pm 1, 0$) combination. The vbL bonding diagram, leading equilibrium MRCI configurations, and Mulliken atomic populations at r_e and r_∞(20 bohr) of the $a\ ^3\Pi$ state are given below

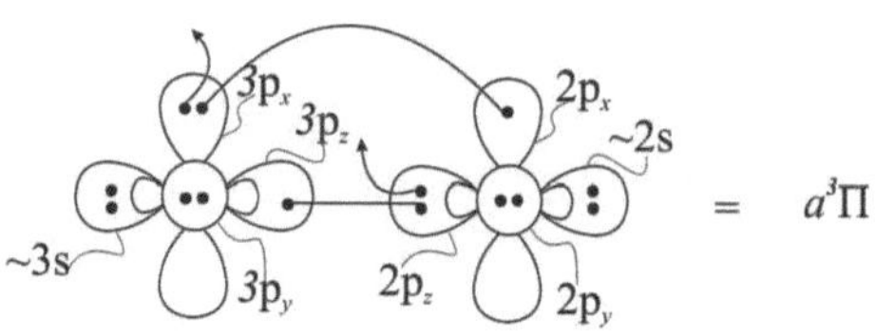

$Cl(^2P;\ M_L = 0)$ $F(^2P;\ M_L = \pm 1)$

$$|a^3\Pi\rangle \approx \left|1\sigma^2 2\sigma^2 3\sigma^2 4\sigma^1 [(0.79) 1\pi_x^2 2\pi_x^1 - (0.32) 1\pi_x^1 2\pi_x^2] 1\pi_y^2 2\pi_y^2\right\rangle - 0.35\left|1\sigma^2 2\sigma^2 3\sigma^1 4\sigma^2 1\pi_x^2 2\pi_x^1 1\pi_y^2 2\pi_y^2\right\rangle$$

$$r_e: 3s^{1.97} 3p_z^{1.50} 3p_x^{1.23} 3p_y^{1.97} /_{Cl} 2s^{1.98} 2p_z^{1.46} 2p_x^{1.73} 2p_y^{1.98} /_F$$

$$r_\infty: 3s^{1.98} 3p_z^{1.03} 3p_x^{1.94} 3p_y^{1.97} /_{Cl} 2s^{1.98} 2p_z^{1.96} 2p_x^{1.03} 2p_y^{1.99} /_F$$

The vbL icon indicates that the $M = 0$ (Cl) + $M = \pm 1$ (F) combination leading to two $3e^- - 2c$(enter) interaction

Table 4 Total energies E (E_h), bond distances r_e (Å), dissociation energies D_e (kcal/mol), harmonic frequencies ω_e (cm^{-1}), anharmonicities $\omega_e x_e$ and $\omega_e y_e$ (cm^{-1}), rotational vibrational coupling constants α_e (cm^{-1}), centrifugal distortions $\bar{D}_e$(cm^{-1}), dipole moments μ_e (D), and energy separations T_e (cm^{-1}) of the $a\ ^3\Pi$ state of $^{35}Cl^{19}F$ at the MRCI + Q method. Previous theoretical and experimental results are given for comparison

Method	$-E$	r_e	D_e^a	ω_e	$\omega_e x_e$	$\omega_e y_e$	$\alpha_e \times 10^3$	$\bar{D}_e \times 10^7$	μ_e^b	T_e
MRCI + Q/A5ζ	559.37191	2.0159	8.70	364.4	9.92	−0.265	6.94	11.5	1.618	18,908
MRCI + Q/A6ζ	559.37720	2.0144	8.70	365.3	10.3	−0.173	7.14	11.5	1.567	18,996
(MRCI/A6ζ)^c	(559.345460)	(2.0293)	(6.91)	(334.7)	(11.8)	(−0.163)	(8.49)	(13.17)	(1.550)	(19,494)
MRCI + DK + Q/A5ζ − rel	560.86954	2.0160	8.75	365.5	10.1	−0.212	7.04	11.5	1.634	18,776
"Best" values^d		2.0145	8.75	366.6	10.5				1.583	18,864
MRD − CI^e		2.064		~350						
MRCI + Q/A5ζ^f	559.36168	2.072	4.73	250						19,393
Expt.^{g}		2.0221	8.802	362.6	8.3					18,825.3

^{a} With respect to the ground-state atoms Cl(^{2}P) + F(^{2}P)

^{b} Calculated by the finite field approach; fixed strength 5 × 10^{-5} a.u.

^{c} MRCI/A6ζ values (no +Q correction) given for comparison

^{d} Corrected MRCI + Q/A6ζ values for core and scalar relativistic effects only; see text

^{e} Ref. [27]

^{f} Ref. [33]

^{g} Results for the $B^3\Pi_{0^+}$ state; see Table 2 and text

is not conducive for strong bonding, rather the opposite, hence the weak binding and considerable bond lengthening as compared to the $X\ ^1\Sigma^+$ state. The population distribution suggests that ~0.5 e$^-$ migrate from F to Cl along the σ frame with a concomitant back transfer from Cl to F of ~0.7 e$^-$ along the π frame, resulting to the Cl^+F^- polarity with the in situ F atom negatively charged by ~0.2 e$^-$.

From Table 4, we can see the excellent agreement at all levels of theory but the MRCI/A6ζ, due to SNE effects (no +Q amendment), and that relativistic effects are very small. Recall that we are working within the Λ − S frame and that the experimental data cited in Table 4 refer to the B ($^3\Pi_{0^+}$) Ω component of the $a\ ^3\Pi$ multiplet (see below). Experiment and theory are in excellent harmony, the absolute differences ΔX from the "best" listed values X_e*, $\Delta X = |X_{expt} - X_e^*|$, being $\Delta r_e = 0.008$ Å, $\Delta D_e = 0.05$ kcal/mol, $\Delta\omega_e = 4.0$ cm^{-1}, $\Delta\omega_e x_e = 2.2$ cm^{-1}, $\Delta\alpha_e$ ~0.3×10^{-3} cm^{-1}, $\Delta\bar{D}_e = 0.5 \times 10^{-7}$ cm^{-1}, and $\Delta T_e = 39$ cm^{-1}. The electric dipole moment of the $a\ ^3\Pi$ state is experimentally unknown, determined theoretically for the first time in the present work; the recommended calculated finite field value is $\mu_e = 1.62 \pm 0.02$ D. Corresponding ΔX differences from Ref. [27] are $\Delta r_e = 0.021$ Å, $\Delta\omega_e = 10.1$ cm^{-1}, and $\Delta T_e = 379$ cm^{-1}; no μ_e or D_e^0 values are given.

Before we turn to the Ω states, a word of caution is needed as to the RCCSD(T) calculations of the $a\ ^3\Pi$ state. Its strong multireference character precludes the use of a single-reference CC method. Indeed, our RCCSD(T)/A5ζ calculations around equilibrium gave results of questionable reliability; that's why they are not included in Table 4. See also the CCSD(T) numbers on the $a\ ^3\Pi$ state of Chen et al. [33].

Taking into account the $\Omega = \Lambda + \Sigma$ coupling, the $a\ ^3\Pi$ state splits into $^3\Pi_2$, $^3\Pi_1$, and $^3\Pi_{0^\pm}$, where to a first approximation the energy separation between the Ω components is given by $\Delta T = A\Lambda\Sigma$, A = constant for a given multiplet. For $\Lambda \neq 0$ as in the present case, however, second-order interactions split slightly the $\Omega = 0$ component into 0^+ and 0^-. In the present case, our calculations do not allow for a 0^+ to 0^- separation; therefore, within our approach, the $^3\Pi_{0^\pm}$ states can be identified with the $^3\Pi_0$.

The A′ $^3\Pi_2$ state is lower in energy (inverted), the calculated MRCI + Q/A5ζ SO $\Delta T(^3\Pi_{0^\pm} - {}^3\Pi_1, {}^3\Pi_1 - {}^3\Pi_2)$ being equidistant with $\Delta T = 260$ cm^{-1}, therefore $A = 260$ cm^{-1}; see the upper inset of Fig. 1. Experimentally, it is found that $\Delta T = 255$ [12, 13, 17] and 313 cm^{-1} (Ref. [16]) for the $\Omega = 0^+ - 1$ and 1–2 splittings, respectively. The 0^+ to 0^- separation has not been determined experimentally, but according to the MRDCI calculations of Buenker and coworkers [27], the 0^- component lies below the 0^+ by ~27 cm^{-1}, a rather large number for this kind of splitting. At the MRCI + Q/A5ζ level of theory, the bond distances of the $^3\Pi_{0^\pm}$, $^3\Pi_1$, and $^3\Pi_2$, states are (experimental values for B, A, and A′ states in parenthesis), $r_e = 2.0245$ (2.0221 or 2.031), 2.0233 (2.0247), and 2.0221 (2.0245) Å, respectively (see Tables 1 and 2). Observe that the absolute differences between experiment and theory are not larger than 0.002 Å.

3.1.3 $1^1\Pi$

The $1^1\Pi$ state is in essence repulsive; its PEC however presents a shallow minimum of less than 0.5 kcal/mol at r_e ~3.3 Å, so it can be characterized as a *van der Waals* (vdW) state; see Fig. 1. It correlates with the $M = 0$ (Cl) + $M = \pm 1$ (F) projections of the ground-state atoms. Around 3.3 Å, there are two leading MRCI configurations of almost equal weight

$$|1^1\Pi\rangle \approx 0.69\left|1\sigma^2 2\sigma^2 3\sigma^1 4\sigma^2 1\pi_x^2 1\pi_y^2 2\bar{\pi}_x^1 2\pi_y^2\right\rangle + 0.64\left|1\sigma^2 2\sigma^2 3\sigma^2 4\sigma^1 1\bar{\pi}_x^1 1\pi_y^2 2\pi_x^2 2\pi_y^2\right\rangle$$

At the MRCI + Q/A6ζ level of theory r(vdW) = 3.31 Å (=6.25 bohr), ΔE(vdW) = −0.43 kcal/mol (= −150 cm^{-1}), $\omega = 44$ cm^{-1}, and $T(1^1\Pi \leftarrow X\ ^1\Sigma^+) = 21{,}795$ cm^{-1}. Interestingly enough the well depth of 150 cm^{-1} can sustain six vibrational levels. At the MRDCI level, Buenker and coworkers predicted a well depth of ~100 cm^{-1} at *r*(vdW) ~6.5 bohr [27].

3.1.4 Repulsive states

Nine more $^{2S+1}\Lambda$ valence states all of strong repulsive character correlate with the ground-state atoms Cl(^{2}P) + F(^{2}P), namely $2^3\Pi$, $2^1\Pi$, $1^3\Sigma^+$, $1^3\Sigma^-$, $1^1\Delta$, $2^1\Sigma^+$, $1^1\Sigma^-$, $1^3\Delta$, and $2^3\Sigma^+$, all calculated at the MRCI + Q/A5ζ level; see lower inset of Fig. 1. The two states of the same spatial–spin symmetry, $1^3\Sigma^+$ and $2^3\Sigma^+$, do not seem to interact at this level, while at distances shorter than 4.5 bohr the energy ordering is the one given above and shown in Fig. 1. Crossings of the pairs $2^1\Sigma^+ - 1^1\Sigma^-$, $2^3\Pi$–$2^1\Pi$, $2^1\Sigma^+ - 1^3\Delta$, and $1^1\Delta$–$1^1\Sigma^-$ are observed at distances ~4.5, 5.0, 5.1, and 5.3 bohr, respectively.

3.2 ClF^+

It is useful to collect at this point all we know experimentally on ClF^+: $D_0^0 = 67.43$ kcal/mol (see the Sect. 1), ω_e ($X\ ^2\Pi_{3/2}$, $^2\Pi_{1/2}$) = 870 or 912 ± 30 cm^{-1}, SO splitting $\Delta E_{SO} = 630$ cm^{-1}, and IE[ClF($X\ ^1\Sigma^+$)→ClF$^+$($X\ ^2\Pi_{3/2}$)] = 12.66 eV [3]; see also Refs. [38, 39]. In comparison with the $X\ ^1\Sigma^+$ state of ClF, the D_0^0 of ClF$^+$ is 7.1 kcal/mol larger.

The IEs of Cl and F are 12.96763 [40] and 17.42282 eV [79], respectively. The ground state of Cl$^+$ is ^{3}P (3s^{2}3p^4) with the first ^{1}D (3$s^2$3p^4), second ^{1}S (3$s^2$3p^4), and third ^{3}P(3$s^1$3p^5) excited states 1.402, 3.414, and 11.57 eV higher, respectively [71]. Thus, the adiabatic dissociation products of the first three channels are Cl$^+$(^{3}P, ^{1}D, ^{1}S) + F(^{2}P). We have constructed all 12 doublets and quartets PECs from the first channel Cl$^+$(^{3}P) + F(^{2}P), ^{2,4}(Σ^+, Σ^-[2], Π[2], Δ), all nine doublets from the second channel Cl$^+$(^{1}D) + F(^{2}P) → ^{2}(Σ^+[2], Σ^-, Π[3], Δ[2], Φ), and the two doublets $^2\Sigma^+$ and $^2\Pi$ correlating with the third channel Cl$^+$(^{1}S) + F(^{2}P), a total of 23 states. Figure 2 displays the 21 PECs correlating adiabatically with the first two channels; the two highest PECs ($^2\Sigma^+$, $^2\Pi$) originating from the third channel are not shown (but see below). Table 5 lists our theoretical results for the $X\ ^2\Pi$, $a\ ^4\Sigma^-$, $2^2\Pi$, and $3^2\Pi$ states along with available experimental and theoretical results by Dunning and coworkers [33] for the first two states. Notice that at the same level of theory, MRCI + Q/A5ζ, the total energies of the $X\ ^2\Pi$ and $a\ ^4\Sigma^-$ states are lower in the present calculations by 6.4 and 8.3 mE_h, respectively, as compared to Ref. [33]; see Table 5. The reason is the extended zero-order space used here; see Sect. 2.

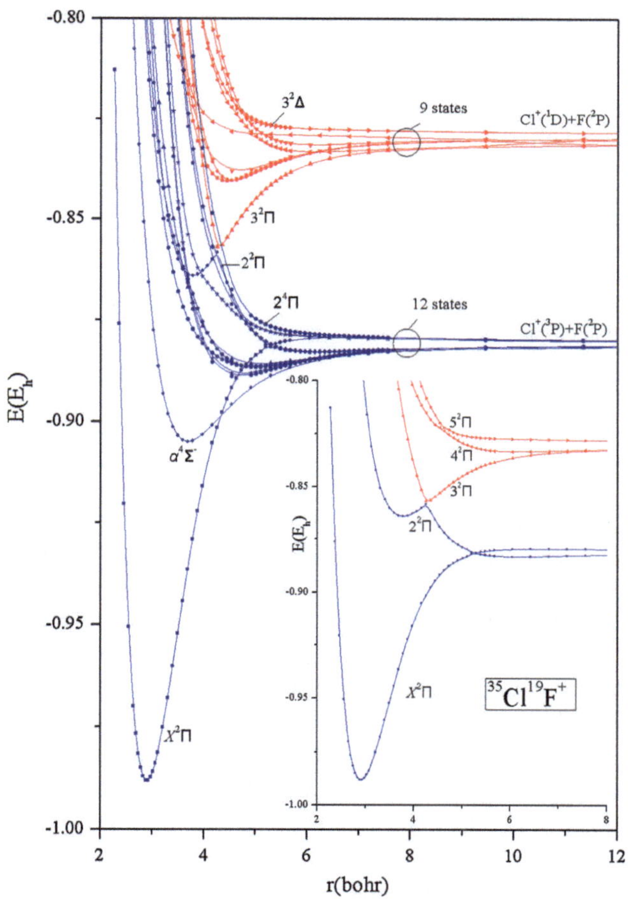

Fig. 2 Relative MRCI + Q/A5ζ PECs of twelve (*first channel*) and nine (*second channel*) states of ClF$^+$. The *inset* displays the avoided crossing among the $^2\Pi$ states

Table 5 Total energies E (E_h), bond distances r_e (Å), dissociation energies D_e and D_0 (kcal/mol), harmonic frequencies ω_e (cm^{-1}) and anharmonicities $\omega_e x_e$ (cm^{-1}), rotational vibrational coupling constants α_e (cm^{-1}), ionization energies IE (eV), and energy separations T_e (cm^{-1}) of the $X\ ^2\Pi$, $a\ ^4\Sigma^-$, $2\ ^2\Pi$ and $3^2\Pi$ bound states of $^{35}Cl^{19}F^+$

Method	$-E$	r_e	D_e^a	D_0^a	ω_e	$\omega_e x_e$	$\alpha_e \times 10^3$	IE	T_e
$X\ ^2\Pi$									
MRCI + Q/A5ζ	558.99504	1.5367	68.35	67.05	911.7	5.78	5.0	12.61	0.0
MRCI + Q/A6ζ	559.00034	1.5357	68.70	67.40	913.7	5.86	5.0	12.61	0.0
RCCSD(T)/A5ζ	559.99515	1.5358	68.10	66.78	929.5	5.58	4.8	12.69	0.0
RCCSD(T)/A6ζ	559.00065	1.5349	68.16	66.83	931.7	5.99	4.8	12.71	0.0
RCCSDT/A5ζ	558.99686	1.5365	68.08		936.6			12.66	
MRCI + Q/A5ζ[b]	558.98865	1.536	67.44	66.14	911.9				
Expt.		1.53[d]		67.43[c]	870[d], 912[e]			12.66[d,e,f]	0.0
$a\ ^4\Sigma^-$									
MRCI + Q/A5ζ	558.91059	1.9491	16.05[a]	15.50[a]	397.1	11.8	2.9		18,535
MRCI + Q/A6ζ	558.91537	1.9475	16.10	15.54	398.5	11.9	2.5		18,649
MRCI + Q/A5ζ[b]	558.90233	1.9745	13.59	13.07	363.2				18,945
$2\ ^2\Pi^g$									
MRCI + Q/A5ζ	558.86428	2.00	20.6	20.0	410	4.5		16.15	28587
Expt.[h]								16.39 (1)	
$3\ ^2\Pi^i$									
	558.8574	~2.27	~16	15	~725			~16	~30,200

[a] Dissociation energies with respect to the adiabatic products $Cl^+(^3P) + F(^2P)$ for the $X\ ^2\Pi$ and $a\ ^4\Sigma^-$ states. $D_0 = D_e - \omega_e/2 + \omega_e x_e/4$

[b] Ref. [33].

[c] See introduction and Ref. [3]. This value refers to $X\ ^2\Pi_{3/2}$

[d] Ref. [39]

[e] Ref. [38]

[f] With respect to $X\ ^2\Pi_{3/2}$

[g] The $2^2\Pi$ is the lowest state correlating diabatically with the second energy channel $Cl^+(^1D) + F(^2P)$. D_e with respect to the diabatic products (second channel)

[h] Ref. [38]; IE = $[E(2^2\Pi) - E(X\ ^1\Sigma^+)] + \Delta\omega_e/2$/MRCI + Q/A5ζ

[i] Pseudostate; see tex

3.2.1 $X\ ^2\Pi$

The lowest state of ClF^+ is certainly of $^2\Pi$ symmetry correlating with $Cl^+(^3P) + F(^2P)$. The vbL icon describing the bonding of ClF^+ is that of the $X\ ^1\Sigma^+$ state of ClF after removing one electron from a $3p_\pi$ atomic (or 2π molecular) orbital of the neutral species and of course with similar leading configurations. The calculated MRCI + Q/A5ζ or A6ζ (RCCSD(T)/A6ζ) ionization energy of ClF($X\ ^1\Sigma^+$) is IE = 12.61 (12.71) eV, in excellent agreement with experiment (IE = 12.66 eV [38, 39]; Table 5). Recall that IE(Cl) = 12.97 eV [40], hence the electron is removed from the chlorine atom in ClF. According to the Mulliken analysis, upon bonding the charge distribution is $^{+0.90}Cl - F^{+0.10}$.

According to Table 5, the bond distance of ClF^+ ranges between 1.5349 (RCCSD(T)/A6ζ) and 1.5367 (MRCI + Q/A5ζ) Å. Core and relativistic effects are quite small changing slightly the bond distance and the binding energy. At the MRCI + Q/A5ζ (RCCSD(T)/A5ζ) core and special relativity are responsible for $\delta r_{core} + \delta r_{rel} = -0.0034$ (-0.0031) + 0.0018 (0.0016) Å = -0.0016 (-0.0015) Å. Therefore, our "best" MRCI + Q(RCCSD(T))/A6ζ bond length is $r_e^* = 1.5357–0.0016\ (1.5349–0.0015) = 1.5341$ (1.5334) Å, both values in agreement with the 1.536 Å of Ref. [33]. The experimental r_e value of ClF^+ has been estimated to be 0.1 Å less than that of ClF($X\ ^1\Sigma^+$) [32], therefore r_e(expt) = 1.628–0.1 ≈ 1.53 Å, in pleasant agreement with the present calculated value(s).

Combined core and relativistic effects increase (decrease) the dissociation energy D_e by 0.49 (0.59) kcal/mol at the MRCI + Q (RCCSD(T))/A5ζ level of theory. Therefore, our "best" MRCI + Q (RCCSD(T))/A6ζ dissociation energy is (see Table 5), $D_e^* = 68.70 + 0.49$ (68.16–0.59) = 69.19 (67.57) kcal/mol, or $D_0^* = 67.89$ (66.21) kcal/mol for the $X\ ^2\Pi$ state of ClF^+. The experimental $D_0 = 67.43$ kcal/mol given in Table 5 refers to the $\Omega = 3/2$ component of the $X\,^2\Pi$ state. Taking into account the SO splitting of both the ClF^+ species and the Cl^+ and F atoms, our final MRCI + Q (RCCSD(T))/A6ζ D_0 value of the $X\,^2\Pi_{3/2}$ state of ClF^+ is $D_0 = D_0^* + 1/2 \times \Delta E_{ClF^+}(^2\Pi_{3/2} - {}^2\Pi_{1/2}) - 1/3 \times \Delta E_F(^2P_{3/2} - {}^2P_{1/2}) - 1/9 \times \Delta E_{Cl^+}(^3P_0 - {}^3P_2) - 3/9 \times \Delta E_{Cl^+}(^3P_1 - {}^3P_2) = 67.89$ (66.21) kcal/mol + $\{1/2 \times 630 - 1/3 \times 404 - 1/9 \times 996 - 3/9 \times 697\}$ cm^{-1} = 67.42 (65.74) kcal/mol, where experimental SO splittings have been used. In particular, the experimental $\Delta E_{ClF^+}(^2\Pi_{3/2} - {}^2\Pi_{1/2})$ SO splitting is 628 ± 30 [38] or 630 ± 30 [39] cm^{-1}, with the $^2\Pi_{3/2}$ being the lower component; the corresponding calculated splitting is 607 cm^{-1}. Obviously the agreement with experiment is excellent, the difference being −0.01 kcal/mol (=−3.5 cm^{-1}) at the MRCI + Q level of theory, but at the RCCSD(T) level the discrepancy is +1.69 kcal/mol (=591 cm^{-1}). As previously reported, RCCSDT/A5ζ results corroborate completely the RCCSD(T)/A5ζ values; see Table 5.

As a final word for the $X\,^2\Pi$ state, it can be said that our results suggest that the experimental harmonic frequency ω_e is closer to 912 cm^{-1} rather than to 870 cm^{-1}; see Table 5.

3.2.2 $a\,^4\Sigma^-$

This is the first excited state of ClF^+, a rather dark state for the experimentalists being spin and Franck–Condon forbidden; no experimental results are available. The $a\,^4\Sigma^-$ state is a relatively weakly bound state, $D_e \approx 16$ kcal/mol, located some 19,000 cm^{-1} above the X state at $r_e = 1.95$ Å (Table 5). It arises by removing a p_π electron from the $a\,^3\Pi$ state of ClF. The bonding interaction between $Cl^+(^3P) + F(^2P)$ is captured by the following vbL icon, analogous to that of the $a\,^3\Pi$ state (vide supra).

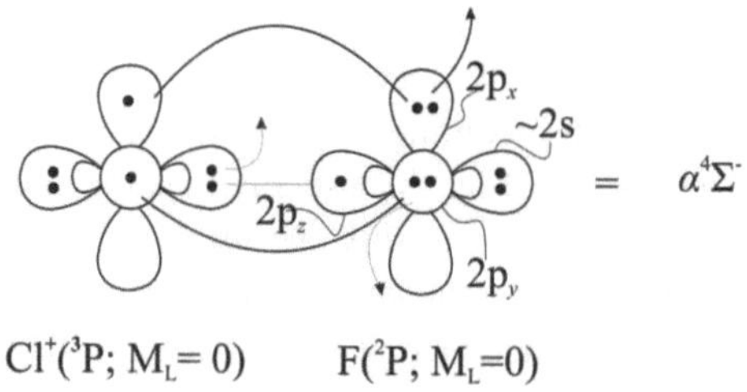

The leading MRCI configurations are

$$|a^4\Sigma^-\rangle \approx \left|1\sigma^2 2\sigma^2 3\sigma^2 4\sigma^1 [(0.82) 1\pi_x^2 2\pi_x^1 - (0.24) 1\pi_x^1 2\pi_x^2] 1\pi_y^2 2\pi_y^1\right\rangle - 0.24\left|1\sigma^2 2\sigma^2 3\sigma^2 4\sigma^1 1\pi_x^2 2\pi_x^1 1\pi_y^1 2\pi_y^2\right\rangle$$

very similar to the leading configurations of the $a\,^3\Pi$ (ClF) state. A total charge of 0.15 e^- is transferred from F to Cl^+ according to the Mulliken analysis. The contrast of the numerical parameters between $a\,^4\Sigma^-$ and $a\,^3\Pi$ (ClF) shows clearly the analogy between these two states. Indeed, at the MRCI + Q/A6ζ level, we have ($a\,^3\Pi$ results in parenthesis): $r_e = 1.9475$ (2.0144) Å, $D_e = 16.10$ (8.70) kcal/mol, $T_e = 18{,}649$ (18,996) cm^{-1}; see Table 5. The almost doubling of D_e of the $a\,^4\Sigma^-$ state as compared to the $a\,^3\Pi$, is due to an extra $3e^- - 2cp_\pi$ "bond" of the former; see the corresponding vbL diagrams. Core and relativistic effects are practically negligible in the $a\,^4\Sigma^-$ state. For instance, the $X\,^2\Pi$–$a\,^4\Sigma^-$ gap (T_e) at the MRCI + Q/A5ζ level increases by 173 and decreases by 175 cm^{-1} due to core and relativistic effects, respectively. Finally, it is interesting to observe that the MRCI + Q/A5ζ r_e and D_e values reported in Ref. [33] are by 0.025 Å larger and 2.5 kcal/mol smaller than the present ones.

Table 6 Total energies E (E_h), bond distances r_e (Å), dissociation energies D_e and D_0 (kcal/mol), harmonic frequencies ω_e (cm^{-1}) and anharmonicities $\omega_e x_e$ (cm^{-1}), rotational vibrational coupling constant α_e (cm^{-1}), and adiabatic electron affinity EA_{ad} (eV) of the $X\,^2\Sigma^+$ state of the $^{35}Cl^{19}F^-$ anion

Method	$-E$	r_e	D_e^a	$D_0^{a,b}$	ω_e	$\omega_e x_e$	$\alpha_e \times 10^3$	EA_{ad}
MRCI + Q/A5ζ	559.53667	2.159	30.32	29.78	375.0	2.40	2.69	2.17
MRCI + Q/A6ζ	559.54233	2.151	30.33	29.80	373.0	3.50	4.06	2.16
RCCSD(T)/A5ζ	559.54378	2.1643	30.14	29.62	364.0	2.62	3.00	2.26
RCCSDT/A5ζ	559.54389	2.1640	29.47					2.23
RCCSD(T)/A6ζ	559.54977	2.1644	29.82	29.30	363.4	2.51	3.08	2.26
CCSD(T)/A5ζ[c]		2.1531			378			2.25
Expt.[d]								2.86 ± 0.2

[a] With respect to $Cl^-(^1S) + F(^2P)$

[b] $D_0 = D_e - \omega_e/2 + \omega_e x_e/4$

[c] Ref. [32]

[d] Ref. [44]

3.2.3 $2^2\Pi$, $3^2\Pi$

These two $^2\Pi$ states correlate adiabatically with the first ($2^2\Pi$) and with the second ($3^2\Pi$) energy channels. The repulsive $2^2\Pi$ state, however, suffers a severe avoided crossing with the incoming $3^2\Pi$ (attractive) state around 4 bohr; see inset of Fig. 2. It seems that the only experimental finding relevant to this energy region is an ionization–excitation energy in very good agreement with the MRCI + Q/A5ζ calculated value, 16.39 ± 0.01 [38] versus 16.15 eV. At this level of theory T_e ($2\ ^2\Pi \leftarrow X\ ^2\Pi$) = 28,587 cm^{-1}, D_e (D_0) = 20.6 (20.0) kcal/mol with respect to the diabatic fragments $Cl^+(^1D) + F(^2P)$, and r_e = 2.00 Å. In Table 5, and with a grain of salt, results are also given for the $3^2\Pi$ (pseudo) state.

3.2.4 Higher states of ClF^+

We discussed three bound states of ClF^+, $X\ ^2\Pi$, $a\ ^4\Sigma^-$, and $2\ ^2\Pi$, all correlating adiabatically with the first channel $Cl^+(^3P) + F(^2P)$. The rest of the states, nine from the first and eight from the second channels, are weakly bound or of repulsive nature never characterized before either experimentally or theoretically; see Fig. 2. The nine states split naturally in two sheafs of five (lower) and four (higher) states each. The symmetries of the first five states are (in parenthesis equilibrium minima r_e in Å and attractive interaction energies $-\Delta E$ in kcal/mol) $1^4\Delta$ (2.48, 4.2), $1^4\Sigma^+$ (2.49, 4.6), $1^2\Sigma^-$ (2.65, 3.50), $1^2\Delta$ (2.61, 3.3), and $1^2\Sigma^+$ (2.68, 2.9) at a mean separation energy of $\bar{T}_e$ = 23,609 cm^{-1}. The next four states of symmetries $1^4\Pi$, $2^2\Sigma^-$, $2^4\Sigma^-$, and $2^4\Pi$ are of repulsive character but the $1^4\Pi$, where r_e = 3.27 Å and ΔE = −1.19 kcal/mol at T_e = 24,568 cm^{-1}; see Fig. 2.

We now turn to the eight doublets which emerge from the second channel. The first three at a mean $\bar{T}_e$ = 34,067 cm^{-1} can be considered as slightly bound, namely $2^2\Delta$ (r_e = 2.35 Å, D_e = 6.9 kcal/mol), $3^2\Sigma^-$ (2.41, 6.8), and $2^2\Sigma^+$(2.47, 5.3). The first of the remaining five states, $4^2\Pi$, shows a well depth of ΔE = −1.3 kcal/mol at r_e = 3.25 Å, while states of symmetries $1^2\Phi$, $3^2\Sigma^+$, $5^2\Pi$, and $3^2\Delta$ are repulsive. The last two PECs calculated at the MRCI + Q/A5ζ level of theory originate from the third channel, $Cl^+(^1S) + F(^2P)$, of $6^2\Pi$ and $4^2\Sigma^+$ symmetries; their PECs are not shown in Fig. 2. The $4^2\Sigma^+$ state is purely repulsive, whereas the $6^2\Pi$ interacts attractively at r_e = 2.71 Å with ΔE = −3.8 kcal/mol and T_e = 50,995 cm^{-1}.

3.3 ClF^-

Molecular anions are not easily tamed species either theoretically or experimentally. For an exhaustive review

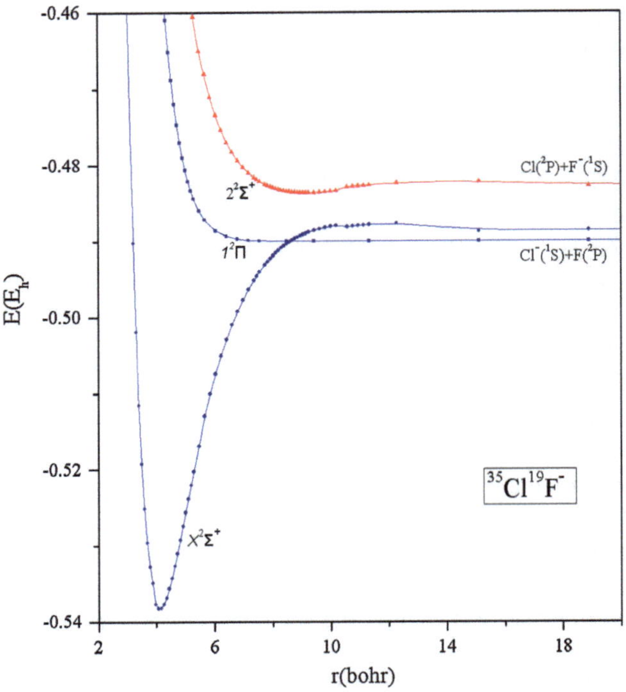

Fig. 3 Relative MRCI + Q/A5ζ PECs of two (*first channel*) and one (*second channel*) states of ClF^-

referring to atomic and molecular electron affinities and the difficulties of obtaining reliable results, see Ref. [45] and references cited therein. As was already discussed in Sect. 1, the only structural parameter that has been measured on ClF^- is its ionization energy, or the adiabatic electron affinity of ClF, the most recent value being EA_{ad} = 2.86 ± 0.2 eV [44], a strongly disputed number (see the Sect. 1). For the best ab initio work so far on ClF^-, we refer to Horný et al. [32]; see also Sect. 1 and Table 6.

The EAs of $Cl(^2P)$ and $F(^2P)$ are 3.612724 ± 0.000027 [80] and 3.4011895 ± 0.0000025 eV [81], respectively. Theoretical EAs at the MRCI + Q (RCCSD(T))/A6ζ level of theory are 3.574 (3.659) and 3.450 (3.414) eV, respectively. It should be said at this place that bound excited states (not resonances) of atomic anions are not common. In particular, it is rather certain that there are no excited states of $Cl^-(^1S)$ and $F^-(^1S)$ [82, 83], despite the high EAs of these atoms, as a matter of fact the highest of all elements [84]. According to the discussion above, one expects four molecular $^{2S+1}\Lambda$ states of $^2\Pi$ and $^2\Sigma^+$ symmetry, related to the channels $Cl^-(^1S) + F(^2P)$, and $Cl(^2P) + F^-(^1S)$, the experimental energy difference between the two adiabatic end products being ΔEA_{ad} = 0.212 eV.

Table 6 lists spectroscopic parameters for the $X\ ^2\Sigma^+$ state of $^{35}ClF^-$, while Fig. 3 displays state-specific MRCI + Q/A5ζ PECs of the $X\ ^2\Sigma^+$, $1^2\Pi$ and $2^2\Sigma^+$ states. The fourth $2^2\Pi$ state, repulsive in character, is not shown because of severe technical problems.

3.3.1 $X\,^2\Sigma^+$

The ground state of ClF^- correlates adiabatically with $Cl^-(^1S) + F(^2P)$. The vbL diagram, MRCI leading equilibrium configuration, and Mulliken population distributions around equilibrium (r_e) and at infinity (r_∞) presented below allow for a better understanding of the considerable binding energy of the $X\,^2\Sigma^+$ state, $D_e \sim 30$ kcal/mol.

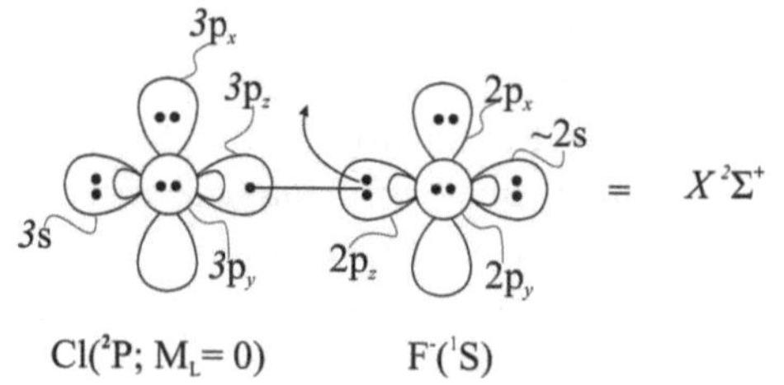

$$|X^2\Sigma^+\rangle \approx 0.93\left|1\sigma^2 2\sigma^2 3\sigma^2 4\sigma^1 1\pi_x^2 1\pi_y^2 2\pi_x^2 2\pi_y^2\right\rangle$$

$$r_e : 3s^{1.98}3p_z^{1.37}3p_x^{1.97}3p_y^{1.97}/_{Cl}2s^{1.98}2p_z^{1.59}2p_x^{1.98}2p_y^{1.98}/_F$$
$$r_\infty : 3s^{1.99}3p_z^{1.95}3p_x^{1.97}3p_y^{1.97}/_{Cl}2s^{1.99}2p_z^{1.02}2p_x^{1.99}2p_y^{1.99}/_F$$

Overall about 0.6 e^- seems to be transferred from Cl^- to F around the equilibrium distance at the MRCI/A5ζ or A6ζ level. However, the morphology of the PECs suggests that due to an interaction at ~9.5 bohr between the $X\,^2\Sigma^+$ and $2^2\Sigma^+$ states, the latter correlating with $F^-(^1S) + Cl(^2P)$, ~0.4 e^- are transferred from F^- to Cl around equilibrium.

According to Table 6, the RCCSD(T)/A6ζ (RCCSDT/A5ζ), $EA_{ad} = 2.26$ (2.23) eV at $r_e = 2.164$ Å, in practical agreement with the EA_{ad} of Horný et al. [32] who at the CCSD(T)/A5ζ level predict $EA_{ad} = 2.25$ eV; MRCI + Q/A5ζ or A6ζ EA_{ad} is by 0.1 eV less. We believe that this settles definitely the question of the EA_{ad} of ClF($X\,^1\Sigma^+$). Using this value and the experimental binding energy of ClF along with the experimental EA of Cl, the "experimental" dissociation energy of ClF^- is $D_0(ClF^-) = D_0(ClF)–[EA(Cl) - EA_{ad}(ClF)] = 60.35$ kcal/mol − (3.6127 − 2.26) eV = 29.16 kcal/mol, completely consistent with the CC results and in excellent agreement with the MRCI + Q results of Table 6. Thus, the recommended D_0 value of the $X\,^2\Sigma^+$ state of ClF^- is 29.5 kcal/mol. It should be added at this point that core and relativistic effects at the RCCSD(T)/A5ζ level reduce the bond distance by 0.002 and 0.001 Å, respectively. Therefore, our "best" bond distance is $r_e{}^* = 2.161$ Å.

3.3.2 $1^2\Pi$

This state correlates with $Cl^-(^1S) + F(^2P;\ M = \pm1)$. The congestion of four electrons in the σ-frame is the cause of a pure repulsive $1^2\Pi$ state the MRCI + Q/A5ζ PEC of which is shown in Fig. 3.

3.3.3 $2^2\Sigma^+$

The MRCI + Q/A5ζ PEC of $2^2\Sigma^+$ state correlates with $F^-(^1S) + Cl(^2P;\ M = 0)$ presenting a well depth of 0.82 kcal/mol at about 9.2 bohr; see Fig. 3.

4 Epitome and remarks

We believe that the present work is a systematic theoretical study of the interhalogen diatomic ClF and its ions $ClF^\pm$ within the $\Lambda - S$ ansatz. Despite the chemical simplicity of the X_2 and XX′ (X, X′ = F, Cl, Br, I) molecules, the determination of reliable properties either experimentally or theoretically is a daunting task. Even for the lightest interhalogen species (ClF), the large number of valence electrons (14), the relative weak bonding, and the intervening spin–orbit effects create a challenging computational *milieu*. Although this investigation leaves much to be done for the $ClF^{0,\pm}$ systems, particularly for the higher states, some of our results are very accurate, some have been calculated for the first time, and in general a wealth of new information is enclosed in this communication useful to workers with some interest in these systems.

Through the use of augmented quintuple and sextuple correlation consistent basis sets, thus in essence removing the error related to the one-electron basis set at least for the lowest states, and multireference CI (MRCI) and single-reference CC (RCCSD(T), RCCSDT) calculations, we have examined a significant number of states of ClF and $ClF^\pm$. In particular, we have constructed full potential $^{2S+1}\Lambda$ energy curves for all states of ClF emanating from the first channel (12), all states of ClF^+ emanating from the first three channels of $Cl^+ + F$ (23), and three states out of four correlating with $Cl^-(^1S) + F(^2P)$ or $Cl(^2P) + F^-(^1S)$ for the ClF^- anion. A number of states have been corrected for core ($2s^22p^6/_{Cl}$), scalar relativistic, and SO effects. Size non-extensivity errors were taken into account through the +Q Davidson correction. The most salient features of the present work are summarized below.

4.1 ClF

The ground state of ClF is of $^1\Sigma^+$ symmetry; MRCI + Q calculated (experimental) values are $D_0^0 = 60.35$ (60.35) kcal/mol, $r_e = 1.6284$ (1.628341(4)) Å, $\omega_e = 786.1$ (783.353) cm^{-1}, $\mu_e = 0.878$ (0.8881) Debye with a charge polarity $^{+q}Cl - F^{-q}$, q = 0.25. Excellent agreement is obtained at the CC level as well. The single σ bond can be adequately described by one configuration function; indeed, $|C_0|^2 \approx 0.9$.

$a\ ^3\Pi$ is the first and only within the $\Lambda - S$ coupling scheme bound state related to ground-state fragments. Its binding energy is about an order of magnitude less than that of the $X\ ^1\Sigma^+$ state. A weak Paschen–Back effect is the cause of the $\Omega = 2, 1, 0^+ (0^-)$ components of the $a\ ^3\Pi$ state, named A′, A, and B, respectively, for historical reasons; no name has been adopted for the 0^- state. Notice that the splitting between the 0^+–0^- components is very small [3] and in the present work $^3\Pi_{0^\pm} = {^3\Pi_0}$. Our "best" MRCI + Q calculated (experimental results referring to $^3\Pi_{0^+}$) values are $D_e^0 = 8.75$ (8.802) kcal/mol, $r_e = 2.0145$ (2.0221) Å, $\omega_e = 366.6$ (362.6) cm^{-1}, $\mu_e = 1.583$ D (no experiment), and $T_e = 18{,}864$ (18,825.3) cm^{-1}. This is a multireference state; therefore, CC calculations are not recommended for its computation. Within the $\Lambda - S$ treatment, there are nine strongly repulsive PECs and a van der Waals $1^1\Pi$ state, the latter with an interaction energy of less than 0.5 kcal/mol and r(vdW) = 3.31 Å at the MRCI + Q level.

4.2 ClF^+

The ground state of ClF^+ is of $^2\Pi$ symmetry. We have detected three $\Lambda - S$ bound states $X\ ^2\Pi$, $a\ ^4\Sigma^-$, and $2^2\Pi$, whereas for the latter two the experimental data are practically none.

For the $X\ ^2\Pi$ state, our MRCI + Q/A6ζ calculated (experimental) values are $D_0^0 = 67.40$ (67.43) kcal/mol, $r_e = 1.5357$ (1.53) Å, $\omega_e = 913.7$ (912) cm^{-1}, and IE = 12.61 (12.66) eV. CC results are in very good agreement as well. The dissociation energy of the $X\ ^2\Pi$ state is by $\Delta D = 7.1$ kcal/mol larger than that of the $X\ ^1\Sigma^+$ of ClF, whereas $\Delta r_e\ (X\ ^2\Pi - X\ ^1\Sigma^+) = -0.093$ Å.

Our MRCI + Q/A6ζ results for the $a\ ^4\Sigma^-$ state, a rather "dark" state for the experimentalists, are $D_0 = 15.54$ kcal/mol, $r_e = 1.9475$ Å, $\omega_e = 398.5$ cm^{-1}, and $T_e = 18{,}649$ cm^{-1}. Notice also that in line with the $a\ ^3\Pi$ state of ClF, $\Delta D = 6.74$ kcal/mol and $\Delta r_e = -0.04$ Å.

For the $2^2\Pi$ state, our MRCI + Q/A5ζ calculations predict $D_0 = 20.0$ kcal/mol, $r_e = 2.00$ Å, $\omega_e = 410$ cm^{-1}, and $T_e = 28{,}587$ cm^{-1}. It is interesting that there is a remarkable agreement with an experimental ionization–excitation energy, 16.39 ± 0.01 eV versus 16.15 eV, proving that the "tentative" assignment of the experimentalists [38] is correct.

4.3 ClF^-

The ground state of ClF^- is of $^2\Sigma^+$ symmetry. The only experimental datum is the (adiabatic) electron affinity EA_{ad} [$ClF\ (X\ ^1\Sigma^+) + e^- \rightarrow ClF^-(X\ ^2\Sigma^+)$] = 2.86 ± 0.2 eV, proved to be wrong by $\sim +0.6$ eV. Our RCCSDT/A5ζ [RCCSD(T)/A6ζ] (MRCI + Q/A6ζ) = 2.23 [2.26] (2.16) eV value, in agreement with previous CC calculations. Clearly, our calculations converge to an EA_{ad} of 2.25 ± 0.01 eV. The MRCI + Q/A6ζ predictions for this state are $D_0^0 = 29.80$ kcal/mol, $r_e = 2.151$ Å, and $\omega_e = 373.0$ cm^{-1}. A charge of 0.4 e^- is transferred from the in situ F^- to Cl around the equilibrium.

The four homonuclear and six heteronuclear dihalogens, XX′ (X, X′ = F, Cl, Br, I) are of emblematic importance for chemistry. We tried to present here a useful theoretical account of the behavior and properties of $ClF^{0,\pm}$, but obviously the subject is far from being closed and further investigation experimental and theoretical as well is clearly needed. We hope that our future work will move toward this direction, particularly toward the excited manifold of these fascinating species.

Acknowledgments One of us (A.V.) expresses his gratitude to Dr. C. N. Sakellaris for helpful discussions.

References

1. Dicciani NK, Burrows C, Greenspan A, Stang PJ (2003) Chem Eng News (number 36), 81, pp. 48(F), *62*(Cl), 96(Br), and 130(I)
2. Greenwood NN, Earnshaw A (1998) Chemistry of the elements, 2nd edn. Butterworth–Heinemann, Oxford, pp 824–825
3. Huber KP, Herzberg G (1979) Molecular spectra and molecular structure. IV constants of diatomic molecules, Van Nostrand Reinhold Co, NY
4. Wahrhaftig AL (1942) J Chem Phys 10:248
5. Gilbert DA, Roberts A, Griswold PA (1949) Phys Rev 76:1723
6. Stricker W, Krauss L (1968) Z Naturforsch 23a:1116
7. Dibeler VH, Walker JA, McCulloh KE (1970) J Chem Phys 53:4414
8. Davis RE, Muenter JS (1972) J Chem Phys 57:2836
9. Nordine PC (1974) J Chem Phys 61:224
10. Combe RD, Pilipovich D, Horne RK (1978) J Phys Chem 82:2484
11. Willis RE Jr, Clark WW III (1980) J Chem Phys 72:4946
12. McDermid IS (1981) J Chem Soc Faraday Trans 2(77):519
13. McDermid IS, Laudenslager JB (1981) Chem Phys Lett 79:370
14. Alekseev VA, Setser DW (1997) J Chem Phys 107:4771
15. Jenkins FA (1953) J Opt Soc Am 43:425
16. Alekseev VA, Setser DW, Tellinghuisen J (1999) J Mol Spectrosc 61:194
17. Alekseev VA, Setser DW, Tellinghuisen J (1999) J Mol Spectrosc 195:162
18. Schmitz H, Schumacher HJ (1947) Z Naturforsch 2a:359
19. Schumacher HJ, Schmitz H, Brodersen PH (1950) Anales Asoc Quim Argent 38:98
20. Straub PA, McLean AD (1974) Theor Chim Acta 32:227
21. Ewig CS, Sur A, Banna MS (1981) J Chem Phys 75:5002
22. Scharf P, Ahlrichs R (1985) Chem Phys 100:237
23. Darvesh KV, Boyd RJ, Peyerimhoff SD (1988) Chem Phys 121:361
24. Peterson KA, Woods RC (1990) J Chem Phys 92:7412
25. Perera SA, Bartlett RJ (1993) Chem Phys Lett 216:606 and references therein
26. deJong WA, Styszynski J, Visscher L, Nieuwpoort WC (1998) J Chem Phys 108:5177 and references therein

27. Alekseyev AB, Liebermann HP, Buenker RJ, Koch KB (2000) J Chem Phys 112:2274 and references therein
28. Buenker RJ, Peyerimhoff SD (1974) Theor Chim Acta 35:33
29. Buenker RJ, Peyerimhoff SD (1975) Theor Chim Acta 39:217
30. Buenker RJ, Peyerimhoff SD, Butsher W (1978) Mol Phys 35:771
31. Ricca A (2000) Chem Phys Lett 323:498
32. Horný L, Sattelmeyer KW, Schaefer HF III (2003) J Chem Phys 119:11615
33. Chen L, Woon DE, Dunning TH Jr (2009) J Phys Chem A 113:12645
34. Chen L, Woon DE, Dunning Jr. TH (2013) J Phys Chem A 117:4251 and references therein.
35. Kalemos A, Mavridis A (2009) J Phys Chem A 113:13972
36. Kalemos A, Mavridis A (2011) J Phys Chem A 115:2378 and references therein
37. Irsa AP, Friedman L (1958) J Inorg Nucl Chem 6:77
38. Anderson CP, Mamantov G, Bull WE, Grimm FA, Carver JC, Carlson TA (1971) Chem Phys Lett 12:137
39. De Kock RL, Higginson BR, Lloyd DR, Breeze A, Cruickshank DWJ, Armstrong DR (1972) Mol Phys 24:1059
40. Radziemski LJ, Kaufmann V (1969) J Opt Soc Am 59:424
41. Harland P, Thynne JCJ (1969) J Phys Chem 73:4031
42. Dispert H, Lacmann K (1978) Int J Mass Spectrom Ion Phys 28:49
43. Illenberger E, Scheunemann HU, Baumgärtel H (1979) Chem Phys 37:21
44. Dudlin AV, Gorokhov LN, Baluev AV (1979) Izv Akad Nauk SSR Ser Khi 11:2408
45. Rienstra-Kiracofe JC, Tschumper GS, Schaefer HF III (2002) Chem Rev 102:231
46. Curtiss LA, Raghavachari K, Redfern PC, Rassolov V, Pople JA (1998) J Chem Phys 109:7764
47. Curtiss LA, Redfern PC, Raghavachari K, Pople JA (2001) J Chem Phys 114:108
48. Law CK, Chien SH, Li WK, Cheung YS (2002) J Phys Chem A 106:11271
49. Dunning TH Jr (1989) J Chem Phys 90:1007
50. Kendall RA, Dunning TH Jr, Harrison RJ (1992) J Chem Phys 96:6796
51. Dunning TH Jr, Peterson KA, Wilson AK (2001) J Chem Phys 114:9244
52. Wilson AK, van Mourik T, Dunning TH Jr (1997) J Mol Struct (THEOCHEM) 388:399
53. Woon DH, Dunning TH Jr (1993) J Chem Phys 98:1358
54. Raghavachari K, Trucks GW, Pople JA, Head-Gordon M (1989) Chem Phys Lett 157:479
55. Watts JD, Gauss J, Bartlett RJ (1993) J Chem Phys 98:8718
56. Knowles PJ, Hampel C, Werner HJ (1993) J Chem Phys 99:5219
57. Knowles PJ, Hampel C, Werner HJ (2000) J Chem Phys 112:3106
58. Werner HJ, Knowles PJ (1988) J Chem Phys 89:5803
59. Knowles PJ, Werner HJ (1988) Chem Phys Lett 145:514
60. Douglas M, Kroll NM (1974) Ann Phys 82:89
61. Hess BA (1985) Phys Rev A 32:756
62. Jansen HB, Ross P (1969) Chem Phys Lett 3:140
63. Boys SF, Bernardi F (1970) Mol Phys 19:553
64. Dunham JL (1932) Phys Rev 41:721
65. Herzberg G (1950) Molecular spectra and molecular structure 1. Spectra of diatomic molecules, 2nd edn. D. van Nostrand Co, Inc., Canada
66. Langhoff RS, Davidson ER (1974) Int J Quantum Chem 8:61
67. Davidson ER, Silver DW (1977) Chem Phys Lett 52:403
68. Sakellaris CN, Mavridis A (2012) J Chem Phys 137:034309
69. CFOUR, a quantum chemical program package written by Stanton JF, Gauss J, Harding ME, Szalay PG with contributions from Auer AA, Bartlett RJ, Benedikt U, Berger C, Bernholdt DE, Bomble YJ, Cheng L, Christiansen O, Heckert M, Heun O, Huber C, Jagau T-C, Jonsson D, Jusélius J, Klein K, Lauderdale WJ, Matthews DA, Metzroth T, Mück LA, O'Neill DP, Price DR, Prochnow E, Puzzarini C, Ruud K, Schiffmann F, Schwalbach W, Stopkowicz S, Tajti A, Vázquez J, Wang F, Watts JD and the integral packages *MOLECULE* (J. Almlöf and P.R. Taylor), *PROPS* (P.R. Taylor), *ABACUS* (Helgaker T, Jensen HJ, Jørgensen P, Olsen J), and ECP routines by Mitin AV, van Wüllen C. For the current version, see http://www.cfour.de
70. MOLPRO is a package of ab initio programs written by Werner H-J, Knowles PJ, Knizia G, Manby FR, Schütz M, Celani P, Korona T, Lindh R, Mitrushenkov A, Rauhut G, Shamasundar KR, Adler TB, Amos RD, Bernhardsson A, Berning A, Cooper DL, Deegan MJO, Dobbyn AJ, Eckert F, Goll E, Hampel C, Hesselmann A, Hetzer G, Hrenar T, Jansen G, Köppl C, Liu Y, Lloyd AW, Mata RA, May AJ, McNicholas SJ, Meyer W, Mura ME, Nicklaß A, O'Neill DP, Palmieri P, Peng D, Pflüger K, Pitzer R, Reiher M, Shiozaki T, Stoll H, Stone AJ, Tarroni R, Thorsteinsson T, Wang M
71. Kramida A, Ralchenko Yu, Reader J, and NIST ASD Team (2012). NIST atomic spectra database (ver. 5.0), [Online]. Available: http://physics.nist.gov/asd [5 June 2013]. National Institute of Standards and Technology, Gaithersburg, MD
72. Tzeli D, Mavridis A (2003) J Chem Phys 118:4984
73. Ewing JJ, Tigelaar HL, Flygare WH (1972) J Chem Phys 56:1957
74. McGurk J, Noris CL, Tigelaar HL, Flygare WH (1973) J Chem Phys 58:3118
75. Green J (1974) Adv Chem Phys 25:179
76. Janda KC, Klemperer W, Novick SE (1976) J Chem Phys 64:2698
77. Fabricant B, Muenter JS (1977) J Chem Phys 66:5274
78. Magoulas I, Kalemos A, Mavridis A (2013) J Chem Phys 138:104312
79. Edlén B (1969) Sol Phys 9:439
80. Berzinsh U, Gustafson M, Hanstorp D, Klinkmüller A, Ljungblad U, Martensson-Pendrill AM (1995) Phys Rev A 51:231
81. Blondel C, Delsart C, Goldfarb F (2001) J Phys B 34:L281
82. Andersen T (1991) Phys Scr 43:23
83. Hogrere H (1998) Phys Scr 58:25
84. Emsley J (1991) The elements, 2nd edn. Clarendon Press, Oxford

Theor Chem Acc (2014) 133:1445
DOI 10.1007/s00214-014-1445-1

REGULAR ARTICLE

Negative electron affinities from conventional electronic structure methods

Kenneth D. Jordan · Vamsee K. Voora · Jack Simons

Received: 22 November 2013 / Accepted: 2 January 2014 / Published online: 14 January 2014

Abstract If the potential V describing the interaction between an excess electron and a ground-state neutral or anionic parent is sufficiently attractive at short range, electron-attached states having positive electron affinities (EAs) can arise. Even if the potential is not attractive enough to produce a bound state, metastable electron-attached states may still occur and have lifetimes long enough to give rise to experimentally detectable signatures. Low-energy metastable states arise when the attractive components of V combine with a longer-range repulsive contribution to produce a barrier behind which the excess electron can be temporarily trapped. These repulsive contributions arise from either the centrifugal potential in the excess electron's angular kinetic energy or long-range Coulomb repulsion in the case of an anionic parent. When there is no barrier, this kind of low-energy metastable state does not arise, but improper theoretical calculations can lead to erroneous predictions of their existence. Conventional electronic structure methods with, at most, minor modifications are described for properly characterizing metastable states and for avoiding incorrectly predicting the existence of metastable states with negative EAs where no barrier is present.

Dedicated to Professor Thom Dunning and published as part of the special collection of articles celebrating his career upon his retirement.

K. D. Jordan · V. K. Voora
Department of Chemistry, University of Pittsburgh, Pittsburgh, PA 15260, USA
e-mail: jordan@pitt.edu

J. Simons (✉)
Department of Chemistry, University of Utah, Salt Lake City, UT 84112, USA
e-mail: jack.simons@utah.edu

Keywords Electron affinities · Metastable anions · Electron–molecule interaction potential

1 Introduction

An electron may interact with the ground state of a parent (neutral, cation, or anion) to produce an electron-attached species having a lower energy than that of the parent. Such species are said to have positive electron affinities (EAs) and occur at discrete (i.e., quantized) energies that can be found using bound-state electronic structure methods. In addition, an electron can interact with the same parent to produce a species having a higher energy than that of the parent. In fact, there exists a continuum of such states having energies $E > 0$, where $E = 0$ is taken to be the energy of the parent plus an electron infinitely far away and having zero kinetic energy.

The full treatment of the states with $E > 0$ lies within the realm of electron–molecule scattering theory. In general, continuum levels cannot be addressed using conventional electronic structure techniques. However, for circumstances discussed in this paper, the density of electron-plus-parent states $\rho(E)$ may be concentrated in certain energy ranges $E \pm \delta E$ to produce states that can be viewed as metastable with lifetimes $\delta t \approx h/\delta E$ related to the energy range δE over which the high state density exists. These states are characterized by energies E that are not rigorously quantized but can be specified within ranges δE, and they can give rise to spectroscopic features that allow them to be experimentally differentiated from the underlying continuum of states. These metastable states are associated with the negative EAs and can be identified using conventional electronic structure codes with, at most, straightforward modifications, but, as we demonstrate here,

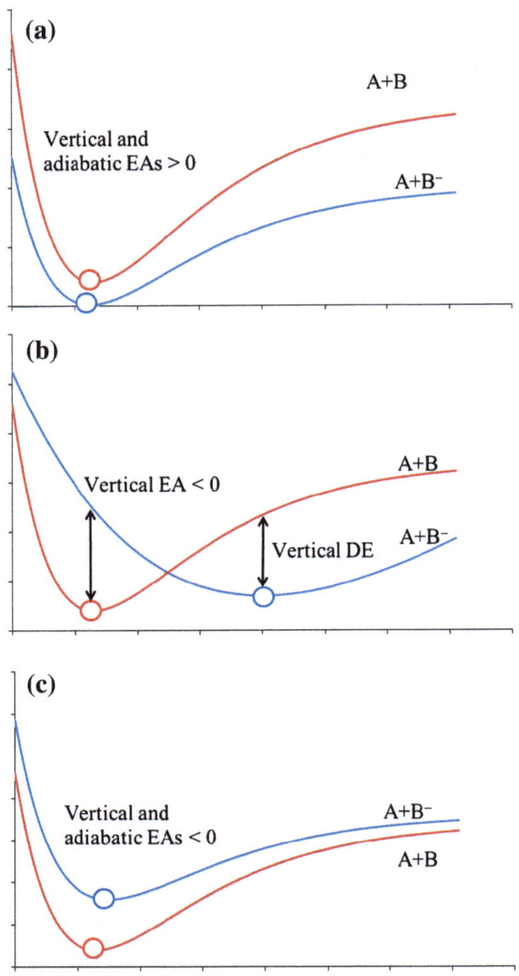

Fig. 1 Qualitative depictions of AB^- anion (*blue*) and AB neutral (*red*) potential energy curves as functions of a geometrical coordinate for three cases discussed in the text

considerable care must be used to distinguish such states with energies lying within the region $E \pm \delta E$ of high state density from states belonging to the underlying continuum whose energy can have any value $E > 0$.

Within the Born–Oppenheimer picture, the electronic stability of molecular anions depends crucially on the relative energies of the anion and neutral potential energy surfaces at various geometries. For example, an electron-attached species may have a positive EA at one geometry yet be metastable and have a negative EA at another geometry. Figure 1a depicts a case in which the anion lies below the neutral at the neutral's equilibrium geometry (which gives a positive vertical EA) and lies below the neutral at the anion's equilibrium geometry [producing a positive vertical detachment energy (DE)]. The adiabatic EA, defined as the energy of the neutral at its equilibrium geometry (the red circles in Fig. 1) minus the energy of the anion at its equilibrium geometry (the blue circles in Fig. 1), is positive for the case shown in Fig. 1a.

Figure 1b depicts a case where the anion lies above the neutral at the minimum of the latter (giving a negative vertical EA) and lies below the neutral at the anion's equilibrium geometry (giving a positive vertical DE). Furthermore, the adiabatic EA is negative if the energy of the anion at its minimum lies above the energy of the neutral at its minimum as shown in Fig. 1b. Alternatively, if the minimum of the anion in Fig. 1b were lowered such that it lies below the energy of the neutral at its minimum, a positive adiabatic EA would result. Finally, Fig. 1c depicts a case in which the anion lies above the neutral at all geometries thus producing negative vertical and adiabatic EAs.

Anions with potential energy surfaces as depicted in Fig. 1b and c are unstable to electron detachment, and the plots of their energies as functions of molecular geometry should be thought of as plots of the center of the range $E \pm \delta E$ of enhanced state density. Also, the lines in such plots should be drawn with Heisenberg widths δE that vary with geometry. However, for visual clarity, the plots of potential energy curves of metastable states will, throughout this paper, be shown as simple lines.

For the case shown in Fig. 1c, the electron will undergo autodetachment at any geometry but it may have a lifetime long enough to allow it to produce a spectroscopic signature that is readily detected in laboratory experiments. For the situation illustrated by Fig. 1b, in the absence of non-Born–Oppenheimer coupling between the electronic and vibration–rotation degrees of freedom, the electron cannot detach at geometries where the anion lies below the neutral. However, if vibrational motion allows the anion to access geometries where its energy lies above the neutral, the anion of Fig. 1b can undergo detachment, in which case the rate of electron loss will depend on the rate at which such geometries are accessed. Finally, even in the case illustrated in Fig. 1a, the anion will become metastable if its level of vibrational or rotational excitation places its total energy above the zero-point level of the neutral molecule. In this case, the magnitude of vibration/rotation–electronic couplings governs the lifetime of the anion. Vibrationally/rotationally excited anions with potentials such as that shown in Fig. 1a usually have much longer lifetimes than anions with potentials as illustrated in Fig. 1c because such non-Born–Oppenheimer coupling is usually weak.

2 The physical content of electron–molecule interaction potentials

To appreciate what determines whether an anion's electronic energy lies above or below that of its parent in its ground electronic state at a particular molecular geometry, it is important to understand the physical content of the interaction potential $V(r)$ experienced by an excess electron interacting with the parent system. A rigorous basis for defining $V(r)$ begins with the Schrödinger equation governing the $(N+1)$-electron wave function of the parent-plus-electron system

$$H(x,r)\Psi(x,r) = E\Psi(x,r) = [H(x) + H(r) + U(x,r)]\Psi(x,r) \quad (1)$$

Here, $H(x)$ is the electronic Hamiltonian containing the kinetic energy, electron–nuclear attractions, and electron–electron interactions for the first N electrons whose coordinates collectively are denoted x; $H(r)$ is the electronic Hamiltonian containing the kinetic energy and electron–nuclear attractions for the $N+1$st electron whose coordinates are denoted r; and $U(x, r)$ is the electron–electron interaction potential between the $N+1$st electron and the parent's N electrons and is of the form $\sum_{i=1}^{N} \frac{e^2}{|r-x_i|}$.

At large values of $|r|$, $V(r)$ should be equal to the electrostatic potential energy of interaction of a negatively charged probe particle at r with the N-electron parent species in its ground electronic state plus the potential energy of interaction of the probe particle with the various moments it induces in the parent. For example, at large $|r|$, $V(r)$ should contain the following terms:

(a) Coulomb attraction (if the parent is a cation) or repulsion (if the parent is an anion),
(b) charge–dipole, charge–quadrupole, and higher charge–multipole interactions,
(c) charge-induced–dipole and higher charge-induced–multipole interactions.

In the present study, we focus on states that arise through electron attachment to the *ground state* of the parent in which case the electronic configuration $\Psi(x,r)$ of the electron-attached species can be qualitatively described as an antisymmetrized product of a wave function $\psi(x,r)$ that is an eigenfunction of $H(x) + U(x,r)$ multiplied by a spin-orbital $\phi(r)$ occupied by the excess electron $\Psi(x,r) = A\psi(x,r)\phi(r)$. The function $\psi(x,r)$ is the ground-state wave function of the parent in the presence of a stationary excess electron whose interaction with the parent is given by $U(x,r)$

$$[H(x) + U(x,r)]\psi(x,r) = E(r)\psi(x,r) \quad (2)$$

$E(r)$ is the energy of the ground-state parent as it interacts with the stationary excess electron located at r, and A is the antisymmetrizer operator.

In addition to the vibration/rotation-excited anions whose electron detachment depends on non-Born–Oppenheimer couplings discussed earlier, there is another class of metastable states that arise in anions but that also is not the focus of the present study. In these states, which are called core-excited, the excess electron is temporarily bound to an excited electronic state of the parent. There are two types of core-excited anion states [1]—those in which the anion lies energetically above its parent and which tend to be short lived, and those that lie energetically below their parent state, and which can be quite long lived. Both types are discussed in Ref. [1] using ethylene as an example, but similar states occur in other olefins and in polyenes. In general, core-excited anions states in which the orbitals involved in the excitation and electron capture are valence, e.g., the $\pi^1\,\pi^{*2}$ state of the ethylene anion, are of the former type, while anions involving excitation to and electron capture into Rydberg orbitals are of the latter type, an example of which is the $\pi^1(3s)^2$ anion of ethylene. The latter anion lies approximately 0.5 eV below the parent $\pi^1 3s^1$ Rydberg state for ethylene, and its autodetachment involves a two-electron process in which one electron is ejected from the 3s orbital, while another electron relaxes from the 3s orbital into the π orbital. The lifetimes for undergoing such two-electron events are often significantly longer than those for tunneling through the barriers arising in the states of primary interest here.

2.1 Electrostatics, polarization, antisymmetry, and orthogonality

It is possible to use perturbation theory to reduce the solution of Eq. (1) to an equation to be solved for the spin-orbital characterizing the excess electron. This results in an equation of the form

$$[H(r) + E(r) - E]\phi(r) - \sum_{i=1}^{N} \int \psi * (x,r)U(x,r)\psi(r_i,r)\phi(x_i)\mathrm{d}x + V_{\mathrm{rep}}(r)\phi(r) = 0 \quad (3)$$

$H(r)$ contains the kinetic energy of the excess electron plus the electron–nuclear Coulomb interactions of this electron. $E(r)$ is, as explained earlier, the energy of the ground-state parent in the presence of a stationary excess electron at r. $E(r)$ can be expressed as the energy E_0 of the parent in the absence of the excess electron plus the sum of electrostatic $V_{\mathrm{es}}(r)$ (e.g., Coulomb, permanent dipole, permanent quadrupole) and induced $V_{\mathrm{ind}}(r)$ (e.g., dipole polarization, quadrupole polarization) interactions between the parent and the stationary electron as well as contributions describing non-adiabatic energies $V_{\text{non-ad}}(r)$ arising from the finite kinetic energy of the excess electron [2–4]. The terms in the sum in Eq. (3) are the exchange contributions, collectively referred to as $V_{\mathrm{exch}}(r)$ arising from the antisymmetry of the $(N + 1)$-electron wave function. Finally, $V_{\mathrm{rep}}(r)$ is a repulsive potential arising from the constraint that the excess electron's spin-orbital be orthogonal to the parent's wave function. In its simplest form in which $E(r)$ is approximated by Hartree–Fock (HF)-level electrostatic interactions, Eq. (3) describes the static-exchange approximation in which $\phi(r)$ is equivalent to an unoccupied (i.e., virtual) HF orbital.

By combining $V_{\text{electrostatic}}(r) + V_{\text{induced}}(r) + V_{\text{non-ad}}(r) + V_{\text{exchange}}(r)$ with $V_{\text{rep}}(r)$, we obtain a one-electron *potential* $V(r)$ that combines with $H(r)$ to yield an equation to be solved for the excess electron's spin-orbital

$$[H(r) + V(r) - E]\phi(r) = 0 \quad (4)$$

The eigenvalues E in Eq. (4) specify the energies of the electron-attached species relative to the parent at a fixed geometry of the molecule. Negative values of E correspond to bound states, and positive values to unbound states. For bound states and metastable states, E is equal to minus the EA. Equation (4) has served as the basis for developing several one-electron models for describing excess electrons interacting with water clusters and other systems [5]. We now turn our attention to the features of the total potential $V(r)$ that combine with $H(r)$ to determine when either a positive EA or a metastable state can be expected.

2.2 Centrifugal potential contributions

For temporary anions with Born–Oppenheimer potential energy surfaces lying above the corresponding neutral's surface (as in Fig. 1c), the attractive components in $V(r)$ are not strong enough to produce a negative eigenvalue E in Eq. (4). In such cases, it is important to consider whether the repulsive angular momentum centrifugal potential arising from the kinetic energy operator $-\frac{\hbar^2}{2m}\nabla_r^2$ can combine with $V(r)$ to produce an *effective radial potential* $V_{\text{eff}}(r)$ having a barrier behind which the excess electron can be temporarily trapped. It is through such a combination of repulsive and attractive potentials that low-energy metastable anion states often arise, with the barrier and well behind it generating a high density of states in certain energy ranges.

To illustrate, consider the specific examples of placing an excess electron into the lowest unfilled valence orbital of N_2, H_2, H_2O, or CO_2. In the case of N_2, the lowest empty valence orbital is of π_g symmetry, which has a dominant component of angular momentum corresponding to the quantum number $L = 2$ (referred to as a *d*-wave in the language of electron scattering theory). The $\frac{L(L+1)\hbar^2}{2mr^2}$ centrifugal potential combines with $V(r)$ to produce an effective potential that has sizable ($\sim$8 eV) barrier, which results in an N_2^- anion that lives long enough to display resolvable vibrational structure in the total cross section for electron scattering even though it is unstable with respect to N_2 plus a free electron [6]. For H_2, the lowest unoccupied valence molecular orbital (LUMO) is of σ_u symmetry and is dominated by an $L = 1$ (*p*-wave) component. This results in a lower angular momentum barrier (through which the electron must tunnel to escape) than in N_2^- and hence a shorter anion lifetime. As a result, H_2^- appears as a broad resonance lacking resolvable vibrational structure in the electron scattering cross section [7].

In the case of H_2O, the lowest antibonding O–H σ_g* valence orbital is of a_1 symmetry and has a large $L = 0$ component. As a result, there is no barrier behind which the excess electron can be trapped, and thus the corresponding $^2A_1$ state of H_2O^- is not detected experimentally. Essentially, the a_1 σ_g* valence orbital has dissolved in the continuum of a free electron plus the neutral molecule. Later, we will demonstrate theoretical tools for handling the metastable states of N_2^- and H_2^- and for avoiding incorrectly predicting a low-energy metastable $^2A_1$ state of H_2O^- to exist. It should be pointed out that there are metastable $^2A_1$ states of H_2O^-, but these states lie at higher energies and have orbital occupancies (e.g., $1a_1^2\,2a_1^2\,1b_2^2\,3a_1^1\,1b_1^2\,4a_1^2$) in which an occupied orbital of H_2O is excited and the excess electron is attached to an excited orbital. That is, they are of the core-excited variety mentioned earlier and are not the kind of low-energy metastable states we are focusing on.

Whether an anion is stable or metastable can also depend on the geometry of the molecule. For example, the lowest-energy unoccupied valence orbital of CO_2 at the linear equilibrium structure is of π_u symmetry and thus has a nonzero centrifugal potential. As a result, a metastable anion results from electron capture into this orbital as can readily be detected in electron scattering measurements (with a peak in the cross section near 3.8 eV) [8]. However, when CO_2 is bent, the in-plane component of the π_u orbital acquires a_1 symmetry, which introduces an $L = 0$ component and lowers the centrifugal barrier and thus the anion's lifetime. The fraction of $L = 0$ character increases with increased bending, and it has been concluded that a metastable state having a lifetime long enough to be detectable occurs only for OCO angles >165° [9].

2.3 Differential electron correlation and problems with Hartree–Fock and DFT

There are often substantial (1–2 eV) differences between the correlation energy of a bound or metastable valence-type anion and its neutral parent. In fact, the correlation contribution to the electron binding is often of the same order of magnitude as the EA itself, so inclusion of electron correlation effects is essential for accurately characterizing these anion states. It is possible for the correlation energy of the parent to be larger than that of the anion, for example, in cases where the excess electron attaches to a vacant valence spin-orbital that contributes strongly to electron correlation in the parent. The occupancy of this spin-orbital in the anion then excludes it from being used to correlate the anion's electrons thus reducing the anion's correlation energy. For example, in olefins, the parent's two electrons occupying π orbitals derive much of their correlation energy through double excitations of the form $\pi^2 \rightarrow \pi^{*2}$. If the π^* orbital is occupied to form the anion, the two electrons occupying the π orbital can no longer use the $\pi^2 \rightarrow \pi^{*2}$ excitation to gain correlation energy. However, it is more common for the anion to have considerably larger correlation energy than its parent neutral. In either case, it is important to include correlation effects because they often contribute a very significant percentage of the electron-binding energy.

Finally, it should be noted that in many cases, density functional theory (DFT) methods using standard GGA functionals such as B3LYP and PBE overbind anion states and can thus predict a stable anion when, in fact, the anion is metastable or there is not even a metastable anion. For example, DFT calculations performed with a flexible basis set incorrectly predict that CO_2 in its linear structure has a bound $^2\Sigma_g^+$ anion. This is less of a problem when using range-separated DFT methods [10] that have the correct long-range exchange interaction.

2.4 Variational collapse is a problem to be overcome

When electron correlation is ignored and the anion is treated within HF theory, a phenomenon known as variational collapse can plague the calculation as we now illustrate. For the reasons explained above, a HF-level treatment of an anion whose parent actually has a positive EA may (incorrectly) predict the anion to lie above the parent if the anion's correlation energy exceeds that of the parent. In such a situation, the HF wave function of the anion will collapse to a function describing the parent plus an approximate continuum function for the excess electron if the atomic orbital basis set has sufficiently diffuse functions in it. This happens because, within the variational HF calculation, a lower energy is achieved by forming a wave function with the $N + 1$st electron in an orbital that is infinitely removed from the parent. Only by properly including the correlation energy of the excess electron will the energy of the anion be adjusted to lie below the energy of the parent.

It is tempting to try to avoid this variational collapse by limiting the basis set's radial extent thus not allowing the $N + 1$st electron to escape. However, as we discuss in great detail in this paper, although this approach produces a HF state in which the excess electron remains attached to the parent, this $(N + 1)$-electron HF wave function cannot be trusted to offer a reasonable zeroth-order function to subsequently use in a correlated calculation. The central issue is how to limit the basis set's radial extent in a way that generates an $N + 1$st orbital that, once electron correlation effects are added to the potential the excess electron experiences, produces a correct description of the bound excess electron. Arbitrarily limiting the basis' radial extent is not a correct approach, but the so-called stabilization methods we discuss later are.

Analogous variational collapse issues also plague calculations when attempting to identify metastable electron-attached states, even when including electron correlation effects. For example, for the case illustrated in Fig. 1c, a straightforward application of traditional electronic structure methods (even correlated methods) to the anion is guaranteed, when using a radially flexible basis set, to collapse onto the neutral plus an approximate continuum function. As discussed later, researchers have developed a variety of methods, including the stabilization method [11, 12] and the coordinate rotation method [13], for avoiding the variational collapse problem when the electron-attached species is metastable, and much of the latter part of this paper is dedicated to illustrating how to use these tools within conventional electronic structure codes.

As noted above, a HF-level description may, because of the differential correlation energy, predict a negative EA for a species that is actually bound. In such cases, if the $(N + 1)$st HF orbital has resulted from variational collapse, using it to form a $(N+1)$-electron wave function as a starting point for subsequent MPn or coupled cluster treatment of electron correlation is problematic. So, even though CCSD(T) often captures a large percentage of the correlation energy, if the initial approximation to the wave function has the variationally collapsed HF orbital occupied, even this powerful technique cannot be trusted. It is essential that the orbital occupied by the excess electron offer a reasonable approximation to the $(N+1)$-electron wave function if correlated methods that use this function as their starting point are to be trusted. If the stabilization methods discussed later are used to form an initial $(N+1)$-electron wave function, then subsequent MPn or coupled cluster treatment of electron correlation will likely be reliable.

There are methods such as Green's function [14] or EOM-CCSD [15] theory that can be especially useful in such problematic cases because their working equations can be cast in a way that does not depend on the qualitative correctness of the HF orbitals. Not only do Green's function and EOM methods provide a route for calculating EAs but they also give descriptions of the electron–parent interaction potential that can supplement the perturbation theory approach with $V(r)$ mentioned earlier. Both theories can be cast in a form [16] in which

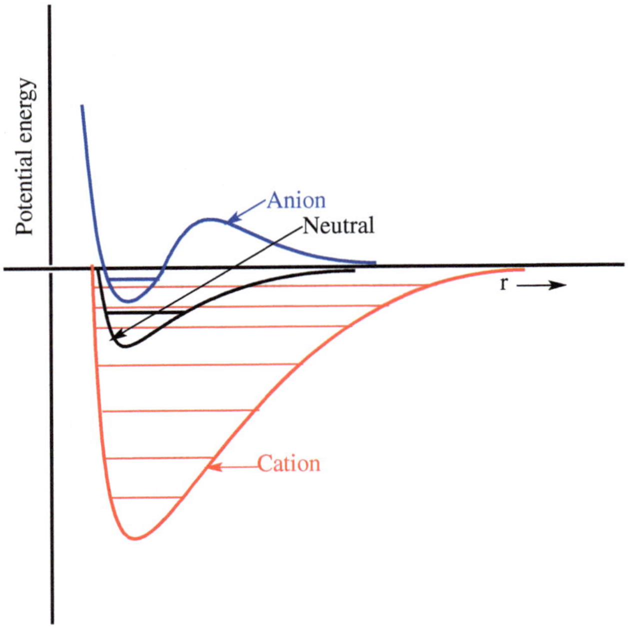

Fig. 2 Qualitative depiction of effective potentials associated with an excess electron interacting with a cation, a neutral, or an anion as a function of the distance between the electron and the parent species

the EA is obtained as an eigenvalue of a one-electron energy-dependent Hamiltonian whose matrix elements account for the excess electron's kinetic energy, its attractions to the parent's nuclei, its electrostatic and induced interactions with the parent's electron density, and the fact that the excess electron is indistinguishable from the parent's electrons (i.e., antisymmetry). The eigenfunctions of these one-electron Hamiltonians are called Dyson orbitals and are closely related to the $\phi(r)$ that we described above.

Having introduced many of the concepts and issues that arise when treating electron-attached species, we move on to address the primary purpose of this paper. It is to present approaches capable of distinguishing between metastable anions arising in regions of enhanced state density and the underlying continuum of states and to offer guidance about how to characterize the former. Before discussing these issues, it will prove useful to examine several illustrative systems to show how the electron–molecule potential and angular momentum play key roles in determining whether the electron-attached state is stable or metastable. In Sect. 3, we describe the potentials $V(r)$ and effective potentials $V_{\mathrm{eff}}(r)$ that govern the interaction between an electron and a cation, neutral, or anion, and we illustrate features of these potentials that produce bound or metastable states. Several examples from the recent literature are used to illustrate the diverse behavior arising in these systems. In Sect. 4, we discuss theoretical tools to focus on metastable electron-attached states. Section 5 provides a summary of the main points.

3 Comparing electron–cation, electron–neutral, and electron–anion interaction potentials

As explained in Sects. 1 and 2, the ability of an atom or molecule to bind an extra electron is governed by the potential $V(r, \theta, \phi)$ between the electron and the atom or molecule. This potential depends on the location (given by the variables r, θ, and ϕ, denoted collectively as $\boldsymbol{r}$) of the electron and on the electrostatic moments and polarizabilities of the underlying parent atomic or molecular species. In Fig. 2, we give qualitative depictions of the radial behavior of $V(r, \theta, \phi)$ for three cases:

1. Interaction of an electron with a cation of charge Z to form a species with charge $Z - 1$, which relates to the formation of a neutral species if $Z = 1$ or to formation of a cation of lower charge if $Z > 1$.
2. Interaction of an electron with a neutral parent to form a singly charged anion.
3. Interaction of an electron with an anion of charge $-Z$ to form an anion of charge $-Z - 1$; if $Z = 1$, this corresponds to forming a dianion from a mono-anion.

3.1 The electron–neutral case

As illustrated in Fig. 2, the depth of the outer region potential well appropriate for an electron interacting with a neutral atom or molecule is less deep than that associated with the electron–cation interaction. The depth of the well depends on the balance between the attractive and repulsive contributions to $V(r, \theta, \phi)$. The spatial range and strength of the repulsive contributions are determined by the sizes and shapes of the parent's occupied orbitals. The relative depths of the potential wells are determined largely by the strength of the long-range attractive components of $V(r, \theta, \phi)$.

The large-r asymptotic form of the polarization potential varies as $-\alpha/2r^4$, where α is the molecular polarizability (assuming that it is isotropic). If the molecule has a permanent dipole, the corresponding electrostatic interactions dominate over the polarization interaction at sufficiently large r. For the range of multipole moments and polarizabilities found in most molecules, the charge–multipole and charge–polarization attractions are not as strong as the Coulomb attraction present in the electron–cation case. This is why the electron–cation well depth almost always exceeds the well depths for electron–neutral or electron–anion systems and is the primary reason behind the observation that electron affinities (EAs) are nearly always smaller than ionization potentials (IPs).

In addition to having a potential well less deep than for electron–cation cases, the fact that the electron–neutral potential varies at large-r as r^{-n} with $n \geq 2$ gives rise to a qualitative change in the pattern of bound quantum states existing in such wells. It can be shown that, within the Born–Oppenheimer approximation, any species having a permanent dipole moment in excess of 1.625 D has an infinite number of bound states into which an excess electron can attach [17–21]. However, in practice, due to Born–Oppenheimer corrections, unless the dipole is very large, only the lowest of these states is significantly bound and then only if the dipole moment is in excess of ca. 2.5 D.

To summarize, the strong electron–cation Coulomb attraction at large-r gives rise to a deep potential well supporting an infinite progression of bound electronic states for an electron interacting with an atomic or molecular cation. These bound states often provide useful spectroscopic fingerprints for studying cation or neutral species because their energy spacings fall in the visible or ultraviolet regions where high-resolution and high-sensitivity light sources and detectors can be employed. In contrast, electron–neutral potentials have shallower wells and produce few, if any, significantly bound states. Moreover, even when bound excited anion states occur, their electronic excitation energies often lie in ranges where light sources and detectors are not routinely available or are of low sensitivity. Therefore, spectroscopic probes based on accessing bound excited states are generally not feasible for anions, although photo-electron spectroscopy, which requires only one bound or long-lived state, is a powerful tool.

3.2 The electron–anion case

For an electron interacting with a negative ion of charge $-Z$, there is a long-range Coulomb repulsive Ze^2/r contribution to $V(r)$. This contribution alters the energy landscape in two ways: (1) it generates the long-range repulsive Coulomb barrier (RCB) shown in Fig. 2, and (2) it shifts the electron–neutral potential upward in energy even within the regions occupied by the parent's valence and inner-shell orbitals. Given what was said earlier about the shallowness of electron–neutral potential wells, it should therefore come as no surprise that multiply charged anions usually have only one or no bound state. As we now illustrate, such species can also have metastable states whose lifetimes depend on the height and thickness of the RCB.

In Fig. 3, we show the molecular structure of copper phthalocyanine, 3,4′,4″,4‴-tetrasulfonate $[CuPc(SO_3)_4]^{4-}$ in which the tetradentate phthalocyanine ligand has four negatively charged sulfonate groups attached to its periphery and a copper atom at its center. To illustrate the range of electron-attached states that can arise in such a species, we discuss the results of photo-electron spectroscopy experiments [22].

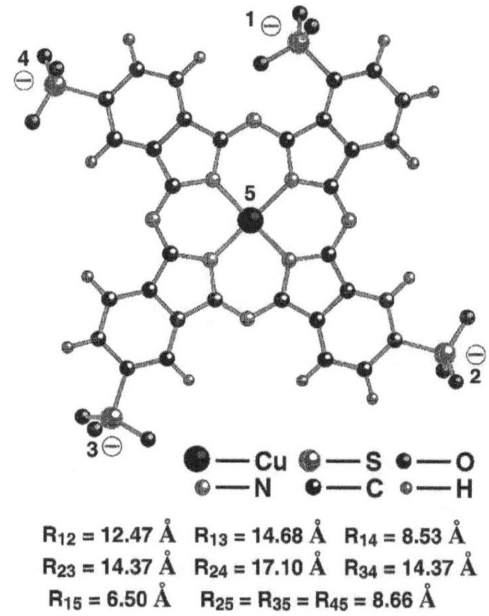

Fig. 3 Copper phthalocyanine complex with four sulfonate groups (from Ref. [22])

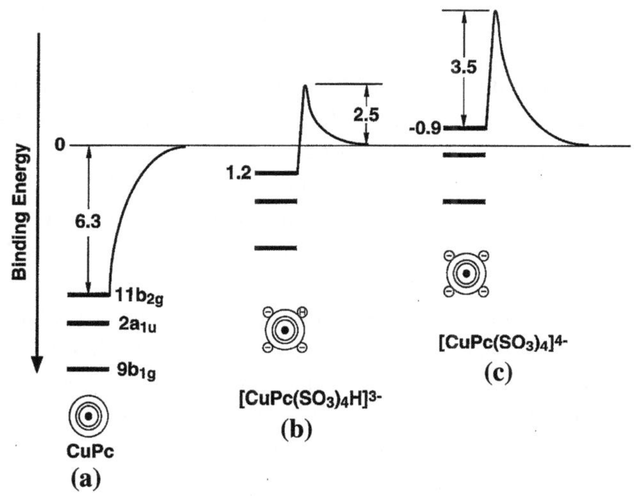

Fig. 4 Qualitative depiction of electron–parent interaction potential showing RCB and electron-binding energies for uncharged (**a**), triply charged (**b**), and quadruply charged (**c**) systems (taken from Ref. [22])

If the four sulfonate groups are rendered neutral and the resulting complex is subjected to photo-electron spectroscopy, it is found that the complex has an IP of 6.3 eV (the Cu-centered electron is detached). In contrast, if one of the four sulfonate groups is neutralized and the resulting triply charged anion is studied, it is found that the system has a DE of 1.2 eV [22]. However, when photons with energy only slightly higher than 1.2 eV are used, essentially no photo-electrons are ejected. Once the photon energy reaches 3.7 eV, ample electron detachment is observed and the kinetic energy of the ejected electrons is found to be ca.

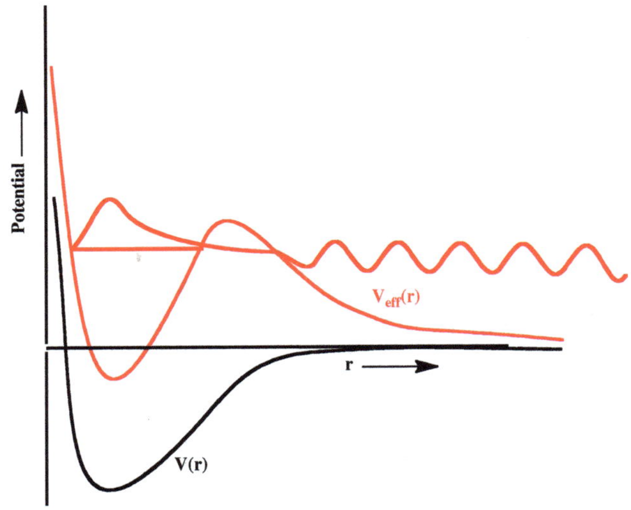

Fig. 5 Electron Mg atom potential $V(r)$ in black and effective potential $V_{eff}(r)$ in red for a nonzero value of L. The radial wave function for the 3p orbital producing the $3s^23p^1$ ^2P temporary anion is also shown

2.5 eV. It is by subtracting the kinetic energy of the electrons from the photon energy that we know the threshold detachment energy is 1.2 eV for this triply charged anion. As suggested in Fig. 4, the height of the RCB for this triply charged anion is 2.5 eV, which is why no photo-electrons are detected at photon energies below 3.7 eV.

When the quadruply charged anion having four sulfonate groups is studied, it is found that very few electrons are detached with photons having energy less than 3.5 eV. However, for photons with energy slightly greater than 3.5 eV, ample electron ejection occurs and electrons having kinetic energies of ca. 4.4 eV are detected. That is, the ejected electrons have higher kinetic energy than the energy of the photons used to eject them! This indicates that the system has a negative detachment energy value of −0.9 eV as shown in Fig. 4.

For the purposes of the present paper, the most important thing to note is that the RCB that arises in such multiply charged anions can not only lower DE values for bound electronic states but can also generate metastable states that have negative DE values, but live long enough to be experimentally detected.

3.3 More examples of metastable states

In the example just discussed, the presence of the RCB generates a potential that can bind the extra electron behind the RCB. As discussed earlier when we treated electrons interacting with N_2, H_2, and CO_2, there are also situations in which the repulsive centrifugal potential generates a barrier in the effective potential that can temporarily bind an electron. To develop these ideas in more detail, let us consider a neutral Mg atom in its ground ^1S state interacting with an extra electron. Here, the electron atom potential $V(r,\theta,\phi)$ depends only on the radial coordinate r describing the distance of the electron from the Mg nucleus although the centrifugal component of the related effective potential V_{eff} also depends on the angular momentum L of the excess electron.

For this electron Mg example, $V(r)$ has a sizeable attractive component because of the significant polarizability of the Mg atom, and, when applied to s or p orbitals, it also has a short-range repulsive contribution to account for orthogonality to the Mg atom's occupied s and p orbitals. Figure 5 depicts $V(r)$ and the effective potential $V_{\text{eff}}(r)$ for a nonzero value of L.

When considering the possibility of an electron binding to the Mg atom, one needs to specify the symmetry of the electron-attached state being studied. For example, we consider the possibilities of forming a $3s^23p^1$ ^2P or a $3s^24s^1$ ^2S state. In the former case, the angular quantum number associated with the extra electron is $L = 1$; in the latter, $L = 0$ applies.

As it turns out, the e^--Mg potential is not sufficiently attractive to support a bound state in any angular momentum channel. Thus, the $3s^24s^1$ ^2S state is not bound, and it is not metastable because there is no angular momentum barrier to trap the electron. However, the combination of the attractive $V(r)$ and the repulsive centrifugal potential $\frac{\hbar^2L(L+1)}{2mr^2}$ produces an effective potential $V_{\text{eff}}(r)$ that can temporarily bind an electron to form a metastable $3s^23p^1$ ^2P state and is readily detected in electron scattering experiments (the anion is only 0.15 eV above the ground state of the neutral [23]). The energy level and the dominant component of the radial wave function of this metastable state are depicted in Fig. 5. In Sect. 4, we will describe how one identifies this metastable state computationally and how one should avoid incorrectly predicting that the $3s^24s^1$ ^2S Mg^- state is metastable.

The wave function of the metastable ^2P Mg^- state has three distinct components: (1) it has a major lobe whose peak is located in the region of the minimum of $V_{\text{eff}}(r)$; (2) in the tunneling region where the energy of this state lies below V_{eff}, it decays exponentially with increasing r; and (3) at larger r, where the energy of the state lies above V_{eff}, it displays sinusoidal variation with a de Broglie wavelength that characterizes the ejected electron's momentum. Of course, the wave function also has smaller-r components but we do not show these in Fig. 5. As detailed in Sect. 4, a proper description of the wave function in all regions is essential to achieving a full description of such metastable-state wave functions and their energies.

Unlike the Mg case just discussed, Ca has a large enough polarizability to cause its electron atom potential $V(r)$ to be sufficiently attractive to produce a $4s^24p^1$

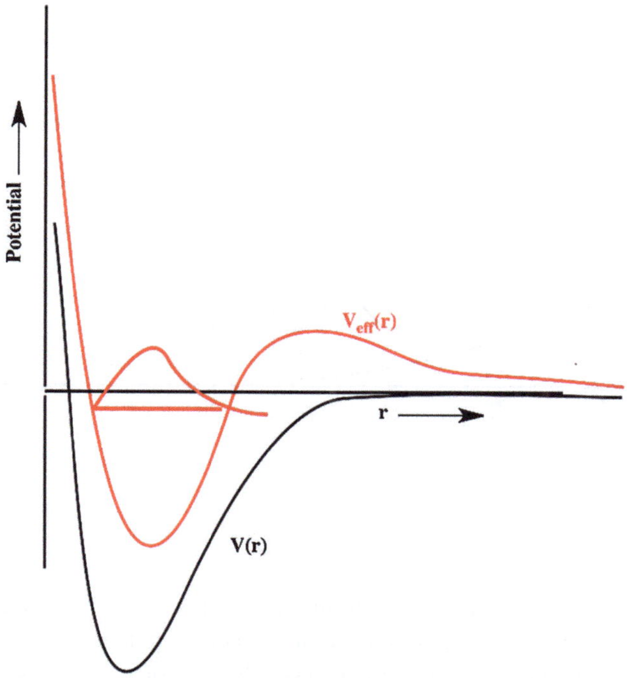

Fig. 6 Schematic of electron Ca atom effective potential (*black*) for $L = 1$ (without the centrifugal term) and the effective potential including a centrifugal potential with $L = 1$ (*red*)

^2P anion state [24] that is bound by 0.025 eV. The relevant effective potential $V_{eff}(r)$ is described qualitatively in Fig. 6. In addition, Ca^- has a metastable ^2D state at 0.7–1.0 eV [25, 26].

In the bound $4s^24p^1$ ^2P state of Ca^-, the wave function has a lobe with its maximum near the minimum of V_{eff}, and an exponentially decaying component in the region where V_{eff} is above the state's energy. However, the wave function of this bound state lacks an asymptotic sinusoidal component, which distinguishes it from the wave function of the ^2P metastable state of Mg^- discussed above.

It is worth noting that the Ca atom does not form a bound $4s^25s^1$ ^2S anion even though it forms a bound $4s^24p^1$ ^2P state. The failure of Ca^- to have a bound ^2S state is due to the fact that the excess electron's 5*s* orbital has to be orthogonal to the 4*s*, 3*s*, 2*s*, and 1*s* orbitals, while the $4s^24p^1$ ^2P state's 4*p* orbital needs to be orthogonal to only the 3*p* and 2*p* orbitals. This offers a nice example of how orthogonality constraints contribute to the total electron–parent potential and thus affect the electron-binding energy.

In summary, an electron interacting with an alkaline earth atom may have a positive EA if V_{eff} has a sufficiently deep well as it does for $4s^24p^1$ ^2P Ca^- but not for $2s^22p$ ^2P Be^-, $3s^23p^1$ ^2P Mg^-, $4s^25s^1$ ^2S Ca^-, or $3s^24s^1$ ^2S Mg^-. Moreover, the $2s^22p^1$ ^2P Be^-, $3s^23p^1$ ^2P Mg^-, and $4s^24d^1$ ^2D Ca^- states exist as observable metastable states; however, the corresponding ^2S anionic states are not metastable because the potentials are not sufficiently attractive to support a bound state and there is no centrifugal barrier to radially constrain the extra electron.

4 Theoretically characterizing anions or multiply charged anions having negative EAs

As explained in Sects. 2 and 3, a ground-state parent's electron-attached state can be metastable when the underlying parent neutral or anion presents to the approaching electron a potential $V(\mathbf{r})$ that is sufficiently attractive to combine with either a nonzero centrifugal potential (in the case of a neutral parent) or the repulsive Coulomb component of $V(\mathbf{r})$ (for an anion parent) to generate an effective potential that can bind the extra electron inside the centrifugal or Coulomb barrier. When there is no barrier present, such metastable states do not arise.

So, how does one find the metastable states and characterize their energies and lifetimes? Let us assume that one is carrying out a conventional electronic structure calculation on a system comprised of an extra electron interacting with either a neutral whose LUMO has nonzero angular momentum (and thus produces a centrifugal barrier) or anionic parent (whose RCB generates the barrier). Furthermore, let us assume that the methods being used allow one to calculate many approximate eigenvalues of the associated Schrödinger equation (e.g., within HF-based Koopmans' theorem [27] by using minus the energies of unoccupied orbitals to approximate the EAs or within configuration interaction (CI) theory by finding several eigenvalues of the $N+1$-electron CI Hamiltonian matrix and subtracting them from the CI energy of the parent). Because such approaches utilize a finite atomic orbital basis set (which we will denote $\{\chi_J(r, \theta, \phi); J = 1, 2,\ldots N\}$), the HF orbital energies $\{\varepsilon_J\}$ or CI eigenvalues $\{E_K\}$ will be finite in number.

Because we anticipate that a true metastable state's wave function must contain a substantial component inside the barrier as well as a tunneling-range part and a large-*r* component that radially oscillates with a de Broglie wavelength indicative of the extra electron's asymptotic momentum, let us assume that the atomic orbital basis set consists of

1. inside-barrier and tunneling-range functions $\{\chi_J(r, \theta, \phi); J = 1, 2,\ldots n\}$ typical of a high-quality electronic structure calculation on the parent neutral or anion, augmented by
2. a set of functions $\{\chi_J(r, \theta, \phi); J = n + 1, n + 2,\ldots N\}$ that are radially more diffuse and are designed to be combined to produce functions that can oscillate radially with de Broglie wavelengths in a range that we anticipate might characterize the extra electron's asymptotic momentum.

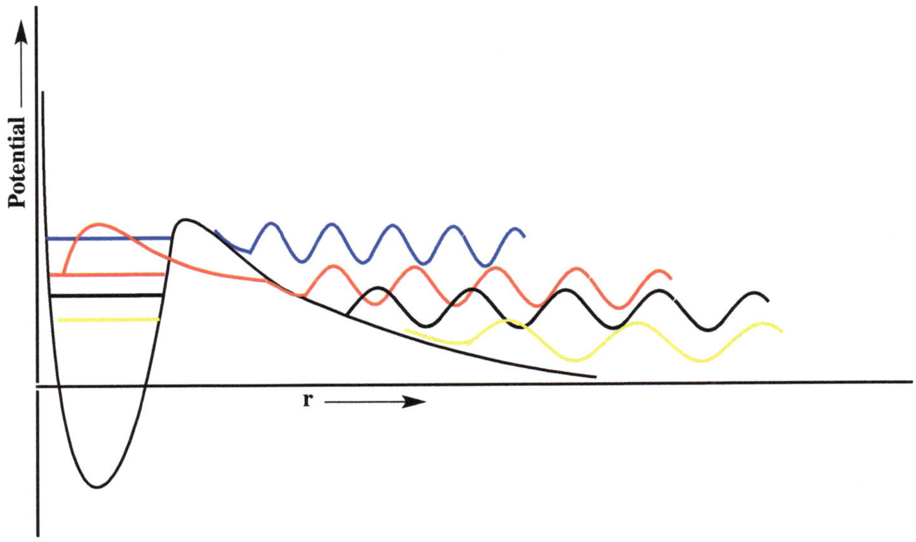

Fig. 7 Effective radial potential (*black*) and wave functions of four states of the electron–parent system including a metastable state whose energy and wave function is shown in *red* and electron–parent collision events whose wave functions are shown in *yellow*, *black*, and *blue*

The primary difficulty that arises when one uses any of the approaches noted above can be illustrated using the plots of Fig. 7 where the inside-barrier and large-r components of the effective radial potential experienced by the extra electron are depicted qualitatively in black for a situation in which the limited basis set generates only a few low-energy states having $E > 0$.

Using a modest basis set, one might find among the four lowest eigenvalues (i.e., by either subtracting CI eigenvalues of the electron–parent system from the CI energy of the parent or using the negative of the HF unoccupied orbital energies) for this simple example

1. A lowest eigenvalue (denoted by the yellow horizontal line in Fig. 7) whose wave function (also in yellow) has little, if any, inside-barrier component and is dominated by a long de Broglie wavelength function existing at large-r but extending outward only as far as the most diffuse atomic basis function in $\{\chi_J(r, \theta, \phi); J = n + 1, n + 2,\ldots N\}$ allows;
2. a higher-energy eigenvalue (black) whose wave function also has little inside-barrier component, exists primarily at large-r, but has a somewhat shorter de Broglie wavelength than the first eigenfunction;
3. an eigenvalue (red) whose wave function has both a large inside-barrier component and an oscillating large-r component whose de Broglie wavelength gives the asymptotic momentum of the extra electron.
4. a fourth eigenvalue (blue) of even higher energy whose wave function is similar in character with the wave functions indicated by yellow and black curves but has a shorter de Broglie wavelength.

The wave function shown in red and its energy correspond to the metastable state. The yellow, black, and blue curves describe the extra electron with various asymptotic kinetic energies encountering and reflecting off of the barrier on the electron–parent effective potential. It is important to note that the metastable state of interest is by no means the lowest-energy state. Moreover, as more and more diffuse basis functions are utilized in the calculation, there is an increasing number of continuum-dominated states lying below (and above) the resonance state.

It might seem possible to simply carry out a conventional quantum chemistry calculation with a reasonable number of diffuse basis functions and visually examine the inside-barrier, tunneling, and asymptotic character of many of the resulting wave functions to find the metastable state. However,

1. when studying electrons interacting with large molecules, it is extremely difficult to examine pictorially a large number of the low-energy wave functions of the electron–parent system, and, more importantly,
2. one often finds that two or more of the eigenfunctions possess substantial components of inside-barrier, tunneling, and asymptotic character. Obviously, a method is needed that can isolate the correct state when mixed states occur.

4.1 The orbital exponent stabilization method

The so-called stabilization method [11, 12] provides a means of identifying the resonances. There are multiple variants of the stabilization method, and here, we consider only the exponent scaling approach where the radial character of the diffuse basis functions is tuned by scaling their orbital exponents. If the (N–n) diffuse functions $\{\chi_J(r,\theta,\phi); J = n + 1, n + 2,\ldots N\}$ are of Gaussian form, the scaling is accomplished by multiplying each of their orbital exponents by a positive scaling factor α

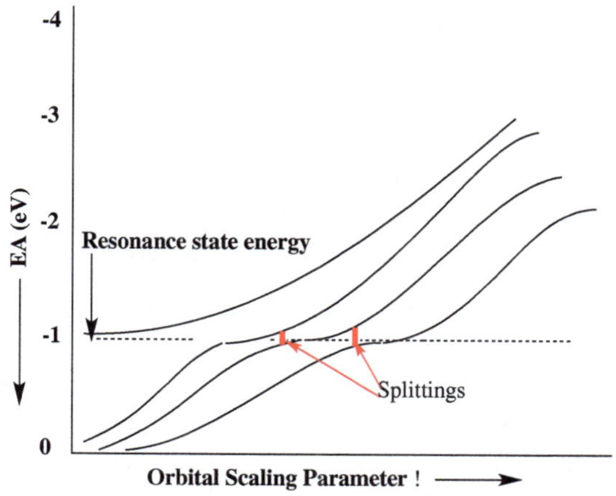

Fig. 8 Example of a stabilization calculation plot showing four $E > 0$ eigenvalues as a function of the orbital scaling parameter α, illustrating the avoided crossings that result due to the presence of a metastable state

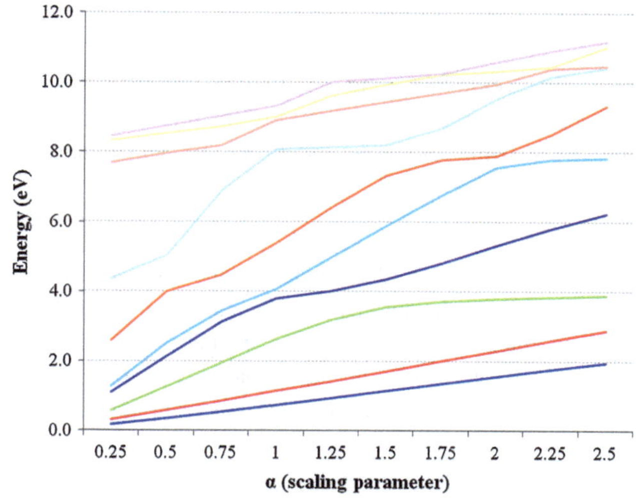

Fig. 9 Stabilization plot of the EA-EOM-CCSD energies (n.b., the EAs are minus the energies at which avoided crossings occur) for ten states of e symmetry ($L = 1$; $M = \pm 1$) for CH_3CN + e; the energies are given relative to that of the parent CH_3CN molecule

(i.e., $\zeta \rightarrow \alpha\zeta$) ranging from less than 1.0 to larger than 1.0 (the range of α-values necessary to realize the avoided crossings discussed below varies from case to case; sometimes it can be from 0.9 to 1.2, and other times it can be from 0.5 to 1.5).

In a stabilization study in search of metastable states for electron–neutral or electron–anion systems, one carries out a series of calculations for different α values for the approximate EAs using a specified set of valence and diffuse atomic basis functions. In each such calculation, one generates approximations to several (four in the example shown in Fig. 7) approximate EAs. One then plots the lowest several approximate EAs as functions of this scaling factor and obtains a stabilization plot such as that depicted in Fig. 8.

The eigenvalues associated with continuum levels move upward in energy as α is increased because increasing α contracts the radial extent of the diffuse functions, in effect reducing the radial "box size" that confine the functions. However, if a metastable state exists, one also finds a series of avoided crossings as seen in Fig. 8. This behavior results from the coupling of (1) one state that, in a diabatic sense, contains primarily inside-barrier and tunneling-range components and whose energy is relatively insensitive to α to (2) other states having nearly the same energy but containing primarily large-r oscillatory character. One can estimate the energy of the metastable anion state as the average of the energies of the two states undergoing avoided crossing at the value of α where their energy gap is smallest. Such metastable states are often called resonance states.

In the example shown in Fig. 8, there is more than one avoided crossing near EA = -1 eV. A series of avoided crossings is generated as increasing α radially constrains all of the diffuse basis functions. To successfully carry out this kind of stabilization calculation, one does not need to use a basis with a large number of diffuse functions that leads to several avoided crossings. In general, it is adequate to employ a basis capable of generating only one large-r-dominated function that undergoes an avoided crossing between the inside-barrier and tunneling-range functions as α is varied.

In Fig. 9, we show a stabilization plot obtained from EA-EOM-CCSD [15] calculations on an excess electron interacting with a H_3C–CN molecule, within C_{3v} symmetry with the two carbon atoms and one nitrogen atom on the z-axis, for states of e symmetry (chosen because the valence π^* orbital is of this symmetry). The basis set used to obtain these data was of 6-31+G* [28] quality with an additional set of diffuse 2s and 2p functions employed on the heavy atoms. The exponents of these supplemental s and p functions (0.0146 on C and 0.0213 on N) and of the most diffuse p functions in the 6-31+G* basis were simultaneously scaled.

In this system, the stabilization plot predicts there to be one metastable state with a negative EA of about -4 eV and another around -8 eV. By examining the character of the occupied orbitals in these two states (i.e., plotting the orbitals), we find the wave function for the lower energy resonance has significant weight from the $C \equiv N$ π^* valence orbital, whereas the higher-energy resonance is associated primarily with the electron being attracted to the dipole moment's positive regions near the methyl group. It should be mentioned that CH_3CN also has a dipole-bound state having $E < 0$ (that we have not treated here), so this species displays both bound and metastable valence anionic states.

4.2 Lifetime estimates

From the avoided crossing in a stabilization plot, it is also possible to estimate the lifetime τ of the metastable state or the rate at which this state undergoes autodetachment

$$\text{Rate} = \frac{1}{\tau} \tag{5}$$

if tunneling through the barrier is the rate-determining factor as it is for the anion states treated in this paper. To do so, one finds the minimum in the energy splitting (see Figs. 8, 9) that occur at the avoided crossings of the stabilization plot and uses the Heisenberg relation

$$\delta E \approx \frac{h}{\tau} \tag{6}$$

with δE taken as 1/2 the minimum splitting, to determine τ. For example, the resonance at ca. 4 eV in Fig. 9 has $\delta E \approx 0.2$ eV, which corresponds to a lifetime of ca. 2×10^{-14} s. It is clear from Fig. 9 that the splitting value is not the same at each avoided crossing of the resonance near 4 eV, so there clearly is uncertainty in the lifetime thus deduced, but this is one of the simplest routes for estimating such decay rates of metastable electron-attached species. A more rigorous approach is to analytically continue the calculated energies into the complex plane and to identify the complex stationary point [29] at which $\partial E/\partial \alpha = 0$. A good example of applying this kind of stabilization method to N_2^- and Mg^- is given if Ref. [30].

4.3 The charge stabilization method

There is another stabilization-based method for estimating the energies of metastable states that is even more straightforward to implement than the exponent scaling stabilization method outlined above. Its potential for efficiency derives from the fact that it does not require the evaluation of two-electron integrals at numerous α-values, which the orbital scaling method does. It involves only modification of the one-electron components of the electronic Hamiltonian. We illustrate this so-called charge stabilization method [31] using the sulfate dianion SO_4^{2-} as an example. Here, we anticipate either being able to bind an electron to the SO_4^{1-} ion (if the attractions are strong enough) or being able to form a metastable dianion in which the electron is trapped behind the RCB of the SO_4^{1-} anion.

Our goal is to compute the EA for binding a second electron to the SO_4^{1-} anion to form the sulfate ion, and let us assume we know the equilibrium S–O bond length of this dianion. A straightforward calculation (e.g., using HF, CI, or other ab initio approach) finds that there is no state of the dianion that lies below the energy of the SO_4^{1-} ion. That is, we find that sulfate is not electronically stable as an isolated species. So, we are faced with finding a metastable state of the sulfate dianion in which the RCB or a centrifugal potential traps the electron.

One could, of course, carry out an exponent scaling stabilization calculation on this system much like we described earlier. Alternatively, one can do the following:

1. Increase the nuclear charge of the sulfur atom from 16 to $16 + \delta q$ using a series of fractional positive values of δq. If δq were allowed to assume the value 1.0, one would be studying the ClO_4^{1-} system (at the geometry of SO_4^{2-}) rather than SO_4^{2-} and ClO_4 is known to have a positive EA. So, for some fractional positive value of δq, the EA must evolve from negative (for SO_4^{1-}) to positive (for ClO_4). Alternatively, one could increase the nuclear charges of the four oxygen atoms each by an identical small amount. Any such change to the attractive potential experienced by the excess electron that preserves the symmetry of the parent can be used.
2. Compute the EA of the fictitious XO_4^{1-} anion (where X is undergoing transmutation from S to Cl) using a good quality inside-barrier and tunneling-range atomic orbital basis and a reliable method, but only use the resultant data for values of δq for which the EA remains positive. For positive values of EA, the results of such standard electronic structure calculations can be trusted, but it is important to keep in mind the warnings issued earlier (e.g., HF may give a negative EA when; in fact, the true EA is positive and included electron correlation is usually important because of the large differential correlation energy that usually occurs).
3. Then, plot the positive EA values as a function of the increased charge δq and extrapolate the data to $\delta q \to 0$, making sure to use only data from δq values for which EA is positive. The value obtained from this extrapolation gives the estimate of the metastable state's negative EA.

An example of such a charge stabilization plot for the SO_4^{1-} system is shown in Fig. 10 with EA data obtained from HF, MP2, MP4, and coupled cluster calculations [32].

Notice that the computed data points corresponding to negative EA values (see those within the red box in Fig. 10) do not fall on the linear fits that are obtained using those data points for which EA is positive, which is why we do not use these data points in making our estimate of the actual negative EA. These energies relate to states that are part of the $SO_4^- + e^-$ continuum and become less reliable for characterizing SO_4^{2-} as δq decreases below ca. 0.25. The HF-level data deviate most strongly from the linear fit, with the MP2 and MP4 data deviating less and the CCSD(T) data deviating only slightly. In these calculations, the radial

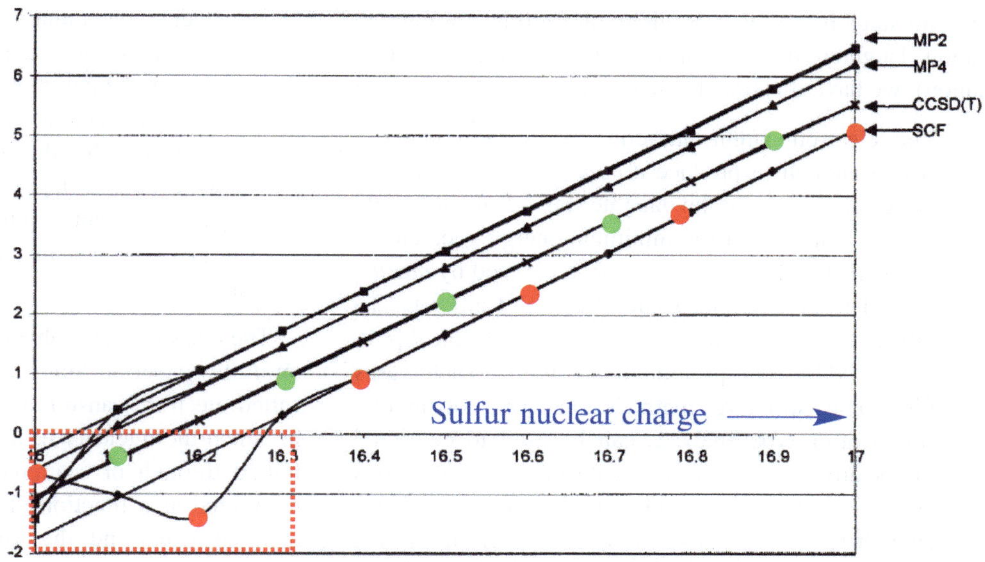

Fig. 10 Charge stabilization plot for the electron SO_4^{1-} system computed at various levels of theory (adapted from Ref. [32]). The HF-level data follow the *red dots*, while the CCSD(T) data follow the *green dots*

extent of the basis set limits the degree of variational collapse that the (N+1)st electron's orbital can undergo. As a result, this orbital retains much valence character, which, in turn, allows the correlation calculations to be reasonably successful even for $\delta q < 0.25$.

The charge stabilization method outlined above does not provide a direct estimate of the lifetime, unlike the exponent scaling stabilization method. However, other workers [33, 34] have extended this kind of approach (i.e., scaling the nuclear charge) employing non-analytic functional forms (e.g., including terms such as $(Z - Z_C)^{3/2}$) for how the EA should scale with nuclear charge Z. In that work, the N+1 to N-electron energy gap ΔE for values of Z for which the energy difference is real was fit to, for example,

$$Z^2\Delta E = a + b(Z - Z_C) + c(Z - Z_C)^2 + d(Z - Z_C)^{3/2} \tag{7}$$

This fit was then used to predict the energy gap for values of $Z < Z_C$ where ΔE has a negative real part (corresponding to a metastable state) and a nonzero imaginary component (inversely related to the lifetime). Such an approach was used [33], for example, to extrapolate from the IP of Ne ($Z = 10$), through the IP of F^- ($Z = 9$), to predict the negative EA of O^- (−5.38 eV) and the lifetime (5×10^{-16} s) of O^{2-}.

It is interesting to point out that the experimental community employs "tricks" similar to the charge stabilization device to estimate negative EAs of metastable anions. For example, we show in Fig. 11 a plot of the measured electron-binding energies for sulfate anions solvated by various numbers of water molecules [35].

When sulfate is solvated by three or more water molecules, the resulting complex has a positive electron-binding energy; when two or fewer water molecules are present, the species is metastable. By plotting the electron-binding energy as shown in Fig. 11 and using only data points for which the complex is stable, the authors of Ref. [35] could extrapolate to predict a negative EA for SO_4^-; doing so, we obtain an EA of −0.5 to −1.0 eV, although the data points in Fig. 11 clearly do not follow a straight line near $n = 3$, so there is considerable uncertainty in this extrapolation. These extrapolated values are in reasonable agreement with the CCSD(T) extrapolation shown in Fig. 10.

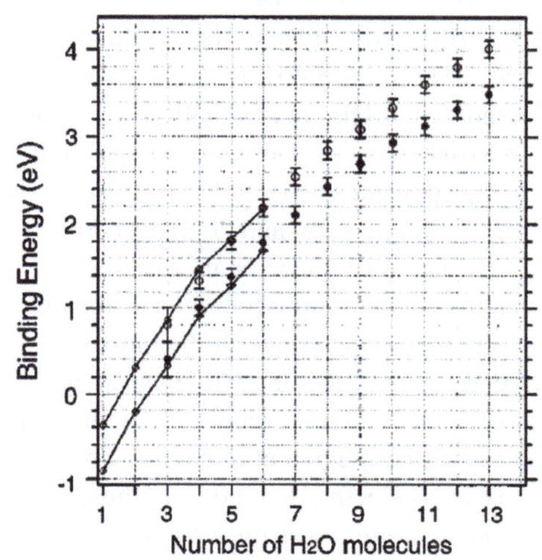

Fig. 11 Plot of the electron-binding energies, determined by photoelectron spectroscopy in Ref. [35] for $(SO_4^{2-})(H_2O)_n$ clusters with $n = 13$. The lower data points relate to adiabatic electron detachment, and the *upper* points to *vertical* detachment

5 Summary

By analyzing the various components contributing to the interaction potential between an excess electron and a ground-state parent that is either a neutral (forming an anion daughter) or an anion (forming a multiply charged anion), we have illustrated that several outcomes can arise.

1. The electron–parent attractive interactions may be strong enough to produce a *bound state* in which the electronic energy of the daughter lies below that of the parent, at least near the minimum energy geometry of the daughter. These cases are characterized by *positive EAs*. The ground electronic states of $Ca^-(^2P)$, Cl^-, OH^-, O_2^- $(H_2O)_4^-$, $[CuPc(SO_3)_4]^{3-}$, and of the dipole-bound $NC–CH_3^-$ provide examples of such cases. Electronic structure methods that do not involve stabilization-type analyses can be used to determine the positive EAs associated with this class of anions. However, one must employ basis functions diffuse enough to accommodate the excess electron and one should use an approach that includes electron correlation corrections because of the large correlation energy difference between the parent and electron-attached systems. Moreover, if the excess electron is not bound in the HF approximation, as is the case for $Ca^-(^2P)$, one must use electronic structure methods that do not depend on the suitability of the HF wave function (or its orbitals and orbital energies) so as to avoid the plague of variational collapse.
2. The electron–parent attractive interactions may not be strong enough to produce a bound state having a positive EA but may, in combination with either a repulsive centrifugal or a repulsive Coulomb potential, produce an effective potential having a barrier that can trap the excess electron behind it. In these cases, a low-energy *metastable electron-attached state* having a *negative EA* can form. The lifetimes of such states are governed primarily by the rate at which the excess electron tunnels through the barrier behind which it is trapped. The low-energy metastable states of $Mg^-(^2P)$, N_2^-, $C_6H_6^-$, SO_4^{2-}, PO_4^{3-}, $[CuPc(SO_3)_4]^{4-}$ give examples of such states. To properly characterize this class of anions, one should employ an approach that avoids the variational collapse that would otherwise occur when using diffuse basis functions capable of describing the long-range character of the metastable state's wave function.
3. If the electron–parent attractive potential is weak and if there is no repulsive Coulomb or centrifugal potential with which it can combine to form a barrier, then no low-energy electron-attached metastable state can be formed. Calculations on such systems using finite basis sets may produce energies for the electron-attached species that lie above the energy of the parent, but these are *not metastable states*. Their nature can be uncovered by improving the basis sets (especially by enhancing the radial range of the diffuse functions) upon which the state will collapse to a function describing the parent plus the excess electron far away (i.e., not attached at all). In a stabilization calculation, one will find that the stabilization plot has no regions of avoided crossing within which inside-barrier and large-*r*-dominant functions are coupled. The $Ca^-(4s^25s^1)$ ^2S state provides an example of such a case.

The class of metastable electron-attached states focused on in this paper are often studied experimentally using so-called electron transmission spectroscopy (ETS) methods in which the intensity of a beam of electrons either transmitted through or scattered from a gas-phase sample is monitored as a function of the kinetic energy of the incident electrons and the kinetic energy of the scattered electrons. Low-energy shape resonance states as studied here are detected, for example, by observation of attenuation of the incident beam within a range of electron kinetic energies $E \pm \delta E$. The center of this range is used to specify the energy of the metastable state and the spread δE over which attenuation is observed is used to specify the state's lifetime via. Eq. (6). Reference [36] provides a good overview of how ETS has been used to characterize many low-energy shape resonance states of organic molecules. Typically, ETS measurements have instrumental resolutions of ca. 0.05 eV. The shape resonance states discussed here often have lifetimes in the 10^{-14} s range, which gives them Heisenberg widths δE of ca. 0.5 eV. The core-excited electron-attached states mentioned earlier in this paper, which require two-electron transitions to undergo electron detachment, have longer lifetimes and thus produce ETS signatures that have much smaller δE values. These order-of-magnitude estimates of the experimental resolution and of lifetimes and widths of metastable states offer some guidance about what accuracy one needs in determining the EAs corresponding to these states.

Acknowledgments K.D.J. acknowledges the support of Grant CHE1111235 from the National Science Foundation.

References

1. Love DE, Jordan KD (1995) Electron impact excitation of the singlet and triplet B_{1u} π π^* states of ethylene near threshold. Chem Phys Lett 235:479–483
2. Callaway J, LaBahn RW, Pu RT, Duxler WM (1968) Extended polarization potential: applications to atomic scattering. Phys Rev 68:12–21
3. Kunc JA (1999) Low-energy electron–atom scattering in a field of model potentials. J Phys B 32:607–619
4. LaBahn RW, Callaway J (1966) Elastic scattering of low-energy electrons from atomic helium. Phys Rev 147:28–40

5. Voora VK, Ding J, Sommerfeld T, Jordan KD (2012) A self-consistent polarization potential model for describing excess electrons interacting with water clusters. J Phys Chem B 117:4365–4370
6. Schulz GJ (1973) Resonances in electron impact on diatomic molecules. Rev Mod Phys 45:423–486
7. Ehrhardt H, Langhans L, Linder F, Taylor HS (1968) Resonance scattering of slow electrons from H_2 and CO angular distributions. Phys Rev 173:222–230
8. Sanche L, Schulz GJ (1973) Electron transmission spectroscopy: resonances in triatomic molecules and hydrocarbons. J Chem Phys 58:479–493
9. Rescigno TN, Byrum DA, Isaacs WA, McCurdy CW (1999) Theoretical studies of low-energy electron-CO_2 scattering: total, elastic, and differential cross sections. Phys Rev A 60:2186–2193
10. Savin A (1996) Recent developments and applications of modern density functional theory. Elsevier, Amsterdam, p 327
11. Taylor HS (1970) Models, interpretations, and calculations concerning resonant electron scattering processes in atoms and molecules. Adv Chem Phys 18:91–147
12. Hazi AU, Taylor HS (1970) Stabilization method of calculating resonance energies: model problem. Phys Rev A 1:1109–1120
13. Moiseyev N, Certain PR, Weinhold F (1978) Complex-coordinate studies of helium autoionizing resonances. Inter J Quantum Chem 14:727–736
14. Schirmer J, Cederbaum LS, Walter O (1983) New approach to the one-particle Green's function for finite Fermi systems. Phys Rev A 28:1237–1259
15. Nooijen M, Bartlett RJ (1995) Equation of motion coupled cluster method for electron attachment. J Chem Phys 102:3629–3647
16. Simons J (2005) Equations of motion (EOM) methods for computing electron affinities and ionization potentials. In: Dykstra CE, Frenking G, Kim KS, Scuseria G (eds) Theory and applications of computational chemistry: the first 40 years, a volume of technical and historical perspectives, chap 17. Elsevier, pp 443–461
17. Jordan KD (1979) The early history of the study of electrons binding to dipoles is overviewed. Negat Ion States Polar Mol Acc Chem Res 12:36–42
18. Wang F, Jordan KD (2003) Theory of dipole-bound anions. Ann Rev Phys Chem 54:367–396
19. Garrett WR (1970) Critical binding of an electron to a non-stationary electric dipole. Chem Phys Lett 5:393–397
20. Garrett WR (1971) Critical binding of an electron to a rotationally excited dipolar system. Phys Rev A 3:961–972
21. Crawford OH (1968) Electron collision frequencies in water vapor. Chem Phys Lett 2:461–463
22. Wang L-S, Wang X-B (2000) Probing free multiply charged anions using photodetachment photoelectron spectroscopy. J Phys Chem A 104:1978–1990
23. Burrow PD, Michejda JA, Comer J (1976) Low-energy electron scattering from Mg, Zn, Cd and Hg: shape resonances and electron affinities. J Phys B Atom Mol Phys 9:3225–3236
24. Petrunin VV, Andersen HH, Balling P, Andersen T (1996) Structural properties of the negative calcium ion: binding energies and fine-structure splitting. Phys Rev Lett 76:744–747
25. Johnston AR, Burrow PD (1979) Temporary negative-ion formation in calcium vapor. Bull Am Phys Soc 24:1189
26. Romanyuk NI, Shpenik OB, Zapesochnyi IP (1980) Cross sections and characteristics of electron scattering by calcium, strontium, and barium atoms. JETP Lett 32:452–455
27. Koopmans T (1934) Über die Zuordnung von Wellenfunktionen und Eigenwerten zu den Einzelnen Elektronen Eines Atoms. Physica 1:104–113
28. Krishnan R, Binkley JS, Seeger R, Pople JA (1980) Self-consistent molecular orbital methods. XX. A basis set for correlated wave functions. J Chem Phys 72:650–654
29. Simons J (1981) Resonance state lifetimes from stabilization graphs. J Chem Phys 75:2465–2467
30. Chao JS-Y, Falcetta MF, Jordan KD (1990) Application of the stabilization method to the $N_2^-(^2\Pi_g)$ and $Mg^-(^2P)$ temporary anion states. J Chem Phys 93:1125–1135
31. Nestmann B, Peyerimhoff SD (1985) Calculation of the discrete component of resonance states in negative ions by variation of nuclear charges. J Phys B 18:615–626
32. Whitehead A, Barrios R, Simons J (2002) Stabilization calculation of the energy and lifetime of metastable SO_4^{2-}. J Chem Phys 116:2848–2851
33. Herrick DR, Stillinger FH (1975) Energy and lifetime of 0^{2-} from analytic continuation of isoelectronic bound states. J Chem Phys 62:436–4365
34. Sergeev A, Kais S (2001) Resonance states of atomic anions. Inter J Quantum Chem 82:255–261
35. Wang X-B, Nicholas JB, Wang L-S (2000) Electronic instability of isolated SO_4^{2-} and its solvation stabilization. J Chem Phys 113:10837
36. Jordan KD, Burrow PD (1987) Temporary anion states of polyatomic hydrocarbons. Chem Rev 87:557–588

Theor Chem Acc (2014) 133:1441
DOI 10.1007/s00214-013-1441-x

REGULAR ARTICLE

The *V* state of ethylene: valence bond theory takes up the challenge

Wei Wu · Huaiyu Zhang · Benoît Braïda · Sason Shaik · Philippe C. Hiberty

Received: 11 September 2013 / Accepted: 12 December 2013 / Published online: 9 January 2014

Abstract The ground state and first singlet excited state of ethylene, so-called *N* and *V* states, respectively, are studied by means of modern valence bond methods. It is found that extremely compact wave functions, made of three VB structures for the *N* state and four structures for the *V* state, provide an $N \rightarrow V$ transition energy of 8.01 eV, in good agreement with experiment (7.88 eV for the $N \rightarrow V$ transition energy estimated from experiments). Further improvement to 7.96/7.93 eV is achieved at the variational and diffusion Monte Carlo (MC) levels, respectively, VMC/DMC, using a Jastrow factor coupled with the same compact VB wave function. Furthermore, the measure of the spatial extension of the *V* state wave function, 19.14 a_0^2, is in the range of accepted values obtained by large-scale state-of-the-art molecular orbital-based methods. The σ response to the fluctuations of the π electrons in the *V* state, known to be a crucial feature of the *V* state, is taken into account using the breathing orbital valence bond method, which allows the VB structures to have different sets of orbitals. Further valence bond calculations in a larger space of configurations, involving explicit participation of the σ response, with 9 VB structures for the *N* state and 14 for the *V* state, confirm the results of the minimal structure set, yielding an $N \rightarrow V$ transition energy of 7.97 eV and a spatial extension of 19.16 a_0^2 for the *V* state. Both types of valence bond calculations show that the *V* state of ethylene is not fully ionic as usually assumed, but involving also a symmetry-adapted combination of VB structures each with asymmetric covalent π bonds. The latter VB structures have cumulated weights of 18–26 % and stabilize the *V* state by about 0.9 eV. It is further shown that these latter VB structures, rather than the commonly considered zwitterionic ones, are the ones responsible for the spatial extension of the *V* state, known to be ca. 50 % larger than the *V* state.

Keywords Valence bond · Quantum Monte Carlo · *V* state of ethylene · Breathing orbitals

Dedicated to Professor Thom Dunning and published as a part of the special collection of articles celebrating his career upon his retirement.

W. Wu (✉) · H. Zhang
The State Key Laboratory of Physical Chemistry of Solid Surfaces, Fujian Provincial Key Laboratory of Theoretical and Computational Chemistry, College of Chemistry and Chemical Engineering, Xiamen University, Xiamen 361005, Fujian, China
e-mail: weiwu@xmu.edu.cn

B. Braïda (✉)
Laboratoire de Chimie Théorique, CNRS, UMR 7616, UPMC Université Paris 06, C. 137, 4 Place Jussieu, 75252 Paris Cedex 05, France
e-mail: benoit.braida@upmc.fr

S. Shaik (✉)
Institute of Chemistry and The Lise Meitner-Minerva Center for Computational Quantum Chemistry, Hebrew University of Jerusalem, 91904 Jerusalem, Israel
e-mail: sason.shaik@gmail.com

P. C. Hiberty (✉)
Laboratoire de Chimie Physique, CNRS UMR8000, Bat. 349, Université de Paris-Sud, 91405 Orsay Cédex, France
e-mail: philippe.hiberty@u-psud.fr

1 Introduction

It often happens that valence bond (VB) theory provides compact and accurate descriptions of difficult test cases which, by contrast, necessitate long configuration expansions in the molecular orbital (MO) framework. As well-known examples, spin-coupled (SC) theory [1, 2] and generalized valence bond (GVB) theory [3, 4] are able to provide the most possible compact wave functions taking care of most of the static correlation energy in a molecule. Moreover, other VB-type methods include also dynamic correlation in a simple and lucid way [5–8]. Usually, dynamic correlation is retrieved in the MO framework by means of complete active space self-consistent-field approach (CASSCF) followed by a perturbative treatment (CASPT2), while VB methods may include dynamic correlation without increasing the number of VB structures, e.g., by using the breathing orbital effect [9–11].

The first singlet excited state of ethylene (so-called the *V* state) is a notoriously difficult test case. Thus, even an extensive CASPT2 calculations, involving an all-valence (12,12) active space, were found to be unsuccessful [12], and the correct description required more elaborate treatments. In the present paper, we pose the question whether VB theory can handle the challenge and still provide a compact but accurate description with a lucid account of electronic correlation? Taking up this challenge is our way of honoring Prof. Dunning who has significantly contributed to the revival of VB theory, among his other numerous achievements.

The *V* state of ethylene, according to the Merer-Mulliken notation [13], is a singlet excited state of B_{1u} symmetry dominated by the $\pi_u \rightarrow \pi_g$ MO configuration, which can be qualitatively represented in VB terms as a resonance between two zwitterionic structures, Scheme 1. The *V* state exhibits a broad UV absorption band with a maximum at 7.66 eV [14]. For a long time now, the *V* state poses an intriguing test case and exceptionally difficult to describe for theoreticians despite the small molecular size. Up to the late seventies, it was not even clear whether this state was of valence or Rydberg nature [15–19]. At the Hartree–Fock level, the *V* state exhibited characteristics of a Rydberg state [15], owing to a very large spatial extent of the wave function as measured by an $\langle x^2 \rangle$ expectation value of 42.1 a_0^2 (*x* being the axis orthogonal to the molecular plane), which is four times larger than the value of 11.9 a_0^2 computed for the ground state (the *N* state) [20]. However, Bender et al. [20] and Buenker and Peyerimhoff [21] showed that wave functions computed with configuration interaction (CI) are considerably more contracted than the

1a 1b

Scheme 1 Traditional representation of the *V* ionic excited state of ethylene

Hartree–Fock one, so that the diffuse character of the state in the latter method is artificial. A few years later, McMurchie and Davidson [22] presented the first definitive theoretical evidence that the *V* state is essentially a valence type. Thus, using a configuration selection scheme by classes of excitations in order to ensure a correct description of all correlation effects, involving the π and π^* electrons, these authors obtained the values 17.8 a_0^2 for $\langle x^2 \rangle$ and 7.96 eV for the $N \rightarrow V$ transition energy. Remarkably, McMurchie and Davidson found that including excitations of $\sigma \rightarrow \sigma^*$ type along with $\pi \rightarrow \pi^*$ ones are essential to give the *V* state its correct spatial extension. This indicates that inclusion of dynamic electron correlation in the π space must by accompanied by a concomitant reoptimization of the σ orbitals [23, 24], a condition that is not met at the Hartree–Fock or CASSCF levels. It follows that even CASPT2 calculations, which are based on these latter levels, overestimate the energy of the *V* state and place it close to the Rydberg $3d\pi_g$ configuration [25]. This energy proximity creates a spurious valence-Rydberg mixing, making the vertical $N \rightarrow V$ transition energy (T_e) artificially correct. As such, getting reasonable T_e values is not sufficient to validate a calculation, since improper orbital optimization can lead to spurious Rydberg character. A valid description of the *V* state must display both an accurate $N \rightarrow V$ transition energy *and* an adequate value of spatial extent of the wave function, as measured by $\langle x^2 \rangle$. Indeed, later computations by Buenker et al. [18, 19] obtained values of 19–20 a_0^2 for $\langle x^2 \rangle$ and 8.05 eV for T_e, employing larger basis sets and MRSDCI configuration selection.

At this point, it is important to note that the calculated T_e value that best matches the experiment is not 7.66 eV, the maximum of the absorption band. The T_e value should be higher, if vibrational corrections were taken into account. On the basis of zero-point energy corrections, Davidson proposed that the best estimate for T_e should be 7.8 eV [26]. More recently, Lasorne et al. [27] found a value of 7.92 eV for T_e at a sophisticated MRCI level and then performed quantum dynamics simulation that reproduced the absorption spectrum with a maximum at 7.70 eV, only 0.04 eV above the experimental value. As such, the theoretical T_e value that best matches experiment lies 0.22 eV above the absorption maximum, giving rise to the value of 7.88 eV as the best theoretical estimate.

Since the difficulty of calculating the T_e and $\langle x^2 \rangle$ values arises from a lack of sufficient electron correlation at the orbital optimization step, various strategies have been devised to optimize the orbitals at a better level than Hartree–Fock or CASSCF. Roos et al. [28] used a multistate CASPT2 scheme, allowing the zero-order states to mix among themselves under the effect of dynamic correlation, and got a T_e value of 7.98 eV, with an $\langle x^2 \rangle$ value in the range 18–20 a_0^2. Krebs and Buenker performed an MRCI calculation in large basis set, including up to more than one million configurations state functions (CSFs), and estimated T_e in the range 7.90–7.95 eV [29]. An even more extensive MRCI calculation was performed by Müller et al. [30], including up to 80 million CSFs, yielding a T_e value of 7.7 eV. Both these MRCI calculations provide $\langle x^2 \rangle$ values in the range 16.5–20 a_0^2. Still another strategy was used by Schautz and Filippi [31]. They performed diffusion Monte Carlo (DMC) calculations after optimization of the orbitals in the presence of the Jastrow factor at the VMC level. Their best T_e value was 7.92 eV, with $\langle x^2 \rangle$ being ~ 20 a_0^2. The quantum Monte Carlo (QMC) method was also used by Anderson and Goddard, in an endeavor to use a GVB determinantal part times a Jastrow factor as a DMC form of trial wave function. For the sake of cost-efficiency, the orbitals were taken from a previous bare GVB calculations and not reoptimized at the DMC level, as orbital reoptimization is a time-consuming part in QMC calculations. This strategy proved very successful for a number of excited states of ethylene and other molecules, but less so for the V state of ethylene, leading to a T_e value of 8.27 eV [32].

Recently, Angeli analyzed in detail the requirements for the correct description of the V state of ethylene and more generally the ionic $\pi \rightarrow \pi^*$ excited states of polyenes, using qualitative VB reasonings [12]. The importance of the σ–π mixing, also called $\sigma\pi$ correlation, was explained in terms of a dynamic response of the σ framework to the field generated by the π electrons in each ionic structure. In short, the σ MOs follow the charge fluctuation in the π system (represented as **1a** $\leftrightarrow$ **1b** in Scheme 1) by polarizing in a dynamic way so as to dampen the net charge on each carbon atom. Importantly, Angeli also showed that taking into account the latter effect induces a spatial contraction of the π atomic orbitals (AOs), thus eliminating the spurious Rydberg character due to incomplete electron correlation. The strategy that was used to get an appropriate set of orbitals for CI was to optimize the orbitals in a self-consistent field manner using a restricted-active-space-SCF (RASSCF) calculation containing all the excitations describing the dynamic σ polarization. Then, these orbitals were used in a CAS-CI(2,2) or CAS-CI(6,6), followed by a multistate perturbative treatment to second order. The procedure proved successful, with $\langle x^2 \rangle$ values in the range 15.8–18.7 a_0^2 and T_e values of 7.65–7.80 eV.

It is clear from the above that the reasons for the early difficulties to describe the V state are now well understood and that there exist well-defined strategies to obtain reasonable values for the diffuseness of this state as well as $N \rightarrow V$ transition energy in the MO-CI framework [12, 22, 33–35]. Still, the corresponding wave functions, which are constructed with large perturbative expansions, lack compactness and lucid interpretability, whereas the compact CASSCF(2,2) wave function is quite inaccurate. However, it has always been our experience that some ab initio VB methods, providing static and dynamic correlations [9–11, 36–38], are able to give quantitative results as well as physical insight by means of very compact wave functions, with a very small number of chemically lucid VB structures. Examples involve dissociation reactions of two-electron bonds [9, 39] and three-electron bonds [40], reaction barriers [41, 42], transition energies [43], and so on [7]. Thus, in view of the difficulties encountered by many researchers and the elaborate strategies that had to be devised to describe the V state of ethylene with MO-CI methods, we decided to address the problem by the use of ab initio VB methods, with the aim of getting accurate values of $N \rightarrow V$ transition energies as well as diffuseness $\langle x^2 \rangle$ values, from compact and insightful VB wave functions. Furthermore, we want the strategy for choosing the relevant VB structures to emerge from clear physical principles, without the need of preliminary large-scale calculations.

There are already some clues that this simple approach should be successful: (1) the effect of paramount importance to the accurate description of the V state is the dynamic response of the σ orbitals to the fluctuations of the π electrons. This effect is very simply described by means of the breathing orbital valence bond (BOVB) method [9–11]. Furthermore, this effect can be easily turned off, allowing one to visualize and assess the effects of dynamic $\sigma\pi$ correlation on the size of orbitals and on the $N \rightarrow V$ transition energy. (2) BOVB, as well as quantum Monte Carlo valence bond methods (QMC-VB) [44–46], includes all the necessary dynamic correlation for both the N and V states, and (3) most of the MO-CI strategies that get an accurate description of the V state are in fact based on qualitative VB reasonings.

Based on these considerations, modern VB methods can take up the challenge of reconciling compactness, accuracy, and physical insight in the calculation of the V state of ethylene. The paper is organized as follows. First, the VB methods are briefly described in a theoretical section. Then, the VB structures are chosen, and finally, the results are presented and discussed.

1.1 Theoretical methods

1.1.1 The VB wave function

A many-electron system wave function, Ψ, is expressed in VB theory as a linear combination of Heitler–London–Slater–Pauling (HLSP) functions, Φ_K in Eq. 1,

$$\Psi = \sum_K C_K \Phi_K \quad (1)$$

where Φ_K corresponds to "classical" VB structures, and C_K is the corresponding structural coefficients. An important feature of our VB calculations is that all the active orbitals, i.e., those that have varying occupancies in the VB structures, are strictly localized on a single atom, like in the classical VB method. This ensures a clear correspondence between the mathematical expressions of the VB structures and their physical meaning, ionic or covalent.

Several definitions exist for the weights of the VB structures [6]. The Coulson–Chirgwin formula [47], Eq. 2, is the equivalent of a Mulliken population analysis in VB theory:

$$W_K = C_K^2 + \sum_{L \neq K} C_K C_L S_{KL} \quad (2)$$

Here, S_{KL} is the overlap integral of two VB structures.

However, as this definition becomes meaningless when the overlaps are large, alternative definitions have also been proposed. Löwdin's scheme for symmetric orthogonalization [48] (Eq. 3)

$$W_K^{Löwdin} = \sum_{IJ} S_{KI}^{1/2} C_I S_{KJ}^{1/2} C_J \quad (3)$$

is an attractive option, which leads to orthogonalized functions deviating as little as possible from the original set.

Another appropriate definition, which also remedies the overlap dependence problem, is the "inverse-overlap" Norbeck-Gallup formula [49] (Eq. 4), which is close to the normalized squares of coefficients:

$$W_K^{Inv} = \frac{NC_K^2}{S_{KK}^{-1}} \quad (4)$$

where N is a normalization factor.

1.1.2 The VB levels

There are several computational approaches for VB theory at the ab initio level [5, 7]. In the VB self-consistent field (VBSCF) procedure [50], both the VB orbitals and structural coefficients are optimized simultaneously to minimize the total energy. The breathing orbital valence bond method (BOVB) [9–11] improves the VBSCF accuracy without increasing the number of VB structures Φ_K. This is achieved by allowing each VB structure to have its own specific set of orbitals during the optimization process, such that the orbitals can be different from one VB structure to the other. In this manner, the orbitals can fluctuate in size and shape so as to fit the instantaneous charges of the atoms on which these orbitals are located. As such, effectively, BOVB covers a much larger variational space than CASSCF/VBSCF, without losing the compactness of the wave function.

The orbitals in either VBSCF or BOVB are not restricted to be orthogonal, and thus, it is not guaranteed that orbitals that are localized on the same fragment will be independent of each other. However, since we always obtain a unique variationally optimized solution, wherein both orbitals and structural coefficients are fully optimized, the nonuniqueness of the orbitals, if happens, does not matter.

The BOVB method has several levels of sophistication [11]. Here, we chose the "SD-BOVB" level, whereby each doubly occupied active orbital is split into a pair of singly occupied orbitals accommodating a spin-pair, so as to account for the radial correlation of the electrons involved in the lone pair. Moreover, as indicated by the letter D (delocalized), the orbitals that do not belong to the active space are allowed to delocalize. The VBSCF and BOVB calculations have been performed using three different basis sets: (1) the standard 6-31G* basis set, (2) the correlation-consistent triple-zeta cc-pVTZ of Dunning et al. [51], and (3) a mixed basis set of triple-zeta + diffuse quality on the carbons, made of the aug-cc-pVTZ basis set [52] for carbons and of the cc-pVDZ basis set [51] for hydrogens. This latter basis set is referred to as aug-VTZ in this work.

1.1.3 The VB-QMC levels

Very recently, a mixed valence bond/quantum Monte Carlo (VB-QMC) method has been proposed which managed to provide high accuracy while keeping the full interpretative capabilities of classical valence bond methods [44–46]. The VB-QMC wave function we utilize here consists of a VBSCF determinant part multiplied by a Jastrow function that is included to account for electronic correlation. The Jastrow function is the same as in previous studies [44–46] and includes explicit two-body electron–electron, electron–nucleus, and also three-body electron–electron–nucleus terms. In this study, the energies of the N and V states are computed by optimizing simultaneously Jastrow parameters, VB structural coefficients, and, in some specified cases, also orbitals. The variational Monte Carlo (VMC) optimization algorithm is used on the multi-structure wave function and separate atoms, respectively, by minimizing the total energy and a small percentage of variance (1 %) using the linear optimization method [46]. These VB-VMC wave functions are then projected onto the ground state using the diffusion Monte Carlo (DMC) algorithm under the fixed-node approximation, a procedure that recovers most of the remaining correlation effects that are missing at the VB-VMC level, while at the same time compensating

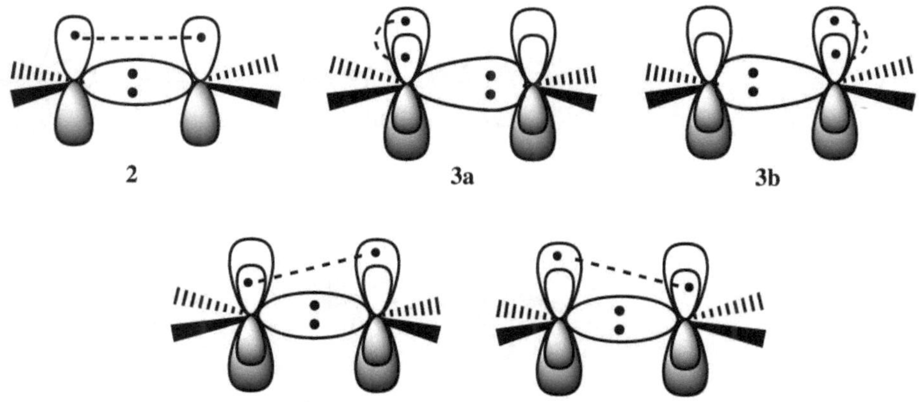

Scheme 2 Set of VB structures for the N and V states in the (2, 2) space. The σ–CC bond is represented as a doubly occupied bonding MO, which is polarized in opposite directions in **3a** and **3b** at the BOVB level

basis set deficiencies, and usually provides very accurate energy differences [44, 53]. This latter method is referred to here as VB-DMC. For all VB-QMC calculations, a systematically convergent triple-zeta polarized basis set of Burkatzki et al. [54] for carbon, supplemented with the s and p diffuse functions taken from the standard aug-cc-pVTZ basis set, and the double-zeta basis set of Burkatzki et al. for the hydrogen atom, was used together with their corresponding pseudo-potentials. The so-constructed basis set will be referred to here as ps-aug-VTZ.

All calculations were performed using the experimental equilibrium geometry of ethylene: $R_{CH} = 1.086$ Å, $R_{CC} = 1.339$ Å, $\angle HCH = 117.6°$. The VBSCF and BOVB calculations were carried out with the Xiamen Valence Bond (XMVB) program [55–57]. All the QMC calculations were carried out using the CHAMP program [58].

1.2 The VB structures set

Restricting ourselves to an MO space of two orbitals and two electrons, so-called (2,2) space, the description of the N state of ethylene becomes straightforward: structure **2**, in which the π bond is fully covalent (Scheme 2), complemented with two ionic structures **3a** and **3b** that we define with split π atomic orbitals (AOs) for full generality. For the V state, the simplest way to get an elementary VB description is to expand the $\pi\pi^*$ MO configuration into VB structures. Doing this in minimal basis set would yield the familiar picture **1a** $\leftrightarrow$ **1b** shown above in Scheme 1. However, the expansion in extended basis sets yields additional terms, because the π and π^* MOs are made from different AOs, p'_a and p'_b in π^* being generally more diffuse than p_a and p_b in π:

$$\pi \propto (p_a + p_b) \tag{5a}$$

$$\pi* \propto (p'_a - p'_b);\ p'_a \neq p_a,\ p'_b \neq p_b \tag{5b}$$

where the labels a and b refer to the left and right carbon atoms, respectively.

Using these orbital expressions and expanding $\pi\pi^*$ give rise to four VB structures instead of two, as shown in Eq. 6 below, dropping normalization constants:

$$\begin{aligned} |\pi\bar{\pi}^* - \bar{\pi}\pi^*| &= |(p_a + p_b)(\bar{p}'_a - \bar{p}'_b) - (\bar{p}_a + \bar{p}_b)(p'_a - p'_b)| \\ &= (|p_a\bar{p}'_a| - |\bar{p}_a p'_a|) - (|p_b\bar{p}'_b| - |\bar{p}_b p'_b|) \\ &\quad - (|p_a\bar{p}'_b| - |\bar{p}_a p'_b|) + (|p_b\bar{p}'_a| - |\bar{p}_b p'_a|) \end{aligned} \tag{6}$$

where the terms in parentheses represent structures **3a**, **3b**, **4a**, and **4b**, respectively, in Scheme 2. In **3a** and **3b**, one recognizes the zwitterionic structures analogous to **1a** and **1b**, with the difference that the electrons pairs are split into different AOs, thus including some radial correlation. On the other hand, **4a** and **4b** are asymmetric covalent types, which have not been considered before. Their negative combination (**4a**–**4b**) has the right symmetry (B_{1u}) to contribute to the V state and has the same constrained coefficient as (**3a**–**3b**) in $\pi\pi^*$ (Eq. 6). The importance or lack thereof of **4a** and **4b** as contributors to the V state will be determined by the quantitative VB calculations below, in which the coefficients of all VB structures will be fully relaxed and optimized.

To calculate the $N \rightarrow V$ transition energy, it is necessary to calculate both the N and V states at a consistent level of accuracy. This will be done by describing the N state with the three structures **2**, **3a**, and **3b** (these two latter ones will combine with a positive sign), while the V state will involve the negative combinations (**3a**–**3b**) and (**4a**–**4b**). This set of five structures will be denoted as VBSCF-5 or BOVB-5, according to the level of VB theory. In all structures **2**–**4** of the (2,2) space, the σ frame, made of the σ–CC and CH bonds, is described as a set of doubly occupied MOs. It should be noted that even this small (2,2) space is sufficient to account for the σ response to the fluctuation of π electrons, provided the underlying σ MOs are allowed to be different for different VB structures, as in the BOVB method. This effect is illustrated (in part) in Scheme 2 where the σ–CC bonding MOs are drawn as polarized in opposite directions in **3a** and **3b**.

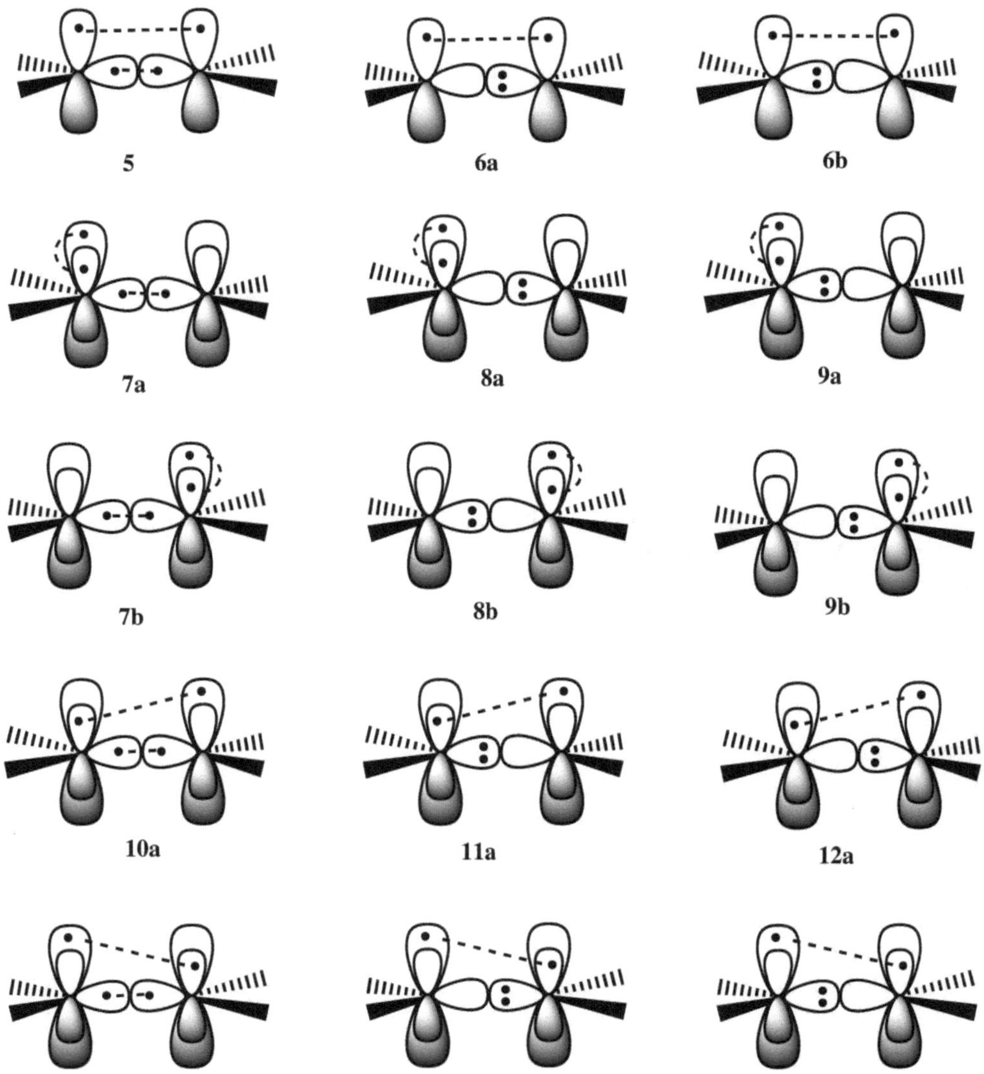

Scheme 3 Set of VB structures for the N and V states in the (4, 4) space

A more detailed way to allow for the σ response to fluctuations of the π electrons is to describe both the π space and the σ–CC bond in a VB fashion. This can be achieved by extending the VB space for the N and V states to the 15 VB structures displayed in Scheme 3. Here, the σ–CC bond is not any more described as a doubly occupied MO, but as a combination of covalent and ionic structures. Hence, the $\sigma\pi$ correlation will be visualized as the difference in weights between structures **8a**, **8b** versus **9a**, **9b**. The VB structure set in Scheme 3 will be denoted as VBSCF-15 or BOVB-15.

2 Results

2.1 VB calculations in the (2, 2) space

Although it is well known that basis sets involving diffuse basis functions are required for the description of states involving a dominant ionic component, it is interesting to start with smaller basis sets and then proceed to the largest.

Table 1 displays the results of calculations in the 6-31G* double-zeta basis set. For the N state, all computational levels provide an $\langle x^2 \rangle$ value in agreement with the accepted value, 11–12 a_0^2 [20]. However, the calculated $\langle x^2 \rangle$ values for the V state appear to be much too small, around 12 a_0^2, showing that the 6-31G* basis set is not appropriate to allow the V state to be 50 % more diffuse than the N state, as it should. It is therefore not surprising that all $N \rightarrow V$ transition energies are largely overestimated in this basis set, even at the BOVB level (by 1.23 eV in BOVB-5).

As can be seen in Table 2, the $\langle x^2 \rangle$ values of the V state are practically not improved in the larger cc-pVTZ basis set, which is of triple-zeta quality but still devoid of diffuse basis functions; however, the transition energies are improved and deserve some comments. The

Table 1 The vertical excitation energy in (2, 2) active space, calculated by HF, CASSCF, and different VB methods with the 6-31G* basis set

	N state		*V* state		T_e (eV)
	E (a.u.)	$\langle x^2\rangle$ (a_0^2)	E (a.u.)	$\langle x^2\rangle$ (a_0^2)	
Hartree–Fock	−78.030983	11.53			
(2, 2) Active space					
CASSCF	−78.059873	11.34	−77.678207	12.08	10.39
VBSCF-3[a]	−78.059878	11.34	−77.657866	12.19	10.94
VBSCF-5[b]	−78.059878	11.34	−77.683298	12.08	10.25
BOVB-3[a]	−78.079418	11.29	−77.718647	11.83	9.82
BOVB-5[b]	−78.079418	11.29	−77.744694	11.80	9.11

[a] Involves structures **2**, **3a** and **3b** for the *N* state, and **3a**, **3b** for the *V* state, as structure **2** cannot contribute to the *V* state for symmetry reasons

[b] Involves structures **2**, **3a** and **3b** for the *N* state, and **3a**, **3b**, **4a**, **4b** for the *V* state. Structures **4a** and **4b** are not included in the *N* state as their symmetric combination would be redundant with the structure **2**

$N \rightarrow V$ transition energy is lower at the VBSCF-5 level than at the CASSCF one, although the latter configuration space contains all VB structures **2–5** and more. This is because the coefficients of the VB structures in the CASSCF space are somewhat constrained, whereas they are fully optimized in VBSCF. Besides, the slight non-equivalence of CASSCF and VBSCF calculations even when both span the same variational space [6] has also been pointed out.[1]

The importance of structures **4a** and **4b** can be appreciated by performing a VB study in a smaller space, restricted to **2**, **3a**, and **3b**. The results at the VBSCF and BOVB levels (referred to as VBSCF-3 and BOVB-3, respectively) show that removing **4a** and **4b** from the VB space causes a considerable increase in the transition energy relative to the reference VBSCF-5 and BOVB-5 levels, in both 6-31G* and cc-pVTZ basis sets, indicating that structures **4a** and **4b** stabilize the *V* state by some 0.7-0.9 eV. Lastly, the effect of dynamic correlation in the π system on the transition energy, as measured by the BOVB-VBSCF difference, is seen to be quite large, ca. 0.9 eV in cc-pVTZ basis set and 1.1 eV in 6-31G*. These differences indicate that allowing the σ response to fluctuations of the π electrons is of importance, as will be analyzed in more details below. The best transition energy in these two basis sets amounts to 8.47 eV, at the BOVB-5/cc-pVTZ level. This last result, which is off by ca. 0.6 eV relative to experiment, as well as the $\langle x^2\rangle$ diffuseness value which is found too small for the *V* state, indicates that basis sets devoid of diffuse basis functions are inappropriate for the problem at hand, as already discovered long ago by Roos et al. [59, 60]. In accord, the rest of the study will be performed in a triple-zeta basis set augmented with diffuse basis functions on the carbon atoms, and a double-zeta basis set on the hydrogen atoms, so-called aug-VTZ (see Theoretical methods).

Table 3 displays results calculated in the aug-VTZ basis set, which in principle combines all the requirements in terms of flexibility and diffuseness to reliably describe an ionic state, which does not have a Rydberg character.

Inspection of Table 3 reveals that the $\langle x^2\rangle$ values of the ground state are practically unchanged relative to smaller basis sets and remain close to the generally accepted value of 12 a_0^2 at all levels. On the other hand, the *V* state is affected by the augmented basis set. Thus, *V* is found much too diffuse at the CASSCF level, as previously found by all researchers, with an $\langle x^2\rangle$ value of 25 a_0^2. Quite expectedly, about the same exaggerated spatial extension is found at the VBSCF-5 level, which lacks dynamic correlation like CASSCF, with the consequences that have been amply discussed in the past literature (vide supra). Thus, the fair agreement between experimental vertical energy and the T_e values that are found at these two levels, 8.25 and 8.26 eV, is artifactual and due to spurious mixing with a higher lying Rydberg configuration.

As shown in the Methods section, dynamic correlation is introduced by going from VBSCF to BOVB levels. As a first step, indicated as BOVB-5(π-only) level, the σ orbitals remain common to all structures and only the π orbitals are allowed to breathe for each of the structures **3a**, **3b**, **4a**, **4b**. It can be seen from Table 3 that there is practically no improvement at this level relative to VBSCF-5: The spatial extension diminishes only slightly, from 24.92 to 24.22 a_0^2, and the $N \rightarrow V$ transition energy is nearly unchanged. In a second step, we allow all orbitals to be different for different structures, at the BOVB-5 level. Here, the σ response to the fluctuation of π electrons in the *V* state is turned on, and the change relative to CASSCF and VBSCF-5 is now clearly significant. The spatial extension of the wave function goes down to 19.14 a_0^2, quite in the range of accepted $\langle x^2\rangle$ values from high-level calculations. This result nicely illustrates *the internal contraction of the wave function under the influence of dynamic correlation* and also shows that essentially *all the dynamic correlation that is attached to the description of the π system in the V state is made of the σ response to the fluctuation of π electrons*. Another way of visualizing the contraction of the wave function is by comparing the different sizes of the π AOs [12], calculated at the CASSCF(2,2) level (devoid of

[1] Slight differences between VBSCF and CASSCF calculations spanning the same variational space are due to the fact that the AOs that compose the MOs in CASSCF may be different in size and shape from one MO to the other, whereas the set of AOs is unique in VBSCF.

Table 2 Vertical excitation energies in (2, 2) active space, calculated by HF, CASSCF, and different VB methods with the cc-pVTZ basis set

	N state		*V* state		T_e (eV)
	E (a.u.)	$\langle x^2 \rangle$ (a_0^2)	E (a.u.)	$\langle x^2 \rangle$ (a_0^2)	
Hartree–Fock	−78.063239	11.84			
(2, 2) Active space					
CASSCF	−78.090945	11.58	−77.741633	13.27	9.51
VBSCF-3[a]	−78.091235	11.59	−77.713055	12.97	10.29
VBSCF-5[b]	−78.091235	11.59	−77.746673	13.32	9.38
BOVB-3[a]	−78.113076	11.71	−77.766555	12.28	9.43
BOVB-5[b]	−78.113076	11.71	−77.801926	12.72	8.47

[a] Involves structures **2**, **3a**, and **3b** for the *N* state, and **3a**, **3b** for the *V* state, as structure **2** cannot contribute to the *V* state for symmetry reasons

[b] Involves structures **2**, **3a**, and **3b** for the N state, and **3a**, **3b**, **4a**, **4b** for the *V* state. Structures **4a** and **4b** are not included in the *N* state as their symmetric combination would be redundant with structure **2**

dynamic correlation), shown in Fig. 1a, versus the dynamically correlated BOVB-5 level, Fig. 1b. First, it can be seen that the p_a and p'_a AOs that are, respectively, part of the π and π^* MOs arising from the CASSCF calculations are very different in size, p'_a being much more diffuse than p_a (Fig. 1a). Second, it clearly appears by comparing Fig. 1a, b that going from CASSCF to BOVB-5 induces a strong contraction of the p'_a AOs. As an effect of $\sigma\pi$ dynamic correlation, the latter AOs have somewhat different shapes in **3a**, **3b** versus **4a**, **4b**, but they have about the same spatial extension, quite smaller than at the CASSCF level.

Another nice feature of the BOVB-5 level is an excellent value of the $N \rightarrow V$ transition energy, 8.01 eV, only 0.13 eV above the recommended value. This good T_e value together with a correct spatial extension of the wave function shows that *compact VB functions with only three configurations for the N state and four ones for the V state are able to capture all the physics of the N → V excitation in ethylene*, whereas many more configurations are required in the MO framework to reproduce the same effects.

It is interesting at this point to estimate the effect of $\sigma\pi$ dynamic correlation or polarization on the T_e value. One could be tempted to simply take it as the VBSCF-5 versus BOVB-5 energy difference, i.e., 0.24 eV (entries 4 vs. 7 in Table 3). However, one must recall that the relatively low T_e value at the VBSCF-5 (or CASSCF) level is artifactual and due to spurious valence-Rydberg mixing. It is therefore

Table 3 The vertical excitation energy in (2, 2) active space, calculated by HF, CASSCF, and different VB methods with the aug-VTZ basis set

	N state		*V* state		T_e (eV)
	E (a.u.)	$\langle x^2 \rangle$ (a_0^2)	E (a.u.)	$\langle x^2 \rangle$ (a_0^2)	
Hartree–Fock	−78.059607	12.01			
(2, 2) Active space					
CASSCF	−78.087096	12.05	−77.783590	25.08	8.26
VBSCF-3[a]	−78.087129	11.68	−77.733637	14.07	9.69
VBSCF-5[b]	−78.087129	11.68	−77.783951	24.92	8.25
BOVB-5 (π-only)[b]	−78.089552	11.73	−77.786128	24.22	8.26
BOVB-3[a]	−78.111200	11.64	−77.782916	13.04	8.93
BOVB-5[b]	−78.111200	11.64	−77.816658	19.14	8.01

[a] Involves structures **2**, **3a** and **3b** for the *N* state, and **3a**, **3b** for the *V* state, as structure **2** cannot contribute to the *V* state for symmetry reasons

[b] Involves structures **2**, **3a** and **3b** for the *N* state, and **3a**, **3b**, **4a**, **4b** for the *V* state. Structures **4a** and **4b** are not included in the *N* state as their symmetric combination would be redundant with the structure **2**

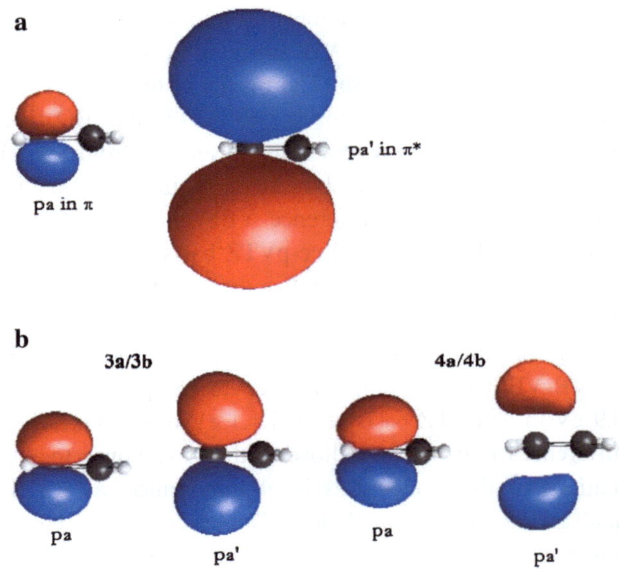

Fig. 1 The π AOs of the *V* state of ethylene, as calculated at the CASSCF(2,2) level (**a** upper figure), and at the dynamically correlated BOVB-5 level (**b** lower figure), showing the contraction of π AOs under the effect of the dynamic σ response to the fluctuation of π electrons. All calculations in aug-VTZ basis set. The isodensity value is 0.045 a.u. in both (**a**) and (**b**)

more instructive to consider this energy difference in a basis set like cc-pVTZ, where the valence-Rydberg mixing does not take place, for lack of diffuse basis functions. Thus, comparing entry 4 versus entry 6 in Table 2 yields the value 0.91 eV, which can be considered as an energetic measure of the importance of σ response to the fluctuation of π electrons, more precisely the greater importance of this σ response in the *V* state relative to the *N* state.

Another outcome of the above VB calculations is the novel finding that structures **4a** and **4b**, which are essentially asymmetric covalent structures, are essential ingredients of the *V* state of ethylene. In terms of energy, their importance can be estimated as the difference between the 2-structure and 4-structure wave functions, and can be seen to be quite large at both the VBSCF and BOVB levels. At the BOVB level, the BOVB-3 versus BOVB-5 energy differences amount to 0.71, 0.96, and 0.92 eV in 6-31G*, cc-pVTZ, and aug-VTZ basis sets, respectively. This significant stabilization can be interpreted as due to two effects: (1) a simple resonance energy, arising from the mixing of **4a**, **4b** with **3a**, **3b**, and (2) the tempering of the charge separation in the π system in **3a** and **3b** when **4a** and **4b** are added. It is also remarkable that without structures **4a** and **4b**, i.e., at the VBSCF-3 and BOVB-3 levels, the spatial extension of the wave function comes out *too small*, even in aug-VTZ basis set, with an $\langle x^2 \rangle$ value of only 13.04 a_0^2 in the *V* state at the BOVB-3 level. Thus, it appears that the zwitterionic structures **3a** and **3b** have no physical reason to be very diffuse because the corresponding VB structures are overall neutral with positive charges in the vicinity of the negative ones. On the other hand, **4a** and **4b** must have diffuse orbitals, so as to clearly differentiate the two π AOs involved in the covalent bond; otherwise, the negative combination **4a–4b** would just vanish (see the above VB expansion). Thus, *the reason for the known fact that the V state is 50 % more diffuse than the N state does not lie in the zwitterionic structures, but in the asymmetrical covalent ones,* **4a** and **4b**.

In view of the essential role played by **4a** and **4b** in the electronic structure of the *V* state, one may anticipate significant weights for these structures. However, the calculation of their weights can only be approximate owing to their large overlaps with **3a** and **3b** (Table 4). These large overlaps originate in the fact that all these structures involve diffuse AOs. For this reason, the usual Coulson–Chirgwin definition of weights [47], which is quite sensitive to overlap [61], is better replaced by the two alternative definitions of the weights, Löwdin's scheme [48] and the "inverse-overlap" formula of Norbeck and Gallup [49]. These two definitions have been applied to calculate the weights of **3a**, **3b**, **4a**, and **4b**, which are displayed in Table 5 together with their coefficients in the BOVB-5 wave function of the *V* state. The Löwdin and inverse-overlap weights are quantitatively different from each other, as always happens when the VB structures have strong mutual overlaps, and therefore, they can only provide orders of magnitude. Still, the calculated weights show that the major structures are **3a** and **3b**, whereas **4a** and **4b** are less important but far from being negligible.

2.2 VB calculations in the (4, 4) space

Extending the space of VB structures so as to include the C–C σ bond into the VB treatment leads to the 15 structures displayed in Scheme 3. Calculations in this full space (**5–12b**) are referred to as VBSCF-15 and BOVB-15. However, being redundant with **5–6b**, structures **10a–12b** are removed from the calculation of the *N* state, while for obvious symmetry, reasons **5** is removed for the calculation of the *V* state. Thus, the calculation of the *N* and *V* states involves 9 structures and 14 structures, respectively.

Since both the π system and the σ C–C bond are now involved in the VB treatment, one may expect the σ response to fluctuation of the electrons to be present already at the VBSCF level, at least to some extent. This σ response will be made apparent by comparing the weights of structures **8a** versus **9a**, or **8b** versus **9b**. If the two π electrons are on the same AO, say the left one as in **7a–9a** (Scheme 3), the σ C–C bond will compensate this charge separation by polarizing itself in the opposite way, through the ionic structures **8a** and **9a**. Therefore, one expects the weight of **8a**, in which both carbon atoms are globally neutral, to be significantly larger that that of **9a**, in which the carbons have formal charges of −2 and +2, respectively. Inspection of Table 6 shows that this is indeed the case, as the weights of **8a** and **9a** are calculated to be 0.129

Table 4 The overlap matrix of *V* state from BOVB-5 calculations with the aug-VTZ basis set in (2, 2) active space

	3a	3b	4a	4b
3a	1.000000			
3b	0.316388	1.000000		
4a	0.845642	0.429452	1.000000	
4b	0.429452	0.845641	0.474833	1.000000

Table 5 The weights and coefficients of VB structures of the *V* state from BOVB-5 calculations with the aug-VTZ basis set in (2, 2) active space

	Löwdin weights	Inverse weights	Coefficients
3a	0.4081	0.3047	1.1882
3b	0.4081	0.3047	−1.1882
4a	0.0919	0.1953	−0.9733
4b	0.0919	0.1953	0.9733

Table 6 The weights and coefficients of VB structures for the V state of ethylene, calculated with the aug-VTZ basis set in (4, 4) active space

	VBSCF-15			BOVB-15		
	Löwdin weights	Inverse weights	Coefficients	Löwdin weights	Inverse weights	Coefficients
6a	0.0003	0.0086	0.0437	0.0000	0.0130	−0.0573
6b	0.0003	0.0086	−0.0437	0.0000	0.0130	−0.0573
7a	0.1786	0.1707	0.9472	0.1984	0.2391	0.8064
8a	0.1281	0.1028	0.4935	0.1149	0.0556	0.2672
9a	0.0656	0.0013	0.0551	0.0915	0.0141	0.1378
7b	0.1788	0.1710	−0.9483	0.1984	0.2391	−0.8064
8b	0.1277	0.1018	−0.4911	0.1149	0.0556	−0.2672
9b	0.0660	0.0014	−0.0567	0.0915	0.0142	−0.1378
10a	0.0624	0.1387	−0.8696	0.0471	0.1376	−0.6180
11a	0.0259	0.0009	−0.0470	0.0248	0.0110	−0.1172
12a	0.0388	0.0768	−0.4400	0.0233	0.0297	−0.1919
10b	0.0624	0.1389	0.8705	0.0471	0.1376	0.6180
11b	0.0260	0.0009	0.0484	0.0248	0.0110	0.1172
12b	0.0390	0.0776	0.4423	0.0233	0.0297	0.1919

Table 7 The vertical excitation energy in (4, 4) active space, as calculated by VB methods with the aug-VTZ basis set

	N state		V state		T_e (eV)
	E (a.u.)	$\langle x^2\rangle$ (a_0^2)	E (a.u.)	$\langle x^2\rangle$ (a_0^2)	
VBSCF-9[a]	−78.110778	11.68	−77.769936	13.44	9.27
VBSCF-15[b]	−78.110778	11.68	−77.811182	22.95	8.15
CASSCF(4, 4)	−78.112095	11.60	−77.800412	22.21	8.48
BOVB-9[a]	−78.121389	11.76	−77.796188	13.07	8.85
BOVB-15[b]	−78.121389	11.76	−77.828470	19.16	7.97

[a] Involves structures **5–9b** for the N state, and **6a–9b** for the V state

[b] Involves structures **5–9b** for the N state, and **6a–12b** for the V state

and 0.066 at the VBSCF-15 level, according to Löwdin's definition of the weights. Moreover, the difference in weights is even more significant with the inverse-overlap weight definition. Of course, similar differences in weights are found between structures **8b** and **9b**.

It should be noted that the σ response to π fluctuation, as it appears from the weights displayed in Table 6, is incomplete at the VBSCF-15 level since it is restricted to the σ C–C bond and does not involve the C–H bonds. Still, one expects some improvement of the description of the V state with respect to the VBSCF calculations in the (2, 2) space. Table 7 reports the absolute energies of the N and V states as calculated in the (4, 4) space, together with their spatial extensions and the $N \rightarrow V$ transition energies T_e. While the CASSCF results are still poor at this level, it can be seen that the VBSCF-15 results are indeed improved relative to the VBSCF-5 results in the (2,2) space (Table 3), as the $\langle x^2\rangle$ value goes down from 24.92 to 22.95 a_0^2, while the transition energy also goes down, from 8.26 to 8.15 eV. However, in spite of these improvements, the results are not fully satisfying at this level, and once again, one must use the BOVB level in the same space to retrieve the missing dynamic σπ correlation. To facilitate convergence, only partial BOVB is performed, in which the VB structures are partitioned into groups: (**5–6b**), (**7a–9a**), (**7b–9b**), (**10a–12a**), and (**10b–12b**). A common set of orbitals is used within each group, but orbitals of one group are allowed to be different from those of other groups. The results are now excellent, with a spatial extension of 19.16 a_0^2 for the V state, and a transition energy of 7.97 eV, only 0.09 eV higher than the reference value of 7.88 eV. Interestingly, these latter results are virtually unchanged relative to the BOVB-5 results in the (2,2) space, showing that all the necessary electron correlation is already taken into account in the smaller VB space.

Since VB structures displaying asymmetrical π covalent bonds proved essential for the description of the V state in the (2,2) space, one may wonder if structures of the same type (**10a–12b**) have a comparable importance in the (4, 4) space. Clearly, the answer is yes: In terms of weights, structures **10a–12b** sum up to 0.190 with the Löwdin definition, and to 0.357 with the inverse-overlap one (Table 5, BOVB-15). This can be compared with the cumulated weights 0.184 and 0.391, respectively, with the same definitions, in the (2,2) space (Table 5). In terms of stabilization energies, the stabilization brought by structures **10a–12b** can be estimated by comparing the calculations of the V state in the 15-structure space (entries 2 and 5 in Table 7), to those using a restricted 9-structure space,

Table 8 Energies of the *N* and *V* states and weights of structures **4a** and **4b**, as calculated at the VMC and DMC levels using VBSCF or BOVB wave functions as trial functions, with the ps-aug-VTZ basis set

QMC method	*N* state		*V* state		T_e (eV)
	Optimized parameter[a]	E (a.u.)	E (a.u.)	Weigths of 4a,4b[b]	
VB-VMC-5[c,d]	j + c	−13.7132(5)	−13.3977(5)	0.138(1)	8.59(2)
BO-VMC-5[c,e]	j + c	−13.7053(5)	−13.4098(5)	0.092(1)	8.04(2)
BO-VMC-5[c,e]	j + c + i	−13.7221(5)	−13.4294(5)	0.074(1)	7.96(2)
BO-VMC-3[c,e,f]	j + c	−13.7053(5)	−13.3751(5)	–	8.98(2)
BO-DMC-5[c,e]	j + c	−13.7382(2)	−13.4455(2)	0.087(1)	7.96(1)
BO-DMC-5[c,e]	j + c + i	−13.7439(2)	−13.4525(2)	0.074(<1) %	7.93(1)

Statistical errors are shown in parenthesis

[a] Parameters optimized at the VMC level: j for Jastrow parameters, c for CSFs coefficients, i for inactive (σ) orbitals

[b] Weights of each of the structures, Löwdin definition, Eq. 3

[c] Involves structures **2**, **3a**, and **3b** for the *N* state, and **3a**, **3b**, **4a**, **4b** for the *V* state

[d] A VBSCF wave function is used as a trial function

[e] A BOVB wave function is used as a trial function

[f] Involves structures **2**, **3a**, and **3b** for the *N* state, and only **3a**, **3b** for the *V* state

involving only structures **5–9b** (entries 1 and 4). The stabilization amounts to 1.12 and 0.88 eV, respectively, at the VBSCF and BOVB level. Moreover, as was also found in the (2,2) space, the spatial extension of the *V* state comes out too small (ca. 13 a_0^2) if structures **10a–12b** are omitted. Thus, from all point of views, it appears that *the newly proposed VB structures displaying asymmetrical π covalent bonds are essential ingredients of the V state of ethylene*, albeit they were never considered before, to our knowledge.

2.3 VB-QMC calculations

Since QMC methods are known to be extremely accurate in principle, provided one uses sufficiently good trial functions, it is interesting to check whether the extra correlation provided by these methods would change or confirm our above results. To this aim, we have used the QMC methods by inputting the above VBSCF and BOVB wave functions as trial wave functions, in so-called VB-QMC methods [44]. Two levels of VB-QMC calculations have been used. At the VMC level, previously optimized VBSCF or BOVB wave functions are supplemented with a Jastrow factor including explicit interparticle correlation (see theoretical section), and wave function parameters are reoptimized. Then, the more accurate DMC level consists of projecting the VMC wave function onto the ground state using the fixed-node DMC algorithm.

Table 8 displays the results obtained at these two levels, where VB-VMC (resp. BO-DMC) corresponds to a wave function that has a VBSCF (resp. BOVB) type of determinantal part. It can be seen that when a simple VBSCF wave function is used as a trial function (VB-VMC, entry 1), the $N \rightarrow V$ transition energy is disappointingly large, 8.59 eV, showing that the Jastrow factor is not able to retrieve the $\sigma\pi$ dynamic polarization that accompanies the fluctuation of the π electrons. On the other hand, using now the BOVB-5 wave function as a trial function, one gets a much better T_e value of 8.04 eV when only the Jastrow parameters and coefficients of the CSFs are optimized (entry 2), and an even better value, 7.96 eV, if a further optimization of the inactive orbitals is performed (entry 3). By contrast, removing structures **4a** and **4b** from the set of VB structures in the trial function, as in the BO-VMC-3 calculation, leads to a much too high $N \rightarrow V$ transition energy, 8.98 eV. This result further confirms that structures **4a** and **4b** are essential components of the *V* state of ethylene, even when a high level of electron correlation is brought by the Jastrow factor.

Fixed-node DMC calculations further improve the BO-VMC-5 results. Using a BOVB-5 trial wave function where only Jastrow and CSFs parameters were reoptimized at the VMC level leads to a DMC transition energy of 7.96 eV, quite close to the experimental value, in excellent agreement with experiment (7.88 eV), showing that the nodal structure of the standard BOVB level is physically correct. This has to be compared with the best DMC value of 8.27 eV (190.8 kcal/mol) obtained by Anderson et al. [32], using a GVB-type trial function. Finally, using the BOVB-5 trial wave function where the σ orbitals were also reoptimized at the VMC level further improves the transition energy to 7.93 eV, identical within the error bars to the DMC value obtained by Schautz and Filippi [31].

3 Conclusions

The V state of ethylene poses a long-standing challenge for theoretical methods, despite the small size of the molecule. The difficulties encountered with standard MO-CI methods are due to (1) the strong charge fluctuation within the π system, which requires a dynamic response of the σ frame already at the orbital optimization step, and (2) a low-lying Rydberg state which may spuriously contaminate the V state. Thus, any meaningful wave function for the V state must satisfy the dual condition of providing an accurate energy relative to the ground state, and a reasonable spatial extension, neither too large nor too small, i.e., being about 50 % larger than that of the ground state but not more.

The strategies of overcoming these difficulties in the MO framework are all based on simple VB reasoning. These methods are generally successful, but lead to long configuration expansions that are treated by multi-reference variational or perturbative methods. This obviously suggests the use of direct VB treatments, which are performed with two different methods, VBSCF that is limited to static electron correlation (like GVB and SC methods), and BOVB that further involves dynamic correlation while keeping the number of VB structures unchanged. At the VBSCF level, the results share the same shortcomings as CASSCF calculations and provide too large $N \rightarrow V$ transition energies and V state wave functions that are too diffuse to qualify as valence state. On the other hand, the BOVB level provides excellent results in terms of both transition energies and spatial extension of the wave functions. The number of configurations amounts to only 14 structures in the (4, 4) VB space, and 4 VB structures in the (2, 2) space. Both VB structure sets lead to practically identical results, showing that 4 VB structures with the BOVB method are already capable of taking care of the σ response to the fluctuations of the π electrons, by means of the breathing orbital effect. Moreover, the contraction of the π AOs under the influence of dynamic electron correlation can be clearly visualized by turning this latter type of correlation on and off, by switching from BOVB to VBSCF.

The BOVB wave functions of the (2, 2) space have also been used as guess functions for diffusion Monte Carlo calculations, yielding results in close agreement with the calculations of Schautz and Filippi [31], even without needing to re-optimize the guess orbitals at the VMC step. This further confirms the adequacy of BOVB wave functions as guesses for DMC calculations, as noted already in previous studies [44, 62].

Another outcome of the present study is the evidence for the importance of VB structures displaying an asymmetric π covalent bond as an essential component of the V state of ethylene. Such VB structures, which have never been considered before in qualitative descriptions of the V state, arise naturally from an expansion of the $\pi\pi^*$ MO configuration into VB structures. These asymmetric π-bonds contribute 18–26 % of the total VB wave function. Moreover, they have quite a significant stabilizing effect, of the order of ca. 0.90 eV, which can be understood as their role in tempering the fluctuations of the π electrons, together with a classical resonance effect due to their mixing with the zwitterionic structures. Lastly, the two asymmetric covalent VB structures are the ones responsible for the spatial extension of the V state being 50 % larger than that of the N state. It must therefore be concluded that the V state of ethylene is not 100 % ionic as usually assumed, but involves a minor though non-negligible covalent component. Incidentally, the disregard of these structures in current qualitative reasoning is due to the usage of a minimal basis set in the decomposition of the $\pi\pi^*$ configuration, in which the asymmetrical covalent structures vanish. Nevertheless, the importance of these structures does not imply that carbon has suddenly five valence orbitals in the V state. The effect of **4a** and **4b** is essentially perturbational, and it dresses the zwitterionic structures with left–right correlation, which tempers the antiresonant interaction (of the **3a**–**3b** combination in the minimal VB set) and the high electron–electron repulsion of the doubly occupied 2p orbitals. It is tempting to assign the diffuse orbitals in **4a** and **4b** as carbon 3p AOs, however, we were not able to verify this character since the orbitals lack the requisite radical node to qualify as 3p.

All in all, the above study shows that performing direct VB calculations in a reasonable basis set may lead to an accurate description of the V state by means of extremely compact wave functions provided the VB treatment involves a built-in dynamic electron correlation. The reason for the success is that VB can achieve this in a direct way, whereas MO-CI methods must do this in a roundabout and sometimes complicated way, owing to orthogonality constraints. By contrast, the choice of VB structures that need be involved in the computations does not require complicated reasoning, but naturally arises from chemical considerations. The additional and sometimes unexpected insight that arises from such calculations is an intrinsic feature of modern ab initio VB methods.

Acknowledgments W. W. is supported by the Natural Science Foundation of China (Nos. 21120102035, 21273176, 21290193). SS thanks the Israel Science Foundation (ISF grant 1183/13). B. B. thanks the IDRIS computational center for an allocation of computer time.

References

1. Cooper DL, Gerratt J, Raimondi M (1991) Chem Rev 91:929
2. Gerratt J, Cooper DL, Karadakov PB, Raimondi M (1998) Spin-coupled theory. Wiley, New York

3. Goddard WA, Dunning TH Jr, Hunt WJ, Hay PJ (1973) Acc Chem Res 6:368
4. Bobrowicz FW, Goddard WA (1977) In: Schaefer III HF (ed) Methods of electronic structure theory, vol 4. Springer, Heidelberg, p 79
5. Hiberty PC, Shaik S (2007) J Comput Chem 28:137
6. Shaik S, Hiberty PC (2008) A chemist's guide to valence bond theory. Wiley, Hoboken
7. Wu W, Su P, Shaik S, Hiberty PC (2011) Chem Rev 111:7557–7593
8. Shaik S, Hiberty PC (2004) Rev Comp Chem 20:1–100
9. Hiberty PC, Humbel S, van Lenthe JH, Byrman C (1994) J Chem Phys 101:5969
10. Hiberty PC, Flament J-P, Noizet E (1992) Chem Phys Lett 189:259
11. Hiberty PC, Shaik S (2002) Theoret Chem Acc 108:255
12. Angeli C (2009) J Comput Chem 30:1319
13. Merer AJ, Mulliken RS (1969) Chem Rev 69:639
14. Wilkinson G, Mulliken RS (1955) J Chem Phys 23:1895
15. Dunning TH Jr, Hunt WJ, Goddard WA (1969) Chem Phys Lett 4:147
16. Brooks BR, Schaefer HF III (1975) Chem Phys 9:75
17. Schaefer HF III (1978) J Chem Phys 68:4839
18. Buenker RJ, Shih S-K, Peyerimhoff SD (1979) Chem Phys 36:97
19. Buenker RJ, Peyerimhoff SD, Shih S-K (1980) Chem Phys Lett 69:7–13
20. Bender CF, Dunning TH Jr, Schaefer HF III, Goddard WA, Hunt WJ (1972) Chem Phys Lett 15:171
21. Buenker RJ, Peyerimhoff SD (1975) Chem Phys 9:75
22. McMurchie LE, Davidson ER (1977) J Chem Phys 66:2959
23. Duben AJ, Goodman L, Pamuk HO, Sinanoglu O (1973) Theoret Chim Acta (Berl) 30:177
24. Sinanoglu O (1969) Adv Chem Phys 14:237
25. Davidson ER (1996) J Phys Chem 100:6161
26. Davidson ER, Jarzecki A (1998) Chem Phys Lett 285:155
27. Lasorne B, Jornet-Somoza J, Meyer H-D, Lauvergnat D, Robb MA, Gatti F (2013) Spect Acta A. http://dx.doi.org/10.1016/j.saa.2013.04.078
28. Finley J, Malmqvist P-Å, Roos BO, Serrano-Andrès L (1998) Chem Phys Lett 288:299
29. Krebs S, Buenker RJ (1997) J Chem Phys 106:7208
30. Müller T, Dallos M, Lischka H (1999) J Chem Phys 110:7176
31. Schautz F, Filippi C (2004) J Chem Phys 120:10931
32. Anderson AG, Goddard WA (2010) J Chem Phys 132:164110
33. Zaitsevskii A, Malrieu J-P (1995) Int J Quantum Chem 55:117–125
34. Garcıa-Cuesta I, Sanchez de Meras AMJ, Koch H (2003) J Chem Phys 118:8216
35. Watts JD, Gwaltney SR, Bartlett RJ (1996) J Chem Phys 105:6979
36. Wu W, Song L, Cao Z, Zhang Q, Shaik S (2002) J Phys Chem A 106:2721
37. Song L, Wu W, Zhang Q, Shaik S (2004) J Comput Chem 25:472
38. Chen Z, Song J, Shaik S, Hiberty PC, Wu W (2009) J Phys Chem A 113:11560
39. Shaik S, Danovich D, Silvi B, Lauvergnat D, Hiberty PC (2005) Chem Eur J 11:6358
40. Hiberty PC, Humbel S, Archirel P (1994) J Phys Chem 98:11697
41. Song L, Wu W, Hiberty PC, Danovich D, Shaik S (2003) Chem Eur J 9:4540
42. Hiberty PC, Megret C, Song L, Wu W, Shaik S (2006) J Am Chem Soc 128:2836
43. Su P, Song L, Wu W, Hiberty PC, Shaik S (2007) J Comput Chem 28:185
44. Braida B, Toulouse J, Caffarel M, Umrigar CJ (2011) J Chem Phys 134:084108
45. Braida B, Hiberty PC (2013) Nat Chem 5:417
46. Toulouse J, Umrigar CJ (2008) J Chem Phys 128:174101
47. Chirgwin HB, Coulson CA (1950) Proc R Soc Lond Ser A 201:196
48. Löwdin PO (1947) Ark Mat Astr Fysik 35A:9
49. Gallup GA, Norbeck JM (1973) Chem Phys Lett 21:495
50. van Lenthe JH, Balint-Kurti GG (1983) J Chem Phys 78:5699
51. Dunning TH Jr (1989) J Chem Phys 90:1007
52. Kendall RA, Dunning TH Jr, Harrison RJ (1992) J Chem Phys 96:6796
53. Petruzielo FR, Toulouse J, Umrigar CJ (2012) J Chem Phys 136:124116
54. Burkatzki M, Filippi C, Dolg M (2007) J Chem Phys 126, 234105. Basis sets and corresponding pseudopotentials are available online: http://burkatzki.com/pseudos/index.2.html
55. Song L, Wu W, Mo Y, Zhang Q (2003) XMVB: an ab initio Non-orthogonal Valence Bond Program, Xiamen University, Xiamen 361005, China
56. Song L, Mo Y, Zhang Q, Wu W (2005) J Comput Chem 26:514
57. Song L, Song J, Mo Y, Wu W (2009) J Comput Chem 30:399
58. Umrigar CJ, Filippi C, Toulouse J. CHAMP, a Quantum Monte Carlo ab initio program, see: http://pages.physics.cornell.edu/~cyrus/champ.html
59. Lindh R, Roos BO (1989) Int J Quantum Chem 35:813
60. Serrano-Andrès L, Merchan M, Nebot-Gil I, Roos BO (1993) J Chem Phys 98:3151
61. Thorsteinsson T, Cooper DL (1998) J Math Chem 23:105–126
62. Domin D, Braida B, Lester WA Jr (2008) J Phys Chem A 112:8964

Theor Chem Acc (2014) 133:1429
DOI 10.1007/s00214-013-1429-6

REGULAR ARTICLE

Comparison of multireference configuration interaction potential energy surfaces for H + O_2 → HO_2: the effect of internal contraction

Lawrence B. Harding · Stephen J. Klippenstein · Hans Lischka · Ron Shepard

Received: 11 September 2013 / Accepted: 20 November 2013 / Published online: 8 December 2013

Abstract A comparison is presented of uncontracted multireference singles and doubles configuration interaction (MRCI) and internally contracted MRCI potential energy surfaces for the reaction $H(^2S) + O_2\left(^3\sum_g^-\right) \rightarrow HO_2(^2A'')$. It is found that internal contraction leads to significant differences in the reaction kinetics relative to the uncontracted calculations.

Keywords MRCI · Internal contraction · H + O_2

1 Introduction

The reaction of hydrogen atom with molecular oxygen

$$H(^2S) + O_2\left(^3\Sigma_g^-\right) \rightarrow HO_2(^2A'') \quad (1)$$

on the electronic ground state potential energy surface (PES) is important in both combustion and atmospheric chemistry. Several experimental [1–4] and theoretical [5–10] studies of the high-pressure limit for this reaction have been reported. All the theoretical studies are based on multireference electronic structure methods. Multireference methods are appropriate here because the reaction results in a change in the number of open-shell orbitals, three for the reactants and just one for the product. Probably the most accurate, global, HO_2 PES currently available was reported in 2005 by Xu et al. [11]. It consists of a spline fit to 15,000 internally contracted multireference configuration interaction (ic-MRCI) energies employing the Dunning, aug-cc-pVQZ basis set and the Davidson correction for higher-order excitations.

The theoretical studies reported in Refs. [5, 10] both employ MRCI calculations with the same, 7-orbital 9-electron, active space and the same orbital basis set (aug-cc-pVTZ). The only difference between the two calculations is that the earlier study is an uncontracted MRCI (uc-MRCI) calculation computed with the COLUMBUS program package [12], while the more recent study is an ic-MRCI calculation computed with the MOLPRO program package [13]. Figure 1 of Ref. [10] shows significant differences between the minimum energy path (MEP) energies from the uncontracted and internally contracted calculations. The purpose of this communication is to explore further the differences between the uc-MRCI and ic-MRCI results for this reaction, and to determine the effect these differences have on the predicted high-pressure-limit reaction rate.

The paper is organized as follows. A brief summary of the uc-MRCI and ic-MRCI methods is presented in Sect. 2. The electronic structure calculations are described in Sect. 3. The transition state theory (TST) calculations of the reaction rates are described in Sect. 4. In Sect. 5, the results of the calculations are presented and discussed. The paper concludes with Sect. 6.

Dedicated to Professor Thom Dunning and published as part of the special collection of articles celebrating his career upon his retirement.

Electronic supplementary material The online version of this article (doi:10.1007/s00214-013-1429-6) contains supplementary material, which is available to authorized users.

L. B. Harding (✉) · S. J. Klippenstein · H. Lischka · R. Shepard
Chemical Sciences and Engineering Division, Argonne National Laboratory, Argonne, IL 60439, USA
e-mail: harding@anl.gov

H. Lischka
Department of Chemistry and Biochemistry, Texas Tech University, Lubbock, TX 79409, USA

2 The uc-MRCI and ic-MRCI formalisms

In a MRCI approach, the wave function is expanded in a basis of reference functions and excitations from those reference functions. In this work, which involves open-shell molecular radical species, the expansion functions are chosen to be spin-adapted configuration state functions (CSF) rather than primitive Slater determinants.

$$\begin{aligned}|\psi^{\rm MRCI}\rangle &= \sum_{n}^{N_{\rm dim}} c_n|n\rangle = \sum_{m}^{N_{\rm ref}} c(m)|m;{\rm ref}\rangle \\ &+ \sum_{m}^{N_{\rm ref}}\sum_{pq} c(m)_q^p\left|m_q^p\right\rangle \\ &+ \sum_{m}^{N_{\rm ref}}\sum_{pqrs} c(m)_{qs}^{pr}\left|m_{qs}^{pr}\right\rangle + \cdots\end{aligned}$$

with

$$\left|m_q^p\right\rangle = E_{pq}|m;{\rm ref}\rangle;\ \left|m_{qs}^{pr}\right\rangle = e_{pqrs}|m;{\rm ref}\rangle;\ldots$$

E_{pq} and e_{pqrs} are the spin-adapted single- and double-excitation operators. The expansion space employed in the present calculations is limited to single and double excitations, which is typical for MRCI applications.

In general, the orbital summations range over all the molecular orbital (MO) basis functions, but when the reference functions are constructed from doubly occupied and active orbital subsets, the allowed combinations of orbital indices in the excitation operators are correspondingly restricted. Furthermore, the above basis functions are typically linearly dependent, which results in further restrictions in the summations. One type of linear dependency occurs when, for example, the molecular orbital φ_s is unoccupied in the reference function $|m;{\rm ref}\rangle$, in which case the basis functions $\left|m_{qs}^{pr}\right\rangle$ should be eliminated from the expansion. A second type of linear dependency occurs when the same expansion function is generated from two different reference functions, $\left|m_{qs}^{pr}\right\rangle = \left|m_{q's'}^{'p'r'}\right\rangle$, in which case only a single instance of the function should be included in the expansion.

An alternative approach to define the expansion space consists of imposing orbital occupation restrictions on subsets of the orbitals to generate the reference functions $|m;{\rm ref}\rangle$ along with a consistent set of orbital occupation restrictions for the single and double excitations. This occupation restriction approach eliminates the complications associated with linear dependence of the basis, but it imposes some restrictions on the general form of the expansion space, and it may also include expansion terms that do not belong to the minimal interacting space [14]. In the COLUMBUS program system, these occupation restrictions are imposed by restricting the allowed nodes of the Shavitt graph in the Graphical Unitary Group Approach (GUGA) (see ref. [12] and references therein).

In the following discussion, the above expansion basis, after the appropriate elimination of linear dependencies, will be called the uc-MRCI space. As discussed in ref. [15], the uc-MRCI expansion dimension $N_{\rm dim}$ increases, approximately linearly, with the reference space dimension $N_{\rm ref}$, resulting in expensive computational procedures and limiting the application of the method generally with respect to molecular system size.

In order to reduce the expansion dimension and the corresponding computational costs of MRCI, the internal contraction approximation was introduced [16, 17]. In this ic-MRCI approximation, the excitation operators are applied to the contracted reference wave function rather than to the individual reference basis functions,

$$|\psi^{ic-{\rm MRCI}}\rangle = \sum_{n}^{N_{\rm dim}} \tilde{c}_n|\tilde{n}\rangle = \tilde{c}_0|\tilde{\psi}_0\rangle + \sum_{pq}\tilde{c}_q^p\left|\tilde{\psi}_q^p\right\rangle + \sum_{pqrs}\tilde{c}_{qs}^{pr}\left|\tilde{\psi}_{qs}^{pr}\right\rangle$$

with

$$|\tilde{\psi}_0\rangle = \sum_{m}^{N_{\rm ref}} c_m^{\rm ref}|m;{\rm ref}\rangle;\ \left|\tilde{\psi}_q^p\right\rangle = E_{pq}|\tilde{\psi}_0\rangle;\ \left|\tilde{\psi}_{qs}^{pr}\right\rangle = e_{pqrs}|\tilde{\psi}_0\rangle.$$

The reference expansion coefficients $c_m^{\rm ref}$ are typically fixed at their optimal MCSCF values and not allowed to vary during the subsequent CI optimization. Thus, the number of variational coefficients $N_{\rm dim}$ in ic-MRCI is comparable to that of a single-reference calculation and is reduced significantly compared to the uc-MRCI dimension. The ic-MRCI expansion spans a subspace of the minimal interacting space [17].

One of the most popular implementations of the ic-MRCI method is the MOLPRO program [13, 18–21]. In this particular implementation, only the double excitations are contracted, the reference functions and the single excitations are left uncontracted. This hybrid approach increases the flexibility of the wave function compared to the fully ic-MRCI expansion while still reducing the computational costs significantly compared to the uc-MRCI expansion.

Both the uc-MRCI and the ic-MRCI energies satisfy the variational bound properties that the computed energies are rigorous upper bounds to the full-CI energies computed with the same MO basis, provided that no further approximations to the Hamiltonian operator or to the computed integrals are imposed. The full-CI energies computed with a given MO basis in turn are upper bounds to the complete-CI energies. The complete-CI energy is the limit of a full-CI energy within a complete MO basis and is the exact solution to the Schrodinger equation within the Born–

Oppenheimer approximation. Because the ic-MRCI expansion spans a subspace of the uc-MRCI expansion, there is also a variational bound relation between these two PESs. Furthermore, these bound relations hold for arbitrary ground and excited electronic states.

$$E_k^{\text{complete-CI}} \leq E_k^{\text{full-CI}} \leq E_k^{\text{uc-MRCI}} \leq E_k^{\text{ic-MRCI}}$$

See ref. [15] for a more complete discussion of the consequences of the ic-MRCI approximation.

Many molecular properties involve energy differences rather than total energies, and in these cases it is possible for a less-flexible method to enjoy a more favorable cancelation of errors. This cancelation of error may be fortuitous, or it may be systematic. In such a situation, the accuracy of the computed properties may not agree with the above ordering of the accuracies of the total energies. The computed reaction rates discussed in the present work depend on such energy differences. These rates depend on the behavior of the PESs, in particular barrier heights and the associated reaction-path partition functions, relative to the asymptotic dissociation limits; these shifted PESs are called *interaction potentials*.

3 Electronic structure calculations

The MRCI calculations employ a 7-orbital 9-electron active space, consisting of all valence orbitals except for the oxygen 2s orbitals that correspond to low-lying lone-pair spectator orbitals for this reaction. The Dunning aug-cc-pVTZ basis set [22] is used for all atoms, resulting in 115 total MOs. Calculations were performed in the standard Jacobi coordinates, R_{OO}, R_{MH} and the HMO angle, where M is the center of mass of the O_2 fragment. The calculations were carried out on a 9 × 31 × 17 grid covering ranges of 1.14–1.38 Å for R_{OO}, 1.3–8.0 Å for R_{MH} and 0°–90° for the HMO angle. The resulting grids of energies were then fit using three-dimensional splines. Both the contracted and the uncontracted calculations were corrected for higher-order excitations with two different approaches. The Davidson correction is an a posteriori correction applied to the MRCI energy, and the averaged quadratic coupled cluster [23] (AQCC) method is an a priori correction in which the CI equation itself is modified. See ref. [15] for a detailed discussion of both approaches. The uc-MRCI and uc-AQCC calculations were computed with the COLUMBUS program system [12], and the ic-MRCI and ic-AQCC calculations were computed with the MOLPRO program [13]. All calculations were performed within the C_s point group. The molecular orbitals were determined with CASSCF calculations, and the four electrons in the O atom 1 s core orbitals were frozen in the subsequent MRCI expansions.

The internal orbital space consists of the active orbitals plus the two spectator oxygen 2 s orbitals. This expansion is denoted CAS+1+2. The uc-MRCI expansions consist of 868 internal CSFs (the 240 reference CSFs of A$''$ symmetry plus excitations within the internal orbital space), 277,727 single excitations, and 27,836,460 double excitations, resulting in $N_{\text{dim}} = 28{,}115{,}055$ total CSFs. In the ic-MRCI calculations, the double excitations are contracted down to 225,117 expansion terms, resulting in an expansion space of dimension $N_{\text{dim}} = 503{,}712$. The ic-MRCI expansion spans a subspace of the minimal interacting space [14, 17], whereas the uc-MRCI expansion includes also terms that interact only indirectly with the references.

Finally, as an additional check on the accuracy of the MRCI calculations, full-CI calculations were carried out along a one-dimensional path. These calculations used the Dunning VDZ basis set [24] (no polarization functions) with the OO distance fixed at R_e and the HOO angle fixed at 115°. These full-CI calculations were performed with MOLPRO [13, 18, 19].

4 Transition state theory calculations

Variable reaction coordinate transition state theory (VRC-TST) [25, 26] was used to predict the H + O_2 high-pressure recombination rate coefficient. The calculations were performed for the four separate analytical potential energy surfaces obtained from fits to the ic-MRCI, ic-MRCI + QC, uc-MRCI and uc-MRCI + QC calculations, where +QC denotes inclusion of the renormalized Davidson correction [15, 27, 28]. The VRC-TST analysis depends on the location of pivot points for the rotational motions of the fragments. For large separations (>4 Å), the O_2 pivot point was placed at its center of mass, while for shorter separations pairs of pivot points displaced along the OO axis were considered. The displacement of these pivot points ranged from 0 to 1 Å. Distances ranging from 1.6 to 8 Å were considered for the H to pivot point separation. Convergence limits of 2 % were employed for the Monte Carlo integrations over the configurational integrals.

5 Results and discussion

MEPs for the title reaction obtained using uc-MRCI and ic-MRCI both with and without the Davidson correction and uc-AQCC and ic-AQCC are all presented in Fig. 1. The results are very similar to those presented in ref. [10]. A key question in this and other radical +O_2 reactions concerns the form of the long-range interaction potential (i.e., for separations ranging from about 2 to 3 Å). The results

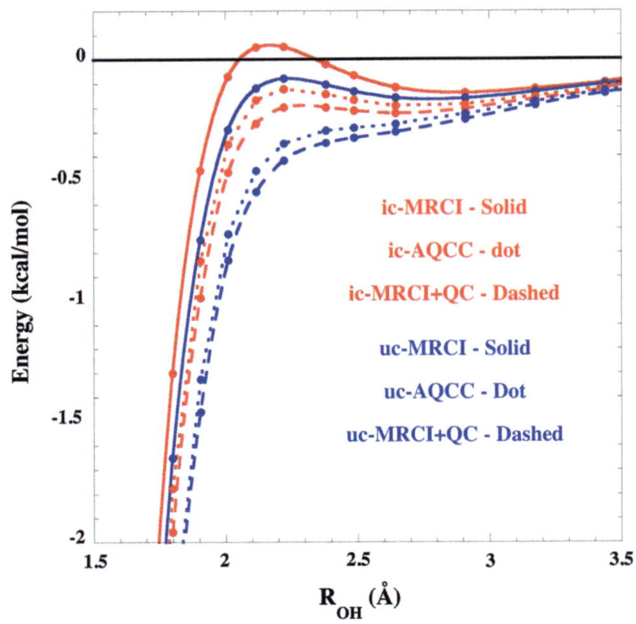

Fig. 1 Minimum energy path energies for the reaction $H + O_2 \rightarrow HO_2$ as a function of the OH bond length. The OO bond length and the HOO angle are optimized at the ic-CAS+1+2/aug-cc-pVTZ level

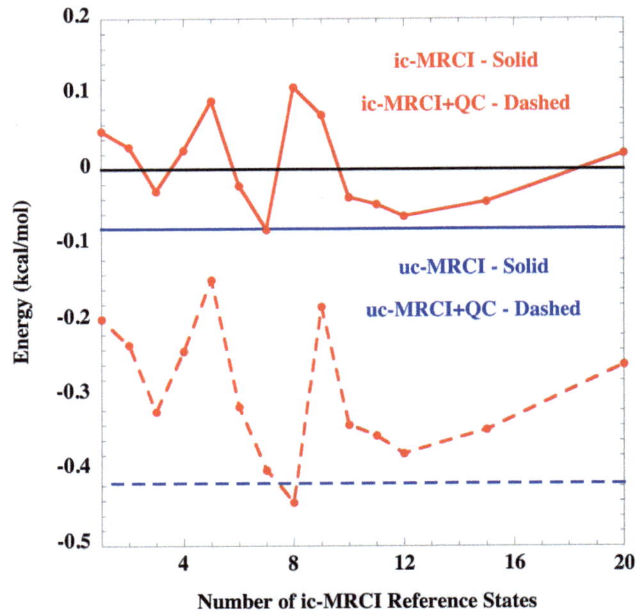

Fig. 2 Comparison of ic-MRCI and uc-MRCI interaction energies as a function of the number of reference states included in the ic-MRCI calculation. All calculations were performed at one intermediate point, $R_{OH} = 4.2$ au (2.22 Å), near the maximum on the ic-MRCI MEP shown in Fig. 1

shown in Fig. 1 illustrate that this long-range potential is very sensitive to the level of theory. The ic-CAS+1+2 calculations exhibit a small, positive barrier. The uc-CAS+1+2, ic-CAS+1+2+QC and ic-AQCC calculations all exhibit submerged barriers. Finally, the uc-CAS+1+2+QC and uc-AQCC calculations exhibit a potential that is monotonically attractive, i.e., with no PES barrier. Interestingly, the difference between the uc-MRCI and ic-MRCI calculations for the overall exothermicity of the reaction is very small, ~0.02 kcal/mol.

To check whether the differences between the ic-MRCI and uc-MRCI energies are primarily due to the contraction or to the interacting space restriction, additional calculations were carried out in which the number of reference states in the ic-MRCI calculations was varied. In principle if the number of references states is increased to equal the number of CSFs in the reference wave function, this should have the effect of fully uncontracting the ic-MRCI calculation within the interacting space. This is not practical for the 9-orbital active space used in the majority of calculations reported here. As a test, a limited number of calculations were performed with a smaller, 5-orbital, 7-electron active space consisting of the O_2 π and π^* orbitals and the hydrogen 1s orbital. For these calculations, the reference wave function consists of 20 CSFs, and when all 20 reference states are included in the ic-MRCI calculation, the resulting interaction energy agrees to within 0.02 kcal/mol with the uc-MRCI calculation. This residual difference is presumably due to the differences in the expansion spaces discussed above. On a cautionary note, we find that the convergence of the ic-MRCI interaction energy as a function of increasing the number of reference states is highly non-monotonic. Although an increment in the number of reference states at a given molecular conformation results in a monotonic decrease in the total energy, the differences of these values can lead to either smaller or larger interaction energies and to non-systematic differences between these ic-MRCI and uc-MRCI energies. This is illustrated in Fig. 2 where the variation in the 9-active orbital ic-MRCI interaction energies (both with and without the Davidson correction) is shown as a function of the number of reference states (1–20) along with the corresponding uc-MRCI energy. Increasing the number of reference states to two or three is seen to result in small improvements in the agreement between the ic-MRCI and uc-MRCI interaction energies. However, other increments can result in significantly worse agreement.

The errors in the MRCI interaction potentials relative to the full-CI are presented in Fig. 3. Without the Davidson correction, the uc-MRCI calculations are in significantly better agreement with the full-CI calculations than are the ic-MRCI. The uc-MRCI calculations are in near perfect agreement with the full-CI for distances greater than 2.5 Å and then become slightly too attractive at shorter distances. The ic-MRCI calculations are not attractive enough in this region. However, inclusion of the Davidson correction changes this picture. With the Davidson correction, the uc-

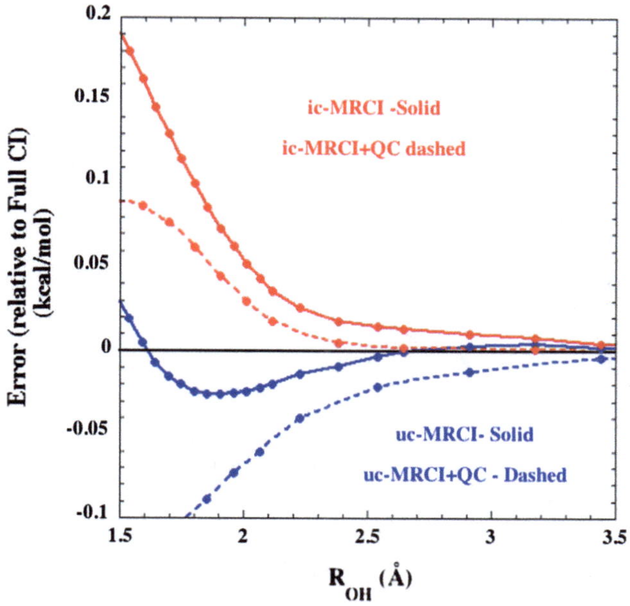

Fig. 3 Errors relative to full-CI in the uc-MRCI/VDZ and ic-MRCI/VDZ energies as a function of the OH bond length. The OO bond length is fixed at R_e, and the HOO bond angle is fixed at 115°

MRCI + QC calculations become too attractive at long range, while the ic-MRCI + QC are in near perfect agreement with full-CI at long range. At shorter distances, the magnitude of the error in the Davidson-corrected uc-MRCI + QC and ic-MRCI + QC calculations is comparable, with the uncontracted calculations being too attractive and the contracted calculation not attractive enough. It should be emphasized that these small-basis VDZ calculations capture only a small percentage of the correlation energy, and it is unclear to what extent these trends would persist with more flexible orbital basis sets. The full-CI total energies are listed in Table 1 of the Supplemental Information.

Contour plots of the uc-MRCI and ic-MRCI PESs for R_{OO} fixed at R_e are given in Fig. 4, and the difference between the two PESs is presented in Fig. 5. The plot shows that the differences between the uc-MRCI and ic-MRCI calculations are largest at H-center-of-mass distances of 1.5–2 Å. The differences also become larger near either the collinear or C_{2v} orientations. In this regard, it may be significant to note that conical intersections exist along both the collinear and C_{2v} approaches.

The results of the TST calculations of the rate are given in Fig. 6 along with experimental results from references [1, 4] and results employed in a recent modeling study of H_2/O_2 flame and ignition data [29]. The latter modeling study showed good agreement across a wide range of observables, including high-pressure flame data [30, 31], for which most other models were deficient. It employed the high-pressure limit obtained in the analysis of Troe [6],

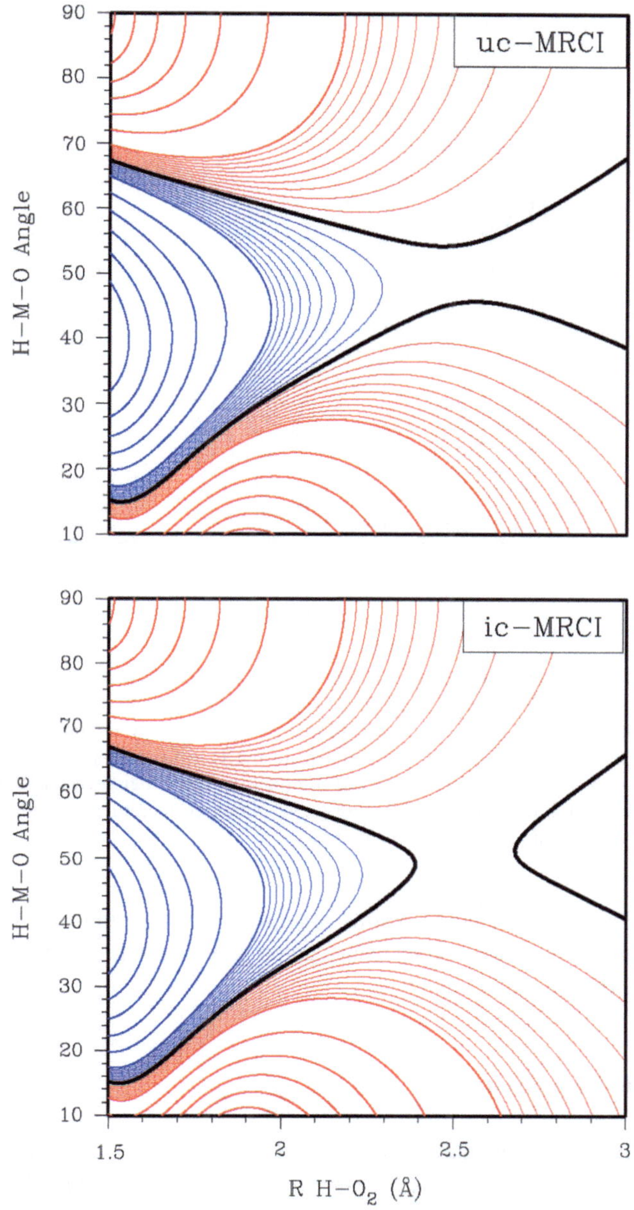

Fig. 4 Comparison of the uc-MRCI and ic-MRCI, CAS+1+2/aug-cc-pVTZ, H + O_2 interaction potentials for R_{OO} fixed at 1.21 Å. Blue contours represent regions lower in energy than the H + O_2 asymptote, while *red* contours represent energies higher than the asymptote, and the *black* contour corresponds to the energy of the asymptote. Contour increments for the heavy (*light*) contours are 5.0 kcal/mol (0.5 kcal/mol)

which in turn was based on the earlier potential energy surface study of Harding et al. [5]. At room temperature, the uc-MRCI-based calculations give rates that are ~30 % larger than the ic-MRCI-based calculations. Inclusion of the Davidson correction increases this difference to ~50 %. At the highest temperatures, the uc-MRCI and ic-MRCI results differ by only 5–10 %. The best agreement between theory and experiment comes from the Davidson-corrected uc-MRCI calculations although both of the

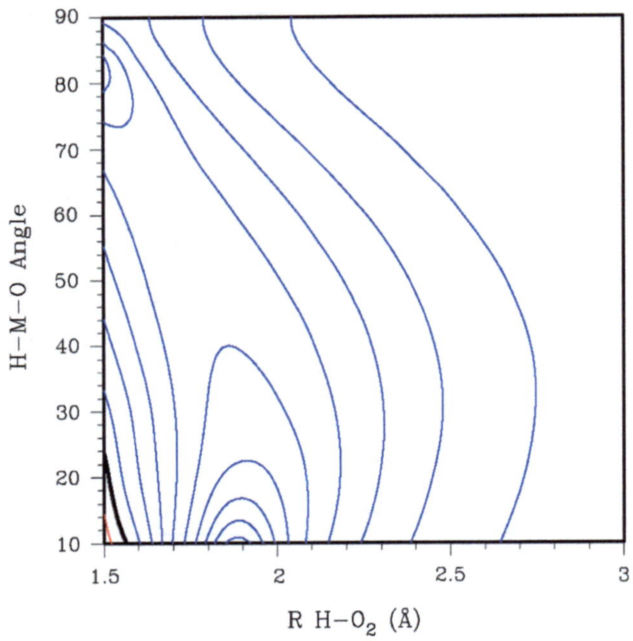

Fig. 5 The difference between the uc-MRCI and ic-MRCI CAS+1+2/aug-cc-pVTZ, H + O_2 interaction potentials for R_{OO} fixed at 1.21 Å. *Blue* contours represent regions where the uc-MRCI potential is more attractive (relative to the H + O_2 asymptote) than the ic-MRCI potential, while *red* contours represent regions where the ic-MRCI potential is more attractive. Contour increments for the heavy contours are 0.1 kcal/mol

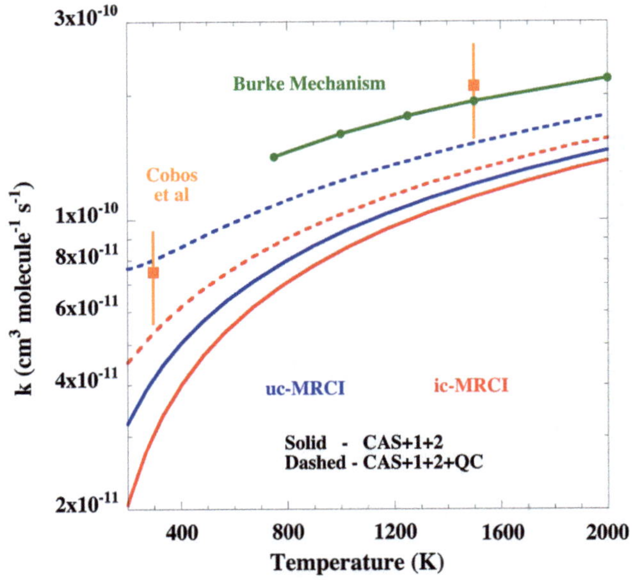

Fig. 6 Results of TST calculations on the high-pressure-limit rate for H + O_2 → HO_2. The results using the uc-MRCI surfaces are shown in *blue* and the ic-MRCI in *red*. *Dashed lines* use PESs that include a Davidson correction. The *orange symbols* are experimental results [1, 4], and the *green line* is a result derived from a recent modeling study [29]

Davidson-corrected results are in reasonably good agreement with the experiment, given the uncertainty in the experimental data.

6 Conclusions

It is demonstrated that ic-MRCI calculations introduce small, but kinetically significant errors in interaction potentials relative to the corresponding uc-MRCI calculations. For the H + O_2 addition reaction, these errors are largest in the vicinity of the dynamical bottleneck. As dynamical bottlenecks often occur in regions where the character of the electronic wave function is changing rapidly (such as near-avoided crossings) and the wave function flexibility is critical, we expect this conclusion to be fairly general.

It is important to note that there is a significant difference in the computational cost of ic-MRCI and uc-MRCI calculations. For the CAS + 1 + 2/aug-cc-pVTZ calculations reported here, an uc-MRCI calculation requires approximately 40–50 times more computer time than the corresponding ic-MRCI calculation. This means that the use of the ic-MRCI approximation enables calculations with larger basis sets and active spaces than is currently feasible with uc-MRCI. When very high accuracy is needed, a hybrid approach might be most efficient in which the contraction error is evaluated with a relatively modest basis set and then applied as a correction to a larger basis set ic-MRCI interaction potential.

Acknowledgments This work was supported by the US Department of Energy, Office of Basic Energy Sciences, Division of Chemical Sciences, Geosciences, and Biosciences, under Contract Numbers DE-AC02-06CH11357. HL was also supported by the National Science Foundation under Project No. CHE-1213263 and by the Robert A.Welch Foundation under Grant No. D-0005.

References

1. Cobos CJ, Hippler H, Troe J (1985) J Phys Chem 89:342
2. Bates RW, Golden DM, Hanson RK, Bowman CT (2001) Phys Chem Chem Phys 3:2337
3. Hahn J, Krasnoperov L, Luther K, Troe J (2004) Phys Chem Chem Phys 6:1997
4. Cobos CJ, Troe J (1985) J Chem Phys 83:1010
5. Harding LB, Troe J, Ushakov VG (2000) Phys Chem Chem Phys 2:631
6. Troe J (2000) Proc Combust Inst 28:1463
7. Marques JMC, Varandas AJC (2001) Phys Chem Chem Phys 3:505
8. Harding LB, Troe J, Ushakov VG (2001) Phys Chem Chem Phys 3:2630
9. Lin SY, Rackham EJ, Guo H (2006) J Phys Chem A 110:1534
10. Sellevag SR, Georgievskii Y, Miller JA (2008) J Phys Chem 112:5085
11. Xu C, Xie D, Zhang DG, Lin SY, Guo H (2005) J Chem Phys 122:244035

12. Lischka H, Müller T, Szalay PG, Shavitt I, Pitzer RM, Shepard R (2011) WIREs Comput Mol Sci 1:191
13. MOLPRO, version 2010.1, a package of ab initio programs, Werner H-J, Knowles PJ, Manby FR, Schütz M, Celani P, Knizia G, Korona T, Lindh R, Mitrushenkov A, Rauhut G, Adler TB, Amos RD, Bernhardsson A, Berning A, Cooper DL, Deegan MJO, Dobbyn AJ, Eckert F, Goll E, Hampel C, Hesselmann A, Hetzer G, Hrenar T, Jansen G, Köppl C, Liu Y, Lloyd AW, Mata RA, May AJ, McNicholas SJ, Meyer W, Mura ME, Nicklass A, Palmieri P, Pflüger K, Pitzer RM, Reiher M, Shiozaka T, Stoll H, Stone AJ, Tarroni R, Thorsteinsson T, Wang M, Wolf A, see http://www.molpro.net
14. McLean AD, Liu B (1973) J Chem Phys 58:1066
15. Szalay PG, Mueller T, Gidofalvi G, Lischka H, Shepard R (2012) Chem Rev 112:108
16. Meyer, W (1977) In: Methods of electronic structure theory. Schaefer III HF (ed), Plenum: New York, pp 413–446
17. Werner H-J, Reinsch E-A (1982) J. Chem Phys 76:3144
18. Knowles PJ, Handy NC (1984) Chem Phys Lett 111:315
19. Knowles PJ, Handy NC (1989) Comp Phys Commun 54:75
20. Werner H-J, Knowles P (1988) J Chem Phys 89:5803
21. Knowles P, Werner H-J (1988) Chem Phys Lett 145:514
22. Kendall RA, Dunning TH Jr, Harrison RJ (1992) J Chem Phys 96:6796
23. Szalay PG, Bartlett RJ (1993) Chem Phys Lett 143:413
24. Dunning TH Jr (1989) J Chem Phys 90:1007
25. Klippenstein SJ (1992) J Chem Phys 96:367
26. Georgievskii Y, Klippenstein SJ (2003) J Phys Chem A 107:9776
27. Langhoff SR, Davidson ER (1974) Int J Quantum Chem 8:61
28. Bruckner KA (1955) Phys Rev 100:36
29. Burke MP, Chaos M, Ju Y, Dryer FL, Klippenstein SJ (2012) Int J Chem Kinet 44:444
30. Burke MP, Chaos M, Dryer FL, Ju Y (2010) Combust Flame 157:618
31. Burke MP, Dryer FL, Ju Y (2010) Proc Combust Inst 33:905

Theor Chem Acc (2014) 133:1421
DOI 10.1007/s00214-013-1421-1

REGULAR ARTICLE

Modern valence-bond description of aromatic annulene ions

Peter B. Karadakov · David L. Cooper

Received: 1 August 2013 / Accepted: 7 November 2013 / Published online: 26 November 2013

Abstract Spin-coupled theory for 'N electrons in M orbitals' active spaces [SC(N,M)], an ab initio valence-bond (VB) approach which uses a compact and easy-to-interpret wave function comparable in quality to a 'N in M' complete-active-space self-consistent field [CASSCF(N,M)] construction, is used to obtain modern VB descriptions of the π-electron systems of the most important annulene rings with $4n + 2$ π electrons: the cyclopropenium ion, the cyclobutadiene dication and dianion, the cyclopentadienide anion, benzene, the cycloheptatrienyl cation, and the cyclooctatetraene dication and dianion in their highest-symmetry nuclear conformations. The SC wave functions for the cyclopropenium ion, cyclopentadienide anion, cycloheptatrienyl cation, cyclooctatetraene dication and dianion are shown to closely resemble the well-known SC model of the classical example of an aromatic system, benzene. The SC orbitals for the cyclobutadiene dication and dianion are more delocalized and demonstrate the ways in which SC wave functions adjust to electron-deficient and electron-rich environments. The high levels of resonance observed in all annulene ions with $4n + 2$ π electrons clearly demonstrate their aromaticity.

Keywords Spin-coupled theory · 'N in M' active space · Valence-bond theory · Aromaticity · Cyclopropenium ion · Cyclobutadiene dication/dianion · Cyclopentadienide anion · Benzene · Cycloheptatrienyl cation · Cyclooctatetraene dication/dianion

1 Introduction

Annulenes and annulene ions with $4n + 2$ π electrons are classical examples of aromatic systems which follow Hückel's $4n + 2$ rule. Using the familiar Frost–Musulin diagrams [1], based on Hückel molecular orbital theory, it is easy to demonstrate that all π electrons in the cyclopropenium ion, $C_3H_3^+$, the cyclobutadiene dication, $C_4H_4^{2+}$, the cyclopentadienide anion, $C_5H_5^-$, benzene, C_6H_6, the cycloheptatrienyl cation (tropylium ion), $C_7H_7^+$, and the cyclooctatetraene dication, $C_8H_8^{2+}$, are placed in doubly occupied bonding MOs, which makes these systems particularly stable. Similarly stable π electron distributions are observed in the dianions of cyclobutadiene and cyclooctatetraene, $C_4H_4^{2-}$ and $C_8H_8^{2-}$, in each of which four additional π electrons (relative to the dications) occupy two doubly degenerate non-bonding orbitals. Another way of explaining the aromaticity of annulenes and annulene ions with $4n + 2$ π electrons is to use the classical valence-bond (VB) concept of resonance: It is possible to draw, for each of these systems, several equivalent stable resonance structures, such as those for the cyclopropenium ion and the cyclopentadienide anion shown below.

Dedicated to Professor Thom Dunning and published as part of the special collection of articles celebrating his career upon his retirement.

P. B. Karadakov (✉)
Department of Chemistry, University of York, Heslington, York YO10 5DD, UK
e-mail: peter.karadakov@york.ac.uk

D. L. Cooper
Department of Chemistry, University of Liverpool, Liverpool L69 7ZD, UK
e-mail: dlc@liverpool.ac.uk

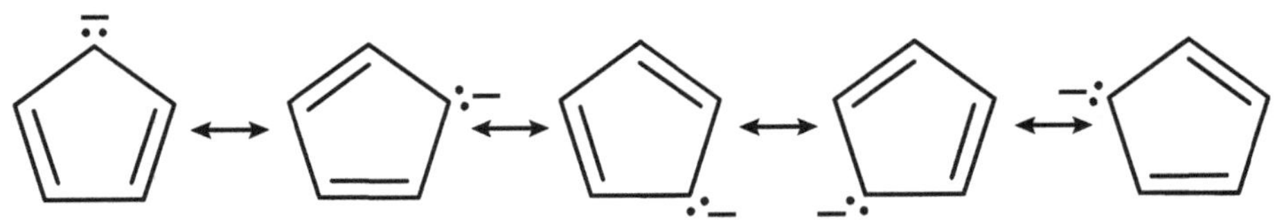

Unfortunately, the only ab initio MO approach which can be used to justify the Frost–Musulin diagrams is the closed-shell Hartree–Fock (HF) method: A HF calculation on a C_mH_m conjugated cycle of D_{mh} symmetry containing an even number of electrons produces, amongst others, m orbitals dominated by carbon $2p_{\pi}$ basis functions; the energies of these orbitals are in a one-to-one correspondence with the orbital energies from a Frost–Musulin diagram with m vertices. The situation with ab initio implementations of classical VB theory is less satisfying: As a rule, the results retain very little of the intuitive clarity of the concept of resonance, as the wavefunctions need to include very large numbers of structures, most of which have next to no chemical relevance.

Spin-coupled (SC) theory, a modern VB approach, uses a post-HF wave function which usually provides a close approximation to a complete-active-space self-consistent field (CASSCF) wave function, but can be interpreted in terms of a small number of VB resonance structures (for a SC review, see Ref. [2]). As a rule, the most important of these VB resonance structures are observed to be the ones which would be suggested by chemical intuition. A good example of SC theory in action is furnished by the SC description of the π-electron system of benzene [3–5]. The six π electrons populate a single product of six non-orthogonal active (or, spin-coupled) orbitals, the spins of which are coupled in all five possible ways leading to an overall singlet. The optimal SC orbitals turn out to be well localized, similar in shape to $C(2p_{\pi})$ atomic orbitals with small symmetrical protrusions toward neighboring carbons, while the optimal spin-coupling pattern, if expressed in terms of Rumer spin functions [6], reveals the presence of two equivalent dominant Kekulé-like structures and three much less important equivalent Dewar-like (or para-bonded) structures. This SC picture reproduces the essential features of the well-known classical VB description of benzene in terms of resonance structures, and yet it comes from a wave function which accounts for close to 90 % of the 'non-dynamic' correlation energy incorporated in a '6 electrons in 6 orbitals' π-space CASSCF [CASSCF(6,6)] wave function [5].

In its original form [8, 9], the SC(N) wave function is equivalent to the full generalized valence-bond (full-GVB) wave function [7] and incorporates a 'N in N' active space, described by means of a single product of N non-orthogonal active orbitals, multiplied by a general N-electron spin function. Clearly, an ansatz of this type cannot accommodate the 'N in M' active spaces with different numbers of active electrons and active orbitals typically employed in CASSCF wave functions. This problem was recently addressed through the development of an extension of SC theory to 'N in M' ($N \neq M$) active spaces [10]. The SC(N,M) wave function retains the essential features of the original SC model: It involves just the products of non-orthogonal orbitals corresponding to all distributions of N electrons between M orbitals in which the smallest number of orbitals possible, $|N - M|$, are doubly occupied (for $N > M$) or omitted (for $N < M$), and all other orbitals are singly occupied; each of these products is combined with a flexible spin function which allows any mode of coupling of the spins of the orbitals within the product. Of course, SC(N,N) signifies precisely the same construction as the original SC(N) wave function.

In the present work, we use appropriate SC(N,M) wave functions to obtain and compare modern VB descriptions of the π-electron systems of the most important annulene rings with $4n + 2$ π electrons: the cyclopropenium ion, the cyclobutadiene dication and dianion, the cyclopentadienide anion, benzene, the cycloheptatrienyl cation, and the cyclooctatetraene dication and dianion. We show that the SC method can produce, at a near-CASSCF level of theory, a convincing general model for the electronic structure of these frequently discussed aromatic systems which justifies the continued use of the popular VB resonance structures.

2 Results and discussion

The highest-symmetry geometries of the cyclopropenium ion, the cyclobutadiene dication and dianion, the cyclopentadienide anion, benzene, the cycloheptatrienyl cation, and the cyclooctatetraene dication and dianion were optimized using π-space CASSCF wave functions within the 6-311G(d,p) basis. These calculations were carried out by means of GAUSSIAN09 [11] under the 'VeryTight'

Table 1 Carbon–carbon and carbon–hydrogen bond lengths (in Å) in the cyclopropenium ion, the cyclobutadiene dication and dianion, the cyclopentadienide anion, benzene, the cycloheptatrienyl cation, and the cyclooctatetraene dication and dianion optimized using π-space CASSCF wave functions within the 6-311G(d,p) basis

System	Point group	Wave function	R_{CC}	R_{CH}
Cyclopropenium ion ($C_3H_3^+$)	D_{3h}	CASSCF(2,3)[a]	1.3540	1.0718
Cyclobutadiene dication ($C_4H_4^{2+}$)	D_{4h}	CASSCF(2,4)[b]	1.4353	1.0838
Cyclobutadiene dianion ($C_4H_4^{2-}$)	D_{4h}	CASSCF(6,4)[c]	1.4639	1.0918
Cyclopentadienide anion ($C_5H_5^-$)	D_{5h}	CASSCF(6,5)[a]	1.4098	1.0794
Benzene (C_6H_6)	D_{6h}	CASSCF(6,6)[a]	1.3952	1.0756
Cycloheptatrienyl cation ($C_7H_7^+$)	D_{7h}	CASSCF(6,7)[a]	1.3945	1.0755
Cyclooctatetraene dication ($C_8H_8^{2+}$)	D_{8h}	CASSCF(6,8)[a]	1.4059	1.0774
Cyclooctatetraene dianion ($C_8H_8^{2-}$)	D_{8h}	CASSCF(10,8)[a]	1.4124	1.0916

[a] Optimized geometry corresponds to a local minimum

[b] Optimized geometry corresponds to a first-order saddle point

[c] Optimized geometry corresponds to a fourth-order saddle point

convergence criteria. The optimized geometries were identified as local minima or saddle points through diagonalizations of the corresponding analytic nuclear Hessians. The optimized carbon–carbon and carbon–hydrogen bond lengths are shown in Table 1. Despite the fact that, for isolated ions, the D_{4h} geometries of the cyclobutadiene dication and dianion correspond to saddle points on the respective potential energy surfaces, both geometries can be stabilized as, for example, in the tetraphenylcyclobutadiene dication reported by Olah and Staral [12] and the tetrakis(trimethylsilyl)cyclobutadiene dianion synthesized by Sekiguchi et al. [13]. While it is possible to obtain local minimum geometries of lower energy and lower symmetry for isolated $C_4H_4^{2+}$ and $C_4H_4^{2-}$, these geometries are most likely to be non-aromatic.

The CASSCF geometry optimizations were followed by fully variational SC(2,3), SC(2,4), SC(6,4), SC(6,5), SC(6,6), SC(6,7), SC(6,8) and SC(10,8) calculations on $C_3H_3^+$, $C_4H_4^{2+}$, $C_4H_4^{2-}$, $C_5H_5^-$, C_6H_6, $C_7H_7^+$, $C_8H_8^{2+}$ and $C_8H_8^{2-}$, respectively. The fully variational SC calculations were carried out at the respective CASSCF/6-311G(d,p) optimized geometries, again within the 6-311G(d,p) basis, by means of the CASVB algorithms [14–17] implemented in MOLPRO [18]. The total energies of the HF, SC and CASSCF wave functions for all cyclic systems studied in this paper and the percentages of CASSCF correlation energy accounted for by the SC wave functions, together with the numbers of configuration state functions (CSFs) included in the CASSCF and SC wave functions, are shown in Table 2. The identical energies of the SC and CASSCF wave functions for the three smallest $4n + 2$ cyclic systems, the cyclopropenium ion and the cyclobutadiene dication and dianion, suggest that these wave functions may coincide. The percentages of CASSCF correlation energy recovered by the SC wave functions for the larger cyclopentadienide anion, cycloheptatrienyl cation and cyclooctatetraene dication and dianion remain very high, well above the 89.6 % obtained for benzene, which indicates that the SC(*N*,*M*) wave functions for these cyclic ions can be viewed as very close approximations to their CASSCF(*N*,*M*) counterparts.

The percentages of CASSCF correlation energy accounted for by the SC wave functions for $C_5H_5^-$, C_6H_6 and $C_7H_7^+$, obtained in the current calculations within the 6-311G(d,p) basis, are very much the same as the corresponding cc-pVDZ and cc-pVTZ numbers reported in Ref. [10]. This is an indication that the relatively small gaps between the energies of the π-space SC and CASSCF wave functions for these systems arise as a consequence of wave function ansatz differences that are largely basis-set-independent.

Each of the SC(2,3), SC(2,4), SC(6,4), SC(6,5), SC(6,6), SC(6,7), SC(6,8) and SC(10,8) calculations on $C_3H_3^+$, $C_4H_4^{2+}$, $C_4H_4^{2-}$, $C_5H_5^-$, C_6H_6, $C_7H_7^+$, $C_8H_8^{2+}$ and $C_8H_8^{2-}$, respectively, produced a set of symmetry-equivalent SC orbitals (see Figs. 1, 2). The SC orbitals for $C_5H_5^-$, C_6H_6 and $C_7H_7^+$ are visually indistinguishable from those calculated within the cc-pVDZ basis and shown in Ref. [10]. These orbitals, as well as the SC orbitals for $C_3H_3^+$, $C_8H_8^{2+}$ and $C_8H_8^{2-}$, resemble distorted $C(2p_\pi)$ atomic orbitals. The SC orbitals for the cyclobutadiene dication and dianion are more delocalized, which is a consequence of the fact that these are the orbitals with the lowest and highest formal occupation numbers within the series, 0.5 and 1.5, respectively.

The overlap integrals between SC orbitals associated with neighboring carbons are listed in Table 3 (for all cyclic systems studied in this paper, we assume that the SC orbitals are numbered clockwise around the ring). The only more significant deviation from the benzene $\langle\psi_\mu|\psi_{\mu+1}\rangle$ overlap integral (ψ_μ and $\psi_{\mu+1}$ are SC orbitals associated with neighboring carbon atoms) is observed for the cyclobutadiene dianion in which each SC orbital has, in addition to the part resembling a distorted $C(2p_\pi)$ atomic orbital, a diagonally opposite component that looks like an inverted-phase reduced-size $C(2p_\pi)$ atomic orbital (see orbital (c) in Figs. 1, 2). The particular shape of the $C_4H_4^{2-}$ SC orbitals is also responsible for the increased magnitude of the 'diagonal' overlap integrals $\langle\psi_\mu|\psi_{\mu+2}\rangle$ (see Table 3).

Table 2 Total HF, SC and CASSCF energies (in a.u.), percentages of the recovered CASSCF correlation energy (in brackets) and numbers of CSFs included in the SC and CASSCF wave functions

System	Wave function	CSFs	Total energy
Cyclopropenium ion ($C_3H_3^+$)	HF	1	−115.031554 (0.0 %)
	SC(2,3)	3	−115.055323 (100.0 %)
	CASSCF(2,3)	6	−115.055323 (100.0 %)
Cyclobutadiene dication ($C_4H_4^{2+}$)	HF	1	−152.936990 (0.0 %)
	SC(2,4)	6	−152.968532 (100.0 %)
	CASSCF(2,4)	10	−152.968532 (100.0 %)
Cyclobutadiene dianion ($C_4H_4^{2-}$)	HF	1	−153.320790 (0.0 %)
	SC(6,4)	6	−153.345554 (100.0 %)
	CASSCF(6,4)	10	−153.345554 (100.0 %)
Cyclopentadienide anion ($C_5H_5^-$)	HF	1	−192.234802 (0.0 %)
	SC(6,5)	10	−192.278712 (97.6 %)
	CASSCF(6,5)	50	−192.279813 (100.0 %)
Benzene (C_6H_6)	HF	1	−230.753520 (0.0 %)
	SC(6,6)	5	−230.818566 (89.6 %)
	CASSCF(6,6)	175	−230.826126 (100.0 %)
Cycloheptatrienyl cation ($C_7H_7^+$)	HF	1	−268.951162 (0.0 %)
	SC(6,7)	35	−269.032346 (95.4 %)
	CASSCF(6,7)	490	−269.036264 (100.0 %)
Cyclooctatetraene dication ($C_8H_8^{2+}$)	HF	1	−306.898421 (0.0 %)
	SC(6,8)	140	−306.999376 (96.9 %)
	CASSCF(6,8)	1,176	−307.002563 (100.0 %)
Cyclooctatetraene dianion ($C_8H_8^{2-}$)	HF	1	−307.371910 (0.0 %)
	SC(10,8)	140	−307.444510 (97.5 %)
	CASSCF(10,8)	1,176	−307.446396 (100.0 %)

The unnormalized SC(2,3) wave function for the cyclopropenium ion involves just three orbital products and the singlet two-electron spin eigenfunction

$$\Psi_{00}(2,3) = \hat{\mathcal{A}}\left[(18)\left(\psi_1\psi_2 + \psi_2\psi_3 + \psi_3\psi_1\right)(\alpha\beta - \beta\alpha)\right] \quad (1)$$

In this equation, '(18)' denotes the 18 core electrons, accommodated in 9 doubly occupied optimized orbitals. Wave function (1) represents the simplest mathematical description of the classical VB resonance in the cyclopropenium ion (see the $C_3H_3^+$ resonance structures in the Introduction and Fig. 3): Each of the products of singlet-coupled overlapping SC orbitals $\psi_1\psi_2$, $\psi_2\psi_3$ and $\psi_3\psi_1$ models the π component of the double bond in one of the three symmetry-equivalent resonance structures (denoted as R in Fig. 3). Due to the fact that the energies of the SC(2,3) and CAS(2,3) wave functions for $C_3H_3^+$ are identical (see Table 2), Eq. (1) can be viewed as an easy-to-interpret alternative form of the CAS(2,3) wave function.

The SC(2,4) and SC(6,4) wavefunctions for the cyclobutadiene dication and dianion can be written down as

$$\Psi_{00}(2,4) = \hat{\mathcal{A}}\left\{(24)\left[C_{a;0}\left(\psi_1\psi_2 + \psi_2\psi_3 + \psi_3\psi_4 + \psi_4\psi_1\right) + C_{b;0}\left(\psi_1\psi_3 + \psi_2\psi_4\right)\right](\alpha\beta - \beta\alpha)\right\} \quad (2)$$

$$\Psi_{00}(6,4) = \hat{\mathcal{A}}\left\{(24)\left[C_{a;0}\left(\psi_3^2\psi_4^2\psi_1\psi_2 + \psi_1^2\psi_4^2\psi_2\psi_3 + \psi_1^2\psi_2^2\psi_3\psi_4 + \psi_2^2\psi_3^2\psi_4\psi_1\right) + C_{b;0}\left(\psi_2^2\psi_4^2\psi_1\psi_3 + \psi_1^2\psi_3^2\psi_2\psi_4\right)\right]\alpha\beta\alpha\beta(\alpha\beta - \beta\alpha)\right\} \quad (3)$$

Each of these wave functions incorporates 24 core electrons, placed in 12 doubly occupied optimized orbitals. The SC orbital products in $\Psi_{00}(2,4)$ and $\Psi_{00}(6,4)$ fall into two groups: In the first group, the two singly occupied orbitals are on neighboring carbons, which gives rise to resonance structures with a single carbon–carbon π bond R_a (see Fig. 3), while in the second group the two singly occupied orbitals are along one of the diagonals of the square formed by the carbons, resulting in diagonally bonded resonance structures R_b. The VB structures for the dianions which are not shown in Fig. 3 can be obtained trivially from those for the dications by replacing the plus signs with minus signs. The relative importance of the R_a and R_b structures can be examined, as is usual in VB theory, by calculating their Chirgwin–Coulson weights [20]. As shown in Fig. 3, the combined Chirgwin–Coulson weight of the R_a structures is larger than that of the R_b structures, but this is actually a consequence of the fact that there are four symmetry-

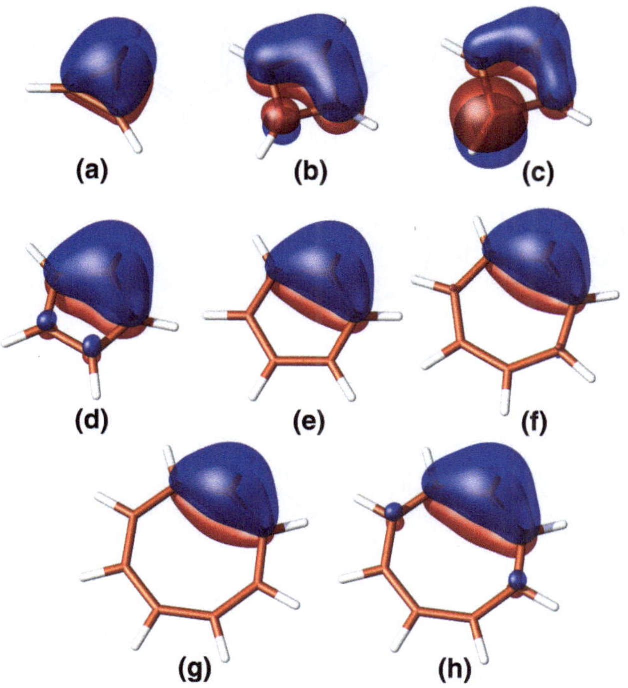

Fig. 1 Active orbitals from the SC wave functions for the cyclopropenium ion (**a**), cyclobutadiene dication (**b**), cyclobutadiene dianion (**c**), cyclopentadienide anion (**d**), benzene (**e**), cycloheptatrienyl cation (**f**), cyclooctatetraene dication (**g**) and cyclooctatetraene dianion (**h**) as isovalue surfaces at $\psi_\mu = \pm 0.05$. All POV-Ray (Persistence of Vision Raytracer) files for the isovalue surfaces were generated by MOLDEN [19]. For further details, see text

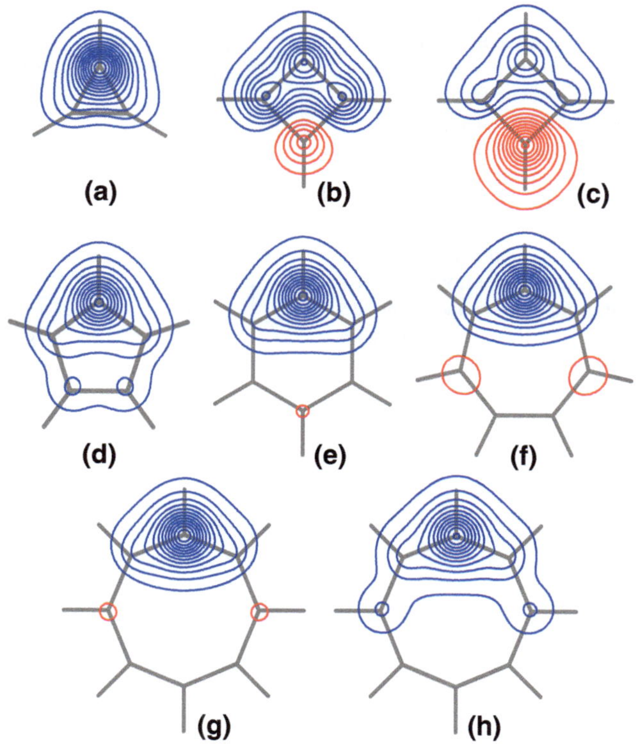

Fig. 2 Active orbitals from the SC wave functions for the cyclopropenium ion (**a**), cyclobutadiene dication (**b**), cyclobutadiene dianion (**c**), cyclopentadienide anion (**d**), benzene (**e**), cycloheptatrienyl cation (**f**), cyclooctatetraene dication (**g**) and cyclooctatetraene dianion (**h**) as contour plots (step 0.025) in a plane 1 Bohr above the respective molecular plane. All contour plots were generated by MOLDEN [19]. For further details, see text

equivalent R_a structures and two symmetry-equivalent R_b structures; if we look at individual structures, the R_b structures, in which the two charges are kept further apart, have larger weights.

The spins of the singly occupied orbitals within the SC wave functions in Eqs. (1), (2) and (3) are coupled by a singlet two-electron spin function which has the same form in any spin basis. The SC wave functions for larger annulene ions incorporate spin functions for more than two electrons and, to make it easier to establish connections between the SC results and classical VB theory, it would be appropriate to make use of the Rumer spin basis [6]. Another option, of which we shall be making use here, is to assemble linear combinations of suitably reordered orbital products with the perfect-pairing spin function which is present in most spin bases, for example, Rumer [6], Kotani [21] and Serber [22, 23].

The SC(6,5), SC(6,6) and SC(6,7) wave functions for $C_5H_5^-$, C_6H_6 and $C_7H_7^+$, respectively, obtained in the current work using the 6-311G(d,p) basis, are very similar to the corresponding wave functions calculated within the cc-pVDZ and cc-pVTZ bases in Ref. [10]. To emphasize the fact that each of the SC(6,5) wave function for the cyclopentadienide anion and the SC(6,6) wave function for benzene contains just two symmetry-unique structures (see Fig. 3), these wave functions can be expressed as (in unnormalized form)

Table 3 Overlaps between neighboring orbitals in the SC wave functions for the cyclopropenium ion, the cyclobutadiene dication and dianion, the cyclopentadienide anion, benzene, the cycloheptatrienyl cation, and the cyclooctatetraene dication and dianion

System	Wave function	$<\psi_\mu\|\psi_{\mu+1}>$
Cyclopropenium ion ($C_3H_3^+$)	SC(2,3)	0.5330
Cyclobutadiene dication ($C_4H_4^{2+}$)	SC(2,4)	0.4661[a]
Cyclobutadiene dianion ($C_4H_4^{2-}$)	SC(6,4)	0.1188[b]
Cyclopentadienide anion ($C_5H_5^-$)	SC(6,5)	0.6812
Benzene (C_6H_6)	SC(6,6)	0.5252
Cycloheptatrienyl cation ($C_7H_7^+$)	SC(6,7)	0.4358
Cyclooctatetraene dication ($C_8H_8^{2+}$)	SC(6,8)	0.4483
Cyclooctatetraene dianion ($C_8H_8^{2-}$)	SC(10,8)	0.6722

[a] 'Diagonal' overlaps $<\psi_\mu|\psi_{\mu+2}> = 0.1871$

[b] 'Diagonal' overlaps $<\psi_\mu|\psi_{\mu+2}> = -0.7261$

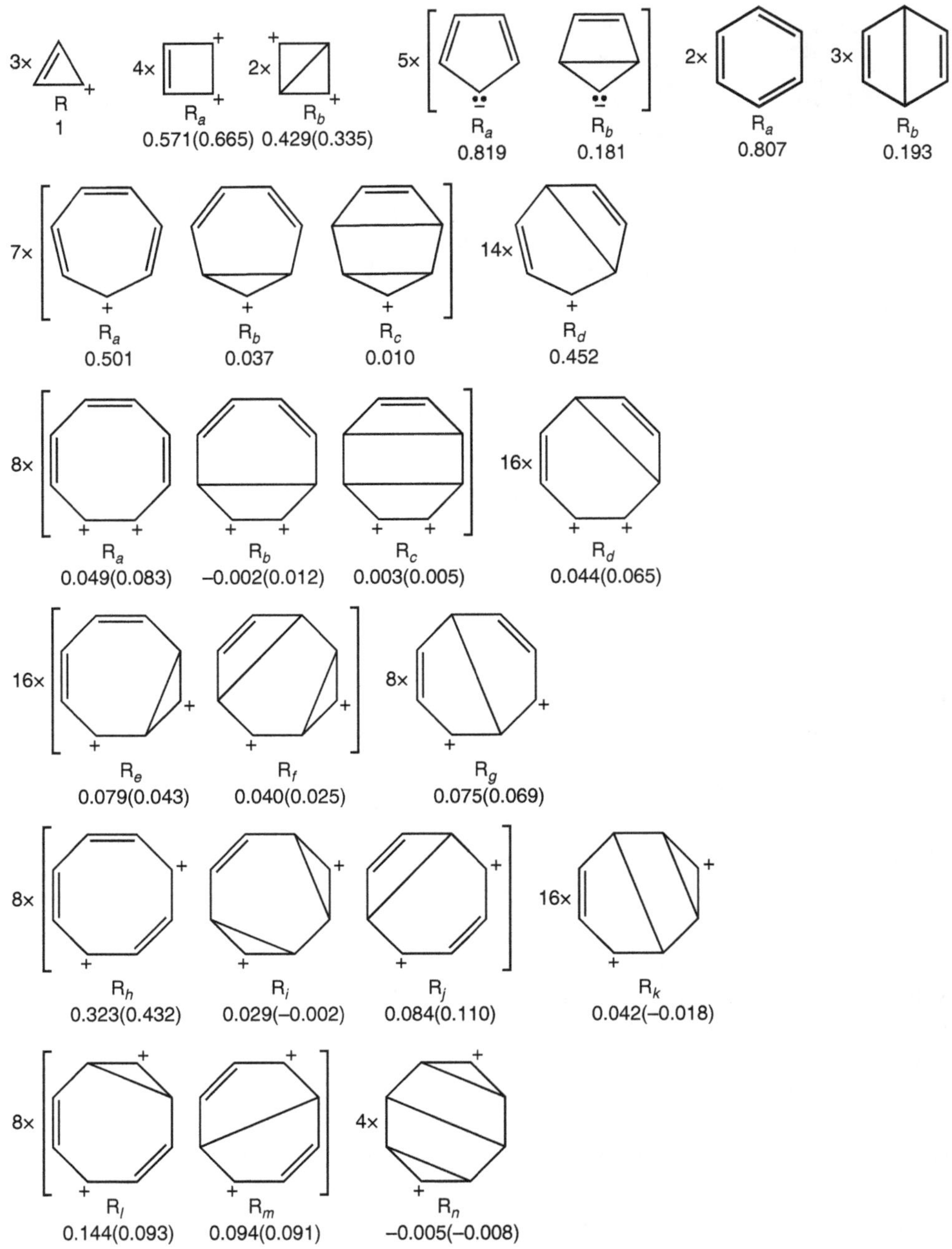

Fig. 3 VB structures included in the SC wave functions for the cyclopropenium ion, cyclobutadiene dication, cyclobutadiene dianion, cyclopentadienide anion, benzene, cycloheptatrienyl cation, cyclooctatetraene dication and cyclooctatetraene dianion, and combined Chirgwin–Coulson weights of the groups of symmetry-unique structures. Results for anions in *brackets*

$$\begin{aligned}\Psi_{00}(6,5) = \hat{\mathcal{A}}\{(30)\big[&C_{a;0}(\psi_1^2\psi_2\psi_3\psi_4\psi_5 + \psi_2^2\psi_3\psi_4\psi_5\psi_1 \\ &+ \psi_3^2\psi_4\psi_5\psi_1\psi_2 + \psi_4^2\psi_5\psi_1\psi_2\psi_3 + \psi_5^2\psi_1\psi_2\psi_3\psi_4) \\ &+ C_{b;0}(\psi_1^2\psi_2\psi_5\psi_3\psi_4 + \psi_2^2\psi_3\psi_1\psi_4\psi_5 \\ &+ \psi_3^2\psi_4\psi_2\psi_5\psi_1 + \psi_4^2\psi_5\psi_3\psi_1\psi_2 \\ &+ \psi_5^2\psi_1\psi_4\psi_2\psi_3)\big]\alpha\beta(\alpha\beta-\beta\alpha)(\alpha\beta-\beta\alpha)\}\end{aligned} \quad (4)$$

$$\begin{aligned}\Psi_{00}(6) = \hat{\mathcal{A}}\{(36)\big[&C_{a;0}(\psi_1\psi_2\psi_3\psi_4\psi_5\psi_6 + \psi_2\psi_3\psi_4\psi_5\psi_6\psi_1) \\ &+ C_{b;0}(\psi_1\psi_4\psi_2\psi_3\psi_5\psi_6 + \psi_2\psi_5\psi_3\psi_4\psi_6\psi_1 \\ &+ \psi_3\psi_6\psi_4\psi_5\psi_1\psi_2)\big](\alpha\beta-\beta\alpha)(\alpha\beta-\beta\alpha)(\alpha\beta-\beta\alpha)\}\end{aligned} \quad (5)$$

If we were to write down the SC(6,7) wave function for $C_7H_7^+$ in a form similar to the SC(6,5) and SC(6,6) wave functions for $C_5H_5^-$ and C_6H_6 respectively, the expression would include four groups of symmetry-equivalent resonance structures: seven R_a, seven R_b, seven R_c and fourteen R_d resonance structures (see the second row of structures in Fig. 3). In contrast to the SC wave functions for the cyclopentadienide anion and benzene, in which the diagonally bonded structures R_b are of relatively low importance, with combined weights of under 20 %, in the case of the tropylium cation structures of this type (R_b, R_c and R_d) are responsible for about 50 % of the wave function.

The SC(6,8) and SC(10,8) wave functions for the cyclooctatetraene dication and dianion contain 140 structures each. There are four groups of symmetry-equivalent structures in which the two charges are on adjacent carbons (see R_a, R_b, R_c and R_d in the third row in Fig. 3), three groups of symmetry-equivalent structures in which the two charges are separated by one carbon (see R_e, R_f and R_g in the fourth row in Fig. 3), four groups of symmetry-equivalent structures in which the two charges are separated by two carbons (see R_h, R_i, R_j and R_k in the fifth row in Fig. 3) and three groups of symmetry-equivalent structures in which the two charges are separated by three carbons (see R_l, R_m and R_n in the bottom row in Fig. 3). The most important structures, both for the cyclooctatetraene dication and dianion, are the eight R_h structures, with a combined weight of 32.3 % in the dication and 43.2 % in the dianion, each of which involves three carbon–carbon π bonds, and the two charges are well separated. The adverse effect of keeping the two charges close together is demonstrated by the low weights of the eight structures R_a, that also involve three carbon–carbon π bonds each, but which show combined weights of just 4.9 % in the dication and 8.3 % in the dianion. The small negative combined weights for some of the cyclooctatetraene dication and dianion structures (for example, R_n) are due to the fact that while the Chirgwin–Coulson weights add to unity, some of them may turn out to be outside the [0, 1] range in the case of strong overlaps between structures.

The SC descriptions of the cyclopropenium ion, cyclopentadienide anion, benzene, cycloheptatrienyl cation, cyclooctatetraene dication and dianion involve remarkably similar active orbitals which are well localized and resemble $C(2p_\pi)$ atomic orbitals with small symmetrical protrusions toward neighboring carbons. All of these systems exhibit extensive resonance between structures constructed from the respective SC orbitals; the dominant resonance structures are similar to the Kekulé-like structures for benzene and make good chemical sense, for example, by keeping multiple charges well separated, as in the cyclooctatetraene dication and dianion. The SC orbitals for the cyclobutadiene dication and dianion demonstrate how the shapes of benzene-like SC orbitals change when placed in electron-deficient or electron-rich environments: In both cases, the orbitals become more delocalized, but resonance remains extensive.

3 Conclusions

SC theory for 'N electrons in M orbitals' active spaces [SC(N,M)] has been used here to obtain modern VB descriptions of the π-electron systems of a series of annulene rings with $4n + 2$ π electrons in their highest-symmetry nuclear conformations. The fully variational SC wave functions for the cyclopropenium ion [SC(2,3)], cyclobutadiene dication [SC(2,4)] and dianion [SC(6,4)], cyclopentadienide anion [SC(6,5)], benzene [SC(6,6)], cycloheptatrienyl cation [SC(6,7)], cyclooctatetraene dication [SC(6,8)] and dianion [SC(10,8)] were shown to recover all ($C_3H_3^+$, $C_4H_4^{2+}$ and $C_4H_4^{2-}$) or almost all ($C_5H_5^-$: 97.6 %, C_6H_6: 89.6 %, $C_7H_7^+$: 95.4 %, $C_8H_8^{2+}$: 96.9 %, $C_8H_8^{2-}$: 97.5 %) of the correlation energy included in their CASSCF counterparts. This indicates that the SC(N,M) wave functions can be regarded as easier-to-interpret, more compact alternatives to CASSCF(N,M) constructions employing analogous active spaces.

The qualitative features of the SC wave functions for the cyclopropenium ion, cyclopentadienide anion, cycloheptatrienyl cation, cyclooctatetraene dication and dianion are very much those that could have been anticipated from the well-known SC description of the π-electron system of benzene [3–5]: The optimized π-space SC orbitals are very similar to those in benzene, while the optimized resonance patterns are fully consistent with the ideas of classical VB theory. The more delocalized nature of the SC orbitals for the cyclobutadiene dication and dianion which emerged from the SC(2,4) and SC(6,4) calculations on these systems can be viewed as a result of the attempts of the respective SC wave functions to adjust to an electron-deficient and an electron-rich environment. The high levels of resonance observed in both $C_4H_4^{2+}$ and $C_4H_4^{2-}$ indicate that these ions are aromatic.

SC theory provides compact VB-style descriptions of the π-electron systems of $C_3H_3^+$, $C_4H_4^{2+}$, $C_4H_4^{2-}$, $C_5H_5^-$, C_6H_6, $C_7H_7^+$, $C_8H_8^{2+}$ and $C_8H_8^{2-}$ in their highest-symmetry nuclear conformations which demonstrate the aromaticity of annulene ions with $4n + 2$ π electrons through the classical VB concept of resonance. While the highest level of theory at which one can justify the familiar Frost–Musulin diagrams, often applied to annulenes and annulene ions with $4n + 2$ π electrons, is closed-shell HF, the equally easy-to-interpret modern VB models reported in this paper emerge from near-CASSCF-quality wave functions.

References

1. Frost AA, Musulin B (1953) J Chem Phys 21:572
2. Cooper DL, Karadakov PB (2009) Int Rev Phys Chem 28:169
3. Cooper DL, Gerratt J, Raimondi M (1986) Nature 323:699
4. Gerratt J (1987) Chem Br 23:327
5. Cooper DL, Wright SC, Gerratt J, Hyams PA, Raimondi M (1989) J Chem Soc Perkin Trans II 719
6. Rumer G (1932) Göttinger Nachr 3:337
7. Ladner RC, Goddard III WA (1969) J Chem Phys 51:1073

8. Gerratt J, Lipscomb WN (1968) Proc Natl Acad Sci USA 59:332
9. Karadakov PB, Gerratt J, Cooper DL, Raimondi M (1992) J Chem Phys 97:7637
10. Karadakov PB, Cooper DL, Duke BJ, Li J (2012) J Phys Chem A 116:7238
11. Frisch MJ, Trucks GW, Schlegel HB, Scuseria GE, Robb MA, Cheeseman JR, Scalmani G, Barone V, Mennucci B, Petersson GA, Nakatsuji H, Caricato M, Li X, Hratchian HP, Izmaylov AF, Bloino J, Zheng G, Sonnenberg JL, Hada M, Ehara M, Toyota K, Fukuda R, Hasegawa J, Ishida M, Nakajima T, Honda Y, Kitao O, Nakai H, Vreven T, Montgomery Jr JA, Peralta JE, Ogliaro F, Bearpark M, Heyd JJ, Brothers E, Kudin KN, Staroverov VN, Kobayashi R, Normand J, Raghavachari K, Rendell A, Burant JC, Iyengar SS, Tomasi J, Cossi M, Rega N, Millam NJ, Klene M, Knox JE, Cross JB, Bakken V, Adamo C, Jaramillo J, Gomperts R, Stratmann RE, Yazyev O, Austin AJ, Cammi R, Pomelli C, Ochterski JW, Martin RL, Morokuma K, Zakrzewski VG, Voth GA, Salvador P, Dannenberg JJ, Dapprich S, Daniels AD, Farkas Ö, Foresman JB, Ortiz JV, Cioslowski J, Fox DJ. Gaussian 09 Revision B.01 (2009) Gaussian Inc, Wallingford CT
12. Olah GA, Staral JS (1976) J Am Chem Soc 98:6290
13. Sekiguchi A, Matsuo T, Watanabe H (2001) Phosphorus Sulfur Silicon Relat Elem 168:51
14. Thorsteinsson T, Cooper DL, Gerratt J, Karadakov PB, Raimondi M (1996) Theor Chim Acta 93:343
15. Cooper DL, Thorsteinsson T, Gerratt J (1998) Adv Quant Chem 32:51
16. Thorsteinsson T, Cooper DL (1998) J Math Chem 23:105
17. Thorsteinsson T, Cooper DL (2000) In: Hernández-Laguna A, Maruani J, McWeeny R, Wilson S (eds) Quantum systems in chemistry and physics. Volume 1: basic problems and model systems. Kluwer, Dordrecht, p 303
18. Werner H-J, Knowles PJ, Knizia G, Manby FR, Schütz M, Celani P, Korona T, Lindh R, Mitrushenkov A, Rauhut G, Shamasundar KR, Adler TB, Amos RD, Bernhardsson A, Berning A, Cooper DL, Deegan MJO, Dobbyn AJ, Eckert F, Goll E, Hampel C, Hesselmann A, Hetzer G, Hrenar T, Jansen G, Köppl C, Liu Y, Lloyd AW, Mata RA, May AJ, McNicholas SJ, Meyer W, Mura ME, Nicklass A, O'Neill DP, Palmieri P, Pflüger K, Pitzer R, Reiher M, Shiozaki T, Stoll H, Stone AJ, Tarroni R, Thorsteinsson T, Wang M, Wolf A (2010) MOLPRO Version 2010.1. A Package of Ab Initio Programs
19. Schaftenaar G, Noordik JH (2000) J Comput-Aided Mol Des 14:123
20. Chirgwin BH, Coulson CA (1950) Proc R Soc Lond Ser A 201:196
21. Kotani M, Amemyia A, Ishiguro E, Kimura T (1963) Tables of molecular integrals. Maruzen, Tokyo
22. Serber R (1934) Phys Rev 45:461
23 Serber R (1934) J Chem Phys 2:697

Theor Chem Acc (2014) 133:1428
DOI 10.1007/s00214-013-1428-7

REGULAR ARTICLE

Bonding in PF_2Cl, PF_3Cl, and PF_4Cl: insight into isomerism and apicophilicity from ab initio calculations and the recoupled pair bonding model

Jeff Leiding · David E. Woon · Thom H. Dunning Jr.

Received: 18 September 2013 / Accepted: 18 November 2013 / Published online: 31 December 2013

Abstract Following previous work on PF_n and SF_nCl compounds, we report high-level ab initio quantum chemical calculations on PF_2Cl, PF_3Cl, and PF_4Cl. Coupled cluster (CCSD(T)) calculations were used to determine the structures and energetics of the various PF_nCl isomers, while generalized valence bond calculations were used to gain a deeper understanding of the factors influencing the structure and energetics of the isomers. Muetterties' rule, which predicts that more electronegative elements occupy the apical positions in pentavalent phosphorus compounds, is examined in the context of the recoupled pair bonding model, which provides invaluable insights into the bonding, structure, and reactivity of hypervalent compounds. While we found more electronegative fluorine is favored in the apical position in closed-shell PF_4Cl, the opposite is true for PF_3Cl and two triplet excited state isomers of PF_2Cl, all of which are open-shell species.

Dedicated to Professor Thom Dunning and published as part of the special collection of articles celebrating his career upon his retirement.

J. Leiding (✉)
Los Alamos National Laboratory, P.O. Box 1663, MS B214, Los Alamos, NM 87545, USA
e-mail: jal@lanl.gov

D. E. Woon · T. H. Dunning Jr.
Department of Chemistry, University of Illinois at Urbana–Champaign, Box 86–6, CLSL, 600 South Mathews, Urbana, IL 61801, USA
e-mail: dewoon@illinois.edu

T. H. Dunning Jr.
e-mail: thdjr@illinois.edu

Keywords Muetterties' rule · Bent's rule · Apicophilicity · Pentavalent phosphorus · Phosphorus chlorofluorides · Hypervalency · Recoupled pair bond

1 Introduction

A well-known guideline of inorganic chemistry, Muetterties' rule [1, 2], predicts that in pentavalent phosphorus compounds, the two apical (or axial) positions will be occupied by the most electronegative ligands in the compound. Muetterties observed this to be the case through NMR studies of a number of phosphorus compounds [1, 2]. It has been argued [3, 4] that Muetterties' rule is a consequence of Bent's rule [5, 6], which states that "atomic s character tends to concentrate in orbitals that are directed toward electropositive groups" [5]. In pentavalent phosphorus compounds, Pauling's hybridization model predicts that the P atom has three sp^2 hybrid orbitals in the equatorial plane and two pd hybrid orbitals along the axial bonds [7, 8]. Thus, according to Bent's rule, the axial positions would be occupied by the most electronegative ligands, as observed by Muetterties. Despite the fact that Pauling's model has been shown to drastically overestimate the importance of d orbitals [7, 9–11], there are theoretical calculations that support the validity of Bent's rule in predicting the isomerism of hypervalent phosphorus and sulfur compounds [3, 4, 12]. Muetterties' rule is directly related to the concept of *apicophilicity* [13], which is defined as the affinity of a given ligand for the apical (or axial) position relative to other ligands. Thus, following Muetterties' rule, greater ligand electronegativity implies greater ligand apicophilicity.

Previously, we reported that generalized valence bond (GVB) theory and the resulting recoupled pair bonding model are able to rationalize the geometric isomerism observed in various ground and excited states of $SF_{n-1}Cl$ (n = 1–6) [14]. To gain a better understanding of the factors influencing the structures and energetics of the analogous phosphorus isomers, we performed high-level CCSD(T) and GVB calculations on PF_2Cl, PF_3Cl, and PF_4Cl. As our previous work has shown [14–21], studying the precursor species and their low-lying excited states provides unique insights into the nature of bonding in a given molecule. Thus, we studied both the closed-shell PF_4Cl isomers and open-shell PF_3Cl isomers, as well as the ground and lowest excited states of PF_2Cl. Experimental studies have focused on the closed-shell trivalent and pentavalent phosphorus chlorofluorides [22–28]. There are previous theoretical studies of PF_4Cl [12, 29–32], most notable of which are DFT and MP2 studies done in the mid-2000s [3, 4]. There are no relevant prior theoretical calculations on either PF_3Cl or the excited states of PF_2Cl, which are also of interest here.

Calculations have shown that a recoupled pair bond is formed when an electron in a singly occupied orbital of an incoming ligand recouples the electrons of a lone pair of another atom, forming a bond pair. The remaining electron from the lone pair resides in an orbital that undercuts the strength of the bond: in the molecular or natural orbital framework [such as in a Hartree–Fock (HF) or complete active space (CAS) wave function], it is an orbital with strong antibonding character in the region of the bond, while in the full GVB framework, it is an orbital that has a large unfavorable overlap with the newly formed bond pair. Regardless of the choice of wave function, the electron left over from recoupling destabilizes the molecule, resulting in a bond energy that is typically much smaller than that of a covalent bond. However, a second bond can be formed with the electron remaining from the recoupling process. When the second ligand is very electronegative, the "left over" orbital polarizes toward and delocalizes onto the incoming ligand during bond formation, thus decreasing the destabilizing effect of the "left over" orbital on the first bond. This results in an unusually strong second bond. This pair of bonds is referred to as a recoupled pair bond dyad. For more information on this process and recoupled pair bonding in general, see any of our previous works [14–19, 21, 33], especially our review article [20].

We note that while Muetterties' rule was originally formulated in the context of closed-shell pentavalent phosphorus compounds, the concept of apicophilicity is more general and is applicable to the open-shell precursor PF_3Cl and, as we will argue below, the open-shell triplet excited states of PF_2Cl. At first glance, one might expect Muetterties' rule to hold for PF_3Cl, for example, given that the geometric and electronic structures of PF_4Cl are derived from PF_3Cl. However, we will demonstrate that Muetterties' rule is not obeyed by PF_3Cl or by the excited states of PF_2Cl. This, however, does not mean that Muetterties' rule is incorrect. We argue instead that Muetterties' rule is not universally applicable to open-shell tri- and tetravalent phosphorus species. This is, of course, not necessarily surprising given that Muetterties' original studies were limited to closed-shell species and that the effect of a singly occupied orbital in the place of a lone pair or bond pair, and ligand can be substantial.

2 Methods

All minimum energy molecular structures were optimized and energies computed with restricted coupled cluster theory with full single and double excitations plus perturbative triples [RCCSD(T)] [34–38]. Large augmented correlation consistent basis sets of triple- or quadruple-ζ quality, with additional tight d functions on Cl and P, i.e., aug-cc-pV(T + d, Q + d)Z, were used for all calculations [39–42]. These basis sets will be referred to as AVTZ and AVQZ. The AVQZ basis set was used for all geometry optimizations of minima and energies used in calculating the bond energies and relative stabilities of the isomers. A transition state (TS) for the $^3A''$ excited electronic state of PF_2Cl was optimized at the PBE0/AVTZ level of theory, followed by a CCSD(T)/AVTZ single-point energy calculation. Spin-coupled valence bond [43–46] (referred to here as GVB) calculations were performed on structures of interest using the AVTZ basis set. Orbitals, orbital overlaps, and spin-coupling coefficients were used to analyze the bonding. All calculations were performed with the Molpro program package [47].

3 Results

3.1 PF_2Cl

Understanding the bonding and isomerism in smaller precursor phosphorus species is helpful for understanding and anticipating the behavior of the larger phosphorus species. Thus, in order to gain insight into PF_3Cl and PF_4Cl, we begin with PF_2Cl.

The ^4S ground state of P has three singly occupied 3p orbitals and a doubly occupied 3s orbital as shown in Fig. 1. As for PF_3 [16], the ground state of PF_2Cl, labeled X^1A', is bound by covalent bonds between the three halogen atoms and phosphorus, i.e., by the formation of three singlet-coupled electron pairs involving the singly occupied halogen p orbitals and the singly occupied 3p orbitals of $P(^4S)$. The calculated geometric structure of PF_2Cl is shown in Fig. 2 and Table 1. Mutual bond pair repulsion

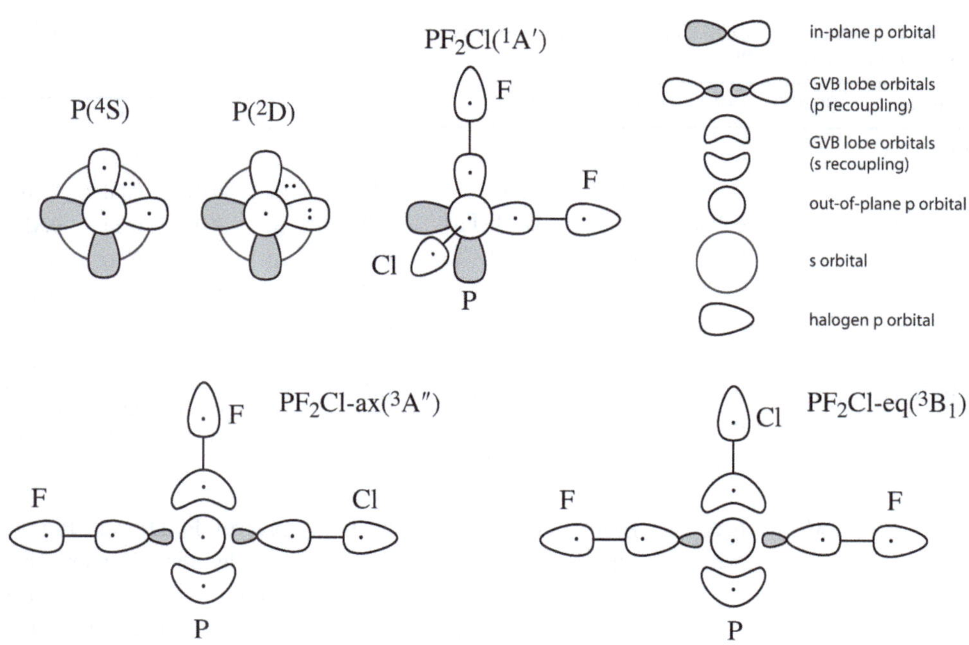

Fig. 1 GVB orbital coupling diagrams for the phosphorus atom and the ground and first two excited states of PF_2Cl. Singlet coupling is indicated by singly occupied orbitals connected by *straight lines*

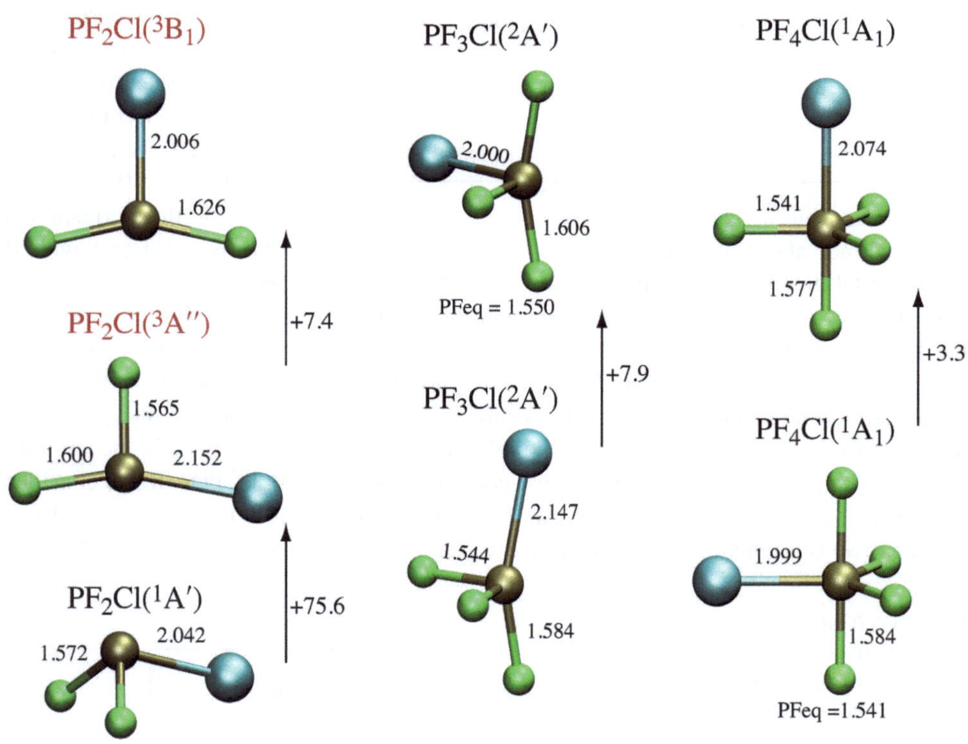

Fig. 2 Geometries for PF_nCl ($n = 2$–4) calculated at the RCCSD(T)/AVQZ level of theory. Bond lengths are in Å. State labels for excited states are shown in *red*. Relative energies in kcal/mol are indicated by *arrows* pointing toward less stable structures. Bond angles are provided in Table 1. (Cl: *blue*, F: *green*, P: *brown*)

between the three phosphorus–halogen bonds accounts for the calculated bond angles, which are about 8° larger than the ideal angle of 90° between the three mutually orthogonal phosphorus 3p orbitals. The calculated RCCSD(T)/AVQZ structure is in excellent agreement with the experimental microwave structure [24]. The experimental P–F and P–Cl bond lengths (r_0) are 1.571 ± 0.003 and 2.030 ± 0.006 Å, respectively, as compared to our calculated (r_e) values of 1.572 and 2.042 Å. Likewise, the experimental F–P–F and F–P–Cl angles are 97.3 ± 0.2° and 99.2 ± 0.3°, respectively, as compared to our calculated values of 97.3° and 98.8°.

PF_2Cl has excited triplet states that arise from the ^2D excited state of P, which lies 32.5 kcal/mol above the ground P(^4S) state (see Fig. 1) [48]. The first excited state of PF_2Cl is the $^3A''$ (PF_2Cl–ax) state, which lies

Table 1 Calculated, RCCSD(T)/AVQZ, bond lengths (in Å) and angles (in degrees) for PF_nCl

Structure	State	Bond	r_e (Å)	Angle	θ_e (°)
PF_2Cl	$^1A'$	P–F	1.572	F–P–F	97.3
		P–Cl	2.042	Cl–P–F	98.8
PF_2Cl–ax	$^3A''$	$P–F_{eq}$	1.565	$F_{eq}–P–F_{ax}$	98.5
		$P–F_{ax}$	1.600	$F_{eq}–P–Cl_{ax}$	101.0
		$P–Cl_{ax}$	2.152	$F_{ax}–P–Cl_{ax}$	160.5
PF_2Cl–eq	$^3B_1$	$P–F_{ax}$	1.626	$Cl_{eq}–P–F_{ax}$	102.3
		$P–Cl_{eq}$	2.006	$F_{ax}–P–F_{ax}$	155.4
PF_3Cl–ax	$^2A'$	$P–F_{eq}$	1.544	$F_{eq}–P–F_{eq}$	102.4
		$P–F_{ax}$	1.584	$Cl_{ax}–P–F_{ax}$	160.2
		$P–Cl_{ax}$	2.147	$Cl_{ax}–P–F_{eq}$	96.33
				$F_{ax}–P–F_{eq}$	96.02
PF_3Cl–eq	$^2A'$	$P–F_{eq}$	1.550	$Cl_{eq}–P–F_{eq}$	105.9
		$P–Cl_{eq}$	2.000	$F_{ax}–P–F_{ax}$	162.6
		$P–F_{ax}$	1.606	$F_{ax}–P–F_{eq}$	92.8
				$F_{ax}–P–Cl_{eq}$	97.2
PF_4Cl–eq	$^1A_1$	$P–F_{eq}$	1.541	$F_{eq}–P–F_{eq}$	117.1
		$P–Cl_{eq}$	1.999	$F_{eq}–P–Cl_{eq}$	121.4
		$P–F_{ax}$	1.584	$F_{ax}–P–F_{eq}$	89.9
				$F_{ax}–P–Cl_{eq}$	90.3
PF_4Cl–ax	$^1A_1$	$P–F_{eq}$	1.541	$F_{eq}–P–F_{eq}$	120.0
		$P–F_{ax}$	1.577	$Cl_{ax}–P–F_{eq}$	90.3
		$P–Cl_{ax}$	2.074	$F_{ax}–P–F_{eq}$	89.7

75.6 kcal/mol above the ground state; the corresponding GVB orbital diagram is shown in Fig. 1. As we previously described for the analogous $^3B_1$ excited state of PF_3 [16], the bonding in this state arises from the bonding of the halogen atoms with the ^2D excited state of phosphorus, where one of the electrons in a singly occupied 3p orbital has been excited into one of the other singly occupied 3p orbitals giving, for example, a $3s^23p_z^23p_x$ configuration. A typical recoupled pair bond dyad can then be formed between two of the halogen ligands (F and Cl) and the $3p_z^2$ pair of $P(^2D)$; see Fig. 1. The third halogen (F) then recouples the $3s^2$ pair. This leaves a singly occupied P $3p_x$ orbital and an sp hybrid-like lobe, the electrons of which are spin-coupled into a triplet. The GVB orbitals of this state are shown in Fig. 3.

Table 2 GVB orbital overlaps for PF_2Cl–ax ($^3A''$)

	φ_1	φ_2	φ_3	φ_4	φ_5	φ_6	φ_7	φ_8
φ_1	1.00							
φ_2	**−0.83**	1.00						
φ_3	−0.01	0.08	1.00					
φ_4	−0.07	0.22	**0.83**	1.00				
φ_5	−0.02	0.10	0.10	0.14	1.00			
φ_6	−0.07	0.22	0.04	0.11	**0.81**	1.00		
φ_7	***0.11***	***−0.32***	***−0.19***	***−0.48***	***−0.12***	***−0.41***	1.00	
φ_8	0.00	0.00	0.00	0.00	0.00	0.00	0.00	1.00

Bond pair overlaps are shown in bold font, and unfavorable overlaps of φ_7 with every other orbital are shown in bold and italic font

The dominant spin-coupling function for the PF_2Cl–ax isomer is the perfect-pairing function with a weight of $w_{PP} = 0.969$. In this spin-coupling, the bond pairs are singlet-coupled (boxed in Fig. 3), and the electrons in the two singly occupied orbitals, φ_7 and φ_8, are triplet-coupled, just as we postulated above. The axial bond lengths of PF_2Cl–ax are 0.028 (P–F) and 0.11 Å (P–Cl) longer than those of the corresponding ground state bonds (Fig. 2). This is consistent with the presence of a recoupled pair bond dyad [20]. Furthermore, the $P–F_{eq}$ bond length is slightly shorter (0.007 Å) than the P–F bonds of the ground state, which is consistent with an increase in the s character of the orbitals participating in the bond. These results support the interpretation of the bonding diagram in Fig. 1. Orbital φ_7 is the orbital "left over" from recoupling the P $3s^2$ pair. It has considerable overlap with the other bond pair orbitals, especially those centered on the P atom, as shown in Table 2. This is also consistent with other

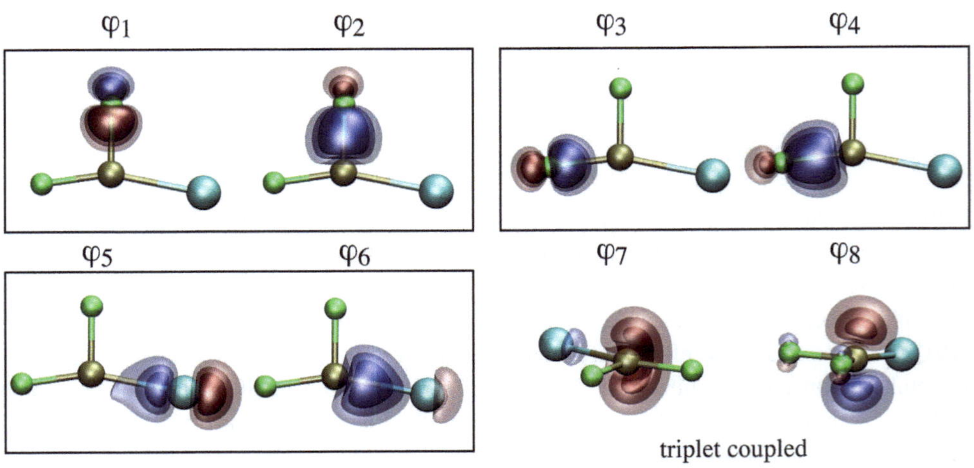

Fig. 3 GVB orbitals for PF_2Cl–ax ($^3A''$). Singlet-coupled bond pairs are *boxed*. The orbital overlap matrix is provided in Table 2. (Cl: *blue*, F: *green*, P: *brown*)

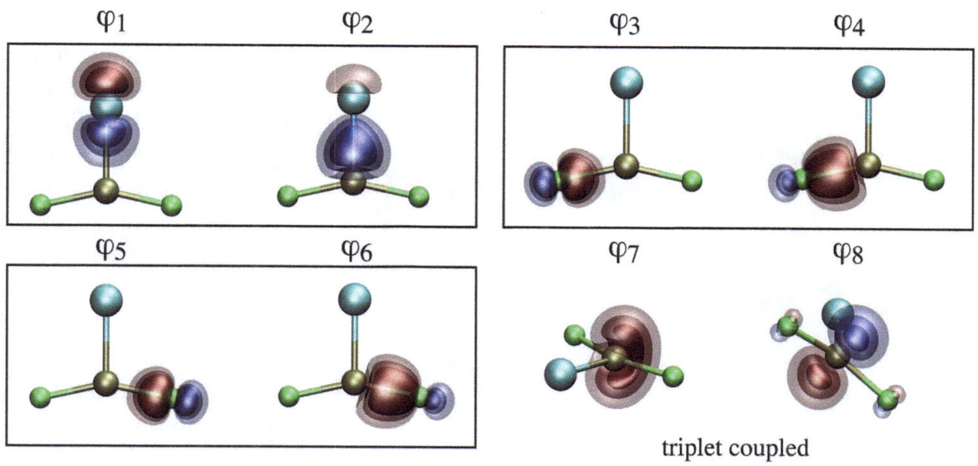

Fig. 4 GVB orbitals for PF_2Cl–eq ($^3B_1$). Singlet-coupled bond pairs are *boxed*. The orbital overlap matrix is provided in Table 3. (Cl: *blue*, F: *green*, P: *brown*)

Table 3 GVB orbital overlaps for PF_2Cl–eq ($^3B_1$)

	φ_1	φ_2	φ_3	φ_4	φ_5	φ_6	φ_7	φ_8
φ_1	1.00							
φ_2	**0.80**	1.00						
φ_3	0.00	−0.09	1.00					
φ_4	−0.08	−0.25	**0.82**	1.00				
φ_5	0.00	−0.09	0.06	0.06	1.00			
φ_6	−0.08	−0.25	0.06	0.12	**0.82**	1.00		
φ_7	***−0.10***	***−0.37***	***0.19***	***0.50***	***0.19***	***0.50***	1.00	
φ_8	0.00	0.00	0.00	0.00	0.00	0.00	0.00	1.00

Bond pair overlaps are shown in bold font, and unfavorable overlaps of φ_7 with every other orbital are shown in bold and italic font

recoupled systems we have studied [19, 20]. We will discuss the importance of these overlaps in Sect. 4.

The second excited state of PF_2Cl is the PF_2Cl–eq ($^3B_1$) state, which lies 7.4 kcal/mol above the PF_2Cl–ax state. This state is analogous to the PF_2Cl–ax ($^3A''$) state discussed above, except the Cl ligand now occupies the equatorial position, i.e., it is an isomer of PF_2Cl–ax. Again, the axial positions are slightly longer (0.054 Å) than a typical P–F covalent bond (as in the PF_2Cl ground state), and the P–Cl bond length is slightly shorter than that of a typical P–Cl covalent bond, indicating increased s character of the bonding orbitals. Furthermore, the P–F bonds are only slightly longer (0.012 Å) than those of the analogous PF_3($^3B_1$) state [16]. The GVB orbitals of this state are shown in Fig. 4. The dominant spin-coupling function is again the perfect-pairing function that couples the bond pairs into singlets and the electrons of φ_7 and φ_8 into a triplet ($w_{PP} = 0.995$). The orbital φ_7 is again the orbital "left over" from forming the P–Cl_{eq} recoupled pair bond, in which Cl recouples the P $3s^2$ pair, and φ_8 is simply a P 3p orbital. Again, φ_7 has considerable overlap with several of the bond pair orbitals as shown in Table 3. These data are consistent with the bonding diagram drawn in Fig. 1.

It should be noted that although PF_2Cl–ax and PF_2Cl–eq have different electronic and spatial symmetries, C_s and C_{2v}, respectively, they both belong to the same C_s symmetry electronic state. The PF_2Cl–eq C_{2v} state collapses to the $^3A''$ C_s symmetry electronic state (of PF_2Cl–ax) by simply breaking its C_{2v} symmetry. Thus PF_2Cl–ax and PF_2Cl–eq are isomers on the same $^3A''$ electronic state. Indeed, we located the barrier separating the two isomers by first optimizing the transition state (TS) at the PBE0/AVTZ level of theory and calculating a CCSD(T)/AVTZ single-point energy of the TS. IRC calculations were performed to verify that this TS connects the PF_2Cl–ax and PF_2Cl–eq isomers. The barrier protecting the PF_2Cl–eq isomer was found to be quite small, just 0.86 kcal/mol (0.52 kcal/mol with ZPE corrections to all structures using PBE0/AVTZ frequencies). However, we have found this to be typical of such isomers. For example, in our studies of SF_nCl, similar isomers were protected by <1.0 kcal/mol [14, 18].

3.2 PF_3Cl

PF_3Cl is an open-shell doublet in its ground electronic state. It has two geometric isomers shown in Fig. 2: one with the Cl in the axial position (PF_3Cl–ax) and the other with Cl in the equatorial position (PF_3Cl–eq). The electronic structure of these isomers is derived from that of the excited triplet isomers of PF_2Cl. Forming a P–F covalent bond with φ_8 of PF_2Cl–ax ($^3A''$) and PF_2Cl–eq ($^3B_1$) yields PF_3Cl–ax and PF_3Cl–eq, respectively. Indeed, the geometric parameters of PF_3Cl–ax and PF_2Cl–ax as well as PF_3Cl–eq and PF_2Cl–eq are quite similar. Again, as in PF_2Cl, the more stable isomer is the one in which Cl occupies the axial position—this time by almost 8 kcal/mol.

The two singly occupied orbitals of the PF_3Cl isomers, which are derived from the φ_7 orbitals of PF_2Cl, are shown in Fig. 5. It is clear that these orbitals are similar to φ_7 of

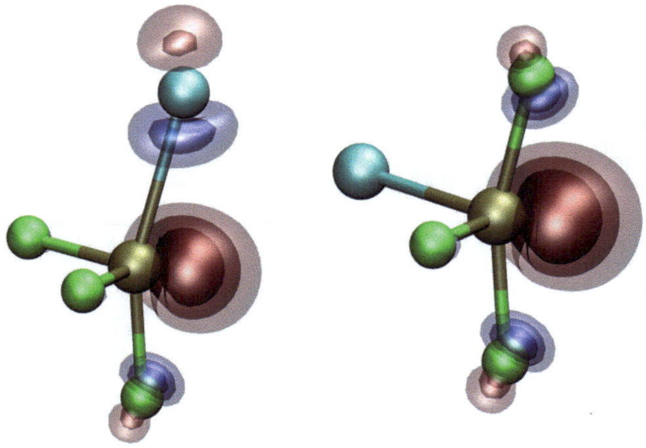

Fig. 5 Singly occupied natural orbitals of PF_3Cl–ax and PF_3Cl–eq. (Cl: *blue*, F: *green*, P: *brown*)

Table 4 Calculated RCCSD(T)/AVQZ sequential bond energies for PF_n and PF_nCl

	D_e (kcal/mol)
P + F → PF	107.1
PF + F → PF_2	120.0
PF_2 + F → PF_3	134.1
PF_3 + F → PF_4	57.4
PF_4 + F → PF_5	136.5
PCl + F → PFCl	113.0
PFCl + F → PF_2Cl	126.2
PF_2Cl + F → PF_3Cl–ax	64.5
PF_2Cl + F → PF_3Cl–eq	56.5
PF_3Cl–eq + F → PF_4Cl–eq	135.0
PF_3Cl–ax + F → PF_4Cl–ax	123.7

PF_n bond energies taken from Woon and Dunning [16]. All species are in their ground electronic states

PF_2Cl, but they have become somewhat distorted in response to the presence of the additional P–F bonds.

The PF_3Cl isomers can also be derived from the ground state of PF_2Cl, by the addition of an F atom opposite either the existing Cl or F ligands. To form the additional P–F bond, the P $3s^2$ pair of electrons must be recoupled, and thus this bond energy is much weaker than a covalent bond. As shown in Table 4, adding F to ground state PF_2Cl yields sequential bond energies of 64.5 and 56.5 kcal/mol for the two isomers, which are much less than the sequential bond energies of PCl + F (113.0 kcal/mol) and PFCl + F (126.2 kcal/mol).

3.3 PF_4Cl

There are two possible geometric isomers for the trigonal bipyramidal structure of PF_4Cl: PF_4Cl–ax and PF_4Cl–eq as shown in Fig. 2. Muetterties' rule correctly predicts the ground state isomer to be PF_4Cl–eq. The PF_4Cl–eq isomer is 3.3 kcal/mol more stable than the PF_4Cl–ax isomer at the RCCSD(T)/AVQZ level of theory. This result is in good agreement with previous theoretical results, despite the fact that our calculations are of much higher quality [3, 4]. The two isomers of PF_4Cl arise from the formation of P–F covalent bonds with PF_3Cl–ax and PF_3Cl–eq. The respective bond energies are shown in Table 4. The bond energies will be discussed in more detail in Sect. 4. The geometric parameters of PF_4Cl–ax and PF_4Cl–eq are similar to those of PF_3Cl–ax and PF_3Cl–eq, with the bond lengths universally shorter in the pentavalent phosphorus species. This is consistent with the behavior we observed in PF_n compounds [16].

The calculated RCCSD(T)/AVQZ PF_4Cl–eq structure is in excellent agreement with the experimental electron diffraction results [25]. The experimental P–F_{eq}, P–F_{ax}, and P–Cl_{eq} bond lengths are 1.535(3), 1.581(4), and 2.000(3) Å, respectively, as compared to our calculated values of 1.541, 1.584, and 1.999 Å. Likewise, the experimental F_{eq}–P–F_{eq} and F_{eq}–P–F_{ax} angles are 117.8(7)° and 90.3(4)°, respectively, as compared to our calculated values of 117.1° and 89.9°. Experimental uncertainties in parentheses represent three standard deviations.

4 Discussion of isomerism

It may be surprising that PF_2Cl–ax and PF_3Cl–ax are the most stable isomers for the excited $^3A''$ electronic state of PF_2Cl and the ground electronic state of PF_3Cl. However, this is consistent with our observations on SF_nCl [14] and, in fact, was the behavior we anticipated based on those observations. After discussing the open-shell cases (PF_2Cl and PF_3Cl) whose more stable structures have Cl occupying the axial position, we will discuss isomerism in the closed-shell pentavalent case, which conforms to Muetterties' rule.

4.1 Open-shell cases

In both excited triplet state isomers of PF_2Cl and the ground state of PF_3Cl, the isomer with Cl occupying the axial position—participating in the recoupled pair bond dyad—is more stable, which is contrary to what one would expect if Muetterties' rule prevailed. The observed isomerism for PF_2Cl can be understood based on the Pauli exchange repulsion between the bond pairs and the electrons in the singly occupied orbitals, as measured by the overlaps of the GVB orbitals. First, consider the more stable PF_2Cl–ax isomer, whose GVB orbital overlaps are shown in Table 2. Although large overlaps between the orbitals within a singlet-coupled pair are favorable, large overlaps between the orbitals in different singlet-coupled pairs or between a singly occupied orbital, and the orbitals in the singlet-coupled pairs are *unfavorable*—they give rise to the exchange repulsions associated with the Pauli principle. In Table 2, the bond pair overlaps are in bold font, and the unfavorable overlaps of the orbital remaining from recoupling the $3s^2$ pair (φ_7) are in bold and italicized font. The sum of the absolute values of the overlaps of φ_7 with every other orbital is 1.63, which is the majority of the sum of every unfavorable overlap (i.e., overlaps between electrons that are not part of the same bond pair). The sum of the overlaps of φ_7 with every other orbital in the PF_2Cl–eq isomer is 1.85, also a majority of the total unfavorable overlaps (the total unfavorable overlap for the PF_2Cl–ax and PF_2Cl–eq isomers is 2.81 and 2.99, respectively). Thus, the more stable isomer is the one with the smallest amount of total unfavorable GVB orbital overlaps. Although summation of the unfavorable overlaps between orbital φ_7 and the other GVB orbitals is an admittedly simple way to estimate the total magnitude of the exchange repulsions, it does appear to be a useful proxy in this case.

The orbitals associated with the axial positions of the PF_2Cl triplet state geometries (φ_3, φ_4, φ_5, φ_6) contribute more to the unfavorable overlaps with φ_7 than the orbitals associated with the equatorial position (φ_1, φ_2). Further, as expected, the phosphorus hybrid GVB orbitals (φ_2, φ_4, φ_6) make up the majority of these overlaps. When Cl occupies the axial position, the naturally longer P–Cl bond length (with respect to P–F) pulls the bond pair orbitals further away from φ_7 and the other bond orbitals, thus reducing unfavorable overlaps. Compare the overlap between orbitals (φ_6, φ_7) in the PF_2Cl–ax and PF_2Cl–eq isomers to observe the largest such difference in overlaps. Thus, the orbitals of the isomers with P–Cl axial bonds have more "breathing room."

It is important at this point to note that this is the same behavior we observed in SF_nCl [14], where we examined the amount of antibonding character present in a given natural orbital remaining from recoupling to explain the stability of the isomers.

We argued in Sect. 3.2 that PF_3Cl derives its electronic structure by F atom addition to the triplet state isomers of PF_2Cl. To further investigate this, we attempted to calculate the GVB wave function for PF_3Cl, but this proved to be very difficult due to technical complications in the orbital optimization process. Instead, we argue by extension that the arguments posed above for PF_2Cl also apply to PF_3Cl. There are good reasons to believe this is the case, as discussed in Sect. 3.2. For example, the nature of the singly occupied orbital remaining from recoupling, the geometric parameters, and the relative energies of the corresponding species are very similar.

The $PF_2X + F \rightarrow PF_3X$ bond energies are reported in Table 4. The bond energy for forming PF_3Cl–eq is similar to that of PF_4, differing by only 0.9 kcal/mol, while the bond energy to form PF_3Cl–ax is about 8.0 kcal/mol larger. This implies that PF_3Cl–eq is as (relatively) stable as PF_4, whereas PF_3Cl–ax is actually *more* stable.

There are some previous relevant studies on the isomerism of other open-shell phosphorus compounds, namely the work by Cramer and co-workers on open-shell fluorophosphoranyl radicals and hydroxylphosphoranyl radicals [49–53]. These authors also found an example of a less electronegative ligand (H) preferring the axial position in $HP(OH)_3$ [51, 53]. In their publications, the authors argue that increased hyperconjugative stabilization present when multiple hydroxyl groups occupy the equatorial positions is enough to overcome the difference in apicophilicity between H and OH. This is arguably not a dominant factor in PF_nCl, given that both F and Cl can contribute to hyperconjugative stabilization (although likely to different extents), whereas H cannot. Although, further calculations would be required to verify this.

Finally, steric interactions likely play a role in the observed isomerism. For example, it could be argued that the steric repulsions are greater for PF_2Cl–eq as compared to PF_2Cl–ax given that the former contains two P–Cl/P–F bond pair interactions at an angle of 102.3° while the latter has one P–Cl/P–F interaction at 101.0° and one P–F/P–F interaction at 98.5°. Similar arguments can be made for PF_3Cl and PF_4Cl as we discuss below; however, such interactions are difficult to quantify and isolate from the other interactions occurring in these systems.

4.2 Closed-shell pentavalent case

Finally, we can discuss the case to which Muetterties' rule applies—PF_4Cl. Bond energies will be helpful in analyzing the isomerism in the PF_4Cl molecule. As discussed in the Introduction, a bond formed between a very electronegative ligand and the orbital left over from the first recoupled pair bond is favorable because it relieves the unfavorable interactions associated with the electron in that orbital. Depending upon the wave function, this occurs either by minimizing the antibonding character of the orbital (HF or CAS wave functions) or by reducing the unfavorable orbital overlaps (GVB wave functions), by drawing that orbital toward the incoming ligand and further from the first (recoupled) bond pair (and any other bond pairs). The PF_3Cl–eq + F → PF_4Cl–eq bond energy is similar to that for the analogous PF_n reaction (135.0 vs. 136.5 kcal/mol). This is consistent with the fact that bond energies for forming PF_3Cl–eq and PF_4 are similar, indicating similar relative stabilities, and equivalent amounts of unfavorable interactions with the orbital remaining from recoupling. On the other hand, the PF_3Cl–ax + F → PF_4Cl–ax bond energy is 11.3 kcal/mol lower than that for forming PF_4Cl–eq. This is consistent with the notion that the orbitals of PF_2Cl–ax and PF_3Cl–ax have less unfavorable overlap with φ_7 than those of PF_2Cl–eq and PF_3Cl–eq, respectively. Thus, we expect the bond energy for forming PF_4Cl–ax to be less than that for forming PF_4Cl–eq. Although this explains why the bond energies would differ by the amount of extra destabilization present in $PF_{2,3}Cl$–eq (~8 kcal/mol), it does not account for the additional 3.3 kcal/mol that causes PF_4Cl–eq to be more stable than PF_4Cl–ax.

Trying to account for such a small energetic difference as 3.3 kcal/mol is difficult, especially since ab initio calculations of the current quality are expected to have an accuracy that is no better than about 1 kcal/mol. It is possible that the more numerous 90° P–Cl/P–F steric interactions in PF_4Cl–ax as compared to PF_4Cl–eq accounts for some of the additional 3.3 kcal/mol. This is the argument often used to explain why groups with bulkier bond/lone pairs occupy equatorial positions in trigonal bipyramids, but this is likely not the only factor important in isomerism. We believe we have provided sufficient evidence to explain the isomerism in excited state PF_2Cl and PF_3Cl, but we have insufficient data to fully explain the energy differences observed in PF_4Cl. Thus, in the future, we plan to explore cases where the energetic differences between the pentavalent isomers are expected to be more dramatic, such as PF_4CH_3 and the multiply chlorine-substituted species, PF_nCl_m $(n + m = 5)$. Preliminary calculations on PF_nCl_m indicate very similar behavior to that observed here.

5 Conclusions and future directions

We have shown that GVB theory and the recoupled pair bond model provide important insights into the bonding and isomerism in PF_2Cl, PF_3Cl, and PF_4Cl. While these insights provide compelling arguments for the observed isomerism in triplet PF_2Cl and ground state PF_3Cl, we currently lack sufficient data to fully explain the relative energetics of the isomers in PF_4Cl. The energies of the excited triplet state isomers of PF_2Cl and ground state PF_3Cl, to which Muetterties' rule does not apply, are readily explained by the differences in the magnitude of the Pauli exchange repulsions in the two isomers, as measured by the differences in the unfavorable overlaps of the orbital "left over" from recoupling the P $3s^2$ pair of electrons, and the other orbitals. We plan to continue this work, and gain a better understanding of the electronic factors important in understanding isomerization in the closed-shell pentavalent phosphorus species, by studying systems where the differences in the isomerization energy are more pronounced, such as $PF_n(CH_3)$ and PF_nCl_m.

Acknowledgments This work was supported by funding from the Distinguished Chair for Research Excellence in Chemistry at the University of Illinois at Urbana–Champaign. This research was supported in part by the National Science Foundation through TeraGrid resources by the National Center for Supercomputing Applications under Grant Number TG-CHE100010. This work was adapted from the first author's PhD dissertation, which is available at http://hdl.handle.net/2142/34409.

References

1. Muetterties EL, Mahler W, Schmutzler R (1963) Inorg Chem 2:613–618
2. Muetterties EL, Mahler W, Packer KJ, Schmutzler R (1964) Inorg Chem 3:1298–1303
3. Noorizadeh S (2005) J Mol Struct Theochem 713:27–32
4. Kandemirli F, Hoscan M, Dimoglo A, Esen S (2008) Phosphorous Sulfur Silicon 183:1954–1967
5. Bent HA (1960) J Chem Educ 37:616–624
6. Bent HA (1961) Chem Rev 61:275–311
7. Cooper DL, Cunningham TP, Gerratt J, Karadakov PB, Raimondi M (1994) J Am Chem Soc 116:4414–4426
8. Pauling L (1931) J Am Chem Soc 53:1367–1400
9. Magnusson E (1990) J Am Chem Soc 112:7940–7951
10. Reed AE, Schleyer PVR (1990) J Am Chem Soc 112:1434–1445
11. Reed AE, Weinhold F (1986) J Am Chem Soc 108:3586–3593
12. Van Der Voorn PC, Drago RS (1966) J Am Chem Soc 88:3255–3260
13. Wang PW, Zhang Y, Glaser R, Reed AE, Schleyer PVR, Streitwieser A (1991) J Am Chem Soc 113:55–64
14. Leiding J, Woon DE, Dunning TH Jr (2011) J Phys Chem A 115:329–341
15. Leiding J, Woon DE, Dunning TH Jr (2011) J Phys Chem A 115:4757–4764
16. Woon DE, Dunning TH Jr (2010) J Phys Chem A 114:8845–8851
17. Chen L, Woon DE, Dunning TH Jr (2009) J Phys Chem A 113:12645–12654
18. Leiding J, Woon DE, Dunning TH Jr (2012) J Phys Chem A 116:1655–1662
19. Leiding J, Woon DE, Dunning TH Jr (2012) J Phys Chem A 116:5247–5255
20. Dunning TH Jr, Woon DE, Leiding J, Chen L (2013) Acc Chem Res 46:359–368
21. Woon DE, Dunning TH (2009) J Phys Chem A 113:7915–7926
22. Cradock S, Rankin DWH (1972) J Chem Soc Faraday Trans 2(68):940–946
23. Morse JG, Parry RW (1972) J Chem Phys 57:5372–5374
24. Brittain AH, Smith JE, Schwendeman RH (1972) Inorg Chem 11:39–42
25. Macho C, Minkwitz R, Rohmann J, Steger B, Wölfel V, Oberhammer H (1986) Inorg Chem 25:2828–2835
26. Eisenhut M, Mitchell HL, Traficante DD, Kaufman RJ, Deutch JM, Whitesides GM (1974) J Am Chem Soc 96:5385–5397
27. Beattie IR, Livingston KMS, Reynolds DJ (1969) J Chem Phys 51:4269–4271
28. Holmes RR (1967) J Chem Phys 46:3718–3723
29. Chesnut DB, Quin LD (2005) Tetrahedron 61:12343–12349
30. Efremov DI, Soifer GB (1997) J Struct Chem 38:901–907

31. Murrell JN, Scollary CE (1976) J Chem Soc Dalton Trans 9:818–822
32. Brun C, Choplin F, Kaufmann G (1972) Inorg Chim Acta 6:77–80
33. Woon DE, Dunning TH Jr (2009) Mol Phys 107:991–998
34. Bartlett RJ (1989) J Phys Chem 93:1697–1708
35. Knowles PJ, Hampel C, Werner H-J (1993) J Chem Phys 99:5219
36. Purvis GD III, Bartlett RJ (1982) J Chem Phys 76:1910–1918
37. Raghavachari K, Trucks GW, Pople JA, Head-Gordon M (1989) Chem Phys Lett 157:479–483
38. Watts JD, Gauss J, Bartlett RJ (1993) J Chem Phys 98:8718–8733
39. Dunning TH Jr (1989) J Chem Phys 90:1007–1023
40. Kendall RA, Dunning TH Jr, Harrison RJ (1992) J Chem Phys 96:6796–6806
41. Wilson AK, van Mourik T, Dunning TH Jr (1996) J Mol Struct Theochem 388:339–349
42. Dunning TH Jr, Peterson KA, Wilson AK (2001) J Chem Phys 114:9244–9253
43. Cooper DL, Thorsteinsson T, Gerratt J (1997) Int J Quantum Chem 65:439–451
44. Cooper DL, Thorsteinsson T, Gerratt J (1998) Adv Quantum Chem 32:51–67
45. Thorsteinsson T, Cooper DL (1998) J Math Chem 23:105–126
46. Thorsteinsson T, Cooper DL, Gerratt J, Karadakov PB, Raimondi M (1996) Theor Chim Acta 93:343–366
47. Werner H-J, Knowles PJ, Manby FR, Schütz M et al (2009) Molpro, Version 2009.1, a package of ab initio programs. Cardiff, UK
48. Martin WC, Zalubas R, Musgrove A (1985) J Phys Chem Ref Data 14:751–802
49. Cramer CJ (1990) J Am Chem Soc 112:7965–7972
50. Cramer CJ (1991) J Am Chem Soc 113:2439–2447
51. Cramer CJ, Gustafson SM (1993) J Am Chem Soc 115:9315–9316
52. Cramer CJ, Gustafson SM (1994) J Am Chem Soc 116:723–734
53. Gustafson SM, Cramer CJ (1995) J Phys Chem 99:2267–2277

Theor Chem Acc (2013) 132:1413
DOI 10.1007/s00214-013-1413-1

REGULAR ARTICLE

MULTIMODE calculations of the infrared spectra of H_7^+ and D_7^+ using ab initio potential energy and dipole moment surfaces

Chen Qu · Rita Prosmiti · Joel M. Bowman

Received: 18 September 2013 / Accepted: 21 October 2013 / Published online: 31 October 2013

Abstract We present a new ab initio potential energy surface (PES) and a dipole moment surface (DMS) for H_7^+ in the bound region. The PES is a linear least-squares fit to 42,525 ab initio points whose energies were computed with CCSD(T)-F12b/cc-pVQZ-F12 theory, and the DMS is a fit to dipole moments calculated at MP2 level of theory. The PES and DMS describe the bound region of H_7^+ precisely. MULTIMODE (MM) calculations of the infrared spectra of H_7^+ and D_7^+ were performed using the new PES and DMS. These calculations were carried out at the lowest three stationary points using the single-reference version of MM, and only the five high-frequency modes were considered. The calculated spectra agree well with the recent experimental predissociation action spectra.

Keywords H_7^+ · Ab initio potential energy surface · Large-amplitude motion · MULTIMODE · Infrared spectrum

Dedicated to Professor Thom Dunning and published as part of the special collection of articles celebrating his career upon his retirement.

Electronic supplementary material The online version of this article (doi:10.1007/s00214-013-1413-1) contains supplementary material, which is available to authorized users.

C. Qu · J. M. Bowman (✉)
Department of Chemistry, Cherry L. Emerson Center for Scientific Computation, Emory University, Atlanta, GA 30322, USA
e-mail: jmbowma@emory.edu

R. Prosmiti
Instituto de Física Fundamental, IFF-CSIC, Serrano 123, 28006 Madrid, Spain
e-mail: rita@iff.csic.es

1 Introduction

The hydrogen cations, H_{2n+1}^+, are believed to be present in the interstellar medium and play an important role in interstellar chemistry [1–3]. However, only H_3^+ has been detected in the interstellar medium [4]. For the spectroscopic detection of the larger clusters, theoretical investigations and predictions of the spectra can be of great importance.

The infrared (IR) spectrum of H_5^+ has been extensively studied recently using ab initio potential energy surfaces (PESs) [5, 6] with various methods such as vibrational self-consistent field and configuration interaction (VSCF/VCI) implemented in the MULTIMODE reaction-path version (MM-RPH) [7, 8], diffusion Monte Carlo (DMC) [7, 9, 10], path-integral Monte Carlo (PIMC) [11, 12], full-dimensional quantum calculations [13], reduced-dimensional simulations [14–16], and multi-configuration time-dependent Hartree [17–19]. These calculations are challenging due to the fluxional nature of the molecule, the high dimensionality, and the strong mode coupling. Nevertheless, these studies have achieved much progress and agree reasonably well with the experimental measurements [7, 8].

However, little work has been done for the H_7^+ cluster due to the difficulty in the experiments and the lack of a high-quality PES for theoretical studies. Okumura et al. [20, 21] reported the first predissociation spectroscopy of this cluster, and the IR photodissociation spectroscopy was revisited by Duncan and coworkers recently [22]. The experimental spectrum of H_7^+ is simpler than that of H_5^+: A weak and broad feature at about 2,200 cm^{-1} and a peak at 3,982 cm^{-1} were reported. Based on DFT/B3LYP/6-311+G** and MP2/6-311+G** calculations of the spectrum with the double-harmonic approximation and the standard scaling method, Duncan and coworkers assigned the feature at about 2,200 cm^{-1} to the asymmetric stretch of the central H_3^+, and the 3,982 cm^{-1} peak to the stretch of the outer H_2. In terms of theoretical studies, calculations using VSCF followed by second-order vibrational Møller–Plesset perturbation theory have been performed for the vibrational analysis of H_n^+ ($n = 5, 7, 9$) with a relatively low-level ab initio method and with only a 2-mode coupling representation of the potential [23]. Just recently, an analytical PES at the MP2/cc-pVQZ level of theory became available [24]; with this PES, DMC and PIMC calculations have been carried out to investigate the dissociation energy of H_7^+ and the vibrational ground-state properties [25]. Both DMC and PIMC results show that the two H_2 almost act as "free" rotors, making this system challenging for quantum vibrational analysis.

Motivated by the latest experimental progress and the availability of the PES, we investigated the spectrum of H_7^+ with MULTIMODE (MM). However, by carrying out DMC calculations, we found that the MP2 PES underestimates the dissociation energy, D_0 of H_7^+ and D_7^+, by roughly 200 cm^{-1}, compared to early experimental enthalpy measurements [26–29]. The PES also underestimates the electronic dissociation energy, D_e, by 65 cm^{-1} compared to the literature CCSD(T)-CBS value [30]. Further, the MP2 normal-mode harmonic frequencies corresponding to the H_3^+ and H_2 stretching modes differ from the CCSD(T) benchmarks by 50 and 100 cm^{-1}, respectively. Thus, if one wants to compare with the recent experimental IR bands, higher-level ab initio data should be employed in the PES parameterization. So, here, we present a new version of the PES at the CCSD(T)-F12b/cc-pVQZ-F12 level of theory that only describes the bound region as well as a dipole moment surface (DMS) based on MP2 theory. MM calculations were performed using the PES and DMS. In Sect. 2, we describe the method of fitting the PES and DMS, and the details of the MM calculations. Then, we present our results and discussions in Sect. 3, and a summary and conclusions are given in Sect. 4.

2 Computational details

2.1 Potential energy and dipole moment surface

As we mentioned above, the MP2 PES is not as accurate as needed for rigorous vibrational analysis, so we considered higher-level ab initio calculations. The coupled-cluster level of theory employing large-size basis sets is computationally expensive as the number of atoms and configurations to be calculated increase. In the last years, significant advances have been made, and the so-called explicitly correlated CCSD(T)-F12 method [31, 32] has been shown to provide energies in impressive agreement with the CCSD(T) complete basis set (CBS) limit values. Thus, we performed different test calculations and found that the CCSD(T)-F12 method presents a computationally efficient way to generate the thousands of data points for the PES. In Table 1, we present the effect of the methods on the total energy E_{opt} and the electronic dissociation energy D_e, as well as the effect of using the cc-pVQZ [33] and cc-pVQZ-F12 [34, 35] basis sets in the calculations. One can see that the difference in D_e is about 65 cm^{-1} for MP2/cc-pVQZ and about 20 cm^{-1} for CCSD(T)-F12b/cc-pVQZ-F12, relative to the benchmark CCSD(T)/CBS value. Overall, we found that the CCSD(T)-F12b energies are in better agreement with the CCSD(T)/CBS limit, and the CCSD(T)-F12a results are lower than the F12b energies. This observed relative performance of the CCSD(T)-F12a/b is consistent with previous investigations [32]. Therefore, as it is also recommended, the F12b energies were considered.

For the new PES, we selected a subset of configurations from the previous MP2 PES and calculated the electronic energies using CCSD(T)-F12b/cc-pVQZ-F12. This subset contains only the configurations in the bound region; thus, the new PES is a local representation. The present PES was generated by a fit to 42,525 ab initio energies calculated with MOLPRO 2010.1 [36], employing the invariant polynomial method [37]. As in our previous work [24], a single expression containing a seven-body term was used to represent the potential function, in permutationally invariant polynomials of Morse-type variables for all internuclear distances. The maximum order of the polynomial was seven, which resulted in 739 terms for the H_7^+ PES.

We also constructed a DMS for H_7^+ using all the 42,525 configurations. The three components of the dipole moment were calculated at MP2 level of theory with the cc-pVQZ-F12 basis set. To represent the dipole moment vector, we used the expression [38]

$$\boldsymbol{\mu}(R) = \sum_{i=1}^{N} w_i(R)\mathbf{r_i} \quad (1)$$

Table 1 Total energies (in a.u.) of the optimal structure of H_7^+ and $H_5^+ + H_2$, together with electronic dissociation energy, D_e, (in cm^{-1}) at the indicated level of theory and basis set

Method/basis set	$E_{opt}^{H_7^+}$	$E_{opt}^{H_5^+} + E_{opt}^{H_2}$	D_e
MP2/cc-pVQZ	−3.69009130	−3.68222854	1,725.7
CCSD(T)/cc-pVQZ	−3.71237896	−3.70571378	1,744.6
CCSD(T)-F12b/cc-pVQZ	−3.71371801	−3.70568117	1,763.9
CCSD(T)-F12a/cc-pVQZ-F12	−3.71539068	−3.70732528	1,770.2
CCSD(T)-F12b/cc-pVQZ-F12	−3.71405842	−3.70598923	1,771.0
CCSD(T)/CBS[a]	−3.71453	−3.70637	1,790.9

[a] From Ref. [24]

where R denotes the molecular configuration and $\mathbf{r}_i$ represents the Cartesian coordinates of the ith nucleus. $w_i(R)$ can be viewed as the effective charge on the ith nucleus, which is a scalar quantity that can be expanded in basis functions that depend on the internuclear distances. The coefficients in the expansion were determined by a linear least-squares fit to ab initio dipole moments. The basis functions transform covariantly under permutations of the like atoms [37]. The sum of the effective partial charges has to be the total charge of the molecule, which is one for H_7^+, and we imposed this as an additional constraint in the fitting procedure.

2.2 MULTIMODE calculations

The code MM [39] was used for vibrational calculations. The code uses the Watson Hamiltonian [40] in mass-scaled normal coordinates. However, the Watson Hamiltonian has difficulty in describing large-amplitude motions such as the torsion; for molecules with low torsional barriers. The MM-RPH code is an extension of MM to include one large-amplitude coordinate in the"reaction-path Hamiltonian" [41]. The MM-RPH version could be employed for H_7^+ since the two H_2 units act almost like free rotors in H_7^+ [25]. However, here, we did not perform MM-RPH calculations; instead, we employed the single-reference version (MM-SR) but limited to the five highest-frequency modes. The reason for choosing MM-SR is that we focus on the experimental part of the spectrum, which is only probing the high-frequency H_3^+ and H_2 internal modes. Therefore, we restricted our calculations to the five high-frequency modes, and these calculations can be conveniently done with MM-SR. Further, based on our previous calculations of the IR spectra of water clusters [42, 43] and $F^-(H_2O)_2$ [44], ignoring the low frequency modes can usually still achieve reasonable agreement with the experiment. However, in recognition of the nearly free-rotor motion of the two H_2 groups, we carried out MM-SR calculations at three configurations on the torsion path, as described below.

In brief, we solved the 5-mode Schrödinger equation

$$\hat{H}\psi(\mathbf{Q}) = E\psi(\mathbf{Q}), \tag{2}$$

where $\mathbf{Q} = [Q_1\ \ Q_2\ \ Q_3\ \ Q_4\ \ Q_5]$ denotes the five modes, and $\hat{H}$ is the Watson Hamiltonian, given by

$$\begin{aligned}\hat{H} = &\frac{1}{2}\sum_{\alpha\beta}(\hat{J}_\alpha - \hat{\pi}_\alpha)\mu_{\alpha\beta}(\hat{J}_\beta - \hat{\pi}_\beta)\\ &-\frac{1}{2}\sum_k \frac{\partial^2}{\partial Q_k^2} - \frac{1}{8}\sum_\alpha \mu_{\alpha\alpha} + V(\mathbf{Q}).\end{aligned} \tag{3}$$

In the Watson Hamiltonian, the $\hat{J}_\alpha$ and $\hat{\pi}_\alpha$ ($\alpha = x, y, z$) are the total and vibrational angular momenta, and $\mu_{\alpha\beta}$ is the inverse of the effective moment of inertia. Q_k ($k = 1, 2, 3, 4, 5$) are the normal coordinates of the five modes, and $V(\mathbf{Q})$ is the full potential with respect to the five modes, while the other modes stay at their minima. In MM, the potential is expanded in a hierarchical N-mode representation [39] and usually is truncated at the 4-mode representation. For high-dimensional problems, this reduces the cost for multi-dimensional integrations. In this work, we mostly used the 4-mode representation, and the 5-mode representation was considered to test the convergence of the calculations.

We solved the Schrödinger equation for zero-total angular momentum using the VSCF/VCI approach [45–47]. In the

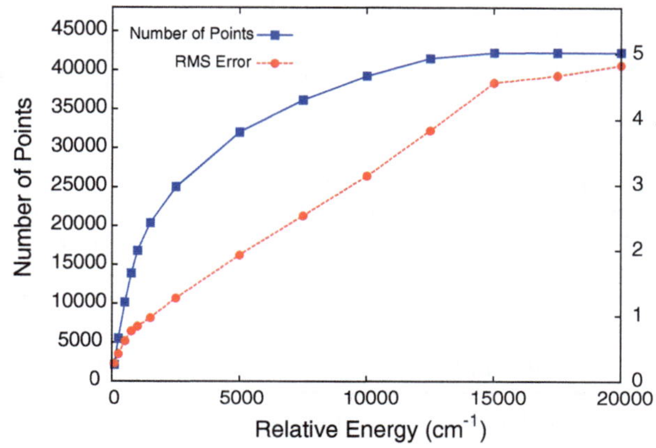

Fig. 1 The number of points and rms error below each energy

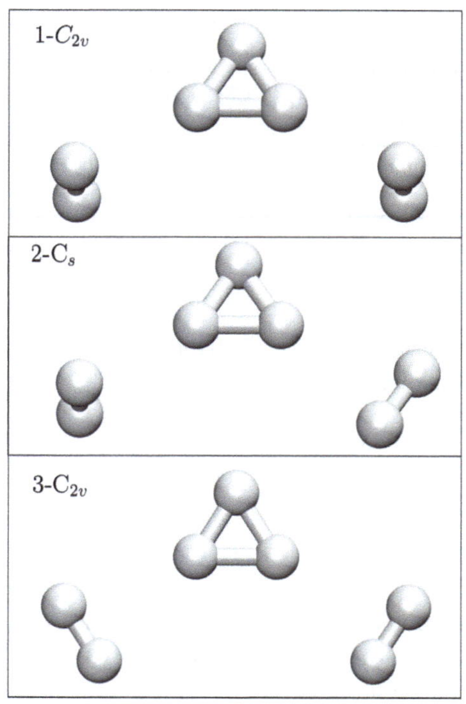

Fig. 2 The structures of the lowest three stationary points

VSCF, step 11 harmonic oscillator wavefunctions for each mode were used to construct the VSCF states. The CI states were expanded using the VSCF ground state, and the virtual states obtained in the VSCF calculation. In the VCI calculations, we allowed the simultaneous excitation of up to five modes, and the largest allowed excitation of an individual mode is 10. We also restricted the sum of total quanta of excitation to be 10. Convergence of the calculations was tested by using the full 5-mode representation of the potential and by changing the size of basis sets (see Sect. 3.2). The sizes of CI matrices were 3,003 and 6,188 for 11 and 13 basis functions, respectively.

The infrared intensities were calculated with the two step "dump-restart" method [48]. Briefly, in the first step, we performed MM calculations as usual, but also recorded the wavefunctions of the states we were interested in. In the second step, we restarted the MM, read in the recorded information of the wavefunctions, and calculated the transition dipole

$$R_{\alpha if} = \int \Psi_i(\mathbf{Q}) D_\alpha(\mathbf{Q}) \Psi_f(\mathbf{Q}) \mathrm{d}\mathbf{Q} \quad (4)$$

where $\alpha = x, y, z$, and $D_\alpha(\mathbf{Q})$ is the DMS. Once the transition dipole $R_{\alpha if}$ has been evaluated, the infrared intensity of the $i \rightarrow f$ transition is given by

$$A_{if}(v) \propto v \sum_\alpha |R_{\alpha if}|^2 \quad (5)$$

where v is the wavenumber of the transition. To finally get the spectra, the stick line shapes were replaced by Gaussian line shapes to achieve the similar line width as the experiment.

Table 2 The energy of the minimum structure (in a.u.) and the relative energies of two saddle points (in cm^{-1}) from CCSD(T)-F12b/cc-pVQZ-F12 calculations and the PES, the difference between them (in cm^{-1}), as well as the values from the previous MP2 PES

Config.	Energy			
	CCSD(T)-F12b	PES	E_{diff}	MP2 PES[a]
1-C_{2v}	−3.71405842	−3.71405830	0.03	−3.69008308
2-C_s	47.3	46.9	−0.4	45.2
3-C_{2v}	113.5	113.9	0.4	122.8

[a] From Ref. [24]

Table 3 Comparison of harmonic frequencies (in cm^{-1}) calculated from the ab initio program at CCSD(T)-F12b/cc-pVQZ-F12 level of theory and the PES

Mode	Description	CCSD(T)-F12b	PES
1	Torsion of H_2	98	96
2	Torsion of H_2	124	122
3		168	167
4	Asymmetric stretch of H_2–H_3^+–H_2	562	565
6	Symmetric stretch of H_2–H_3^+–H_2	572	571
7		695	694
8		732	733
9		797	798
10		907	908
11	Bending of H_3^+	2,294	2,293
12	Asymmetric stretch of H_3^+	2,525	2,527
13	Symmetric stretch of H_3^+	3,288	3,286
14	Out-of-phase stretch of two H_2	4,237	4,235
15	In-phase stretch of two H_2	4,239	4,237
ZPE		10,821	10,818

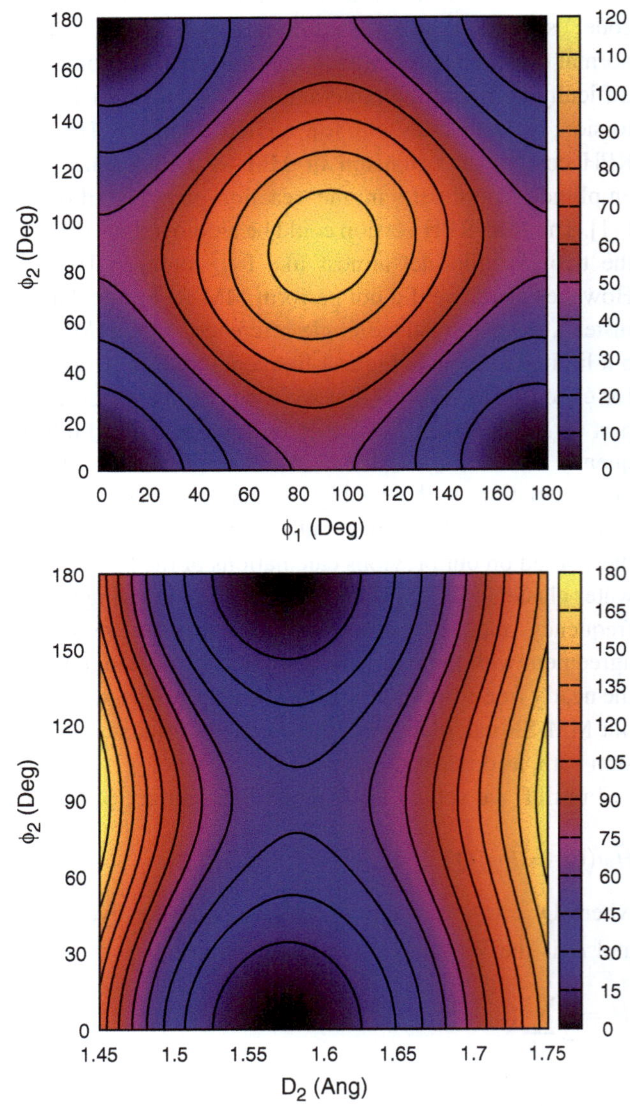

Fig. 3 Contour plots of the PES. In the *upper panel*, H_7^+ is fixed at 1-C_{2v}, and ϕ_1 and ϕ_2 are torsional angles of two H_2; in the lower panel, H_7^+ is also fixed at the minimum. ϕ_2 is the torsional angle of H_2 and D_2 is the H_3^+–H_2 distance

3 Results and discussion

3.1 Properties of the PES and DMS

The overall root-mean-square (rms) error of the fit is 17.2 cm^{-1}, which is an order of magnitude smaller than that of the previous MP2 PES. Figure 1 shows the number of points and the rms error below each energy in the PES. As is shown in the figure, most of the configurations in the data set are below 15,000 cm^{-1}, and the rms error for this region is as small as 5 cm^{-1}. Thus, this fit provides an accurate representation of H_7^+ surface in the bound region up to energies of about 15,000 cm^{-1} above the global minimum. Figure 2 shows the structures of the lowest three stationary points that are used as the reference geometries in the MM calculations.

In Table 2, we list the energies of the three optimized stationary points on the fitted PES and their comparison with the previous MP2 PES and the CCSD(T)-F12b ab initio energies. One can see that the difference between the energy from the PES and the ab initio value for each stationary point is less than 1 cm^{-1}. Normal-mode analysis is also carried out for each stationary point, and the harmonic frequencies predicted by the PES for the minimum are listed in Table 3. The difference between the PES and the ab initio value is less than 5 cm^{-1} for each mode. The energies of other stationary points and the frequencies of 2-C_s and 3-C_{2v} are provided in the supplementary material.

In Fig. 3, we present contour plots of the present PES of H_7^+. In the upper panel of the figure, we show the contour lines of the potential with the H_3^+ core fixed, while the two H_2 units are both rotating in the ϕ_1 and ϕ_2 coordinates, where ϕ_1 and ϕ_2 are the torsional angles of the two H_2 units. The torsional angle is zero for the minimum configuration and is 90 degree for the saddle point. One can see the behavior of the PES around the 4 symmetric 1-C_{2v} minima, the 4 equivalent 2-C_s saddle points connecting them, and the 3-C_{2v} stationary point. In the lower panel, the H_7^+ is kept fixed at the 1-C_{2v} configuration and the contour plot is presented as a function of the D_2, which is the H_3^+–H_2 distance, and ϕ_2 coordinates. The region around the two symmetric 1-C_{2v} minima and their 2-C_s barrier are shown up to energies of 180 cm^{-1}.

The rms fitting error of the dipole moments is 0.0006 a.u. The *x*-, *y*-, and *z*-components of the dipole moments along specified normal modes (see the descriptions of the normal modes in Table 3) at the minimum are plotted in Fig. 4. As one can see, among the five high-frequency modes, the dipole moments along the bending of H_3^+ (mode 11), the asymmetric stretch of H_3^+ (mode 12), and out-of-phase stretch of two H_2 (mode 14) have significant changes; thus, these modes may play an important role in the IR spectrum of H_7^+. Here, we should note that this DMS is only valid in the bound region, and it is not correct when H_7^+ dissociates to H_5^+ and H_2.

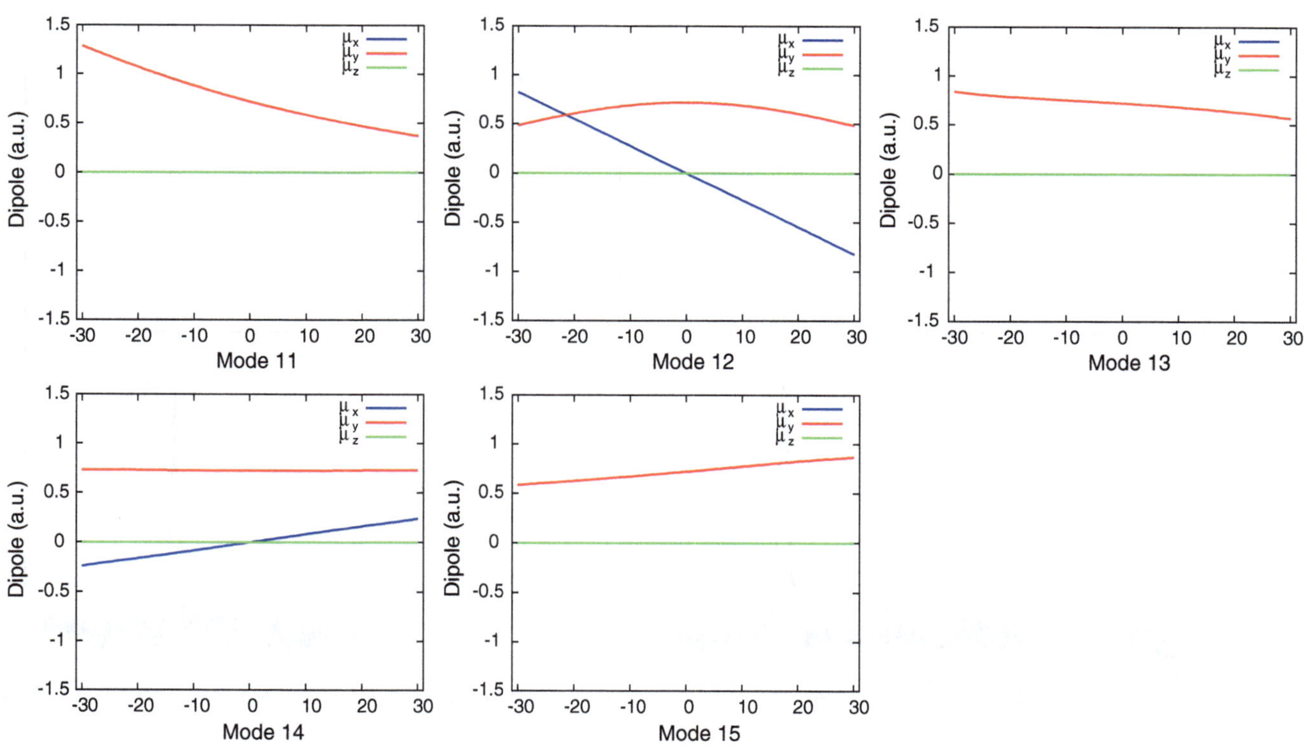

Fig. 4 One-dimensional cuts of the DMS along the specified normal modes at the minimum configuration. (See Table 3 for the description of the normal modes.) In mode 11, 13, and 15, μ_x and μ_z overlap

Table 4 MULTIMODE results from 4- and 5-mode representations of the potential and different numbers of basis functions

Config.	Mode	4-Mode representation		5-Mode representation	
		11 basis	13 basis	11 basis	13 basis
1-C_{2v}					
	v_{11}	1,909.7	1,909.1	1,909.7	1,909.1
	v_{12}	2,172.9	2,172.7	2,172.9	2,172.7
	v_{13}	2,907.0	2,906.5	2,907.0	2,906.5
	$v_{11}+v_{12}$, v_{14}	3,941.1	3,939.5	3,941.3	3,939.6
	v_{15}	3,983.9	3,983.8	3,983.9	3,983.8
	v_{14}, $v_{11}+v_{12}$	4,004.2	4,003.5	4,004.0	4,003.2
2-C_s					
	v_{11}	1,928.1	1,927.6	1,928.1	1,927.6
	v_{12}	2,174.3	2,174.1	2,174.3	2,174.1
	v_{13}	2,908.9	2,908.4	2,908.9	2,908.4
	$v_{11}+v_{12}$, v_{14}, v_{15}	3,952.8	3,951.7	3,952.8	3,951.6
	v_{15}, v_{14}, $v_{11}+v_{12}$	3,983.9	3,983.8	3,983.9	3,983.8
	v_{15}, $v_{11}+v_{12}$, v_{14}	4,012.0	4,011.0	4,012.1	4,011.1
3-C_{2v}					
	v_{11}	1,928.2	1,927.7	1,928.2	1,927.7
	v_{12}	2,185.1	2,184.8	2,185.0	2,184.8
	v_{13}	2,912.8	2,912.3	2,912.8	2,912.3
	v_{14}, $v_{11}+v_{12}$	3,960.7	3,959.6	3,960.9	3,959.8
	v_{15}	3,989.2	3,989.1	3,989.2	3,989.1
	v_{14}, $v_{11}+v_{12}$	4,018.6	4,017.5	4,018.4	4,017.2

3.2 IR spectra

The frequencies of the five fundamentals and a combination band from the MM calculations are listed in Table 4. We used both 4- and 5-mode representations for the PES and expanded the wavefunctions with different number of basis functions. As one can see in the Table 4, the differences between these calculations are mostly less than 1 cm^{-1}. Therefore, the 4-mode representation with 11 basis functions is adequate to achieve good convergence for this problem, and thus, it is employed for the calculations of the spectra.

In the MM calculations, we find that for the 1-C_{2v} and 3-C_{2v} reference configurations, two states show strong mixing: (1) the out-of-phase stretch of the two H_2; (2) the combination band of the H_3^+ bending and H_3^+ asymmetric stretch. While for 2-C_s, three states, namely two stretches of the two H_2 and the combination band of H_3^+ bending and asymmetric stretch, have strong coupling. Comparing the results from three stationary points, we find the frequencies vary only about 20 cm^{-1}.

Since the previous DMC and PIMC simulations suggest that the two H_2 almost rotate freely at vibrational ground state, we combine the three spectra to obtain an overall spectrum for H_7^+, which is shown in Fig. 5. In the range from 2,000 to 4,500 cm^{-1}, our calculated spectrum has two

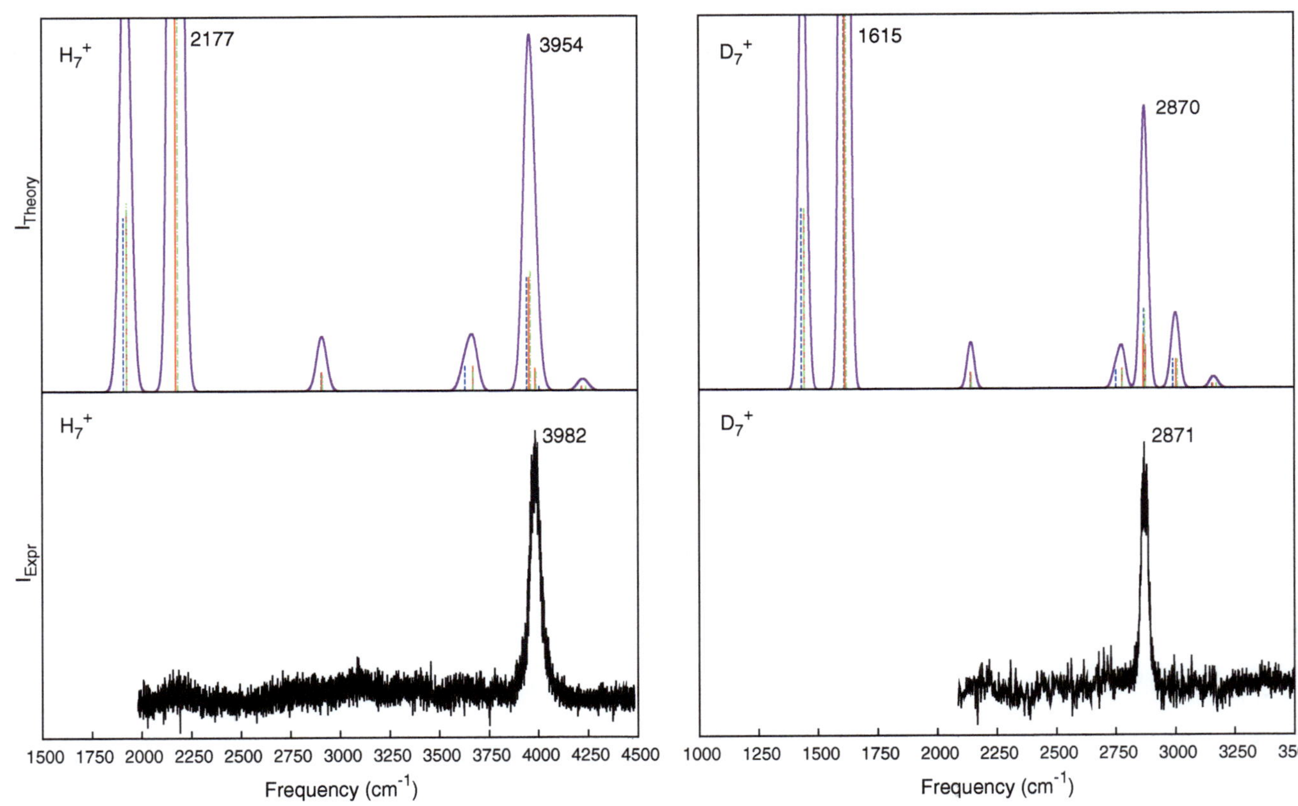

Fig. 5 Comparison of the calculated and experimental spectra of H_7^+ and D_7^+, respectively. We combined the spectra of 1-C_{2v} (*blue sticks*), 2-C_s (*red sticks*), and 3-C_{2v} (*green sticks*) in one figure, and replaced the stick line shapes with Gaussian line shapes to obtain the calculated spectrum (*purple lines*)

relatively intense peaks at 2,177 and 3,954 cm^{-1}, respectively, which are in good agreement with the experimental ones at about 2,200 and 3,982 cm^{-1}. As is explained in Ref. [22], the relative intensities of the two experimental bands can be attributed to the lower laser power at low frequency and the lower photodissociation yield with lower power. Therefore, the differences of the intensities between our computations and the experiment are understandable. Comparing with the scaled DFT spectrum of H_7^+, which still overestimates the peak at 3,982 cm^{-1} by 100 cm^{-1} [22], our spectrum from the VCI calculations is in better agreement with the experimental action spectrum. In addition, we assign the peak at 3,982 cm^{-1} to the mixed state mentioned above, instead of the stretch of H_2. The spectrum from the three stationary points for D_7^+ is also shown in Fig. 5. As is shown, it does not exhibit significant differences compared to that of H_7^+, except that the positions of the peaks are shifted by a factor of roughly $\sqrt{2}$.

4 Summary and conclusions

We constructed a new potential energy surface and a dipole moment surface for H_7^+ cluster, which is very precise, to describe the bound region of it. The PES was used in MM calculations, in which we only considered the five high-frequency internal modes of H_3^+ and H_2. These reduced-dimensional calculations resulted in spectra of H_7^+ and D_7^+ that are in good agreement with the experiment. More rigorous calculations need to couple more modes and to correctly treat the two torsional motions of the H_2 groups, which will be the goal in the future.

Acknowledgments C.Q. and J.M.B. thank NASA for financial support through Grant No. 370NNX12AF42G from the NASA Astrophysics Research and Analysis program. R.P. thanks the Centro de Calculo (IFF-CSIC) and SGAI (CSIC) for allocation of computer time. Supports from MICINN, Spain, Grant No. FIS2011-29596-C02-01, Consolider-Ingenio 2010 Programme CSD2009-00038 (MICINN), and COST Action CM1002 (CODECS) are gratefully acknowledged by R.P. We thank Mike Duncan for sending his experimental spectra.

References

1. Duley WW (1996) Astrophys J 471:L57
2. Petrie S, Bohme DK (2007) Mass Spectrom Rev 26:258
3. Snow TP, Bierbaum VM (2008) Annu Rev Anal Chem 1:229
4. Geballe TR, Oka T (1996) Nature 384:334
5. Xie Z, Braams BJ, Bowman JM (2005) J Chem Phys 122:224307
6. Aguado A, Barragán P, Prosmiti R, and Delgado-Barrio G, Villarreal P, Roncero O (2010) J Chem Phys 133:024306
7. Cheng TC, Bandyopadyay B, Wang Y, Carter S, Braams BJ, Bowman JM, Duncan MA (2010) J Phys Chem Lett 1:758
8. Cheng TC, Jiang L, Asmis KR, Wang Y, Bowman JM, Ricks AM, Duncan MA (2012) J Phys Chem Lett 3:3160
9. Lin Z, McCoy AB (2012) J Phys Chem Lett 3:3690
10. Lin Z, McCoy AB (2013). J Phys Chem A. doi:10.1021/jp4014652
11. Pérez de Tudela R, Barragán P, Prosmiti R, Villarreal P, Delgado-Barrio G (2011) J Phys Chem A 115:2483
12. Barragán P, Pérez de Tudela R, Prosmiti R, Villarreal P, Delgado-Barrio G (2011) Phys Scr 84:028109
13. Song H, Lee SY, Yang M, Lu Y (2013) J Chem Phys 138:124309
14. Sanz-Sanz C, Roncero O, Valdés A, Prosmiti R, Delgado-Barrio G, Villarreal P, Barragán P, Aguado A (2011) Phys Rev A 84:060502
15. Aguado A, Sanz-Sanz C, Villarreal P, Roncero O (2012) Phys Rev A 85:032514
16. Valdés A, Barragán P, Sanz-Sanz C, Prosmiti R, Villarreal P, Delgado-Barrio G (2012a) Theor Chem Acc 131:1210
17. Valdés A, Prosmiti R, Delgado-Barrio G (2012b) J Chem Phys 136:104302
18. Valdés A, Prosmiti R, Delgado-Barrio G (2012c) J Chem Phys 137:214308
19. Valdés A, Prosmiti R (2013) J Phys Chem A. doi:10.1021/jp3121947
20. Okumura M, Yeh LI, Lee YT (1985) J Chem Phys 83:3705
21. Okumura M, Yeh LI, Lee YT (1988) J Chem Phys 88:79
22. Young JW, Cheng TC, Bandyopadhyay B, Duncan MA (2013) J Phys Chem A 117:6984
23. Barbatti M, Nascimento MAC (2003) J Chem Phys 119:5444
24. Barragán P, Prosmiti R, Wang Y, Bowman JM (2012) J Chem Phys 136:224302
25. Barragán P, Pérez de Tudela R, Qu C, Prosmiti R, Bowman JM (2013) J Chem Phys 139:024308
26. Bennett SL, Field FH (1972) J Am Chem Soc 94:8669
27. Hiraoka K, Kebarle P (1975) J Chem Phys 62:2267
28. Beuhler RJ, Ehrenson S, Friedman L (1983) J Chem Phys 79:5982
29. Hiraoka K (1987) J Chem Phys 87:4048
30. Prosmiti R, Villarreal P, Delgado-Barrio G (2003) J Phys Chem A 107:4768
31. Adler TB, Knizia G, Werner HJ (2007) J Chem Phys 127:221106
32. Knizia G, Adler TB, Werner HJ (2009) J Chem Phys 130:054104
33. Dunning TH Jr (1989) J Chem Phys 90:1007
34. Peterson KA, Adler TB, Werner HJ (2008) J Chem Phys 128:084102
35. Yousaf KE, Peterson KA (2009) J Chem Phys 129:184108
36. Werner HJ, Knowles PJ, Knizia G et al (2010) MOLPRO, version 2010.1, a package of ab initio programs. http://www.molpro.net
37. Braams BJ, Bowman JM (2009) Int Rev Phys Chem 28:577
38. Wang Y, Huang X, Shepler BC, Braams BJ, Bowman JM (2011) J Chem Phys 134:094509
39. Bowman JM, Carter S, Huang X (2003) Int Rev Phys Chem 22:533
40. Watson JKG (1968) Mol Phys 15:479
41. Miller WH, Handy NC, Adams JE (1980) J Chem Phys 72:99
42. Wang Y, Bowman JM (2011) J Chem Phys 134:154510
43. Liu H, Wang YM, Bowman JM (2012) J Phys Chem Lett 3:3671
44. Kamarchik E, Bowman JM (2013) J Phys Chem Lett 4:2964
45. Bowman JM (1978) J Chem Phys 68:608
46. Bowman JM, Christoffel K, Tobin F (1979) J Phys Chem 83:905
47. Christoffel KM, Bowman JM (1982) Chem Phys Lett 85:220
48. Burcl R, Carter S, Handy NC (2003) Chem Phys Lett 380:237

Theor Chem Acc (2014) 133:1451
DOI 10.1007/s00214-014-1451-3

REGULAR ARTICLE

Properties of local vibrational modes: the infrared intensity

Wenli Zou · Dieter Cremer

Received: 12 September 2013 / Accepted: 13 January 2014 / Published online: 1 February 2014

Abstract For the local (adiabatic) vibrational modes of Konkoli and Cremer (Int J Quantum Chem 67:29–40, 1998), infrared intensities are derived by setting up the appropriate adiabatic conditions. It is shown that the local mode intensities are independent of the coordinates used to describe a molecule and correctly reflect the molecular symmetry and isotope composition. Normal mode intensities are related to local mode intensities via an adiabatic connection scheme (ACS). The ACS reveals intensity changes due to local mode mixing and avoided crossings, which are easily identified and quantified. The infrared intensities of simple molecules such as H_2O, CH_4, O_3, HOOH, CH_3OH, and the water dimer are discussed, and the influence of isotopes is quantified.

Keywords Local vibrational modes · Local stretching force constant · Infrared intensities · Local mode intensities · Adiabatic connection scheme · Isotope effects

1 Introduction

One of the primary objectives in chemistry is to determine the properties of the chemical bond [1]. Chemists have collected bond dissociation energies (BDE), bond lengths, stretching force constants, and other properties to derive suitable bond strength descriptors [2–4]. Although BDE values may be useful in a qualitative sense, they fail to be bond strength descriptors in a quantitative way because they depend on both the strength of the bond to be broken and the stabilization of the dissociation fragments caused by electron density redistribution, geometry relaxation, and avoided crossings between electronic states [5, 6]. The bond length has been used as bond strength descriptor for small, nonpolar molecules however becomes problematic for molecules with strongly polar bonds as is documented in the literature [7]. More suitable as bond strength descriptors are the stretching force constants of a vibrating molecule, which are obtained with the help of vibrational spectroscopy [4, 8, 9].

The use of stretching force constants to describe the chemical bond dates back to the 20s and 30s of the last century when Badger [10] found a relationship between force constant and bond length for diatomic molecules [11]. The extension of the Badger relationship to polyatomic molecules turned out to be difficult because spectroscopically derived stretching force constants are not unique, reflect coupling between the vibrational modes, and depend on the internal coordinates used for the description of the molecule in question [11]. Repeated attempts have been made to use stretching force constants by assuming that the bond stretching frequencies of certain functional groups are less effected by mode–mode coupling and therefore provide at least approximate measures for the bond strength via the associated force constants [12, 13]. These attempts are based on the general understanding that the stretching force constants of a molecule in its equilibrium geometry are the appropriate measures of the bond strength. Vibrational force constants are related to the curvature of the Born–Oppenheimer potential energy surface (PES) E(**q**) spanned by the internal coordinates q_n of

Dedicated to Professor Thom Dunning and published as part of the special collection of articles celebrating his career upon his retirement.

W. Zou · D. Cremer (✉)
Computational and Theoretical Chemistry Group (CATCO), Department of Chemistry, Southern Methodist University, 3215 Daniel Ave, Dallas, TX 75275-0314, USA
e-mail: dieter.cremer@gmail.com; dcremer@gmail.com

the molecule in question. They can be obtained by calculating the Hessian of E($\mathbf{q}$), which collects all second derivatives of the molecular energy with regard to the q_n-coordinates and is identical (apart from some conversion factors) to the force constants matrix $\mathbf{F}^q$ expressed in terms of internal coordinates.

The stretching force constant corresponds to an infinitesimally small change of the bond, and therefore, it is an ideal dynamic measure of the bond strength, which is no longer influenced by electronic structure reorganization or geometry relaxation effects. However, the stretching force constants obtained for a polyatomic molecule by either directly calculating the Hessian matrix or, alternatively, deriving them from measured stretching frequencies by solving the basic equation of vibrational spectroscopy [8] can because of coupling effects, no longer be related to individual bonds. Therefore, vibrational spectroscopists have pursued various ways of obtaining local mode stretching force constants.

Already in the 60s, Decius [14] suggested to solve the force constant problem by reverting to the inverse force constant matrix $\mathbf{\Gamma} = (\mathbf{F}^q)^{-1}$ and introducing the compliance constants Γ_{nn} as local bond strength descriptors. Ample work has been carried out with the compliance constants to describe the properties of chemical bonds [15–19] although their physical meaning and relationship to the normal vibrational modes remained unclear. McKean [20–22] solved the problem of obtaining local XH stretching force constants by synthesizing isotopomers of a given molecule where all H atoms but the target hydrogen were replaced by deuterium. By measuring then the *isolated* XH stretching frequency, a reasonable approximation for a local mode frequency was obtained. Henry [23] obtained local mode information on CH-stretching vibrations from overtone spectra. Apart from this, there were numerous attempts to set up relationships between stretching force constants or frequencies and bond strength descriptors such as BDE values, bond orders, bond lengths, etc., which are discussed in a 2010 review article that underlines the necessity of obtaining local mode information from normal vibrational modes [11].

Konkoli and Cremer [24] determined for the first time local vibrational modes directly from normal vibrational modes by solving the mass-decoupled Euler–Lagrange equations. Each local mode is associated with an internal coordinate q_n ($n = 1,\ldots,N_{vib}$ with $N_{vib} = 3N - \Sigma$; N number of atoms; Σ number of translations and rotations), which drives the local mode [24]. These authors also demonstrated that each normal vibrational mode can be characterized in terms of local vibrational modes, where their characterization of normal mode (CNM) method is superior to the potential energy distribution analysis [11, 25]. Cremer et al. [26] developed a way of calculating from a complete set of $3N - \Sigma$ measured fundamental frequencies the corresponding local mode frequencies. In this way, one can distinguish between calculated harmonic local mode frequencies (force constants) and experimentally based local mode frequencies (force constants), which differ by anharmonicity effects [27, 28]. Larsson and Cremer [29] showed that McKean's isolated stretching frequencies are equal to the local mode frequencies if there is a complete decoupling of the CH-stretching modes in a deuterium isotopomer. Zou et al. [30] proved that the reciprocal of the compliance constant of Decius is identical with the local force constant of Konkoli and Cremer. Furthermore, they proved that the local vibrational modes of Konkoli and Cremer are the only modes, which directly relate to the normal vibrational modes.

A local mode depends only on the internal coordinate it is associated with (*leading parameter principle* [24]) and is independent of all other internal coordinates used to describe the geometry of a molecule. Accordingly, it is also independent of using redundant or non-redundant coordinate sets. The number of local vibrational modes can be larger than N_{vib}, and therefore, it is important to determine those local modes, which are essential for the reproduction of the normal modes. They can be determined with the help of an adiabatic connection scheme (ACS), which relates local vibrational frequencies to normal vibrational frequencies by increasing a scaling factor λ from 0 (local frequencies) to 1 (normal frequencies). For a set of redundant internal coordinates and their associated local modes, all those frequencies converge to zero for $\lambda \rightarrow 1$, which do not contribute to the normal modes so that a set of N_{vib} dominant local modes remains [30, 31].

The infrared intensities of vibrational modes have been used to determine effective atomic charges of a molecule [32–34]. The measured intensities are associated with the atomic polar tensor (APT), which is the matrix of dipole moment derivatives with regard to the geometrical parameters of a molecule. If it is possible to obtain the APT from measured infrared intensities and if in addition the geometry of a molecule is known, one can directly determine effective atomic charges from measured infrared intensities. Much work has been done in this direction [32–35] where however all attempts so far have been based on normal rather than local vibrational modes. Since each normal mode is delocalized because of mode–mode coupling, it is questionable whether reliable charge information can be obtained from normal mode intensities. It is much more likely that in these cases, as in the case of the bond strength description, local mode rather than normal mode information is needed.

In this work, we will make the first and necessary step for obtaining effective atomic and bond charges from infrared intensities and APT by deriving the local mode intensity. Furthermore, we will relate the local mode intensities to those of the normal modes utilizing an intensity ACS as was recently done for the frequency ACS [30]. Equipped with these theoretical tools, we will be able to analyze normal mode intensities and discuss them in terms of local mode intensities, mode–mode coupling, isotope-dependence, and symmetry.

Before doing so it is necessary to clarify the term *local mode* because it is used in the literature in at least four different ways. (i) In computational chemistry, the normal modes are calculated using the classical description of a vibrating molecule introduced by Wilson et al. [8]. In this description, normal modes are delocalized because of mode coupling and their counterparts (derived by Konkoli and Cremer [24]) are the *local modes* of this work. Other terms such as *isolated* [20] or *intrinsic* [36] have been used in connection with local mode descriptions, but these latter terms refer to normal vibrational modes, which are *local* only in an approximate sense. (ii) Henry et al. [23, 37–40] have developed local mode (an)harmonic oscillator models to quantum mechanically calculate the overtones of XH stretching modes. The higher overtone modes ($n = 5$ or 6) for isolated XH groups are largely decoupled, which justifies speaking of local modes. Contrary to the normal and local modes of classical physics (see i), the local modes of the oscillator models and their frequencies are true eigenfunctions and eigenvalues of a quantum mechanical Hamiltonian acting on the vibrational wave function. (iii) Reiher et al. [41–43] calculate unitarily transformed normal modes associated with a given band in the vibrational spectrum of a polymer where the criteria for the transformation are inspired by those applied for the localization of molecular orbitals. The authors speak in this case of local vibrational modes because the modes are localized in just a few units of a polymer. Nevertheless, Reihers local modes are still delocalized within the polymer units. (iv) Yet, another use of the term local modes is made in solid-state physics where it refers to the vibrational mode(s) of an impurity in a solid material [44, 45].

The results of this work will be presented in three sections. In Sect. 2, the theory of the local mode intensities and the intensity ACS will be developed. Local mode intensities are analyzed and discussed for some small molecules in Sect. 3. It is shown how the normal mode intensities can be stepwise converted into local mode intensities and vice versa. In Sect. 4, the chemical relevance of the local mode intensities is discussed. In the final section, conclusions are drawn and an outlook is presented.

2 Theory of local vibrational modes

The vibrational secular equation expressed in Cartesian coordinates is given by Eq. (1): [8, 9, 46]

$$\mathbf{F}^x\tilde{\mathbf{L}} = \mathbf{M}\tilde{\mathbf{L}}\boldsymbol{\Lambda} \tag{1}$$

where $\mathbf{F}^x$ is the force constant matrix, $\mathbf{M}$ the mass matrix, matrix $\tilde{\mathbf{L}}$ collects the vibrational eigenvectors $\tilde{\mathbf{l}}_\mu$ in its columns, and $\boldsymbol{\Lambda}$ is a diagonal matrix with the eigenvalues λ_μ, which leads to the (harmonic) vibrational frequencies ω_μ according to $\lambda_\mu = 4\pi^2c^2\omega_\mu^2$. In Eq. (1), the number of vibrational modes is given by N_{vib}, i.e., Σ translational and rotational motions of the molecule are already eliminated. Here and in the following, a tilde above a vector or matrix symbol indicates mass weighting. Matrix $\tilde{\mathbf{L}}$ has the following properties

$$\tilde{\mathbf{L}}^\dagger\mathbf{M}\tilde{\mathbf{L}} = \mathbf{I} \tag{2}$$

$$\tilde{\mathbf{L}}^\dagger\mathbf{F}^x\tilde{\mathbf{L}} = \boldsymbol{\Lambda} \tag{3}$$

i.e., matrix $\tilde{\mathbf{L}}$ and eigenvalue matrix $\boldsymbol{\Lambda}$ are obtained by diagonalization of the mass-weighted force constant matrix. Usually, the normal mode vectors $\tilde{\mathbf{l}}_\mu$ are renormalized according to

$$\mathbf{l}_\mu = \frac{1}{\sqrt{\tilde{\mathbf{l}}_\mu^\dagger\tilde{\mathbf{l}}_\mu}}\tilde{\mathbf{l}}_\mu = \sqrt{m_\mu^R}\tilde{\mathbf{l}}_\mu \tag{4}$$

or

$$\mathbf{L} = \tilde{\mathbf{L}}(\mathbf{M}^R)^{1/2} \tag{5}$$

where $m_\mu^R = \left(\tilde{\mathbf{l}}_\mu^\dagger\tilde{\mathbf{l}}_\mu\right)^{-1}$ is the reduced mass of mode μ.

Matrix $\mathbf{L}$ also satisfies Eq. (1) in the form

$$\mathbf{F}^x\mathbf{L} = \mathbf{M}\mathbf{L}\boldsymbol{\Lambda} \tag{6}$$

which leads to

$$\mathbf{L}^\dagger\mathbf{F}^x\mathbf{L} = \mathbf{K} \tag{7}$$

$$\mathbf{L}^\dagger\mathbf{M}\mathbf{L} = \mathbf{M}^R \tag{8}$$

Equations (7) and (8) define the diagonal normal force constant matrix $\mathbf{K}$ and the reduced mass matrix $\mathbf{M}^R$ (with elements m_μ^R), respectively.

The vibrational secular equation expressed in internal coordinates q_n is given by Wilson et al. [8]

$$\mathbf{F}^q\tilde{\mathbf{D}} = \mathbf{G}^{-1}\tilde{\mathbf{D}}\boldsymbol{\Lambda} \tag{9}$$

Here, $\tilde{\mathbf{D}}$ contains the normal mode vectors $\tilde{\mathbf{d}}_\mu(\mu = 1,\ldots,N_{vib})$, and matrix $\mathbf{G} = \mathbf{B}\mathbf{M}^{-1}\mathbf{B}^\dagger$ (Wilson matrix) gives the kinetic energy in terms of internal coordinates [8]. The eigenvector matrix $\tilde{\mathbf{D}}$ has the properties

$$\tilde{\mathbf{D}}^{\dagger}\mathbf{G}^{-1}\tilde{\mathbf{D}} = \mathbf{I} \quad (10)$$

$$\tilde{\mathbf{D}}^{\dagger}\mathbf{F}^{q}\tilde{\mathbf{D}} = \mathbf{\Lambda} \quad (11)$$

Renormalization of $\tilde{\mathbf{D}}$ according to

$$\mathbf{D} = \tilde{\mathbf{D}}(\mathbf{M}^{R})^{1/2} \quad (12)$$

leads to

$$\mathbf{F}^{q}\mathbf{D} = \mathbf{G}^{-1}\mathbf{D}\mathbf{\Lambda} \quad (13)$$

and

$$\mathbf{D}^{\dagger}\mathbf{F}^{q}\mathbf{D} = \mathbf{K} \quad (14)$$

$$\mathbf{D}^{\dagger}\mathbf{G}^{-1}\mathbf{D} = \mathbf{M}^{R} \quad (15)$$

The relationship between $\mathbf{D}(\tilde{\mathbf{D}})$ and $\mathbf{L}(\tilde{\mathbf{L}})$ is given by Zou et al. [31]

$$\mathbf{L} = \mathbf{C}\mathbf{D} \quad (16)$$

$$\tilde{\mathbf{L}} = \mathbf{C}\tilde{\mathbf{D}} \quad (17)$$

Matrix $\mathbf{C}$ is the pseudo-inverse matrix of $\mathbf{B}$, where the latter is a rectangular ($N_{\text{vib}} \times 3N$) matrix containing the first derivatives of the internal coordinates q_n with regard to the Cartesian coordinates.

$$\mathbf{C} = \mathbf{M}^{-1}\mathbf{B}^{\dagger}\mathbf{G}^{-1} \quad (18)$$

Equations (1)–(18) are needed to present and derive in the following the properties of the local vibrational modes.

2.1 Properties of a local mode

The local vibrational modes of Konkoli and Cremer [24] can be directly determined from the normal vibrational modes. The local mode vector $\mathbf{a}_n$ associated with $q_n (n = 1, \ldots, N_{\text{para}}$ with N_{para} being the number of internal coordinates to specify the molecular geometry) is given by

$$\mathbf{a}_n = \frac{\mathbf{K}^{-1}\mathbf{d}_n^{\dagger}}{\mathbf{d}_n\mathbf{K}^{-1}\mathbf{d}_n^{\dagger}} \quad (19)$$

where the local mode is expressed in terms of normal coordinates Q_μ associated with force constant matrix $\mathbf{K}$. Here, $\mathbf{d}_n$ denotes a row vector of the matrix $\mathbf{D}$. The local mode force constant k_n^a of mode n (superscript a denotes an adiabatically relaxed, i.e., local mode) is obtained with Eq. (20):

$$k_n^a = \mathbf{a}_n^{\dagger}\mathbf{K}\mathbf{a}_n = (\mathbf{d}_n\mathbf{K}^{-1}\mathbf{d}_n^{\dagger})^{-1} \quad (20)$$

Local mode force constants, contrary to normal mode force constants, have the advantage of being independent of the choice of the coordinates to describe the molecule in question [24, 26]. In recent work, Zou et al. [30, 31] proved that the compliance constants Γ_{nn} of Decius [14] are simply the reciprocal of the local mode force constants: $k_n^a = 1/\Gamma_{nn}$.

The reduced mass of the local mode $\mathbf{a}_n$ is given by the reciprocal diagonal element G_{nn} of the $\mathbf{G}$-matrix [24]. Local mode force constant and mass are sufficient to determine the local mode frequency ω_n^a

$$(\omega_n^a)^2 = \frac{1}{4\pi^2c^2}k_n^a G_{nn} \quad (21)$$

2.2 Adiabatic connection scheme (ACS) relating local to normal mode frequencies

With the help of the compliance matrix $\mathbf{\Gamma}^q = (\mathbf{F}^q)^{-1}$, the vibrational eigenvalue Eq. (9) can be expressed as [30]

$$(\mathbf{\Gamma}^q)^{-1}\tilde{\mathbf{D}} = \mathbf{G}^{-1}\tilde{\mathbf{D}}\mathbf{\Lambda} \quad (22)$$

or

$$\mathbf{G}\tilde{\mathbf{R}} = \mathbf{\Gamma}^q\tilde{\mathbf{R}}\mathbf{\Lambda} \quad (23)$$

where a new eigenvector matrix $\tilde{\mathbf{R}}$ is given by

$$\tilde{\mathbf{R}} = (\mathbf{\Gamma}^q)^{-1}\tilde{\mathbf{D}} = \mathbf{F}^q\tilde{\mathbf{D}} = (\tilde{\mathbf{D}}^{-1})^{\dagger}\mathbf{K} \quad (24)$$

Next, the matrices $\mathbf{\Gamma}^q$ and $\mathbf{G}$ are partitioned into diagonal ($\mathbf{\Gamma}_d^q$ and $\mathbf{G}_{\text{d}}$) and off-diagonal ($\mathbf{\Gamma}_{od}^q$ and $\mathbf{G}_{\text{od}}$) parts: [30]

$$(\mathbf{G}_d + \lambda\mathbf{G}_{od})\tilde{\mathbf{R}}_\lambda = (\mathbf{\Gamma}_d^q + \lambda\mathbf{\Gamma}_{od}^q)\tilde{\mathbf{R}}_\lambda\mathbf{\Lambda}_\lambda \quad (25)$$

where the off-diagonal parts can be successively switched on with a scaling factor λ ($0 \leq \lambda \leq 1$), so that the local mode description given by the diagonal parts ($\lambda = 0$) is stepwise converted into the normal mode description obtained for $\lambda = 1$. For each value of λ a specific set of eigenvectors and eigenvalues collected in $\tilde{\mathbf{R}}_\lambda$ and $\mathbf{\Lambda}_\lambda$, respectively, is obtained. Equation (25) is the basis for the ACS.

2.3 Infrared intensity of a normal mode

The infrared intensity of normal mode μ is determined by [34, 47, 48]

$$I_\mu^{nm} = (\boldsymbol{\delta}_\mu^{nm})^{\dagger}\boldsymbol{\delta}_\mu^{nm} \quad (26)$$

where superscript nm denotes a normal mode and the dipole derivative vectors $\boldsymbol{\delta}_\mu^{nm}$ are collected in a matrix $\boldsymbol{\delta}^{nm}$ given by Eq. (27):

$$\boldsymbol{\delta}^{nm} = \mathscr{C}\mathbf{\Delta}\tilde{\mathbf{L}} = \mathscr{C}\mathbf{\Delta}\mathbf{L}(\mathbf{M}^R)^{-1/2} \quad (27)$$

The APT matrix $\mathbf{\Delta}$ is of dimension $3 \times 3N$ and contains the dipole moment derivatives with regard to Cartesian coordinates [32, 34]. If the normal mode intensity I_μ^{nm} is given in km/mol and $\mathbf{\Delta}$ and $\mathbf{M}^R$ in atomic units, the conversion factor $\mathscr{C}$ in Eq. (27) is 31.22307.

2.4 Infrared intensity of a local mode

The intensity I_n^a of a local mode $\mathbf{a}_n$ associated with internal coordinate q_n has to fulfill a number of requirements: (i) I_n^a must be characteristic of the local mode in question (and the associated displacement coordinate), however independent of any other internal coordinate used for the description of the molecular geometry. (ii) It must be characteristic of the masses of the atoms participating in the local vibration, but at the same time it must be independent of any other atomic masses in the molecule. (iii) In case of symmetry, symmetry-equivalent local modes must possess identical intensities. (iv) For diatomic molecules, the local mode intensity must be identical with the normal mode intensity.

For the derivation of the local mode intensity, Eq. (27) is re-written in terms of internal coordinates utilizing Eqs. (10), (17), and (18):

$$\delta^{nm} = \mathscr{C}\mathbf{\Delta}\left(\mathbf{M}^{-1}\mathbf{B}^{\dagger}\mathbf{G}^{-1}\right)\tilde{\mathbf{D}} \tag{28}$$

$$= \mathscr{C}\mathbf{\Delta}\mathbf{M}^{-1}\mathbf{B}^{\dagger}(\tilde{\mathbf{D}}^{\dagger})^{-1} \tag{29}$$

If $N_{\text{para}} = N_{\text{vib}}$, the inverse of $\tilde{\mathbf{D}}$ exists.

For the adiabatic situation with $\lambda = 0$, Eq. (22) becomes

$$(\mathbf{\Gamma}^q)_d^{-1}\tilde{\mathbf{D}}_0 = \mathbf{G}_d^{-1}\tilde{\mathbf{D}}_0\mathbf{\Lambda} \tag{30}$$

and the normalization condition (10) takes the form

$$\tilde{\mathbf{D}}_0^{\dagger}\mathbf{G}_d^{-1}\tilde{\mathbf{D}}_0 = \mathbf{I} \tag{31}$$

where the subscript 0 denotes $\lambda = 0$. If the local modes are ordered according to increasing frequencies ω_{μ}^2, then matrix $\tilde{\mathbf{D}_0}$ will be diagonal. Hence, matrix $\mathbf{D}_0 = \mathbf{G}_d^{-1/2}\tilde{\mathbf{D}}_0$ is also diagonal where $\mathbf{M}_0^R = \mathbf{G}_d^{-1}$. Accordingly, it holds that

$$\mathbf{D}_0^{\dagger}\mathbf{D}_0 = \tilde{\mathbf{D}}_0^{\dagger}\mathbf{G}_d^{-1}\tilde{\mathbf{D}}_0 = \mathbf{I} \tag{32}$$

or

$$\sum_{\nu}^{N_{\text{vib}}}(\mathbf{D}_0)_{\mu\nu}(\mathbf{D}_0)_{\nu\mu} = (\mathbf{D}_0)_{\mu\mu}^2 = 1 \tag{33}$$

which implies that $(\mathbf{D}_0)_{\mu\mu} = \pm 1$ where only the positive value is used for reasons of simplicity. We conclude that matrix $\mathbf{D}_0$ is the unit matrix.

Hence, the local mode condition with $\lambda = 0$ implies that (i) $\mathbf{D}_0 = \mathbf{I}$, (ii) $\mathbf{M}_0^R = \mathbf{G}_d^{-1}$, and (iii) $\tilde{\mathbf{D}}_0 = \mathbf{G}_d^{1/2}$. Equation (29) takes for local mode $\mathbf{a}_n$ the form

$$\delta^a = \mathscr{C}\mathbf{\Delta}\mathbf{M}^{-1}\mathbf{B}^{\dagger}\mathbf{G}_d^{-1/2} \tag{34}$$

which leads to the local mode intensity

Table 1 Local and normal mode intensities I_n^a and I_{μ}^{nm} of H_2O and HDO

Molecule	Parameter	I_n^a (km/mol)	Mode μ	I_{μ}^{nm} (km/mol)
H_2O	H–O	23.4868	1	40.8595
	H–O	23.4868	2	3.2361
	H–O–H	69.1712	3	69.5078
	H·H	11.3848		
HDO	H–O	23.4868	1	24.7171
	D–O	14.9527	2	11.2932
	H–O–D	59.8634	3	59.5745
	D–O	14.9527		
	H·D	16.9479		

B3LYP/cc–pVTZ calculations based on Eqs. (26) and (34)

$$I_n^a = (\delta_n^a)^{\dagger}\delta_n^a \tag{35}$$

It can be easily proved that Eq. (34) is both isotope-independent and parameter-independent and, in addition, leads to symmetry-equivalent intensities thus fulfilling the requirements (i), (ii), and (iii) for local mode intensities. For a diatomic molecule, Eq. (34) collapses to Eq. (27). The fulfillment of (i), (ii), and (iii) is demonstrated for the water molecule (see Table 1).

The water example shows that the bending intensity is relatively large and literally identical for normal and local mode where the difference results from a small coupling between bending and symmetric stretching mode. The local OH stretching intensities are identical and of medium size, which changes as a consequence of coupling between the two OH stretching modes. The symmetric stretching mode has only a small intensity because it leads to a relatively small change in the charge distribution. It is noteworthy that the measured intensity of the symmetric stretching mode is 2.2 km/mol [33], which is a result of a small mixing of stretching and bending vibration. The asymmetric stretching mode causes a large change in the charge distribution and accordingly has a large normal mode intensity.

If the bending parameter is replaced by the distance between the H atoms, a much lower intensity is obtained because the charge changes become smaller during this non-bonded stretching motion. However, none of the OH stretching vibrations changes its intensity value confirming that the local modes are independent of the other internal coordinates used for determining the molecular geometry. In this connection, it is important to note that the HH distance is used here only for test purposes. If the frequency ACS of the water molecule is set up with a redundant set of six internal coordinates (three distances and three angles in a hypothetical triangle), three local mode frequencies associated with HH distance and the two OHH angles converge to zero.

If one H is replaced by D, the local DO stretching intensity adopts a lower value because of the large mass of the D atom. Again, this does not lead to a change in the local OH stretching intensity. Also, there is no change when the bending angle is replaced by the H · D non-bonded distance.

Because of the parameter-independence, Eq. (34) can be formulated for an individual local mode associated with q_n as

$$\boldsymbol{\delta}_n^a = \frac{\mathscr{C}}{\sqrt{G_{nn}}}\boldsymbol{\Delta}\mathbf{M}^{-1}\mathbf{b}_n^\dagger \qquad (36)$$

where $\mathbf{b}_n$ is a vector of matrix $\mathbf{B}$. Since $\mathbf{b}_n^\dagger/\sqrt{G_{nn}}$ has the unit of amu$^{1/2}$, the conversion factor $\mathscr{C}$ in Eq. (36) does not depend on $\mathbf{b}_n$, i.e., the conversion factors for bond stretching, bending, or torsion are the same, which is different from the case of local mode force constants [31].

2.5 Adiabatic connection scheme for intensities

Equation (29) can be used as a starting point for deriving an ACS for infrared intensities.

$$\boldsymbol{\delta}_\lambda = \mathscr{C}\boldsymbol{\Delta}\mathbf{M}^{-1}\mathbf{B}^\dagger(\tilde{\mathbf{D}}_\lambda^\dagger)^{-1} \qquad (37)$$

i.e., the intensity I_λ depends on the eigenvector matrix $\tilde{\mathbf{D}}_\lambda$ obtained from Eq. (38):

$$\boldsymbol{\Gamma}_\lambda^{-1}\tilde{\mathbf{D}}_\lambda = \mathbf{G}_\lambda^{-1}\tilde{\mathbf{D}}_\lambda\boldsymbol{\Lambda}_\lambda \qquad (38)$$

or

$$\left(\boldsymbol{\Gamma}_d^q + \lambda\boldsymbol{\Gamma}_{od}^q\right)^{-1}\tilde{\mathbf{D}}_\lambda = (\mathbf{G}_d + \lambda\mathbf{G}_{od})^{-1}\tilde{\mathbf{D}}_\lambda\boldsymbol{\Lambda}_\lambda \qquad (39)$$

with the scaling factor λ increasing from 0 (local modes; $\tilde{\mathbf{D}}_{\lambda=0} = \mathbf{G}_d^{1/2}$ and $\boldsymbol{\delta}_{\lambda=0} = \boldsymbol{\delta}^a$) to 1 (normal modes; $\tilde{\mathbf{D}}_{\lambda=1} = \tilde{\mathbf{D}}$ and $\boldsymbol{\delta}_{\lambda=1} = \boldsymbol{\delta}^{nm}$).

Equation (39) has a number of advantages compared with Eq. (25). The latter becomes unstable in the case of small normal mode frequencies associated with small normal mode force constants because $\tilde{\mathbf{R}}$ of Eq. (24) becomes also small. This can no longer happen if one works with the inverse of matrices $\mathbf{G}_\lambda$ and $\boldsymbol{\Gamma}_\lambda$.

In the following, we give some application examples, which illustrate the method described. Also, the usefulness of local mode intensities is discussed.

3 Local mode intensities and ACS for infrared intensities

In Figs. 1a–d, 2a, b and 3a, b, intensity and frequency ACS diagrams are shown for H_2O, CH_4, O_3, HOOH, CH_3OH, and the water dimer, $(H_2O)_2$, as obtained (if not otherwise noted) at the ωB97X-D/aug-cc-pVTZ level of theory [49, 50]. For ozone, CCSD/aug-cc-pVTZ calculations were carried out to get more reliable results. The calculated geometries and the notation of the atoms are given in Fig. 4. In Table 2, calculated normal and local mode frequencies and intensities are compared. In the following, we will discuss interesting features of the intensity ACS, which help to understand the magnitude of the normal mode intensity.

3.1 Water

The two local OH stretching modes are equivalent and, accordingly, the two local OH stretching frequencies are identical as are the corresponding intensities (see Fig. 1a). These identities reflect the symmetry of the molecule although none of the two local OH stretching vectors can by classified to have a symmetry of the C_{2v} point group. However, an infinitesimal increase of λ by ϵ leads to an large change in the OH stretching intensities caused by the fact that the symmetry of the molecule is switched on. In this sense, $\lambda = \varepsilon$ ($\varepsilon \to 0$) denotes a catastrophe point [51] in the ACS diagram for intensities (according to Thom's theory of catastrophes) [52].

Contrary to the vibrational frequencies, the λ-dependent intensities explicitly depend on the mode vectors [see Eq. (37)]. For a given normal mode μ, one can define the mode dependent part as

$$\mathbf{t}_\mu(\lambda) = [(\tilde{\mathbf{D}}_\lambda^\dagger)^{-1}]_\mu = [(\mathbf{G}_\lambda)^{-1}\tilde{\mathbf{D}}_\lambda]_\mu \qquad (40)$$

which in the case of the local mode ($\lambda = 0$) becomes

$$\mathbf{t}_n(\lambda = 0) = [(\tilde{\mathbf{D}}_{\lambda=0}^\dagger)^{-1}]_n = [\mathbf{G}_d^{-1/2}]_n \qquad (41)$$

i.e., the local mode vector $\mathbf{t}_n$ contains only zeroes with the exception of position n.

This may be demonstrated for H_2O at $\lambda = 0$ where the following matrix contains the three column vectors $\mathbf{t}_n$ corresponding to internal coordinates H–O–H, O–H1, and O–H2:

$$\begin{pmatrix} 0.0000 & 0.0000 & 0.9737 \\ 0.0000 & 0.9737 & 0.0000 \\ 1.2416 & 0.0000 & 0.0000 \end{pmatrix} \qquad (42)$$

At $\lambda = \varepsilon = 10^{-4}$, the column vectors change strongly

$$\begin{pmatrix} -0.0001 & 0.6886 & 0.6885 \\ -0.0001 & 0.6886 & -0.6885 \\ 1.2416 & 0.0008 & 0.0000 \end{pmatrix} \qquad (43)$$

indicating that there is an intensity catastrophe leading to the splitting into a large asymmetric OH stretching intensity and a small symmetric OH stretching intensity.

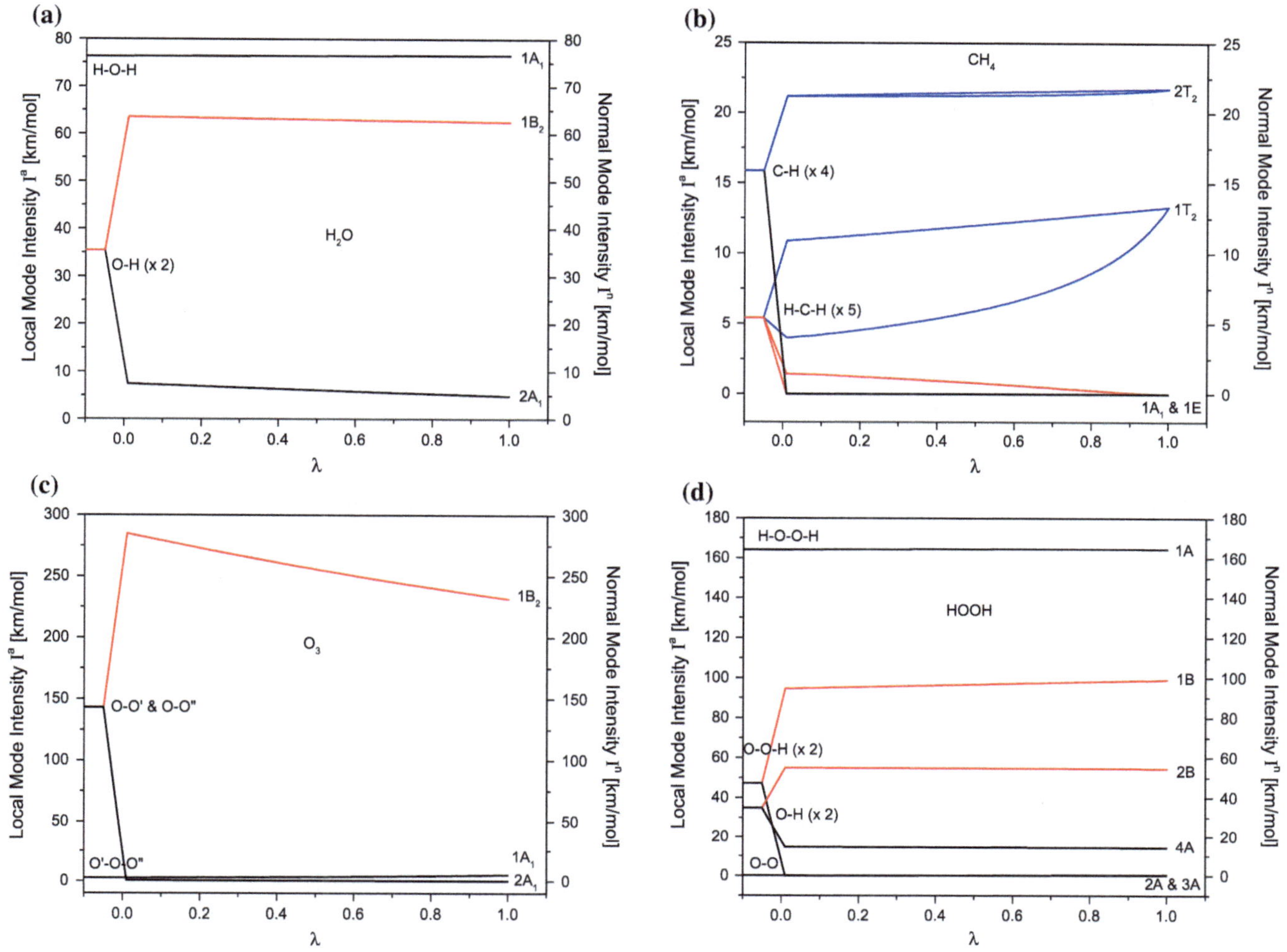

Fig. 1 Intensity ACS for **a** water, **b** methane, **c** ozone, and **d** hydrogen peroxide. The different mode symmetries are indicated by *different colors*. For the purpose of identifying intensity catastrophes, the $I(\lambda)$ curves are started with an arbitrary negative λ value so that the splitting at $\lambda = 0$ can be made visible. For the numbering of atoms, see Fig. 4. ωB97X-D/aug–cc–pVTZ or CCSD/aug–cc–VTZ (ozone) calculations

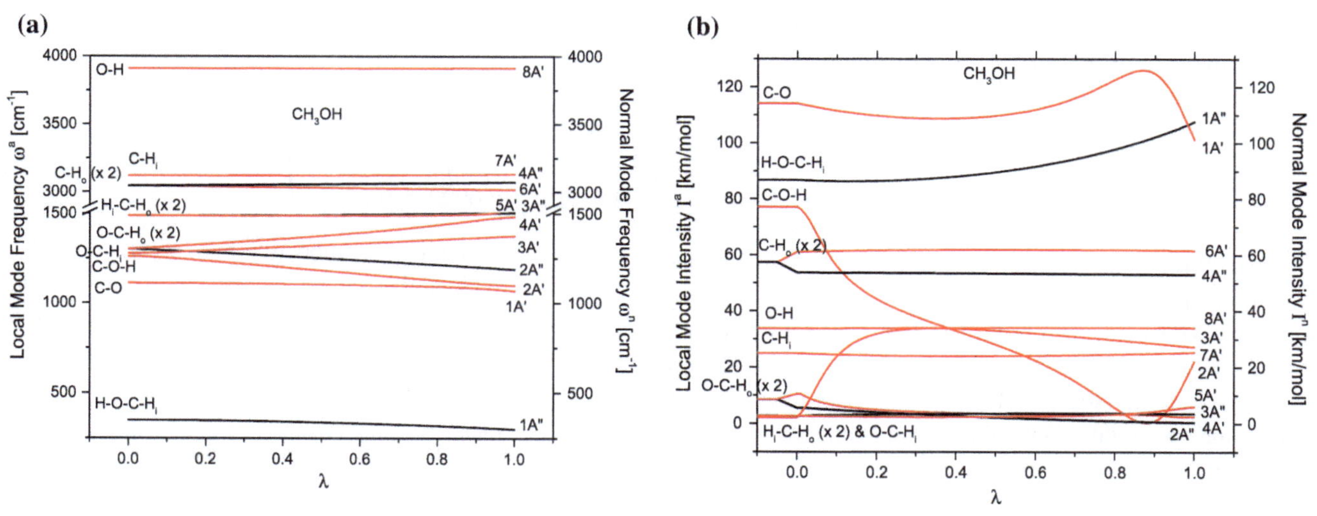

Fig. 2 **a** Frequency and **b** intensity ACS for methanol, CH_3OH. The different mode symmetries are indicated by *different colors*. For the purpose of identifying intensity catastrophes, the $I(\lambda)$ curves are started with an arbitrary negative λ value so that the splitting at $\lambda = 0$ can be made visible. For the numbering of atoms, see Fig. 4 . ωB97X-D/aug–cc–pVTZ calculations

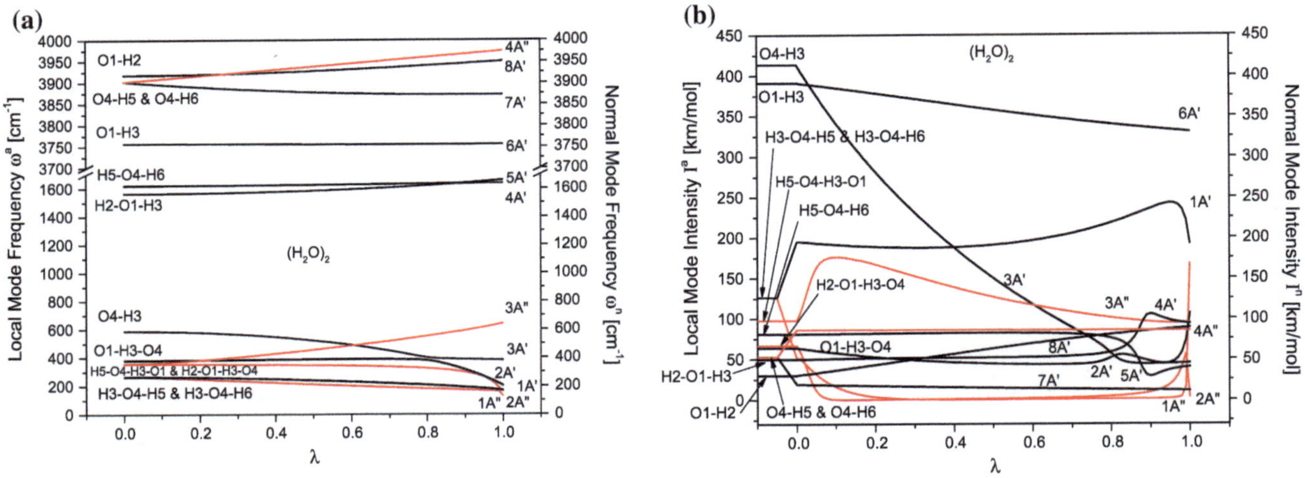

Fig. 3 **a** Frequency and **b** intensity ACS for the water dimer, $(H_2O)_2$. The different mode symmetries are indicated by *different colors*. For the purpose of identifying intensity catastrophes, the $I(\lambda)$ curves are started with an arbitrary negative λ value so that the splitting at $\lambda = 0$ can be made visible. For the numbering of atoms, see Fig. 4. ωB97X-D/aug–cc–pVTZ calculations

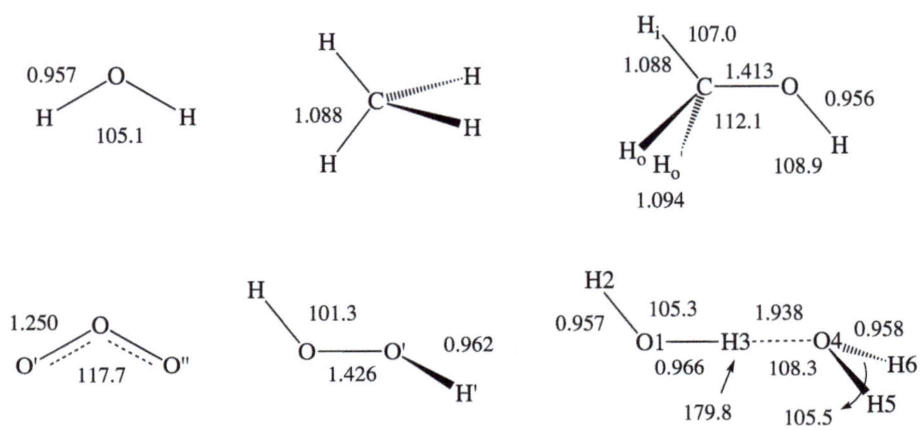

Fig. 4 ωB97X-D/aug–cc–pVTZ or CCSD/aug–cc–VTZ (ozone) geometries of the molecules investigated

The magnitude of the intensity splitting for equivalent modes depends on the corresponding values of the APT (the change in the charge distribution caused by the local mode), the ratio of the masses involved and the coupling of the mode vectors. In the case of H_2O, the two stretching mode vectors involve as a common atom the O atom, which is a prerequisite for a large coupling (proximity effect). However, the angle between the mode vectors is with 105° (Fig. 4) close to 90° where the coupling of the OH stretching modes vanishes. Also the light–heavy–light situation of the three atoms involved leads to a smaller coupling. However, the changes in the charge distribution upon OH stretching are significant so that the splitting values ΔI of the local OH stretching intensities at the catastrophe point takes a medium-seized value of ±28 km/mol (see Table 3; Fig. 1a).

An intensity catastrophe is not found in the case of isotopomer HOD because of the difference in the OH and OD stretching modes. At $\lambda = 0$, three different $\mathbf{t}_n$ vectors are determined:

$$\begin{pmatrix} 0.0000 & 0.0000 & 0.9737 \\ 0.0000 & 1.3375 & 0.0000 \\ 1.4164 & 0.0000 & 0.0000 \end{pmatrix} \tag{44}$$

and at $\lambda = \varepsilon$,

$$\begin{pmatrix} 0.0001 & 0.0007 & 0.9737 \\ 0.0001 & 1.3375 & -0.0007 \\ 1.4164 & 0.0008 & 0.0006 \end{pmatrix} \tag{45}$$

for which the changes are moderate in comparison to (44). It can be concluded that *sudden changes* in the intensity are connected with catastrophe points caused by a switching on of the molecular symmetry as a result of an infinitesimal change of $\lambda = 0$ to $\lambda = \varepsilon$.

3.2 Other examples with catastrophe points

CH_4, O_3, and HOOH. For methane, $N_{vib} = N_{para} = 9$ where the four CH bond lengths and five of the six H–C–H bending angles are used. There are just two local mode

Table 2 Comparison of normal mode and local mode frequencies and intensities obtained by ωB97X-D/aug-cc-pVTZ or CCSD/aug-cc-VTZ (ozone) calculations

Molecule mode μ	Sym.	ω_μ^{nm} (cm^{-1})	I_μ^{nm} (km/mol)	Local mode parameter n	ω_n^a (cm^{-1})	I_n^a (km/mol)
H_2O	C_{2v}					
1	A_1	1,634.8	76.6	HOH	1,634.8	76.4
2	A_1	3,878.3	4.9	OH	3,913.2	35.5
3	B_2	3,985.4	62.6	OH′	3,913.2	35.5
CH_4	T_d					
1, 2, 3	$1T_2$	1,360.4	13.3	HCH (×3)	1,450.1	5.4
4, 5	1E	1,577.0	0	HCH (×2)	1,450.1	5.4
6	$1A_1$	3,041.7	0	CH	3,126.2	15.9
7, 8, 9	$2T_2$	3,159.8	21.7	CH (×3)	3,126.2	15.9
O_3	C_{2v}					
1	$1A_1$	761.8	5.2	O′OO″	866.5	2.5
2	$1B_2$	1,253.7	231.8	OO′	1,249.7	143.0
3	$2A_1$	1,272.9	0.2	OO″	1,249.7	143.0
HOOH	C_2					
1	1A	3,93.5	164.5	HOO′H′	394.0	164.0
2	2A	1,025.6	0.4	OO′	1,020.5	0.0
3	1B	1,361.6	99.0	OO′H′	1,365.2	47.3
4	3A	1,468.8	0.4	O′OH	1,365.2	47.3
5	2B	3,839.7	54.6	OH	3,836.2	34.9
6	4A	3,841.1	14.3	O′H′	3,836.2	34.9
CH_3OH	C_s					
1	1A″	300.4	107.7	$HOCH_i$	348.9	86.8
2	1A′	1,069.6	101.4	CO	1,111.7	114.2
3	2A′	1,100.0	22.0	COH	1,260.4	77.2
4	2A″	1,189.0	0.5	OCH_o	1,302.2	8.6
5	3A′	1,377.8	27.2	OCH_i	1,277.5	2.3
6	4A′	1,485.0	2.6	OCH_o'	1,302.2	8.6
7	3A″	1,505.3	3.5	H_iCH_o	1,487.8	2.7
8	5A′	1,523.2	6.0	H_iCH_o'	1,487.8	2.7
9	6A′	3,019.2	61.6	CH_o	3,044.3	57.4
10	4A″	3,073.3	53.1	CH_o'	3,044.3	57.4
11	7A′	3,134.5	25.3	CH_i	3,120.3	25.0
12	8A′	3,916.3	34.1	OH	3,911.3	33.8
$(H_2O)_2$	C_s					
1	1 A″	129.8	167.2	H3O4H5	264.7	126.3
2	2A″	166.8	3.4	H5O4H3O1	354.7	66.5
3	1A′	168.2	192.3	H3O4H6	264.7	126.3
4	2A′	203.3	106.2	O4H3⋆	587.6	413.4
5	3A′	381.3	45.0	O1H3O4⋆	380.9	63.9
6	3A″	643.2	95.4	H2O1H3O4	357.0	98.0
7	4A′	1,637.8	93.4	H5O4H6⋆	1,621.7	81.0
8	5A′	1,659.2	40.0	H2O1H3⋆	1,564.4	49.7
9	6A′	3,755.4	330.4	O1H3	3,757.2	390.9
10	7A′	3,871.7	11.1	O4H5	3,902.3	52.8
11	8A′	3,950.4	88.7	O1H2	3,918.1	30.1
12	4A″	3,974.8	85.0	O4H6	3,902.3	52.8

A star indicates that a change in ordering occurred due to an avoided crossing. For a notation of atoms, see Fig. 4

frequencies and two local mode intensities, which in two separate catastrophes (catastrophe points both at $\lambda = \varepsilon$) slit up into five different intensities with weight factors 3:2:1:2:1 (stretching, bending, bending, bending, stretching; see Fig. 1b). In the case of the CH-stretching modes, an intensity value of 22 km/mol associated with the $2T_2$-symmetrical (triply degenerate) asymmetric CH-stretching modes and an intensity of 0 associated with the A_1-symmetrical CH-stretching mode is obtained. It is noteworthy that the intensities of the CH-stretching modes fulfill a sum rule

$$\sum_k^{Nd}(\Delta I)_k = \sum_k^{Nd}(I_{\lambda=\epsilon})_k - (I_{\lambda=0})_k = 0 \qquad (46)$$

The *intensity sum rule* is a result of the molecular symmetry and is fulfilled provided that the members of a set of equivalent local modes (i.e., all members of the set have identical local mode frequencies) cannot mix with other modes possessing the same symmetry. Hence, the deviation from the zero value given by the intensity sum rule is a measure for the degree of mode mixing. In Table 3, the sum rule is tested for the infrared intensities of the molecules investigated in this work.

The sum rule of the local mode intensities is nicely fulfilled for the set of local CH-stretching modes (ω_n^a = 3,126 cm^{-1}, Table 2) and the set of local HCH bending modes (ω_n^a = 1,450 cm^{-1}, Table 2). For the former, the positive ΔI values of the three asymmetric CH-stretching intensities ($2T_2$ symmetry) is balanced by the strong decrease in the symmetric CH-stretching intensity ($1A_1$) leading to a sum of just −0.02 km/mol. For the HCH bending intensities, the situation is different as the intensities associated with the $1T_2$- and $1E$-symmetrical mode sets split up into three parts (2:1:2). This is a result of the fact that from six possible HCH bending angles only five are relevant thus fulfilling the requirement $N_{vib} = N_{para}$. The sum rule leads to a value of 0.02, which is indicative for some residual mixing of modes of the same symmetry.

In the case of ozone (see Fig. 1c), there is a deviation from the intensity sum rule by 0.67 km/mol (see Table 3), which results from the mixing of the symmetric OO stretching mode with the bending mode. This is stronger

Table 3 Splittings ΔI of intensities of equivalent local modes at the catastrophe point $\varepsilon = 0.01$

Molecule parameter	Sym.	$I_{\lambda=0}$ (km/mol)	$I_{\lambda=\varepsilon}$ (km/mol)	ΔI (km/mol)
H_2O				
OH	$2A_1$	35.49	7.43	−28.06
OH	$1B_2$	35.49	63.52	28.03
Sum				−0.03
CH_4				
HCH	$1T_2$ (×2)	5.43	10.87	5.44
HCH	$1T_2$ (×1)	5.43	3.93	−1.50
HCH	1E (×1)	5.43	1.50	−3.93
HCH	1E (×1)	5.43	0.00	−5.43
Sum				0.02
CH	$1A_1$	15.89	0.00	−15.89
CH	$2T_2$	15.89	21.18	5.29
Sum				−0.02
O_3				
OO	$2A_1$	143.02	0.45	−141.90
OO	$1B_2$	143.02	284.92	142.57
Sum				0.67
HOOH				
OOH	3A	47.31	0.00	−47.31
OOH	1B	47.31	94.67	47.36
Sum				0.05
OH	4A	34.89	14.71	−20.18
OH	2B	34.89	55.07	20.18
Sum				0
CH_3OH				
OCH_o	4A′	8.58	5.53	−3.05
OCH_o	2A″	8.58	10.59	2.01
Sum				−1.04
CH_o	6A′	57.37	53.67	−3.70
CH_o	4A″	57.37	61.11	3.74
Sum				0.04
$(H_2O)_2$				
HOH	1A′	126.29	51.13	−75.16
HOH	1A″	126.29	195.00	68.71
Sum				−6.45
OH	7A′	52.82	19.05	−33.77
OH	4A″	52.82	86.47	33.65
Sum				−0.12

Sum denotes the value according to the sum rule of Eq. (46). For a notation of atoms, see Fig. 4

than in the case of H_2O because of an increase in the bending angle from 105° to 118° (Fig. 4) and a mass ratio equal to 1, which both facilitate mode mixing.

The intensity ACS in Fig. 1c reveals that the intensity of the $1B_2$-symmetrical mode decreases from 284.9 to 231.8 km/mol. This is a direct effect of mass coupling, as reflected by the increase in the negative off-diagonal element of matrix $\mathbf{G}^{-1}$, which connects the asymmetric with the symmetric OO stretching mode. Actually, the same effects can be found for the intensities of the asymmetric XH stretching modes in H_2O and CH_4. However, the decrease in the local mode intensities is much smaller in these cases because mass coupling is smaller for these light–heavy–light situations.

It is interesting to note that the intensity of the local OO stretching modes are high (143 km/mol, Table 2), which is predominantly due to a large ATP element, which in turn is in line with a relatively large change in charge upon OO stretching in a molecule with high biradical character.

For hydrogen peroxide, the sum rule is exactly fulfilled for the OH stretching intensities and approximately for the OOH bending intensities (Table 3). Mass coupling has a relatively small effect on the intensities as can be seen in Fig. 1d. The largest element in the APT is found for the torsional motion as a result of the relatively large change in the charge distribution upon HOOH torsion.

3.3 Methanol, CH_3OH

There are 12 normal and 12 local vibrational modes, which are shown in the frequency ACS of Fig. 2a. The corresponding intensity ACS is given in Fig. 2b. There are familiar features such as the intensity splitting of the two CH_o stretching intensities and that of the two O–C–H_o bending intensities where only the first fulfill the intensity sum rule and the latter deviate because of coupling with other modes of the same symmetry (see Table 3). The CO stretching mode and the H–O–C–H_i torsional mode possess the largest intensities which is due the polarity of the CO bond and the relatively large changes in the charge distribution accompanying these vibrational modes as is confirmed by the corresponding elements of the APT.

Contrary to the intensity ACS shown in Fig. 1a–d, in which the intensity lines mostly change almost linearly (after a possible catastrophe point) from the local mode to the normal mode intensities for $\lambda = 1$, there are strong variations in the intensity of the CO stretching, the C–O–H bending, the O–C–H_o bending, and the O–C–H_i bending modes (Fig. 2b). These variations in the intensity are the results of avoided crossings between these modes. For example, there is an avoided crossing at $\lambda = 0.87$ between modes $1A'$ and $2A'$, which are related to the local CO stretching and C–O–H bending modes (see Fig. 2a). At the avoided crossing, there is a strong mode mixing accompanied by the exchange of mode character and mode energy. Figure 2b reveals that as a consequence also the mode intensities change in the sense that the CO stretching intensity is enhanced and that of the C–O–H bending intensity by about the same amount decreased. In the region of the avoided crossings, the two intensity curves change in a complementary fashion.

Avoided crossings are also found at $\lambda = 0.02$ between modes $2A'$ and $3A'$, which are related to the local C–O–H and O–C–H_i bending modes (Fig. 2a) and at $\lambda = 0.98$ between modes $4A'$ and $5A'$, which are related to the local O–C–H_o and H–C–H bending modes (Fig. 2a). These avoided crossings are responsible for the steep complementary changes in the C–O–H and O–C–H_i bending intensities I (λ) for small λ (i.e., on the local mode side). Involved is also the local O–C–H_i intensity due to an avoided crossing between modes $3A'$ and $4A'$ (Fig. 2a). The avoided crossing at $\lambda = 0.98$ is however too late to have a large impact on the $4A'$ and $5A'$ intensities (Fig. 2b)

3.4 H_2O dimer, $(H_2O)_2$

The intensity ACS of the water dimer (Fig. 3b) is characterized by a large intensity change of the H-bond (O4 $\cdots$ H3) stretching intensity from 588 to 106 km/mol of the 2A′ mode, which due to a transfer of the mode character from the 3A′ to the 2A′ mode at the avoided crossing at $\lambda = 0.8$ (Fig. 3a) where the latter becomes the H-bond stretching mode (mixed with O1 − H3 $\cdots$ O4 bending character) and, because of the 2A′ − 1A′ avoided crossing at $\lambda = 0.98$ (Fig. 3a), an addition of H3 $\cdots$ O4−H5 bending character [27]. Again at the two avoided crossings, the changes of the 3A′ and 2A′ (2A′ and 1A′) intensity curves are complementary (Fig. 3b).

This observation leads to the important conclusion that the local H-bond stretching intensity has a large value because of a large change in charge accompanying the stretching motion. This effects the polarization of the charge distribution in the two water molecules since this is determined by H-bonding. However, mass coupling (caused by stepwise switching on of the masses of the other four atoms in the water dimer; H3 and O4 have already their correct masses) leads to a significant decrease in the intensity. Therefore, the measured normal mode intensity is no longer a reliable descriptor of the charge distribution caused by H-bonding in the dimer.

As noted before, there are no avoided crossings in the intensities. Hence, one must follow the frequency ACS, which is obtained by solving the vibrational eigenvalue problem in dependence of λ to determine that normal mode, which is dominated by H-bond stretching character. This is the 2A′ mode, which can be confirmed by the analysis of normal modes in terms of local modes [25, 53].

It is noteworthy that the local O1–H3 stretching mode has also a relatively large intensity because it is directly involved in the charge polarization caused by H-bonding. However, this mode (6A′) does not experience any avoided crossings with other A′ modes (Fig. 3a) and its mass dependence is smaller than that of O4 $\cdots$ H3. Therefore, the decrease in the intensity is just from 391 to 330 km/mol (Table 2).

There are also jumps in the intensity ACS curve of the local H3 $\cdots$ O4−H5 bending mode (converting to the $1A''$ normal mode, which starts at 126.3 km/mol for $\lambda = 0$ drops down to 51 km/mol because of a catastrophe point (mixing with the H3 $\cdots$ O4−H6 bending), then continues to decrease to 0 km/mol because of avoided crossings with the torsional modes H2−O1−H3 $\cdots$ O4 and H5−O4 $\cdots$ H3−O1 (converting into $3A''$ and $2A''$), and finally experiencing a steep increase to 167 km/mol because of an avoided crossing with the $2A''$ mode at $\lambda = 0.98$, which makes the intensity of the $3A''$ mode drop down to 3 km/mol. Other changes in the intensity curves of Fig. 3b can also be explained by identifying the avoided crossings in the frequency ACS. Conversely, an avoided crossing in the frequency ACS can be confirmed by inspection of the intensity ACS and identifying then the complementary changes in the intensity lines of the modes involved.

4 Chemical relevance of the local mode intensities

As mentioned in the introduction, local mode intensities are derived to get a direct insight into the charge distribution of a molecule. Secondly, the local mode intensities together with the local mode frequencies provide the basis for analyzing infrared spectra. Also, the local mode intensities are tools for a better quantum chemical calculation of infrared intensities. Finally, local mode intensities can be used for the calibration of weakly coupled or completely uncoupled anharmonic oscillator models.

4.1 Infrared intensities and the molecular charge distribution

The derivation of atomic charges from measured quantities has been a major effort in chemistry. Promising in this respect is the determination of effective atomic charges from infrared intensities, [32] which was strongly advocated by Person and Zerbi [33], Galabov and Dudev [34] and their co-workers. The approach had limited success although the line of action was well-defined. This had to do with the fact that for the determination of effective atomic charges, the APT is needed, which could not be obtained from intensities without addition information from quantum chemistry. However, if the APT is fully known, effective atomic charges, which reasonably correlate with natural bond orbital (NBO) charges, [54] can be determined as was demonstrated by Milani et al. [35].

In our work, we pursue a different approach. We will use the local bond stretching intensities to determine effective bond charges, [34] which determine the charge transfer between two bonded atoms and by this the bond polarity. The exact quantum chemical calculation of the bond polarity is only possible by using highly correlated coupled cluster methods because this requires a well-balanced description of covalent and ionic states in a correlated wave function. For the understanding of bonding, one needs to know the covalent and the ionic (polar) contributions to the bond strength.

The new method of calculating effective bond charges from local mode intensities will comprise the following steps: (i) Calculation of the APT of a target molecule; (ii) Improvement in the APT with the help of measured normal mode intensities; (iii) Use of Eq. (35) to obtain local mode intensities; (iv) Determination of bond charges Δp_n using the relationship $\Delta p_n = \sqrt{I_n^a / G_{nn}}$; and (v) Calculation of bond dipole moments from the known molecular geometry and the Δp_n values. If only relative intensities are measured, which is mostly the case, a reliable quantum chemical calculation of the infrared intensities is needed to convert them into absolute intensities. For quantum chemically calculated intensities, it will be interesting to see how the effective bond polarity and the bond dipole moment derived from local mode intensities differ from those obtained by a population analysis (often derived in a somewhat arbitrary way).

4.2 Analysis of infrared spectra

In this work, we have shown that the normal mode intensities are the result of mode mixing. Therefore, they are not associated with a specific structural unit. By determining the local mode frequency and intensity, the consequences of mode coupling for normal mode frequencies and intensities can be given in detail by the ACS diagrams such as those shown in Figs. 1, 2 and 3. This is particularly interesting when local mode properties are determined on the basis of measured vibrational data [27, 28]. Then, it is possible to determine the strength of a bonding interaction from the local stretching force constant and the bond polarity from the local intensity. Since this information would be extracted from experiment rather than quantum chemical data, the shortcomings of method, basis set, or harmonic approximation used in a quantum chemical calculation would not need to be discussed. Such an analysis would show that H-bonding is combined with a large charge separation contrary to the small intensity of the H-bond stretching band at 143 cm^{-1} [27]. The combination of APT and local mode intensity would provide the effective atomic charges and the magnitude of the effective bond charge equal to the charge transfer.

4.3 Improved scaling procedures

The local mode frequencies can be used for a superior scaling of quantum chemical frequencies calculated for large molecules utilizing the harmonic approximation. Since a local mode is associated with a given internal coordinates, local mode frequencies for molecular units such as XH, AB, ABH, ABC, etc. can be determined from measured [26] and calculated frequencies. The determination of scaling factors for well-defined structural units is straightforward and can be used to an individual normal mode frequency. Each normal mode can be decomposed into local mode contributions, i.e., for each mode the percentage of XH stretching, ABC bending, etc. can be determined. Then, each mode contribution is assigned the appropriate local mode scaling factor and an individual frequency scaling factor is calculated from the properly weighted local scaling factors of the local modes contributing to the normal mode in question. This dynamic scaling approach is superior to previous static scaling procedures, which could not consider the effect of mode coupling. A stretching mode may have a significant contribution from bending, and therefore, significant down-scaling of the frequency as needed for pure stretching frequencies is inappropriate because harmonic bending frequencies have to be less reduced than harmonic stretching frequencies.

The need for individual intensity scaling is even larger than for frequency scaling as was emphasized in various articles [55, 56]. The use of an individual scaling scheme as described in the case of the normal mode frequencies is only possible with the help of the local mode intensities. For example, in a recent investigation, the cyano-stretching intensities were scaled with a common factor leading to an improved but not exact agreement with experiment [56]. This we see as a result of different coupling situations for the CN stretching vibration with other modes, thus triggering stronger changes in the intensity ACS. Another source of error results from solvent influences, which are different for polar and nonpolar groups in a molecule. The calculation of local mode intensities can provide exact information with regard to solvent dependable intensity changes.

4.4 Calibration of harmonically coupled oscillator models (HOM)

The local mode model of molecular vibrations developed in the 70s by Henry et al. [37] has been used in the description of CH-stretching overtone spectra [38–40, 57,

58]. In this approach, harmonically coupled Morse oscillators (including anharmonicity effects) are used to quantum mechanically determine the vibrational frequencies and wave functions of the HOM. By employing quantum chemical methods to calculate the dipole moment function and then the oscillator strengths in the CH-stretching overtone spectra of various small and medium-seized organic molecules (alkanes, alkenes, alcohols, H-bonded complexes, etc.) a useful account of frequencies and intensities of overtones up to $v = 6$ could be determined [23].

In previous work, Cremer and co-workers demonstrated that the local mode frequencies of Konkoli and Cremer [24] linearly correlate ($R^2 = 0.990$) with the overtone frequencies of Henry for $v = 6$ [11]. Also, it could be shown by these authors that residual couplings lead to deviations from the ideal local oscillator model. This opens up the possibility of providing suitable local force constants for the parametrization of the HOM. This can be done for anharmonically corrected force constants where the latter are derived from measured vibrational frequencies using the Konkoli–Cremer approach [26]. Rong et al. [40] observed that the harmonically coupled anharmonic oscillator model leads to exaggerated intensities depending on the quantum chemical method and basis set used for calculating the dipole moment function. Clearly, the prediction of overtone intensities can be improved by proper scaling where again local mode intensities derived from experiment and from calculations would lead to suitable scaling factors as described in the previous subsection.

5 Conclusions

In this work, we have introduced the local mode intensities and the intensity ACS for the purpose of analyzing normal mode intensities. A number of conclusions can be drawn from the work presented here.

1. The local mode intensity has been derived by expressing the normal mode intensity in internal coordinates and then applying the adiabatic condition ($\lambda = 0$), which leads to three requirements defining $\mathbf{D}_0, \mathbf{M}_0^R, \tilde{\mathbf{D}}_0$, and by this also the local mode intensity I_n^a associated with mode $\mathbf{a}_n$ and the internal coordinate q_n. Local mode intensities I_μ^{nm} are independent of the coordinates q_m, $(m \neq n)$ used for the description of the molecular geometry. They are also independent of any variation in the isotope distribution in other parts of the molecule but the target fragment described by q_n. The local mode intensity values of a molecule comply with the molecular symmetry and become identical for diatomic molecules to the normal mode intensity.
2. The intensity ACS has been derived in this work by exploiting the dependence of the normalized internal coordinate mode vectors on λ, which can be expressed in the form $(\mathbf{G}_\lambda)^{-1}\tilde{\mathbf{D}}_\lambda$. The first term of the matrix product reflects the effect of mass coupling on the intensity, which increases as the off-diagonal elements of the $\mathbf{G}$-matrix increase with λ. The second product term describes the effect of mode mixing on the intensity I_λ.
3. Mass coupling leads to an essentially linear decrease or increase in the local to the normal mode intensity. This can be anticipated by an analysis of the matrix $\mathbf{G}^{-1}$.
4. Symmetry-equivalent local modes strongly couple, which leads to a large change in the corresponding intensities upon an infinitesimal increase of λ from 0 to ϵ, for which we have coined the term *intensity catastrophe* because the sudden change in the local mode intensities complies with the mathematical definition of a catastrophe [51]. The changes in the intensities accompanying an intensity catastrophe depend on the position of the local mode units in the molecule (proximity effect), the alignment of the local mode vectors, the ratio of the masses involved and the change in the charge distribution caused by the local mode vibrations.
5. All members of a group of equivalent local modes are involved in an intensity catastrophe and if there is no coupling with modes outside the group, the intensity changes ΔI caused by the catastrophe sum to zero (*intensity sum rule*). The deviation from the sum rule is a quantitative measure for the coupling with other modes of the same symmetry.
6. Avoided crossings in the frequency ACS between modes of the same symmetry can lead to strong nonlinear changes in the intensity curves I_λ, which are largely complementary for the vibrational modes involved. Depending on the type of avoided crossing, the mode character is transferred from one mode to the other or the original mode changes its character by absorbing additional local mode contributions. This has to be considered when analyzing intensity changes from local to normal modes.
7. There are no avoided crossings in an intensity ACS. However avoided crossings in the frequency ACS can be detected by identifying modes with strong, but complementary changes in the intensities.
8. In each case investigated, one obtains detailed physical explanations why a normal mode intensity adopts a particular value. This is facilitated by evaluating matrices $\mathbf{\Delta}, \mathbf{M}^{-1}, \mathbf{B}^\dagger, (\mathbf{G}_\lambda)^{-1}$, and $\tilde{\mathbf{D}}_\lambda$ for

specific values of λ between 0 and 1 where the first three matrices do not change with λ.

9. There have been attempts to relate normal mode intensities to bond charges [34, 47]. On the background of this work, we can say that these attempts are only meaningful in cases of minimal local mode coupling and large mass ratios. In general, this approach is useless as long as it does not start from local mode intensities.
10. We have shown the chemical relevance of local mode intensities with regard to the determination of bond charges and bond polarity, for the analysis of infrared spectra, for the development of a dynamic scaling method of calculated harmonic infrared intensities, and for the calibration of harmonic and anharmonic oscillator models.

In this work, we have not put emphases on the correct calculation of the infrared intensities because the derivation of the basic concept of local mode intensities and the analysis of normal mode intensities in terms of the former is at the primary focus of this work. However, in future work we will focus on the determination of local mode intensities from experimental intensities or coupled cluster intensities. Also, we will derive effective bond charges from local mode intensities, which are no longer contaminated by mass- or mode-coupling.

Acknowledgments This work was financially supported by the National Science Foundation, Grant CHE 1152357. We thank SMU for providing computational resources.

References

1. Pauling L (1960) The nature of the chemical bond. Cornell University Press, Ithaca
2. Lide DR (2009) CRC handbook of chemistry and physics, 90th edn. CRC, Boca Raton, FL
3. Luo Y-R (2007) Comprehensive handbook of chemical bond energies. Taylor and Francis, Boca Raton
4. Herzberg G, Huber KP (1979) Spectra and molecular structure. IV. Constants of diatomic molecules. Van Nostrand, Reinhold, NY
5. Cremer D, Wu A, Larsson A, Kraka E (2000) Some thoughts about bond energies, bond lengths, and force constants. J Mol Model 6:396–412
6. Kalescky R, Kraka E, Cremer D (2013) Identification of the strongest bonds in chemistry. J Phys Chem A 117:8981–8995
7. Cremer D, Kraka E (2010) From molecular vibrations to bonding, chemical reactions, and reaction mechanism. Curr Org Chem 14:1524–1560
8. Wilson EB, Decius JC, Cross PC (1955) Molecular vibrations. The theory of infrared and Raman vibrational spectra. McGraw-Hill, New York
9. Woodward LA (1972) Introduction to the theory of molecular vibrations and vibrational spectroscopy. Oxford University Press, Oxford
10. Badger RM (1934) A relation between internuclear distances and bond force constants. J Chem Phys 2:128–131
11. Kraka E, Larsson A, Cremer D (2010) Generalization of the badger rule based on the use of adiabatic vibrational modes in vibrational modes in computational IR spectroscopy. In: Grunenberg J (ed) Computational spectroscopy: methods, experiments and applications, Wiley, New York, p 105
12. Silverstein RM, Webster FX, Kiemle D (2005) Spectrometric identification of organic compounds. Wiley, Hobroken
13. Smith B (1999) Infrared spectral interpretation: a systematic approach. CRC Press, Boca Raton
14. Decius JC (1963) Compliance matrix and molecular vibrations. J Chem Phys 38:241–248
15. Grunenberg J, Goldberg N (2000) How strong is the Gallium≡Gallium triple bond? Theoretical compliance matrices as a probe for intrinsic bond strengths. J Am Chem Soc 122:6045–6047
16. Brandhorst K, Grunenberg J (2008) How strong is it? The interpretation of force and compliance constants as bond strength descriptors. Chem Soc Rev 37(8):1558–1567
17. Brandhorst K, Grunenberg J (2010) Efficient computation of compliance matrices in redundant internal coordinates from Cartesian Hessians for nonstationary points. J Chem Phys 132:184101–184107
18. Madhav MV, Manogaran S (2009) A relook at the compliance constants in redundant internal coordinates and some new insights. J Chem Phys 131:174112–174116
19. Espinosa A, Streubel R (2011) Computational studies on azaphosphiridines, or how to effect ring-opening processes through selective bond activation. Chem Eur J 17:3166–3178
20. McKean DC (1978) Individual CH bond strengths in simple organic compounds: effects of conformation and substitution. Chem Soc Rev 7:399–422
21. Duncan JL, Harvie JL, McKean DC, Cradock C (1986) The ground-state structures of disilane, methyl silane and the silyl halides, and an SiH bond length correlation with stretching frequency. J Mol Struct 145:225–242
22. Murphy WF, Zerbetto F, Duncan JL, McKean DC (1993) Vibrational-spectrum and harmonic force-field of trimethylamine. J Phys Chem 97:581–595
23. Henry BR (1987) The local mode and overtone spectra: a probe of molecular structure and conformation. Acc Chem Res 20:429–435
24. Konkoli Z, Cremer D (1998) A new way of analyzing vibrational spectra. I. Derivation of adiabatic internal modes. Int J Quantum Chem 67:1–9
25. Konkoli Z, Cremer D (1998) A new way of analyzing vibrational spectra. II. Characterization of normal vibrational modes in terms of internal vibrational modes. Int J Quantum Chem 67:29–40
26. Cremer D, Larsson JA, Kraka E (1998) New developments in the analysis of vibrational spectra: on the use of adiabatic internal, vibrational modes. In: Parkanyi C (ed) Theoretical and computational chemistry, volume 5, theoretical organic chemistry. Elsevier, Amsterdam, p 259
27. Kalescky R, Zou W, Kraka E, Cremer D (2012) Local vibrational modes of the water dimer—comparison of theory and experiment. Chem Phys Lett 554:243–247
28. Kalescky R, Kraka E, Cremer D (2013) Local vibrational modes of the formic acid dimer—the strength of the double hydrogen bond. Mol Phys 111:1497–1510
29. Larsson A, Cremer D (1999) Theoretical verification and extension of the McKean relationship between bond lengths and stretching frequencies. J Mol Struct 485:385–407
30. Zou W, Kalescky R, Kraka E, Cremer D (2012) Relating normal vibrational modes: to local vibrational modes with the help of an adiabatic connection scheme. J ChemPhys 137:084114
31. Zou W, Kalescky R, Kraka E, Cremer D (2012) Relating normal vibrational modes to local vibrational modes: benzene and naphthalene. J Mol Model 19:2865–2877
32. Person WB, Newton JH (1974) Dipole moment derivatives and infrared intensities. I. Polar tensors. J Chem Phys 61:1040–1049

33. Person WB, Zerbi G (1982) Vibrational Intensities in infrared and Raman spectroscopy. Elsevier, Amsterdam
34. Galabov B, Dudev T (1996) Vibrational intensities. Elsevier, Amsterdam
35. Milani A, Tommasini M, Castiglioni C (2012) Atomic charges from ir intensity parameters: theory, implementation and application. Theor Chem Acc 131:1–17
36. Boatz JA, Gordon MS (1989) Decomposition of normal-coordinate vibrational frequencies. J Phys Chem 93:1819, 1989
37. Hayward RJ, Henry BR (1975) A general local-mode theory for high energy polyatomic overtone spectra and application to dichloromethane. J Mol Spectrosc 57:221–235
38. Kjaergaard HG, Yu H, Schattka BJ, Henry BR, Tarr AW (1990) Intensities in local mode overtone spectra: propane. J Chem Phys 93:6239–6248
39. Kjaergaard HG, Turnbull DM, Henry BR (1993) Intensities of CH- and CD-stretching overtones in 1,3-butadiene and 1,3-butadiene-d_6. J Chem Phys 99:9438–9452
40. Rong Z, Henry BR, Robinson TW, Kjaergaard HG (2005) Absolute intensities of CH stretching overtones in alkenes. J Phys Chem A 109:1033–1041
41. Jacob CR, Luber S, Reiher M (2009) Understanding the signatures of secondary-structure elements in proteins via Raman optical activity spectroscopy. Chem Eur J 15:13491–13508
42. Jacob CR, Reiher M (2009) Localizing normal modes in large molecules. J Chem Phys 130:084106
43. Liegeois V, Jacob CR, Champagne B, Reiher M (2010) Analysis of vibrational Raman optical activity signatures of the $(TG)_N$ and $(GG)_N$ conformations of isotactic polypropylene chains in terms of localized modes. J Phys Chem A 114:7198–7212
44. Sokolov VI, Grudzev NB, Farina IA (2003) Local vibrational mode in zinc telluride associated with a charged nickel impurity. Phys. Solid State 45:1638–1643
45. Sangster MJL, Harding JH (1986) Calculation of local and gab mode frequencies from impurities in alkali halide crystals. J Phys C Solid State Phys 19:6153–6158
46. Califano S (1976) Vibrational states. Wiley, London
47. Zerbi G (1982) Introduction to the theory of vibrational frequencies and vibrational intensities. In: Person W and Zerbi G (eds) Vibrational intensities in infrared and Raman spectroscopy, Elsevier, Amsterdam, p 23
48. Yamaguchi Y, Goddard JD, Osamura Y, Schaefer HFS (1994) A new dimension to quantum chemistry: analytic derivative methods in ab initio molecular electronic structure theory. Oxford University Press, Oxford
49. Chai J-D, Head-Gordon M (2008) Long-range corrected hybrid density functionals with damped atomatom dispersion corrections. Phys Chem Chem Phys 10:6615–6620
50. Dunning TH (1989) Gaussian basis sets for use in correlated molecular calculations. J Chem Phys 90:1007–1023
51. Thom R (1975) Structural stability and morphogenesis. Benjamin, Reading
52. Poston T, Stewart I (1981) Catastrophe theory and its applications. Pitma, Boston
53. Konkoli Z, Larsson JA, Cremer D (1998) A new way of analyzing vibrational spectra IV. Application and testing of adiabatic modes within the concept of the characterization of normal modess. Int J Quantum Chem 67:41
54. Reed AE, Curtiss LA, Weinhold F (1988) Intermolecular interactions from a natural bond orbital, donor–acceptor viewpoint. Chem Rev 88:899
55. Hallas MD, Schlegel HB (1998) Comparison of the performance of local, gradient-corrected, and hybrid density functional models in predicting infrared intensities. J Chem Phys 109:10587–10593
56. Stoyanov SS (2010) Scaling of computed cyano-stretching frequencies and IR intensities of nitriles, their anions, and radicals. J Phys Chem A 114:5149–5161
57. John U, Nair KPR (2005) Near ir overtone spectral investigations of cyclohexanol using local mode model-evidence for variation of anharmonicity with concentration due to hydrogen bonding. Spectrochim Acta A 61:2555–2559
58. Morita M, Takahashi K (2013) Multidimensional oh local mode calculations for $OH^{-}(H_2O)_3$-importance of intermode anharmonicity. Phys Chem Chem Phys 15:114–124

Theor Chem Acc (2014) 133:1454
DOI 10.1007/s00214-014-1454-0

REGULAR ARTICLE

The infrared spectra of $C_{96}H_{25}$ compared with that of $C_{96}H_{24}$

Charles W. Bauschlicher Jr. · Alessandra Ricca

Received: 6 August 2013 / Accepted: 18 January 2014 / Published online: 8 February 2014

Abstract The addition of one H atom to $C_{96}H_{24}$ has been studied for the neutral, cation, and anion. Hydrogen atom binding at the solo site is the most favorable for all three charge states. The solo and duo sites are significantly more strongly bound than the endo positions. One extra hydrogen atom has very little effect on the infrared spectra. It is unlikely that species with one extra hydrogen could be identified from the astronomical emission spectra.

Keywords Astrochemistry · Molecular data · Infrared bands · Interstellar medium · DFT

1 Introduction

The observed spectra of many astronomical objects show emission features at 3.3, 6.2, 7.7, 8.6, and 11.2 μm, which have been attributed [1, 2] to polycyclic aromatic hydrocarbons (PAHs). A detailed analysis of these spectra reveals that the PAH features vary between different classes of objects and even spatially within extended objects (e.g., [3–8]). This suggests that the observed emission features depend on the specific PAH molecules present and the conditions within the emission zones. One puzzle that remains today is the cause of the variation in the band that falls near 6.2 μm. Peeters et al. [3] analyzed the spectra of 57 sources and found three classes of objects, which they denoted as Class A, B, and C. Class A peaks between 6.19 and 6.23 μm, Class B peaks between 6.235 and 6.28 μm, and Class C peaks at 6.29 μm and is more symmetric than the other two classes. As expected, the three classes are representative of different sources [9]. Class A sources consist of interstellar material illuminated by a star, including HII regions, reflection nebulae, and the general interstellar medium (ISM) of the Milky Way and other galaxies. Class B sources are associated with circumstellar material, such as planetary nebulae, a variety of post-asymptotic giant branch (post-AGB) objects and Herbig Ae/Be stars. Class C sources are limited to a few extreme carbon-rich post-AGB objects. Classes A and B are more common than Class C and therefore have attracted more attention. For simplicity, the Class A and Class B bands in the 6.2–6.3 μm region are commonly referred to as the 6.2 and 6.3 μm bands, respectively. The Class B bands are consistent with the C–C stretching frequency of typical PAH molecules, with the intensity showing that many of species are ions. The origin of the Class A species is far from obvious, and some candidates have been proposed to explain the Class A emission, for example, hydrogenated amorphous carbon [10, 11], PAHs where inner carbons have been replaced by nitrogen atoms (denoted endo-PANHs) [12] and singly protonated PAHs (denote as $HPAH^+$) [13]. While experiments [13] show that small $HPAH^+$s have a band near 6.2 μm, the overall features of their spectra are inconsistent with astronomical observations. We studied [14] protonated PAH species computationally and found that the spectra of small $HPAH^+$s agreed with experiment. However, we found that

Dedicated to Professor Thom Dunning and published as part of the special collection of articles celebrating his career upon his retirement.

C. W. Bauschlicher Jr. (✉)
Entry Systems and Technology Division Mail Stop 230-3, NASA Ames Research Center, Moffett Field, CA 94035, USA
e-mail: Charles.W.Bauschlicher@nasa.gov

A. Ricca
Carl Sagan Center, SETI Institute, 189 Bernardo Ave., Suite 100, Mountain View, CA 94043, USA
e-mail: Alessandra.Ricca-1@nasa.gov

the large $HPAH^+$s, which have an overall spectra consistent with observations, have a C–C stretching band that is consistent with Class B. Since large $HPAH^+$s are more likely to represent the species emitting in space, we concluded that it is very unlikely that the $HPAH^+$ species are responsible for the Class A emission.

In our previous study of the $HPAH^+$ species, we only considered protonation at the edge of the PAH leading to the formation of a CH_2 group. This edge binding site yields the largest binding energies and is consistent with the bonding position in the small $HPAH^+$s studied in experiment. In our study of endo-PANHs [12], we observed that the substitution of a carbon atom with a nitrogen atom shifted the 6.3 μm band to shorter wave length and that this shift increased as the substitution moved toward the center of the PAH molecule. This implies that the effect on the IR spectra depends on the substitution site. A similar effect could occur for hydrogen addition; therefore, our decision to study only $HPAH^+$s with hydrogen atoms added to edge sites, based on binding energy, may have been too restrictive. In this manuscript, we report on adding H atoms to all possible nonequivalent carbon atoms in $C_{96}H_{24}$ and compare the binding energy and IR spectra as a function of hydrogen binding site.

2 Model and methods

The parent molecule, circumcircumcoronene ($C_{96}H_{24}$), is shown in Fig. 1. The hydrogen atom can add to ten unique sites, two are at the edge of the molecule, and 8 are

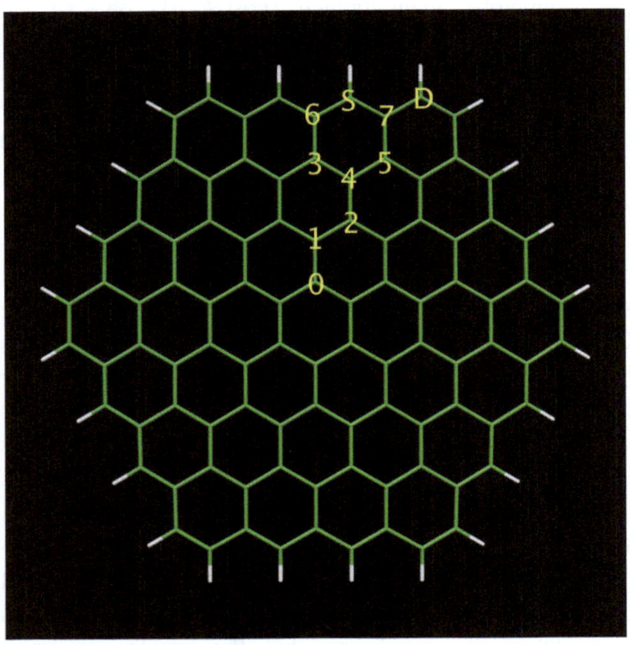

Fig. 1 Carbon sites for the hydrogen addition

internal. The edge sites are classified as solo (S) or duo (D) by the number of hydrogens on the same six-membered ring. The solo site has one hydrogen on the ring, and the duo has two hydrogens. The eight internal or endo sites are numbered from 0 to 7 starting from the center of the PAH molecule out to the edge, see Fig. 1. The hydrogen addition at the solo site and the duo site has been reported previously for the $C_{96}H_{25}^+$ cation [14], and in this work, we extend the study to include the neutral and anion species. The structures were fully optimized and the harmonic frequencies computed using density functional theory (DFT). We used the hybrid B3LYP [15, 16] functional in conjunction with the 4-31G basis set [17]. Previous work [18] has shown that the computed B3LYP/4-31G harmonic frequencies, scaled by a single scale factor of 0.958, are in excellent agreement with the matrix isolation mid-IR fundamental frequencies of the PAH molecules. However, for some PAH molecules, the results obtained with the B3LYP functional are inconsistent with experiment. We have found [19] that switching to the BP86 functional [20, 21] can be used to evaluate the validity of the B3LYP results. If the B3LYP and BP86 results are in agreement, then B3LYP spectra are likely to be reliable. If the two functionals yield different results, then additional information is required to determine if either spectra is correct. This problem arises for one molecule in this work and is discussed in more detail below.

While experience has shown that the B3LYP/4-31G and BP86/4-31G approaches yield reliable IR spectra for PAH molecules, bond energies have not been as extensively calibrated. Therefore, in this work, we perform some calibration calculations for the C–H bond energies of 1-hydro, 2-hydro, and 9-hydronaphthalene. B3LYP calculations are preformed using the 4-31G and 6-31G** basis sets [17]. Coupled cluster singles and doubles calculations [22], including the effect of connected triples determined using perturbation theory [23], CCSD(T), are also performed. The CCSD(T) calculations use the correlation consistent polarized triple zeta (cc-pVTZ) set of Dunning [24].

The computational studies yield integrated band intensities in km/mol, which we broaden in wavenumber space because it is linear in energy. After broadening, the *y* axis units are those of a cross section and reported in $10^5 \times cm^2$/mol. The *x* axis units are converted from cm^{-1} to μm to compare with observational results, which are commonly reported in μm.

The B3LYP spectra are broadened by a linewidth of 30 cm^{-1} for the bands shortward of 9 μm, while a linewidth of 10 cm^{-1} is used for the bands longward of 10 μm; these values are consistent with current observational and theoretical modeling, see discussion in Bauschlicher et al. [25]. For the 9–10 μm region, the full width at half maximum

Table 1 The C–H binding energy (in kcal/mol) for the "extra" hydrogen in naphthalene (in kcal/mol) as a function of the level of theory

Site	Cation			Neutral			Anion		
	B3LYP		CCSD(T)	B3LYP		CCSD(T)	B3LYP		CCSD(T)
	4-31G	6-31G**	cc-pVTZ	4-31G	6-31G**	cc-pVTZ	4-31G	6-31G**	cc-pVTZ
1-Hydro	69.0	70.1	71.6	34.7	36.7	31.4	57.9	59.5	62.2
2-Hydro	65.5	66.9	68.9	29.6	31.9	26.7	53.1	55.0	56.8
9-Hydro	47.4	49.8	53.1	7.6	10.1	6.7	36.0	38.0	43.5

These are D_e values

(FWHM) is scaled in a linear fashion (in wavenumber space) from 30 to 10 cm^{-1}. In addition to ignoring any further variations of linewidth as a function of mode, Fermi resonances are not taken into account. Despite these limitations, these idealized spectra can be useful in better understanding the astronomical spectra.

Typically, the astronomical PAHs are observed as the emission from highly vibrationally excited molecules. Hence, when comparing with observations, our computed 0 K absorption spectra should be shifted to the red to account for the difference between absorption and emission from vibrationally excited molecules. The size of this shift depends on many factors such as the size of the molecule, the anharmonicity of modes, and temperature of the emitting species. In this work, we focus on changes in the spectra with hydrogen position, so the shift is ignored; however, some consideration of the shift should be made before comparing these spectra with observation.

All of the calculations are performed using Gaussian 09 [26] or earlier versions of this code. The interactive molecular graphics tool MOLEKEL [27] was used to aid the analysis of the vibrational modes.

3 Results and discussion

The binding energy for the aliphatic C–H bond in 1-hydro, 2-hydro, and 9-hydronaphthalene is shown in Table 1 as function of the level of theory for the cation, neutral, and anion species. Overall, there is good agreement between the different levels of theory. The B3LYP/6-31G** values are 2–3 kcal/mol larger than the B3LYP/4-31G values for all species. The CCSD(T) values are larger than the B3LYP/4-31G binding energies for the cation and anion but smaller for the neutral. The 1-hydro and 2-hydro have errors in the 2.6 to 4.3 kcal/mol range, while the errors for the 9-hydro are 5.7 and 7.5 kcal/mol for the cation and anion. While there are errors associated with the B3LYP/4-31G approach, it appears to be sufficiently accurate that we can evaluate the trends in the binding energies for the larger $C_{96}H_{24}$ system.

The B3LYP/4-31G C–H binding energies of the added hydrogen atoms for the sites considered are summarized in Table 2. These binding energies have been corrected for zero-point energy using our computed harmonic frequencies. The cation and anion have larger binding energies than the neutral as they both have an open-shell electron for the incoming hydrogen to bond with. The cation binding is slightly larger than the anion. Adding the H atom to sites 0–7 causes a puckering of the carbon framework because the hybridization of the carbon at the binding site changes from sp^2 to sp^3. This leads to a disruption of the π bonding network and therefore to a smaller binding energy than bonding of the hydrogen at the edge, where the π bonding is less disrupted by the addition of an extra hydrogen. As shown in Table 2, this favoring of the solo and duo sites is rather significant. For the cation, we were able to find the transition state for the hydrogen transfer from site 0 to site 1. The barrier is almost equal to the binding energy. An inspection of the transition state geometry shows that the carbons are almost planar; that is, the π bonding is essentially restored to that of the parent, with the hydrogen only weakly interacting with the parent PAH.

Typically, the band positions and intensities, as well as, the total intensities are similar for all the PAH species of

Table 2 The C–H binding energy (in kcal/mol) for the "extra" hydrogen in $C_{96}H_{25}$ and the transition state barrier (in kcal/mol) for the hydrogen transfer between sites 0 and 1 (TS 0→1)

Site	Cation	Neutral	Anion
0	19.3	9.0	16.1
1	20.6	9.0	17.4
2	20.8	10.1	17.6
3	20.5	10.8	15.9
4	16.1	8.6	12.8
5	25.0	12.4	21.3
6	16.0	3.3	11.2
7	18.5	10.4	14.9
D	35.8	32.0	32.2
S	53.9	41.4	50.0
TS 0→1	18.2		

The binding energies have been corrected for zero-point energy approximately the same size and with the same charge state, see Figs. 2, 3, and 4. This is true for all species considered in this work, except for site 3, where the neutral has two very strong bands at 9.33 and 10.47 μm, the cation has a band at 9.17 μm that is 3 times the cross section of the bands for the other $C_{96}H_{25}^{+}$ species, and the anion has a very large band (4 times the cross section of any other $C_{96}H_{25}^{-}$ species) at 9.71 μm. Since the spectra

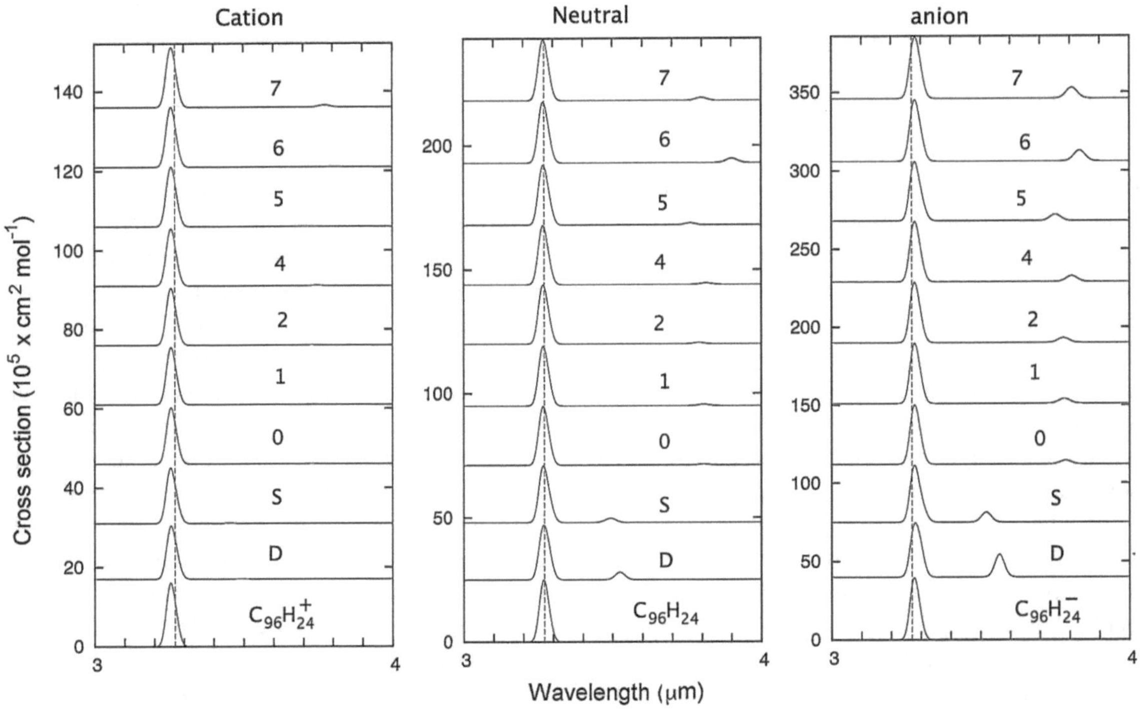

Fig. 2 The C–H stretching region of the computed spectra of the $C_{96}H_{25}$ species. The spectra of the parents, $C_{96}H_{24}$, $C_{96}H_{24}^{+}$, and $C_{96}H_{24}^{-}$, are also shown. A *dashed line* has been added at 3.27 μm to help compare the spectra

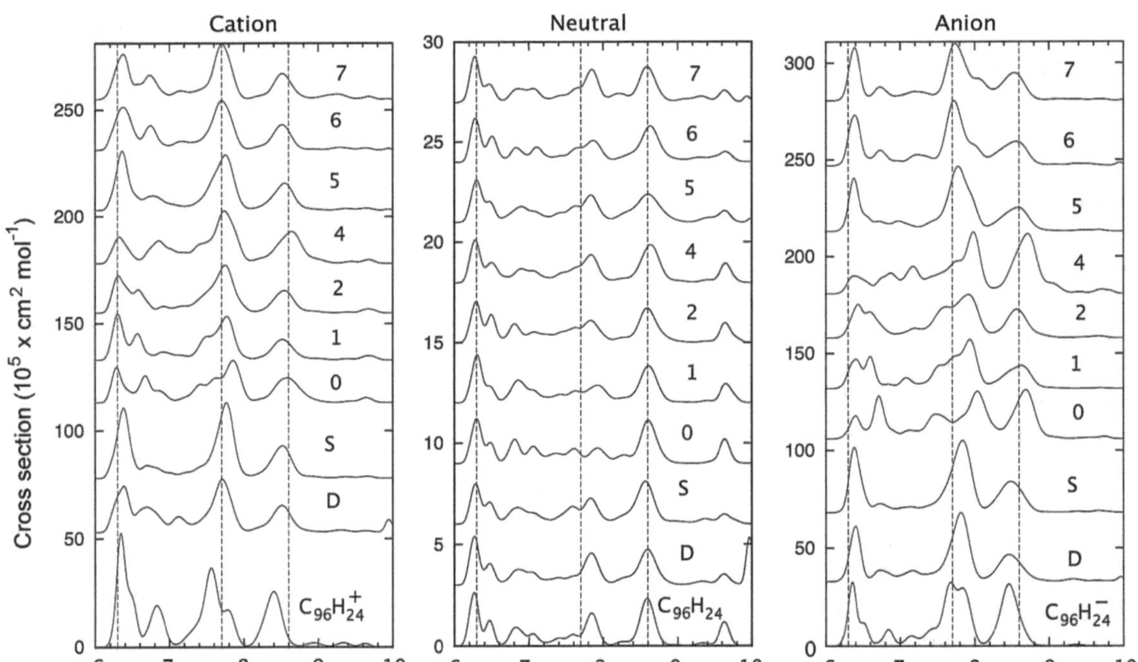

Fig. 3 The 6–10 μm region of the computed spectra of the $C_{96}H_{25}$ species. The spectra of the parents, $C_{96}H_{24}$, $C_{96}H_{24}^{+}$, and $C_{96}H_{24}^{-}$, are also shown. *Dashed lines* have been added at 6.3, 7.7, and 8.6 μm to help compare the spectra

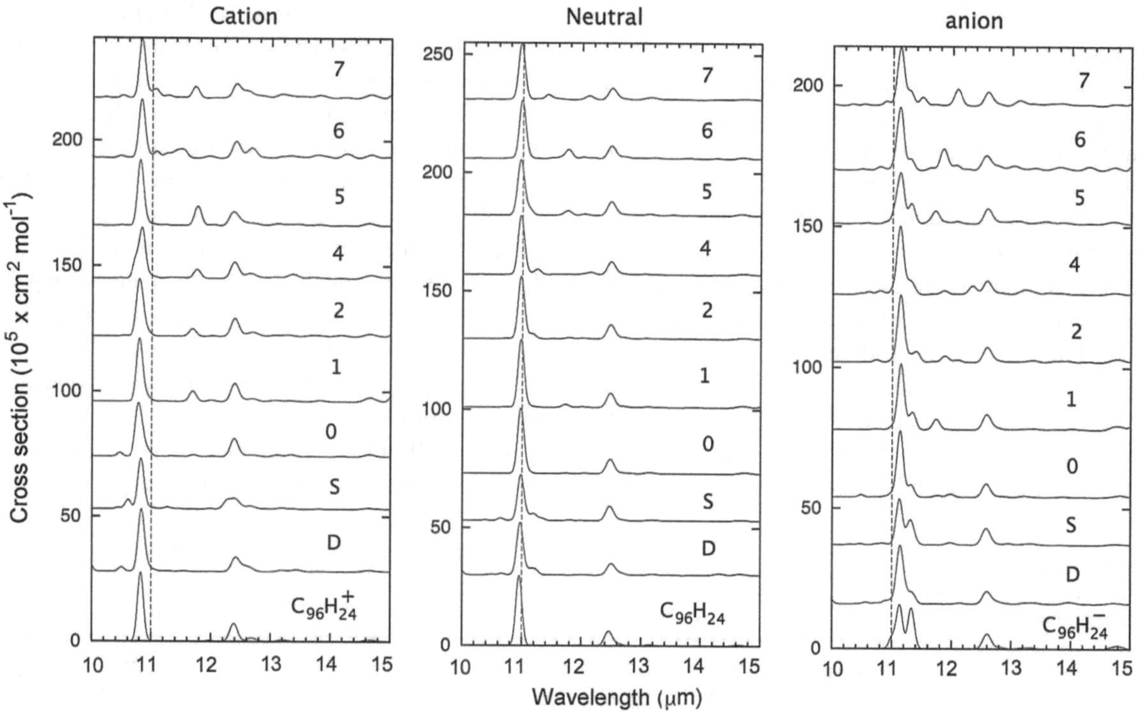

Fig. 4 The 10–15 μm region of the computed spectra of the $C_{96}H_{25}$ species. The spectra of the parents, $C_{96}H_{24}$, $C_{96}H_{24}^+$, and $C_{96}H_{24}^-$, are also shown. A *dashed line* has been added at 11.0 μm to help compare the spectra

for site 3 are clearly inconsistent with those of the other PAH spectra, we repeated the calculations using the BP86 functional for site 3 and several other sites. We first note that for all sites considered, excluding site 3, the BP86 and B3LYP spectra looked very similar, thus confirming the reliability of the B3LYP functional for these sites. For the neutral and the cation, the BP86 spectra for site 3 did not look like the B3LYP spectra, but rather looked consistent with the spectra of the other sites. For the anion, the BP86 spectrum for site 3 did not look like the B3LYP spectrum, and it also looked different from the other sites as well, namely the large band shifted to 9.0 μm. The fact that the B3LYP and BP86 results are inconsistent with each other and inconsistent with the anion results at the other sites suggests that neither the B3LYP nor BP86 is reliable for the anion of site 3. Because BP86 appears to fail for this system, we also tested a second nonhybrid functional, namely the exchange–correlation function of Perdew et al. [28], which is denoted as PW91PW91 in the Gaussian program. The PW91PW91 spectra for the ions of site 3 are very similar to those obtained using the BP86 functional. While we cannot rule out finding a functional that will work for this site, it is beyond the scope of this work to try all possible functionals. The BP86 results for the neutral and cation are consistent with our observation that similar-sized species with the same charge have similar spectra, which suggests that the BP86 results for the neutral and cation are reasonable. However, we adopt

Table 3 The strongest bands in the C–H stretching region for $C_{96}H_{24}$ and $C_{96}H_{25}$

Molecule	Cation		Neutral		Anion	
	λ	I	λ	I	λ	I
$C_{96}H_{24}$	3.255	630.6	3.268	931.3	3.279	1479.9
D	3.256	542.3	3.269	892.4	3.281	1459.9
	3.495	2.9	3.524	97.6	3.563	464.1
S	3.255	558.5	3.267	910.9	3.279	1488.0
	3.452	4.7	3.494	63.3	3.519	213.7
0	3.256	554.8	3.267	921.9	3.279	1468.2
	3.730	2.2	3.806	15.3	3.786	75.9
1	3.256	561.4	3.267	931.0	3.279	1477.4
	3.728	1.5	3.806	19.3	3.782	95.8
2	3.256	558.6	3.267	923.9	3.279	1491.8
	3.729	2.9	3.790	17.3	3.779	95.8
4	3.256	563.0	3.266	918.2	3.280	1523.3
	3.747	4.0	3.814	18.8	3.807	119.5
5	3.257	588.4	3.267	963.5	3.280	1564.7
	3.708	3.3	3.761	25.5	3.752	129.3
6	3.256	586.6	3.267	958.1	3.280	1579.5
	3.805	2.9	3.902	55.2	3.834	217.0
7	3.257	583.6	3.268	949.3	3.281	1574.3
	3.774	16.7	3.799	37.1	3.808	220.5

The band position (λ) and the intensity (I) are given in μm and km/mol, respectively. Only bands with an intensity larger than 50 km/mol for any one charge state are reported a conservative approach and do not report any infrared spectra for site 3.

The C–H stretching region of the computed spectra is shown in Fig. 2, and the band positions and intensities for the strong bands are summarized in Table 3. (N.b. All of the spectral information will be included in the next update of our PAH database [29] at http://www.astrochem.org/pahdb.) Overall, the spectra of all of the $C_{96}H_{25}$ species look very similar, where there is one main peak, with the position varying slightly with charge state and a second, weaker band that is red shifted with a peak position that varies somewhat with site. The position of the strong peak is very similar to that in the parent, where the maximum difference between the $C_{96}H_{24}$ and the $C_{96}H_{25}$ species is 0.002 μm for all three charge states. The intensity of the main peak is also similar for the parent and $C_{96}H_{25}$ species. The weaker band arises from the C–H stretching involving the "extra" hydrogen. For the edge sites, there are actually two bands, which correspond to the symmetric and asymmetric stretching of the CH_2 group, but these two bands fall close to each other and for a FWHM of 30 cm^{-1} they do not appear as separate bands. For the cations and many of the neutral species, these bands are much weaker than the strong band, see Table 3. For the anions, these bands are more intense, but still much weaker than the main band. The 0–7 positions show a larger shift away from the main band than do the solo and duo sites.

The 6–10 μm region of the computed spectra is shown in Fig. 3. We first note that the range of cross section

Table 4 The strongest bands in the C–C stretching and C–H in-plane bending region for $C_{96}H_{24}$ and $C_{96}H_{25}$

Molecule	Cation		Anion		Molecule	Cation		Anion	
	λ	I	λ	I		λ	I	λ	I
$C_{96}H_{24}$	8.401	1023.3	8.462	1354.0	D	8.509	589.0	8.464	744.6
	7.782	469.8	7.834	920.8		7.710	1387.7	7.814	1866.8
	7.563	1816.1	7.676	1461.4					
			7.151	193.5		7.122	335.5	7.166	309.1
	6.827	997.8	6.842	312.0		6.694	876.4	6.730	334.0
			6.499	296.4					
	6.345	2718.2	6.354	1251.3		6.384	1265.5	6.399	1272.4
S	8.517	722.9	8.487	886.9	0	8.578	683.4	8.690	1139.0
	7.765	1893.8	7.836	2194.0		7.854	850.7	8.036	1370.8
						7.632	352.0		
			7.136	165.6		7.420	351.5	7.478	783.3
	6.697	449.2	6.730	234.5		6.845	207.2	7.126	180.9
						6.672	570.1	6.714	1001.8
	6.380	1797.2	6.381	1643.0		6.288	850.3	6.398	555.7
1	8.516	479.4	8.633	693.7	2	8.542	512.3	8.568	743.7
	7.770	1340.0	7.935	1534.7				7.915	1327.9
			7.548	379.8		7.749	1530.9	7.617	755.4
	7.130	56.6	7.081	232.2		7.198	106.1	7.197	251.8
	6.913	212.6	6.841	68.0		6.918	126.6		
	6.568	539.4	6.597	610.6		6.579	493.7	6.591	701.0
	6.303	997.9	6.405	755.5		6.311	1042.7	6.434	879.9
4	8.642	865.1	8.715	1634.5	5	8.546	610.6	8.589	752.5
	7.736	1722.0	7.982	2213.4		7.754	1587.7	7.777	2172.8
	7.187	114.2	7.173	659.5				6.977	328.7
	6.854	790.4	6.873	643.9		6.774	548.7	6.715	202.3
	6.323	755.8	6.382	678.0		6.363	1584.4	6.377	1420.2
6	8.512	527.6	8.567	759.7	7	8.516	552.7	8.536	704.5
								8.033	295.8
	7.702	1384.8	7.723	1853.2		7.705	1401.6	7.736	1565.2
	7.167	203.4	7.230	355.8		7.155	190.1	7.194	357.9
	6.743	569.1	6.732	432.5		6.733	722.9	6.730	348.7
	6.376	1412.4	6.389	1318.7		6.378	1242.7	6.386	1211.2

The band position (λ) and the intensity (I) are given in μm and km/mol, respectively. Only bands with an intensity larger than 150 km/mol are reported values for the neutrals is about ten times smaller than for the cation or anion, which is typical of PAH molecules. Since the observed astronomical emission in the 6–10 μm region has about the same intensity as that at about 11 μm, it is clear that the observed emission at 6–10 μm arises from the ions. The bands with intensity greater than 150 km/mol are given in Table 4. Note that only ion bands are shown as all of the neutral bands are smaller than the cutoff. The absence of any strong neutral bands in this region of the spectra is consistent with other neutral PAHs and shows that adding one H atom does not significantly change the intensities in this region. In fact, a detailed examination of the spectra shows that the addition of one hydrogen makes only extremely small changes in the spectra. Focusing on the 6.2–6.3 μm region shows that adding a hydrogen atom to the inner most rings causes a shift to shorter wave length; the shifts relative to the parent are −0.057, −0.042, −0.034, and −0.022 μm for the 0, 1, 2, and 4 sites, respectively. For the remaining sites, there is a small positive shift from 0.018 for site 5 to 0.035 μm for site S. Thus, we see no evidence that these endo species can account for the observed variation in this 6.2–6.3 μm region as a function of astronomical source. These small shifts in the peak position are of the same magnitude as the variation in band position for PAH ions with different numbers of rings or with a different arrangement of rings. That is, it would be impossible to separate the addition of H atoms from some variation in the PAH population in the emission zone.

The 10–15 μm region of the spectra is shown in Fig. 4. The band positions and intensities of the strong bands are summarized in Table 5. For the cation, the main peak is at approximately 10.8 μm and is attributed to the solo out-of-plane hydrogen motion. This band looks very similar for all species, the largest shift is −0.046 μm for site 0. The duo band at approximately 12.4 μm is also very similar for all species. The only difference between the species are some weak bands, most notably the band between 11 and 12 μm. The neutral spectra look similar to those of the cation, with the two main band positions slightly shifted to the red (approximately at 11 and 12.45 μm) and the weak bands are slightly less noticeable. The variation in the strongest peak with hydrogen site is even smaller for the neutral than for the cation, with a shift of 0.025 μm for site 6. The anions show slightly more variation between the species. The main solo peak is further to the red (11.4 μm) as does the duo peak (12.75 μm). However, the solo peak is split for some of the species, while others have some weak peaks between the solo and duo peaks, as found for the cation and neutral. The strongest anion band in this region shows even smaller changes with hydrogen addition than the cation or neutral, with a maximum shift of −0.015 μm for site 6. All of these changes are smaller [30] than the difference between $C_{54}H_{18}$ and $C_{96}H_{24}$, and it would be difficult to differentiate between an added hydrogen atom or a variation in PAH population in an observed astronomical spectra.

Table 5 The strongest bands in the C–H out-of-plane bending region for $C_{96}H_{24}$ and $C_{96}H_{25}$

Molecule	Cation		Neutral		Anion	
	λ	I	λ	I	λ	I
$C_{96}H_{24}$	12.385	76.9	12.464	73.8	12.602	65.9
					11.329	155.3
	10.825	299.0	10.957	317.1	11.131	207.1
D	12.415	92.8	12.502	63.1	12.598	62.5
	10.833	304.1	10.971	262.5	11.138	329.8
S	12.379	84.9	12.475	73.6	12.582	71.2
					11.308	107.7
	10.824	236.3	10.979	231.3	11.123	184.9
0	12.382	82.3	12.480	71.5	12.582	58.2
	10.779	306.7	10.977	314.8	11.133	283.3
1	12.389	83.8	12.483	71.7	12.587	69.5
					11.335	63.1
	10.792	307.7	10.977	315.9	11.142	276.3
2	12.387	80.9	12.484	73.6	12.591	69.3
	10.791	305.1	10.975	312.4	11.135	290.8
4	12.378	77.1	12.491	65.9	12.588	63.2
	10.828	311.0	10.971	282.0	11.122	344.8
5	12.362	99.9	12.488	74.8	12.594	67.8
	11.761	84.3			11.711	63.5
					11.308	73.7
	10.798	313.8	10.964	309.5	11.123	253.4
6	12.661	53.0				
	12.401	77.8	12.489	72.5	12.568	82.6
	11.494	89.0	11.766	52.3	11.846	91.9
	10.817	274.0	10.982	286.8	11.116	281.8
7	12.407	117.1	12.498	59.7	12.594	60.5
					12.080	65.6
	10.818	275.0	10.969	267.7	11.121	299.0

The band position (λ) and the intensity (I) are given in μm and km/mol, respectively. Only bands with an intensity larger than 50 km/mol are reported

4 Conclusions

The addition of an extra hydrogen to $C_{96}H_{24}$ has very little effect on the infrared spectra. This includes adding hydrogen atoms to either the edge or central carbons. Clearly, it would be very difficult to identify large PAHs with a single hydrogen added in an astronomical spectra. Binding hydrogens to the edge carbons is much more favorable than adding them to central carbons. The edge solo site is favored over the edge duo site. The barrier for hydrogen migration is a sizable fraction of the total binding energy. This suggests that adding hydrogen atoms to the center of PAHs followed by hydrogen migration to the edge will not be a very efficient process, since the energy for hydrogen loss and hydrogen migration is so similar. It seems more likely that H atoms will add and detach from various sites with the solo hydrogenated species being the longest lived followed by the duo hydrogenated species.

Acknowledgments AR thanks the NASA's Astrophysics Theory and Fundamental Physics (ATFP) (NNX09AD18G) program for its generous support of this work.

References

1. Allamandola LJ, Tielens AGGM, Barker JR (1989) Astrophys J S 71:733
2. Puget JL, Léger A (1989) Ann Rev Astro Astrophys 27:161
3. Peeters E, Hony S, Van Kerkhoven C, Tielens AGGM, Allamandola LJ, Hudgins DM, Bauschlicher CW (2002) Astro Astrophys 390:1089
4. van Diedenhoven B et al (2004) Astrophys J 611:928
5. Joblin C, Abergel A, Bregman J, et al. (2000) ISO beyond the peaks: The 2nd ISO workshop on analytical spectroscopy. Eds. Salama A, Kessler MF, Leech K, Schulz B, ESA-SP, 456:49
6. Bregman J, Temi P (2005) Astrophys J 621:831
7. Sloan GC, Keller LD, Forrest WJ et al (2005) Astrophys J 632:956
8. Compiègne M, Abergel A, Verstraete L et al (2007) Astro Astrophys 471:205
9. Tielens AGGM (2008) Ann Rev Astron Astrophys 46:289
10. Hu A, Alkhesho I, Duley WW (2006) Astrophys J 653:L157
11. Pino T, Dartois E, Cao A-T, Carpentier Y, Chamaillé Th, Vasquez R, Jones AP, d'Hendecourt L, Bréchignac Ph (2008) Astro Astrophys 490:665
12. Hudgins DM, Bauschlicher CW, Allamandola LJ (2005) Astrophys J 632:316
13. Ricks AM, Douberly GE, Duncan MA (2009) Astrophys J 702:301
14. Ricca A, Bauschlicher CW, Allamandola LJ (2011) Astrophys J 727:128
15. Becke AD (1993) J Chem Phys 98:5648
16. Stephens PJ, Devlin FJ, Chabalowski CF, Frisch MJ (1994) J Phys Chem 98:11623
17. Frisch MJ, Pople JA, Binkley JS (1984) J Chem Phys 80:3265, and references therein
18. Bauschlicher CW, Langhoff SR (1997) Spectrochimica Acta A 53:1225
19. Bauschlicher CW, Ricca A (2010) Mol Phys 108:2647
20. Becke AD (1988) Phys Rev A 38:3098
21. Perdew JP (1986) Phys Rev B 33:8822
22. Bartlett RJ (1981) Ann Rev Phys Chem 32:359
23. Raghavachari K, Trucks GW, Pople JA, Head-Gordon M (1989) Chem Phys Lett 157:479
24. Dunning TH (1989) J Chem Phys 90:1007
25. Bauschlicher CW, Peeters E, Allamandola LJ (2009) Astrophys J 697:311
26. Frisch MJ et al. (2013) Gaussian 09, Revision D.01, Gaussian, Inc., Wallingford CT
27. Flükiger P, Lüthi HP, Portmann S, Weber J (2000) MOLEKEL 4.2; Swiss Center for Scientific Computing: Manno, Switzerland, 2000-2002
28. Perdew JP, Chevary JA, Vosko SH, Jackson KA, Pederson MR, Singh DJ, Fiolhais C (1992) Phys Rev B, 46:6671–6687
29. Bauschlicher CW, Boersma C, Ricca A, Mattioda AL, Cami J, Peeters E, Sanchez de Armas F, Saborido GP, Hudgins DM, Allamandola LJ (2010) Astrophys J Suppl 189:341
30. Ricca A, Bauschlicher CW, Boersma C, Tielens AGGM, Allamandola LJ (2012) Astrophys J 754:75

Theor Chem Acc (2014) 133:1423
DOI 10.1007/s00214-013-1423-z

REGULAR ARTICLE

The mechanism of the cycloaddition reaction of 1,3-dipole molecules with acetylene: an investigation with the unified reaction valley approach

Marek Freindorf · Thomas Sexton · Elfi Kraka · Dieter Cremer

Received: 10 September 2013 / Accepted: 12 November 2013 / Published online: 4 December 2013

Abstract The unified reaction valley approach (URVA) is used in connection with a dual-level approach to describe the mechanism of ten different cycloadditions of 1,3-dipoles XYZ (diazonium betaines, nitrilium betaines, azomethines, and nitryl hydride) to acetylene utilizing density functional theory for the URVA calculations and CCSD(T)-F12/aug-cc-pVTZ for the determination of the reaction energetics. The URVA results reveal that the mechanism of the 1,3-dipolar cycloadditions is determined early in the van der Waals range where the mutual orientation of the reactants (resulting from the shape of the enveloping exchange repulsion spheres, electrostatic attraction, and dispersion forces) decides on charge transfer, charge polarization, the formation of radicaloid centers, and the asynchronicity of bond formation. All cycloadditions investigated are driven by charge transfer to the acetylene LUMO irrespective of the electrophilic/nucleophilic character of the 1,3-dipole. However, an insufficient charge transfer typical of an electrophilic 1,3-dipole leads to a higher barrier. Energy transfer and energy dissipation as a result of curvature and Coriolis couplings between vibrational modes lead to an unusual energy exchange between just those bending modes that facilitate the formation of radicaloid centers. The relative magnitude of the reaction barriers and reaction energies is rationalized by determining reactant properties, which are responsible for the mutual polarization of the reactants and the stability of the bonds to be broken or formed.

Keywords Unified reaction valley approach · 1,3-Dipolar cycloadditions · Reaction mechanism · Mutual polarization · Energy transfer and dissipation

Dedicated to Professor Thom Dunning and published as part of the special collection of articles celebrating his career upon his retirement.

Electronic supplementary material The online version of this article (doi:10.1007/s00214-013-1423-z) contains supplementary material, which is available to authorized users.

M. Freindorf · T. Sexton · E. Kraka (✉) · D. Cremer (✉)
Computational and Theoretical Chemistry Group (CATCO), Department of Chemistry, Southern Methodist University, 3215 Daniel Ave, Dallas, TX 75275-0314, USA
e-mail: ekraka@gmail.com

D. Cremer
e-mail: dieter.cremer@gmail.com

1 Introduction

1,3-Dipolar cycloadditions are pericyclic reactions leading to five-membered rings with heteroatoms [1–5]. The reactions involve a 1,3-dipole molecule and a reaction partner with a multiple bond as a dipolarophile. The reaction mechanism is in most cases concerted and stereo- and regiospecific with regard to both 1,3-dipole and dipolarophile. The interest in these cycloadditions results from the fact that a large variety of 1,3-dipole molecules XYZ rapidly reacts in an exothermic fashion to the desired cycloaddition product yielding a wealth of heterocycles. A suitable combination of groups X, Y, and Z containing C, N, or O leads to 18 different 1,3-dipole molecules, 12 of which are N-centered and 6 are O-centered. The XYZ framework of a 1,3-dipole molecule can contain 16 valence electrons as in the case of the diazonium betaines **1**, **2**, and **3** or the nitrilium betaines **4**, **5**, and **6** of Fig. 1. Alternatively, XYZ may have a total of 18 valence electrons as in the case of the azomethines **7**, **8**, and **9** of Fig. 1.

1,3-Dipole molecules are better described by resonance structures rather than simple Lewis structural formulas

Fig. 1 1,3-Dipolar cycloaddition systems **1–9** investigated in this work. The name of the 1,3-dipole and the five-membered ring formed is given

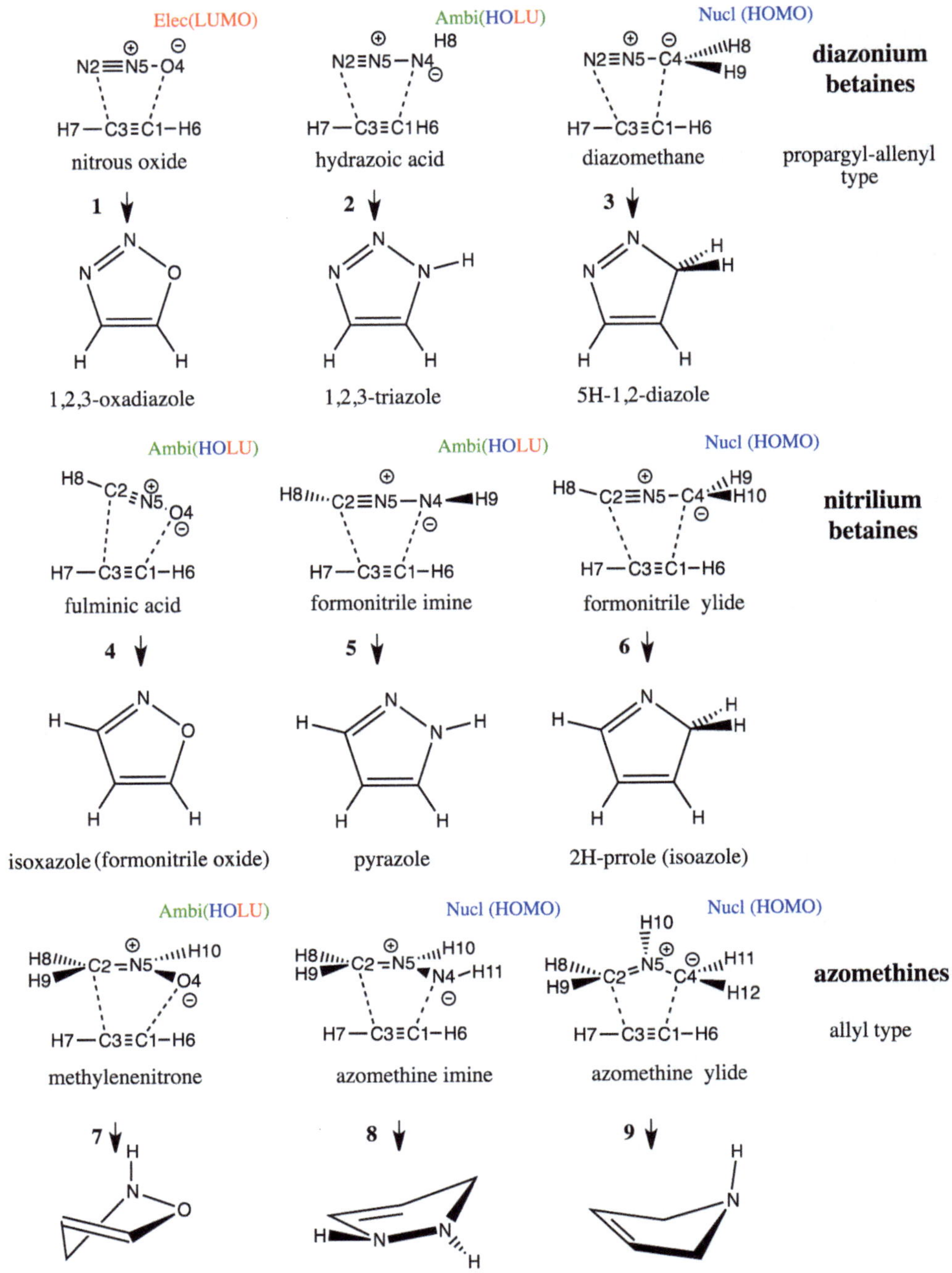

with standard single, double, or triple bonds. The propargyl-allenyl-type resonance structures (only the former is shown in Fig. 1) are typical of the 16 valence electrons systems **1–6**, and the allyl-type resonance structures (only one shown) are typical of the N-centered 18 valence electron systems **7–9**. Although the electronic structures of the 1,3-dipole molecules are different, the mechanism of their cycloadditions to dipolarophiles is similar as reflected by their exothermicity, similar reaction barriers, and a similar stereochemistry [1–5]. This has been explained to some part by applying the Woodward–Hoffmann rules [6] and identifying 1,3-dipolar cycloadditions as concerted, symmetry-allowed pericyclic $\pi^4 s + \pi^2 s$ reactions involving 4 π-electrons of the 1,3-dipole and 2 π-electrons of the double or triple bond of the dipolarophile leading in this way to a Hückel-aromatic, stabilized transition state (TS) according to the Evans–Dewar–Zimmerman rules [7]. The state- and orbital-symmetry analysis can explain the relatively low barriers of 1,3-dipolar cycloadditions and their stereochemistry. Detailed trends in reaction rates and changes in the stereochemistry have been explained, often with success, on the basis of frontier orbital theory [8]. An even more detailed insight has been provided by a large number of quantum chemical investigations [9–21] where especially the thorough and extensive work of Houk et al. [22–37] has to be mentioned.

1,3-Dipolar cycloadditions are versatile tools for the synthesis of heterocyclic molecules [1–5]. Applications range from asymmetric organic synthesis [38, 39], catalysis [40–43], materials science [44], and drug design [45] to chemical biology [46, 47]. The mechanism of the 1,3-dipolar cycloaddition reactions had been controversially debated starting in the late 1960s when Huisgen [48] postulated a concerted mechanism and Firestone [49] advocated a stepwise biradical mechanism. Numerous quantum chemical investigations have been performed since then to shed light on the mechanism of 1,3-dipolar cycloadditions [33, 50, 51], to provide accurate energetic data [29, 34–36] and kinetic data [24], and to investigate the regioselectivity of these reactions [9, 11, 17, 28]. Calculations could solve the Huisgen–Firestone controversy; the concerted mechanism is favored for the reactions of the unsubstituted 1,3-dipoles with ethylene and acetylene [21, 26, 27], whereas the stepwise mechanism becomes likely when the 1,3-dipoles and dipolarophiles are substituted by radical-stabilizing groups [25, 31, 52].

Still, open mechanistic questions concern the driving force of the cyclization and the electronic factor(s) that determine(s) the barrier height. Currently, three different rate-determining factors are discussed in the literature: (1) the *distortion energy* of dipole and dipolarophile as defined by Houk et al. [22, 23, 30, 32]; (2) reaction acceleration by an excitation of the *bending vibrations* of the 1,3-dipole [13, 14, 26, 27]; and (3) the *biradical character* of the 1,3-dipole molecule [10, 12].

Most of the previous work on the 1,3-dipolar cycloadditions has focused on finding a rationale for the relatively low reaction barriers because they predominantly determine the reaction rate and the yield of a given product. A general assumption in this connection is that the similarity of the activation energies indicates a similarity of the reaction mechanism.

In this work, we will show that comparable reaction barriers do not necessarily imply a similarity in the reaction mechanism. Conversely, two reactions may have a similar mechanism although they have different barriers. We approach the question of the mechanism of the 1,3-dipolar cycloadditions in a different way than previous investigations by utilizing the unified reaction valley approach (URVA) of Kraka and Cremer [53–57]. URVA explores the reaction path and the reaction valley via its direction and curvature, using a number of tools that provide direct information about the reaction mechanism. According to a definition two of us have given recently [53], the mechanism of a reaction is determined by the nature and sequence of elementary electronic events that take place in each phase of the reaction starting from the first long-range interactions between the reactants in the entrance channel and terminating with the finalization of the products in the exit channel.

We will show that the fate of the reaction is determined early on in the van der Waals region of the entrance channel. Furthermore, we will reveal that all 1,3-dipolar cycloadditions investigated in this work pass through the same reaction phases; however, there are distinct differences between different classes of cycloadditions. Also, it will become apparent that there is a difference between reactions with radicaloid and biradicaloid character and that the dipolarophile rather than the 1,3-dipole initiates bond formation.

The current investigation is based on a detailed analysis of the reaction path and the reaction valley, which will lead to a better understanding of the mechanism than it is possible with a conventional mechanistic analysis based on the features of the stationary points along the reaction path. Our focus will be on well-defined transient points along the reaction path corresponding to bond breaking and bond formation, rehybridization, and (bi)radicaloid formation.

The results of this work are presented in the following way. In Sect. 2, the computational methods used in this work are described. The results of the quantum chemical investigation of ten 1,3-dipolar cycloadditions are presented and discussed in Sect. 3. Conclusions will be drawn in the final section.

2 Computational methods

The current investigation is based on URVA [53–57], the reaction path Hamiltonian of Miller, Handy, and Adams [58], and the local vibrational mode description of Konkoli and Cremer [59–62] recently theoretically justified by Zou and Cremer [63, 64]. URVA has been repeatedly described in previous publications [55, 65–69] as well as several review articles [53, 54, 56, 57]. Therefore, we refrain from a detailed account of the theory of URVA and the use of local vibrational modes and instead outline the methods used in a more qualitative fashion.

Each 1,3-dipolar cycloaddition system is investigated with a dual-level approach: (1) A high level such as CCSD(T)-F12/aug-cc-pVTZ [70–72] is used to determine the energetics of the reaction from the relative energies (enthalpies) of the stationary points along the reaction path. (2) The reaction valley and the reaction path embedded in the reaction valley are calculated at a lower level using density functional theory (DFT). The primary objective of URVA is to determine those locations along the reaction path at which the chemical processes of bond breaking and forming take place, where the first van der Waals interactions lead to chemical change, where rehybridization leads to the formation of radicaloid and biradicaloid structures, and where, after bond formation, the electrons are reorganized to form delocalized π-electron ensembles. To get these insights, a paradigm shift with regard to conventional mechanistic studies is made: The primary concerns are no longer the energy or geometry changes of the reaction complex, but rather the features of the reaction path, which can be directly related to changes in the electronic structure [53, 54, 56].

Chemical change and chemical processes always imply a curving of the reaction path [53, 54, 56]. If the reaction path becomes a straight line, physical processes driven exclusively by electrostatic forces take place with no or just a steady change in the electronic structure. Chemical processes are characterized by catastrophe points in the electron density distribution of the reaction complex (infinitesimal changes in the density lead, e.g., to singularity catastrophes in its topological structure reflecting sudden changes in bonding), and this in turn is connected with directional changes (curving) of the reaction path (which must not be confused with the energy curvature of the potential energy surface). Changes in the reaction path can be described with the help of the two Frenet vectors: path tangent (direction) and path curvature (curving). A third Frenet vector leading to the path torsion is only of relevance in case of repeated curving of the path over a larger area, which seldom occurs. The reaction complex and the reaction path are intimately related, and analyzing the latter on the basis of just two quantities (direction and curvature vector) solves the problem of monitoring of the 3N-L (N: number of atoms; L: number of translations and rotations) degrees of freedom of the reaction complex during the reaction.

The curvature vector *kappa* has the dimension of 3N–L. In URVA, the scalar curvature (the length of the curvature vector) is investigated to obtain an easy-to-analyze function $\kappa(s)$ (scalar curvature as a function of the mass-weighted arc length s; $s < 0$: entrance channel of the reaction; $s = 0$: TS; $s > 0$: exit channel). The square of the scalar curvature is equal to the sum of the squares of the normal mode curvature coupling coefficients $B_{\mu,s}$, which describe the coupling between the translational mode along the reaction path and the normal vibrational modes $\mathbf{l}_\mu$ spanning the reaction valley [58]. The normal vibrational modes, in turn, can be decomposed into the local vibrational modes $\mathbf{a}_n$ of Konkoli and Cremer [59, 60]. Each local mode is driven by an internal coordinate q_n of the reaction complex. By decomposing the scalar curvature into local mode curvature coupling coefficients $A_{n,s}$, those parts of the reaction complex can be directly identified, which are related to the curving of the reaction path and enhancements of the scalar curvature.

In this way, all electronic structure changes of the reaction complex along the reaction path can be monitored. They present the basis of the reaction mechanism, which develops in the form of a sequence of reaction phases along the reaction path. Each reaction phase is characterized by a curvature enhancement or peak enclosed by locations of minimal or even zero curvature [53, 54, 56]. The positions of minimal curvature correspond to transient structures of the reaction complex with interesting electronic structure features. Suitable changes of the reaction complex and/or the reaction environment can transform these transient points into stationary points, and therefore, the terms *hidden intermediate* and *hidden transition state* have been coined to describe the mechanistic relevance of these curvature minima [53, 65, 67].

There are three quantities, which are used in this work to describe the reaction mechanism as it develops via the scalar curvature function $\kappa(s)$: (1) local mode curvature coupling coefficients $A_{n,s}$ to describe the mechanism; (2) normal mode curvature coupling coefficients $B_{\mu,s}$ to describe energy transfer from the vibrational modes into the translational motion along the path; and (3) normal mode Coriolis (mode–mode) coupling coefficients $B_{\mu,\nu}$ to describe energy dissipation between the modes. Apart from this, URVA calculates charge transfer and charge polarization as a function of s and determines local mode force constants and frequencies of the reaction complex [63, 64].

For URVA, a representative reaction path must be chosen, which can be the intrinsic reaction coordinate (IRC) path of Fukui [73], a reasonable down-hill path, or any other representative path [68, 74]. In this work, we use the IRC path and follow it with an algorithm recently suggested by Hratchian and Kraka [75] where B3LYP/6-31G(d,p) [76–78] is used to describe the reaction valley. For each cycloaddition path, between 630 and 930 path points were calculated. At each second path point, generalized harmonic vibrational frequencies of the reaction complex were determined. Curvature, curvature coupling coefficients, and all other URVA-related quantities were calculated with the *ab initio* package COLOGNE13 [79]. The analysis of the charge distribution (charge transfer and charge polarization) was based on the natural bond orbital (NBO) method of Weinhold et al. [80, 81]. For the CCSD(T)-F12 calculations [71], the program package MOLPRO [82] was used. Reaction enthalpies $\Delta_R H(298)$, activation enthalpies $\Delta H^a(298)$, reaction free energies $\Delta_R G(298)$, and $\Delta G^a(298)$ were calculated using CCSD(T)-F12/aug-cc-pVTZ energies in connection with B3LYP/6-31G(d,p) or ωB97X-D/aug-cc-pVTZ geometries and vibrational frequencies [83, 84]. The latter functional is known to provide an improved description of dispersion interactions [83, 85]. Basis set superposition error (BSSE) corrections were obtained using the counterpoise method [86].

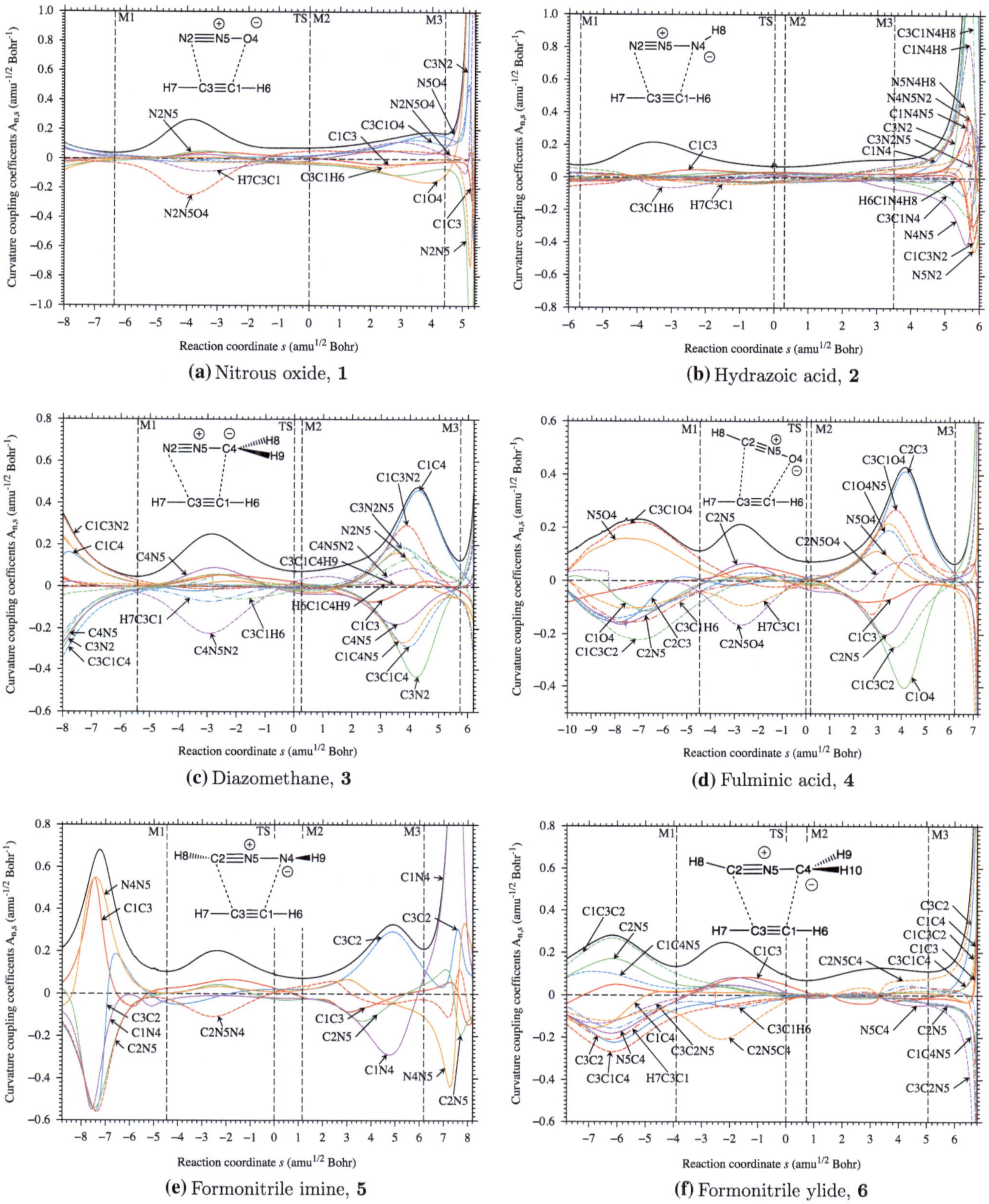

(a) Nitrous oxide, **1**

(b) Hydrazoic acid, **2**

(c) Diazomethane, **3**

(d) Fulminic acid, **4**

(e) Formonitrile imine, **5**

(f) Formonitrile ylide, **6**

Fig. 2 Scalar curvature as a function of the reaction path parameter s (*solid black line*) for 1,3-dipolar cycloaddition systems **1–10**. Local mode curvature coupling coefficients are given in *color*. The borders of the reaction phases are indicated by *vertical dashed lines* at curvature points M1, M2, and M3. The TS at $s = 0$ amu$^{1/2}$ Bohr is also indicated by a *vertical dashed line*. **a** Nitrous oxide, **1**, **b** hydrazoic acid, **2**, **c** diazomethane, **3**, **d** fulminic acid, **4**, **e** formonitrile imine, **5**, **f** formonitrile ylide, **6**, **g** methylene nitrone, **7**, **h** azomethine imine, **8**, **i** azomethine ylide, **9**, **j** nitryl hydride, **10**

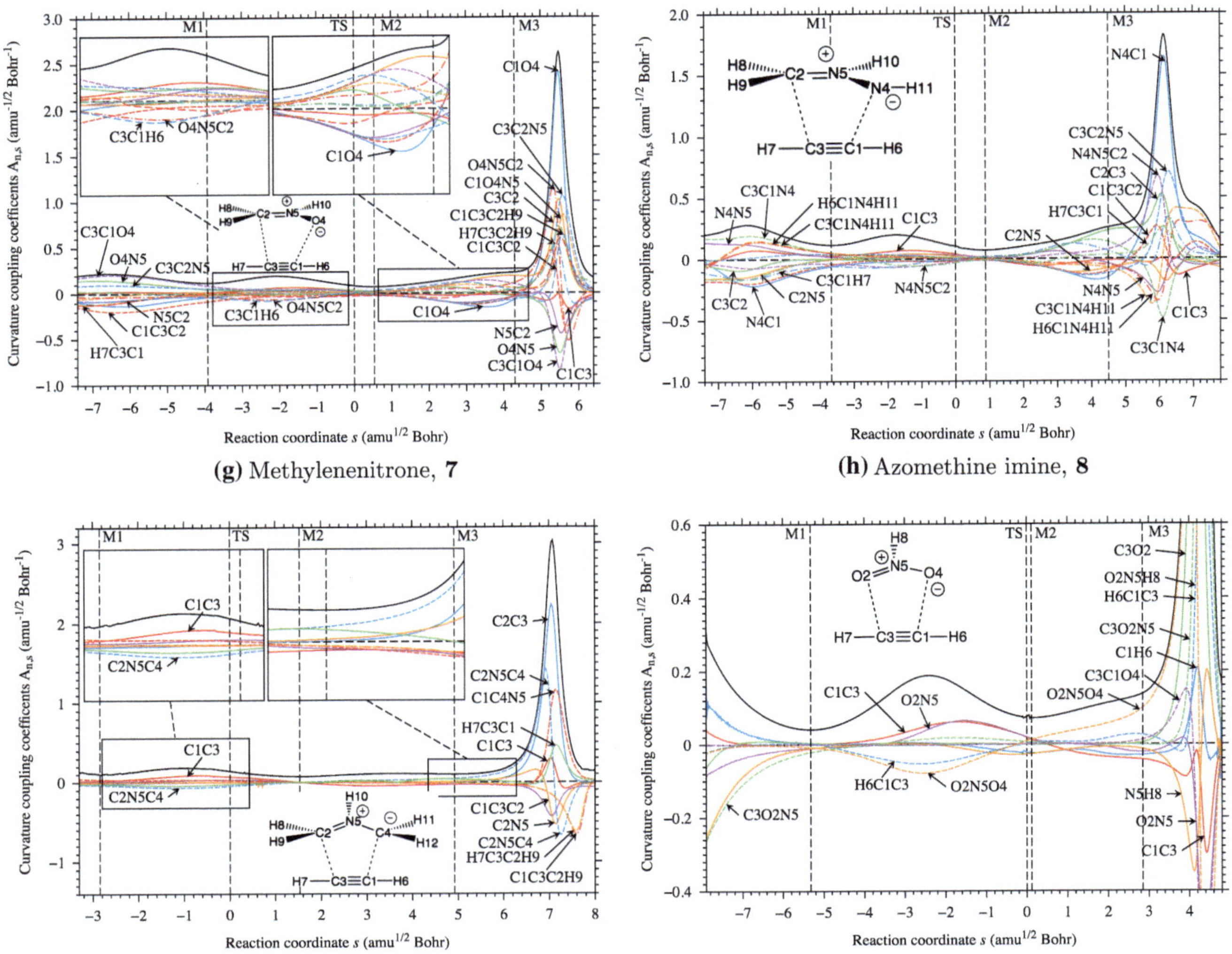

(g) Methylenenitrone, **7**

(h) Azomethine imine, **8**

(i) Azomethine ylide, **9**

(j) Nitryl hydride, **10**

Fig. 2 continued

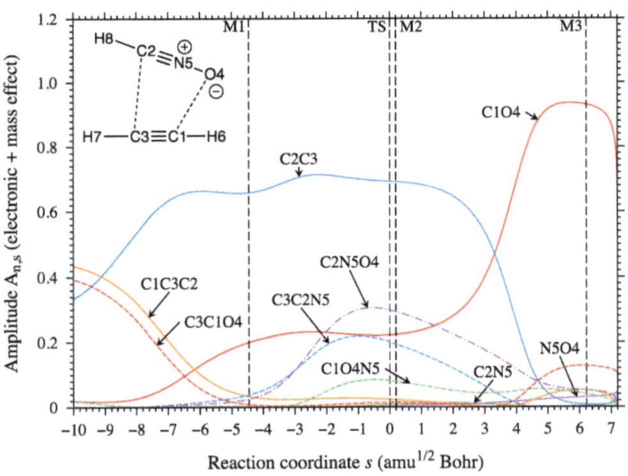

Fig. 3 Components of the reaction path direction given as a function of the reaction parameter s for reaction system **4**

3 Results and discussion

In Fig. 2, curvature diagrams $\kappa(s)$ of the 1,3-dipolar cycloaddition systems **1–9** listed in Fig. 1 and the nitryl hydride system **10** are shown, where each curvature is decomposed in terms of local mode curvature coupling coefficients $A_{n,s}(s)$. An example for the composition of the reaction path direction in terms of local mode contributions is given for reaction system **4** in Fig. 3 and the charge transfer from the 1,3-dipole to acetylene in Fig. 4a and b. Calculated energy barriers and reaction energies are analyzed in Figs. 5 and 6. An example of the dependence of the vibrational frequencies on the path parameter s is given in Fig. 7a and b for system **4**.

The energy data for the ten cycloaddition reactions investigated in this work are summarized in Table 1. The reaction and activation enthalpies obtained at the

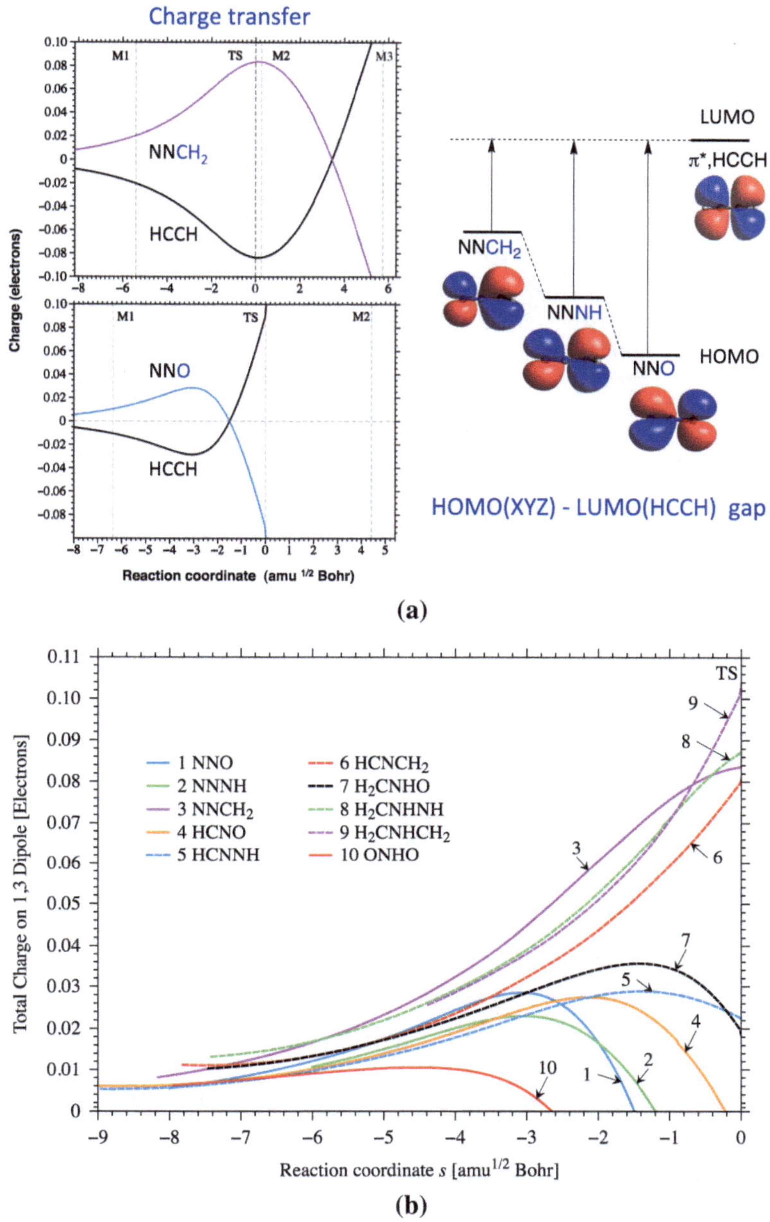

Fig. 4 **a** Charge transfer from the HOMO of the 1,3-dipole (schematically drawn for NNZ) to the LUMO of acetylene in the case of reaction systems **1**, **2**, and **3**. **b** Charge transfer from the 1,3-dipole to acetylene in the entrance channel (TS at $s = 0$ amu$^{1/2}$ Bohr) given as a function of the reaction parameter s

CCSD(T)-F12a/aug-cc-pVTZ level of theory compare well with the corresponding CBS-QB3 values of Houk et al. [30, 50]. The corresponding B3LYP and ωB97X-D results (after BSSE corrections) show a reasonable agreement with the CCSD(T)-F12a data, which makes them both suitable for the description of the reaction valley. To reduce the computational costs, B3LYP/6-31G(d,p) was used for the URVA analysis as described above. It is interesting to note that the entropy contributions $\Delta_R S(298)$ to the free reaction energies $\Delta_R G(298)$ are fairly constant (−36 to −41 entropy units, Table 1) as are the contributions $\Delta S^a(298)$ to the free activation energies, $\Delta G^a(298)$, which vary from −32 to −37 entropy units. The values collected in Table 1 suggest that entropy effects can be excluded from the discussion of the reaction mechanism.

All ten cycloaddition reactions follow a four-phase mechanism irrespective of the nature of the 1,3-dipole reacting with acetylene. These phases will be discussed below by making reference to the curvature diagrams presented in Fig. 2.

Reaction phase 1 (entrance channel from start to curvature point M1): There are four properties of the 1,3-dipole molecule, which are decisive in the first phase of the cycloaddition reaction:

1. The first property is the exchange repulsion envelope around the reactants, which decides on the approach mode. If one terminal group of the 1,3-dipole is more electronegative than the other, the exchange repulsion envelope at this end (e.g., an O atom) will be more

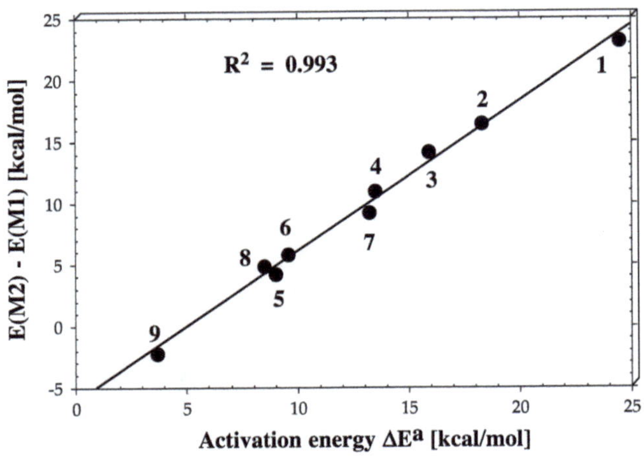

Fig. 5 Comparison of the energy change in reaction phase 2, E(M2) − E(M1), with the activation energy ΔE^a for the nine 1,3-dipolar cycloaddition reactions (for numbering of reaction complexes, see Fig. 1)

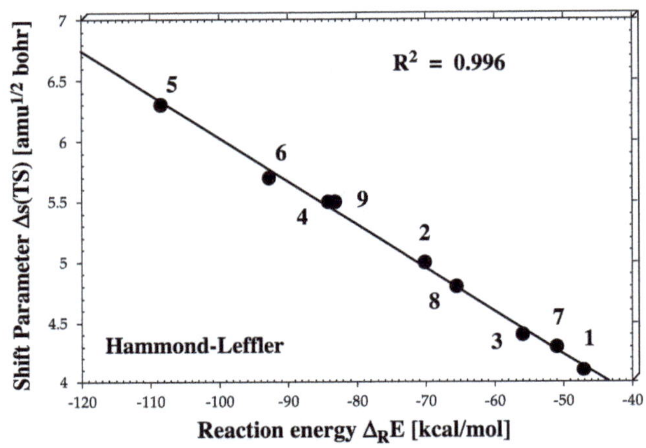

Fig. 6 Comparison of the parameter $\Delta s(TS)$, which determines the shift of the TS relative to the center M3 of the bond-forming processes, with the reaction energy $\Delta_R E$ for the nine 1,3-dipolar cycloaddition reactions (for numbering of reaction complexes, see Fig. 1). A positive $\Delta s(TS)$ value indicates an early TS according to the Hammond–Leffler postulate

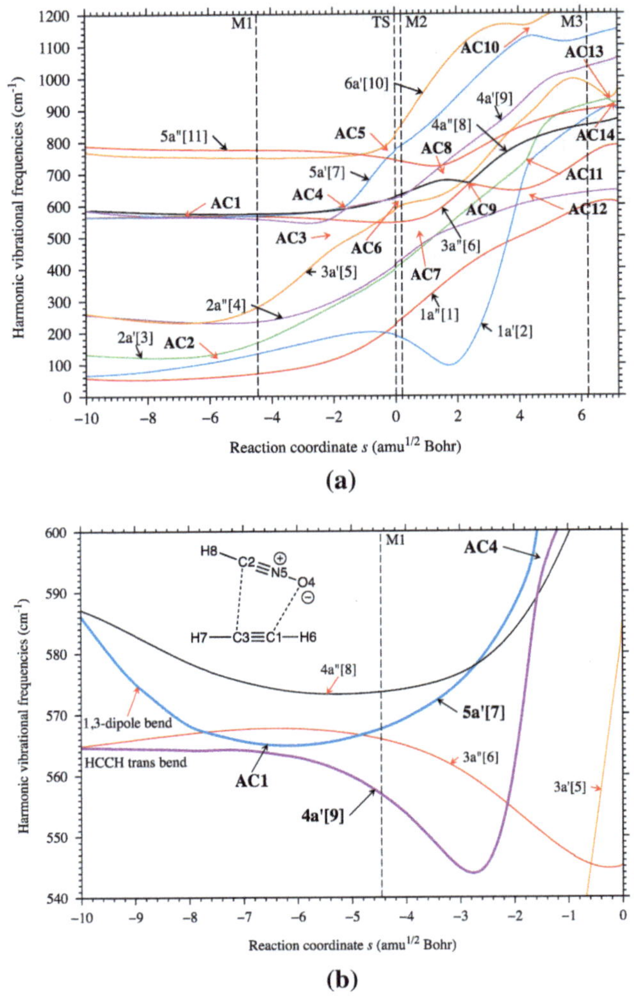

Fig. 7 Vibrational frequencies of reaction system **4** given as a function of the reaction parameter *s*. The avoided crossings (AC) are the positions of mode–mode (Coriolis) coupling leading to energy dissipation. **a** Framework vibrations. **b** 1,3-Dipole and acetylene bending

contracted than at the other terminal group. The reactants will approach each other in such a way that stabilizing dispersion interactions and electrostatic attractions are maximized while minimizing the short-range exchange repulsion. The approach distance will be shorter for the more electronegative group than for the other group. Accordingly, there is a larger or smaller inclination angle of the axis of the 1,3-dipole molecule (when it is linear, otherwise the 1,3 connection axis) relative to the axis of acetylene. These are given in Table 2 for the first point in the entrance channel where interaction energies are smaller than just a few tenth of a kcal/mol.

One can see that the eletronegativity difference between X2 and Z4 determines sign and magnitude of the inclination angle (Table 2). Systems with an electronegative group Z (**1**, **4**, **5**, **7**, and **8**; compared with Fig. 1) have a positive inclination angle, systems **3** and **6** have a negative inclination angle (i.e., the distance C3X2 is shorter than the distance C1Z4), whereas systems **2**, **9**, and **10** approach each other in a parallel fashion because of symmetry and/or similar (identical) electronegativities of X2 and Z4. The mode of approach has direct consequences for the sequence of bond formation as will be shown below.

2. The second property relevant to the mechanism is the dipole moment of the 1,3-dipole molecule, where only its component parallel to the axis of the approaching acetylene molecule matters. This has been determined

Table 1 Energetics of cycloaddition reactions **1–10**

System	$\Delta_R E$			ΔE^a			CCSD(T)-F12a		CBS-QB3		ωB97X-D		CCSD(T)-F12a	
	A	B	C	A	B	C	$\Delta_R H(298)$	$\Delta H^a(298)$	$\Delta_R H(298)$	$\Delta H^a(298)$	$\Delta_R S(298)$	$\Delta S^a(298)$	$\Delta_R G(298)$	$\Delta G^a(298)$
1 (NNO)	−44.1	−47.1	−42.3	30.2	24.5	27.5	−39.5	27.3	−38.8	26.7	−35.7	−31.6	−28.8	36.7
2 (NNNH)	−70.6	−70.1	−67.]	22.5	18.5	19.7	−63.1	20.0	−63.7	19.0	−40.2	−34.0	−51.1	30.2
3 ($NNCH_2$)	−56.9	−56.0	−56.8	17.4	15.9	14.3	−52.6	14.9	−51.2	14.0	−39.7	−34.0	−40.8	25.0
4 (HCNO)	−83.7	−84.2	−82.0	16.7	13.5	12.9	−77.5	13.0	−75.8	13.3	−38.5	−31.9	−66.0	22.5
5 (HCNNH)	−112.1	−108.4	−107.9	10.2	9.0	7.7	−102.1	8.1	−102.5	7.5	−41.2	−33.6	−89.8	18.1
6 ($HCNCH_2$)	−96.2	−92.8	−95.3	9.2	9.6	6.3	−90.3	6.8	−88.8	6.5	−41.4	−33.9	−77.9	16.9
7 (H_2CNHO)	−52.0	−51.0	−50.3	14.7	13.2	13.0	−46.5	13.8	−45.6	12.9	−39.8	−36.4	−34.7	24.7
8 (H_2CNHNH)	−69.6	−65.6	−66.4	8.4	8.6	1.0	−62.2	7.7	−61.3	6.5	−40.0	−36.2	−50.3	18.5
9 (H_2CNHCH_2)	−87.0	−83.2	−84.7	1.9	3.8)	1.1	−79.9	1.8	−79.0	0.4	−38.5	−34.3	−68.4	12.0
10 (ONHO)	−22.4	−23.8	−22.1	32.9	27.3	31.1	−19.3	31.4	−18.8	28.4	−38.3	−35.9	−7.9	42.1

Reaction energies and reaction barriers are given for A: ωB97X-D, B: B3LYP, and C: CCSD(T)-F12a. B3LYP calculations carried out with the 6-31G(d,p) basis set. All other calculations were carried out with the aug-cc-pVTZ basis set. Enthalpies and free energies were obtained with ωB97X-D geometries and thermochemical corrections. Energy differences are given in kcal/mol and entropy differences in cal/(mol K). CBS-QB3 reference data from Ref. [50]. Basis set superposition error corrections were obtained using the counterpoise method

for the first point in the entrance channel and is given in Table 2 as μ_z (the direction z is that defined by the acetylene CC axis) together with the total dipole moment of the 1,3-dipole. The magnitude of the dipole component μ_z reflects the ability of the 1,3-dipole to polarize the acetylene molecule. It has a direct influence on the barrier of the reaction.

3. The third property is the polarizability of the reactants. If one compares different dipolarophiles, one would have to compare their polarizabilities α where the magnitude of the α_{zz} component would be the most important. Since this is not the case, we consider here only the fact that both reactants "communicate" via their polarizing power and their polarizability. Polarization of the charge distribution of acetylene leads to an induced dipole moment, which in turn polarizes the charge distribution of the 1,3-dipole molecule. The latter process depends on its polarizability where again the zz-component of the polarizability tensor rather than the isotropic polarizability will be important. A large α_{zz} facilitates the reaction and leads to a lower reaction barrier.
4. The fourth property, which strongly influences reaction rate and reaction mechanism, is the ability of both partners of accepting or donating charge. Previous studies have discussed the charge transfer between dipolarophile and 1,3 dipole within the framework of the frontier molecular orbital (FMO) theory. Using this concept, the 1,3-dipoles were categorized as being nucleophilic [Nucl(HOMO); Fig. 1] donating charge from their HOMO to the LUMO of the dipolarophile, electrophilic [Elec(LUMO); Fig. 1] accepting charge from the dipolarophile in their LUMO, or amphiphilic [Amphi(HOLU), Fig. 1] reacting either way depending on the character of the dipolarophile [1–5]. We have investigated the charge transfer between the reactants and found, contrary to the qualitative description of FMO theory, that the charge transfer starts already in the van der Waals region, i.e., in phase 1. Then, it is directed from the 1,3-dipole to the acetylene molecule and continues to develop to its full strength in phase 2 (compared with Fig. 4a and b).

On the basis of these facts, the curvature of the reaction path, and its decomposition into local mode curvature coupling coefficients (see Fig. 2a–j), one can describe the events in phase 1 as follows where system **4** (Fig. 2d) is used as an example. Guided by the form of its exchange repulsion envelope, HCNO approaches acetylene as much as possible at its O terminus (Z4), which leads to an inclination angle of 41°. Accordingly, the local mode curvature coupling coefficients associated with the orientation angles C3C1O4 and C1C3C2 become relatively

Table 2 Properties of 1,3-dipoles and reaction complexes **1–10**: Dipole moment component μ_z and total dipole moment μ_{total} (in parenthesis), polarizability component α_{zz} and isotropic polarizability α_{iso} (in parenthesis) of the reaction complex calculated for the complex geometry at the first point of the reaction path in the entrance channel

System	μ_z (μ_{total}) [Debye]	α_{zz} (α_{iso}) (Å^3)	k^a(XYZ) (mdyn Å/rad^2)	ω^a(XYZ) (cm^{-1})	ω_μ(XYZ) (cm^{-1})	k^a(C3X2) (mdyn/Å)	k^a(C1Z4) (mdyn/Å)	ω_μ(C3X2) (cm^{-1})	ω_μ(C1Z4) (cm^{-1})	Orientation (°)	Charge pol. (me)	Δ s(TS) [amu$^{1/2}$ Bohr]
1 (NNO)	−0.01 (0.01)	4.1 (2.1)	0.693	604	604 (599)	5.506	5.766	1,113	1,166	5.4	8	4.1
2 (NNNH)	1.03 (1.95)	5.4 (2.8)	0.618	563	530 (520)	5.853	6.802	1,185	1,469	0.9	−23	5.0
3 ($NNCH_2$)	1.69 (1.72)	6.3 (3.4)	0.600	549	590 (582)	3.959	4.509	964	992	−10.3	−14	4.4
4 (HCNO)	−2.43 (3.16)	3.7 (2.7)	0.513	517	569 (565)	5.789	5.879	1,051	1,132	41.2	55	5.5
5 (HCNNH)	−0.20 (2.08)	6.5 (3.7)	0.598	546	535 (531)	6.039	6.838	1,064	1,153	21.0	22	6.3
6 ($HCNCH_2$)	0.74 (1.66)	8.3 (4.5)	0.605	544	494 (492)	4.222	4.699	936	1,011	−18.0	−47	5.7
7 (H_2CNHO)	−2.85 (3.44)	5.2 (3.4)	1.490	761	572 (570)	4.121	5.130	943	1,072	17.1	103	4.3
8 (H_2CNHNH)	−1.60 (2.37)	4.9 (4.1)	1.417	744	544 (545)	4.092	4.534	946	1,095	36.5	38	4.8
9 (H_2CNHCH_2)	0.00 (1.36)	9.0 (5.0)	1.173	675	477 (479)	3.941	3.941	967	967	0	0	5.5
10 (ONHO)	0.00 (2.59)	3.5 (2.2)	2.061	914	783 (782)	4.535	4.535	976	976	0	0	5.0

Direction z is defined by the acetylene CC axis. Local mode force constant $k^a(XYZ)$, local mode frequency $\omega^a(XYZ)$, and normal mode bending frequency $\omega_\mu(XYZ)$ of the isolated 1,3-dipole molecule (values in parentheses correspond to normal mode XYZ bending frequencies of the reaction complex calculated for the complex geometry at the first point of the reaction path in the entrance channel); local mode force constants $k^a(C3X2)$ and $k^a(C1Z4)$, normal mode frequencies $\omega_\mu(C3X2)$ and $\omega_\mu(C1Z4)$ of the product; orientation angle of the 1,3-dipole (X2Z4 axis) relative to the CC axis of acetylene and charge polarization in acetylene (surplus of negative charge at C1H compared to HC3) calculated for the complex geometry at the first point of the reaction path in the entrance channel; shift parameter $\Delta s(TS)$ of the TS relative to the center M3 of bond formation

large and they are responsible (via their squares) for a significant curvature enhancement in phase 1. Since in this phase, C1Z4 and C3X2 are approach parameters measuring the degree of exchange repulsion, their curvature coupling coefficients can also make a contribution to the scalar curvature; however, they will always be negative, i.e., the reaction complex resists a shortening of the approach distance (a closer approach requires energy).

Because of a substantial magnitude of the μ_z-component (−2.43 D), the charge of acetylene is polarized so that the C3H end becomes negative and the C1H end positive (difference in local charges 55 me). At the same time as charge polarization starts, also the charge transfer to acetylene begins (see Fig. 4b). Hence, phase 1 can be characterized as the orientation, polarization, and charge transfer phase, in which the electronic structure of the reactants is not significantly altered. In this sense, one can also speak of a van der Waals phase. Phase 1 will be well developed in the curvature diagram if the orientation of the reactants to each other is important as reflected by the orientation angle of Table 2. Reaction systems **1**, **2**, and **9** have small or zero orientation angles, and therefore, phase 1 is rather small (see Fig. 2a, b, i).

Reaction Phase 2 (from curvature point M1 to curvature point M2): As Fig. 4b reveals, charge transfer fully develops in phase 2, which we call the rehybridization phase as will be explained below. Charge transfer to acetylene and the simultaneous polarization of the acetylene density leads to a labilization that invokes bending and the formation of a radicaloid center. If acetylene accepts negative charge, it will try to adopt a bent structure. The amount of bending at each terminus depends on whether the charge is equally distributed in the corresponding $\pi^\star$ MO (see **9**) or, via charge polarization, shifted to the CH group farther away from the more electronegative terminal group of the 1,3-dipole. When free acetylene accepts negative charge, a trans form is adopted, which is also formed in phase 1 (supported by attraction with an electronegative terminal group Z4 (X2)), however with bending angles deviating by just a couple of degree from the linear form (revealed by the local mode curvature coupling coefficient of the HCC bending type in phase 1). In phase 2, however, acetylene is forced into a cis-bended form (e.g., **9**) or a form significantly bent at just one center whereas the other CCH unit remains quasi-linear.

We have investigated the local vibrational modes of a charge-polarized acetylene molecule and find a smaller bending force constant for the more negatively charged acetylene terminus. Hence, charge transfer and charge polarization are the driving forces for bending and rehybridization of acetylene. A radicaloid or biradicaloid center(s) with fractional unpaired electrons protruding in the direction of the attacking 1,3-dipole is (are) generated. The

HCC local mode curvature coupling coefficients clarify which HC group bends first. In all those cases where an electronegative group in the 1,3-dipole leads to a larger orientation angle, the not directly attacked CCH group bends more and becomes the location for the first bond formation. This means that a close attack on one CH terminus of acetylene causes the other terminus to prepare first for bonding.

Charge loss leads to a labilization of the rigid (multiple) bond structure of the 1,3-dipole so that bending becomes possible. The curvature enhancement in phase 2 is dominated by the XYZ and CCH local mode curvature coupling coefficients. These are negative, thus indicating that energy is required to enforce rehybridization and bending of the two reactants. Both reactants change at the same time so that there is no way of saying that the one or the other reactant is leading the reaction. However, snapshots of the geometry and the curvature coupling coefficients reveal that in asymmetric cases with a sizable orientation angle, bending of one HCC acetylene group is clearly ahead of the other CCH group.

The charge transfer curves shown in Fig. 4b reveal that there is a clear distinction between the 1,3-dipole molecules with regard to charge-donating capacities. This is large for nucleophilic 1,3-dipoles (see Fig. 1) and extends beyond the TS, thus leading to relatively low energy barriers (compared to all other cycloaddition systems or within a given group as, e.g., in the group of the diazonium betaines in the case of reaction system **3**). For electrophilic 1,3-dipoles such as **1**, the labilization of acetylene is not sufficient for the formation of a radicaloid center and therefore a large energy barrier results (Table 1). For the purpose of verifying this hypothesis, we investigated also the cycloaddition of the strongly electrophilic 1,3-dipole molecule nitryl hydride to acetylene (system **10**, Fig. 2j) and found the predicted reaction behavior: Insufficient charge transfer (see Fig. 4b) is coupled with a relatively large reaction barrier of 29.4 kcal/mol (Table 2).

In view of the nature of the curvature enhancement dominating reaction phase 2, we speak of the rehybridization and bending phase, which leads to the formation of (bi)radicaloid centers.

Reaction Phase 3 (from curvature point M2 to curvature point M3): Reaction phase 3 can be distinguished from phase 4 only in the case of reaction systems **3**, **4**, and **5** where a curvature peak of medium size results from the formation of the first bond between the reactants. In all other cases, the scalar curvature of phase 3 appears as a shoulder of a large curvature peak developing in phase 4. Nevertheless, a distinction between phases 3 and 4 becomes possible when one investigates the local mode curvature coupling coefficients of the two bond formation processes. In phase 3, one is positive (bond formation supporting) and the other negative (bond formation resisting). Therefore, it is justified to characterize phases 3 and 4 as the first and second bond formation phases.

In phase 3, the C3X2 bond curvature coupling coefficient is positive (when Z4 corresponds to the electronegative terminus of the 1,3-dipole), whereas the C1Z4 interaction (characterized by a negative curvature coupling coefficient) is still resisting bond formation. This corresponds to the fact that bond C3X2 is formed before bond C1Z4. In phase 1, the electronegative end of the 1,3-dipole is much closer to acetylene, suggesting that bond C1Z4 is formed first. However, charge transfer and charge polarization facilitate the formation of a radicaloid center at the other terminus. Radicaloid centers with lower exchange repulsion between them can bridge much easier a larger distance so that bond formation, contrary to general expectations, takes place earlier between the atoms farther apart.

In the cases of systems **2** and **3** (N2 is now the electronegative end), the situation is reversed and the C1Z4 curvature coupling coefficient is ahead of the resisting C3X2 curvature coupling coefficient, indicating that the C1Z4 bond is formed now before the C3X2 bond. In case of symmetry, phase 3 is less pronounced and is dominated by C1C3 and C2N5C4 adjustments. This is also true for system **6** for which both termini of the 1,3-dipole have similar electronegativities.

Reaction Phase 4 (from curvature point M3 to the end in exit channel): The question, which bond is formed first when generating the five-membered ring, can be easily answered by inspection of the components dominating the reaction path direction. Reaction path direction and reaction path curvature are complementary in so far as a local mode, when making a large contribution to one of the path quantities, can only make a small contribution to the other. As shown in Fig. 3 for the reaction path direction of system **4** (HCNO), the approach distances C2C3 and C1O4 dominate the path direction for most of its range. The amplitude of the mode associated with parameter C2C3 drops to small values between $s = 3$ and $s = 5$ amu$^{1/2}$ Bohr in phase 3, which is exactly that location where the C2C3 bond is formed. A similar decrease in the amplitude of the mode associated with C1O4 occurs much later in phase 4 when the C1O4 bond is finalized. Hence, the formation of the five-membered ring bonds is asynchronous with the less polar bond C2C3 being formed first in phase 3 (when C1O4 is still resisting bond formation) and the more polar bond C1O4 later in phase 4.

The less polar bond in **4** is formed by electron pairing in a *soft*, homolytic fashion leading to a smaller curvature enhancement or just a shoulder. For the formation of the more polar bond, exchange repulsion has to be overcome and a stronger bond is formed. This leads to a large curvature peak, which is enhanced by contributions from the adjustment steps of the other bonds in the five-membered ring being formed. In the case of the symmetric systems **9** and **10**, the simultaneous formation of the two new bonds leads to a large curvature.

The delay between the formation of X2C3 and C1Z4 is proportional to the electronegativity difference between termini X and Z. The new bonds are formed at the same time for systems **9** and **10** (because of symmetry). In the case of systems **2** and **6**, formation takes place almost at the same time because the electronegativities of groups N and NH or HC and CH_2 are almost the same.

There is a basic difference between reaction systems **1–6** and systems **7–9** because in the former cases a planar five-membered ring is formed characterized by electron delocalization, i.e., the new bonds adopt some double-bond characters whereas in the latter cases puckered rings are generated with little electron delocalization and the new bonds being essentially single bonds. Although the degree of puckering is small (puckering amplitudes [87–89] q are 0.187 (**7**), 0.098 (**8**), and 0.146 Å (**9**), which correspond to maximal dihedral angles of 19.0, 9.9, and 14.7°, respectively), it leads to unique ring conformations (a twist form close to pseudorotation phase angle [87–89] $\phi = 342°$ for **7**, an envelope form close to $\phi = 324°$ for **8**, and a pure envelope form with an axial NH bond with $\phi = 180°$ for **9**; see bottom of Fig. 1), which are optimal with regard to anomeric stabilization and bond staggering.

When these single bonds are finalized, the five-membered ring adjusts to its equilibrium conformation. This adjustment does not require significant electronic structure changes, and therefore, the curvature is small in the last part of phase 4 of the corresponding cycloaddition reactions (see Fig. 2g–i). In the other six systems (including also system **10**), the last stage of phase 4 corresponds to π-delocalization, which is connected with a relatively large curvature. This is also relevant for the reverse reaction: The splitting of the two bonds requires initially major electron reorganization, and the scalar curvature starts with a high value.

Discussion of the reaction energetics The discussion of the reaction mechanism of cycloaddition systems **1–9** reveals a close mechanistic relationship, which should make it possible to explain the energetics of these reactions in a consistent way. Clearly, the fate of the reaction complex with regard to stereochemistry or the sequence of the bond formation steps is early decided in phase 1 (van der Waals phase). Orientation of the reactants, charge transfer to acetylene, and mutual polarization in phase 1, although decisive, involve only minor energy changes. The major changes in the electronic structure of the reactants occur in phase 2 (rehybridization and bending phase), in which via bending of the XYZ-dipole molecule, the pyramidalization of the terminal $-CH_2$ groups (if existent), and bending of one or both CCH units of acetylene, the radicaloid centers are prepared so that bond formation becomes possible.

In all cases investigated, the TS is located in phase 2 always before (or at) the end of this phase given by M2 and thereby always after the curvature peak. Since at this curvature peak the largest change in energy takes place (strongest change in the electronic structure) in the entrance channel, it is reasonable to assume that the energy difference $\Delta E(P2) = E(M2) - E(M1)$ must be proportional to the activation energy. This hypothesis is confirmed by Fig. 5 where calculated $\Delta E(P2)$ values are correlated with the energy barriers. The energy changes in phase 2 determine the reaction barrier; however, this does not say that they determine the mechanistic sequence, which obviously is determined already in phase 1.

There are three electronic factors determining the barrier.

1. The reaction barriers decrease linearly with (absolutely seen) increasing dipole moment μ_z (parallel to the CC axis of acetylene, see Table 2) where the correlation is more significant with the more accurate CCSD(T)-F12 barriers. The polarizing power of the 1,3-dipole expressed in this way has a larger effect for the more electrophilic dipoles **1**, **2**, and **3** than for the more nucleophilic dipoles **7**, **8**, and **9**.
2. The reaction barriers also decrease with increasing polarizability component α_{zz} (see Table 2). This effect is best developed for the 1,3-dipoles with linear XYZ unit (**1–6**) and less for the bent dipoles (the azomethines, see Fig. 1) with a three-dimensional structure. The larger the polarizability component α_{zz} is, the easier can be the density of the 1,3-dipole molecule polarized so that radicaloid centers are formed.
3. Also shown in Table 2 are the local mode XYZ bending force constants k^a, which are a measure for the ease of bending of the heavy atom framework of XYZ in phase 2. For the diazonium betaines and the azomethines investigated, a larger bending force constant (in line with a larger electronegativity difference between X and Z) leads to a larger barrier. This effect is, however, outweighed by the charge polarization and polarizability effects in the case of the nitrilium betaines.

In Table 2, the calculated local mode stretching force constants k^a of bonds C3X2 and C1Z4 of the newly formed five-membered rings are also given. They confirm that the more polar bond (formed after the less polar bond) is always the somewhat stronger (corresponding to a larger k^a value). The sum of these k^a values reflects whether a planar ring with a delocalized 6π-system is formed (large sum) or a (non)planar ring with limited π-delocalization. In the former case, the barrier for cycloreversion is high.

Differences between the various cycloaddition reactions Although the reaction mechanisms of the ten cycloadditions investigated in this work are very similar, there are also remarkable differences, which become apparent from the URVA analysis.

1. Differences between the reaction of an electrophilic and a nucleophilic 1,3-dipole (for the former the charge transfer ceases already in the entrance channel, whereas for the latter it has it maximum in the exit channel);
2. Differences between symmetric and asymmetric 1,3-dipoles (for the former, phase 3 is a phase to prepare the reaction complex for bond formation, whereas for all other reaction systems, phase 3 is the phase for forming the soft bond via long-range interactions between radicaloid centers);
3. Similarly, there is a difference between 1,3-dipoles with termini of similar or strongly different electronegativities;
4. Diazonium and nitrilium betaine reactions lead to a different phase 4 compared to that of the azomethine cycloadditions (the product of the former reactions is a planar ring with electron delocalization, and therefore, phase 4 encompasses this process indicated by a large scalar curvature at the end of the phase. The latter reaction systems lead to a puckered ring, and therefore, phase 4 is characterized by a small curvature peak and small curvature values at the end reflecting a conformational adjustment);
5. If the orientation phase includes conformational adjustments of the 1,3-dipole, it is characterized by a larger curvature enhancement. Such a case is given by system **5** where the orientation leads to a rotational adjustment of the 1,3-dipole in the reaction complex (see Fig. 2e);
6. There is a fine distinction between different reaction systems depending on whether the trans-bending mode of acetylene is strongly, just weakly, or not at all excited in the entrance channel so that the radicaloid centers of acetylene develop with a very different pace (see below);
7. Systems such as **6** and **8**, which possess identical reaction barriers within calculational accuracy, have different reaction mechanisms as can be seen from the curvature diagrams in phase 4 (see Fig. 2f, h). This reflects that the TS and energy barrier are just cumulative descriptors of a reaction, which result from different electronic effects and their changes, but which do not provide a detailed insight into the mechanism.

Comparison with other mechanistic studies In the "Introduction," three mechanistic hypotheses are mentioned, which are currently used to explain mechanistic details of 1,3-dipolar cycloaddition reactions. They are discussed below.

The distortion energy description Houk et al. [30, 32] introduced the *distortion energy* as the energy required to distort the dipole and the dipolarophile into the TS geometries without allowing any interaction between them. The total activation energy is the sum of the distortion and interaction energy of the two distorted fragments frozen in the TS geometry. These authors found a strong correlation between the activation energy and the distortion energy for the reaction of dipoles **1–9** with ethylene and acetylene. Based on these findings, they concluded that the concerted mechanism leads to a TS geometry at which the overlap between the orbitals of the termini of the cycloaddends directly leads to the cycloadduct without further geometry distortion. To achieve this, the TS geometry requires distortion of the dipole, and the distortion is related to the dipole stability.

These observations are in line with the URVA results in so far as the TSs of the 1,3-dipolar cycloadditions are located close to the end of the rehybridization phase, which is the phase with the energy-consuming electronic structure changes. In many other cases investigated so far, the TS occurs at a less prominent position, which would make the mechanistic interpretation of the distortion energy more difficult [53, 65–67, 90].

Conventional investigations of chemical reactions focus on the explanation of the height of the reaction barrier. This is important, however, that the reaction barrier alone cannot provide a detailed insight into the reaction mechanism as presented in this work. As shown above, similar reaction barriers do present neither a necessary nor sufficient condition for similarity of the reaction mechanism. Likewise, two reactions may have similar mechanisms although they have different barriers. The energy barrier of a reaction is just a cumulative measure of all electronic structure changes taking place at the position of the TS, and therefore, it can only provide indirect and remote insight into the electronic structure changes taking place during the reaction.

The bending hypothesis Houk and coworkers concluded that the correlation of the activation energy with the distortion energy implies that the vibrational excitation of the reactants is an important feature of the mechanism [28]. When analyzing the transition vector for the reactions of the diazonium betaines **1–3** with acetylene or ethylene, these authors found that the translational vector associated with the imaginary frequency is dominated by the bending of the 1,3-dipole (up to 80 %). Classical trajectory calculations confirmed the dominance of the dipole bending excitation, which parallels the trend in the distortion energies. Barnes and Hase performed variational transition state calculations for the cycloaddition of diazonium betaines **1–3** and ethylene [14]. They concluded that the reaction rate for a fixed temperature will be increased by selective excitation of the diazonium betaines bending vibrational modes in line with Houk's findings.

The analysis of the reaction path direction reveals that in none of the reaction systems investigated does bending dominate the path direction. However, the curvature enhancements shown in phase 2 are dominated by the normal mode curvature coupling coefficient associated with the XYZ and acetylene bending modes. Hence, pumping up these modes will lead to an acceleration of the reaction, provided the energy channeled into the bending modes is not dissipated by Coriolis couplings with other vibrational modes.

In Fig. 7a, the normal mode frequencies of reaction complex **4** are given as a function of the reaction parameter s in the range up to 1200 cm^{-1}, which includes all bending and torsional modes. Also shown are the avoided crossings (ACs) between vibrational states of the same symmetry AC1 to AC14, which are the positions of strong mode–mode couplings. Clearly, there is the possibility of a continuous exchange of energy between the vibrational modes along the reaction path, which should lead to energy dissipation and thereby a reduction in mode-selective rate enhancement.

If energy dissipation, involving key vibrational modes, takes place before the TS, mode-selective rate enhancement is quenched. Figure 7a and b shows that in the entrance channel the XYZ bending mode, # 7, undergoes Coriolis coupling with the trans-HCCH bending mode, # 9, at $s = -6.85$ amu$^{1/2}$ Bohr (AC1) and at $s = -1.80$ amu$^{1/2}$ Bohr (AC4), which leads to a mixing of their characters and an exchange of energy. Shortly before the TS, there is another avoided crossing (AC5) involving modes #7 and #10. This will effectively dissipate any energy pumped into the XYZ bending mode # 7, which is in contrast to the hypothesis of Houk [28] and Hase et al. [14].

The analysis of normal modes #7, #9, and #10 reveals that energy stored in the first two modes can be transferred to the cis-bending mode of HCCH (see Fig. 7a, b). This mode must be triggered to establish radical centers at C1 and/or C3. Any energy stored in the XYZ bending mode #7 will activate cis-HCCH bending, which may accelerate the reaction as long as XYZ bending takes place at the same time. However, this cannot be guaranteed on the basis of the mode–mode coupling situation before the TS.

All ten reaction systems investigated exhibit a similar mode–mode coupling pattern, which leads to an energy dissipation between those bending modes, which must be excited for rate acceleration. Hence, the previous observations of Houk and Hase are confirmed even though complicated Coriolis coupling patterns as shown in Fig. 7a normally make mode-selective rate enhancements ineffective. It will be interesting to see whether other dipolarophiles than acetylene lead to the same effective mechanism despite a multitude of dissipation possibilities.

The biradical hypothesis Hiberty et al. [12] found a correlation between the biradical character of 1,3-dipoles **1-9** and their reactivity toward ethylene or acetylene using the breathing-orbital valence bond ab initio method [20]. Each 1,3-dipole is described as a linear combination of three valence bond structures, two zwitterions and one biradical, for which the weights in the total wave function can be quantitatively estimated.

A reaction mechanism was proposed [12], in which the 1,3-dipole first distorts so as to reach a reactive electronic state that has a significant biradical character, which then adds with little or no barrier to the dipolarophile. By determining the biradical character of the 1,3-dipole either as the weight of the biradical structure at equilibrium or as the energy gap between the ground state of the 1,3-dipole and its biradical diabatic state, they found a useful correlation between the biradical character and the barrier heights. In line with Houk′s energy distortion/interaction model, the barrier height for the cycloaddition of a given 1,3-dipole to ethylene or acetylene is dominated by biradical energy, which rationalizes that for both ethylene and acetylene addition, the barrier heights are nearly the same despite significant differences in the exothermicity of these reactions.

Again, this approach provides an energy-counting argument, which does not necessarily lead to a mechanistic insight. As shown in this work, the reaction is initiated by a charge transfer to the dipolarophile and polarization of the density of the latter. The generation of (a) (bi)radicaloid center(s) at acetylene is as important as that at the 1,3-dipole molecule and precedes or is parallel to the bending of the 1,3-dipole. Also as noted above the major part of the reaction barrier is due to the rehybridization and (bi)radicaloid formation of the 1,3-dipole. However, mechanistically more important are the charge transfer, the population of a $\pi^{\star}$(HCCH) orbital, and the *softening* of the bending motion, which is the prerequisite for radicaloid formation. In the first phase, the 1,3-dipole polarizes the acetylene molecule and not vice versa. The difference in electronegativities of the 1,3-dipole termini X and Z decides on the amount of charge transfer and the polarization.

We can distinguish between 1,3-dipolar cycloadditions, which start with just one radicaloid center (bending of one CCH group: e.g., system **4**) and those which, because of symmetry or similar electronegativities of X and Z, require the bending of both CCH groups in acetylene and the formation of a biradicaloid (e.g., systems **9** or **6**). This mechanistic differentiation is supported by a sophisticated energy exchange between 1,3-dipole bending, HCCH trans-bending, and HCCH cis-bending motions before the TS. In view of the URVA analysis presented in this work, it is appropriate to speak in this connection of radicaloid- and biradicaloid-driven cycloadditions. However, it would be a simplification to consider just the biradical formation of the 1,3-dipole molecule excluding the role of the dipolarophile.

The Hammond–Leffler postulate The Hammond–Leffler postulate, although not directly formulated in this way [91, 92], derives from the exothermic or endothermic character of a reaction a characterization of the TS as being early (in the first case) and late (in the second case). A quantification of this postulate has turned out to be difficult until recently. We showed that the TS of a reaction is shifted against the center of the bond-breaking/forming processes (identified via scalar curvature and local mode curvature coupling coefficients) by an amount $\Delta s(TS)$ along the reaction path, which is proportional to the reaction energy, $\Delta_R E$ [69, 93]. The larger the shift parameter is, the more exothermic the reaction is, and the earlier the TS is located in the entrance channel. For endothermic reactions, the shift parameter becomes negative [69, 93].

In Fig. 6, the $\Delta s(TS)$ values obtained in this work (see also Table 2) are given in dependence of the calculated reaction energies $\Delta_R E$. A linear relationship results, which identifies reaction systems **5**, **6**, **4,** and **9** as having an early TS. This is a result of forming a stable five-membered ring (a pyrazole, isoazole, or isoxazole with a delocalized 6π-electron system) or converting an unstable 1,3-dipole such as the azomethine ylide of reaction system **9** with high biradical character into a more stable 2,5-dihydro-pyrrole. The scalar curvature presents an important tool to verify and quantify the Hammond–Leffler postulate.

4 Conclusions

This investigation confirms that chemical reactivity is the result of the mutual polarization of the reactants (proportional to their polarizing power and polarizability) and charge transfer between them, which leads to a change in their electronic structure and prepares them for the reaction.

1. All ten 1,3-dipolar cycloadditions investigated in this work with URVA follow a 4-phase reaction mechanism with well-defined reaction phases defined by the scalar curvature of the reaction path and features of the unified reaction valley. These phases are identified and characterized in their chemical nature with the help of the local mode curvature coupling coefficients.
   (1) *Reaction phase 1* is the orientation phase where charge transfer and charge polarization are initiated;
   (2) *Reaction phase 2* is the rehybridization and bending phase where radicaloid or biradicaloid structures are formed;
   (3) *Reaction phase 3* is the phase where the homolytic formation of the *softer* bond starts;
   (4) In *Reaction phase 4* being characterized by a large curvature peak, the formation of the *harder* (more polar) bond and the finalization of the new five-membered ring take place, accompanied by π-delocalization (systems **1–6** and **10**) or ring puckering (systems **7–9**).
2. The orientation of the 1,3-dipole XYZ relative to acetylene is a consequence of the shape of the exchange repulsion envelope around the 1,3-dipole. This in turn depends on the electronegativity difference between X and Z. The orientation of the 1,3-dipole determines the degree of charge transfer and charge polarization and decides on the asynchronicity of the formation of the two five-membered ring bonds. Hence, the mechanism of the cycloaddition reaction is decided early in the van der Waals range (phase 1). It is important to note that *the bond with larger CX or CZ distance in the orientation phase is formed first*, which has to do with the fact that two radical centers can undergo long-range interactions.
3. The driving force of all ten cycloaddition reactions is the charge transfer from the 1,3-dipole to the acetylene molecule. This is also true for the electrophilic 1,3-dipole molecules, which differ from the nucleophilic dipoles by a reduced charge transfer ability ceasing before the TS (see Fig. 4b). Consequently, a much higher energy barrier results. The amount of charge transfer can be qualitatively predicted by investigating the frontier orbitals of 1,3-dipole and dipolarophile, which explains the success of FMO theory for describing these types of reactions. However, we also note that the detailed information provided by the URVA analysis of a reaction (see, e.g., Fig. 4b) cannot be expected from FMO theory.
4. In separate calculations, we have shown that any transfer of negative charge into the $\pi^\star$ MO of acetylene leads to trans-bending, which is initiated in phase 1, although it is in conflict with the actual cis-bending required later in the reaction. Charge polarization in acetylene as induced by the approaching 1,3-dipole triggers CCH bending at the more negatively charged acetylene terminus as could be shown with the help of the local mode bending force constants.

5. The 1,3-dipolar cycloaddition reactions are the first reactions investigated so far [55, 65–69, 74, 90] for which the mechanistically important electronic structure changes take place in the phase (here phase 2) containing the TS so that the reaction barrier is almost identical to the energy changes in this phase (phase 2; see Fig. 5). In most other reactions, the location of the TS is not significant for the individual mechanistic changes because it is just an *accountant* for the sum of all energy changes rather than a descriptor of specific energy changes associated with individual mechanistic steps.
6. The energy barrier is determined by mutual charge polarization of the reactants and therefore directly related to the polarizing power and the polarizability of the 1,3-dipole molecule as reflected by its dipole moment component in the direction of the acetylene molecule and its polarizability component in the same direction.
7. Another factor influencing the magnitude of the reaction barrier is the stiffness of the XYZ-dipole as measured by the local XYZ bending force constant. A large (small) force constant implies a large (small) barrier apart from the nitrilium betaines where mutual polarization effects dominate the reaction barrier. The local stretching force constants of the five-membered rings formed reflect their degree of electron delocalization and they are measures for the height of the barrier of the cycloreversion reaction.
8. Previous work emphasizing the importance of an excitation of the XYZ bending vibration for a mode-specific acceleration of the reaction is confirmed, however, with some important restrictions. There is a multitude of Coriolis couplings, which in general dissipate any mode-specific energy surplus. In the case of the cycloaddition reactions **1–10**, the peculiar situation exists that XYZ bending couples with trans- and cis-bending of HCCH in the entrance channel so that any energy excess in the XYZ bending mode facilitates the formation of radicaloid centers in the dipolarophile without guaranteeing that XYZ bending is enhanced. Future investigations have to clarify whether this is a more general phenomenon.
9. The center of the chemical processes of bond forming leading to the five-membered ring is located in the exit channel. The shift parameter $\Delta s(TS)$ measures the earliness of the TS according to the Hammond–Leffler postulate and correlates linearly with the exothermicity of the reaction as described by $\Delta_R E$, thus providing a quantitative confirmation of this postulate.

The chemical reaction mechanism of the 1,3-dipolar cycloadditions is determined in the van der Waals range far off the TS. It is typical of a symmetry-allowed pericyclic reaction that the energy needed to overcome the reaction barrier is used to *prepare* the reactants for the reaction and that the actual bond formation takes place after the TS in the exit channel. The stability of the bonds being formed and the degree of electron delocalization in the five-membered ring determine the exothermicity of the reaction.

The many mechanistic details, which the URVA analysis of the 1,3-dipolar cycloadditions unravels, provide a solid basis for predictions concerning changes in the reaction mechanism and the energetics upon replacing the dipolarophile, substituting 1,3-dipole and/or dipolarophile, or using metal catalysis to accelerate the cycloaddition reaction. Any means that (1) increase the charge transfer to the dipolarophile and/or (2) facilitate its charge polarization will lower the barrier of the cycloaddition. Substituents of the 1,3-dipole will lower the barrier if they increase its polarizability and charge polarization in such a way that the asynchronicity of bond formation is supported rather than hindered. The same applies also with regard to steric effects. In view of the lengths of this work, the detailed discussion of these effects will be the topic of a forthcoming publication.

Acknowledgments This work was financially supported by the National Science Foundation, Grant CHE 1152357. We thank SMU for providing computational resources.

References

1. Padwa A, Pearson WH (2002) Synthetic applications of 1,3-dipolar cycloaddition chemistry toward heterocycles and natural products. Wiley, New York
2. Nair V, Suja TD (2007) Intramolecular 1,3-dipolar cycloaddition reactions in targeted syntheses. Tetrahedron 63:12247–12275
3. Padwa A, Bur SK (2007) The domino way to heterocycles. Tetrahedron 63:5341–5378
4. Rane D, Sibi M (2011) Recent advances in nitrile oxide cycloadditions. Synthesis of isoxazolines. Curr Org Synth 8:616–627
5. Kaur J (2011) Azomethines and 1,3 dipoles—leading to new heterocycles: studies on 1,3 dipolar cycloadditions. LAP LAMBERT Academic Publishing, New York
6. Woodward RB, Hoffmann R (1969) The conservation of orbital symmetry. Angew Chem Int Ed Engl 8:781–853
7. Dewar MJS (1971) Aromaticity and pericyclic reactions. Angew Chem Int Ed Engl 10:761–776
8. Nguyen TA (2007) Frontier orbitals: a practical manual. Wiley, New York
9. Gold B, Dudley GB, Alabugin IV (2013) Moderating strain without sacrificing reactivity: design of fast and tunable noncatalyzed alkyne-azide cycloadditions via stereoelectronically controlled transition state stabilization. J Am Chem Soc 135:1558–1569
10. Barbosa AGH, Monteiro JGS (2012) On the electronic structure of the diazomethane molecule. Theor Chem Acc 131:1297–1323
11. Das TK, Salampuria S, Banerjee M (2010) Computational DFT study of the 1,3-dipolar cycloadditions of 1-phenylethyl- trans-2-

methyl nitrone to styrene and 1-phenylethyl nitrone to allyl alcohol. J Mol Struc Theochem 959:22–29
12. Braida B, Walter C, Engels B, Hiberty PC (2010) A clear correlation between the diradical character of 1,3-dipoles and their reactivity toward ethylene or acetylene. J. Am. Chem. Soc. 132:7631–7637
13. Engels B, Marian CM (2009) What controls the reactivity of 1,3-dipolar cycloadditions? Angew Chem Int Ed Engl 48:7968–7970
14. Barnes GL, Hase WL (2009) Bent out of shape. Nat Chem 1:103–104
15. Mladenovic M, Elhiyani M, Lewerenz M (2009) Electric and magnetic properties of the four most stable CHNO isomers from ab initio CCSD(T) studies. J Chem Phys 131:034302–1–034302–14
16. Mladenovic M, Lewerenz M, McCarthy MC, Thaddeus P (2009) Isofulminic acid, HONC: ab initio theory and microwave spectroscopy. J Chem Phys 131:174308–117430810
17. Benchouk W, Mekelleche SM (2008) Theoretical analysis of the regioselectivity of 1,3-dipolar cycloaddition of C-(methoxycarbonyl)-N-methyl with methyl acrylate and vinyl acetate. J Mol Struc Theochem 852:46–53
18. Merino P, Tejero T, Chiacchio U, Romeoc G, Rescifina A (2007) A DFT study on the 1,3-dipolar cycloaddition reactions of C-(hetaryl) nitrones with methyl acrylate and vinyl acetate. Tetrahedron 63:1448–1458
19. Sakai S, Nguyen MT (2004) Theoretical determination of the electronic mechanisms of 1,3-dipolar cycloaddition reactions of fulminic acid and diazomethane. J Phys Chem A 108:9169–9179
20. Hiberty PC, Shaik S (2002) A clear correlation between the diradical character of 1,3-dipoles and their reactivity toward ethylene or acetylene. Theor Chem Acc 108:255–272
21. Nguyen MT, Chandra AK, Sakai S, Morokuma K (1999) Another look at the mechanism of the concerted 1,3-dipolar cycloaddition of fulminic acid to acetylene. J Org Chem 64:65–69
22. Lopez SA, Munk ME, Houk KN (2013) Mechanisms and transition states of 1,3-dipolar cycloadditions of phenyl azide with enamines: a computational analysis. J Org Chem 78:1576–1582
23. Lan Y, Wheeler SE, Houk KN (2012) Extraordinary difference in reactivity of ozone (OOO) and sulfur dioxide (OSO): a theoretical study. J Chem Theory Comput 7:2104–2111
24. Krenske EH, Houk KN, Holmes AB, Thompson J (2011) Entropy versus tether strain effects on rates of intramolecular 1,3-dipolar cycloadditions of N-alkenylnitrones. Tetrahedron Lett 52:2181–2184
25. Lan Y, Houk KN (2010) Mechanism and stereoselectivity of the stepwise 1,3-dipolar cycloadditions between a thiocarbonyl ylide and electron-deficient dipolarophiles: a computational investigation. J Am Chem Soc 132:17921–17927
26. Xu CE, Doubleday L, Houk KN (2010) Dynamics of 1,3-dipolar cycloadditions: energy partitioning of reactants and quantitation of synchronicity. J Am Chem Soc 132:3029–3037
27. Xu CE, Doubleday L, Houk KN (2009) Dynamics of 1,3-dipolar cycloaddition reactions of diazonium betaines to acetylene and ethylene: bending vibrations facilitate reaction. Angew Chem Int Engl 48:2746–2748
28. Schoenebeck F, Ess DH, Jones GO, Houk KN (2009) Reactivity and regioselectivity in 1,3-dipolar cycloadditions of azides to strained alkynes and alkenes: a computational study. J Am Chem Soc 131:8121–8133
29. Ess DH, Jones GO, Houk NK (2008) Transition states of strain-promoted metal-free click chemistry: 1,3-dipolar cycloadditions of phenyl azide and cyclooctynes. Org Lett 10:1633–1636
30. Ess DH, Houk KN (2008) Theory of 1,3-dipolar cycloadditions—distortion/interaction and frontier molecular orbital models. J Am Chem Soc 130:10187–10198
31. Jones GO, Houk KN (2008) Predictions of substituent effects in thermal azide 1,3-dipolar cycloadditions: implications for dynamic combinatorial (reversible) and click (irreversible) chemistry. J Org Chem 73:1333–1342
32. Ess DH, Houk KN (2007) Distortion/interaction energy control of 1,3-dipolar cycloaddition reactivity. J Am Chem Soc 129:10646–10647
33. Ess DH, Jones GO, Houk NK (2006) Conceptual, qualitative, and quantitative theories of 1,3-dipolar and diels-alder cycloadditions used in synthesis. Adv Synth Catal 348:2337–2361
34. Jones GO, Ess DH, Houk KN (2005) Activation energies and reaction energetics for 1,3-dipolar cycloadditions of hydrazoic acid with C–C and C–N multiple bonds from high-accuracy and density functional quantum mechanical calculations. Helvetica Chimica Acta 88:1702–1710
35. Ess DH, Houk KN (2005) Activation energies of pericyclic reactions: performance of DFT, MP2, and CBS-QB3 methods for the prediction of activation barriers and reaction energetics of 1,3-dipolar cycloadditions, and revised activation enthalpies for a standard set of hydrocarbon pericyclic reactions. J Phys Chem A 109:9542–9553
36. Jung ME, Min S, Houk KN, Ess D (2004) Synthesis and relative stability of 3,5-diacyl-4,5-dihydro-1h-pyrazoles prepared by dipolar cycloaddition of enones and α-diazoketones. J Org Chem 69:9085–9089
37. Houk KN, Gonzalez J, Li Y (1995) Pericyclic reaction transition states: passions and punctilios, 1935–1995. Acc Chem Res 28:81–90
38. Gothelf KV, Jorgensen KA (1998) Asymmetric 1,3-dipolar cycloaddition reactions. Chem Rev 98:863–909
39. Kissane M, Maguire AR (2010) Asymmetric 1,3-dipolar cycloadditions of acrylamides. Chem Soc Rev 39:845–883
40. Xing Y, Wang N-X (2012) Organocatalytic and metal-mediated asymmetric (3 + 2) cycloaddition reactions. Coord. Chem. Rev. 256:938–952
41. Amblard F, Cho JH, Schinazi RF (2009) Cu(I)-catalyzed Huisgen azide-alkyne 1,3-dipolar cycloaddition reaction in nucleoside, nucleotide, and oligonucleotide chemistry. Chem Rev 109: 4207–4220
42. Naodovi M, Yamamoto H (2008) Asymmetric silver-catalyzed reactions. Chem Rev 108:3132–3148
43. Stanley LM, Sibi MP (2008) Enantioselective copper-catalyzed 1,3-dipolar cycloadditions. Chem Rev 108:2887–2902
44. Lutz JF (2007) 1,3-dipolar cycloadditions of azides and alkynes: a universal ligation tool in polymer and materials science. Angew Chem Int Ed Engl 46:1018–1025
45. Tron GC, Pirali T, Billington RA, Canonico PL, Sorba G, Genazzani AA (2007) Click chemistry reactions in medicinal chemistry: applications of the 1,3-dipolar cycloaddition between azides and alkynes. Med Res Rev 28:278–308
46. Jewett J, Bertozzi CR (2010) Cu-free click cycloaddition reactions in chemical biology. Chem Soc Rev 39:1272–1279
47. Pieters RJ, Rijkers RMJ, Liskamp DTS (2007) Application of the 1,3-dipolar cycloaddition reaction in chemical biology: approaches toward multivalent carbohydrates and peptides and peptide-based polymers. QSAR Comb Sci 26:1181–1190
48. Huisgen R (1968) On the mechanism of 1,3-dipolar cycloadditions. A reply. J Org Chem 33:2291–2297
49. Firestone R (1968) On the mechanism of 1,3-dipolar cycloadditions. J Org Chem 33:2285–2290
50. Lan Y, Zou L, Cao Y, Houk KN (2011) Computational methods to calculate accurate activation and reaction energies of 1,3-dipolar cycloadditions of 24 1,3-dipoles. J Phys Chem A 115:13906–13920
51. Tantillo DJ, Lee JK (2011) Reaction mechanisms: pericyclic reactions. Annu Rep Prog Chem Sect B 107:266–286
52. Domingo LR, Picher T (2004) A DFT study of the Huisgen 1,3-dipolar cycloaddition between hindered thiocarbonyl ylides and tetracyanoethylene. Tetrahedron 60:5053–5058

53. Kraka E, Cremer D (2010) Computational analysis of the mechanism of chemical reactions in terms of reaction phases: hidden intermediates and hidden transition state. Acc Chem Res 43:591–601
54. Cremer D, Kraka E (2010) From molecular vibrations to bonding, chemical reactions, and reaction mechanism. Curr Org Chem 14:1524–1560
55. Konkoli Z, Cremer D, Kraka E (1997) Unified reaction valley approach: mechanism of the reaction $CH_3 + H_2 \rightarrow CH_4 + H$. J Phys Chem A 101:1742–1757
56. Kraka E (2011) Reaction path Hamiltonian and the unified reaction valley approach. In: Allen W, Schreiner PR (eds) Wiley interdisciplinary reviews: computational molecular science. Wiley, New York, pp 531–556
57. Kraka E (1998) Reaction path Hamiltonian and its use for investigating reaction mechanism. In: Schleyer PR, Allinger NL, Clark T, Gasteiger J, Kollman P, Schaefer HF, Schreiner PR (eds) Encyclopedia of computational chemistry, volume 4. Wiley, Chichester, p 2437
58. Miller WH, Handy NC, Adams JE (1980) Reaction path Hamiltonian for polyatomic molecules. J Chem Phys 72:99–112
59. Konkoli Z, Cremer D (1998) A new way of analyzing vibrational spectra I. Derivation of adiabatic internal modes. Int J Quant Chem 67:1–11
60. Konkoli Z, Cremer D (1998) A new way of analyzing vibrational spectra III. Characterization of normal vibrational modes in terms of internal vibrational modes. Int J Quant Chem 67:29–41
61. Kraka E, Larsson JA, Cremer D (1998) New developments in the analysis of vibrational spectra: on the use of adiabatic internal, vibrational modes. In: Parkanyi C (eds) Theoretical and computational chemistry, volume 5, theoretical organic chemistry, C. Elsevier, Amsterdam, p 259
62. Kraka E, Larsson JA, Cremer D (2010) Generalization of the badger rule based on the use of adiabatic vibrational modes. In: Grunenberg J (eds) Vibrational modes in computational IR spectroscopy. Wiley, New York, pp 105–149
63. Zou W, Kalesky R, Kraka E, Cremer D (2012) Relating normal vibrational modes to local vibrational modes with the help of an adiabatic connection scheme. J Chem Phys 137:084114–1–084114–11
64. Zou W, Kalescky R, Kraka E, Cremer D (2012) Relating normal vibrational modes to local vibrational modes benzene and naphthalene. J Mol Model 19:2865–2877
65. Cremer D, Wu A, Kraka E (2001) The mechanism of the reaction $FH + H_2C{=}CH_2 \rightarrow H_3C\text{-}CFH_2$. Investigation of hidden intermediates with the unified reaction valley approach. Phys Chem Chem Phys 3:674–687
66. Kraka E, Wu A, Cremer D (2003) Mechanism of the Diels–Alder reaction studied with the united reaction valley approach: mechanistic differences between symmetry-allowed and symmetry-forbidden reactions. J Phys Chem A 107:9008–9021
67. Kraka E, Joo H, Cremer D (2010) A stunning example for a spontaneous reaction with a complex mechanism: the vinylidene–acetylene cycloaddition reaction. Mol Phys 19(20):2667–2685
68. Joo H, Kraka E, Quapp W, Cremer D (2007) The mechanism of a barrierless reaction: hidden transition state and hidden intermediate in the reaction of methylene with ethene. Mol Phys 105:2697–2717
69. Kraka E, Zou W, Freindorf M, Cremer D (2012) Energetics and mechanism of the hydrogenation of XH_n for group IV to group VII elements X. J Chem Theory Comput 8:4931–4943
70. Raghavachari K, Trucks GW, Pople JA, Head-Gordon M (1989) A fifth-order perturbation comparison of electron correlation theories. Chem Phys Lett 157:479–483
71. Adler TB, Knizia G, Werner H-J (2007) A simple and efficient CCSD(T)-F12 approximation. J Chem Phys 127:221106–221110
72. Dunning T Jr (1989) Gaussian basis sets for use in correlated molecular calculations I. The atoms boron through neon and hydrogen. J Chem Phys 90:1007–1023
73. Fukui K (1981) The path of chemical reactions—the IRC approach. Acc Chem Res 14:363–368
74. Quapp W, Kraka E, Cremer D (2007) Finding the transition state of quasi-barrierless reactions by a growing string method for Newton trajectories: application to the dissociation of methylenecyclopropene and cyclopropane. J Chem Phys A 111:11287–11293
75. Hratchian HP, Kraka E (2013) Improved predictor-corrector integrators for evaluating reaction path curvature. J Chem Theor Comput 9:1481–1488
76. Becke AD (1993) Density-functional thermochemistry. III. The role of exact exchange. J Chem Phys 98:5648–5652
77. Stevens PJ, Devlin FJ, Chabalowski CF, Frisch MJ (1994) Ab initio calculation of vibrational absorption and circular dichroism spectra using density functional force fields. J Phys Chem 98:11623–11627
78. Hariharan PC, Pople JA (1973) The influence of polarization functions on molecular orbital hydrogenation energies. Theor Chimica Acta 28:213–222
79. Kraka E, Filatov M, Zou W, Gräfenstein J, Izotov D, Gauss J, He Y, Wu A, Polo V, Olsson L, Konkoli Z, He Z, Cremer D (2013) COLOGNE13
80. Reed AE, Curtiss LA, Weinhold F (1988) Intermolecular interactions from a natural bond orbital, donor-acceptor viewpoint. Chem Rev 88:899–926
81. Weinhold F, Landis CR (2003) Valency and bonding: a natural bond orbital donor–acceptor perspective. Cambridge University Press, Cambridge
82. Werner HJ, Knowles PJ, Kniza G, Manby FR, Schütz M, and others v MOLPRO, version 2010.1, a package of ab initio programs. see http://www.molpro.net
83. Chai J-D, Head-Gordon M (2008) Systematic optimization of long-range corrected hybrid density functionals. J Chem Phys 128:084106–1–084106–14
84. Chai J-D, Head-Gordon M (2008) Long-range corrected hybrid density functionals with damped atom–atom dispersion corrections. Phys Chem Chem Phys 10:6615–6620
85. Thanthiriwatte KS, Hohenstein EG, Burns LA, Sherrill CD (2011) Assessment of the performance of DFT and DFT-D methods for describing distance dependence of hydrogen-bonded interactions. J Chem Theor Comp 7:88–96
86. Boys SF, Bernardi F (1970) The calculation of small molecular interactions by the differences of separate total energies. Some procedures with reduced errors. Mol Phys 19:553–566
87. Cremer D, Pople JA (1975) A general definition of ring puckering coordinates. J Am Chem Soc 97:1354–1358
88. Cremer D, Szabo KJ (1995) Ab initio studies of six-membered rings, present status and future developments. In: Juaristi E (eds) Methods in stereochemical analysis, conformational behavior of six-membered rings, analysis, dynamics, and stereoelectronic effects. VCH Publishers, USA, pp 59–134
89. Cremer D, Izotov D, Zou W, Kraka E (2011) Southern Methodist University, Dallas, TX
90. Kraka E, Cremer D (2002) Mechanism and dynamics of organic reactions: 1,2-H shift in methylchlorocarbene. J Phys Org Chem 15:431–447
91. Hammond GS (1955) A correlation of reaction rates. J Am Chem. Soc 77:334–338
92. Leffler JE (1953) Parameters for the description of transition states. Science 117:340–341
93. Cremer D, Kraka E (2012) Verification and quantification of the Hammond–Leffler postulate. Rev Proc Quim: 27–30

Theor Chem Acc (2014) 133:1415
DOI 10.1007/s00214-013-1415-z

REGULAR ARTICLE

Active Thermochemical Tables: dissociation energies of several homonuclear first-row diatomics and related thermochemical values

Branko Ruscic · David Feller · Kirk A. Peterson

Received: 5 September 2013 / Accepted: 23 October 2013 / Published online: 22 November 2013

Abstract The current Active Thermochemical Tables (ATcT) results for the bond dissociation energies of the homonuclear diatomics H_2, C_2, N_2, O_2, and F_2 are reported and discussed. The role and origin of the distributed provenance of ATcT values is analyzed. Ramifications in terms of the enthalpies of formation of H, C, N, O, and F atoms, which are fundamental thermochemical quantities, are presented. In addition, the current ATcT bond dissociation energies and enthalpies of formation of HF, CH, CO, CN, NO, OH, CO_2, H_2O, and triplet and singlet CH_2 are also reported.

Keywords Enthalpy of formation · Thermochemical network · Theory–experiment interface · Benchmark values

1 Introduction

Active Thermochemical Tables (ATcT) [1, 2] are a novel paradigm for obtaining accurate, reliable, and internally consistent thermochemical values for a broad range of chemical species, accompanied by statistically sound uncertainties that conform to the accepted standard in thermochemistry (95 % confidence intervals). These characteristics of the ATcT thermochemical values make them very desirable for developing and benchmarking highly accurate state-of-the-art electronic structure approaches [3–6].

In a companion paper [7], the Feller–Peterson–Dixon (FPD) procedure at the highest currently feasible computational level is applied to H_2 and 13 small first- and second-row molecules, and the results are compared to dissociation energies obtained from ATcT. The aim of the current paper is to report the current ATcT values for the bond dissociation energies of H_2, C_2, N_2, O_2, and F_2 and elucidate their provenances.

Gas-phase homonuclear diatomics of hydrogen, nitrogen, oxygen, and fluorine are the thermodynamic reference states for the corresponding elements. Their dissociation energies are tightly related to the enthalpies of formation of the corresponding atoms, which are fundamental thermochemical quantities that belong to the CODATA 'key' set [8]. Since the CODATA evaluation, these quantities have gained additional relevance: the availability of accurate and reliable values for enthalpies of formation of atoms has become *a sine qua non* for electronic structure methods that use the total atomization energy route to obtain practical enthalpies of formation.

2 The approach of Active Thermochemical Tables

As opposed to traditional thermochemistry, which uses a sequential approach to build a set of thermochemical values (A begets B, B begets C, and so on), ATcT are based

Dedicated to Professor Thom Dunning and published as part of the special collection of articles celebrating his career upon his retirement.

B. Ruscic (✉)
Chemical Sciences and Engineering Division, Argonne National Laboratory, Argonne, IL 60439, USA
e-mail: ruscic@anl.gov

B. Ruscic
Computation Institute, University of Chicago, Chicago, IL 60637, USA

D. Feller · K. A. Peterson
Department of Chemistry, Washington State University, Pullman, WA 99164, USA

on constructing, analyzing, correcting, and solving a thermochemical network (TN) [1, 2]. The TN contains the available experimental determinations that thermochemically interconnect the targeted chemical species (such as bond dissociation energies, reaction enthalpies, constants of equilibria, ionization energies, and electron affinities), complemented by similar determinations extracted from high-level theoretical calculations (such as energies of various isodesmic reactions, bond dissociation energies, and total atomization energies). Each determination in the TN has an initially assigned uncertainty, reflecting its perceived 95 % confidence interval. Uncertainties are an important integral component of the TN: they determine the weight by which each determination contributes to the overall knowledge content of the TN and thus regulate the uncertainties of the final results. After the TN is constructed and checked for stoichiometric correctness, ATcT proceed with an iterative statistical analysis that exploits the available thermochemical cycles in the TN and evaluates all determinations for mutual consistency, producing a ranked list of potential 'offenders,' i.e., determinations with too optimistic uncertainties, which would unduly skew the final results if left uncorrected. The uncertainty of the most likely 'offender' is then augmented by a small increment, and the statistical analysis is repeated until internal consistency is achieved across the entire TN. Once the TN is internally consistent, ATcT obtain the final results by solving it simultaneously for all included chemical species.

The ATcT TN approach provides a number of significant advantages over the traditional sequential approach, described in more detail elsewhere [1, 2, 9, 10]. Because the results are obtained by simultaneously satisfying all relevant determinations present in an internally consistent TN, the ATcT values are not only more accurate,[1] but also more robust[2] than those obtained from a sequential approach. A typical sequentially obtained thermochemical value has a critical dependency on the reliability and integrity of the particular determination that was selected for its derivation. In contrast, ATcT values typically have a distributed provenance and thus do not critically depend on the accuracy and reliability of a single determination. In rare cases when the ATcT thermochemistry of some species displays a strong dependency on a single thermochemical determination, the latter is marked as a 'weak link,' and the corresponding section of the TN is earmarked for fortification by acquiring new experimental and/or theoretical determinations. The distributed nature of the provenance,[3] together with the fact that all provenance contributors are brought into mutual consistency before computing the final solution, are very important aspects of the ATcT approach that enable ATcT to produce results of superior accuracy and robustness.

The current ATcT TN (ver. 1.118) spans over a thousand chemical species and contains nearly 16,000 determinations. Describing even a modest fraction of these determinations is obviously well outside the scope of any single paper. However, the degree to which each of these determinations contributes to the enthalpy of formation of a given species (or to some other related thermochemical quantity, such as a particular bond dissociation energy) can be evaluated by performing a variance/covariance decomposition. This enables us to focus the discussion only on those determinations that make a prominent contribution to the provenance of the targeted thermochemical quantity.

3 ATcT results and their discussion

Since gas-phase H_2, N_2, O_2, and F_2 are the thermodynamic reference states for the corresponding elements (having by definition an enthalpy of formation of zero at all temperatures), the enthalpies of formation of the corresponding atoms are exactly equal to half the bond dissociation enthalpy of the homonuclear diatomic molecule. The same, of course, does not hold for C_2. For carbon, the thermodynamic reference state is graphite, and thus, the enthalpy of formation of the carbon atom corresponds to the enthalpy of vaporization of graphite to monatomic carbon, while the enthalpy of formation of C_2 corresponds to the enthalpy of vaporization of graphite to diatomic carbon. The two are related via the dissociation energy of diatomic carbon. While C_2 will be discussed here in pertinent detail, describing the genesis of the ATcT value for carbon atom would require a separate discussion that is outside the scope of the present paper. Nevertheless, for the sake of completeness, the current ATcT enthalpy of formation of C atom is also reported here.

Each of the cases discussed below in more detail is interesting in its own way: (1) H_2 is a relatively simple case devoid of complications that allows us to illustrate some of

[1] The term 'precision' is normally reserved to describe the spread of values (i.e., the width of the distribution), without taking into account the possible bias between the true value and the central value of the distribution, also known as 'trueness.' Occasionally, the term 'accuracy' is used as a synonym for 'trueness.' In the present paper, the term 'accuracy' encompasses the best estimate of both 'trueness' and 'precision,' in keeping with ISO 5725 standard.

[2] In the present paper, the term 'robustness' is used only in a qualitative sense and implicates a resistance to change if one of the relevant determinations is removed, or if a new determination of similar quality is added.

[3] 'Provenance' is in the present paper used as a quantitative descriptor; the contribution of an arbitrary determination to the provenance of an arbitrary thermochemical quantity is defined numerically as the relative contribution of that determination to the final value of the variance of the thermochemical quantity in question.

the important characteristics of ATcT results; (2) C_2 illustrates a case that suffers from a dearth of experimental data; (3) N_2 illustrates a case that involves the resolution of a 'weak link' before reliable thermochemistry can be obtained; (4) O_2 illustrates a case where one needs to include an often 'forgotten' correction to experimental data in order to achieve the final accuracy; (5) F_2 provides a case where the ATcT must use arbitration to resolve inconsistencies between relevant determinations.

3.1 ATcT values for $D_0(H_2)$ and $\Delta_f H°(H)$

The current ATcT value for $D_0(H_2)$ is $432.068_{06} \pm 0.000_{02}$ kJ/mol (see Table 1), or, equivalently, $\Delta_f H°_0(H) = 216.034_{03} \pm 0.000_{01}$ kJ/mol (217.997_{88} kJ/mol at 298.15 K; see Table 2).[4] Other than gradually gaining in accuracy as the TN was expanded over time, these ATcT values have not changed significantly since the previously reported versions [1, 3, 4, 9–11]. The CODATA value [8], subsequently adopted in the popular tables of Gurvich et al. [12, 13] and in the JANAF Tables[5] [14, 15], was already quite accurate, $\Delta_f H°_{298}(H) = 217.998 \pm 0.006$ kJ/mol, and thus in this particular case, the improvement in accuracy by almost three orders of magnitude is probably of no practical thermochemical consequence.

The CODATA value for the enthalpy of formation of H relies entirely on a single measurement: the H_2 dissociation energy of Herzberg [16] (see additional discussion of the CODATA value in Ruscic et al. [1]). In contrast to this, the provenance of the ATcT value is significantly distributed, reflecting a typical outcome of the ATcT TN approach. The first 90 % of the provenance of $\Delta_f H°(H)$ and/or $D_0(H_2)$ includes no less than 9 different determinations, and in order to pedigree 99.9 % of the provenance, one needs to include 30 different determinations. The most prominent experimental contributors to the provenance are the determination of the ionization energy of *ortho*-H_2 of Liu et al. [17], their reevaluation of the ionization energy of H and of the dissociation energy of H_2^+ using the CODATA 2006 constants [18], the evaluation of the ionization energy of H by Erickson [19] (additionally rescaled to current values of natural constants), the determination of $D_0(H_2)$ by Zhang et al. [20], the evaluation of the same quantity by Stoicheff [21], the determination of *para–ortho* separation in H_2 by Jennings et al. [22], and the determination of the ionization energy of H_2^+ by de Lange et al. [23]. With respect to theoretical contributors, the most prominent contributors are the determinations of $D_0(H_2)$ and of *para–ortho* separation in H_2 by Piszczatowski et al. [24], of the Lamb shift in H by Johnson and Soff [25], of $D_0(H_2^+)$ by Moss [26], of *para–ortho* separation in H_2 by Schwartz and Le Roy [27], and of $D_0(H_2)$ by Wolniewicz [28] and by Kolos and Rychlewski [29].

The list of provenance contributors attests to the fact that ATcT utilize the knowledge content of available determinations, irrespective of whether they are of experimental or theoretical origin. One also easily notes that in addition to being highly distributed, the provenance does not correspond to a straightforward collection of competing determinations of $D_0(H_2)$, but it includes such species as *ortho*-H_2, H_2^+, H^+. These two observations illustrate an important general characteristic of the ATcT approach: the final values have distributed provenances, reflecting the fact that they have been obtained by simultaneously satisfying *all statistically relevant determinations* through *all statistically relevant thermochemical cycles* available in the TN.

The determination of $D_0(H_2)$ by Herzberg [16], which provided the CODATA/JANAF/Gurvich value for the enthalpy of formation of H, is included in the TN, but ends up contributing rather negligibly to the final result. The primary reason for this is that it seriously lags in accuracy when compared to the current list of prominent provenance contributors. Not surprisingly, this is a rather typical situation for many determinations that might have been the best available (and hence preferred) at the time of the CODATA/JANAF/Gurvich evaluations, but have since become overshadowed by newer and more accurate data. Within the ATcT TN, the transition of less accurate determinations toward obsolescence is governed automatically and usually happens gradually: as more accurate determinations are added to the TN, the less accurate determinations—though still contributing to the overall knowledge content of the TN—are slowly eased toward lower provenance ranks.

3.2 ATcT values for $D_0(C_2)$ and $\Delta_f H°(C_2)$

CODATA [8] did not evaluate C_2, ostensibly because the experimental data available at the time were lacking sufficient accuracy and consistency. JANAF Tables [14, 15] grounded the thermochemistry of this species on the spectroscopically based extrapolation of Messerle and

[4] The significant digits in the values given in the text and in Tables 1 and 2 are un-subscripted and correspond either to a maximum of three significant digits after the decimal point or to two significant digits in the related uncertainty (whichever is less); when given, extra digits are subscripted.

[5] As opposed to Gurvich et al. [12, 13], who adopted the values from the final CODATA report [8], the third edition of JANAF [14] has adopted a slightly different set of values from an interim report of the CODATA Task Group. The fourth edition of JANAF [15] retains the interim values adopted in the third edition. For H atom, the value adopted by JANAF is larger by 0.001 kJ/mol, for O atom is lower by 0.010 kJ/mol, for F atom is larger by 0.010 kJ/mol.

Table 1 ATcT values for the 0-K dissociation energies, D_0, and 298.15 bond dissociation enthalpies, BDE_{298}, of H_2, C_2, N_2, O_2, F_2, HF, CH, CO, CN, NO, OH, CO_2, H_2O, and triplet and singlet CH_2 (in kJ/mol)

Species	D_0	BDE_{298}	Uncertainty
H_2	432.068_{06}	435.995_{75}	$\pm 0.000_{02}$
C_2	602.52_7	605.03_1	$\pm 0.27_8$
N_2	941.146	944.870	±0.047
O_2	493.687_8	498.458_3	$\pm 0.004_2$
F_2	154.57_5	158.78_7	$\pm 0.10_8$
HF	565.966	570.082	±0.008
CH	334.60_2	338.71_7	$\pm 0.11_4$
CO	1072.041	1,076.631	±0.055
CN	745.25_3	749.31_4	$\pm 0.14_1$
NO	626.830	630.574	±0.060
OH	425.625	429.735	±0.026
CO_2[b]	526.150	532.182	±0.025
H_2O[c]	492.215	497.321	±0.002
CH_2[d] (triplet)	417.90_0	422.64_1	$\pm 0.14_0$
CH_2[e] (singlet)	380.23_3	385.11_8	$\pm 0.14_8$

The listed values are from ver. 1.118a of the ATcT thermochemical network

[a] The conversion of $\Delta_f H°(C_2)$ from 0 K to 298.15 K uses a newly computed ATcT partition function for C_2; with the prior partition function for C_2 (from Gurvich et al. [12, 13]), BDE_{298} value would be 605.43_0 kJ/mol

[b] The current total atomization energy of CO_2 is $TAE_0(CO_2) = 1{,}598.191 \pm 0.054$ kJ/mol (1,608.812 kJ/mol at 298.15 K)

[c] The current total atomization energy of H_2O is $TAE_0(H_2O) = 917.840 \pm 0.026$ kJ/mol (927.056 kJ/mol at 298.15 K)

[d] The current total atomization energy of triplet CH_2 is $TAE_0(^3CH_2) = 752.50_2 \pm 0.12_7$ kJ/mol (761.35_8 kJ/mol at 298.15 K)

[e] The current total atomization energy of singlet CH_2 is $TAE_0(^1CH_2) = 714.83_5 \pm 0.13_5$ kJ/mol (723.83_6 kJ/mol at 298.15 K)

Kraus [30], taking $D_0(C_2) = 589.7 \pm 3.8$ kJ/mol.[6] In contrast to this, Huber and Herzberg [31] termed the determination of Messerle and Krauss [30] as 'somewhat doubtful' and listed $D_0(C_2) = 599$ kJ/mol (with an implied uncertainty of about ±9 kJ/mol) based on high-temperature measurements of Brewer et al. [32] and of Kordis and Gingerich [33]. Gurvich et al. [12, 13] examined nearly all high-temperature measurements that were available at the time [33–40], and ultimately anchored the thermochemistry of C_2 on the same spectroscopic measurement as JANAF, but reinterpreted it as implying $D_0(C_2) = 600 \pm 10$ kJ/mol.

The current ATcT value for $D_0(C_2)$ is $602.52_7 \pm 0.27_8$ kJ/mol (Table 1). The related 0-K ATcT enthalpy of formation of C_2 is $820.26_3 \pm 0.29_0$ kJ/mol (828.72_9 kJ/mol at 298.15 K, Table 2). The Gurvich et al. [12, 13] value $\Delta_f H°_0(C_2) = 822.4 \pm 10$ kJ/mol (830.5 kJ/mol at 298.15 K) and their $D_0(C_2) = 600 \pm 10$ kJ/mol, as well as the nearly identical $D_0(C_2)$ value of Huber and Herzberg [31], are clearly consistent with the ATcT value, though less accurate by one-and-a-half orders of magnitude. The often quoted and used JANAF [14, 15] value $\Delta_f H°_0(C_2) = 829.3 \pm 3.8$ kJ/mol (837.7 kJ/mol at 298.15 K) is evidently inconsistent with the current ATcT value. Table 2 also lists the current ATcT enthalpy of formation of carbon atom, $\Delta_f H°_0(C) = 711.395 \pm 0.054$ kJ/mol (716.880 kJ/mol at 298.15 K). The latter value has changed very slightly since the previously reported [10] interim ATcT value.

The provenance of the ATcT value for $D_0(C_2)$ is quite distributed: 90 % of the provenance contribution is due to 23 determinations, and in order to account for 99.9 % of the provenance, it takes no less than 715 determinations. In spite of the fact that the presence of C_2 is clearly visible in nearly every hydrocarbon flame (quoting Hoffmann [41]: 'the lovely blue color of hot hydrocarbon flames is due in large part to emission from excited C_2 molecules on their way to soot or CO_2'), and in spite of numerous detailed spectroscopic studies of this molecule [42–51], there is very little in terms of accurate experimental determinations that could help define $D_0(C_2)$. Among the top fifty provenance contributors to the ATcT value, only four are experimental

[6] Note that there is an inconsistency in the JANAF Tables [14, 15]: the listed enthalpy of formation of C_2 was derived in the third edition [14] by combining the assumed bond dissociation energy of 589.7 ± 3.8 kJ/mol with an older value for the enthalpy of formation of C atom, which is lower by 1.7 kJ/mol than the listed value. Thus, from the listed enthalpies of formation for C_2 and C, one nominally obtains $D_0(C_2) = 593.1 \pm 3.9$, or 3.4 kJ/mol higher than their original assumption. The inconsistency stays uncorrected in the fourth edition [15].

Table 2 ATcT enthalpies of formation, $\Delta_f H°$, of H, C, N, O, F, C_2, HF, CH, CO, CN, NO, OH, CO_2, H_2O, and triplet and singlet CH_2 at 0 K and 298.15 K (in kJ/mol)

Species	0 K	298.15 K	Uncertainty
H	216.034_{03}	217.997_{88}	$\pm 0.000_{01}$
C	711.395	716.880	±0.054
N	470.573	472.435	±0.023
O	246.843_9	249.229_2	$\pm 0.002_1$
F	77.287	79.393	±0.053
C_2	820.26_3	828.72_9^a	$\pm 0.29_0$
HF	−272.644	−272.691	±0.053
CH	592.82_7	596.16_1	$\pm 0.12_3$
CO	−113.802	−110.522	±0.026
CN	436.71_5	440.00_1	$\pm 0.14_5$
NO	90.586	91.090	±0.062
OH	37.253	37.492	±0.026
CO_2	−393.108	−393.474	±0.014
H_2O	−238.928	−241.831	±0.026
CH_2 (triplet)	390.96_0	391.51_8	$\pm 0.13_4$
CH_2 (singlet)	428.628	429.040	$\pm 0.14_2$

The listed values are from ver. 1.118a of the ATcT thermochemical network

[a] The conversion of $\Delta_f H°(C_2)$ from 0 K to 298.15 K uses a newly computed ATcT partition function for C_2; with the prior partition function for C_2 (from Gurvich et al. [12, 13]), the 298.15 K value would be 828.330 kJ/mol

determinations: the Gibbs energy of vaporization of graphite to produce C_2 as extracted by a third law treatment of the measurements of Kordis and Gingerich [33], the energy of dissociation of acetylene to C_2 and two H atoms extracted from the measurements of Urdahl et al. [52], the high-temperature equilibrium constant between C_2 and C atoms determined by Wachi and Gilmatrin [38], and a reinterpretation of the bond dissociation measurements by Messerle and Kraus [30] that is very similar to that given by Gurvich et al. [12, 13]. Each of these four experimental determinations individually contributes less than 1 % to the provenance. The other experimental determinations mentioned at the beginning of the C_2 discussion are also included in the TN, but they occupy even lower provenance ranks. The primary reasons for a low contribution of experimental determinations to the provenance are their insufficient accuracy and poor mutual consistency. Rather than experiment, the relevant provenance contributors in the current version of ATcT derive from theory: the bond dissociation energy and the energy for loss of two hydrogen atoms from acetylene based on W4.4a calculations of Karton et al. [53–55] and on the FPD method results of Feller et al. [56], together with the analogous quantities based on slightly lower levels of theory, such as W4.3, W4.2, and W4 [53, 55], and on an earlier version of the FPD method [57], as well as the C–H bond dissociation energy of C_2H from an approach that utilized explicitly correlated coupled-cluster methods [6], as well as the same bond energy from a CCSD(T)/CBS-based approach [58].

Although the current ATcT value for $D_0(C_2)$ is dominated by virtual (i.e., computational) determinations, rather than actual (i.e., experimental) determinations, this was not necessarily the case in earlier versions of the ATcT TN. The section of TN relevant to C_2 initially contained all the available experimental data mentioned above, but very few high-accuracy computational results. The resulting values for $D_0(C_2)$ and $\Delta_f H°(C_2)$ were rather inaccurate, partly because of the relative paucity of data and partly because of their limited accuracy. In addition, some of the high-temperature determinations had a tendency to skew the resulting bond dissociation energy to higher values, because at that point, the TN contained a rather limited set of thermochemical cycles involving C_2 and thus ATcT had no good thermodynamic pathways to check the involved determinations for consistency. As high-accuracy computational results have gradually appeared and were inserted in the TN, they incrementally boosted the TN knowledge content, allowing ATcT to perform a more rigorous evaluation of the relevant TN section for internal consistency. The resulting $D_0(C_2)$ and $\Delta_f H°(C_2)$ were gradually gaining in accuracy and converging toward the current value, but at the same time, the newer and increasingly accurate theoretical results began pushing the less accurate experiments down the rank list of provenance contributors.

The thermochemistry of C_2 is clearly a case where the ATcT results benefit immensely from the availability of high-accuracy state-of-the-art electronic structure methods. However, it is worth stressing here that although the current ATcT thermochemistry of C_2 is dominated by theoretical determinations, the results presented here are superior to any of the included computational determinations taken alone. By virtue of the underlying TN approach, the ATcT results reflect a statistically weighted summary consensus of the determinations contributing to their pedigree—after they were brought into mutual consistency by statistically evaluating all available thermochemical cycles—and are thus superior to any of their constituent determinations taken alone.

3.3 ATcT values for $D_0(N_2)$ and $\Delta_f H°(N)$

CODATA [8] used a bond dissociation energy of N_2 of 941.64 ± 0.60 kJ/mol from Büttenbender and Herzberg [59] to derive the 298.15 K enthalpy of formation of nitrogen atom of 472.68 ± 0.40 kJ/mol, which was then adopted by Gurvich et al. [12, 13] and JANAF[7] [14, 15]. In the very early stages of the development of the ATcT TN, when the network was much smaller, the resulting ATcT enthalpy of formation of N atom was strongly dominated by the same spectroscopic determination by Büttenbender and Herzberg [59]. Consequently, this determination was marked as a 'weak link' in the TN, and steps were undertaken to enhance the TN section surrounding N. More exhaustive literature searches have subsequently unearthed the determinations of Carroll and Mulliken [60] and Roncin et al. [61], and the analyses of Lofthus [62] and Le Roy et al. [63]. However, while their addition to the TN introduced competing values for $D_0(N_2)$, this enhancement did not create new thermochemical cycles that would allow ATcT to perform more exhaustive checks for consistency in the relevant portion of the TN. A crucial breakthrough toward alleviating the 'weak link' symptom occurred in the form of a tailored collaborative effort with the group of Ng [64], where accurate synchrotron-based photoionization measurements have been used to determine in three separate ways the onset of photodissociative ionization of N_2 (corresponding to the process $N_2 \rightarrow N^+ + N + e^-$).

The current ATcT value is $D_0(N_2) = 941.146 \pm 0.047$ kJ/mol. In contrast to the very early versions of the TN that suffered from the 'weak link' situation, the provenance of $D_0(N_2)$—and hence also the provenance of $\Delta_f H°(N)$—is currently quite distributed, with a distribution size not dissimilar to that found for $D_0(C_2)$: it takes 17 determinations to elucidate 90 % of provenance, and as many as 1,036 determinations to account for 99.9 % of the provenance. However, as opposed to the situation for C_2, the primary contributors to the current value of $D_0(N_2)$ are entirely experimental. The most relevant contributors are the three collaborative photoionization measurements mentioned above [64], the measurements of Roncin et al. [61], the measurement of Carroll and Mulliken [60], as well as the original determination by Büttenbender and Herzberg [59].

The corresponding ATcT value for $\Delta_f H°_0(N) = 470.573 \pm 0.023$ kJ/mol (472.435 kJ/mol at 298.15 K, Table 2) coincides with the previously reported [10] interim value. The CODATA [8] value is more than an order of magnitude less accurate and is 0.24 kJ/mol higher; perhaps a not entirely unrelated detail is that the bond dissociation energy explicitly quoted by CODATA should have produced an uncertainty for the enthalpy of formation of nitrogen atom of $\pm 0.3_0$ kJ/mol, but in the final analysis, the CODATA Task Group has decided for some reason to increase it to $\pm 0.4_0$ kJ/mol. One suspects that the undocumented addition of a safety margin to the uncertainty by CODATA is related to the fact that $D_0(N_2)$ was involved in a very lively controversy during the first half of the twentieth century (see Gaydon's book [65] for a historical overview of this subject).

While the uncertainty of the CODATA value is large enough to accommodate the current ATcT value in the lower section of the bound, the actual CODATA value is *outside* the ATcT error bar by a substantial margin.[8] Both the shift in the value and the increased accuracy of the ATcT value are of relevance to high-accuracy theoretical approaches that utilize the total atomization energy route to obtain practical enthalpies of formation, particularly for chemical species that have multiple nitrogen atoms.

3.4 ATcT values for $D_0(O_2)$ and $\Delta_f H°(O)$

The current ATcT value for $D_0(O_2) = 493.687_8 \pm 0.004_2$ kJ/mol (Table 1), and the corresponding $\Delta_f H°_0(O) = 246.843_9 \pm 0.002_1$ kJ/mol (249.229_2 kJ/mol at 298.15 K, Table 2); these have not changed from the previous ATcT version [10].

The CODATA [8] value for the enthalpy of formation of oxygen atom is based on $D_0(O_2) = 493.58 \pm 0.18$ kJ/mol obtained by Brix and Herzberg [66], and the resulting 298.15 K enthalpy of formation for oxygen atom of 249.18 ± 0.10 kJ/mol was adopted by Gurvich et al. [12, 13] and JANAF[5] [14, 15]. The provenance of the ATcT value is reasonably distributed: 90 % of the provenance can be attributed to 4 determinations, and 99.9 % of the provenance to 106 determinations. The prominent contributors are the bond dissociation energy determination of Lewis et al. [67], the ion-pair threshold determination of Martin and Hepburn [68], and the bond dissociation energies determined by Gibson et al. [69] and Cosby and Huestis [70].

Some of the (otherwise very accurate) spectroscopic values for $D_0(O_2)$ that are found in the literature are—in a strict thermodynamic sense—incorrect and require an additional small correction before they can be used for high-accuracy thermochemical purposes. Namely, the thermodynamic definition of dissociation energy is that it corresponds to the energy difference between the lowest *existing* rovibrational level of a molecule and the lowest *existing* level of the dissociation asymptote. Both Herzberg [71] and Huber and Herzberg [31] quoted this definition,

[7] JANAF [14] explicitly declares that it adopts the CODATA value for the enthalpy of formation of N, but lists an uncertainty that is smaller than CODATA's by a factor of 4, as if the authors failed to convert it from kcal/mol to kJ/mol. The problem is uncorrected in the fourth edition [15].

[8] While a detailed comparison of quantities with vastly different accuracies is nontrivial, particularly if the more accurate value is within the uncertainty bounds of the less accurate value, but the converse is not true, if one anchors the reference frame to the ATcT value for $\Delta_f H°(N)$, the corresponding CODATA [8] value is too high by more than 20 standard deviations.

and yet the dissociation energy of Brix and Herzberg [66] was given with respect to the nonexistent $v = 0$, $N = J = 0$ level of the $X\,^3\Sigma_g^-$ state. Prima facie, the lowest existing level in the ground state of O_2 would be 1.783 cm^{-1} lower and would correspond to $N = 0$, $J = 1$ (F_1), but that level is wiped out by nuclear spin statistics in the dominant isotopic variant $^{16}O_2$. The lowest actual level in $X\,^3\Sigma_g^-$ of $^{16}O_2$ is $N = 1$, $J = 0$ (F_3), which is 1.086 cm^{-1} lower than the nonexistent reference level. CODATA [8] apparently entirely ignored this (admittedly small) correction to the Brix and Herzberg $D_0(O_2)$ value [66], presumably considering it superfluous in view of the ±15 cm^{-1} overall uncertainty of that determination. However, Gibson et al. [69] and Lewis and al. [67] follow suit and also refer their D_0 to the same nonexistent level; in their case, the correction is relevant since they both quote sub-cm^{-1} uncertainties. In a similar vein, Martin and Hepburn [68] explicitly report their ion-pair formation threshold relative to the $N = 1$, $J = 2$ level of the ground state of O_2, which is 2.1 cm^{-1} above the lowest existing level of $X\,^3\Sigma_g^-$ of O_2—again a relevant correction in view of their uncertainty of ±0.7 cm^{-1}.

The ATcT value for $\Delta_f H°(O)$ is significantly more accurate than the value selected by CODATA [8] (by more than two orders of magnitude) and slightly higher (by 0.05 kJ/mol). As in the case of N atom, the current value for the O atom is of relevance to high-accuracy theoretical approaches that utilize the total atomization energy route to obtain enthalpies of formation.

3.5 ATcT values for $D_0(F_2)$ and $\Delta_f H°(F)$

The current ATcT value for $D_0(F_2) = 154.57_5 \pm 0.10_8$ kJ/mol (Table 1), or, equivalently, $\Delta_f H°_0(F) = 77.287 \pm 0.053$ kJ/mol (79.393 kJ/mol at 298.15 K, Table 2). The ATcT value for the enthalpy of formation of F atom is nearly six times more accurate, but otherwise numerically very similar to the CODATA [8] value of $\Delta_f H°_{298}(F) = 79.38 \pm 0.30$ kJ/mol.

The CODATA value for $\Delta_f H°(F)$ was obtained by adopting $D_0(F_2) = 154.56 \pm 0.60$ kJ/mol from Colbourn et al. [72]. The latter determination requires additional discussion. Though often referred to as a spectroscopic measurement, it is, in fact, an educated estimate based on spectroscopic observations on F_2, rather than a direct measurement of $D_0(F_2)$. Colbourn et al. have spectroscopically determined the B_v and G_v values from $v = 0$ to $v = 22$ of the $X\,^1\Sigma_g^+$ state of F_2. They concluded that the dissociation limit must be higher than the highest observed vibrational level. The inference that the $v = 22$ vibrational level must be bound was rationalized by noting that the ground state curve is unlikely to have a barrier on the basis that both the $C_6 r^{-6}$ and the $C_8 r^{-8}$ terms in the long-range potential correspond to attraction between the atoms and thus have the same sign. Colbourn et al. further remarked that an extrapolation of the vibrational levels beyond $v = 22$ to the dissociation limit would be difficult because the intervals between the high vibrational levels vary rapidly and follow no obvious pattern (a situation not dissimilar to the case of $D_0(OH)$ [73, 74], now apparently considered to be a classic [75]). Colbourn et al. [72] estimate that the dissociation limit of F_2 is 90 cm^{-1} higher than the $v = 22$ vibrational level, with an uncertainty of ±50 cm^{-1}, or slightly more than half of the incremental interval. The exact procedure by which Colbourn et al. arrived at this estimate is not given in their paper, but one suspects that the expertise of this world-renowned spectroscopic group must have played an important role during the process.

Several years ago, Bytautas and Ruedenberg [76] carefully analyzed the long-range region of the ground state potential energy function of F_2 (obtained by their CEEIS method). They found that while the London dispersion interaction, which scales as r^{-6}, is attractive, the interaction between quadrupoles of the fluorine atoms, which scales as r^{-5}, is repulsive because of coaxial alignment. There is also an additional repulsive force due to loss of spin–orbit coupling upon the bond formation. Consequently, their potential energy curve has a small ($\sim$9 cm^{-1}) dissociation barrier at about 4 Å. The quandary that has arisen immediately after the work of Bytautas and Ruedenberg was published, concerns the issue of whether the 9 cm^{-1} barrier affects the D_0 estimate of Colbourn et al. [72] or not. Clearly, the barrier is much too small to vitiate the main stepping stone in the estimate of Colbourn et al. [72], which is the assumption that the highest observed vibrational level, 90 cm^{-1} below the presumed dissociation limit, is fully bound. Would the knowledge of the barrier have resulted in an estimate of D_0 that was lower by 9 cm^{-1}, or would it have perhaps resulted in an estimated error bar larger than ±50 cm^{-1}? It is unclear that it would cause either of the two, particularly since the estimate of Colbourn et al. explicitly did not rely on a straightforward extrapolation of the observed vibrational levels, and likely relied heavily on spectroscopic experience.

An additional and more serious problem surrounding the correct value of $D_0(F_2)$ has arisen even before the discovery of the dissociation barrier by Bytautas and Ruedenberg [76]. Yang et al. [77] have reported the ion-pair formation threshold of F_2 (corresponding to the process $F_2 \rightarrow F^+ + F^-$), which was obtained by velocity mapping the F^- fragment. When combined with the generally accepted values for the ionization energy and electron affinity of F atom, it produced $D_0(F_2) = 154.96 \pm 0.10$ kJ/mol; this value was subsequently slightly lowered in an erratum [78] to 154.92 kJ/mol. The value of Yang et al. appears six times more accurate than the estimate of Colbourn et al. but is 0.36 kJ/mol higher.

From the perspective of traditional sequential thermochemistry, establishing whether the Colbourn et al. estimate (either uncorrected or lowered by 0.11 kJ/mol on account of the dissociation barrier) or the higher determination of Yang et al. is correct is a crucial issue: once one of the conflicting determinations is selected as the preferred value, it immediately implies a particular value for $\Delta_f H°(F)$.

However, from the perspective of the ATcT approach, establishing beforehand which of the possibilities is correct is not a crucial issue at all. Arbitrating between mutually inconsistent determinations is one of the things that the ATcT approach is explicitly designed to do: as long as there are sufficient alternate thermochemical cycles in the TN, errant determination(s) will be identified during the ATcT analysis and brought into consistency with the rest of the TN by augmenting their uncertainties. As we shall show in a moment, the TN contains a truly large number of determinations that can help establish the dissociation energy of F_2 both directly and through alternate thermochemical cycles, and thus ATcT should have no problem with such arbitration.

In general, the ATcT analysis typically finds that the optimal solutions are outside the initially assigned uncertainties for roughly 10 % of the determinations (or about double the expected number if all initial uncertainties truly corresponded to probabilistically correct 95 % confidence intervals). Since the introduction of the ion-pair threshold of Yang et al. [77, 78] into the TN a number of versions ago (using their original uncertainty of ± 8 cm^{-1} or $\pm 0.09_6$ kJ/mol), the ATcT analysis consistently finds this determination to be problematic, unless the original uncertainty is increased by approximately a factor of 3.6, to $\pm$ 0.34_4 kJ/mol. Other than it being inconsistent with the remaining knowledge content of the TN (where the latter consists of the cumulative knowledge of well over a thousand related determinations, vide infra), it is unclear at this point what exactly may be the problem with the threshold of Yang et al. or with its original uncertainty.

As hinted to earlier, the provenance of the ATcT value for $D_0(F_2)$ is extremely distributed—even more so than the other species examined in this report: the top 90 % provenance is spread out over 78 determinations, while accounting for 99.9 % of the provenance involves 1,197 determinations! The prominent contributors to the provenance are a mix of experiment and theory. They include the experimental ion-pair formation threshold by Yang et al. [77, 78] (with an uncertainty augmented during the ATcT analysis), the bond dissociation determination of Colbourn et al. [72], together with the theoretical bond dissociation energies of Bytautas et al. [79], Csontos et al. [80], Harding et al. [4], Karton et al. [5], Feller et al. [56], as well as a large number of other experimental and theoretical determinations. These include additional species such as ClF, CF_4, Cl, FOF, HF, HCl, CF, CF_2, CF_3, COF_2, C_2F_4, CO_2, CH_4, CH, H_2O, H_2, and graphite, through alternative thermochemical cycles that ATcT find to be of some relevance to establishing the final result. At first blush, some of these additional species involved in the provenance may appear surprising, though they can be all rationalized. For example—to provide just a few random examples—the enthalpy of reaction of F_2 with Cl atoms to produce ClF and F atoms corresponds to the difference in the bond dissociation energies of ClF and F_2. When combined with the equilibrium constant for the reaction of F_2 and Cl_2 to produce ClF, it corresponds to the difference in the bond dissociation energies of Cl_2 and F_2. Or, the calorimetric combustion of graphite in F_2 to produce CF_4, when combined with the vaporization enthalpy of graphite and either with the successive bond dissociation energies of CF_4 or with the total atomization energy of CF_4, produces a net cycle corresponding to two F_2 molecules dissociating to constituent atoms. Taken alone—such as would be the case in sequential thermochemistry—none of these alternate thermochemical cycles would be able to sufficiently accurately establish the bond dissociation energy of F_2. However, within the ATcT TN approach, which attempts to satisfy all available thermochemical cycles simultaneously, the additional thermochemical cycles synergistically enhance the overall knowledge about a satisfactory value for $D_0(F_2)$ in a way that becomes relevant to the final result.

As mentioned at the beginning of the subsection on F_2, while the ATcT value and the CODATA value for the enthalpy of formation of F atom differ in their accuracy, their nominal values are numerically nearly identical—a fortuitous coincidence in view of the fact that the ATcT value is highly independent from the $D_0(F_2)$ value of Colbourn et al. [72]; the latter determination contributes only about 3 % to the provenance of the ATcT value. From the ATcT value for $D_0(F_2)$, one can infer that the dissociation limit is located 91 ± 9 cm^{-1} beyond the last vibrational level observed by Colbourn et al. Their estimate was 90 ± 50 cm^{-1}. The two values are fortuitously in near-perfect agreement, much better than statistically justified by the two uncertainties.

Csontos et al. [80] have quite recently confirmed the existence of the dissociation barrier of Bytautas and Ruedenberg [76] by using several multi-reference methods, and have computed $D_0(F_2) = 154.95 \pm 0.48$ kJ/mol using an enhanced HEAT-like approach. Csontos et al. [80] noted that the value of Yang et al. [78] of $D_0(F_2) = 154.92 \pm 0.10$ kJ/mol and the value of $D_0(F_2) = 154.52 \pm 0.12$ kJ/mol based on a previously reported [10] interim ATcT value for $\Delta_f H°_0(F) = 77.26 \pm 0.06$ kJ/mol are outside each other's error bars, but the uncertainty of their computed bond dissociation energy, although remarkably low, was not low enough for them to be able to make a distinction between the two values. However, they remarked that their computed value 'almost precisely agrees' with the value of Yang et al. [78], and indicated support for the latter. Csontos et al. [80] emphasized the fact that neither the experimental ion-pair threshold nor the computed value would suffer from the existence of the dissociation barrier in the ground state of F_2, while direct experimental measurements of the dissociation energy which would. Finally, they conclude that "the ATcT approach uses $D_0(F_2)$ reported by Colbourn and co-workers, which is derived from extrapolation of the vibrational levels for the ground state F_2 molecule to the dissociation limit" and that because of the neglect of the dissociation barrier "the uncertainties [of the ATcT values for $D_0(F_2)$ and $\Delta_f H°(F)$] … should be increased." The quoted conclusions of Csontos et al. [80] are incorrect; these authors assumed that the ATcT result is derived directly from the value of Colbourn et al. [72], while in fact, as we have shown above, the provenance of the ATcT values for $D_0(F_2)$ and $\Delta_f H°(F)$ is highly distributed and essentially (in fact, 97 %) independent of the value of Colbourn et al. As discussed above, the numerical near-coincidence between the estimate of Colbourn et al. and the roughly five times more accurate ATcT value is fortuitous. Furthermore, the possible influence of the dissociation barrier on the ATcT result was checked immediately after the result of Bytautas and Ruedenberg [79] became public. With the possible exception of the estimate of Colbourn et al. [72], none of the other ~1,200 determinations contributing to the provenance of the thermochemistry of F is affected by the dissociation barrier in the ground state of F_2. Furthermore, irrespective of whether the estimate of Colbourn et al. requires an adjustment or not, its contribution to the final ATcT result is quite small, and a small change in that value would negligibly affect the ATcT value and/or its uncertainty.

3.6 Current ATcT values for $D_0(HF)$, $D_0(CH)$, $D_0(CO)$, $D_0(CN)$, $D_0(NO)$, $D_0(OH)$, $D_0(CO_2)$, $D_0(H_2O)$, and $D_0(CH_2)$

The ATcT bond dissociation energies of HF, CH, CO, CN, NO, OH, CO_2, H_2O, and triplet and singlet CH_2 are given in Table 1, and the corresponding enthalpies of formation are listed in Table 2. The essential ATcT thermochemistry for H_2O has been published recently [11]. Strictly speaking, the ATcT results for the other species represent interim values, since the corresponding sections of the TN have not yet been finalized and analyzed in earnest. However, it would be quite surprising if these interim values underwent a substantial change during the final analysis.

3.7 Benchmarking the benchmarks

The D_0 values given in Table 1 are used in the companion paper [7] for benchmarking highly accurate computational results obtained by the FPD procedure. These theoretical results are benchmarks in their own right: the underlying computations have been carried out at the highest level currently feasible by the available hardware, and the individual sources of error were carefully monitored and minimized during each of the computational steps. It goes without saying that in order to keep the benchmarking procedure as objective and independent as possible, the FPD computational results from the companion paper have not been included in the current TN. The agreement between the ATcT results and the computed D_0 values is very good indeed: in all cases, the two sets display an abundant overlap within their uncertainties, except for C_2, where the overlap is rather marginal. The latter, however, does not come as a great surprise: the ATcT result for C_2 suffers from a paucity of good experimental data as well as from the fact that the corresponding TN section is somewhat underdeveloped in terms of available thermochemical cycles, and, at the same time, the computational efforts are rendered complex by the multi-configurational character of C_2.

From the viewpoint of ATcT, benchmarking state-of-the-art theory is a two-way street. Once the fidelity level of the benchmarked theoretical results is understood, agreement between the computed results and the ATcT values indirectly also validates the latter. Conversely, if disagreement is found for a particular chemical species, it signals a problem either with the computation or with the TN section related to the species in question, or perhaps both. Furthermore, upon conclusion of the benchmarking procedure, the ATcT TN is ready to incorporate the just benchmarked theoretical results, as well as additional computations that use the newly benchmarked theoretical approach and target the introduction of new chemical species into the TN or fortification of underdeveloped sections of the TN.

4 Conclusions

The current ATcT results for the bond dissociation energies of the homonuclear diatomics H_2, C_2, N_2, O_2, and F_2 are reported and compared to values found in traditional thermochemical tabulations. The provenance of the ATcT values, obtained by examining the results of the variance decomposition approach, is discussed. The results that are presented here illustrate, inter alia, that the underlying TN approach produces values that generally have a significantly distributed provenance and thus, as opposed to values obtained by traditional sequential thermochemistry, do not critically depend on the reliability of a single measurement. In addition, the bond dissociation energy of N_2 illustrates a case that involves the resolution of a 'weak link' in the TN by acquiring additional experimental data. The bond dissociation energy of O_2 illustrates a case where the reported measurements are very accurate, but need additional corrections in order to conform to the thermochemical definition of a bond dissociation energy. The thermochemistry of C_2 illustrates a case where ATcT rely on state-of-the-art electronic structure methods because of a paucity of sufficiently accurate and mutually consistent experimental measurements, while the bond dissociation energy of F_2 illustrates a case where ATcT use the cumulative knowledge content of the TN to identify and resolve an inconsistent piece of information. Ramifications in terms of enthalpies of formation of atoms, which are fundamental thermochemical quantities, are also discussed. The current ATcT bond dissociation energies and enthalpies of formation of HF, CH, CO, CN, NO, OH, CO_2, H_2O, and triplet and singlet CH_2 are also given.

Acknowledgments This work was supported by the US Department of Energy, Office of Science, Office of Basic Energy Sciences, Division of Chemical Sciences, Geosciences and Biosciences under Contract No. DE-AC02-06CH11357.

References

1. Ruscic B, Pinzon RE, Morton ML, von Laszewski G, Bittner SJ, Nijsure SG, Amin KA, Minkoff M, Wagner AF (2004) Introduction to Active Thermochemical Tables: several "key" enthalpies of formation revisited. J Phys Chem A 108:9979–9997
2. Ruscic B, Pinzon RE, von Laszewski G, Kodeboyina D, Burcat A, Leahy D, Montoya D, Wagner AF (2005) Active Thermochemical Tables: thermochemistry for the 21st century. J Phys: Conf Ser 16:561–570
3. Klopper W, Ruscic B, Tew DP, Bischoff FA, Wolfsegger S (2009) Atomization energies from coupled-cluster calculations augmented with explicitly-correlated perturbation theory. Chem Phys 356:14–24
4. Harding ME, Vázquez J, Ruscic B, Wilson AK, Gauss J, Stanton JF (2008) HEAT: high accuracy extrapolated ab initio thermochemistry. III. Additional improvements and overview. J Chem Phys 128:114111
5. Karton A, Rabinovich E, Martin JML, Ruscic B (2006) W4 theory for computational thermochemistry: in pursuit of confident sub-kJ/mol predictions. J Chem Phys 125:144108
6. Aguilera-Iparraguirre J, Boese AD, Klopper W, Ruscic B (2008) Accurate ab initio computation of thermochemical data for C_3H_x ($x = 0, \ldots, 4$) species. Chem Phys 346:56–68
7. Feller D, Peterson KA, Ruscic B (2013) Improved accuracy benchmarks of small molecules using correlation consistent basis sets. Theor Chem Acc. doi:10.1007/s00214-013-1407-z
8. Cox JD, Wagman DD, Medvedev VA (1989) CODATA key values for thermodynamics. Hemisphere, New York
9. Ruscic B, Pinzon RE, Morton ML, Srinivasan NK, Su M-C, Sutherland JW, Michael JV (2006) Active Thermochemical Tables: accurate enthalpy of formation of hydroperoxyl radical, HO_2. J Phys Chem A 110:6592–6601
10. Stevens WR, Ruscic B, Baer T (2010) The heats of formation of C_6H_5, $C_6H_5^+$, and C_6H_5NO by TPEPICO and Active Thermochemical Tables analysis. J Phys Chem A 114:13134–13145
11. Ruscic B (2013) Active Thermochemical Tables: water and water dimer. J Phys Chem A. doi:10.1021/jp403197t
12. Gurvich LV, Veyts IV, Alcock CB (1989) Thermodynamic properties of individual substances, 4th edn, vol 1, parts 1 and 2. Hemisphere, New York
13. Gurvich LV, Veyts IV, Alcock CB (1991) Thermodynamic Properties of Individual Substances, 4th edn, vol 2, parts 1 and 2. Hemisphere, New York
14. Chase MW, Davies CA, Downey Jr JR, Frurip DJ, McDonald RA, Syverud AN (1985) JANAF thermochemical tables, vol 14, 3rd edn. J Phys Chem Ref Data (Suppl 1)
15. Chase Jr MW (ed) (1998) NIST-JANAF Thermochemical Tables, 4th edn. J Phys Chem Ref Data, Monograph 9
16. Herzberg G (1970) The dissociation energy of the hydrogen molecule. J Mol Spectrosc 33:147–168
17. Liu J, Salumbides EJ, Hollenstein U, Koelemeij JCJ, Eikema KSE, Ubachs W, Merkt F (2009) Determination of the ionization and dissociation energies of the hydrogen molecule. J Chem Phys 130:174306
18. Mohr PJ, Taylor BN, Newell DB (2008) CODATA recommended values of the fundamental physical constants: 2006. J Phys Chem Ref Data 37:1187–1284
19. Erickson GW (1977) Energy levels of one-electron atoms. J Phys Chem Ref Data 6:831–869
20. Zhang YP, Cheng CH, Kim JT, Stanojevic J, Eyler EE (2004) Dissociation energies of molecular hydrogen and the hydrogen molecular ion. Phys Rev Lett 92:203003
21. Stoicheff BP (2001) On the dissociation energy of molecular hydrogen. Can J Phys 79:165–172
22. Jennings DE, Bragg SL, Brault JW (1984) The $v = 0$–0 spectrum of H_2. Astrophys J 282:L85–L88
23. de Lange A, Reinhold E, Ubachs W (2002) Spectroscopy on some g symmetry states in H_2 and determination of the ionization potential. Phys Rev A 65:064501
24. Piszczatowski K, Lach G, Przybytek M, Komasa J, Pachucki K, Jeziorski B (2009) Theoretical determination of the dissociation energy of molecular hydrogen. J Chem Theory Comput 5:3039–3048
25. Johnson WR, Soff G (1985) The Lamb shift in hydrogen-like atoms, $1 <= Z <= 110$. Atomic Data Nucl Data Tables 33:405–446
26. Moss RE (1993) Calculations for the vibration-rotation levels of H_2^+ in its ground and first excited electronic states. Mol Phys 80:1541–1554
27. Schwartz C, Le Roy RJ (1987) Nonadiabatic eigenvalues and adiabatic matrix elements for all isotopes of diatomic hydrogen. J Mol Spectrosc 121:420–439
28. Wolniewicz L (1995) Nonadiabatic energies of the ground state of the hydrogen molecule. J Chem Phys 103:1792–1799
29. Kolos W, Rychlewski J (1993) Improved theoretical dissociation energy and ionization potential for the ground state of the hydrogen molecule. J Chem Phys 98:3960–3967
30. Messerle G, Krauss LZ (1967) Die Dissoziation des C_2-Moleküls durch Rotation. Z Naturforsch A 22:2023–2026a
31. Huber KP, Herzberg G (1979) Molecular spectra and molecular structure. IV. Constants of diatomic molecules. Van Nostrand Reinhold, New York

32. Brewer L, Hicks WR, Krikorian OH (1962) Heat of sublimation and dissociation energy of gaseous C_2. J Chem Phys 36:182–188
33. Kordis J, Gingerich KA (1973) Gaseous phosphorus compounds. IX. Mass spectrometric studies of equilibria in the carbon-phosphorus system. J Chem Phys 58:5058–5066
34. Chupka WA, Inghram MG (1955) Direct determination of the heat of sublimation of carbon with the mass spectrometer. J Phys Chem 59:100–104
35. Drowart J, Bums RP, DeMaria G, Inghram MG (1959) Mass spectrometric study of carbon vapor. J Chem Phys 31:1131–1132
36. Burns RP, Jason AJ (1964) Evaporation coefficient of graphite. J Chem Phys 40:1161
37. Zavitsanos PD (1970) The vaporisation of pyrolytic graphite. In: Price P, Williams JE (eds) 2nd European symposium on the time-of-flight mass spectrometer, dynamics mass spectrometry series. Heyden, London
38. Wachi FM, Gilmartin DE (1972) Heat of formation and entropy of C_3 molecule. High Temp Sci 4:423–431
39. Wyatt JR, Stafford FE (1972) Mass spectrometric determination of the heat of formation of ethynyl radical, C_2H, and some related species. J Phys Chem 76:1913–1918
40. Zavitsanos PD, Carlson GA (1973) Experimental study of the sublimation of graphite at high temperatures. J Chem Phys 59:2966–2973
41. Shaik S, Rzepa HS, Hoffmann R (2013) One molecule, two atoms, three views, four bonds? Angew Chem Int Ed 52: 3020–3033
42. Brooke JSA, Bernath PF, Schmidt TW, Bacskay GB (2013) Line strengths and updated molecular constants for the C2 Swan system. J Quant Spectrosc Rad Trans 124:11–20
43. Bornhauser P, Sych Y, Knopp G, Gerber T, Radi PP (2011) Shedding light on a dark state: the energetically lowest quintet state of C_2. J Chem Phys 134:044302
44. Nakajima M, Joester JA, Page NI, Reilly NJ, Bacskay GB, Schmidt TW, Kable SH (2009) Quantum chemical study and experimental observation of a new band system of C_2, e $^3\Pi_g$ - c $^3\Sigma_u^+$. J Chem Phys 131:044301
45. Tanabashi A, Hirao T, Amano T, Bernath PF (2007) The Swan system of C_2: a global analysis of Fourier transform emission spectra. Astrophys J Suppl Ser 169:472–484
46. Kokkin DL, Bacskay GB, Schmidt TW (2007) Oscillator strengths and radiative lifetimes for C_2: Swan, Ballik–Ramsay, Phillips, and d $^3\Pi_g$–c $^3\Sigma_u^+$ systems. J Chem Phys 126:084302
47. Joester JA, Nakajima M, Reilly NJ, Kokkin DL, Nauta K, Kable SH, Schmidt TW (2007) The d $^3\Pi_g$–c $^3\Sigma_u^+$ band system of C_2. J Chem Phys 127:214303
48. Wakabayashi T, Ong A-L, Krätschmer W (2002) Laser induced fluorescence spectra of the D $^1\Sigma_u^+$–B' $^1\Sigma_g^+$ and C $^1\Pi_g$–A $^1\Pi_u$ systems in solid Ne. J Chem Phys 116:5996–6001
49. Lloyd GM, Ewart P (1999) High resolution spectroscopy and spectral simulation of C_2 using degenerate four-wave mixing. J Chem Phys 110:385–392
50. Prasad CVV, Bernath PF (1994) Fourier transform spectroscopy of the Swan (d $^3\Pi_g$–a $^3\Pi_u$) system of the jet-cooled C_2 molecule. Astrophys J 426:812–821
51. Martin M (1992) C_2 spectroscopy and kinetics. J Photochem Photobiol, A 66:263–289
52. Urdahl RS, Bao Y, Jackson WM (1991) An experimental determination of the heat of formation of C_2 and the C–H bond dissociation energy in C_2H. Chem Phys Lett 178:425–428
53. Karton A, Taylor PR, Martin JML (2007) Basis set convergence of post-CCSD contributions to molecular atomization energies. J Chem Phys 127:064104
54. Karton A, Tarnopolsky A, Lamere J-F, Schatz GC, Martin JML (2008) Highly accurate first-principles benchmark data sets for the parametrization and validation of density functional and other approximate methods. Derivation of a robust, generally applicable, double-hybrid functional for thermochemistry and thermochemical kinetics. J Phys Chem A 112:12868–12886
55. Karton A, Tarnopolsky A, Martin JML (2009) Atomization energies of the carbon clusters C_n (n = 2–10) revisited by means of W4 theory as well as density functional, Gn, and CBS methods. Mol Phys 107:977–990
56. Feller D, Peterson KA, Dixon DA (2008) A survey of factors contributing to accurate theoretical predictions of atomization energies and molecular structures. J Chem Phys 129:204105
57. Feller D, Peterson KA (2007) Probing the limits of accuracy in electronic structure calculations: is theory capable of results uniformly better than "chemical accuracy"? J Chem Phys 126:114105
58. Peterson KA, Dunning TH Jr (1997) Benchmark calculations with correlated molecular wave functions. VIII. Bond energies and equilibrium geometries of the CH_n and C_2H_n (n = 1–4) series. J Chem Phys 106:4119–4140
59. Büttenbender G, Herzberg G (1935) Über die Struktur der zweiten positiven Stickstoffgruppe und die Prädisssoziation des N_2-Moleküls. Ann Phys 413:577–610
60. Carroll PK, Mulliken RS (1965) $^3\Pi$ levels and predissociations of N_2 near the 12.135 eV dissociation limit. J Chem Phys 43:2170–2179
61. Roncin J-Y, Launay F, Larzilliere M (1984) Evidence for predissociation of N_2 into N (^{2}D) + N (^{2}D) from new high-resolution vacuum-ultraviolet emission bands. Phys Rev Lett 53:159–162
62. Lofthus A, Krupenie PH (1977) The spectrum of molecular nitrogen. J Phys Chem Ref Data 6:113–307
63. Le Roy RJ, Huang Y, Jary C (2006) An accurate analytic potential function for ground-state N_2 from a direct-potential-fit analysis of spectroscopic data. J Chem Phys 125:164310
64. Tang X, Hou Y, Ng CY, Ruscic B (2005) Pulsed field-ionization photoelectron-photoion coincidence study of the process $N_2 + h\nu \rightarrow N^+ + N + e^-$: bond dissociation energies of N_2 and $N_2{}^+$. J Chem Phys 123:074330
65. Gaydon AG (1968) Dissociation energies and spectra of diatomic molecules, 3rd edn. Chapman and Hall, London, pp 184–196
66. Brix P, Herzberg G (1954) Fine structure of the Schumann–Runge bands near the convergence limit and the dissociation energy of the oxygen molecule. Can J Phys 32:110–135
67. Lewis BR, Berzins L, Carver JH, Gibson ST (1985) Decomposition of the photo absorption continuum underlying the Schumann–Runge bands of $^{16}O_2$. I. Role of the B $^3\Sigma_u^-$ state: a new dissociation limit. J Quant Spectrosc Radiat Transfer 33:627–643
68. Martin JDD, Hepburn JW (1997) Electric field induced dissociation of molecules in Rydberg-like highly vibrationally excited ion-pair states. Phys Rev Lett 79:3154–3157
69. Gibson ST, Lewis BR, Baldwin KG, Carver JH (1991) Rotational features in the fluorescence excitation spectrum of O ($^1D_2$) from vacuum ultraviolet laser photodissociation of O_2. J Chem Phys 94:1060–1068
70. Cosby PC, Huestis DL (1992) On the dissociation energy of O_2 and the energy of the $O_2{}^+$ b $^4\Sigma_g^-$ state. J Chem Phys 97:6108–6112
71. Herzberg G (1950) Molecular spectra and molecular structure. I. Spectra of diatomic molecules, 2nd edn. Van Nostrand, Princeton, NJ
72. Colbourn EA, Dagenais M, Douglas AE, Raymonda JW (1976) The electronic spectrum of F_2. Can J Phys 54:1343–1359
73. Ruscic B, Feller D, Dixon DA, Peterson KA, Harding LB, Asher RL, Wagner AF (2001) Evidence for a lower enthalpy of formation of hydroxyl radical and a lower gas-phase bond dissociation energy of water. J Phys Chem A 105:1–4
74. Ruscic B, Wagner AF, Harding LB, Asher RL, Feller D, Dixon DA, Peterson KA, Song Y, Qian X, Ng C-Y, Liu J, Chen W, Schwenke DW (2002) On the enthalpy of formation of hydroxyl

radical and gas-phase bond dissociation energies of water and hydroxyl. J Phys Chem A 106:2727–2747
75. Barna D, Nagy B, Csontos J, Császár AG, Tasi G (2009) Benchmarking experimental and computational thermochemical data: a case study of the butane conformers. J Chem Theor Comput 8:479–486
76. Bytautas L, Ruedenberg K (2009) Ab initio potential energy curve of F_2. IV. Transition from the covalent to the van der Waals region: competition between multipolar and correlation forces. J Chem Phys 130:204101
77. Yang J, Hao Y, Li J, Zhou C, Mo Y (2005) A combined zero electronic kinetic energy spectroscopy and ion-pair dissociation imaging study of the F_2^+ (X $^2\Pi_g$) structure. J Chem Phys 122:134308
78. Yang J, Hao Y, Li J, Zhou C, Mo Y (2007) Erratum: a combined zero electronic kinetic energy spectroscopy and ion-pair dissociation imaging study of the F_2^+ (X $^2\Pi_g$) structure. J Chem Phys 127:209901
79. Bytautas L, Matsunaga N, Nagata T, Gordon MS, Ruedenberg K (2007) Accurate ab initio potential energy curve of F_2. III. The vibration rotation spectrum. J Chem Phys 127:204313
80. Csontos B, Nagy B, Csontos J, Kállay M (2013) Dissociation of the fluorine molecule. J Phys Chem A 117:5518–5528

Theor Chem Acc (2014) 133:1515
DOI 10.1007/s00214-014-1515-4

REGULAR ARTICLE

Transition metal atomic multiplet states through the lens of single-reference coupled-cluster and the equation-of-motion coupled-cluster methods

Varun Rishi · Ajith Perera · Rodney Bartlett

Received: 24 February 2014 / Accepted: 2 June 2014 / Published online: 22 June 2014

Abstract Transition metal atoms require a correct description of quasi-degeneracy and spin states, which lead to many closely lying multiplets. Before embarking on the theoretical description of transition metal complexes, the problems encountered in the atoms should be shown to be amenable to a chosen ab initio electronic structure method. It is usually thought that multi-reference methods should be a necessity for the correct description of transition metal multiplets. To the contrary, this paper explores the use of single-reference coupled-cluster theory in most of its manifestations. These include a variety of orbital choices from variationally optimal restricted open-shell references, to Brueckner orbital references, fractionally occupied references, quasi-restricted Hartree–Fock and other choices. In addition, the equation-of-motion coupled-cluster method for target multiplet states is considered. Relativistic corrections are obtained from a Douglas–Kroll–Hess fifth-order approximation which is found to be superior to effective core potentials, although there is little coupling between the relativistic effects and electron correlation.

Keywords Transition metals · Coupled cluster · Multi-reference · Douglas–Kroll–Hess · Equation-of motion coupled-cluster

Dedicated to Professor Thom Dunning and published as part of the special collection of articles celebrating his career upon his retirement.

V. Rishi · A. Perera (✉) · R. Bartlett
Quantum Theory Project, Department of Chemistry, University of Florida, Gainesville, FL, USA
e-mail: perera.ajith@gmail.com

1 Introduction

Transition metal (TM) atoms represent the prototype electronic structure problem that exhibits quasi-degeneracy, due to their five degenerate d orbitals and nearby s level that can be occupied in multiple ways by different numbers of electrons. The ground states and the low-lying excited states of the TM atoms have been the subject of various past [1, 2] and more recent studies [3, 4] as they pose a unique set of challenges to *quantum chemical* methods: multiple spin states and the potential inadequacy of single-reference-based correlation methods [5] plus non-negligible relativistic effects. This is also true for larger TM systems and consequently recent studies have been carried out to establish benchmarks [6–8] in order to judge the efficacy of the current approaches. Because of their quasi-degeneracy, the 'knee-jerk' reaction from a quantum chemist is that these systems require multi-reference techniques for their correct description. It is argued that only then, can the 'static' correlation be correctly accommodated. However, the great power of single-reference coupled-cluster (SR-CC) theory is that it converges to the full CI so efficiently when starting from even a poor reference determinant, that for many apparent multi-reference problems SR-CC provides the best ab initio correlated results obtainable at the basis set limit. When augmented with the equation-of-motion CC method which retains the simplicity and 'black-box' nature of SR-CC but now for an expanded number of problems including excited, ionized, doubly ionized, electron attached states, many apparent MR problems can be easily handled within the SR-CC/EOM framework. In this paper, we address this question for TM multiplets.

2 Synopsis of theory

2.1 SR-CC

The equations of SR-CC are now well known [9, 10],

$$E = \langle 0|e^{-T}He^{T}|0\rangle = \langle 0|\bar{H}|0\rangle$$

$$Q\bar{H}|0\rangle = 0$$

where Q indicates the orthogonal complement to the single-determinant reference function, $|0\rangle$ and can be separated into different excitation levels such as singles, doubles, triples (Q_1, Q_2, Q_3). The cluster operator,

$$T = T_1 + T_2 + T_3$$

is composed of single, double, triple, and higher 'connected' cluster excitation operators. At the T_n cluster limit for n electrons, the exact result is obtained in a given basis set (full CI). This result is size-extensive and gives a rigorous upper bound to the exact energy. It is also completely invariant to orbital choice. Short of the full CI, CC theory does not guarantee an upper bound to the energy, but is always extensive, which is guaranteed by the exponential wave function ansatz. This property, which is not shared by CI methods short of the full CI limit, is why SR-CC can rapidly exhaust most electron correlation effects while truncated CI methods, even built upon a MR framework, have zero overlap with the exact correlated wavefunction for an infinite number of electrons. To the degree that SR-CC at a given level, CCSD [11], CCSD(T) [12–15], ΛCCSD(T) [16], CCSDT [17], approaches the exact result, further MR extensions are not necessary. Where this point occurs will vary with the problem addressed. All the CC methods mentioned are also invariant to occupied–occupied and virtual–virtual orbital transformations, but not to transformations that mix occupied and virtual orbital space. However, they are nevertheless 'insensitive' to those as well as we will see later.

Properties in CC theory depend upon generalized expectation values which are calculated using the 1- and 2-body density matrices,

$$(\gamma_N)_{qp} = \langle 0|(1+\Lambda)e^{-T}\{p^{+}q\}e^{T}|0\rangle$$

where the de-excitation operator Λ serves as the left-hand eigenvector of the non-Hermitian Hamiltonian, $\bar{H}$,

$$\langle 0|(1+\Lambda)\bar{H} = \langle 0|E$$

and is related to the resolvent of the $\bar{H}$ operator,

$$\langle 0|\Lambda = \langle 0|\bar{H}Q(E-\bar{H})^{-1}Q$$

Besides properties, Λ has a role in the generalized ΛCCSD(T) method that is used here.

2.2 CCSDT-3

For the purposes of this paper, the highest computational level we will consider is CCSDT-3. This means that subject to projection only onto Q_1, Q_2, and Q_3,

$$\bar{H} = e^{(-T_1-T_2-T_3)}He^{(T_1+T_2+T_3)} = \tilde{H} + [\tilde{H}, T_3]$$

$$\tilde{H} = e^{(-T_1-T_2)}He^{(T_1+T_2)}$$

Then, the CCSDT-3 equations [18] are

$$Q_1\bar{H}|0\rangle = 0$$

$$Q_2\bar{H}|0\rangle = 0$$

$$Q_3\tilde{H}|0\rangle = 0$$

These are then iterated to convergence. By avoiding the T_3 term in the Q_3 projection, the method is an iterative $\sim n^7$ instead of the $\sim n^8$ for the full CCSDT. We have seen few examples where the differences between CCSDT-3 and CCSDT are significant, and even for those, CCSDT-3 might provide a better approximation due to error cancelation involving T_4 than does the full CCSDT method.

We explore the significance of triple excitations in the CC theory included iteratively or non-iteratively in the specific context of the energy separation between the closely spaced TM atomic states.

3 Results and discussion

3.1 Non-relativistic multiplet splitting

Table 1 lists the splitting energies for late first series TM atoms (Fe-Cu) obtained with non-relativistic CC calculations at various levels of correlation using a ROHF reference. They are compared with experimental values corrected for relativistic contributions as provided in a previous study [19].

Correlation effects are enormously important, changing the SCF results by more than a factor of 2. The next significant change occurs while moving from CCSD to a method that includes effects of triple excitations. This amounts to more than $\sim$0.2 eV improvement in the splitting values. In general, good accuracy can be achieved with coupled-cluster methods with triples corrections. Specifically, the multiplet splitting values obtained with CCSDT-3 are within 0.1–0.2 eV of the non-relativistic experimental measurements except for the '^{3}F–^{1}S' splitting of Ni, which will be a continuing theme. ΛCCSD(T) tends to improve the agreement to experiment as compared to CCSD(T) except for the above mentioned case for Ni atom where even CCSDT-3 performs poorly.

Table 1 Non-relativistic ROHF-CC calculations on multiplet states of Fe, Co, Ni, and Cu in cc-pwCVQZ (NR) basis set

	Excitation	DFT (TPSS)	DFT (B3LYP)	SCF	CCSD	CCSD(T)	ΛCCSD(T)	CCSDT-3	Exp (NR)
Fe	$^5F(d^6s^2)$–^5D (d^7s^1)	−0.46	−0.18	1.838	0.837	0.732	0.714	0.695	0.65
Co	$^4F(d^7s^2)$–$^4F(d^8s^1)$	−0.47	−0.09	1.552	0.368	0.224	0.198	0.174	0.17
Ni	$^3F(d^8s^2)$–$^3D(d^9s^1)$	−0.67	−0.31	0.833	−0.409	−0.526	−0.566	−0.486	−0.33
	$^3F(d^8s^2)$–^1S (d^{10})	1.49	1.73	5.117	1.576	1.064	0.896	0.742	1.57
Cu	$^2D(d^9s^2)$–$^2S(d^{10}s^1)$	−2.23	−1.86	−0.27	−1.638	−1.807	−1.856	−1.878	−1.85

The DFT results are taken from a study by Furche and Perdew[6]

Table 2 Relativistic ROHF-CC (DKH5) calculations on multiplet states of Fe, Co, Ni, and Cu in cc-pwCVQZ-DK basis set

	Excitation	SCF	CCSD	CCSD(T)	ΛCCSD(T)	CCSDT-3	Exp (eV)
Fe	$^5F(d^6s^2)$–^5D (d^7s^1)	2.106	1.103	0.993	0.975	0.954	0.87
Co	$^4F(d^7s^2)$–$^4F(d^8s^1)$	1.864	0.680	0.532	0.505	0.479	0.42
Ni	$^3F(d^8s^2)$–^3D (d^9s^1)	1.201	−0.044	−0.168	−0.210	−0.135	−0.03
	$^3F(d^8s^2)$–^1S (d^{10})	5.694	2.180	1.658	1.487	1.299	1.71
Cu	$^2D(d^9s^2)$–$^2S(d^{10}s^1)$	0.153	−1.208	−1.383	−1.434	−1.459	−1.49

As the procedure used to obtain these non-relativistic experimental numbers tends to be arbitrary, a better approach is to introduce relativistic effects directly into the ab initio calculation.

3.2 Relativistic corrections

First row TM are not usually thought to manifest very large relativistic corrections, but for high-accuracy quantum chemical calculations it remains a necessity [20]. The easiest way to introduce essential effects would be through relativistic based effective core potentials (ECP) [21, 22]. This simply removes the inner shell electrons from the calculation via introducing an ECP. The latter can be obtained from doing a true Dirac–Fock calculation for the TM, and generating the pseudo-potential by insisting that for the remaining electrons their description remains the same as obtained via the Dirac–Fock calculation. The ECP used here, Stuttgart 1997 relativistic small core (RSC), has pseudo-potential parameters adjusted to reproduce atomic valence-energy spectra obtained from Dirac–Fock calculations. Of course, the potential limitation of this procedure would be that only non-correlated theory is employed to generate the ECP. If the same is done following a 4-component correlated calculation, then one might expect that the ECP would be better suited to use in highly correlated CC calculations, but we know of no such ECP generation.

Alternatively, there are ways to simplify the full 4-component calculation such as the Douglas–Kroll–Hess (DKH) procedure [23–25] where by a series of unitary transformations of the Dirac Hamiltonian, the large (electronic) and small (positronic) component of the wavefunction are uncoupled and a block-diagonal Hamiltonian is obtained. This block-diagonal Hamiltonian can be expressed as a sum of a series of 'even' terms of orders in the external potential. The truncation of this series at nth order defines the DKHn Hamiltonian [26]. Here, we use a fifth-order approximation that has been implemented [27] in ACESII [28].

The results of the ROHF-CC calculations with the Douglas–Kroll–Hess method (DKH5) are shown in Table 2.

Above all-electron CC calculations with inclusion of relativistic effect through the DKH5 procedure tend to approach experimental J-averaged splittings [29] with CCSDT-3 results being the most accurate. As compared to the non-relativistic results (Table 1), different correlated methods display similar general features. One improvement, though, is seen for the hard to describe '^3F–^1S' splitting of Ni which is now in better agreement with experiment. The cc-pwCVQZ-DK basis set [30] optimized for valence 3d and 4s as well as semi-core 3 s and 3p electrons has been used and was obtained from the EMSL basis set library [31, 32]. Although we report all-electron results, no core addition to these bases that are mostly meant to describe the valence electrons is made.

Similarly, we obtain the following results for ROHF-CC calculations with relativistic ECP's (Stuttgart RSC (Relativistic Small Core) ECP) which include relativistic effects primarily as a core effect (Table 3)

When we compare the two relativistic results, it is apparent that the all-electron CC calculations with inclusion of scalar relativistic effects through the

Table 3 Relativistic ROHF-CC calculations on multiplet states of Fe, Co, Ni, and Cu in 'Stuttgart 1997 RSC (Relativistic Small Core) ECP basis set

	Excitation	CCSD(T)	ΛCCSD(T)	CCSDT-3	Exp(eV)
Fe	$^5F(d^6s^2)$–^5D (d^7s^1)	1.413	1.353	1.333	0.87
Co	$^4F(d^7s^2)$–$^4F(d^8s^1)$	0.911	0.827	0.805	0.42
Ni	$^3F(d^8s^2)$–^3D (d^9s^1)	0.202	0.082	0.132	−0.03
	$^3F(d^8s^2)$–^1S (d^{10})	2.325	1.856	1.479	1.71
Cu	$^2D(d^9s^2)$–$^2S(d^{10}s^1)$	−1.285	−1.438	−1.442	−1.49

Douglas-Kroll-Hess method performs better compared to the valence CC calculations using relativistic ECP's particularly for Fe and Co states. The quality of the basis set for the treatment of valence electrons in the 'Stuttgart' set might contribute to the overall poor performance as compared to the all-electron calculations since fewer functions are allocated to the correlation treatment.

3.3 Choice of vacuum state

Common to all the SR-CC methods is the vacuum state $|0\rangle$. In a properly written CC program, it does not matter what orbitals are used in the vacuum state, so we have many choices within SR-CC. If the particular choice buys anything, then SR-CC will benefit from that. That, and the chosen approximation for the T amplitudes, are really the only degrees of freedom in a CC calculation. Of course, other pragmatic simplifications can be made, like dropping core orbitals or only allowing active orbitals to be involved in the evaluation of T_3 and higher amplitudes to avoid the excessive scaling that would otherwise ensue, but for this paper, neither of those choices will be made. Instead, this will address the differences obtained for different choices of the vacuum state, $|0\rangle$.

The obvious choice is to use some flavor of Hartree–Fock theory. For closed shells that would be spin and spatially restricted (RHF). For open shells, we can choose unrestricted (UHF) or if we insist upon maximum double occupancy up to the open-shell orbitals, we have restricted open-shell HF (ROHF). Unlike the former, the latter is a spin eigenfunction. However, once either is used in a SR-CC calculation the final result is not a spin eigenfunction. We monitor the spin state from the generalized expectation value,

$$\langle S^2 \rangle = \langle 0|(1+\Lambda)e^{-T}S^2e^{T}|0\rangle = \bar{S}(\bar{S}+1)$$

to define the average multiplicity, M.

$$M = 2\bar{S} + 1 = \sqrt{(1 + 4\langle S^2 \rangle)}$$

For UHF cases, the projected value is often used,

$$\langle 0|e^{-T}S^2e^{T}|0\rangle = \bar{S}(\bar{S}+1)$$

as that is consistent with the 'projected' CC theory itself. But for an ROHF reference for $|0\rangle$, it is trivially true that S would be the eigenvalue for the exact spin state since the $\langle 0|$ would project out any contaminant. The expectation value does not have that limitation, so its value is more indicative of the contamination of the actual CC spin state. In Table 4 we list all the excited and ionized states studied here with relative energies and spin-expectation values.

For all of these examples, the spin state is essentially perfect within the error bars of the other components of the calculation such as the basis set and level of correlation. So, SRCC, with ROHF or even a UHF Ref. [33], seldom suffers from large spin contamination [34], especially for high spin states. Low spin states are another matter.

Because of the Thouless theorem that says any single determinant of orbitals can be rotated to any other single determinant up to a normalization constant via $\exp(T_1)$, CCSD and beyond have a very convenient insensitivity to orbital choice. In other words, rather than worrying about the optimization of the reference determinant, once it is determined that calculations like CCSD and beyond are to be done, the CCSD step is expected to automatically fix most remaining deficiencies in the reference determinant. This provides flexibility in the choice of the vacuum state that can be exploited in a wide variety of applications.

Examples might include using a Kohn–Sham determinant [35] when there is a known deficiency in the HF solution, as occurs frequently for metal systems. Perhaps we would like to use the first natural determinant (N) composed of natural orbitals because that determinant provides the density matrix with maximum overlap to the exact, correlated density matrix. Maybe we would like to use the Brueckner determinant [36] (B), which has maximum overlap with the exact correlated wavefunction.

Another choice pertinent to TM is to use orbitals that are obtained from a closed-shell calculation (QRHF) [37]. These orbitals are taken from a HF calculation of a related neutral or ionized state and are not variationally optimal, but should be readily accommodated by Coupled-cluster methods. To see how well this works, consider the results obtained (Table 5)

On the scale of the expected accuracy of these calculations (~0.2 eV), most of these results are close to those obtained using the ROHF reference with the differences being shown in '()' in Table 5. Clearly, the smallest differences are for CCSDT-3 results. This is expected as the

Table 4 ROHF-CC calculations on neutral and cationic states of Fe, Co, Ni, Zn, and Cu in cc-pwCVQZ-DK basis set

	Avg. Mult.	CCSD	CCSD(T)	ΛCCSD(T)	CCSDT-3	EXP (eV)
Fe						
$^5D(d^6s^2)$	4.99996	0.00	0.00	0.00	0.00	0.00
$^5F(d^7s^1)$	5.00007	1.103	0.993	0.975	0.954	0.87
Fe$^+$						
$^6D(d^6s^1)$	5.99996	7.691	7.840	7.828	7.838	7.90
$^4F(d^7)$	4.00009	8.174	8.209	8.192	8.193	8.15
Co						
$^4F(d^7s^2)$	3.99995	0.00	0.00	0.00	0.00	0.00
$^4F(d^8s^1)$	4.00003	0.681	0.532	0.505	0.479	0.42
Co$^+$						
$^5F(d^7s^1)$	5.00002	8.076	8.229	8.218	8.228	8.28
$^3F(d^8)$	2.99988	7.897	7.905	7.884	7.885	7.85
Ni						
$^3D(d^9s^1)$	2.99999	−0.0439	−0.168	−0.210	−0.135	−0.03
$^3F\ (d^8s^2)$	2.99997	0.00	0.00	0.00	0.00	0.00
$^1S\ (d^{10})$	1.00000	2.180	1.658	1.487	1.299	1.71
Ni$^+$						
$^4F\ (sd^8)$	3.99997	8.137	8.350	8.337	8.451	8.67
$^2D(d^9)$	1.99978	7.311	7.353	7.322	7.427	7.59
Cu						
$^2S(d^{10}s^1)$	1.99985	−1.208	−1.383	−1.434	−1.459	−1.49
$^2D(d^9s^2)$	1.99996	0.000	0.000	0.000	0.000	0.000
$^2P(d^{10}p^1)$	1.99076	2.538	2.437	2.398	2.315	2.295
Cu$^+$						
$^3D(d^9s^1)$	2.99998	8.815	8.974	8.967	8.976	9.04
$^1S\ (d^{10})$	1.00000	6.249	6.237	6.202	6.206	6.23
Zn						
$^1S(d^{10}s^2)$	1.00000	0.000	0.000	0.000	0.000	0.000
Zn$^+$						
$^2S(d^{10}s^1)$	1.99996	9.147	9.305	9.301	9.308	9.39
$^2D(d^9s^2)$	1.99998	16.903	17.112	17.128	17.130	17.30

Relative energies (in eV) are calculated w.r.t one of the indicated neutral states

higher the cluster operator, the closer to the full CI results which would manifest full orbital invariance. When compared directly to experiment, the actual answer using the closed-shell QRHF orbitals tends to be a bit better in an absolute sense when there are significant changes from the ROHF-based calculation. Perhaps this effect can be ascribed to the fact that the target states are now expressed relative to exactly the same set of orbitals which might benefit taking relative energy differences.

For TM in particular, or other highly degenerate systems, another choice for the QRHF orbitals is to generate them via a HF calculation on an open-shell system by insisting upon fractional occupations of the valence *d* orbitals (FON). Such a calculation provides a *spherically symmetric* state with exactly five degenerate d orbitals unlike any normal UHF or ROHF calculation which would lift the degeneracy. For this set of orbitals, as was the case in the prior QRHF reference calculations, the individual multiplets for different occupancies of the 5 degenerate d orbitals are readily apparent, instead of being complicated by which of the non-degenerate orbitals are occupied in a given multiplet. Hence, one set of orbitals can be used to generate all possible multiplet occupations. One difference that this procedure allows compared to the prior QRHF case based upon a nearby closed shell is the possibility of generating multiple sets of reference orbitals corresponding to the different choices of the occupancies of the d orbitals. Again the fact the orbitals are not variationally optimal in the fractional occupation case is expected to be accommodated by CCSD and beyond so that there is no remaining significant error from using these orbitals for a subsequent treatment of TM multiplets (see Table 6).

Here too, most results show modest changes, but for those that are >0.1 eV, the absolute answers are generally improved. An advantage here is the access to the states of Fe and Co atoms. Neither are close to a closed-shell

Table 5 QRHF-CC calculations in calculations on neutral and cationic states of Ni, Zn, and Cu in a cc-pwCVQZ-DK basis set

QRHF		CCSD	CCSD(T)	ΛCCSD(T)	CCSDT-3	EXP
Ni	$^3D\ (d^9s^1)$	0.1520 (−0.1959)	−0.652 (0.484)	−0.5491 (0.3391)	−0.0748 (−0.0602)	−0.03
	$^3F\ (d^8s^2)$	0.00	0.00	0.00	0.00	0.00
	$^1S\ (d^{10})$	2.242 (−0.062)	1.244 (0.414)	1.005 (0.482)	1.326 (−0.027)	1.71
Ni^+	$^4F\ (sd^8)$	8.514 (−0.377)	8.361 (−0.011)	8.450 (−0.113)	8.597 (−0.146)	8.67
	$^2D(d^9)$	7.493 (−0.182)	6.892 (0.461)	6.934 (0.388)	7.462 (−0.035)	7.59
Cu	$^2S\ (d^{10}s^1)$	−1.356 (0.148)	−1.320 (−0.063)	−1.460 (0.026)	−1.450 (−0.009)	−1.49
	$^2D(d^9s^2)$	0.000	0.000	0.000	0.000	0.000
	$^2P(d^{10}p^1)$	2.425 (0.113)	2.475 (−0.038)	2.352 (0.046)	2.311 (0.004)	2.295
Cu^+	$^3D\ (d^9s^1)$	8.809 (0.006)	8.968 (0.006)	8.962 (0.005)	8.972 (0.004)	9.04
	$^1S\ (d^{10})$	6.150 (0.099)	6.263 (−0.026)	6.1486 (0.0534)	6.201 (0.005)	6.23
Zn	$^1S(d^{10}s^2)$	0.000	0.000	0.000	0.000	0.000
Zn^+	$^2S\ (d^{10}s^1)$	9.171 (−0.024)	9.286 (0.019)	9.293 (0.008)	9.305 (0.003)	9.39
	$^2D(d^9s^2)$	16.970 (−0.067)	17.097 (0.015)	17.192 (−0.064)	17.129 (0.001)	17.30

Relative energies (in eV) are calculated w.r.t. one of the indicated neutral states (d^ns^2) of the Ni, Cu, and Zn atoms. The numerical difference with the variationally optimal ROHF reference are indicated in '( )'

reference where the prior QRHF procedure could be used, but with orbitals generated using fractional occupation HF calculation, these high spin open-shell multiplets can be accessed and the results are quite accurate.

Another choice is the Brueckner set of orbitals in which the virtual orbitals are rotated in such a way as to make the single amplitudes zero i.e., $T_1 = 0$. This condition ensures that overlap between the reference function and the exact wavefunction is a maximum i.e.,

$$|\langle \Phi_{\text{Brueckner}} | \Psi \rangle| = \max$$

Again we see modest changes relative to ROHF-based SRCC calculations (Table 7).

3.4 Relativistic corrections and orbital invariance

An additional aspect of orbital invariance pertains to the nature of the relativistic correlation corrections. The DKH5 procedure changes the mean-field Hamiltonian and orbitals for the subsequent CC correlated calculation. So all coupling between relativity and correlation in this approximation comes from these two effects. Hence, one might ask does the orbital invariance of CCSD and beyond still apply even though the Hamiltonian has changed? If it did, then a non-relativistic calculation can be directly added to the DKH5 mean-field and the results will be essentially the same. In Table 8, we show results obtained in that manner.

The results show that the amount of correlation energy using a particular correlated method with a non-relativistic Hamiltonian or a relativistic (DKH transformed) Hamiltonian is almost the same, and thus, we have orbital invariance in CC theory even when terms have been added to the non-relativistic Hamiltonian.

3.5 Equation-of-motion CC

One more route to obtaining multiplets of TMs is to use indirect means based upon the EOM-CC method [38, 39]. Then, depending upon the choice of starting point, these states can be excited states, ionized, or electron attached states. Using indirect EOM-CC approaches is particularly helpful if the reference state is described by a SR-CC for a closed shell, because then it follows that all accessible open-shell states are spin eigenfunctions. This is obviously especially useful for TM multiplets since much of their complexity arises from their spin properties. It is also unlike other SR-CC results where $\langle S^2 \rangle$ should be monitored.

The EOM-CC equations are well known. Once a SR-CC reference solution is obtained, then all sectors of Fock space can be assessed by solving

$$\bar{H}R_k|0\rangle = R_k\omega_k|0\rangle$$

where ω_k is the difference in energy between the reference state and the target excited, or ionized, or attached state. Furthermore, there is a whole spectrum of target states that are accessible via diagonalization of the matrix equation. Like T, for any k,

$$R = r_0 + R_1 + R_2 + R_3 + \cdots$$

The particular sector of Fock space to be studied is governed by the meaning of R_p. For excited electronic states, R_1 would correspond to single excitations or 1-particle, 1-hole operators (1p1h) $\{a^+i\}$, with R_2 being 2p2h $\{a^+i\ b^+j\}$, etc. Only for electronic excited states of the same symmetry as the reference state does r_0 have meaning. For ionized states, the excitation into orbital a would

Table 6 QRHF-CC calculations based on reference wave functions composed of orbitals obtained from 'fractional occupation' HF calculations (FON)

	FON	CCSD	CCSD(T)	ΛCCSD(T)	CCSDT-3	EXP
Fe	^{5}D(d^6s^2)	0.00	0.00	0.00	0.00	0.00
	^{5}F(d^7s^1)	0.972 (0.131)	1.068 (−0.075)	1.019 (−0.044)	0.949 (0.005)	0.87
Fe^+	^{6}D(d^6s^1)	7.673 (0.018)	7.846 (−0.006)	7.841 (−0.013)	7.839 (−0.001)	7.90
	^{4}F(d^7)	8.103 (0.071)	8.365 (−0.156)	8.336 (−0.144)	8.194 (−0.001)	8.15
Co	^{4}F(d^7s^2)	0.00	0.00	0.00	0.00	0.00
	^{4}F(d^8s^1)	0.480 (0.201)	0.688 (−0.154)	0.616 (−0.111)	0.493 (−0.014)	0.42
Co^+	^{5}F(d^7s^1)	8.066 (0.01)	8.316 (−0.087)	8.325 (−0.107)	8.230 (−0.002)	8.28
	^{3}F(d^8)	8.055 (−0.158)	8.210 (−0.305)	8.167 (−0.283)	8.023 (−0.138)	7.85
Ni	^{3}D (d^9s^1)	−0.319 (0.2751)	0.110 (−0.278)	0.001 (−0.211)	−0.076 (−0.059)	−0.03
	^{3}F (d^8s^2)	0.00	0.00	0.00	0.00	0.00
Ni^+	^{4}F (sd^8)	8.143 (−0.006)	8.349 (0.001)	8.346 (−0.009)	8.460 (−0.009)	8.67
	^{2}D(d^9)	7.149 (0.162)	7.504 (−0.151)	7.437 (−0.115)	7.455 (−0.028)	7.59
Cu	^{2}S ($d^{10}s^1$)	−1.554 (0.346)	−0.966 (−0.417)	−1.112 (−0.322)	−1.355 (−0.104)	−1.49
	^{2}D(d^9s^2)	0.000	0.000	0.000	0.000	0.000
	^{2}P($d^{10}p^1$)	2.277 (0.261)	2.711 (−0.274)	2.625 (−0.227)	2.315 (0.00)	2.295
Cu^+	^{3}D (d^9s^1)	8.828 (−0.013)	8.963 (0.011)	8.967 (0.00)	8.975 (0.001)	9.04
	^{1}S (d^{10})	6.036 (0.213)	6.476 (−0.239)	6.392 (−0.19)	6.255 (−0.049)	6.23

Neutral and cationic states of Fe, Co, Ni, Zn, and Cu in the cc-pwCVQZ-DK basis set are studied. Relative energies (in eV) are calculated w.r.t neutral states (d^ns^2) of Fe, Co, Ni, Cu, and Zn atoms. The numerical difference with the variationally optimal ROHF reference CC results are indicated in '( )'

Table 7 Brueckner-CC calculations on neutral and cationic states of Ni, Zn, and Cu in a cc-pwCVQZ-DK basis set

Brueckner		CCSD(ROHF)	CCSD	CCSD(T)	CCSDT-3	EXP
Fe	^{5}D(d^6s^2)	0.00	0.00	0.00	0.00	0.00
	^{5}F(d^7s^1)	1.103	1.071 (0.032)	0.977 (0.016)	0.949 (0.005)	0.87
Fe^+	^{6}D(d^6s^1)	7.691	7.637 (0.054)	7.838 (0.002)	7.838 (0.00)	7.90
	^{4}F(d^7)	8.174	8.131 (0.043)	8.202 (0.007)	8.193 (0.000)	8.15
Co	^{4}F(d^7s^2)	0.00	0.00	0.00	0.00	0.00
	^{4}F(d^8s^1)	0.681	0.704 (−0.023)	0.516 (0.016)	0.476 (0.003)	0.42
Co^+	^{5}F(d^7s^1)	8.076	8.068 (0.008)	8.235 (−0.006)	8.232 (−0.004)	8.28
	^{3}F(d^8)	7.897	7.905 (−0.008)	7.904 (0.001)	7.889 (−0.004)	7.85
Ni	^{3}D (d^9s^1)	−0.0439	−0.010 (−0.0339)	−0.190 (0.022)	−0.137 (0.002)	−0.03
	^{3}F (d^8s^2)	0.00	0.00	0.00	0.00	0.00
	^{1}S (d^{10})	2.180	2.274 (−0.094)	1.540 (0.118)	1.303 (−0.004)	1.71
Ni^+	^{4}F (sd^8)	8.137	8.128 (0.009)	8.358 (−0.008)	8.456 (−0.005)	8.67
	^{2}D(d^9)	7.311	7.329 (−0.018)	7.353 (0.000)	7.437 (−0.01)	7.59
Cu	^{2}S ($d^{10}s^1$)	−1.208	−1.168 (−0.04)	−1.412 (0.029)	−1.460 (0.001)	−1.49
	^{2}D(d^9s^2)	0.000	0.000	0.000	0.000	0.000
	^{2}P($d^{10}p^1$)	2.538	2.578 (−0.04)	2.424 (0.013)	2.331 (−0.016)	2.295
Cu^+	^{3}D (d^9s^1)	8.815	8.806 (0.009)	8.987 (−0.013)	8.985 (−0.009)	9.04
	^{1}S (d^{10})	6.249	6.272 (−0.023)	6.236 (0.001)	6.220 (−0.014)	6.23
Zn	^{1}S ($d^{10}s^2$)	0.000	0.000	0.000	0.000	0.000
Zn^+	^{2}S ($d^{10}s^1$)	9.147	9.140 (0.007)	9.321 (−0.016)	9.320 (−0.012)	9.39
	^{2}D(d^9s^2)	16.903	16.863 (0.04)	17.122 (−0.01)	17.127 (0.003)	17.30

Relative energies (in eV) are calculated w.r.t one of the neutral states of Ni, Cu, and Zn atoms. For iron's ^{5}D state, there was difficulty in converging the Brueckner iteration process, so the final rotated t_1 amplitudes are <0.08. The numerical difference with the variationally optimal ROHF reference are indicated in '()'

correspond to the continuum, so R_1 is limited to just one-hole states, $\{i\}|0\rangle$ and R_2 to 2h1p, $\{ija^+\}|0\rangle$, etc. For double ionization, DIP, R_2 is the first contributing operator, $\{ij\}|0\rangle$, with R_3 being 3h1p operators, $\{ijka^+\}|0\rangle$. For properties in target EOM-CC states, the left-hand eigenvector solution is required,

Table 8 The correlation energy contribution obtained using non-relativistic CCSD and CCSD(T) calculations are added, respectively, to SCF energy (using DKH5 transformed Hamiltonian) and compared to the relativistic total energy separation at these correlated levels

	Excitation	SCF (DKH5)	E_{cor}(NR) (CCSD level)	E_{SCF}(DKH5) + E_{cor}(NR)	CCSD (DKH5)	E_{cor}(NR) CCSD(T)	E_{SCF}(DKH5) + E_{cor}(NR)	CCSD(T) (DKH5)
Fe	$^5F(d^6s^2)$–^5D (d^7s^1)	2.106	−1.001	1.105	1.103	−1.106	1.000	0.993
Co	$^4F(d^7s^2)$–$^4F(d^8s^1)$	1.864	−1.184	0.680	0.680	−1.328	0.536	0.532
Ni	$^3F(d^8s^2)$–^3D (d^9s^1)	1.201	−1.242	−0.041	−0.044	−1.359	−0.158	−0.168
	$^3F(d^8s^2)$–^1S (d^{10})	5.694	−3.541	2.153	2.180	−4.053	1.641	1.658
Cu	$^2D(d^9s^2)$–$^2S(d^{10}s^1)$	0.153	−1.368	−1.215	−1.208	−1.537	−1.384	−1.383

See Tables 1 and 2

$$\langle 0|L_k\bar{H} = \langle 0|L_k\omega_k$$

$$(\gamma_N)^{lk}_{qp} = \langle 0|L_k e^{-T}\{p^+q\}e^T R_l|0\rangle$$

as was Λ for the reference state.

Below (Table 9) are the energies for the 'd^{n-2} $s^{2-} > d^{n-1}s^{1}$' excitation calculation using various methods such as CIS, EOM-MBPT(2), EOM-CCSD, and EOM-CCSD(T) approaches with the relativistic effects included at the level of Hamiltonian through the DKH5 transformation.

The EOM-CCSD and EOM-CCSD(T) results are less accurate as compared to the reference CCSDT-3 values with error of 0.1–0.3 eV in the excitations studied above. When we add triples to get EOM-CCSD(T), we normally expect the excitation energies to improve by $\sim$0.2 eV [40] and it definitely seems to be the case here. A possible reason for this large correction is the fact that these are transitions from an open-shell to an open-shell system. Generally, EOM-CC calculations calculate transition energies from a closed-shell system to open-shell systems. Such EOM calculations provide spin eigenfunctions for the target states which is clearly not the case in the above UHF-reference EOM calculations where sorting out spin states can be difficult even with spin-expectation values. On the other hand, the inclusion of (T) into the EOM calculation indicates that iterative triples methods will certainly help to improve these multiplet energies.

A superior approach for EE-EOM-CC for such problems is the ROHF-based scheme [41] proposed by Szalay and Gauss, where the target state is guaranteed to be a spin eigenfunction. Also, the spin-flip variant of EOM-CC of Krylov [42] could be used for some states.

Other variants of the EOM-CCSD method, IP-EOM-CCSD, EA-EOM-CCSD, and DIP-EOM-CCSD can be used to describe selected multiplet states for the copper and zinc atoms (Table 10) where the transition is from closed shell to open shells. This can be done in multiple ways depending upon the chosen reference state and the particular variant of EOM.

As described in Table 10, the neutral and singly ionized states of copper are not obtained via a single IP-EOM or DIP-EOM or an EA-EOM calculation but by a combination of all three which gives us access to all the studied states from the starting closed shell (either a Cu^- $(d^{10}s^2)$ or Cu^+ (d^{10})). A comment on the quality of results is due as we can see the splittings between the neutral states is not as numerically precise as compared to the SRCC or the EE-EOM results. A possible explanation might lie in the use of non-optimal set of orbitals to describe the states and the comparative 'sensitivity' of EOM results to the choice of orbitals used. Unlike the SR-CC, the R_1 operator in EOM

Table 9 $d^{n-2}s^2$- > $d^{n-1}s^1$ excitation energies calculated by EE-EOM-CCSD method for Fe, Co, Ni, and Cu atoms in UHF reference, inclusion of relativistic effects by DKH5 transformation

	Excitation (reference-excited)	CIS	EOM-MBPT(2)	EOM-CCSD	EOM-CCSD(T)	CCSDT-3 (ROHF)	Exp(eV)
Fe	$^5F(d^6s^2)$–^5D (d^7s^1)	4.148	1.406	1.260	1.234	0.954	0.87
Co	$^4F(d^7s^2)$–$^4F(d^8s^1)$	−1.695	−0.258	0.173	0.580	0.479	0.42
Ni	$^3F(d^8s^2)$–$^3D(d^9s^1)$	−0.414	−0.095	−0.288	−0.110	−0.135	−0.03
Cu	$^2D(d^9s^2)$–$^2S(d^{10}s^1)$	−3.735	−2.065	−1.363	−1.720	−1.459	−1.49

The 'cc-pwCVQZ-DK' (uncontracted) basis is used

Table 10 The neutral and cationic states of Cu and Zn atoms obtained using IP/EA variants of EOM-CCSD method in cc-pwCVQZ-DK (uncontracted) basis set, and the inclusion of relativistic effects using DKH5

	Copper	IP/DIP/EA-EOM-CCSD	Exp (eV)
Cu	^2S $(d^{10}s^1)$	−2.177	−1.49
	$^2D(d^9s^2)$	0.00	0.000
	$^2P(d^{10}p^1)$	1.457	2.295
Cu^+	^3D (d^9s^1)	9.062	9.04
	^1S (d^{10})	5.127	6.23
Zn	^1S $d(^{10}s^2)$	0.000	0.000
Zn^+	^2S $(d^{10}s^1)$	9.244	9.39
	$^2D(d^9s^2)$	17.269	17.30

In the case of copper atom, we obtain the neutral states ^2S $(d^{10}s^1)$ and ^2D (d^9s^2) as single ionizations from the copper anionic state (Cu^-–^1S $(d^{10}s^2)$) and ^2P $(d^{10}p^1)$ as an electron addition to the copper cationic state (Cu^+ (d^{10})) using EA-EOM-CCSD. The two cationic states, ^3D (d^9s^1) and ^1S (d^{10}), are obtained as double ionizations from copper anion (Cu^-–^1S $(d^{10}s^2)$) using DIP-EOM-CCSD. Similarly in the case of Zinc, we obtain the cationic states as ionizations from the closed-shell ground ^1S state using IP-EOM-CCSD

that mostly accounts for orbital relaxation is linear instead of exponential, which causes the orbital choice to be more important.

The ionization potentials obtained through IP-EOM-CCSD are quite accurate as compared to the experimental result with an accuracy of 0.1 eV with the exception of the ^2S state of Cu and the ^1S state of Cu^+.

4 Conclusions

A great deal of important chemistry involves TM systems. Besides the large number of electrons that have to be described, there are critical degeneracies, spin effects and relativistic corrections. One objective of this work is to assess the suitability of SR-CC methods to be able to correctly describe these systems. Though many of the complications that are found in smaller systems might be present to a lesser extent in larger TM systems with attached ligands that will quench some of these degeneracies and spin complications, the prototype for all TM systems are the TM atoms themselves. Since these atomic multiplets are fundamental to the chemistry of TM's, they need to be accurately described by quantum chemical methods themselves. We demonstrate the wealth of different SR-CC methods that could be used. In general, the CCSDT-3 results obtained here provide multiplet splittings to an accuracy of ~0.1 eV with a few exceptions. Also it does not seem to matter significantly whether ROHF, QRHF, UHF, FON, or Brueckner orbital-based references are used. The use of orbitals generated by fractional occupation-based HF calculations further expands the accessibility of closed-shell-based QRHF references to any d occupations as demanded in the case of the Fe and Co multiplets.

A comparison of all-electron calculations with inclusion of relativistic effects through DKH5 methods and valence-electron calculation with relativistic ECP basis is assessed, with the DKH5 procedure preferred. For a discussion on the infinite-order generalization of DKH and exact two-component theories see Barsyz and Sadlej [43]. Their approach would be the obvious generalization of DKH.

In all these calculations, iterative triple excitations via CCSDT-3 are found to be essential to obtain accurate results (~0.1 eV). Even though the SR-CC is not a strict spin eigenfunction, the computed spin multiplicities for the SR-CC calculations are quite accurate.

The most difficult multiplets to describe with QRHF are those that involve d^3 to d^7 since these are not close to a suitable closed-shell reference state. For such cases, FON orbitals offer an attractive alternative that provides a spherically symmetric set of orbitals which leads to unambiguous orbital occupations. Finally, the EE-EOM-CCSD method is used to calculate the energies for $d^{n-2}s^2$-> $d^{n-1}s^1$ excitations for the hard to access states. At the level of EOM-CCSD without triples, these were found to be not as accurate as the ones calculated through other approaches using SRCC methods, as the EE-EOM method would require open-shell to open-shell excitations, which are less recommended. However, once the non-iterative triples correction is added, the results are much better.

Iterative triples models for EOM-CC, analogous to CCSDT-3 for the single state, would be recommended.

Closed-shell-based EOM is used in the IP/EA-EOM-CC methods to obtain the ionization potentials for Cu and Zn which were generally found to be accurate as compared to the experimental values.

In summary, for TM multiplets, SR-CC methods such as CCSDT-3 provide highly accurate results despite their extensive degeneracy and spin state complications. Once ligands are added, much of the complexity encountered in the atoms would be alleviated. Hence, there is little reason to automatically assume that any TM complex demands a multi-reference description. All indications are that such systems can be described as well as other molecules composed of light atoms, as long as the large number of electrons is under control. With regard to that number, the extensivity of CC methods becomes especially pertinent for large number of electrons. Size-extensive methods follow a path that is correct from the limit of N non-interacting molecular units to the electron gas of infinite electrons, guaranteeing that the linked diagram energy is linear in either N. CC theory adheres to this all important path unlike CI methods, or even DFT ones.

Acknowledgments This work was sponsored by the U S Army Research Office. It is our pleasure to contribute this paper to the special issue in honor of Dr. Thom Dunning, whose work has been instrumental in the development of quantum chemistry over the last 45 years. We also thank Peter Schwerdtfeger and Markus Reiher for their role in instructing Tom Watson in the implementation of DKH5 into ACES II and III.

References

1. Langhoff SR, Bauschlicher CW (1988) Annu Rev Phys Chem 39:181 and references therein
2. Harrison JF (2000) Chem Rev 100:679 and references therein
3. Balabanov NB, Peterson KA (2006) J Chem Phys 125:074110
4. Luo S, Averkiev B, Yang KR, Xu X, Truhlar DG (2014) J Chem Theory Comput 10(1):102–121
5. Jiang W, DeYonker NJ, Wilson AK (2012) J Chem Theory Comput 8(2):460
6. Furche F, Perdew J (2006) J Chem Phys 124:044103
7. Schultz NE, Zhao Y, Truhlar DG (2005) J Phys Chem A 109:11127
8. DeYonker NJ, Peterson KA, Steyl G, Wilson AK, Cundari TR (2007) J Phys Chem A 111(44):11269
9. Bartlett RJ, Musial M (2007) Rev Mod Phys 79:1
10. Shavitt I, Bartlett RJ (2009) Many body methods in chemistry and physics. Cambridge Uni. Press, Cambridge
11. Purvis GD, Bartlett RJ (1982) J Chem Phys 76:1910
12. Urban M, Noga J, Cole SJ, Bartlett RJ (1985) J Chem Phys 83:4041
13. Raghavachari K, Trucks GW, Pople J, Head-Gordon M (1989) Chem Phys Lett 157
14. Watts JD, Gauss J, Bartlett RJ (1993) J Chem Phys 98:8718
15. Bartlett RJ, Watts J, Kucharski SA, Noga J (1990) Chem Phys Lett 165:513
16. Kucharski SA, Bartlett RJ (1998) J Chem Phys 108:5243
17. Noga J, Bartlett RJ (1987) J Chem Phys 86:7041
18. Noga J, Bartlett RJ, Urban M (1987) Chem Phys Lett 134:126
19. Raghavachari K, Trucks GW (1989) J Chem Phys 91:1062
20. Martin RL, Hay PJ (1981) J Chem Phys 75:4539
21. Dolg M, Cao X (2011) Chem Rev 112:403
22. Schwerdtfeger P (2011) ChemPhysChem 12(17):3143
23. Douglas M, Kroll NM (1974) Ann Phys 82:89
24. Hess BA (1986) Phys Rev A 33:3742
25. Reiher M, Wolf A (2009) Relativistic quantum chemistry. Wiley-Vch, Hoboken
26. Reiher M (2012) WIREs Comput Mol Sci 2:139
27. Watson TJ (2012) Ph.D. dissertation, Univ. of Florida, Gainesville
28. Stanton JF, Gauss J, Perera SA, Watts JD, Yau AD, Nooijen M, Oliphant N, Szalay PG, Lauderdale WJ, Gwaltney SR, Beck S, Balkova A, Bernholdt DE, Baeck KK, Rozyczko P, Sekino H, Huber C, Pittner J, Cencek W, Taylor D, Bartlett RJ. ACES II is a program product of the quantum theory project, University of Florida, Gainesville. Integral packages included are VMOL (J. Almlof and P.R. Taylor); VPROPS (P. Taylor); ABACUS (T. Helgaker, H.J. Aa. Jensen, P. Jørgensen, J. Olsen, and P.R. Taylor)
29. Moore CE (1971) Atomic energy levels. National Bureau of Standards, Washington
30. Balabanov NB, Peterson KA (2005) J Chem Phys 123:064107
31. Schuchardt KL, Didier BT, Elsethagen T, Sun L, Gurumoorthi V, Chase J, Li J, Windus TL (2007) J Chem Inf Model 47(3):1045
32. Feller D (1996) J Comp Chem 17(13):1571
33. Purvis GD, Sekino H, Bartlett RJ (1988) Coll Czech Chem Commun 53:2203
34. Stanton JF (1994) J Chem Phys 101:371
35. Villaume S, Daniel C, Strich A, Perera A, Bartlett RJ (2005) J Chem Phys 122:044313
36. Handy NC, Pople J, Head-Gordon M, Ragavachari K, Trucks GW (1989) Chem Phys Lett 164:185
37. Rittby M, Bartlett RJ (1988) J Chem Phys 76:3033
38. Watts JD (2008) In: Shukla MK, Leszczynski J (eds) Radiation induced molecular phenomenon in nucleic acids, Springer
39. Bartlett RJ (2012) WIREs Comput Mol Sci 2:126
40. Watson TJ, Lotrich VF, Szalay PG, Perera A, Bartlett RJ (2013) J Phys Chem A 117(12):2569
41. Szalay PG, Gauss J (2000) J Chem Phys 112:4027
42. Krylov A (2006) Acc Chem Res 39(2):83
43. Barysz M, Sadlej AJ (2001) J Mol Struct (THEOCHEM) 573(1):181

Theor Chem Acc (2014) 133:1459
DOI 10.1007/s00214-014-1459-8

REGULAR ARTICLE

Molecular orbital interpretation of the metal–metal multiple bonding in coaxial dibenzene dimetal compounds of iron, manganese, and chromium

Hui Wang · Dong Die · Hongyan Wang · Yaoming Xie · R. Bruce King · Henry F. Schaefer III

Received: 4 November 2013 / Accepted: 31 January 2014 / Published online: 26 February 2014

Abstract Both coaxial and perpendicular singlet spin state structures are found for the dibenzene dimetal complexes $(C_6H_6)_2M_2$ (M = Fe, Mn, and Cr) using density functional theory. For $(C_6H_6)_2M_2$ (M = Fe, Mn), the coaxial structure is the lower energy structure, whereas for $(C_6H_6)_2Cr_2$ the perpendicular structure is the lower energy structure. These coaxial structures are predicted to have very short M–M distances of ~1.98 Å for $(C_6H_6)_2Fe_2$, ~1.75 Å for $(C_6H_6)_2Mn_2$, and ~1.68 Å for $(C_6H_6)_2Cr_2$. Investigation into the frontier molecular orbitals suggests a formal 2π Fe=Fe double bond in $(C_6H_6)_2Fe_2$, a $\sigma + 2\pi$ Mn≡Mn triple bond in $(C_6H_6)_2Mn_2$, and a $\sigma + 2\pi + \delta$ quadruple bond in $(C_6H_6)_2Cr_2$. This gives each metal atom in these coaxial $(C_6H_6)_2M_2$ (M = Fe, Mn, Cr) derivatives a 16-electron configuration suggesting an 8-orbital d^5p^3 metal valence orbital manifold without the involvement of the s orbital. The coaxial $(C_6H_6)_2M_2$ (M = Fe, Mn) derivatives have ideal sixfold D_{6h} symmetry. However, distortion of coaxial $(C_6H_6)_2Cr_2$ from D_{6h} symmetry to D_{2h} symmetry is observed because of involvement of only one orbital from the $\{d(xy), d(x^2 - y^2)\}$ set of δ symmetry of each chromium atom in the Cr≣Cr formal quadruple bond.

Keywords Dibenzene dimetal compounds · Metal–metal multiple bonding · Molecular orbitals · Density functional theory

Dedicated to Professor Thom Dunning in recognition of his seminal contributions to theoretical chemistry and published as part of the special collection of articles celebrating his career upon his retirement.

Electronic supplementary material The online version of this article (doi:10.1007/s00214-014-1459-8) contains supplementary material, which is available to authorized users.

H. Wang · H. Wang (✉)
School of Physical Science and Technology, Southwest Jiaotong University, Chengdu 610031, China
e-mail: wanghyxx@yahoo.com

D. Die
Research Center for Advanced Computation, School of Physics and Chemistry, Xihua University, Chengdu, 610039, China

Y. Xie · R. B. King (✉) · H. F. Schaefer III
Department of Chemistry and the Center for Computational Quantum Chemistry, University of Georgia, Athens, GA 30602, USA
e-mail: rbking@chem.uga.edu

1 Introduction

The chemistry of metal–metal multiple bonding [1, 2] dates back nearly a half century to the seminal 1965 work of Cotton and Harris [3, 4] on the rhenium–rhenium quadruple bond in the octachlorodirhenate(III) dianion $Re_2Cl_8^{2-}$. Structure determination of this anion in its dipotassium salt showed four terminal chlorine atoms bonded to each rhenium atom and an unusually short rhenium–rhenium distance of 2.24 Å suggesting a high-order metal–metal multiple bond (Fig. 1). A more detailed analysis of the rhenium–rhenium bonding in $Re_2Cl_8^{2-}$ led to the postulation of a formal quadruple bond, thereby giving each rhenium atom only a 16-electron configuration in contrast to the favored 18-electron configuration in transition metal organometallic chemistry and related areas of coordination chemistry [5–9]. The rhenium–rhenium bond in $Re_2Cl_8^{2-}$ represented not only the first example of a metal–metal multiple bond but also the first example of a quadruple bond of any type.

Shortly after the identification of the Re–Re quadruple bond in the purely inorganic anion $Re_2Cl_8^{2-}$, the first

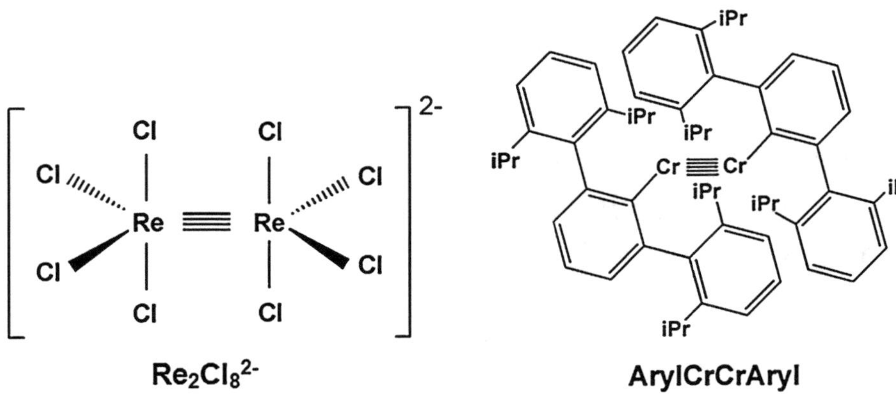

Fig. 1 First species with metal–metal quadruple and quintuple bonds to be synthesized and characterized structurally by X-ray crystallography

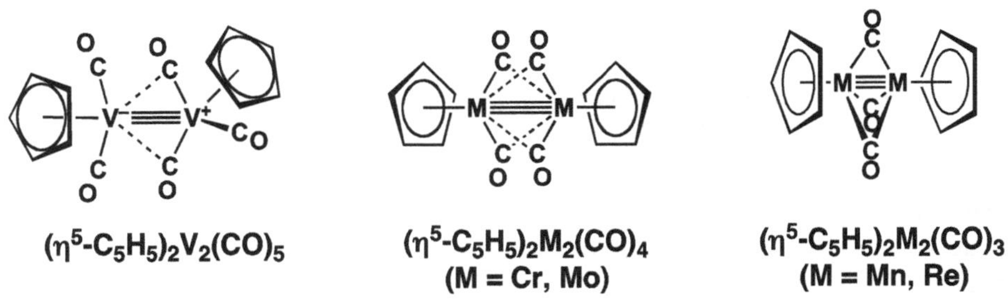

Fig. 2 Binuclear cyclopentadienyl metal carbonyls with formal M≡M triple bonds that have been structurally characterized by X-ray crystallography

example of metal–metal multiple bonding in an organometallic system was discovered. Thus, in 1967, King and Bisnette [10] reported the thermal reaction of $Mo(CO)_6$ with pentamethylcyclopentadiene to give the tetracarbonyl $(\eta^5\text{-}Me_5C_5)_2Mo_2(CO)_4$ rather than the expected hexacarbonyl $(\eta^5\text{-}Me_5C_5)_2Mo_2(CO)_6$ analogous to the unsubstituted cyclopentadienyl derivative. Formulation of the metal–metal bond in $(\eta^5\text{-}Me_5C_5)_2Mo_2(CO)_4$ as the Mo≡Mo triple bond suggested by the 18-electron rule was later supported by an X-ray structural determination indicating an unusually short metal–metal distance consistent with a triple bond (Fig. 2). Thus, the metal–metal distance in $(\eta^5\text{-}Me_5C_5)_2Mo_2(CO)_4$ was found to be 2.488 Å as compared to 3.235 Å for $(\eta^5\text{-}C_5H_5)_2Mo_2(CO)_6$, which clearly has a metal–metal single bond [11]. Subsequently, the analogous chromium compounds $(\eta^5\text{-}R_5C_5)_2Cr_2(CO)_4$ (R=H, [12] Me [13, 14]) were also synthesized and likewise found to have relatively short Cr≡Cr distances around 2.24 Å suggesting formal triple bonds. Further work indicated related formal metal–metal triple bonds to be stable central units in the binuclear cyclopentadienylmetal carbonyls $(\eta^5\text{-}C_5R_5)_2V_2(CO)_5$ [15, 16] and $(\eta^5\text{-}C_5R_5)_2M'_2(CO)_3$ (M′=Mn, [17] Re [18]) (Fig. 2). In all of these binuclear cyclopentadienyl metal carbonyl derivatives containing postulated formal M≡M triple bonds, each metal atom has the favored 18-electron configuration.

The highest known formal metal–metal bond order in a stable molecule then remained four for 40 years until the 2005 discovery by Power et al. [19] of a binuclear chromium(I) aryl of the type ArylCrCrAryl (Aryl=C_6H_3-2,6(C_6H_3-2,6-Pr^i_2)$_2$), with an extremely short metal–metal distance suggesting a formal quintuple bond (Fig. 1). This seminal discovery stimulated numerous theoretical studies on high-order metal–metal bonds [20–25]. In addition, various research groups reported further experimental work on low oxidation state transition metal aryls of the type ArylMMAryl [26, 27] as well as chromium(I) amidinate, [28, 29] 2-amino-pyridine, [30], and diazadiene [31] complexes, apparently containing formal metal–metal quintuple bonds. Note that simple electron counting in the binuclear chromium(I) aryls RCrCrR gives each chromium atom only a 12-electron configuration, even with a formal chromium–chromium quintuple bond. Chromium is also a favorable metal to form binuclear derivatives with formal quadruple bonds such as in the chromium(II) carboxylates, $Cr_2(O_2CR)_4{\cdot}2L$, which typically have short Cr≡Cr distances around 2.3 Å [32].

The available experimental results have led to the discovery of a number of binuclear transition metal derivatives having formal quadruple or quintuple bonds between the pair of transition metals as noted above. However, in all such stable compounds, each transition metal atom has less than the favored 18-electron configuration. In recent years, we have undertaken a number of theoretical studies in order to suggest possible systems for the synthesis of stable binuclear molecules containing metal–metal bonds of orders four and larger and the generally favored 18-electron configurations for the metal atoms forming such multiple bonds. An investigation of the unsaturated niobium carbonyls $Nb_2(CO)_9$ and $Nb_2(CO)_8$ predicted short Nb–Nb distances and high WBI values, suggesting the formal Nb–Nb quadruple and quintuple bonds required to

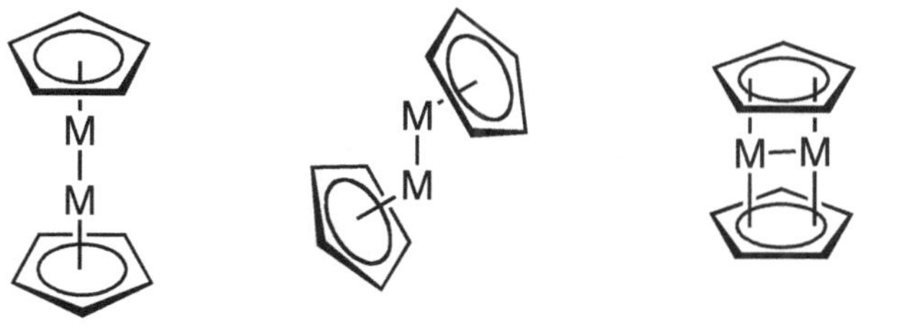

Fig. 3 Coaxial, bent, and perpendicular dimetallocenes

give each niobium atom the favored configuration [33]. However, an analysis of the highest lying bonding MOs suggested formal $\sigma + 2\pi$ formal Nb≡Nb triple bonds and 16-electron niobium configurations in these species. Thus, the occupied δ bonding orbitals in these species appeared to be canceled out by filled δ^* antibonding orbitals so that the net δ bonding in the Nb–Nb bond becomes zero. The metal d orbitals that would otherwise be used for δ bonding in the Nb–Nb multiple bonds are diverted away from the Nb–Nb bond for $d\pi \rightarrow p\pi$ back-bonding to the carbonyl groups. A related situation appears to occur in the singlet $(C_6H_6)_2Cr_2(CO)_2$ and $(C_6H_6)_2Cr_2(CO)$ structures [34].

Dimetallocenes, in which a pair of metal atoms is sandwiched between two parallel planar carbocyclic rings, provide additional examples of interesting metal–metal bonded organometallic systems. Such dimetallocenes can have coaxial, bent, or perpendicular structures (Fig. 3). The first coaxial dimetallocene, namely decamethyl-dizincocene $(\eta^5\text{-}Me_5C_5)_2Zn_2$, was synthesized by Carmona et al. in 2004 [35]. This dizincocene derivative can be regarded as a formal zinc(I) derivative with a formal Zn–Zn single bond leading to the favored 18-electron configuration for each zinc atom. In order to explore possible metal–metal multiple bonding in dimetallocenes, the third-row transition metal derivatives Cp_2M_2 (Cp = $\eta^5\text{-}C_5H_5$; M = Os, Re, W, Ta) [36] were investigated in view of the expected tendencies for such metals to form metal–metal multiple bonds as indicated by the original rhenium–rhenium quadruple bond in $Re_2Cl_8^{2-}$ discussed above [3]. Bent dimetallocene structures were predicted for all of these systems with short metal–metal distances suggesting high-order metal–metal multiple bonds. Of particular interest are the singlet Cp_2M_2 (M = Os, Re) systems for which analysis of the frontier bonding molecular orbitals suggests a formal Os–Os quintuple bond ($\sigma + 2\pi + 2\delta$) in singlet Cp_2Os_2 and a formal Re–Re sextuple bond ($2\sigma + 2\pi + 2\delta$) in singlet Cp_2Re_2. Such high formal metal–metal bond orders give each metal atom in both molecules the favored 18-electron metal configuration.

These Cp_2M_2 (M = Os, Re, W, Ta) dimetallocenes with apparent high-order metal–metal multiple bonds all have bent dimetallocene structures. In order to find linear dimetallocene structures with possible high-order metal–metal multiple bonds, the dibenzene dimetal derivatives $(C_6H_6)_2M_2$ of the first-row transition metals (M = Ti, V, Cr, Mn, Fe, Co, Ni) were investigated by density functional theory. Perpendicular dimetallocene structures were found for all seven $(C_6H_6)_2M_2$ derivatives with bridging benzene ligands related to the known dibenzene dipalladium dication $(C_6H_6)_2Pd_2^{2+}$ in $(C_6H_6)_2Pd_2(AlCl_4)_2$ and $(C_6H_6)_2Pd_2(Al_2Cl_7)_2$, first reported in 1970 [37]. However, only the chromium, manganese, and iron derivatives $(C_6H_6)_2M_2$ (M = Cr, Mn, Fe) were found to form coaxial isomers with terminal benzene ligands and the metal–metal axes collinear with the C_6 axes of the benzene rings. The $(C_6H_6)_2M_2$ isomers of the remaining four metals (M = Ti, V, Co, Ni) with terminal benzene ligands were predicted to have bent dimetallocene structures similar to those of the bent dimetallocenes Cp_2M_2 (M = Os, Re, W, Ta), which were objects of the previous theoretical study on metal–metal multiple bonding [36].

Application of the 18-electron rule [5–9] predicts formal high metal–metal bond orders in the coaxial $(C_6H_6)_2M_2$ (M = Cr, Mn, Fe) derivatives ranging from six for the chromium derivative to four for the iron derivative. The prospect of synthesizing $(C_6H_6)_2Cr_2$ or other $(arene)_2Cr_2$ derivatives is particularly interesting since a formal bond order of six for the chromium–chromium bond would exceed the formal bond order of five in the Power Aryl-CrCrAryl derivatives (Fig. 1). In order to gain insight into the nature of the metal–metal multiple bonding in the coaxial $(C_6H_6)_2M_2$ (M = Cr, Mn, Fe) derivatives, we have investigated the frontier molecular orbitals in these systems. We come to the somewhat disappointing conclusion that these $(C_6H_6)_2M_2$ complexes are best regarded as systems with 16-electron metal configurations rather than 18-electron metal configurations having formal bond orders ranging from four for the chromium derivative to two for the iron derivative. The metal d orbitals that could otherwise be used for components of the metal–metal bond appear to be preempted for back-bonding into the δ antibonding orbitals of the benzene rings.

2 Theoretical methods

The density functional theory (DFT) methods B3LYP [38, 39] and BP86 [40, 41] with double-ζ plus polarization (DZP) basis sets were selected for this study similar to our numerous previous theoretical studies on first-row transition metal complexes. Thus, for carbon and hydrogen, the DZP basis set used here adds one set of pure spherical harmonic d functions with orbital exponents $\alpha_d(C) = 0.75$ and $\alpha_p(H) = 0.75$ to the Huzinaga–Dunning standard contracted DZ sets designated as $(9s5p1d/4s2p1d)$ for

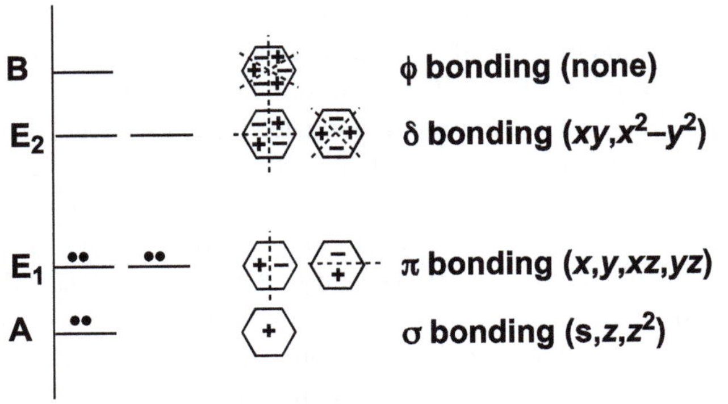

Fig. 4 Orbitals involved in the metal-ring bonding in the coaxial $(C_6H_6)_2M_2$ structures of the d-block transition metals with the metal–metal bond as the z axis

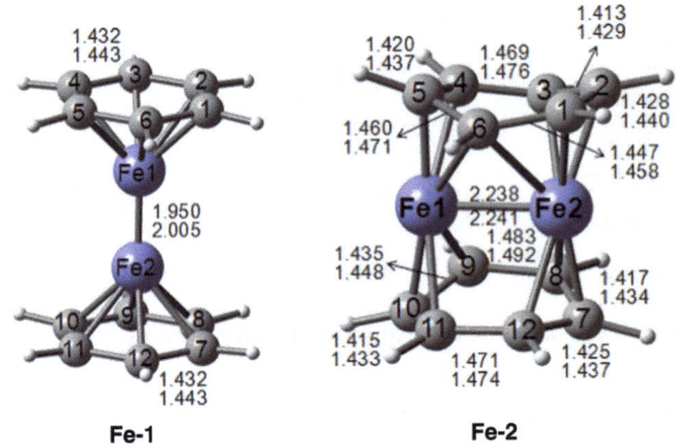

Fig. 5 Coaxial and perpendicular singlet structures for $(C_6H_6)_2Fe_2$. In Figs. 5, 6, and 7, the upper distances were obtained by the B3LYP method and the lower distances by the BP86 method (all bond distances are in angstroms); the carbon atoms in both C_6H_6 rings are anticlockwise labeled

carbon and (4*s*1*p*/2*s*1*p*) for hydrogen [42, 43]. For the first-row transition metals, in our loosely contracted DZP basis set, the Wachters' primitive sets were used, but augmented by two sets of p functions and one set of d functions and contracted following Hood et al. [44] and designated as (14*s*11*p*6*d*/10*s*8*p*3*d*).

The geometries of all of the structures were fully optimized using both the DZP B3LYP and DZP BP86 methods. All of the computations were carried out with the Gaussian 09 program package [45], in which an ultrafine integration grid (99, 590) was used. Imaginary vibrational frequencies of magnitude less than $100i$ cm^{-1} are questionable, because of the limitation of the numerical integration for the DFT methods [46]. In general, we do not follow the imaginary eigenvector in search of the minimum in such cases.

3 Results and discussion

3.1 Metal-ring bonding in benzene metal complexes

Figure 4 shows the benzene frontier molecular orbitals and the available atomic orbitals of a *d*-block transition metal in a coaxial $(C_6H_6)_2M_2$ derivative with which they can overlap. The coordinate system is chosen so that the z axis is collinear with the metal–metal bond.

The six π electrons of a neutral benzene ligand in a coaxial $(C_6H_6)_2M_2$ derivative fill pairwise the lowest lying non-degenerate A molecular orbital and the immediately higher lying degenerate E_1 molecular orbital pair (Fig. 4). A benzene–metal bond in such a molecule using these six π electrons is effectively a formal triple bond with one σ component and two orthogonal π components. The benzene–metal σ bond can effectively use a linear hybrid of the $p(z)$ and $d(z^2)$ metal orbitals leaving the other hybrid for a σ component of the metal–metal bond. The metal s orbital, although of suitable symmetry, is less likely to play a role in this metal–metal σ bond because of its non-directional nature. Similarly, one of the two orthogonal benzene–metal π bonds can use a linear hybrid of the $p(x)$ and $d(xz)$ metal orbitals leaving the other hybrid for a π component of a metal–metal bond. An second orthogonal benzene–metal π bond can similarly use a linear hybrid of the $p(y)$ and $d(yz)$ metal orbitals leaving the other such hybrid for a second orthogonal π component of a metal–metal bond. Thus, a d-block transition metal has enough atomic orbitals in a coaxial $(C_6H_6)_2M_2$ derivative to form a benzene–metal $\sigma + 2\pi$ triple bond and a metal–metal $\sigma + 2\pi$ triple bond.

The $\sigma + 2\pi$ forward benzene–metal bonding can be supplemented by two δ metal–benzene back bonds at 45° angles into the empty E_2 benzene antibonding orbitals of δ symmetry using the metal $d(xy)$ and $d(x^2 - y^2)$ orbitals (Fig. 4). However, use of these metal orbitals for such back-bonding makes them unavailable for δ components of quadruple and quintuple metal–metal multiple bonds. Thus metal-ring back-bonding into the empty E_2 benzene orbitals of δ symmetry is directly competitive with increasing the metal–metal formal bond order beyond three by addition of δ components to a $\sigma + 2\pi$ formal triple bond.

In summary, the two critical points for understanding the interplay between metal–benzene bonding and metal–metal multiple bonding in coaxial $(C_6H_6)_2M_2$ derivatives of d-block transition metals are the following:

1. The metal $d(xy)$ and $d(x^2 - y^2)$ orbitals are available for either metal-ring δ back-bonding or the δ components of formal metal–metal quadruple and quintuple bonds but not both.
2. The non-directionality of metal s orbitals can remove them from the metal bonding orbital manifold leaving an eight-orbital p^3d^5 manifold corresponding to a

Table 1 Fe–Fe distances (in Å), total energies (E, in Hartree), and relative energies (ΔE, in kcal/mol) for the two $(C_6H_6)_2Fe_2$ structures

	Fe-1 (D_{6h})		Fe-2 (C_1)	
	B3LYP	BP86	B3LYP	BP86
Fe-1–Fe-2	1.950	2.005	2.238	2.241
−Energy	2,991.93527	2,992.39065	2,991.91996	2,992.38003
ΔE	0.0	0.0	9.6	6.7

None of these structures has any imaginary vibrational frequencies

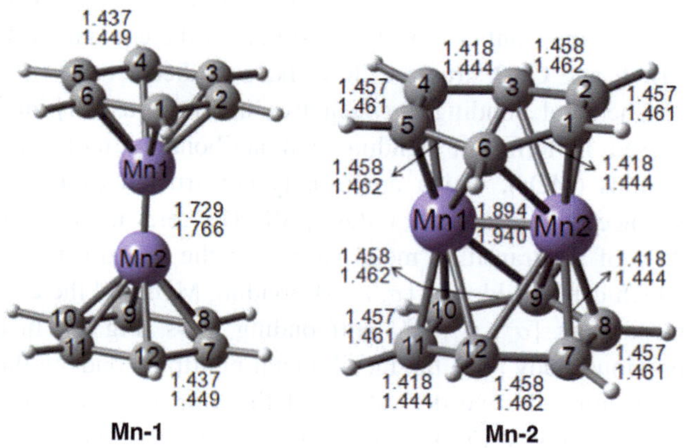

Fig. 6 Coaxial and perpendicular singlet structures for $(C_6H_6)_2Mn_2$

favored 16-electron rather than a favored 18-electron configuration.

3.2 Molecular structures

3.2.1 $(C_6H_6)_2Fe_2$

A D_{6h} coaxial $(C_6H_6)_2Fe_2$ structure **Fe-1** with two hexahapto η^6-C_6H_6 rings was found with a very short Fe–Fe distance of 1.950 Å (B3LYP) or 2.005 Å (BP86) (Fig. 5; Table 1). The perpendicular $(C_6H_6)_2Fe_2$ structure **Fe-2** with a longer Fe–Fe distance of 2.238 Å (B3LYP) or 2.241 Å (BP86) is predicted to lie 9.6 kcal/mol (B3LYP) or 6.7 kcal/mol (BP86) above **Fe-1**. In **Fe-2** the central Fe_2 unit is bridged by a η^3,η^3-C_6H_6 and a η^4,η^3-C_6H_6 ring with one of the carbon atoms in the latter ring being shared by both iron atoms.

3.2.2 $(C_6H_6)_2Mn_2$

Similar computations were performed on the singlet $(C_6H_6)_2Mn_2$ derivatives. A D_{6h} coaxial structure **Mn-1** was found with an ultrashort Mn–Mn distance of 1.729 Å (B3LYP) or 1.766 Å (BP86) (Fig. 6; Table 2), which is comparable to the Cr–Mn bond distance of 1.82 Å in a recently reported heteronuclear compound [47]. In **Mn-1**, both η^6-C_6H_6 rings are hexahapto ligands. The perpendicular

Table 2 Mn–Mn distances (in Å), total energies (E, in Hartree), and relative energies (ΔE, in kcal/mol) for the $(C_6H_6)_2Mn_2$ structures

	Mn-1 (D_{6h})		Mn-2 (D_2)	
	B3LYP	BP86	B3LYP	BP86
Mn-1–Mn-2	1.729	1.766	1.894	1.940
−Energy	2,766.48242	2,766.91561	2,766.46220	2,766.90093
ΔE	0.0	0.0	12.7	9.2

None of these structures has any imaginary vibrational frequencies

structure **Mn-2** with a somewhat longer Mn–Mn distance of 1.894 Å (B3LYP) or 1.940 Å (BP86) is predicted to lie 12.7 kcal/mol (B3LYP) or 9.2 kcal/mol (BP86) above **Mn-1**. In **Mn-2** (Fig. 2), each benzene ring bridges the central Mn_2 unit as an η^4,η^4-C_6H_6 ligand.

3.2.3 $(C_6H_6)_2Cr_2$

The DFT computations on $(C_6H_6)_2Cr_2$ predict the perpendicular structure **Cr-1** to lie 13.9 kcal/mol (B3LYP) or 23.5 kcal/mol (BP86) in energy below the coaxial structure **Cr-2** (Fig. 7; Table 3). The relative energies between the two $(C_6H_6)_2Cr_2$ structures have been confirmed by a single-point DFT calculation using a def2-TZVPP basis set, which gives the same energy ordering as that from the DZP basis set. Thus, this chromium derivative $(C_6H_6)_2Cr_2$ differs from $(C_6H_6)_2Fe_2$ and $(C_6H_6)_2Mn_2$ for which the coaxial structures are the lower energy structures (Tables 1, 2). Also the coaxial $(C_6H_6)_2Cr_2$ structure **Cr-2** has only D_{2h} symmetry in contrast to the D_{6h} symmetry of the coaxial $(C_6H_6)_2M_2$ (M = Fe, Mn) structures **Mn-1** and **Fe-1**. Thus, in each benzene ring of **Cr-2**, the lengths of four of the six C–C bonds in each benzene ring are ~1.45 Å, whereas those of the remaining two C–C bonds are only ~1.41 Å. This differs from **Mn-1** and **Fe-1** for which all of the benzene C–C bonds are predicted to be ~1.44 Å. The perpendicular structure **Cr-1** with C_{2v} symmetry has a small imaginary frequency of 58i cm^{-1} using the B3LYP/DZP method. However, vibrational frequency analysis with the def2-TZVPP basis set gives all real frequencies for **Cr-1**. The central Cr_2 unit is bridged by the two η^3,η^3-C_6H_6 rings.

An important feature of each $(C_6H_6)_2Cr_2$ structure is a very short Cr–Cr distance. The coaxial structure **Cr-2** has a particularly short Cr–Cr distance of 1.668 Å (B3LYP) or 1.698 Å (BP86), which is very close to the experimental value for diatomic Cr_2 (1.679 Å) [48]. The same Cr–Cr distance of 1.668 Å (B3LYP) or 1.698 Å (BP86) for D_{2h} singlet coaxial $(C_6H_6)_2Cr_2$ is also given by Liu et al. [49] in recent work. In addition, in our previous work, it is shown that the predicted Cr–Cr distances for the singlet structures $(\eta^6$-$C_6H_6)_2Cr_2(CO)_n$ decrease monotonically as carbonyl groups are lost starting from 2.954 Å (BP86) in

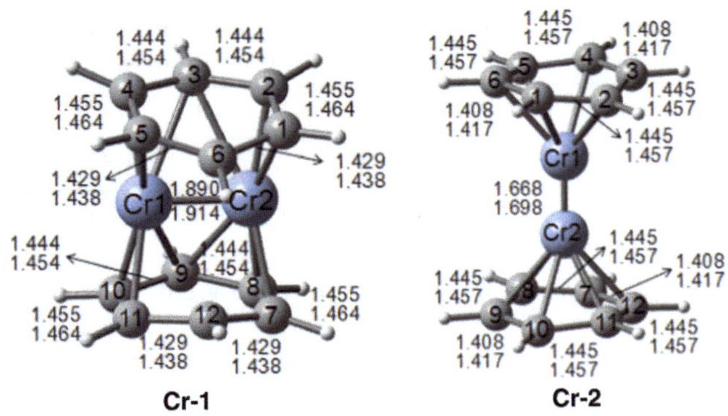

Fig. 7 Coaxial and perpendicular singlet structures for $(C_6H_6)_2Cr_2$

Table 3 Cr–Cr distances (in Å), total energies (*E*, in Hartree), and relative energies (Δ*E*, in kcal/mol) for the $(C_6H_6)_2Cr_2$ structures

	Cr-1 (C_{2v})		Cr-2 (D_{2h})	
	B3LYP	BP86	B3LYP	BP86
Cr-1–Cr-2	1.890	1.914	1.668	1.698
−Energy	2,553.45219	2,553.84721	2,553.43010	2,553.80983
Δ*E*	0.0	0.0	13.9	23.5

$(C_6H_6)_2Cr_2(CO)_5$ to 1.947 Å (BP86) in $(C_6H_6)_2Cr_2(CO)$ [34]. The short Cr–Cr distance in **Cr-2** without carbonyl groups is in accord with the trend. The Cr–Cr distance in the perpendicular structure **Cr-1** of 1.890 Å (B3LYP) or 1.914 Å (BP86), although still extremely short, is ~0.2 Å longer than that in the coaxial structure **Cr-2**.

3.3 Natural bond orbital (NBO) analysis

Table 4 lists the M–M bond distances, Wiberg's bond indices (WBIs), and electron delocalization indices DI(M,M) [50] using the BP86 method. The WBI value for the Cr–Cr bond in the coaxial **Cr-2** structure has the extraordinarily high value of 3.40, suggesting a high-order multiple bond. The WBIs for the metal–metal bonds in the coaxial structures $(C_6H_6)_2Mn_2$ (**Mn-1**) and $(C_6H_6)_2Fe_2$ (**Fe-1**) are progressively lower at 2.41 and 1.46, respectively, suggesting decreasing formal orders of the metal–metal bonds in the sequence Cr>Mn>Fe in accordance with expectation based on the numbers of metal valence electrons. The DI(M,M) values [50] are in essential agreement with the WBIs.

3.4 Molecular orbitals in the singlet coaxial $(C_6H_6)_2M_2$ structures (M = Fe, Mn, and Cr)

The short M–M distances and the 18-electron rule [5–9] suggest formal sextuple, quintuple, and quadruple M–M bonds in the singlet coaxial $(C_6H_6)_2Cr_2$, $(C_6H_6)_2Mn_2$, and $(C_6H_6)_2Fe_2$ structures, respectively. In order to gain some insight into the metal–metal multiple bonding in these species, the frontier molecular orbitals of the coaxial **Fe-1**, **Mn-1**, and **Cr-2** structures have been studied.

Table 4 WBIs and DI(M,M)s for the M–M bonds in the singlet coaxial $(C_6H_6)_2M_2$ (M = Cr, Mn, and Fe) structures by BP86

	Structure	M–M WBI	DI (M,M)	M–M distance, Å	Formal M–M bond order
$(C_6H_6)_2Cr_2$	**Cr-2**	3.40	3.38	1.698	4
$(C_6H_6)_2Mn_2$	**Mn-1**	2.41	2.71	1.766	3
$(C_6H_6)_2Fe_2$	**Fe-1**	1.46	1.98	2.005	2

Chemical bonding between two atoms is usually discussed in terms of bonding and antibonding molecular orbitals (MOs), which are usually constructed using the valence atomic orbitals (AOs) [20]. The relevant valence AOs of the transition metal atoms are the five d orbitals, which can provide the {σ, π, δ} bonding MOs and the corresponding {σ^*, π^*, δ^*} antibonding MOs (Fig. 4). In a previous study [36], the metal–metal multiple bonds in the bent dimetallocene derivatives of the third-row transition metals, namely Cp_2M_2 (M = Os, Re, W, Ta), were successfully characterized in terms of the frontier MOs. This approach was used to identify the $\sigma + 2\pi + 2\delta$ components of a formal quintuple bond in a singlet bent Cp_2Os_2 structure and the $2\sigma + 2\pi + 2\delta$ components of a formal sextuple bond in a singlet bent Cp_2Re_2 structure. Here, we use a similar approach to study M–M bonding in the likewise singlet coaxial $(C_6H_6)_2M_2$ structures (M = Fe, Mn, and Cr).

3.4.1 Molecular orbitals in the singlet coaxial $(C_6H_6)_2Fe_2$ structure (Fig. 8)

The HOMO-7 in the coaxial $(C_6H_6)_2Fe_2$ structure **Fe-1** corresponds to the σ component of the Fe–Fe bond using a combination of the metal {$d(z^2)$, $p(z)$} orbitals. The shape of HOMO-7 suggests that the iron $d(z^2)$ atomic orbital is a major contributor to HOMO-7. However, the HOMO of **Fe-1** is the σ^* antibonding orbital, which cancels the corresponding σ bonding orbital (HOMO-7) so that the Fe–Fe multiple bond in **Fe-1** does not have a σ component.

The degenerate HOMO-3/HOMO-4 molecular orbital pair in **Fe-1** (Fig. 8) corresponds to two orthogonal π bonding components of the Fe–Fe multiple bond. These orthogonal Fe–Fe π bonding components use iron {$d(xz)$, $p(x)$} hybrid orbitals for one of the π components and the iron {$d(yz)$, $p(y)$} hybrid orbitals for the other π component. The degenerate HOMO-8/HOMO-9 molecular orbital pair corresponds to the benzene–iron π bonding components, also using iron {$d(xz)$, $p(x)$} and {$d(yz)$, $p(y)$} hybrid orbitals. The remaining two iron d orbitals, namely the $d(xy)$ and $d(x^2 - y^2)$ orbitals, are used for the degenerate δ

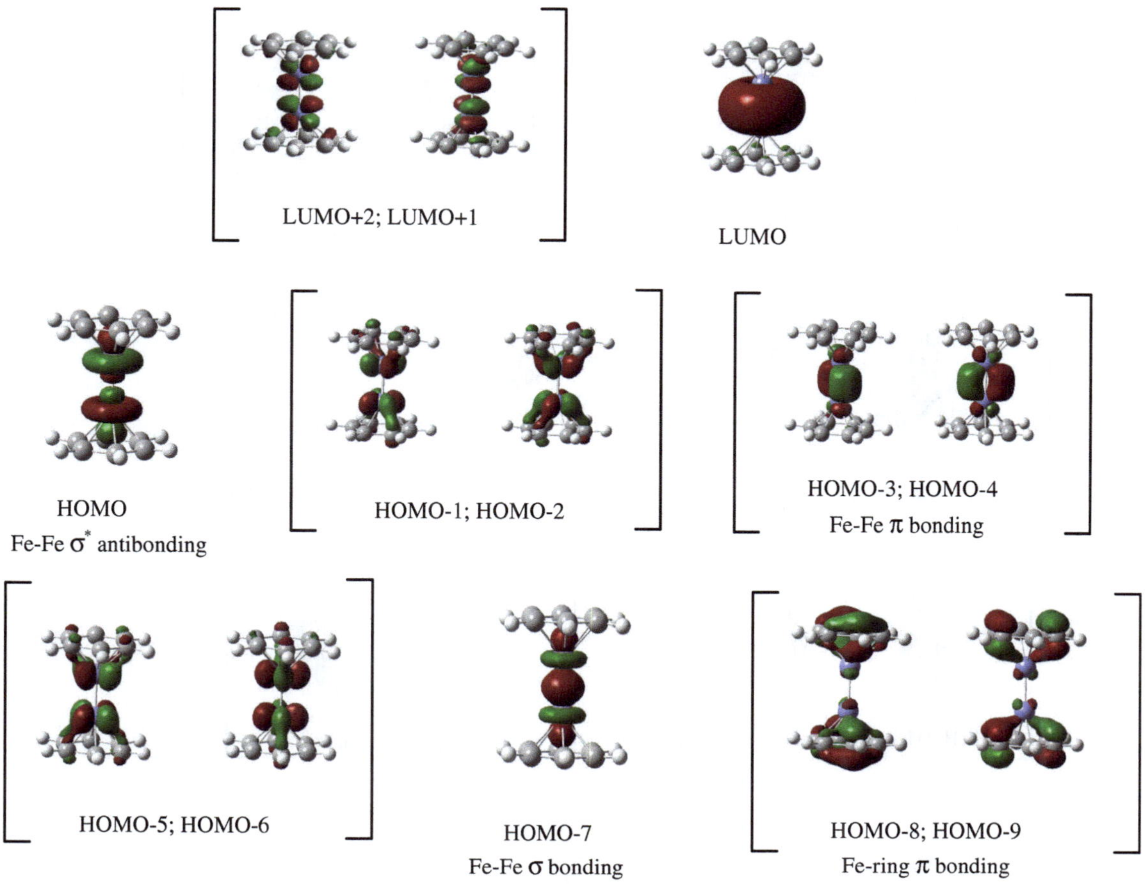

Fig. 8 Active MOs related to the two iron atoms in the coaxial $(C_6H_6)_2Fe_2$ structure **Fe-1**. Pairs of degenerate orbitals are enclosed in brackets

bonding orbitals (HOMO-5/HOMO-6). However, they are canceled by the degenerate δ^* antibonding orbitals (HOMO-1/HOMO-2) so that the Fe–Fe bond in **Fe-1** has no δ component (Fig. 8).

This MO analysis thus suggests a formal Fe=Fe double bond 2π bond in **Fe-1** rather than the quadruple bond expected from the 18-electron rule. This Fe=Fe double bond is seen to have no components of σ and δ symmetries. A formal Fe=Fe double bond in **Fe-1** corresponds to a 16-electron configuration for each iron atom rather than the normally favored 18-electron configuration. The LUMO of **Fe-1** appears to correspond to the iron s orbitals suggesting that the s orbitals are not involved in either the iron–iron bonding or the benzene–iron bonding. This is consistent with an eight valence orbital d^5p^3 manifold corresponding to a 16-electron configuration of each iron atom.

3.4.2 *Molecular orbitals in the singlet coaxial $(C_6H_6)_2Mn_2$ structure (Fig. 9)*

For the coaxial $(C_6H_6)_2Mn_2$ structure **Mn-1**, the degenerate HOMO-2/HOMO-3 pair corresponds to the two δ bonding components of the Mn–Mn bond. However, these bonding interactions are canceled by the degenerate HOMO/HOMO-1 pair corresponding to the two δ^* antibonding components of the Mn–Mn bond so that the Mn–Mn multiple bond in **Mn-1** has no δ components. The degenerate HOMO-4/HOMO-5 molecular orbital pair in **Mn-1** (Fig. 9) corresponds to two orthogonal π bonding components of the Mn–Mn multiple bond. These orthogonal Mn–Mn π bonding components use manganese $\{d(xz), p(x)\}$ hybrid orbitals for one of the π components and manganese $\{d(yz), p(y)\}$ hybrid orbitals for the other π component. The degenerate HOMO-7/HOMO-8 molecular orbital pair corresponds to the benzene–manganese π bonding components, also using manganese $\{d(xz), p(x)\}$ and $\{d(yz), p(y)\}$ hybrid orbitals.

The MO pattern for the π and δ bonding in the coaxial $(C_6H_6)_2Mn_2$ structure **Mn-1** thus is similar to that in the coaxial $(C_6H_6)_2Fe_2$ structure **Fe-1**. However, a significant difference is found in the σ bonding in the two structures. For **Mn-1**, HOMO-6 corresponds to the σ component of the Mn–Mn bond using the metal $\{d(z^2), p(z)\}$ orbitals with a shape suggesting major involvement of the manganese $d(z^2)$ orbital (Fig. 9). However, for **Mn-1**, unlike **Fe-1**, the corresponding σ^* antibonding orbital is the empty LUMO

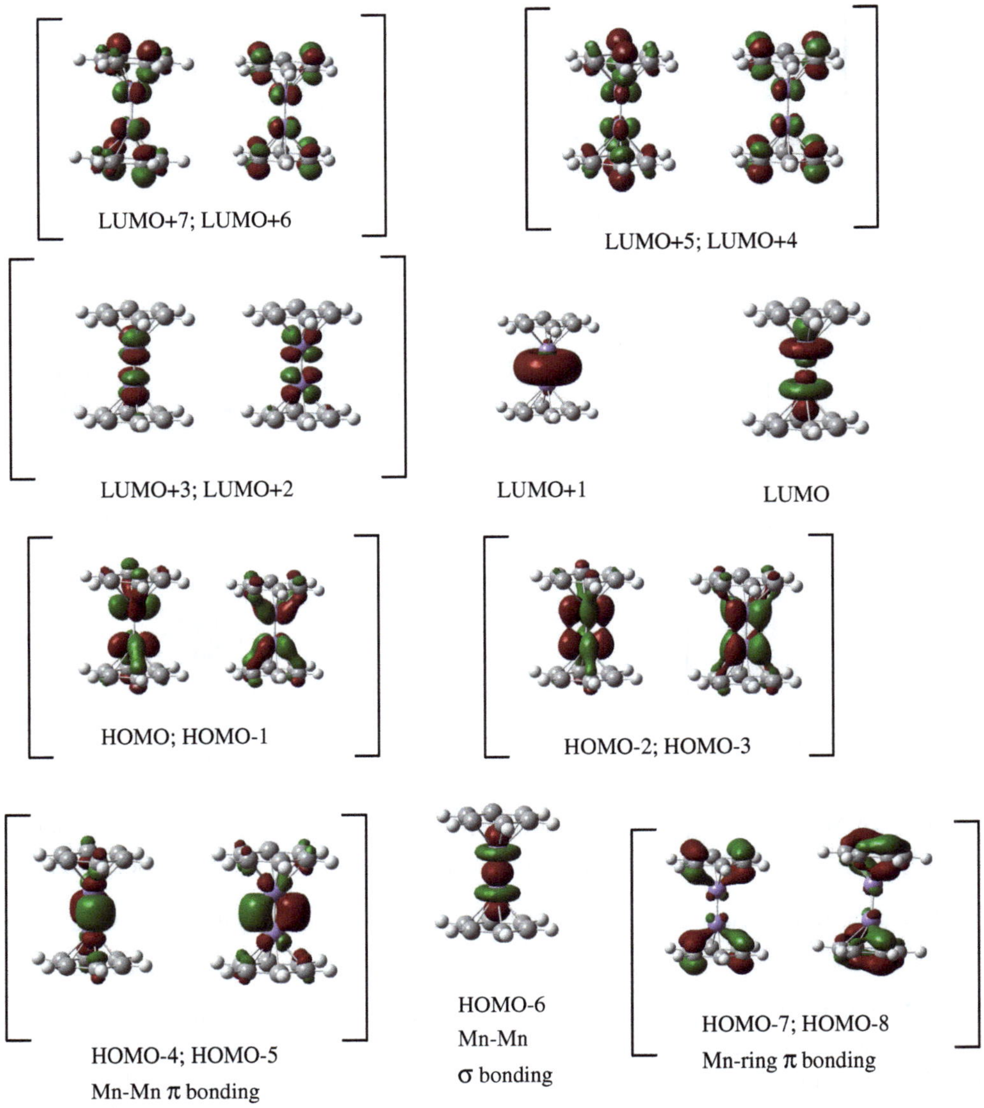

Fig. 9 Active MOs related to the two manganese atoms in the coaxial $(C_6H_6)_2Mn_2$ structure **Mn-1**. Pairs of degenerate orbitals are enclosed in brackets

orbital so that there are no σ^* antibonding electrons to cancel the σ bonding component of the Mn–Mn bond. This analysis of the frontier MOs thus suggest a $\sigma + 2\pi$ formal Mn$\equiv$Mn triple bond in **Mn-1**, thereby giving each metal atom a 16-electron configuration similar to that in **Fe-1**. The empty LUMO + 1 of **Mn-1** appears to correspond to the manganese s orbitals suggesting that the s orbitals are not involved in either the manganese–manganese or the benzene–manganese bonding. This again is consistent with an eight valence orbital d^5p^3 manifold corresponding to a 16-electron configuration of each iron atom. In fact, the differences between the frontier MOs of the coaxial $(C_6H_6)_2M_2$ (M = Mn, Fe) relate completely to the presence of two fewer valence electrons in the manganese system relative to the iron system.

3.4.3 Molecular orbitals in the singlet coaxial $(C_6H_6)_2Cr_2$ structure (Fig. 10)

The analysis of the chromium–chromium bonding in the singlet coaxial $(C_6H_6)_2Cr_2$ structure **Cr-2** is somewhat more complicated than that for **Fe-1** and **Mn-1** owing to the lower D_{2h} symmetry of **Cr-2** relative to the D_{6h} symmetry of **Fe-1** and **Mn-1**. The HOMO-5 in the coaxial $(C_6H_6)_2Cr_2$ structure **Cr-2** corresponds to the σ component of the Cr–Cr bond using the metal $\{d(z^2), p(z)\}$ orbitals (Fig. 10). The degenerate HOMO-3/HOMO-4 molecular orbital pair in **Cr-2** (Fig. 9) corresponds to two orthogonal π bonding components of the Cr–Cr multiple bond. These orthogonal Cr–Cr π bonding components use chromium $\{d(xz), p(x)\}$ hybrid orbitals for one of the π components

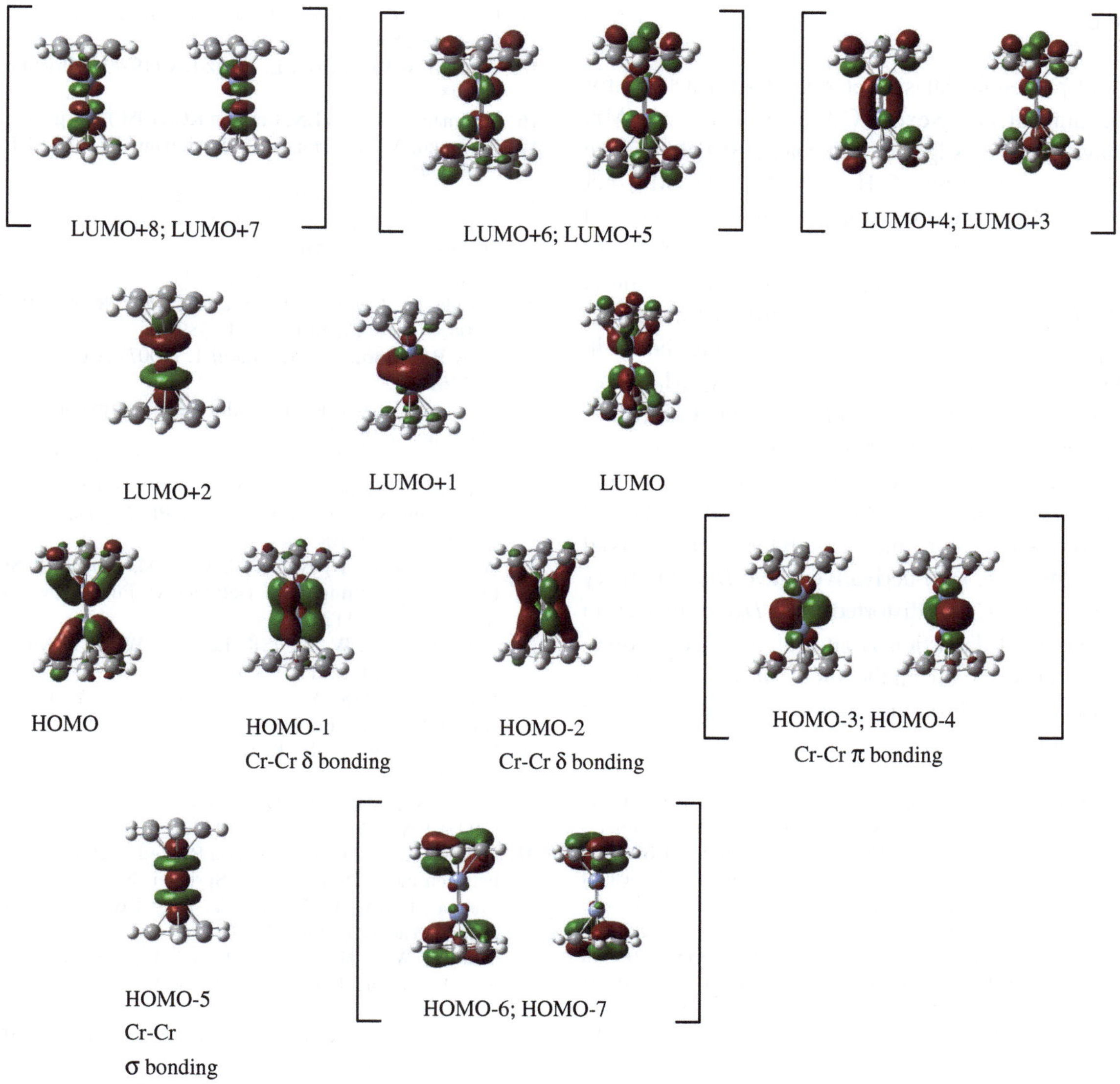

Fig. 10 Active MOs related to the two chromium atoms in the coaxial $(C_6H_6)_2Cr_2$ structure **Cr-2**. Pairs of degenerate orbitals are enclosed in brackets

and chromium $\{d(yz), p(y)\}$ hybrid orbitals for the other π component. The degenerate HOMO-6/HOMO-7 molecular orbital pair corresponds to the benzene–chromium π bonding components, also using chromium $\{d(xz), p(x)\}$ and $\{d(yz), p(y)\}$ hybrid orbitals.

The δ components HOMO-1 and HOMO-2 of the Cr–Cr bond in the coaxial $(C_6H_6)_2Cr_2$ structure **Cr-2** (Fig. 10) are distinctly different from those in the coaxial $(C_6H_6)_2M_2$ (M = Fe, Mn) structures discussed above. Firstly, they are non-degenerate because of the lower D_{2h} symmetry of coaxial $(C_6H_6)_2Cr_2$ **Cr-2** relative to the coaxial $(C_6H_6)_2M_2$ (M = Fe, Mn) structures with higher D_{6h} symmetry. The HOMO is a δ^* Cr–Cr antibonding orbital canceling out one of the two δ components of the Cr–Cr bond. This leaves one δ component to supplement the $\sigma + 2\pi$ components of the Cr–Cr bond leading to a formal $\sigma + 2\pi + \delta$ quadruple bond. This gives each chromium atom in **Cr-2** a 16-electron configuration similar to the metal atom configurations in the other coaxial $(C_6H_6)_2M_2$ (M = Fe, Mn) derivatives **Fe-1** and **Mn-1**. The diversion of one of the chromium $\{d(xy), d(x^2 - y^2)\}$ orbitals from metal-ring back-bonding to a single δ component of the Cr–Cr quadruple bond in **Cr-2** has the following two effects:

1. The benzene–metal bond is weakened as reflected by an increase in the M-C (benzene) distances from ~2.04 to ~2.08 Å in **Fe-1** and **Mn-1**, respectively, to ~2.13 and ~2.22 Å in the $(C_6H_6)_2Cr_2$ structure **Cr-2**.
2. The use of one of the metal δ orbitals of the $\{d(xy), d(x^2 - y^2)\}$ for metal–benzene back-bonding and the other metal δ orbital for the Cr–Cr quadruple bond in **Cr-2** lifts their degeneracy, thereby reducing the symmetry from D_{6h} in **Fe-1** and **Mn-1** to D_{2h} in **Cr-2**.

4 Summary

Coaxial and perpendicular singlet spin state structures for dibenzene dimetal complexes $(C_6H_6)_2M_2$ (M = Fe, Mn, and Cr) have been investigated. The coaxial structure is the lower energy structure for $(C_6H_6)_2M_2$ (M = Fe, Mn) but not for $(C_6H_6)_2Cr_2$. These coaxial structures are predicted to have very short M–M distances of $\sim$1.98 Å for $(C_6H_6)_2Fe_2$, $\sim$1.75 Å for $(C_6H_6)_2Mn_2$, and $\sim$1.68 Å for $(C_6H_6)_2Cr_2$. Investigation into the frontier molecular orbitals suggests a formal 2π Fe=Fe double bond in $(C_6H_6)_2Fe_2$, a $\sigma + 2\pi$ Mn≡Mn triple bond in $(C_6H_6)_2Mn_2$, and a $\sigma + 2\pi + \delta$ Cr–Cr quadruple bond in $(C_6H_6)_2Cr_2$. This gives each metal atom in these coaxial $(C_6H_6)_2M_2$ derivatives (M = Fe, Mn, Cr) a 16-electron configuration suggesting an 8-orbital d^5p^3 metal valence orbital manifold without involvement of the s orbital. The coaxial $(C_6H_6)_2M_2$ (M = Fe, Mn) derivatives have D_{6h} symmetry but coaxial $(C_6H_6)_2Cr_2$ is distorted from D_{6h} symmetry to D_{2h}. This observed distortion is attributed to the involvement of only one orbital from the $\{d(xy), d(x^2 - y^2)\}$ set of δ symmetry of each chromium atom in the formal Cr–Cr quadruple bond.

Acknowledgments We are grateful for financial support from the China Scholarship Council, and hospitality of Center for Computational Quantum Chemistry of the University of Georgia, USA. We also acknowledge financial support from the Fundamental Research Funds for the Central Universities (Grant SWJTU12CX084), the China National Science Foundation (Grant 11174237), the Sichuan Province, Applied Science and Technology Project (Grant 2013JY0035), the open research fund of the Key Laboratory of Advanced Scientific Computation, Xihua University (Grant: szjj2012-035), and the U.S. National Science Foundation (Grants CHE-1057466 and CHE-1054286).

References

1. Cotton FA, Walton RA (1993) Multiple bonds between metal atoms, 1–27, 2nd edn. Clarendon, Oxford
2. Radius U, Breher F (2006) Angew Chem Int Ed 45:3006
3. Cotton FA, Harris CB (1965) Inorg Chem 4:330
4. Cotton FA (1965) Inorg Chem 4:334
5. Langmuir I (1921) Science 54:59
6. Bose DM (1926) Z Phys 219
7. Reiff F (1931) Z Anorg Allg Chem 202:375
8. Sidgwick NV, Bailey RW (1934) Proc Roy Soc London A144:521
9. Pyykkö P (2006) J Organomet Chem 691:4336
10. King RB, Bisnette MB (1967) J Organomet Chem 8:287
11. Huang JS, Dahl LF (1983) J Organomet Chem 243:57
12. Curtis MD, Butler WM (1978) J Organomet Chem 155:131
13. King RB, Efraty A, Douglas WM (1973) J Organomet Chem 60:125
14. Potenza J, Giordano P, Mastropaolo D, Efraty A (1974) Inorg Chem 13:2540
15. Cotton FA, Kruczynski L, Frenz BA (1978) J Organomet Chem 160:93
16. Huffman JC, Lewis LN, Caulton KG (1980) Inorg Chem 19:2755
17. Herrmann WA, Serrano R, Weichmann J (1983) J Organomet Chem 246:C57
18. Hoyano JK, Graham WAG (1982) Chem Comm 27
19. Nguyen T, Sutton AD, Brynda M, Fettinger JC, Long GJ, Power PP (2005) Science 310:844
20. Frenking G (2005) Science 310:796
21. Brynda M, Gagliardi L, Widmark PO, Power PP, Roos BO (2006) Angew Chem Int Ed 45:3804
22. Roos BO, Borin AC, Gagliardi L (2007) Angew Chem Int Ed 46:1469
23. Merino G, Donald KJ, D'Acchioli JS, Hoffmann R (2007) J Am Chem Soc 129:15295
24. Brynda M, Gagliardi L, Roos BO (2009) Chem Phys Lett 471:1
25. Wagner FR, Noor A, Kempe R (2009) Nat Chem 1:529
26. La Macchia G, Gagliardi L, Power PP, Brynda M (2008) J Am Chem Soc 130:5104
27. Wolf R, Ni C, Nguyen T, Brynda M, Long GJ, Sutton AD, Fischer RC, Fettinger JC, Hellman M, Pu L, Power PP (2007) Inorg Chem 46:11277
28. Tsai YC, Hsu CW, Yu JSK, Lee GH, Wang Y, Kuo TS (2008) Angew Chem Int Ed 47:7250
29. Hsu CW, Yu JSK, Yen CH, Lee GH, Wang Y, Tsai YC (2008) Angew Chem Int Ed 47:9933
30. Noor A, Wagner FR, Kempe R (2008) Angew Chem Int Ed 47:7246
31. Kreisel KA, Yap GPA, Dmitrenko O, Landis CR, Theopold KH (2007) J Am Chem Soc 129:14162
32. Cotton FA, Murillo CA, Walton RA (eds) (2005) Multiple bonds between metal atoms, 35–68. Springer, New York
33. Tang L, Luo Q, Li QS, Xie Y, King RB, Schaefer HF (2012) J Chem Theory Comput 8:862
34. Wang H, Wang HY, Sun ZH, King RB (2013) Chem Phys 421:49
35. Resa I, Carmona E, Gutierrez-Puebla E, Monge A (2004) Science 305:1136
36. Xu B, Li QS, Xie Y, King RB, Schaefer HF (2010) J Chem Theory Comput 6:735
37. Allegra A, Tettamanti Casagrande G, Immirzi A, Porri L, Vitulli G (1970) J Am Chem Soc 92:289
38. Becke AD (1993) J Chem Phys 98:5648
39. Lee C, Yang W, Parr RG (1988) Phys Rev B 37:785
40. Becke AD (1988) Phys Rev A 38:3098
41. Perdew JP (1986) Phys Rev B 33:8822
42. Dunning TH (1970) J Chem Phys 53:2823
43. Huzinaga S (1965) J Chem Phys 42:1293
44. Hood DM, Pitzer RM, Schaefer HF (1979) J Chem Phys 71:705
45. Frisch MJ et al (2009) Gaussian 09. Gaussian Inc., Wallingford, CT
46. Xie Y, Schaefer HF, King RB (2000) J Am Chem Soc 122:8746
47. Clouston LJ, Siedschlag RB, Rudd PA, Planas N, Hu S, Miller AD, Gagliardi L, Lu CC (2013) J Am Chem Soc 135:13142
48. Bondybey VE, English JH (1983) Chem Phys Lett 94:443
49. Sun Z, Schaefer HF, Xie Y, Liu Y, Zhong R (2013) Mol Phys 111:2523
50. Keith TA (2013) AIMAII (Version 13.11.04) TK Gristmill Software, Overland Park, KS, USA (aim.tkgristmill.com)

Theor Chem Acc (2013) 132:1408
DOI 10.1007/s00214-013-1408-y

REGULAR ARTICLE

All electron ab initio calculations on the ScTi molecule: a really hard nut to crack

Apostolos Kalemos · Aristides Mavridis

Received: 4 September 2013 / Accepted: 14 October 2013 / Published online: 5 November 2013

Abstract The lightest and yet completely unexplored intermetallic ScTi system has been studied by high-level ab initio methods and quantitative basis sets. We have studied in all 21 $^{2S+1}\Lambda$ states and constructed potential energy curves at the MRCI/cc–pvQZ computational level. The ground state of the system is of $^6\Delta$ symmetry with $r_e = 2.65$ Å and $D_e = 32.4$ kcal/mol with respect to the adiabatic fragments or $D_e^0 = 10.4$ kcal/mol with respect to the ground state atoms while its first excited state seems to be of $^4\Delta$ symmetry with similar bonding features lying only 3 kcal/mol higher. An exceptional feature is the rather high Mulliken charge (average value of $\sim$0.5 e$^-$) transferred to Sc but the small calculated dipole moments.

Keywords ScTi · Ab initio · Multireference

1 Introduction

The nine 3*d*-transition metal elements (M) from Sc ($Z = 21$) to Cu ($Z = 29$) comprise the most emblematic group of metals on Earth. As a matter of fact, Cu and Fe mark the dawn of human civilization, dubbed the Copper and Iron Ages, and perhaps the beginning of what we call today "Modern Chemistry." Ironically, and despite the explosive progress in the physical sciences, we cannot claim that we understand satisfactorily certain classes of "small" molecular systems. Prominent among them are the 3*d*-MM′ diatomics (M, M′ = Sc–Cu). Since the early 90 s, our research group has been engaged in an in-depth theoretical study of all 3*d*-M diatomics, $MX^{0,\pm}$, where X = H [1–3], B [4–8], C [9–27], O [28–33], F [34–36], Al [37], P [38–40], and Cl [41] through all electron high-level ab initio calculations. One of the most interesting features of the MX species is the extremely rich variety of the binding modes entailed due to the metals' 3*d* electrons, which result in a remarkable density of high spin and orbital angular momenta atomic ^{2S+1}L states. This large density of states is the driving force behind the wealthy chemistry of such systems and is also the root of all computational problems encountered, which still challenge all existing state-of-the art theoretical methods. Both features are enhanced in the 3*d*-dimers of either homo-(M_2) or hetero-(MM′) nuclear nature due to the simultaneous presence of two capricious atoms. There are $\binom{9}{2} = 36$ 3*d* intermetallic (MM′) and nine homonuclear diatomics (M_2), a total of 45 molecules each one with a personality of its own.

In the recent past, we have carefully examined through multireference variational and single reference methods, three homonuclear and one heteronuclear diatomics, namely Sc_2 [42–47],[1] Ti_2 [48], Mn_2 [49], and TiFe [50]. Our computational and conceptual difficulties with these systems are well described in the above four references, so

Dedicated to Professor Thom Dunning and published as part of the special collection of articles celebrating his career upon his retirement.

A. Kalemos (✉) · A. Mavridis
Department of Chemistry, Laboratory of Physical Chemistry, National and Kapodistrian University of Athens, Panepistimiopolis, 157 71 Athens, Greece
e-mail: kalemos@chem.uoa.gr

[1] It is interesting to follow the discussion on the ground state of Sc_2, $^5\Sigma_u^-$ versus $^3\Sigma_u^-$, and the theoretical methods involved. The dispute was triggered by Maxtain et al. [47] who erroneously suggested that the ground state of Sc_2 is $^3\Sigma_u^-$. The controversy was settled in reference [42], see also [44].

there is no need for repetition. One thing, however, is clear: the M_2 (or MM′) class is an unusual set of molecules, challenging, therefore interesting, and undisciplined to the conventional chemical wisdom. While the Cr (4 $s^1 3d^5$, ^{7}S) and Cu ($4s^1 3d^{10}$, ^{2}S) atoms whose open structure shows a proclivity for M_2 or MM′ bonding, the 4 $s^2 3d^n$ (n = 1–3 and 5–8) ground state configurations of the remaining 3d metals with ratios of mean 4s and 3d radii $\frac{\langle r_{4s}\rangle}{\langle r_{3d}\rangle}$ ranging from 2.36 to 3.36 [51] do not allow for direct, non-van der Waals bonding [42, 48–50]. Hence, covalent interaction can only be expected from at least a $4s^2$–$4s^1$ distribution, i.e., an M′–M* interaction where the star (*) refers to an excited state atom.

Presently, we have examined the first that is the lightest intermetallic MM′ diatomic ScTi. We are not aware of any experimental or ab initio investigation on ScTi but a DFT study by Gutsev et al. [52] who studied all ScM (M = Sc–Zn) dimers and their anions. Based on a [10s7p4d1f] Gaussian basis set and through the generalized gradient approximation for the exchange correlation potential, they found a $^6\Sigma^+$ (r_e = 2.290 Å, D_e = 2.22 eV, ω_e = 272 cm^{-1}, μ = 0.57 D) ground state with a $^2\Sigma^+$ (r_e = 2.065 Å, ω_e = 382 cm^{-1}, μ = 1.07 D) state lying just 0.16 eV higher. An NBO analysis revealed that in both states, the Sc and Ti atoms are in situ excited in their $4s^1 3d^2$ and $4s^1 3d^3$ electronic configurations, respectively.

The interaction of the ground state fragments Sc ($4s^2 3d^1$, ^{2}D) + Ti ($4s^2 3d^2$, ^{3}F) results in 40 molecular states, that is, ^{2,4}(Σ^+[2], Σ^-[3], Π[5], Δ[4], Φ[3], Γ[2], H[1]) doublets and quartets, expected to be of van der Waals type [42, 48, 49]. The first excited dissociation channel Sc ($4s^2 3d^1$, ^{2}D) + Ti ($4s^1 3d^3$, ^{5}F) located 0.806 eV higher gives rise also to 40 molecular $^{2S+1}\Lambda$ states of the same spatial symmetry as before but of quartet and sextet spin multiplicities. Finally, the interaction of two excited atoms, i.e., Sc ($4s^1 3d^2$, ^{4}F) + Ti ($4s^1 3d^3$, ^{5}F) at 0.806 [ΔE_{Ti}(^{5}F ← ^{3}F)] + 1.427 [ΔE_{Sc}(^{4}F ← ^{2}D)] = 2.233 eV [53] above the ground state fragments, results in 112 molecular $^{2S+1}\Lambda$ states of ^{2,4,6,8}(Σ^+[4], Σ^-[3], Π[6], Δ[5], Φ[4], Γ[3], H[2], I) symmetry. Between the first dissociation channel [Sc ($4s^2 3d^1$, ^{2}D) + Ti ($4s^2 3d^2$, ^{3}F)] and the one that combines two excited fragments [Sc ($4s^1 3d^2$, ^{4}F) + Ti ($4s^1 3d^3$, ^{5}F)], there are in total 17 channels producing hundreds of molecular states within an asymptotic energy range of only 2.233 eV [53].

Based on preliminary calculations at the complete active space self-consistent field (CASSCF) computational level, we decided to study states of quartet and sextet spin multiplicity with spatial symmetry ranging from Σ^+ to H (Λ = 5). For both metal atoms, the Dunning [54] correlation-consistent polarized valence quadruple zeta (cc–pVQZ) basis set was used generally contracted to [8s7p5d3f2g1h] [55], consisting of 104 × 2 = 208 spherical Gaussian functions. The internally contracted (*ic*) variant of the multireference CISD (MRCI = CASSCF + 1 + 2) method was used in all our calculations as implemented in the MOLPRO 2010.1 code [56]. The one electron basis set was issued from state average CASSCF wave functions that in turn resulted by distributing all 7 valence electrons to 15 orbitals correlating with the (4s + 3d) × 2 + p'(3) valence space of the fragments supplemented by three p type functions to "angularly" correlate the "4s" electrons. During the CASSCF optimization, a number of states from all four irreducible representations (A_1, A_2, B_1, B_2) of the C_{2v} point group were averaged in order to assure a smooth evolution along the inter nuclear distance. For the quartet spin multiplicity, the reference space consisted of ~60,000 configuration functions (CF) giving rise to ~255 × 10^6 CFs at the MRCI level, internally contracted to ~12 × 10^6 CFs. For the sextet spin multiplicities, the corresponding spaces contain ~17,000 (CASSCF), ~91 × 10^6 (MRCI), and ~8 × 10^6 (*ic*MRCI) CFs. Due to severe technical complications, only parts of the potential energy curves (PEC) could be calculated. No core–subvalence ($3s^2 3p^6$), relativistic, or spin–orbit effects were taken into account. Technical difficulties in the calculation only permitted the Davidson +Q correction to be applied to the $X\,^6\Delta$ ground state.

2 Results and discussion

Table 1 collects the numerical data for 21 states, while Figs. 1, 2, 3, and 4 display the corresponding PECs. In what follows, we discuss in some detail the bonding characteristics of the states studied grouped according to their symmetry.

X $^6\Delta$, 6 $^6\Sigma^+$, 10 $^6\Gamma$, 16 $^6\Delta$, 19 $^6\Delta$, 20 $^6\Sigma^+$, and 21 $^6\Gamma$. The ground state of the ScTi system seems to be of $^6\Delta$ symmetry. Its most important equilibrium configurations at the CASSCF level are (only the seven valence electrons appear in what follows).

$$\begin{aligned}|X\,^6\Delta(^6A_1)\rangle \approx\ & 0.76\left|1\sigma^2 2\sigma^1 3\sigma^1 1\pi_x^1 1\pi_y^1 1\delta_-^1\right\rangle \\ & + 0.21\left|1\sigma^2 2\sigma^1 5\sigma^1 (1\pi_x^1 2\pi_y^1 - 2\pi_x^1 1\pi_y^1) 1\delta_-^1\right\rangle \\ & + 0.18\left|1\sigma^2 2\sigma^1 5\sigma^1 1\pi_x^1 1\pi_y^1 1\delta_-^1\right\rangle\end{aligned}$$

with 1σ ≈ 0.75[4s(Sc)] + 0.71[4s(Ti)], 2σ ≈ − 0.66 [4s(Sc)] + 0.38[4p_z(Sc)] + 0.66[4s(Ti)] + 0.32[4p_z(Ti)], 3σ ≈ 0.39[3d_σ(Sc)] + 0.86[3d_σ(Ti)], 5σ ≈ 0.87[3d_σ(Sc)] − 0.48[3d_σ(Ti)], 1π ≈ 0.46[3d_π(Sc)] − 0.64[3d_π(Ti)], 2π ≈

Table 1 Total energies E (hartree), equilibrium distances r_e (Å), dissociation energies D_e (kcal/mol), harmonic frequencies ω_e and anharmonic corrections $\omega_e x_e$ (cm^{-1}), dipole moment μ (Debye), and energy separations T_e (kcal/mol) of 21 molecular states of $^{45}Sc^{48}Ti$ at the MRCI(+Q) level of theory

State	−E	r_e	D_e	ω_e	$\omega_e x_e$	μ	T_e
$X^6\Delta$	1,608.259 250	2.647	32.4	218.3	0.685	0.05	0.0
	(1,608.267 82)	(2.659)	36.8	(223.2)	(3.816)		(0.0)
$2\ ^4\Delta$	1,608.254 387	2.637	7.4	220.3	−2.458	0.65	3.05
$3\ ^4\Pi$	1,608.246 750	2.587	2.6	159.6	−0.955	0.16	7.84
$4\ ^4\Phi$	1,608.242 781	2.773		192.5	8.681		10.33
$5\ ^4\Phi$	1,608.240 477	2.943		239.4	0.825		11.78
$6\ ^6\Sigma^+$	1,608.239 6	2.96	20.1			0.10	12.3
(Local minimum)							
	1,608.236 578	2.440		238.2	−0.812		
$7\ ^4\Gamma$	1,608.239 276	2.787		196.2	3.643	0.24	12.53
$8\ ^4\Delta$	1,608.238 994	2.927		231.8	5.292	0.27	12.71
$9\ ^4\Pi$	1,608.238 172	2.351		300.6	0.131	−0.06	13.23
(Local minimum)							
	1,608.238 054	2.708		201.8	0.994		
$10\ ^6\Gamma$	1,608.237 719	3.005	18.9	233.1	6.845	0.09	13.51
$11\ ^6\Pi$	1,608.237 008	2.940	18.4	180.0	3.299	0.08	13.96
$12\ ^6H$	1,608.236 484	2.981	18.1	213.5	2.307	−0.14	14.29
$13\ ^4\Sigma^+$	1,608.236 155	2.775		195.6	2.764	0.29	14.49
$14\ ^6\Phi$	1,608.235 912	2.969	17.7	221.1	7.237	0.28	14.64
$15\ ^6\Pi$	1,608.235 722	2.960	17.6	211.3	2.353	0.19	14.76
$16\ ^6\Delta$	1,608.233 835	2.969	16.4	211.0	4.012	0.21	15.95
$17\ ^6\Pi$	1,608.232 718	2.993	15.7	200.1	−3.053	0.05	16.65
$18\ ^6\Phi$	1,608.232 450	2.985	15.6	219.0	2.935	0.01	16.82
$19\ ^6\Delta$	1,608.230 819	2.893	14.6	239.3	5.049	0.00	17.84
$20\ ^6\Sigma^+$	1,608.227 748	3.031	12.6	269.7	7.965		19.77
$21\ ^6\Gamma$	1,608.226 659	2.992	11.9				20.45

$[3d_\pi(Sc)] + [3d_\pi(Ti)]$, $1\delta_- \approx [3d_{xy}(Ti)]$. The MRCI Mulliken atomic distributions.

$$4s^{2.09}4p_z^{0.43}3d_{z^2}^{0.40}4p_x^{0.09}3d_{xz}^{0.40}4p_y^{0.09}3d_{yz}^{0.40}3d_{xy}^{0.05}/_{Sc},\ q_{Sc} = -0.98$$

$$4s^{0.29}4p_z^{0.07}3d_{z^2}^{0.60}4p_x^{0.08}3d_{xz}^{0.47}4p_y^{0.08}3d_{yz}^{0.47}3d_{xy}^{0.94}/_{Ti},\ q_{Ti} = +0.98$$

hint to an in situ Sc atom in its $4s^2 3d_{z^2}^1$ (^{2}D, $M_L = 0$) state and a Ti atom in its $4s^1 3d_{xz}^1 3d_{yz}^1 3d_{xy}^1$ (^{5}F, $M_L = \pm 2$) state, whereas the two atoms are bound together by two half π bonds due to a Ti-to-Sc charge transfer of 1 e^-, and a σ interaction of four electrons coupled into a triplet. The above bonding situation can be summarized pictorially by the following valence bond Lewis (vbL) scheme.

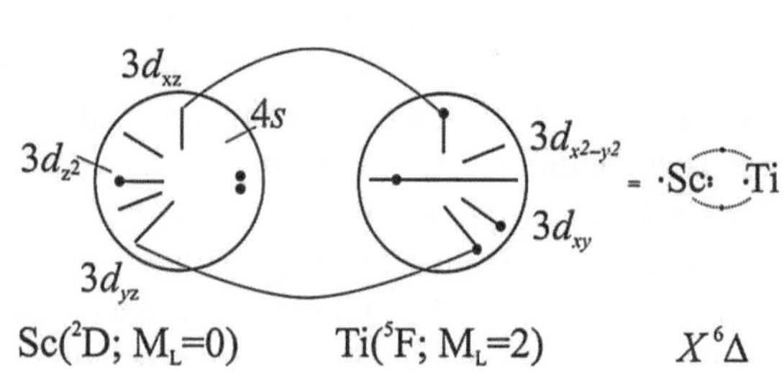

The MRCI(+Q) equilibrium distance and bond energy are $r_e = 2.647$ (2.659) Å and $D_e = 32.4$ (36.8) kcal/mol with respect to Sc($4s^2 3d^1$, ^{2}D) + Ti($4s^1 3d^3$, ^{5}F); see Table 1. Its MRCI (+Q) binding energy with respect to the ground state atoms is $D_e^0 = 10.4$ (15.3) kcal/mol. Its permanent electric dipole moment calculated as an expectation value is close to zero, $\mu = 0.052$ D, surprisingly small

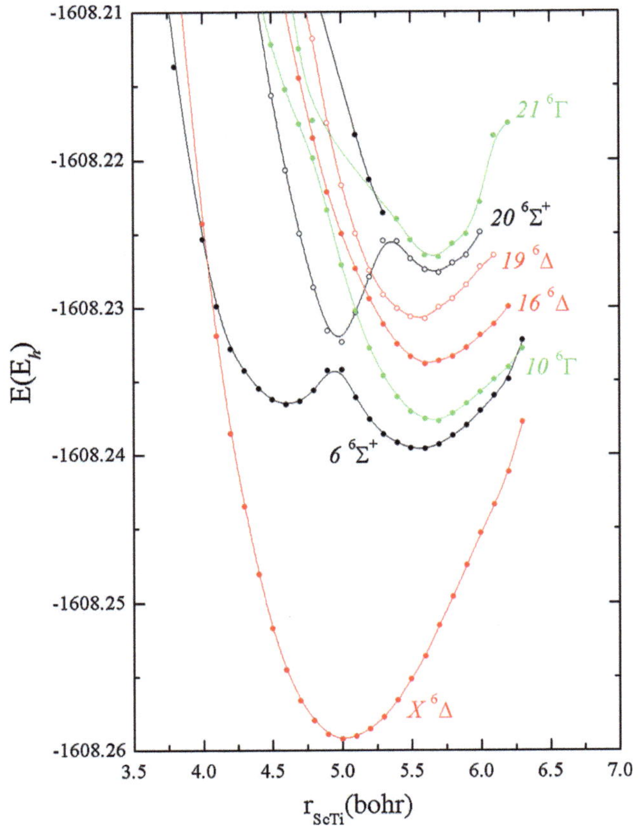

Fig. 1 MRCI/cc–pVQZ PECs of the $X\ ^6\Delta$, 6 $^6\Sigma^+$, 10 $^6\Gamma$, 16 $^6\Delta$, 19 $^6\Delta$, 20 $^6\Sigma^+$, and 21 $^6\Gamma$ states

if one considers the 1e$^-$ charge transfer from Ti to Sc. Perhaps the symmetry defining electron on the $3d_{xy}$ (Ti) orbital compensates for the centroid of charge on the Sc atom created by the charge transfer along the π-frame. It is interesting at this point to compare the ScTi $X\ ^6\Delta$ state to the Sc_2 $X\ ^5\Sigma_u^-$ [42] state. As was discussed in reference 42, the Sc_2 ground state results from a ground Sc ($4s^2 3d^1$, ^{2}D; $M_L = 0$) and an excited Sc* ($4s^1 3d^2$, ^{4}F; $M_L = 0$) atoms held together by a 3e$^-$-2c (enter) σ interaction and two half π delocalizations due to the inversion symmetry. When a $3d_{xy}$ electron is attached to the Sc* (^{4}F) state, the ScTi $X\ ^6\Delta$ state is obtained with similar bonding features.

Two more states of $^6\Delta$ symmetry have been studied, namely the 16 $^6\Delta$ and 19 $^6\Delta$ ones, both dissociating adiabatically to the Sc(^{2}D) + Ti(^{5}F) fragments. Their equilibrium CASSCF wave functions and corresponding MRCI Mulliken populations read.

$$|16^6\Delta(^6A_1)\rangle \approx \left|1\sigma^2 2\sigma^1[(0.52)5\sigma^1-(0.39)7\sigma^1]1\pi_x^1 1\pi_y^1 1\delta_-^1\right\rangle - 0.31\left|1\sigma^2 2\sigma^1 3\sigma^1(2\pi_x^1 1\pi_y - 1\pi_x^1 2\pi_y^1)1\delta_-^1\right\rangle$$

$$4s^{1.57}4p_z^{0.33}3d_{z^2}^{0.53}4p_x^{0.11}3d_{xz}^{0.31}4p_y^{0.11}3d_{yz}^{0.31}3d_{xy}^{0.10}/_{Sc},\ q_{Sc} = -0.39$$

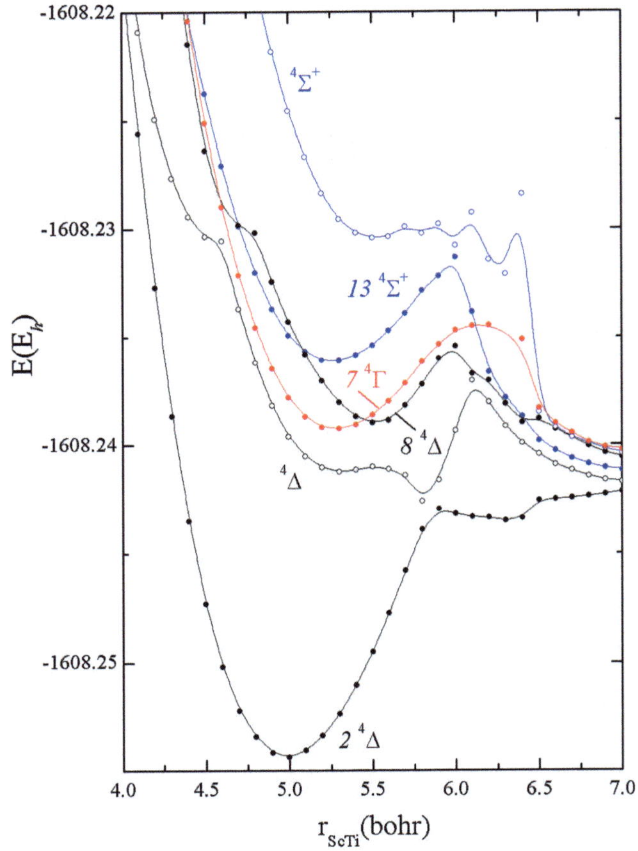

Fig. 2 MRCI/cc–pVQZ PECs of the 2 $^4\Delta$, 7 $^4\Gamma$, 8 $^4\Delta$, and 13 $^4\Sigma^+$ states

$$4s^{0.86}4p_z^{0.13}3d_{z^2}^{0.50}4p_x^{0.10}3d_{xz}^{0.50}4p_y^{0.10}3d_{yz}^{0.50}3d_{xy}^{0.90}/_{Ti},\ q_{Ti} = +0.39$$

and

$$|19^6\Delta(^6A_1)\rangle \approx 0.68\left|1\sigma^2 2\sigma^1 3\sigma^1 1\pi_x^1 1\pi_y^1 2\delta_-^1\right\rangle + 0.25\left|1\sigma^2 2\sigma^1 3\sigma^1(1\pi_x^1 2\pi_y^1 - 2\pi_x^1 1\pi_y^1)2\delta_-^1\right\rangle$$

$$4s^{1.46}4p_z^{0.37}3d_{z^2}^{0.20}4p_x^{0.10}3d_{xz}^{0.11}4p_y^{0.10}3d_{yz}^{0.11}3d_{xy}^{0.86}/_{Sc},\ q_{Sc} = -0.33$$

$$4s^{0.94}4p_z^{0.13}3d_{z^2}^{0.70}4p_x^{0.07}3d_{xz}^{0.68}4p_y^{0.07}3d_{yz}^{0.68}3d_{xy}^{0.22}/_{Ti},\ q_{Ti} = +0.33$$

The bonding character in both the 16 and 19 $^6\Delta$ states remains the same as in the $X\ ^6\Delta$ state; however, their orthogonality imposes a number of changes. The main CASSCF configuration of the $X\ ^6\Delta$ state ($C_0 = 0.76$) is the $\left|1\sigma^2 2\sigma^1 3\sigma^1 1\pi_x^1 1\pi_y^1 1\delta_-^1\right\rangle$. In 16 $^6\Delta$, the 3σ orbital has been replaced by the [(0.52) 5σ–(0.39) 7σ] orthogonal orbital (5σ ≈ 0.63[$3d_\sigma$(Sc)] + 0.24[$4p_z$(Sc)] − 0.55[$3d_\sigma$(Ti)] − 0.26[$4p_z$(Ti)] and 7σ ≈ 0.67[$3d_\sigma$(Sc)] − 0.53[$4p_z$(Sc)] +

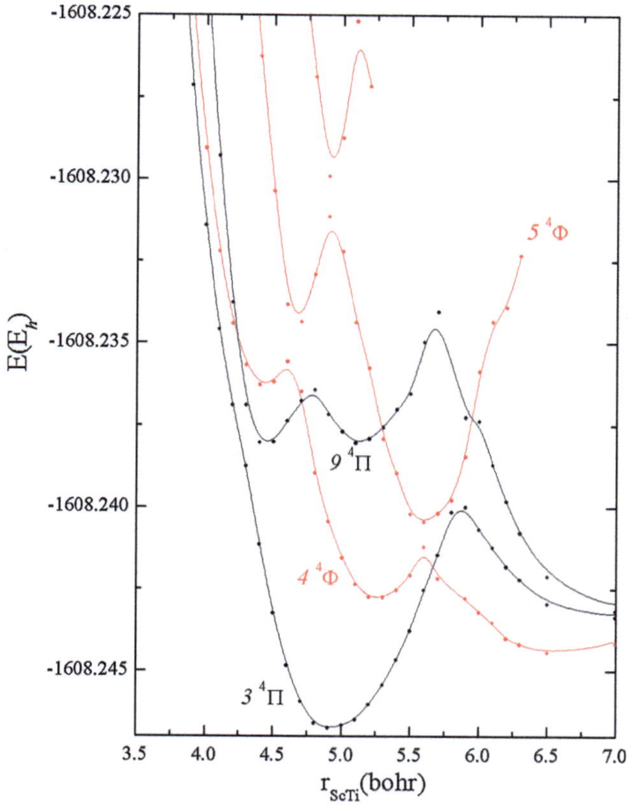

Fig. 3 MRCI/cc-pVQZ PECs of the 3 $^4\Pi$, 4 $^4\Phi$, 5 $^4\Phi$, and 9 $^4\Pi$ states

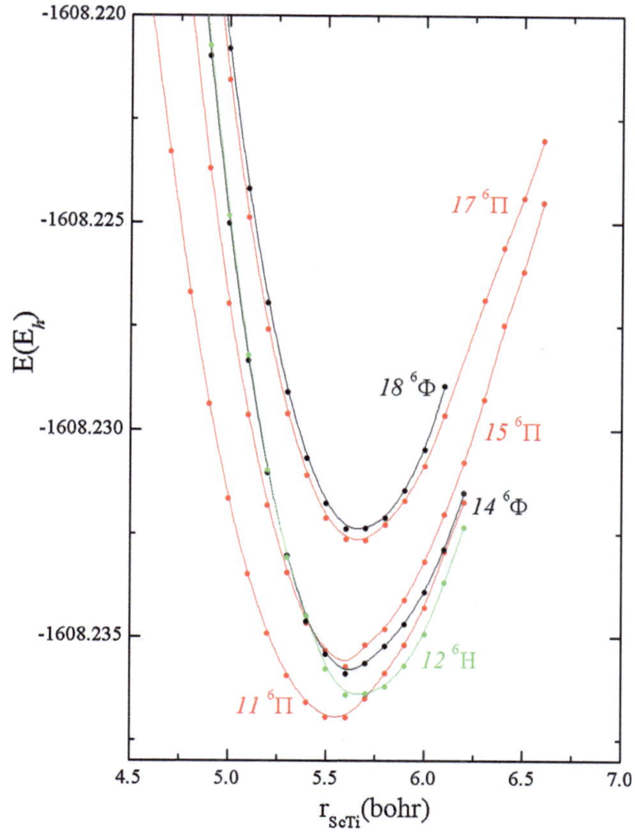

Fig. 4 MRCI/cc-pVQZ PECs of the 11 $^6\Pi$, *12* ^{6}H, 14 $^6\Phi$, 15 $^6\Pi$, 17 $^6\Pi$, and 18 $^6\Phi$ states

0.32[$4p_z$(Ti)]), while in the 19 $^6\Delta$ state, the $2\delta_-$ orbital is found localized on the Sc atom instead of the Ti ($X\,^6\Delta$) one, or in other words, the transition $X\,^6\Delta \rightarrow 19\,^6\Delta$ is made possible through the transfer of the symmetry defining δ e^- from the Ti to the Sc atom.

In the 16 $^6\Delta$ state, two half π bonds are created through the transfer of $0.40 \times 2 = 0.80$ e^- from Ti to Sc, while 0.43 e^- are back transferred through the σ frame. The Mulliken populations reveal an in situ Sc atom in its $4s^23d^1$ (^{2}D; $M_L = 0$) electronic configuration as was also found in the X state, while the Ti atom defines the $\Lambda = 2$ symmetry through its $4s^13d^1_{xz}3d^1_{yz}3d^1_{xy}$ configuration. This character resemblance should be the product of an avoided crossing since the 16 $^6\Delta$ state should correlate adiabatically with a different combination of fragments, see Table 1 for numerical results.

The main characteristic of the 19 $^6\Delta$ state is the different location of the symmetry defining electron as contrasted to both the X and 16 $^6\Delta$ states. The Mulliken populations show an in situ Sc atom in its ground $4s^23d^1$ (^{2}D; $M_L = \pm 2$) configuration, while the Ti atom is found in its $4s^13d^1_{xz}3d^1_{yz}3d^1_{\sigma}$ configuration, a 20 %|^{5}F> +80 %|^{5}P> mixture. The π Ti-to-Sc migration is limited to only $0.20 \times 2 = 0.40$ e^-, while a weak δ-bond is created through a 0.2 e^- transfer from Sc to Ti. A $(4s4p_z3d_{z^2})$Sc hybrid helps to relieve the congestion of four σ electrons. Both states appear to have similar equilibrium distances and bond strengths: $r_e = 2.969$ (16 $^6\Delta$) and 2.893 (19 $^6\Delta$) Å, $D_e = 16.4$ (16 $^6\Delta$) and 14.6 (19 $^6\Delta$) kcal/mol with respect to Sc ($4s^23d^1$, ^{2}D) + Ti ($4s^13d^3$, ^{5}F).

The next states to be discussed are of $^6\Sigma^+$ and $^6\Gamma$ symmetry. The 6 $^6\Sigma^+$ state appears to have two minima of completely different electronic character, Fig. 1. The adiabatic PECs of the two $^6\Sigma^+$ states when properly rearranged define two diabatic PECs that should be considered simultaneously by a 2×2 post-Born–Oppenheimer treatment in order to extract the rovibrational levels of the system. The CASSCF equilibrium configurations of the global 6 $^6\Sigma^+$ minimum are

$$|6^6\Sigma^+\rangle_g \approx 0.54\left|1\sigma^22\sigma^1(1\delta^1_+1\pi^1_x1\pi^1_y2\delta^1_- + 2\delta^1_+1\pi^1_x1\pi^1_y1\delta^1_-)\right\rangle$$

which in connection to its MRCI atomic distributions

$$4s^{1.48}4p_z^{0.34}3d_{z^2}^{0.07}3d_{x^2-y^2}^{0.44}4p_x^{0.09}3d_{xz}^{0.16}4p_y^{0.09}3d_{yz}^{0.16}3d_{xy}^{0.47}/_{Sc},\quad q_{Sc} = -0.34$$

$$4s^{0.85}4p_z^{0.13}3d_{z^2}^{0.09}3d_{x^2-y^2}^{0.53}4p_x^{0.08}3d_{xz}^{0.70}4p_y^{0.08}3d_{yz}^{0.70}3d_{xy}^{0.50}/_{Ti},\quad q_{Ti} = +0.34$$

suggest a $|^2D,\ M_L = \pm 2\rangle_{Sc} \otimes |^5F,\ M_L = \mp 2\rangle_{Ti}$ asymptotic combination. The two atoms interact attractively by a

$3e^{-}$–2c σ interaction between a $(4s4p_z3d_{z^2})^{1.9}_{Sc}$ and a $(4s4p_z3d_{z^2})^{1.07}_{Ti}$ hybrids and a 0.25 × 2 = 0.50 e^- Ti-to-Sc π migration, while two δ electrons, one at each atom, maintain the Σ character of the $1\sigma^2 2\sigma^1 1\pi_x^1 1\pi_y^1$ structure.

The 10 $^6\Gamma$ state is the orthogonal complement of the $^6\Sigma^+$ global minimum, their only difference being a sign change "−" ($^6\Gamma$) instead of a "+" ($^6\Sigma^+$) in their main CASSCF configurations and with identical Mulliken populations. Their asymptotic channel is evidently the same as before but with equal M_L projections, i.e., $|^2D,\ M_L = \pm 2\rangle_{Sc} \otimes |^5F,\ M_L = \pm 2\rangle_{Ti}$, so as to result in $\Lambda = 4$.

The leading equilibrium configurations of the local 6 $^6\Sigma^+$ minimum are

$$|6^6\Sigma^+\rangle_l \approx \left|1\sigma^2[(0.11)2\sigma^1 + (0.62)3\sigma^1]1\delta_+^1 1\pi_x^1 1\pi_y^1 1\delta_-^1\right\rangle - 0.35\left|1\sigma^2 3\sigma^1(1\delta_+^1 1\pi_x^1 1\pi_y^1 2\delta_-^1 + 2\delta_+^1 1\pi_x^1 1\pi_y^1 1\delta_-^1)\right\rangle$$

The "0.35" component is similar to the electronic character of the global 6 $^6\Sigma^+$ minimum, while the "first" component results from that of the $X\,^6\Delta$ state by replacing a σ electron by a δ_+ one. This local minimum is a "resonant" hybrid between two structures whose main difference is the location of the $\delta_\pm$ electrons. In the "first" component, both are placed on Ti while in the second ("0.35") a δ electron is on Sc and the other one on Ti.

Two more states of $^6\Sigma^+$ and $^6\Gamma$ symmetry separated by ~20 kcal/mol from the $X\,^6\Delta$ state have also been found. They are of intense multi configuration character that precludes an easy VB transcription of their bonding characteristics.

2 $^4\Delta$, 7 $^4\Gamma$, 8 $^4\Delta$, and 13 $^4\Sigma^+$. The first excited state of the ScTi system is of $^4\Delta$ symmetry with bonding features practically identical to those of the $X\,^6\Delta$ state lying just 3 kcal/mol higher. The main CASSCF configurations read as follows

$$\begin{aligned}|2\,^4\Delta(^4A_1)\rangle \approx\ & 0.56\left|1\sigma^2 2\sigma^1 3\bar{\sigma}^1 1\pi_x^1 1\pi_y^1 1\delta_-^1\right\rangle \\ & + 0.45\left|1\sigma^2 2\sigma^1 3\sigma^1 1\bar{\pi}_x^1 1\pi_y^1 1\delta_-^1\right\rangle \\ & + 0.32\left|1\sigma^2 2\sigma^1 3\sigma^1 1\pi_x^1 1\bar{\pi}_y^1 1\delta_-^1\right\rangle \\ & + 0.14\left|1\sigma^2 2\sigma^1 3\sigma^1 1\pi_x^1 1\pi_y^1 1\bar{\delta}_-^1\right\rangle.\end{aligned}$$

All four configurations differ from the "0.76" CF of the $X\,^6\Delta$ state only in the way spins are coupled, both molecular orbitals and atomic distributions/charges being the same. So, it is certainly not a surprise that the $^4\Delta$ state is the first excited state of this diatomic. Both r_e (2.637 vs 2.647 (X) Å) and ω_e (220.3 vs 218.3 (X) cm^{-1}) values are quite similar, an indication of the same binding mode. The only stunning difference is the dipole moment value $\mu = 0.65$ D versus $\mu = 0.05$ D of the X state, an amazing side effect of the quartet spin multiplicity.

In Fig. 2, the adiabatic potential energy curve of a second $^4\Delta$ state is shown but with an irregular topology. This "fuzziness" will play certainly a role in the complexity of the spectrum, but offers little in the discussion of the chemical bond of the system. One of its most important characteristics is the avoided crossing with the 2 $^4\Delta$ state at 5.9 bohr (Fig. 2), responsible for the excited in situ Ti character. The last state of $(8)^4\Delta$ symmetry presently studied is unbound with respect to its adiabatic dissociation limit, i.e., Sc (^{2}D) + Ti (^{3}F), but features a potential well of 2.2 kcal/mol, sustaining some vibrational levels. Its electronic character is of extreme multiconfigurational character, thus forbiding any attempt to transcribe its bonding nature into simple vbL structures.

Two more states of quartet spin multiplicity of Γ and Σ^+ spatial symmetry have been studied. Their PECs shown in Fig. 2 reveal a parallel evolution within the minimum energy region. This is also corroborated by the similarity of the r_e (2.787 ($^4\Gamma$) versus 2.775 ($^4\Sigma^+$) Å) and ω_e (196.2 ($^4\Gamma$) versus 195.6 ($^4\Sigma^+$) cm^{-1}) values. Their main CASSCF equilibrium CFs are

$$\begin{aligned}|7^4\Gamma(^4A_1)\rangle \approx\ & 0.38\left|1\sigma^2 2\sigma^1 3\sigma^1 1\delta_+^1(1\pi_x^2 - 1\pi_y^2)\right\rangle \\ & + 0.31\left|1\sigma^2 2\sigma^1 3\sigma^1 1\bar{\pi}_x^1 1\pi_y^1 1\delta_-^1\right\rangle \\ & - 0.44\left|1\sigma^2 2\sigma^1 3\sigma^1 1\pi_x^1 1\bar{\pi}_y^1 1\delta_-^1\right\rangle\end{aligned}$$

and

$$\begin{aligned}|13^4\Sigma^+\rangle \approx\ & 0.36\left|1\sigma^2 2\sigma^1 3\sigma^1 1\delta_+^1(1\pi_x^2 - 1\pi_y^2)\right\rangle \\ & - 0.29\left|1\sigma^2 2\sigma^1 3\sigma^1 1\bar{\pi}_x^1 1\pi_y^1 1\delta_-^1\right\rangle \\ & + 0.42\left|1\sigma^2 2\sigma^1 3\sigma^1 1\pi_x^1 1\bar{\pi}_y^1 1\delta_-^1\right\rangle\end{aligned}$$

while their MRCI atomic populations (similar for both states)

$$4s^{1.75}4p_z^{0.38}3d_{z^2}^{0.28}4p_x^{0.08}3d_{xz}^{0.48}4p_y^{0.08}3d_{yz}^{0.48}/_{Sc},\ q_{Sc} = -0.58$$

$$4s^{0.66}4p_z^{0.10}3d_{z^2}^{0.73}3d_{x^2-y^2}^{0.47}4p_x^{0.08}3d_{xz}^{0.39}4p_y^{0.08}3d_{yz}^{0.39}3d_{xy}^{0.52}/_{Ti},\ q_{Ti} = +0.58$$

point to an excited Ti atom creating a σ and π interaction with a ground state Sc atom.

3 $^4\Pi(1)$, 4 $^4\Phi(1)$, 5 $^4\Phi(2)$, and 9 $^4\Pi(2)$. These four quartet states correlate adiabatically with the ground state fragments Sc(^{2}D) + Ti(^{3}F) although it is evident from their PECs (see Fig. 3) that their minima are due to excited atomic states that interact with lower states via numerous avoided curve crossings.

The equilibrium character of the $3^4\Pi$ state is a mixture of two different structures as can be seen by its main CASSCF configurations

$$|3^4\Pi(^4B_1)\rangle \approx 0.45\left|1\sigma^2 2\sigma^1 3\sigma^1 1\pi_x^1 1\pi_y^2\right\rangle + 0.41\left|1\sigma^2 2\sigma^1 (1\delta_+^1 1\pi_x^1 1\pi_y^2 - 1\pi_x^2 1\pi_y^1 1\delta_-^1)\right\rangle.$$

The MRCI Mulliken populations are also consistent with the above structural mixture

$$4s^{2.0}4p_z^{0.49}3d_{z^2}^{0.21}4p_x^{0.09}3d_{xz}^{0.45}4p_y^{0.11}3d_{yz}^{0.62}/_{Sc},\ q_{Sc} = -1.11$$

$$4s^{0.18}4p_z^{0.08}3d_{z^2}^{0.47}3d_{x^2-y^2}^{0.22}4p_x^{0.08}3d_{xz}^{0.58}4p_y^{0.10}3d_{yz}^{0.96}3d_{xy}^{0.22}/_{Ti},\ q_{Ti} = +1.11.$$

Both of the above data clearly show the difficulty to draw a simple vbL bonding scheme. Note the remarkable Ti-to-Sc charge transfer of more than one electron and $\mu = 0.16$ D.

The 4 $^4\Phi$ state relates to the "0.41" component of the previously discussed 3 $^4\Pi$ state. Its electronic character at the global minimum is $|4^4\Phi(^4B_1)\rangle_g \approx 0.49|1\sigma^2 2\sigma^1 (1\delta_+^1 1\pi_x^1 1\pi_y^2 + 1\pi_x^2 1\pi_y^1 1\delta_-^1)\rangle$.

Its PEC presents two avoided crossings at 5.6 and 4.7 bohr with a second $(5)^4\Phi$ state, where the latter avoided crossing is due to an incoming third $^4\Phi$ state. The local minimum appears to be of the same electronic nature.

The $5^4\Phi$ state appears to have two potential minima, both of them due to avoided crossings with the $4^4\Phi$ state at the distances 5.6 and 4.7 bohr mentioned above. The main CASSCF configurations are as follows

$$|5^4\Phi(^4B_1), 5.6\text{ bohr}\rangle \approx \left|1\sigma^2 2\sigma^1 [(0.55)1\bar{\pi}_x^1 2\pi_x^1 1\pi_y^1 + (0.40)1\pi_x^1 2\pi_x^1 1\bar{\pi}_y^1]\right\rangle$$

and $|5^4\Phi(^4B_1), 4.7\text{ bohr}\rangle \approx 0.54\left|1\sigma^2 3\sigma^1 (1\delta_+^1 1\pi_x^1 1\pi_y^2 + 1\pi_x^2 1\pi_y^1 1\delta_-^1)\right\rangle$.

The 9 $^4\Pi$ state shows two quasi-degenerate minima both of which result from avoided crossings with nearby curves of the same symmetry. The nominally global potential minimum appears at 4.4 bohr and is due to the avoided crossing with an incoming 3rd $^4\Pi$ state at 4.8 bohr and with the $3^4\Pi$ state at 4.4 bohr, while the second potential minimum is due to the avoided crossing with a 3rd $^4\Pi$ state at 5.7 bohr with the repulsive part of the PEC of 3 $^4\Pi$. Equilibrium configurations are similar at both minima, i.e.,

$$|9^4\Pi(^4B_1)\rangle_g \approx 0.49\left|1\sigma^2 3\sigma^1 (1\delta_+^1 1\pi_x^1 1\pi_y^2 - 1\pi_x^2 1\pi_y^1 1\delta_-^1)\right\rangle + 0.32\left|1\sigma^2 2\sigma^1 3\sigma^1 1\pi_x^1 1\pi_y^2\right\rangle$$

and

$$|9^4\Pi(^4B_1)\rangle_l \approx 0.55\left|1\sigma^2 2\sigma^1 3\sigma^1 1\pi_x^1 1\pi_y^2\right\rangle - 0.23\left|1\sigma^2 2\sigma^1 (1\delta_+^1 1\pi_x^1 1\pi_y^2 - 1\pi_x^2 1\pi_y^1 1\delta_-^1)\right\rangle$$

revealing a resonance between two vbL bonding schemes.

11 $^6\Pi$, 12 ^{6}H, 14 $^6\Phi$, 15 $^6\Pi$, 17 $^6\Pi$, and 18 $^6\Phi$. We have studied three $^6\Pi$ (11, 15, 17), two $^6\Phi$ (14, 18), and one ^{6}H ($\Lambda = 5$) states all of which correlate adiabatically with Sc (^{2}D) + Ti (^{5}F).

The three $^6\Pi$ states are energetically close with T_e values of 13.96, 14.76, and 16.65 kcal/mol and similar equilibrium bond distances, $r_e = 2.940$ (11 $^6\Pi$), 2.960 (15 $^6\Pi$), and 2.993 (17 $^6\Pi$) Å. At their equilibrium distance, the main CASSCF configurations are

$$|11^6\Pi(^6B_1)\rangle \approx 0.49\left|1\sigma^2 2\sigma^1 3\sigma^1 (1\delta_+^1 1\pi_y^1 2\delta_-^1 + 2\delta_+^1 1\pi_y^1 1\delta_-^1)\right\rangle - 0.32\left|1\sigma^2 2\sigma^1 3\sigma^1 1\pi_x^1 (1\delta_+^1 2\delta_+^1 + 1\delta_-^1 2\delta_-^1)\right\rangle$$

$$|15^6\Pi(^6B_1)\rangle \approx 0.45\left|1\sigma^2 2\sigma^1 (1\delta_+^1 1\pi_x^1 1\pi_y^1 2\pi_y^1 - 1\pi_x^1 2\pi_x^1 1\pi_y^1 1\delta_-^1)\right\rangle$$

and

$$|17^6\Pi(^6B_1)\rangle \approx 0.48|1\sigma^2 2\sigma^1 3\sigma^1 1\pi_x^1 (1\delta_+^1 2\delta_+^1 + 1\delta_-^1 2\delta_-^1)\rangle + 0.33\left|1\sigma^2 2\sigma^1 3\sigma^1 1\pi_y^1 (1\delta_+^1 2\delta_-^1 + 1\delta_-^1 2\delta_+^1)\right\rangle$$

They are of multi configuration nature reflected as well in the Mulliken populations; thus, a simple vbL picture cannot be drawn. Along the same lines is the nature of the $^6\Phi$ and ^{6}H states. Their most important equilibrium CFs are

$$|14^6\Phi(^6B_1)\rangle \approx 0.54\left|1\sigma^2 2\sigma^1 (1\delta_+^1 1\pi_x^1 1\pi_y^1 2\pi_y^1 - 1\pi_x^1 2\pi_x^1 1\pi_y^1 1\delta_-^1)\right\rangle$$

$$|18^6\Phi(^6B_1)\rangle \approx 0.32\left|1\sigma^2 2\sigma^1 3\sigma^1 5\sigma^1 (1\pi_y^1 1\delta_-^1 - 1\delta_+^1 1\pi_x^1)\right\rangle - 0.31\left|1\sigma^2 2\sigma^1 3\sigma^1 2\delta_+^1 (1\pi_y^1 1\delta_-^1 + 1\delta_+^1 1\pi_x^1)\right\rangle + 0.28\left|1\sigma^2 2\sigma^1 3\sigma^1 (1\delta_+^1 1\pi_y^1 2\delta_-^1 + 1\pi_x^1 1\delta_-^1 2\delta_-^1)\right\rangle$$

and

$$|12^6\text{H}(^6B_1)\rangle \approx 0.43|1\sigma^2 2\sigma^1 3\sigma^1 (1\pi_x^1 1\delta_-^1 2\delta_-^1 - 1\delta_+^1 2\delta_+^1 1\pi_x^1 - 1\delta_+^1 1\pi_y^1 2\delta_-^1 + 2\delta_+^1 1\pi_y^1 1\delta_-^1)\rangle$$

thus rendering very difficult a vbL transcription of the wave functions. One major characteristic of these states is their spectroscopic similarities due to the sharing of the same configurations but with different signs.

3 Conclusions

The ScTi intermetallic diatomic molecule has been studied by high-level all electron ab initio MRCI calculations, in conjunction with correlation-consistent basis sets of quadruple cardinality. No previous experimental or ab initio data are available in the literature. We have constructed PECs for 21 states at the MRCI/cc–pVQZ level; no core, relativistic, or other corrections have been applied. Due to severe technical difficulties, our results are of semi-quantitative nature. Our general conclusions are the following.

(a) All 21 examined states (13 sextets and 8 quartets) are of multi reference character and exhibit weak bonding energies.

(b) The ground state seems to be of $^6\Delta$ symmetry with $r_e(+Q) = 2.647$ (2.66) Å, $D_e(+Q) = 32.4$ (36.8) kcal/mol, and $\omega_e(+Q) = 218.3$ (223) cm^{-1} featuring two half π bonds and a four electron σ interaction. The first excited state is of $^4\Delta$ symmetry just 3 kcal/mol higher and with similar binding characteristics to that of the X $^6\Delta$ state. It is interesting to note that DFT (GGA) calculations predict totally different results: a ground state of $^6\Sigma^+$ symmetry at $r_e = 2.290$ Å and $D_e = 51.2$ kcal/mol, and an excited state ($^2\Sigma^+$) 3.7 kcal/mol higher at $r_e = 2.065$ Å [52].

(c) In most of the states, a ~0.4 e$^-$ Ti-to-Sc charge transfer is recorded. For three of the states, however, X $^6\Delta$, 2 $^4\Delta$, and 3 $^4\Pi$, a Mulliken Ti-to-Sc charge transfer of one electron is observed. Interestingly, the corresponding electric dipole moments are surprisingly small, that is, $\mu = 0.05$, 0.65, and 0.16 D, respectively.

(d) In all states presently studied, the in situ Ti atom is found in its first excited ^{5}F ($4s^13d^3$) state, a prerequisite for the development of covalent bonding between the two atoms.

References

1. Mavridis A, Harrison JF (1989) J Chem Soc Faraday Trans II(85):1391
2. Mavridis A, Harrison JF (1990) J Chem Soc Faraday Trans II(86):1625
3. Sakellaris CN, Mavridis A (2012) J Chem Phys 137:034309
4. Kalemos A, Mavridis A (1998) J Phys Chem A 102:5982
5. Kalemos A, Mavridis A (1999) Adv Quantum Chem 32:69
6. Kalemos A, Mavridis A (1999) J Phys Chem A 103:3336
7. Kalemos A, Mavridis A (2000) J Chem Phys 113:2270
8. Tzeli D, Mavridis A (2008) J Chem Phys 128:034309
9. Kerkines ISK, Mavridis A (2000) J Phys Chem A 104:11777
10. Kalemos A, Mavridis A, Harrison JF (2001) J Phys Chem A 105:755
11. Kalemos A, Mavridis A (2002) J Phys Chem A 106:3905
12. Tzeli D, Mavridis A (2002) J Chem Phys 116:4901
13. Kerkines ISK, Pittner J, Čársky P, Mavridis A, Hubač I (2002) J Chem Phys 117:9733
14. Tzeli D, Mavridis A (2003) J Chem Phys 118:4984
15. Kerkines ISK, Mavridis A (2003) Collect Czech Chem Commun 68:387
16. Tsouloucha A, Kerkines ISK, Mavridis A (2003) J Phys Chem A 107:6062
17. Kerkines ISK, Mavridis A (2004) Mol Phys 102:2451
18. Tzeli D, Mavridis A (2005) J Chem Phys 122:056101
19. Kalemos A, Dunning TH Jr, Mavridis A (2005) J Chem Phys 123:014301
20. Kalemos A, Dunning TH Jr, Mavridis A (2005) J Chem Phys 123:014302
21. Tzeli D, Mavridis A (2005) J Phys Chem A 109:9249
22. Kalemos A, Dunning TH Jr, Mavridis A (2006) J Chem Phys 124:154308
23. Tzeli D, Mavridis A (2006) J Phys Chem A 110:8952
24. Tzeli D, Mavridis A (2007) J Chem Phys 126:194304
25. Kalemos A, Dunning TH Jr, Mavridis A (2008) J Chem Phys 129:174306
26. Tzeli D, Mavridis A (2010) J Chem Phys 132:194312
27. Brugh DJ, Morse MD, Kalemos A, Mavridis A (2010) J Chem Phys 133:034303
28. Miliordos E, Mavridis A (2007) J Phys Chem A 111:1953
29. Miliordos E, Mavridis A (2010) J Phys Chem A 114:8536
30. Sakellaris CN, Papakondylis A, Mavridis A (2010) J Phys Chem A 114:9333
31. Sakellaris CN, Miliordos E, Mavridis A (2011) J Chem Phys 134:234308
32. Sakellaris CN, Mavridis A (2012) J Phys Chem A 116:6935
33. Sakellaris CN, Mavridis A (2013) J Chem Phys 138:054308
34. Koukounas C, Kardahakis S, Mavridis A (2004) J Chem Phys 120:11500
35. Kardahakis S, Koukounas C, Mavridis A (2005) J Chem Phys 122:054312
36. Koukounas C, Mavridis A (2008) J Phys Chem A 112:11235
37. Tzeli D, Mavridis A (2000) J Phys Chem A 104:6861
38. Mavridis A, Metropoulos A (1993) J Phys Chem 97:10955
39. Glezakou VA, Harrison JF, Mavridis A (1996) J Phys Chem 100:13971
40. Tzeli D, Mavridis A (2004) J Chem Phys 121:2646
41. Kardahakis S, Mavridis A (2009) J Phys Chem A 113:6818
42. Kalemos A, Kaplan IG, Mavridis A (2010) J Chem Phys 132:024309
43. Camacho C, Cimiraglia R, Witek HA (2010) Phys Chem Chem Phys 12:5058
44. Camacho C, Witek HA, Cimiraglia P (2010) J Chem Phys 132:224306
45. Soto J, Avila FJ, Otero JC, Arenas JF (2011) Phys Chem Chem Phys 13:7230
46. Camacho C, Cimiraglia R, Witek HA (2011) Phys Chem Chem Phys 13:7232
47. Maxtain JM, Rezabal E, Lopez X, Ugalde JM, Gagliardi L (2008) J Chem Phys 128:194315
48. Kalemos A, Mavridis A (2011) J Chem Phys 135:134302
49. Tzeli D, Miranda U, Kaplan IG, Mavridis A (2008) J Chem Phys 129:154310
50. Krechkivska O, Morse MD, Kalemos A, Mavridis A (2012) J Chem Phys 137:054302
51. Bunge CE, Barrientos JA, Bunge AV (1993) At Data Nucleic Data Tables 53:113
52. Gutsev GL, Jena P, Rao BK, Khanna SN (2001) J Chem Phys 114:10738
53. Kramida A, Ralchenko Yu, Reader J, NIST ASD Team (2012). *NIST Atomic Spectra Database* (ver. 5.0), [Online]. Available: http://physics.nist.gov/asd2013, August 15]. National Institute of Standards and Technology, Gaithersburg, MD
54. Dunning TH Jr (1989) J Chem Phys 90:1007
55. Balabanov N, Peterson KA (2005) J Chem Phys 123:064107

56. MOLPRO is a package of ab initio programs written by Werner H-J, Knowles PJ, Knizia G, Manby FR, Schütz M, Celani P, Korona T, Lindh R, Mitrushenkov A, Rauhut G, Shamasundar KR, Adler TB, Amos RD, Bernhardsson A, Berning A, Cooper DL, Deegan MJO, Dobbyn AJ, Eckert F, Goll E, Hampel C, Hesselmann A, Hetzer G, Hrenar T, Jansen G, Köppl C, Liu Y, Lloyd AW, Mata RA, May AJ, McNicholas SJ, Meyer W, Mura ME, Nicklaß A, O'Neill DP, Palmieri P, Peng D, Pflüger K, Pitzer R, Reiher M, Shiozaki T, Stoll H, Stone A, Tarroni RJ, Thorsteinsson T, Wang M

Theor Chem Acc (2014) 133:1452
DOI 10.1007/s00214-014-1452-2

REGULAR ARTICLE

Explicitly correlated coupled cluster benchmarks with realistic-sized ligands for some late-transition metal reactions: basis sets convergence and performance of more approximate methods

Manoj K. Kesharwani · Jan M. L. Martin

Received: 9 October 2013 / Accepted: 15 January 2014 / Published online: 5 February 2014

Abstract CCSD(T)-F12b benchmark calculations have been performed for the energetics and barrier heights of three late-transition metal systems, in increasing order of size: oxidative additions at bare Pd, a model for the Grubbs catalyst, and competitive CC/CH activation by a Rh(PCP) pincer complex. The results depend weakly on the basis set on the main-group atoms but are rather more sensitive to the basis set on the metal. An aug-cc-pwCVTZ-PP set on the metal combined with cc-pVTZ-F12 on the main-group elements yields barriers that are effectively converged in the basis set, but even the combination with aug-cc-pwCVTZ-PP on the metal and cc-pVDZ-F12 on the main group, or of def2-TZVPP on the metal and def2-TZVP on the main group, works well enough for most benchmark purposes. Inner-shell correlation cannot be neglected for even semi-accurate work. Simple nonempirical (meta-)GGAs with D3BJ dispersion work quite well for the Grubbs and pincer cases but break down for the Pd example, which requires exact exchange. Hybrids of these same functionals, such as PBE0, TPSS0, and B3PW91, are among the best performers through rung four on Perdew's ladder. For the Grubbs case, dispersion is very important and D3BJ clearly is superior over D2. Only the DSD double hybrids consistently perform well in the absence of dispersion corrections.

Keywords Explicitly correlated methods · Catalysis · Late-transition metals · Density functional theory · CCSD(T)-F12 · Basis set convergence

Dedicated to Professor Thom Dunning and published as part of the special collection of articles celebrating his career upon his retirement.

Electronic supplementary material The online version of this article (doi:10.1007/s00214-014-1452-2) contains supplementary material, which is available to authorized users.

M. K. Kesharwani · J. M. L. Martin (✉)
Department of Organic Chemistry, Weizmann Institute of Science, 76100 Reḥovot, Israel
e-mail: gershom@weizmann.ac.il

1 Introduction

Homogenous catalysis by late-transition metals is of great industrial importance [1–11]. Because of the size of typical ligands involved (dozens to hundreds of atoms), computational mechanistic studies on such reactions (e.g., olefin metathesis reactions [12, 13], Heck reactions [14, 15], and pincer-ligated metal complex catalyzed reactions [16–18]) have traditionally relied on the density functional theory, usually with a semi-arbitrarily chosen exchange–correlation functional.

Some papers do explore multiple basis sets and exchange–correlation functionals. Ideally, however, ab initio benchmarks should be available: since most late-transition metal complexes of interest in this area are relatively free of static correlation effects, CCSD(T) [19, 20] can be considered "the gold standard[1]." However, such studies are very rare (see, e.g., Refs. [21–29] for some examples) because, with basis sets large enough to yield meaningful results, the computational cost of CCSD(T) (not to mention the storage and RAM requirements) quickly becomes prohibitive.

One way out of this predicament is offered by explicitly correlated methods (see Refs. [30, 31] for recent reviews),

[1] The use of this term for CCSD(T) originated with Thom. H. Dunning, who is being honored by this collection.

which exhibit dramatically accelerated basis set convergence. For basis sets saturated in the angular momenta present, explicitly correlated methods should have [32] leading $O(L^{-7})$ basis set convergence (with L being the highest angular momentum). In practice, with real-life basis sets such as the correlation consistent sets pioneered by Dunning [33, 34], we can expect a gain of 2–3 angular momentum steps [35]: for example, a CCSD-F12/aug-cc-pVTZ calculation can be expected to be similar to a CCSD/aug-cc-pV5Z or even aug-cc-pV6Z calculation in quality. (See Ref. [36] for a recent example application.)

This means that, using explicitly correlated methods, we might be able to obtain near-basis set limit benchmark data for realistically sized catalyst model systems and then assess the performance of more approximate methods against these.

In this paper, we will present such a study. In the first stage, we will consider some small prototypes for C–C, C–H, H–H, and C–X insertion by atomic Pd. These systems, which were the subject of a previous benchmark study [37] using conventional CCSD(T) with basis set extrapolations [38, 39], are small enough that we can establish convergence to the basis set limit to within better than 0.1 kcal/mol.

At the second stage, we will consider a model system (Fig. 2) for the Grubbs catalyst that was previously the subject of a benchmark study [40] by a layered MP2:CCSD(T) strategy.

At the third stage, we will consider competitive CC and CH activation in a Rh(I) PCP pincer complex (Fig. 3) [41]. The explicitly correlated data on this latter system will be published separately as part of a paper on pincer complexes [42].

Finally, we will evaluate the performance of a variety of DFT functionals for a test set consisting of the above three problems as well as for each problem individually.

The overall purpose of the present paper is twofold. The first is to establish the smallest basis set that can be used for reliable CCSD(T)-F12 benchmarks on reactions that are actually representative of the ones that we face in homogenous catalysis applications (e.g., [43–45]). The second is to establish, for systems beyond the grasp of CCSD(T)-F12, whether commonly used DFT functionals such as B3LYP [46, 47][2] or PBE0 [48] yield acceptable accuracy, and whether any of the more recent alternatives (such as dispersion-corrected functionals [49] and double hybrids [50]) represent an improvement.

2 Methods

All explicitly correlated coupled cluster calculations were carried out using MOLPRO [51] 2010.1 and 2012.1 running on the Faculty of Chemistry cluster at the Weizmann Institute of Science. The density functional calculations were carried out using either Gaussian [52] 09 or ORCA [53] running on the same hardware.

The orbital basis sets used include the Weigend–Ahlrichs [54] def2-SVP, def2-TZVP, def2-TZVPP, and def2-QZVPP basis sets, as well as augmented variants thereof such as def2-ATZVPP as defined in the MOLPRO basis set library. These latter basis sets are defined as having one diffuse function of each angular momentum added with the exponent obtained by even-tempered extrapolation.

For the Coulomb fitting and the RI-MP2, the program defaults (Hättig [55, 56], Weigend [57]) were employed unless noted otherwise. (Details of all auxiliary basis sets used with any specific orbital basis set can be found in Table S1 in the ESI).

In addition, for the main-group elements, we employed the cc-pVnZ-F12 basis sets of Peterson et al. [58] together with the associated auxiliary basis sets [59, 60]. cc-pVnZ-F12 are variants, optimized for explicit correlated calculations, of the venerable correlation consistent basis sets introduced and developed by Dunning [33, 34, 61–65].

For the second-row transition metal atoms Ru, Rh, and Pd, we also applied the cc-pwCVTZ-PP and aug-cc-pwCVTZ-PP orbital basis sets [66]. (It is perhaps worth noting that both the Weigend–Ahlrichs basis sets and the (aug-)cc-pwCVnZ-PP basis sets use the same Stuttgart–Cologne-type relativistic energy-consistent pseudopotentials [67] on the metal atoms.) The MP2 and CABS basis sets for these elements were both taken from the very recent paper by Hill [68].

The geminal exponent was left at its default value of 1.0 throughout.

The CCSD-F12b ansatz [69, 70] and the 3C(Fix) approximation [71] were used throughout for the F12 calculations. For the (T) term, we employed the Marchetti–Werner approximation [72, 73].

Aside from CCSD-F12 and CCSD(T)-F12, we obtained MP2-F12, SCS-MP2-F12 [74–76],[3] S2-MP2-F12 [77], SCS-CCSD-F12 [78],[4] and SCS(MI)CCSD-F12 [79][5] results as by-products of the CCSD(T)-F12b calculations.

As all species studied in the present paper are closed-shell singlets, first-order spin–orbit coupling does not affect any energetics. We made no attempt to account for second-order spin–orbit coupling.

In the DFT calculations, we employed the Weigend–Ahlrichs basis sets exclusively, using no auxiliary basis

[2] This paper defines the B3 hybrid exchange functional; the first published reference to the "B3LYP" combination appears to be in [47].

[3] For a physical interpretation of the SCS-MP2 method in terms of Feenberg scaling, see [75]; for a review, see [76].

[4] Coefficients: cAB = 1.27, cSS = 1.13 for general-purpose applications.

[5] Coefficients: cAB = 1.11, cSS = 1.28, optimized for weak interactions.

sets at all in Gaussian and the associated RI-MP2 auxiliary basis sets [55] for the double-hybrid calculations in ORCA.

The DFT functionals evaluated include (grouped by rung in the "Jacob's Ladder" [80]):

- GGAs (second rung): BP86 [81, 82], BLYP [81, 83], PBE [84], and B97-D [85]
- meta-GGAs (third rung): TPSS [86], M06L [87], and mPWB95 [88]
- hybrid GGAs (incomplete fourth rung): B3LYP, B3PW91 [89, 90], PBE0, MPW1K [91], PBEh1PBE [92], B97-1 [93], and B97-2 [94]
- hybrid meta-GGAs (complete fourth rung): B1B95 [95], BB1K [96], BMK [97], M06 [98], M06-2X [98], MPW1B95 [99], MPWB1K [99], TPSS0 [100],[6] and TPSS1KCIS [101]
- double hybrids (fifth rung): B2GP-PLYP [102], DSD-PBEP86 [103], DSD-PBEPBE, and DSD-PBEB95 [104]. These latter three are spin-component-scaled double hybrids, which come in different parameterizations for no dispersion correction ("noD"), D2, and D3BJ dispersion corrections.

Finally, we included two range-separated functionals, namely ωB97X-D [105] and M11 [106].

Empirical dispersion corrections were computed either directly in Gaussian or ORCA as appropriate, or using the DFTD3 external program by Grimme [107, 108].

Reference geometries for the Pd model systems and for the Grubbs catalyst model were the same as used in the Refs. [37, 40].

Our largest available nodes have 32 Intel E5-4640 CPU cores, 512 GB of RAM, and 4 TB of solid-state storage. Even so, we were unable to run the largest calculations fully distributed-parallel due to memory requirements and I/O congestion: hence, we resorted to mixed parallelism in which we ran 8 or 16 MPI processes on 4 and 2 OpenMP threads, respectively.

3 Results and discussion

In order to ensure that CCSD(T)-F12b is indeed an adequate level of theory for these problems, we calculated a number of diagnostics for nondynamical correlation (see Ref. [109] for a recent overview) such as the T1 diagnostic [110], the largest T2 amplitudes, and the percentages of SCF and (T) in the CCSD(T)-F12b atomization energies, %TAE[SCF] and %TAE[CCSD(T)-F12b] [111]. Particularly, the latter is known to be [109, 111] a reliable predictor for the importance of post-CCSD(T) correlation effects. The indices can be found in the Electronic Supporting Information (Table S2, ESI): all of the systems occurring in the present work appear to be dominated by dynamical correlation, meaning that CCSD(T) is a "gold standard" in the Dunning sense.

Our "best estimate" energetics for all three systems can be found in Table 1.

3.1 The Pd prototype insertion systems

A general representation of these oxidative additions and the stationary points involved is given in Fig. 1.

For these systems, we were able to consider fairly large basis sets that would be approaching the basis set limit even in a conventional calculation. The largest basis sets we did consider for CCSD(T)-F12b were def2-AQZVPP and awCVTZ-F12. The latter consists of aug-cc-pwCVTZ-PP basis set on the transition metal combined with cc-pVTZ-F12 on the main-group elements.

All valence electrons were correlated on the main-group elements; for the transition metal, in addition to the usual valence orbitals, it can be argued that the (4*s*,4*p*) subvalence orbitals are so close to the valence shell that correlation from them must be included. We thus compared results including these orbitals (labeled "Pd(4*s*,4*p*)") and excluding them (labeled "default core"). Full results can be found in the ESI: error statistics can be found in Table 2.

In order to remove a potential ambiguity, we should emphasize that error statistics for "Default core" and "Pd(4*s*,4*p*)" are relative to the CCSD(T)-F12b result with the largest available basis set (in this section, def2-AQZVPP) and the same selection of frozen-core orbitals, i.e., "Default core" or "Pd(4*s*,4*p*)". In other words, they should reflect nothing but basis set incompleteness for either case.

First of all, both with and without Pd(4*s*,4*p*) correlations, the two largest basis sets yield results very close to each other, differing by 0.10 and 0.08 kcal/mol RMS, respectively. It should be noted that only the aug-cc-pwCVTZ-PP basis set is explicitly designed to accommodate subvalence correlation on the metal: yet, the def2-AQZVPP is apparently flexible enough in the relevant region that it does not suffer from basis set incompleteness error in this respect.

Second, as we obtain essentially equivalent results with two very different large basis sets, we can feel confident that either set is adequate for reference purposes.

Third, while these systems are perhaps too small for a truly meaningful CPU time comparison, the awCVTZ-F12 basis set is about three times less expensive than def2-AQZVPP on our hardware (not surprising when basis set sizes are compared), and in light of the very small differences between the results, awCVTZ-F12 should be quite adequate for benchmarking purposes for reactions of this type.

[6] This functional, which uses TPSS correlation with a 25:75 mixture of Hartree–Fock like and TPSS exchange (rather than the 10:90 mixture in TPSSh) appears to have been introduced independently in [100] and Ref. [37], where it was called TPSS25TPSS.

Table 1 Best CCSD(T*)-F12 energetics (kcal/mol) for the Pd prototype reactions, the Grubbs catalyst, and the Rh(PCP) benchmark

	Pd + H_2	Pd + CH_4	Pd + C_2H_6 (CH ins.)	Pd + C_2H_6 (CC ins.)	Pd + CH_3Cl	
Pd prototype insertion systems[a]						
Default core						
D_e (complex)	19.01	8.24	8.60	8.60	12.89	
$\Delta E_{forward}$	5.60	14.58	14.84	27.84	14.36	
Reaction energy	6.29	10.92	10.29	6.98	−15.00	
Pd(4*s*,4*p*)						
D_e (complex)	20.75	9.47	9.87	9.87	14.33	
$\Delta E_{forward}$	4.57	12.99	13.08	26.38	13.93	
Reaction energy	4.88	8.46	7.84	4.22	−16.85	
	1 − PH_3 + C_2H_4	3-*cis*	3-*trans*	TS-*cis*	TS-*trans*	4-*cis*	4-*trans*
Grubbs catalyst model system[b]							
Default core	−24.26	−5.76	−16.29	−5.52	−7.67	−18.49	−17.66
Ru(4*s*,4*p*)	−24.83	−6.78	−16.94	−6.49	−8.38	−19.68	−18.75
	2	3	4	2-TS	3-TS		
Rh-PCP pincer complexes[c]							
Default core	−8.66	−9.42	0.43	0.72	9.16		
Rh(4*s*,4*p*)	−8.51	−9.40	0.31	0.72	9.09		

[a] Calculated with basis set def2-AQZVPP, and all energies are relative to the corresponding reactant complex

[b] Calculated with basis set awCVTZ-F12, and all energies are relative to (**2** + C_2H_4)

[c] Calculated with basis set def2-TZVP(P), and all energies are relative to species **1**

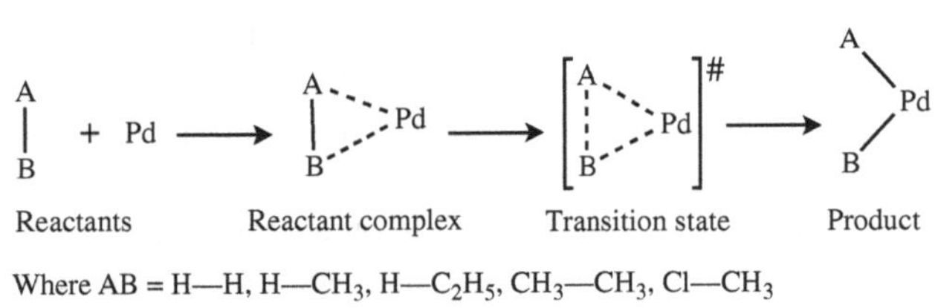

Fig. 1 Schematic representation of the prototype oxidative additions at Pd

Fourth, at 1.7 kcal/mol RMS (reaching as large as 2.8 kcal/mol for the CH_3–Pd–CH_3 insertion product), the Pd(4*s*,4*p*) contributions to the barrier heights and reaction energies are clearly too large to be ignored and well exceed basis set incompleteness errors (see below) for many basis sets.

We will now consider the latter. Removal of the diffuse functions from def2-AQZVPP only has a very modest effect (0.07 kcal/mol RMS), since the underlying basis set is fairly "diffuse" already. In stark contrast, removing the diffuse functions from just the Pd atom in awCVTZ-F12 causes an error of 1.25 kcal/mol RMS. Quintal et al. [37] did find them to be quite important in conventional calculations as well, but clearly their importance is amplified in explicitly correlated ones.

An ordinary def2-TZVPP basis set yields an unacceptable basis set incompleteness error of 1.23 kcal/mol, which drops to 1.09 kcal/mol when Pd(4*s*,4*p*) correlation is excluded. Yet, def2-ATZVPP reproduces the reference data to 0.10 kcal/mol RMS! While the CPU time cost of def2-ATZVPP is already only 2/5 of awCVTZ-F12 and 1/8 of def2-AQZVPP, even this basis set may prove too costly for more realistic-sized transition metal complexes and thus we considered further economies. Adding diffuse functions to def2-TZVPP on just the angular momenta that are occupied in valence states (which we denoted def2-TZVPP+A) only modestly reduces RMSD from 1.23 to 1.02 kcal/mol. If an additional angular momentum layer is added, however (denoted def2-TZVPP+AA in the Table 2), RMSD drops to just 0.16 kcal/mol.

We applied the same "calendar basis set" [112] approach also for def2-QZVPP, but there the underlying basis set is extended enough already that it does not matter much.

The def2-TZVPPD basis set [113] yields a disappointing RMSD = 1.53 kcal/mol: we should note that the standard definition does not include diffuse functions on any transition metals, including Pd.

Can we reduce the basis set further? We attempted def2-TZVP variants, but even def2-ATZVP already has an RMSD

Table 2 CCSD(T*)-F12b RMSD in kcal/mol for Pd-catalyzed prototype oxidative addition reactions

	def2-AQZVPP	def2-QZVPP	def2-QZVPP+A[a]	def2-QZVPP+AA[b]	def2-QZVPPD	awCVTZ-F12[c]	pwCVTZ-F12[d]	awCVTZ[e]	pwCVTZ[f]	awCVDZ-F12[g]	awCVD(T)Z-F12[h]
RMSD											
Default core	0.00	0.08	0.04	0.05	0.10	0.08	1.16	0.10	0.75	0.42	0.27
Pd(4*s*,4*p*)	0.00	0.07	0.07	0.06	0.11	0.10	1.25	0.14	0.80	0.71	0.28
No. of orbitals	457	382	399	416	412	336	311	298	241	184	236
CPU time (min)	422.1	225.8	264.5	301.6	292.7	134.0	105.8	79.9	41.4	15.6	38.0

	def2-ATZVPP	def2-TZVPP	def2-TZVPP+A	def2-TZVPP+AA	def2-TZVPPD	def2-TZVP	def2-ATZVP	def2-SVP	def2-ASVP	SVP-TZVPP+AA[i]
RMSD										
Default core	0.13	1.09	0.92	0.17	1.35	1.61	0.98	0.79	0.90	0.99
Pd(4*s*,4*p*)	0.10	1.23	1.03	0.17	1.53	2.27	1.23	2.32	2.94	0.96
No. of orbitals	259	202	219	236	232	138	186	89	123	130
CPU time (min)[j]	51.3	36.4	27.4	36.3	35.3	7.3	21.4	2.4	6.8	6.3

[a] "A" represents the addition of diffuse functions by even-tempered expansion on valence angular momenta only

[b] "AA" represents the addition of diffuse functions by even-tempered expansion on *s*, *p*, *d* angular momenta of main-group elements and *s*, *p*, *d*, *f* angular momenta of transition metal

[c] cc-pVTZ-F12 basis set on main-group elements and aug-cc-pwCVTZ-PP basis set on transition metal

[d] cc-pVTZ-F12 basis set on main-group elements and cc-pwCVTZ-PP basis set on transition metal

[e] aug-cc-pVTZ basis set on main-group elements, cc-pVTZ basis set on H, and aug-cc-pwCVTZ-PP basis set on transition metal

[f] cc-pVTZ basis set on main-group elements and cc-pwCVTZ-PP basis set on transition metal

[g] cc-pVDZ-F12 basis set on main-group elements and aug-cc-pwCVDZ-PP basis set on transition metal

[h] cc-pVDZ-F12 basis set on main-group elements and aug-cc-pwCVTZ-PP basis set on transition metal

[i] def2-SVP basis set on main-group elements and def2-TZVPP+AA basis set on transition metal

[j] Given CPU times are for complex (ethane + Pd) calculated on our hardware (Intel Xeon E5-4640, 32 cores, 512 GB RAM, 8 solid-state disks in a 4 TB RAID-0 array) using two-layer parallelism over 4 MPI processes with 2 OpenMP threads each

of 1.2 kcal/mol with Pd(4*s*,4*p*) correlation included, 1.0 kcal/mol without. Deleting the diffuse functions increases these numbers to 2.3 and 1.6 kcal/mol, respectively.

The very simplest and least expensive basis set we tested, namely def2-SVP, actually yields a surprisingly reasonable performance if Pd(4*s*,4*p*) correlation is excluded (RMSD = 0.8 kcal/mol). Even for F12 methods, however, such good agreement with the basis set limit for such small basis sets is almost certainly due to error compensation between basis set incompleteness error and BSSE (basis set superposition error). Note that the error shoots up to 2.3 kcal/mol when Pd(4*s*,4*p*) correlation is included. If anything, including diffuse functions makes things worse, RMSD = 2.9 kcal/mol. Clearly, the metal basis set is intrinsically too inflexible to deal with the important subvalence correlation in this system.

Finally, combining the cc-pVDZ-F12 basis set on main-group elements with aug-cc-pwCVDZ-PP on the transition metals yields error statistics (0.42 kcal/mol default core and 0.71 kcal/mol with Pd subvalence orbitals included) markedly superior to def2-ASVP. If in addition just the metal basis set is enlarged to aug-cc-pwCVTZ, the error relative to the respective basis set limits is reduced below 0.3 kcal/mol both with and without Pd(4*s*,4*p*) correlation. This latter awCVD(T)Z-F12 option thus seems promising for larger systems (see below).

Considering that, in most transition metal complexes, the ratio of the number of transition metal atoms to the number of main-group atoms is quite small, it may be worthwhile to expand the basis set on just the metal while retaining a fairly small basis set on the main-group elements. Calculations with def2-TZVPP+AA basis set on metal and def2-SVP on other elements yield about 1.0 kcal/mol RMS error with both included and excluded Pd(4*s*,4*p*) correlation.

How well do our best explicitly correlated results compare with the previous orbital-based benchmark data of Quintal et al. [37]. The RMSD between both sets of results is 0.6 kcal/mol—this may sound quite small, but is comparable to the deviation from the reference data that Quintal et al. found for some DFT functionals! In light of this, revisiting this dataset with large basis set explicitly correlated calculations, as done presently, turned out to be worthwhile.

Let us now consider lower-level ab initio electron correlation methods (Table S3, ESI). Our reference here, again, is the CCSD(T*)-F12/def2-AQZVPP data with Pd(4*s*,4*p*) correlated. Omitting the (T) entirely causes an RMSD of 4.0 kcal/mol—which is actually worse than MP2-F12 (3.6 kcal/mol) and especially SCS-MP2-F12 (2.2 kcal/mol). If we forgo the Marchetti–Werner [72] scaling of the (T) contribution, the results are affected by 0.3 kcal/mol RMS. For the sake of completeness, RMSD = 20.7 kcal/mol for SCF.

We will now assess the DFT functionals. Double-hybrid DFT calculations were performed in the def2-QZVPP basis set using ORCA; all other DFT calculations were performed using Gaussian 09 in a def2-ATZVPP basis set.

RMS errors with respect to the reference values (CCSD(T*)-F12/def2-AQZVPP) are given in Table 3. Most pure GGAs and meta-GGAs assessed here (BP86, BLYP, PBE, TPSS) perform quite poorly (RMSD >4 kcal/mol), M06L and mPWB95 being the exceptions to that rule. Hybrids perform much better, and PBE0 is in fact the best performer of all with RMSD ~0.7 kcal/mol. The stellar performance of PBE0 for this problem had also been noted by Quintal et al. [37].

"Kinetics-friendly" functionals (i.e., hybrid meta-GGA functionals with high percentages of HF exchange, such as BMK, M06-2X, MPWB1K, and BB1K) perform badly with >4.5 kcal/mol RMS error, which is in agreement with prior observations on the performance of high-Hartree–Fock functionals for transition metal compounds [37, 98, 114–116]. Hybrid mGGAs with more typical percentages of HF exchange (B1B95, M06, MPW1B95, mPW25B95, mPW28B95, TPSS0, TPSS1KCIS, TPSS21KCIS, and TPSS25KCIS) perform more reasonably, with <2.5 kcal/mol RMSD. mPW25B95 stands out with an RMS error 0.8 kcal/mol.

Among the double hybrids, B2GP-PLYP and DSD-PBEP86 perform especially well at ~1.0 kcal/mol RMSD. DSD-PBEhB95 yields a subpar ~2.2 kcal/mol RMSD for this particular problem.

The two range-separated hybrids (M11 and wB97X) yield very poor results (RMSD ~5.0 kcal/mol).

Furthermore, we have evaluated Grimme D2 and D3BJ dispersion corrections to some of the studied DFT functionals. For the problem at hand, the effect of dispersion corrections is small; however, for some DFT functionals, the RMSD decrease is significant, particularly with D3BJ. These functionals include B3LYP and B3PW91, but also the already very good RMSD for PBE0 drops below 0.5 kcal/mol, and B2GP-PLYP likewise receives a boost.

Inspection of individual errors reveals that, in general, functionals yielding RMS errors >4.0 kcal/mol have comparatively larger errors for the reaction barriers and reaction energies.

3.2 Grubbs catalyst model system

Let us now consider a somewhat more realistic-sized problem, namely the Grubbs catalyst model system (Fig. 2), that was the subject of an earlier benchmark study by Zhao and Truhlar [40] using conventional orbital-based correlation methods and an additivity approximation.

The largest basis set we were able to apply in the present study is awCVTZ-F12, as defined in the previous section.

Table 3 RMSD in kcal/mol for various DFT functionals relative to CCSD(T*)-F12b/def2-AQZVPP for Pd-catalyzed prototype oxidative addition reactions

	B2GP-PLYP	DSD-PBEB95	DSD-PBEP86	DSD-PBEPBE	BP86	BLYP	PBE	M06L	mPWB95	TPSS	B97-1
No disp.	1.07	2.17	0.98	1.37	6.15	4.33	5.94	1.79	2.20	5.53	1.52
DFT-D2	0.73	1.91	1.00	0.98	7.09	4.94	6.55	1.87		6.28	1.32
DFT-D3BJ	0.71	1.95	0.99	0.83	6.81	4.58	6.30			5.86	

	B97-2	B97D	B3LYP	B3PW91	MPW1K	PBE0	PBEh1PBE	B1B95	BB1K	BMK	M06
No disp.	2.78		2.20	1.54	4.98	0.74	0.70	1.84	4.81	5.33	6.48
DFT-D2	1.79	3.97[a]	1.39	1.46		0.44		1.17		5.57	6.33
DFT-D3BJ		3.21[b]	0.90	1.00		0.52				5.51	

	M06-2X	MPW 1B95	mPW 25B95	mPW 28 B95	MPW B1K	TPSS0	TPSS 1KCIS	TPSS 21KCIS	TPSS 25KCIS	ωB97-X	M11
No disp.	10.14	2.01	0.83	1.36	5.00	1.53	1.74	1.79	2.44	5.32	4.93
DFT-D2	10.10					0.98	1.96			3.76[c]	4.78
DFT-D3BJ		1.70			4.71	0.65					

[a] Calculations were performed with the B97D functional and its built-in dispersion correction

[b] Calculations were performed with the B97D3 functional and its built-in dispersion correction

[c] Calculations were performed with the ωB97X-D functional and its built-in dispersion correction

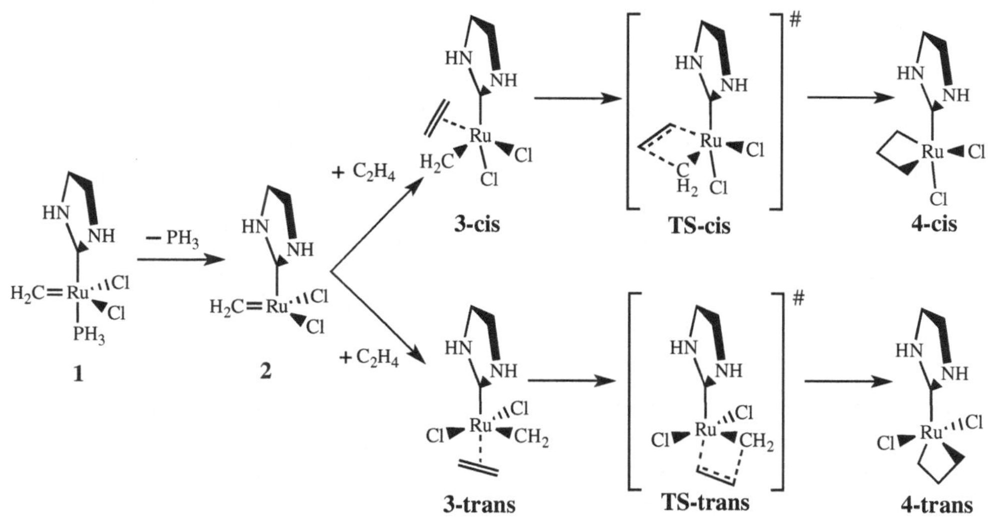

Fig. 2 Schematic representation of species involved in the catalytic cycle of the model Grubbs catalyst

These calculations took about 6 days wall-clock (for the case of 3-*cis*) on our hardware (using two-layer parallelism over 8 MPI processes with 4 OpenMP threads each). CCSD(T*)-F12b/awCVTZ-F12 calculated relative energies are given in Table 1. Calculated results show that the *trans*-bounded π complex (3-*trans*) is energetically ∼10 kcal/mol more stable than the *cis*-bounded π complex (3-*cis*). Furthermore, the transition state for the *trans*-pathway (TS-*trans*) is energetically more favorable than the *cis*-pathway (TS-*cis*). However, *cis*-Ru-cyclobutane (4-*cis*) is energetically more stable than 4-*trans*. From these results, one can say that the *trans*-pathway is kinetically more favorable; however, thermodynamically *cis*-pathway is favorable. These observations agree with those of Zhao and Truhlar [40].

Here, Ru(4*s*,4*p*) core correlation cannot be ignored for accurate work either (0.91 kcal/mol RMS), but the contribution is, counterintuitively, actually smaller than for the Pd system. However, it may be that the Pd model systems are simply too "bare" to be very representative of real-life situations.

Employing regular cc-pVTZ rather than cc-pVTZ-F12 (which is about the same size as aug-cc-pVTZ) on the main-group elements has only a small effect (0.18 kcal/mol RMS; Table 4); a similar further error is incurred by stripping the diffuse functions off the metal. They are

Table 4 CCSD(T*)-F12b RMSD in kcal/mol for the Grubbs model system

	awCVTZ-F12	awCVTZ	pwCVTZ	awCVD(T)Z-F12[a]	def2-TZVPP	def2-TZVPP+A
RMSD						
Default core	0.00	0.24	0.40	0.33	0.35	0.38
Ru(4*s*,4*p*)	0.00	0.18	0.37	0.34	0.36	0.36
No. of orbitals	886	758	573	548	556	605

	def2-TZVPP+AA	def2-TZVP(P)[b]	def2-TZVP(P)+AA	def2-SVP	def2-ASVP
RMSD					
Default core	0.23	0.33	0.27	1.19	0.92
Ru(4*s*,4*p*)	0.21	0.32	0.22	1.18	0.56
No. of orbitals	662	450	556	239	345

[a] aug-cc-pwCVTZ-PP basis set on Ru atom and cc-pVDZ-F12 on all other atoms

[b] def2-TZVPP basis set on Ru atom and def2-TZVP on all other atoms

clearly much less important for Ru in this complex than for the previous problem—where the metal is both more electron-rich and unrealistically poor in ligands, so their importance was exacerbated. def2-TZVPP yields an error of 0.36 kcal/mol, which can be reduced to 0.21 kcal/mol by adding diffuse functions to *s*, *p*, *d* angular momenta of main-group elements and *s*, *p*, *d*, *f* angular momenta of the metal. As it turns out, reducing the basis set on the main-group elements to def2-TZVP+AA while leaving the metal basis set at def2-TZVPP+AA yields comparable errors of 0.22 kcal/mol. This is quite useful, as even a reduction in the basis set by about 20 % translates into substantial cost and resource savings in n^2N^4 methods like CCSD. The computational cost with def2-TZVP(P)+AA basis set was about 37 h on 16 CPU cores, which is about one-eighths the aggregate CPU time of the awCVTZ-F12 basis set (6 days on 32 cores, see above).

Reducing the basis set still further, we find that def2-ASVP works surprisingly well—but even the simple def2-SVP has a relatively mild basis set error of about 1.2 kcal/mol in this case, and the awCVD(T)Z-F12 combination discussed above actually outperforms pwCVTZ, at RMSD = 0.34 kcal/mol with Ru(4*s*,4*p*) correlation included. (At the other end, we attempted def2-ATZVPP calculations but those yielded erratic results due to a numerical problem involving the CABS basis set.)

The RMS difference between our best results and the previous MP2:CCSD(T) study [40] is 0.64 kcal/mol, on the one hand validating these earlier results and on the other hand showing that explicitly correlated methods can yield results of similar quality using relatively small basis sets (such as def2-TZVP(P)) and modest computational resources.

In the spirit of Ref. [40] and of the ccCA approach [117, 118], we attempted layered schemes in which the system is treated at the MP2-F12 level with a larger basis set and the CCSD(T*)-F12b—MP2-F12 difference was treated as an additive correction with a smaller basis set. (Such layered approximations, while a key ingredient of the "focal point analysis" [119], arguably go back to the historic work of Pople et al. [120].) This approach slightly improves on CCSD(T*)-F12b calculations with the smaller basis set only (Table 5) but does appear to entail a sacrifice in accuracy. Combination of def2-SVP calculated CCSD(T*)-F12b results with MP2-F12/def2-QZVPP, awCVTZ-F12, and def2-TZVPP yields RMS errors of 0.73, 0.83, and 0.66 kcal/mol, respectively.

As for lower-level ab initio methods, while MP2-F12 yields RMSD = 6.54 kcal/mol and CCSD-F12b 3.53 kcal/mol, the corresponding SCS methods have much lower RMSD of 1.96 and 1.52 kcal/mol, respectively (Table S3, ESI).

Turning now to Table 6 for the DFT performance, the great importance of dispersion for this particular problem is immediately apparent. Only the DSD double hybrids perform well without any dispersion correction, as does (to a lesser extent) M06. With the D3BJ correction, TPSS-D3BJ outshines the other functionals, but PBE-D3BJ and B3PW91-D3BJ also perform well, as do, to a lesser extent, BP86-D3BJ, PBE0-D3BJ, and TPSS0-D3BJ. Among DSD double-hybrid functionals, with the D3BJ correction, DSD-PBEB95 performs excellently with 0.79 kcal/mol RMS error. Other double hybrids also acquit themselves well.

With the D2 correction, B2GP-PLYP-D2 renders an excellent performance with 0.77 kcal/mol RMS error. Except for double-hybrid methods, all other functionals with D2 correction yield RMS error more than 1.5 kcal/mol. It should also be noted that D3BJ is markedly superior to D2 for this problem, much more so than for the two other benchmarks here.

Table 5 RMSD (kcal/mol) of MP2-F12:CCSD(T*)-F12b two-layer additivity approximations for Grubbs model systems

Higher Lower	def2-QZVPP def2-SVP	awCVTZ-F12 def2-SVP	def2-TZVPP def2-SVP	awCVTZ-F12 def2-TZVPP	def2-QZVPP def2-TZVPP	def2-QZVPP def2-TZVP(P)	def2-QZVPP def2-TZVP(P)+AA
Default core	0.96	0.95	1.08	0.28	0.19	0.22	0.15
Ru(4*s*,4*p*)	0.73	0.82	0.66	0.28	0.20	0.22	0.16

RMSD relative to CCSD(T*)-F12b/awCVTZ-F12

Approximations were done by using equation $E = E(\text{MP2-F12/Higher}) + E(\text{CCSD}(T*)\text{-F12b/Lower}) - E(\text{MP2-F12/Lower})$

Table 6 RMSD in kcal/mol for various DFT functionals relative to CCSD(T*)-F12b/awCVTZ-F12 for the Grubbs model system

	B2GP-PLYP	DSD-PBEB95	DSD-PBEP86	DSD-PBEPBE	BP86	BLYP	PBE	M06L	mPWB95	TPSS	B97-1
No disp.	3.14	1.18	1.24	1.26	6.89	14.72	4.40	3.65	4.50	4.50	6.39
DFT-D2	0.77	1.10	1.48	1.12	2.41	5.80	2.61	2.54		3.77	1.98
DFT-D3BJ	1.56	0.79	1.37	1.19	1.56	7.97	1.31			0.78	

	B97-2	B97D	B3LYP	B3PW91	MPW1K	PBE0	PBEh1PBE	B1B95	BB1K	BMK	M06
No disp.	9.43		11.30	6.98	4.34	3.56	3.43	4.27	3.80	5.54	1.74
DFT-D2	2.29	6.93[a]	3.35	2.72		3.28		3.57		7.41	1.56
DFT-D3BJ		8.42[b]	5.56	1.22		2.00				6.42	

	M06-2X	MPW1B95	mPW25B95	mPW28B95	MPWB1K	TPSS0	TPSS1KCIS	TPSS21KCIS	TPSS25KCIS	ωB97X	M11
No disp.	5.06	3.05	3.10	3.06	3.47	3.58	6.07	5.71	5.56	3.46	3.77
DFT-D2	4.71					4.52	1.75			3.16[c]	3.40
DFT-D3BJ		2.22			3.62	2.04					

[a] Calculations were performed with the B97D functional and its built-in dispersion correction

[b] Calculations were performed with the B97D3 functional and its built-in dispersion correction

[c] Calculations were performed with the ωB97XD functional and its built-in dispersion correction

3.3 Rh-PCP pincer complexes

Complexes of late-transition metals and pincer ligands show remarkable versatility as selective catalysts: for more details, see the reviews by Rybtchinski and Milstein [121, 122] and by van der Boom and Milstein [123].

Mechanisms, kinetics, and the competition between CC and CH activation pathways have in the past been studied computationally in our group [41, 124, 125] using the B3LYP density functional and rather meager basis sets. (The relevant chemistry is discussed at length in these references and in the review articles of Milstein and coworkers, and this discussion will not be repeated here.) A detailed re-investigation using explicitly correlated methods will be published elsewhere: here, we will only present results for the Rh(PCP) case (Fig. 3) [41].

Initial geometries were taken from the supporting information and re-optimized at the PBE0/def2-TZVPP level. At those geometries, single-point energy calculations were performed at the CCSD(T)-F12b level with various basis sets.

As the ligand system here is much larger than that for the Grubbs catalyst model system, let alone for the bare-Pd prototype reactions, hardware limitations (particularly storage space and central memory) precluded even the def2-TZVPP+AA basis set. However, since in the previous section we found an RMSD of just 0.32 kcal/mol for the smaller def2-TZVP(P) set—i.e., the combination of def2-TZVPP on the metal atom and def2-TZVP on the remainder—we attempted this approach here as well. These calculations ran to completion uneventfully. In addition, we applied the awCVD(T)Z-F12 combination that proved successful in the two previous sections: the RMSD between both sets of results is just 0.23 kcal/mol with Rh(4*s*,4*p*) correlation included. As it is not a priori clear which is the more reliable set of data, we have given RMSD relative to both basis set combos in Table 7.

We shall first briefly turn to n-particle convergence. In this case, omitting (T) causes a relatively small error of about 1.7 kcal/mol RMSD, compared to 4 kcal/mol for the bare-Pd systems and 3.5 kcal/mol for Grubbs (Table S3,

ESI). Unlike the latter system, SCS-MP2-F12 does quite poorly here, with RMSD = 4.6 kcal/mol. SCS-CCSD-F12 however does have an error just below 1 kcal/mol.

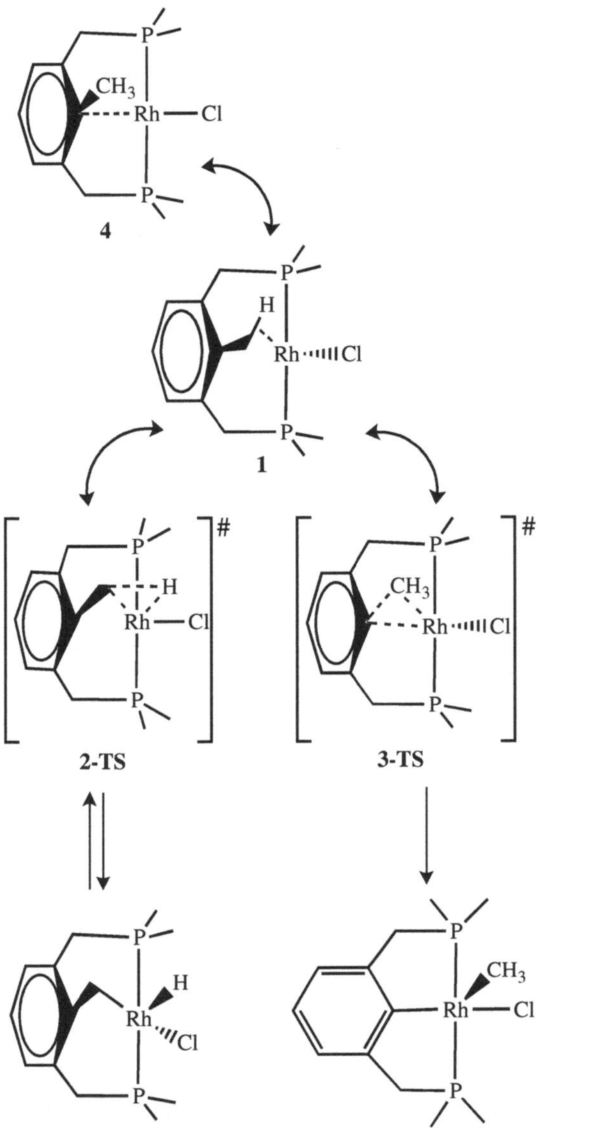

Fig. 3 Schematic representation of the stationary points involved in the CH/CC activation starting from one common intermediate structure

Simply reducing the metal basis set to def2-TZVP as well causes an error of 0.8 kcal/mol RMS, as does stripping the entire basis set down to def2-ASVP. As clearly the metal basis set is so important, we tried the combination of def2-TZVPP on the single metal atom with the simple def2-SVP basis set on everything else and found that this causes a surprisingly low error of 0.38 kcal/mol, presumably again due to error compensation between BSSE and intrinsic basis set incompleteness. For comparison, using def2-SVP on the metal as well puts one in 1 kcal/mol basis set error territory.

Yet, it seems that basis set dependence is actually the weakest of the three problem sets studied here, and interestingly enough, the same is true for the importance of Rh(4*s*,4*p*) correlation. If the latter were excluded, the def2-ASVP would actually cause a basis set incompleteness error of just 0.2 kcal/mol and def2-SVP of only 0.5 kcal/mol.

We can now proceed to benchmarking the DFT treatments (for which the def2-TZVPP basis set was used throughout). The results can be found in Table 8. Here, dispersion corrections have some effect, but nothing like what is seen in the Grubbs case. The performance of TPSS-D3BJ, at RMSD = 0.57 kcal/mol, can only be described as stunning: without dispersion, this only deteriorates slightly to 1.1 kcal/mol. But even the uncorrected PBE and BP86 functionals turn in errors below 1 kcal/mol, as do B97-1 and TPSS1KCIS. The double hybrids B2GP-PLYP and DSD-PBEB95 again turn in very good performances with or without dispersion corrections. B97D and B97D3 actually perform very well for this problem, unlike the mediocre performance for bare Pd (where it suffers from not being a hybrid) and the very poor one for Grubbs.

3.4 DFT functionals assessed jointly for all three problems

The average RMSD of all three problems combined is probably a more useful metric of performance for late-transition metal catalysis than each individually. Table 9 presents these statistics.

Table 7 CCSD(T*)-F12b RMSD in kcal/mol for Rh-PCP pincer complexes

	def2-TZVP(P)	awCVD(T)Z-F12[a]	def2-TZVP	SVP-TZVPP[b]	def2-SVP	def2-ASVP
RMSD w.r.t. CCSD(T*)-F12b/def2-TZVP(P)						
Default core	0.00	0.31	0.46	0.38	0.51	0.20
Rh(4*s*,4*p*)	0.00	0.23	0.78	0.38	0.98	0.84
RMSD w.r.t. CCSD(T*)-F12b/awCVD(T)Z-F12						
Default core	0.31	0.00	0.67	0.28	0.50	0.31
Rh(4*s*,4*p*)	0.23	0.00	0.92	0.32	0.84	0.73

[a] cc-pVDZ-F12 basis set on main-group elements and aug-cc-pwCVTZ-PP basis set on transition metal

[b] def2-SVP basis set on main-group elements and def2-TZVPP basis set on transition metal

Table 8 RMSD in kcal/mol for various DFT functionals relative to CCSD(T*)-F12b/def2-TZVP(P) for Rh-PCP pincer complexes

	B2GP-PLYP	DSD-PBEB95	DSD-PBEP86	DSD-PBEPBE	BP86	BLYP	PBE	M06L	mPWB95	TPSS	B97-1
No disp.	1.07	0.83	1.99	1.73	0.76	1.55	0.89	2.80	1.13	1.09	0.86
DFT-D2	0.71	0.67	1.56	1.13	1.71	1.14	1.49	2.97		0.45	1.19
DFT-D3BJ	0.92	0.56	1.71	1.38	1.18	1.11	1.23			0.57	

	B97-2	B97D	B3LYP	B3PW91	MPW1K	PBE0	PBEh1PBE	B1B95	BB1K	BMK	M06
No disp.	1.34		1.52	1.08	2.98	1.72	2.06	2.30	3.52	4.12	5.78
DFT-D2	2.38	1.22[a]	2.30	2.42		2.49		3.30		3.68	5.99
DFT-D3BJ		1.31[b]	1.97	1.97		2.21					

	M06-2X	MPW 1B95	mPW25 B95	mPW 28B95	MPWB1K	TPSS0	TPSS 1KCIS	TPSS21KCIS	TPSS25KCIS	ωB97x	M11
No disp.	4.21	2.75	2.30	2.52	3.88	0.91	0.65	1.07	1.36	3.03	2.46
DFT-D2	4.28					2.00	1.43			2.59[c]	2.64
DFT-D3BJ		3.02			4.15	1.54					

[a] Calculations were performed with the B97D functional and its built-in dispersion correction

[b] Calculations were performed with the B97D3 functional and its built-in dispersion correction

[c] Calculations were performed with the ωB97X-D functional and its built-in dispersion correction

Table 9 RMSD in kcal/mol for various DFT functionals averaged over the Pd reactions, the Grubbs catalyst, and the Rh-PCP C–C and C–H activation pathways (see text)

	B2GP-PLYP	DSD-PBEB95	DSD-PBEP86	DSD-PBEPBE	BP86	BLYP	PBE	M06L	mPWB95	TPSS	B97-1
No disp.	1.76	1.39	1.41	1.45	4.60	6.87	3.74	2.75	2.61	3.71	2.92
DFT-D2	0.74	1.23	1.35	1.08	3.74	3.96	3.55	2.46		3.50	1.50
DFT-D3BJ	1.06	1.10	1.36	1.13	3.19	4.55	2.95			2.40	

	B97-2	B97D	B3LYP	B3PW91	MPW1K	PBE0	PBEh1PBE	B1B95	BB1K	BMK	M06
No disp.	4.52		5.01	3.20	4.10	2.01	2.07	2.80	4.04	5.00	4.66
DFT-D2	2.15	4.04	2.35	2.20		2.07		2.68		5.55	4.63
DFT-D3BJ		4.31	2.81	1.39		1.58				5.97	

	M06-2X	MPW1B95	mPW 25B95	mPW28B95	MPWB1K	TPSS0	TPSS1KCIS	TPSS21KCIS	TPSS25 KCIS	ωB97x	M11
No disp.	6.47	2.60	2.08	2.31	4.12	2.01	2.82	2.86	3.12	3.94	3.72
DFT-D2	6.36					2.50	1.71			3.17	3.61
DFT-D3BJ		2.31			4.16	1.41					

The "workhorse" functionals BP86 and PBE, combined with D3BJ dispersion, yield values of about 3 kcal/mol—in both cases, the average of excellent performance for Grubbs and Rh(PCP) and apparent inability to cope with the bare-Pd systems. The same holds true for TPSS. The hybrids do not suffer from this problem, with PBE0-D3BJ, TPSS0-D3BJ, and B3PW91-D3BJ clocking in at respectable values of 1.6, 1.4, and 1.4 kcal/mol. All three functionals are clearly superior to B3LYP.

With the D2 correction, B97-1 and TPSS1KCIS performed surprisingly well with 1.5 and 1.7 kcal/mol RMS error, respectively. Only the double hybrids do still better, the best overall being B2GP-PLYP-D2 at 0.7 kcal/mol.

Heavily parametrized kinetic functionals such as BMK, M06, and M06-2X clearly are found wanting: the two range-separated hybrids M11 and wB97X-D do better, but still are not "magic bullets" for this class of problems. Functionals with B95 correlation all yield similar RMSDs of 2–3 kcal/mol.

As the present paper was being finalized for publication, a study by Chen and coworkers [126] appeared in which they propose ab initio benchmarks for Zr-mediated reactions. Basis set extrapolation [127, 128] of conventional ab initio data with comparatively small basis sets was used there in conjunction with core–valence corrections obtained from explicitly correlated calculations. For these early-transition metal systems, they found M06-2X to be the best performer [126]: the same group had argued earlier [129–131] that different systems required different DFT functionals.

4 Conclusions

We have performed CCSD(T*)-F12b benchmark calculations for the energetics and barrier heights of three late-transition metal systems, in increasing order of size: oxidative additions at bare Pd, a model for the Grubbs catalyst, and competition between CC and CH activation by a Rh(PCP) pincer complex. The results depend weakly on the basis set on the main-group atoms but are rather more sensitive to the basis set on the metal. An aug-cc-pwCVTZ-PP basis set on the metal combined with cc-pVTZ-F12 on the main-group elements yields barriers that are converged in terms of the basis set, but two smaller basis set combinations, namely def2-TZVP(P) and awCVD(T)Z-F12, appear to work well enough for most benchmark purposes. The former refers to the combination of def2-TZVP on the main group with def2-TZVPP on the transition metal, the latter to cc-pVDZ-F12 on the main group combined with aug-cc-pwCVTZ-PP on the metal. ($4s$,$4p$) inner-shell correlation cannot be neglected for even semi-accurate work.

While SCS-CCSD-F12 can approach CCSD(T)-F12b to within about 1.5 kcal/mol, this is a similar accuracy as can be obtained from the best DFT functionals, despite the vast difference in computational cost. Lower-level ab initio methods are generally not worthwhile.

For the Grubbs case, dispersion is very important and D3BJ clearly is superior over D2. Only the DSD-"noD" double hybrids perform consistently well in the absence of dispersion corrections.

Simple nonempirical (meta) GGAs with D3BJ dispersion work quite well for the Grubbs and pincer cases but break down for the Pd example: apparently, this latter situation (low oxidation state, no ligands) requires exact exchange. Hybrids of these same functionals, such as PBE0, TPSS0, and B3PW91, are among the best performers. If good performance for main-group barriers and general thermochemistry are also important, that narrows the choice to the double hybrids.

Acknowledgments J.M.L.M. is the Thatcher Professor of Chemistry at the Weizmann Institute of Science and a member *ad personam* of the Lise Meitner-Minerva Center for Computational Quantum Chemistry. The authors would like to thank Dr. Brina Brauer for assistance with the M11 calculations, Drs. Irena Efremenko (Weizmann) and Sebastian Kozuch (U. of North Texas) for helpful discussions, and Dr. Mark Vilensky for technical assistance with the Faculty of Chemistry computing cluster. This research was partially supported by the Lise Meitner-Minerva Center for Computational Quantum Chemistry as well as by the Helen and Martin A. Kimmel Center for Molecular Design.

References

1. Garralda MA (2009) Dalton Trans 3635–3645. doi: 10.1039/B817263C
2. Troegel D, Stohrer J (2011) Coord Chem Rev 255:1440–1459
3. Bahuleyan BK, Park D-W, Ha C-S, Kim I (2006) Catal Surv Asia 10:65–73
4. Michalak A, Ziegler T (2005) Top Organomet Chem 12:145–186
5. Cabri W, Candiani I (1995) Acc Chem Res 28:2–7
6. Beletskaya IP, Cheprakov AV (2000) Chem Rev 100:3009–3066
7. Poverenov E, Milstein D (2013) Top Organomet Chem 40:21
8. Gelman D, Romm R (2013) Top Organomet Chem 40:289
9. Nishina N, Yamamoto Y (2013) Top Organomet Chem 43:115
10. Zhang Q, Wu C, Zhou L, Li J (2013) Organometallics 32:415–426
11. Vougioukalakis GC, Grubbs RH (2010) Chem Rev 110:1746–1787
12. Martinez H, Miró P, Charbonneau P, Hillmyer MA, Cramer CJ (2012) ACS Catal 2:2547–2556
13. Minenkov Y, Occhipinti G, Jensen VR (2013) Organometallics 32:2099–2111
14. Geoghegan K, Evans P, Rozas I, Alkorta I (2012) Chem Eur J 18:13379–13387
15. Ray D, Nasima Y, Sajal MK, Ray P, Urinda S, Anoop A, Ray JK (2013) Synthesis 45:1261–1269
16. Nicolas E, Martin-Vaca B, Mézailles N, Bourissou D, Maron L (2013) Eur J Inorg Chem 2013:4068–4076
17. Chang Y-H, Nakajima Y, Tanaka H, Yoshizawa K, Ozawa F (2013) J Am Chem Soc 135:11791–11794
18. Cho D, Ko KC, Lee JY (2013) Organometallics 32:4571–4576
19. Raghavachari K, Trucks GW, Pople JA, Head-Gordon M (1989) Chem Phys Lett 157:479–483
20. Watts JD, Gauss J, Bartlett RJ (1993) J Chem Phys 98:8718–8733
21. Mutter ST, Platts JA (2010) Chem Eur J 16:5391–5399
22. Steinmetz M, Grimme S (2013) ChemistryOpen 2:115–124
23. Granatier J, Lazar P, Otyepka M, Hobza P (2011) J Chem Theory Comput 7:3743–3755
24. Craciun R, Picone D, Long RT, Li S, Dixon DA, Peterson KA, Christe KO (2010) Inorg Chem 49:1056–1070
25. Craciun R, Vincent AJ, Shaughnessy KH, Dixon DA (2010) Inorg Chem 49:5546–5553; erratum (2011) 50:5307
26. Craciun R, Long RT, Dixon DA, Christe KO (2010) J Phys Chem A 114:7571–7582
27. Chen M, Craciun R, Hoffman N, Dixon DA (2012) Inorg Chem 51:13195–13203
28. Li S, Hennigan JM, Dixon DA, Peterson KA (2009) J Phys Chem A 113:7861–7877
29. Vasiliu M, Li S, Arduengo AJ, Dixon DA (2011) J Phys Chem C 115:12106–12120
30. Kong L, Bischoff F, Valeev E (2012) Chem Rev 112:75–107
31. Hättig C, Klopper W, Köhn A, Tew DP (2012) Chem Rev 112:4–74

32. Kutzelnigg W, Morgan III JD (1992) J Chem Phys 96:4484–4508; erratum 97:8821
33. Dunning TH (1989) J Chem Phys 90:1007–1023
34. Dunning TH, Peterson KA, Woon DE (1998) Basis sets: correlation consistent sets. In: Schleyer PVR (ed) Encyclopedia of computational chemistry, vol 5. Wiley, Chichester. doi:10.1002/0470845015.cca053
35. Hill JG, Peterson KA, Knizia G, Werner HJ (2009) J Chem Phys 131:194105
36. Peterson KA, Dixon DA, Stoll H (2012) J Phys Chem A 116:9777–9782
37. Quintal MM, Karton A, Iron MA, Boese AD, Martin JML (2006) J Phys Chem A 110:709–716
38. Halkier A, Helgaker T, Jørgensen P, Klopper W, Koch H, Olsen J, Wilson AK (1998) Chem Phys Lett 286:243–252
39. Martin JML, de Oliveira G (1999) J Chem Phys 111:1843–1856
40. Zhao Y, Truhlar DG (2009) J Chem Theory Comput 5:324–333
41. Sundermann A, Uzan O, Milstein D, Martin JML (2000) J Am Chem Soc 122:7095–7104
42. Kesharwani MK, Efremenko I, Martin JML, to be published
43. Poverenov E, Efremenko I, Frenkel AI, Ben-David Y, Shimon LJW, Konstantinovski L, Martin JML, Milstein D (2008) Nature 455:1093–1096
44. Efremenko I, Poverenov E, Martin JML, Milstein D (2010) J Am Chem Soc 132:14886–14900
45. Montag M, Efremenko I, Cohen R, Shimon LJW, Leitus G, Diskin-Posner Y, Ben-David Y, Salem H, Martin JML, Milstein D (2010) Chem Eur J 16:328–353
46. Becke AD (1993) J Chem Phys 98:5648
47. Stephens PJ, Devlin FJ, Chabalowski CF, Frisch MJ (1994) J Phys Chem 98:11623
48. Adamo C, Barone V (1999) J Chem Phys 110:6158
49. Grimme S (2011) WIREs Comput Mol Sci 1:211–228
50. Grimme S (2006) J Chem Phys 124:034108
51. MOLPRO is a package of ab initio programs written by Werner H-J, Knowles PJ, Knizia G, Manby FR, Schütz M, Celani P, Korona T, Lindh R, Mitrushenkov A, Rauhut G, Shamasundar KR, Adler TB, Amos RD, Bernhardsson A, Berning A, Cooper DL, Deegan MJO, Dobbyn AJ, Eckert F, Goll E, Hampel C, Hesselmann A, Hetzer G, Hrenar T, Jansen G, Köppl C, Liu Y, Lloyd AW, Mata RA, May AJ, McNicholas SJ, Meyer W, Mura ME, Nicklaß A, O'Neill DP, Palmieri P, Pflüger K, Pitzer R, Reiher M, Shiozaki T, Stoll H, Stone AJ, Tarroni R, Thorsteinsson T, Wang M, Wolf A
52. Frisch MJ, Trucks GW, Schlegel HB, Scuseria GE, Robb MA, Cheeseman JR, Scalmani G, Barone V, Mennucci B, Petersson GA, Nakatsuji H, Caricato M, Li X, Hratchian HP, Izmaylov AF, Bloino J, Zheng G, Sonnenberg JL, Hada M, Ehara M, Toyota K, Fukuda R, Hasegawa J, Ishida M, Nakajima T, Honda Y, Kitao O, Nakai H, Vreven T, Montgomery Jr, JA, Peralta JE, Ogliaro F, Bearpark M, Heyd JJ, Brothers E, Kudin KN, Staroverov VN, Kobayashi R, Normand J, Raghavachari K, Rendell A, Burant JC, Iyengar SS, Tomasi J, Cossi M, Rega N, Millam NJ, Klene M, Knox JE, Cross JB, Bakken V, Adamo C, Jaramillo J, Gomperts R, Stratmann RE, Yazyev O, Austin AJ, Cammi R, Pomelli C, Ochterski JW, Martin RL, Morokuma K, Zakrzewski VG, Voth GA, Salvador P, Dannenberg JJ, Dapprich S, Daniels AD, Farkas Ö, Foresman JB, Ortiz JV, Cioslowski J, Fox DJ (2009) Gaussian 09, Revision D1. Gaussian, Inc, Wallingford CT
53. Neese F (2012) WIREs Comput Mol Sci 2:73–78
54. Weigend F, Ahlrichs R (2005) Phys Chem Chem Phys 7:3297–3305
55. Hättig C (2005) Phys Chem Chem Phys 7:59–66
56. Hellweg A, Hättig C, Höfener S, Klopper W (2007) Theor Chem Acc 117:587–597
57. Weigend F (2006) Phys Chem Chem Phys 8:1057–1065
58. Peterson KA, Adler TB, Werner H-J (2008) J Chem Phys 128:084102
59. Yousaf KE, Peterson KA (2008) J Chem Phys 129:184108
60. Yousaf KE, Peterson KA (2009) Chem Phys Lett 476:303–307
61. Kendall RA, Dunning TH, Harrison RJ (1992) J Chem Phys 96:6796–6806
62. Woon DE, Dunning TH (1994) J Chem Phys 100:2975–2988
63. Dunning TH, Peterson KA, Wilson AK (2001) J Chem Phys 114:9244–9253
64. Wilson AK, Woon DE, Peterson KA, Dunning TH (1999) J Chem Phys 110:7667–7676
65. Balabanov NB, Peterson KA (2005) J Chem Phys 123:064107
66. Peterson KA, Figgen D, Dolg M, Stoll H (2007) J Chem Phys 126:124101
67. Cao X, Dolg M (2011) WIREs Comp Mol Sci 1:200–210
68. Hill JG (2013) J Comput Chem 34:2168–2177
69. Adler TB, Knizia G, Werner H-J (2007) J Chem Phys 127:221106
70. Knizia G, Adler TB, Werner H-J (2009) J Chem Phys 130:054104
71. Ten-no S (2004) Chem Phys Lett 398:56–61
72. Marchetti O, Werner HJ (2008) Phys Chem Chem Phys 10:3400–3409
73. Marchetti O, Werner HJ (2009) J Phys Chem A 113:11580–11585, specifically eq (8) there
74. Grimme S (2003) J Chem Phys 118:9095
75. Szabados A (2006) J Chem Phys 125:214105
76. Grimme S, Goerigk L, Fink RF (2012) WIREs Comput Mol Sci 2:886
77. Fink RF (2010) J Chem Phys 133:174113
78. Takatani T, Hohenstein EG, Sherrill CD (2008) J Chem Phys 128:124111
79. Pitoňák M, Rezac J, Hobza P (2010) Phys Chem Chem Phys 12:9611
80. Perdew JP, Schmidt K (2001) AIP Conf Proc 577:1
81. Becke AD (1988) Phys Rev A 38:3098
82. Perdew JP (1986) Phys Rev B 33:8822–8824
83. Lee C, Yang W, Parr RG (1988) Phys Rev B 37:785
84. Perdew JP, Burke K, Ernzerhof M (1996) Phys Rev Lett 77:3865; erratum (1997) Phys Rev Lett 78:1396
85. Grimme S (2006) J Comp Chem 27:1787–1799
86. Tao J, Perdew JP, Staroverov VN, Scuseria GE (2003) Phys Rev Lett 91:146401
87. Zhao Y, Truhlar DG (2006) J Chem Phys 125:194101
88. Adamo C, Barone V (1998) J Chem Phys 108:664–675
89. Perdew JP (1991) In: Ziesche P, Eschrig H (eds) Electronic structure of solids'91. Akademie, Berlin, p 11
90. Perdew JP, Chevary JA, Vosko SH, Jackson KA, Pederson MR, Singh DJ, Fiolhais C (1992) Phys Rev B 46:6671–6687; erratum (1993) 48:4978
91. Lynch BJ, Fast PL, Harris M, Truhlar DG (2000) J Phys Chem A 104:4811–4815
92. Ernzerhof M, Perdew JP (1998) J Chem Phys 109:3313–3320
93. Hamprecht FA, Cohen AJ, Tozer DJ, Handy NC (1998) J Chem Phys 109:6264–6271
94. Wilson PJ, Bradley TJ, Tozer DJ (2001) J Chem Phys 115:9233–9242
95. Becke AD (1996) J Chem Phys 104:1040–1046
96. Zhao Y, Lynch BJ, Truhlar DG (2004) J Phys Chem A 108:2715–2719
97. Boese AD, Martin JML (2004) J Chem Phys 121:3405–3416
98. Zhao Y, Truhlar DG (2008) Theor Chem Acc 120:215–241
99. Zhao Y, Truhlar DG (2004) J Phys Chem A 108:6908
100. Grimme S (2005) J Phys Chem A 109:3067
101. Zhao Y, Lynch BJ, Truhlar DG (2005) Phys Chem Chem Phys 7:43–52

102. Karton A, Tarnopolsky A, Lamère JF, Schatz GC, Martin JML (2008) J Phys Chem A 112:12868
103. Kozuch S, Martin JML (2011) Phys Chem Chem Phys 13:20104–20107
104. Kozuch S, Martin JML (2013) J Comput Chem 34:2327–2344
105. Chai J-D, Head-Gordon M (2008) J Chem Phys 128:84106
106. Peverati R, Truhlar DG (2011) J Phys Chem Lett 2:2810–2817
107. Grimme S, Antony J, Ehrlich S, Krieg H (2010) J Chem Phys 132:154104
108. Grimme S, Ehrlich S, Goerigk L (2011) J Comput Chem 32:1456–1465
109. Fogueri UR, Kozuch S, Karton A, Martin JML (2013) Theor Chem Acc 132:1291
110. Lee TJ, Taylor PR (1989) J Quant Chem Symp S23:199–207
111. Karton A, Rabinovich E, Martin JML, Ruscic B (2006) J Chem Phys 125:144108
112. Papajak E, Zheng J, Xu X, Leverentz HR, Truhlar DG (2011) J Chem Theory Comput 7:3027–3034
113. Rappoport D, Furche F (2010) J Chem Phys 133:134105
114. Schultz N, Zhao Y, Truhlar DG (2005) J Phys Chem A 109:4388–4403
115. Schultz N, Zhao Y, Truhlar DG (2005) J Phys Chem A 109:11127–11143
116. Zhao Y, Schultz NE, Truhlar DG (2006) J Chem Theory Comput 2:364–382
117. DeYonker NJ, Cundari TR, Wilson AK (2006) J Chem Phys 124:114104
118. Mahler A, Wilson AK (2013) J Chem Theory Comput 9:1402–1407
119. Allen WD, East ALL, Császár AG (1993) In: Laane J, Dakkouri M, van der Veken B, Oberhammer H (eds) Structures and conformations of non-rigid molecules. Kluwer, Dordrecht, p 343
120. Curtiss LA, Raghavachari K, Trucks GW, Pople JA (1991) J Chem Phys 94:7221–7230 and references therein
121. Rybtchinski B, Milstein D (2004) ACS Symp Ser 885:70
122. Rybtchinski B, Milstein D (1999) Angew Chem Int Ed 38:870–883
123. van der Boom ME, Milstein D (2003) Chem Rev 103:1759–1792
124. Rybtchinski B, Oevers S, Montag M, Vigalok A, Rozenberg H, Martin JML, Milstein D (2001) J Am Chem Soc 123:9064–9077
125. Sundermann A, Uzan O, Martin JML (2001) Organometallics 20:1783–1791
126. Sun Y, Chen H (2013) J Chem Theor Comput 9:4735–4743
127. Karton A, Martin JML (2006) Theor Chem Acc 115:330–333
128. Truhlar DG (1998) Chem Phys Lett 294:45–48
129. Kang RH, Lai WZ, Yao JN, Shaik S, Chen H (2012) J Chem Theory Comput 8:3119–3127
130. Lai WZ, Yao JN, Shaik S, Chen H (2012) J Chem Theory Comput 8:2991–2996
131. Kang RH, Yao JN, Chen H (2013) J Chem Theory Comput 9:1872–1879

Theor Chem Acc (2014) 133:1463
DOI 10.1007/s00214-014-1463-z

REGULAR ARTICLE

Simulating Cl K-edge X-ray absorption spectroscopy in MCl_6^{2-} (M = U, Np, Pu) complexes and $UOCl_5^{-}$ using time-dependent density functional theory

Niranjan Govind · Wibe A. de Jong

Received: 23 September 2013 / Accepted: 7 February 2014 / Published online: 21 February 2014

Abstract We report simulations of the X-ray absorption near edge structure at the Cl K-edge of actinide hexahalides MCl_6^{2-} (M = U, Np, Pu) and the $UOCl_5^{-}$ complex using linear response time-dependent density functional theory extended for core excitations. To the best of our knowledge, these are the first calculations of the Cl K-edge spectra of $NpCl_6^{2-}$ and $PuCl_6^{2-}$. In addition, the spectra are simulated with and without the environmental effects of the host crystal as well as ab initio molecular dynamics to capture the dynamical effects due to atomic motion. The calculated spectra are compared with experimental results, where available and the observed trends are discussed.

Keywords Actinides · Plutonium · Uranium · Neptunium · K-edge · Chlorine · X-ray absorption · Spectroscopy · XAS · XANES

Dedicated to Professor Thom Dunning and published as part of the special collection of articles celebrating his career upon his retirement.

N. Govind (✉) · W. A. de Jong (✉)
Environmental Molecular Sciences Laboratory,
Pacific Northwest National Laboratory,
Richland, WA 99352, USA
e-mail: niri.govind@pnl.gov

W. A. de Jong
e-mail: wadejong@lbl.gov

Present Address:
W. A. de Jong
Scientific Computing Group, Lawrence Berkeley National Laboratory, Berkeley, CA 94720, USA

1 Introduction

X-ray absorption spectroscopy (XAS) is a powerful and widely used technique for probing the local electronic and geometrical structure properties of matter at the atomic scale. In XAS, the electrons are excited from an atomic core state to bound valence or continuum states, and the spectrum is divided into energetically separate edges. By convention, the edges are labeled according to the core electron that is excited (for example, K-edge for $1s$, L_1-edge for $2s$, L_2-edge for $2p_{1/2}$ and L_3-edge for $2p_{3/2}$). Absorptions near the rising edge, also referred to as X-ray absorption near edge spectroscopy or X-ray absorption near edge structure (XANES), involve excitations from the relevant core state to increasingly high-energy virtual states. This gives insight into the chemical state of the atom (for example, coordination, bonding and oxidation state). Higher energy absorption sufficient to result in photoelectron ejection is called extended X-ray absorption fine structure or EXAFS. This yields information about atoms in the vicinity, atomic identity, coordination and solvation structure of charged ions in solution [1–3].

Since EXAFS involves backscattering from nearby atoms, the physics is essentially captured by the phase shifts and scattering amplitudes from which the spectrum can be calculated using a damped spherical photoelectron wavefunction approximation. Nowadays, these are routinely computed using Green's function-based approaches [4, 5], and the approach is especially powerful when combined with ab initio molecular dynamics simulations (AIMD) [6, 7]. XANES, on the other hand, requires a full electronic structure treatment of the absorbing center and neighboring atoms constituting the chemical environment. Over the years, various approaches have been pioneered like the static exchange approximation (STEX) and static approaches [8, 9]. All these approaches incorporate the relaxation of the core hole and valence orbitals indirectly by manual preparation of the core hole state and/or use of modified core potentials that mimic the core relaxation when a core electron is excited [10–13]. The absorption spectra are computed by calculating the transition dipole elements of the initial and final states of interest. In a nutshell, these methods are very sensitive to the choice of initial and final states.

Recently, linear response time-dependent density functional theory (LR-TDDFT) has been extended to treat core-level excitations [14–18]. The advantage of this approach is that no manual preparation of the core hole state is needed, and peak separations and trends are properly captured. However, the quality of the results is dependent on the exchange-correlation functional choice [19]. Another approach is real-time TDDFT (RT-TDDFT) [20]. We have recently applied our LR-TDDFT approach to several studies including the K-edges of oxygen, carbon, fluorine in a number of molecular systems (CO, H2O and fluorobenzenes) [18], ruthenium L_3-edge in$[Ru(NH_3)_6]^{3+}$, ruthenium L_3-edge in a series of model Ru(II) and Ru(III) complexes and mixed-valence metal (Ru/Fe) dimers [19], the K-edges spectra of oxygen, nitrogen and sulfur in cysteine [21], and the entangled valence electron-hole dynamics in stimulated attosecond X-ray Raman scattering [22]. Approaches combining DFT and the restricted open-shell configuration interaction method have also been recently reported [23, 24].

Here, we apply our approach to theoretically study the Cl K-edge spectra of the closed-shell $UOCl_5^-$ complex and the open-shell tetravalent actinide hexahalides series MCl_6^{2-} (M = U, Np, Pu) formally containing 2, 3 and 4 *5f*-electrons. Cl K-edge spectroscopy has been used to study the electronic structure of actinide complexes and to probe the role of *5f* and *6d* orbitals in the formation of covalent bonds [25, 26]. Across the actinides series, the *5f*-orbitals become more contracted and their role in bonding changes, which we expect to be reflected in the Cl K-edge spectra [27]. Our study specifically probes the role of the environment of the host crystal and the dynamical effects due to atomic motion on the shape of the spectra. The computed spectra are compared with recent gas-phase computational results as well as available experiments [25]. The rest of this paper is organized as follows: We first summarize our computational approach followed by an analysis and discussion of our results.

2 Computational approach

Within LR-TDDFT, the excitation energies are obtained as solutions to the frequency domain time-dependent Kohn–Sham equations (TDKS). The implementation in the NWChem quantum chemistry program [28, 29] is based on the iterative subspace Davidson algorithm, which is an effective approach for valence excitations [30–34]. However, this approach becomes expensive for core excitations because the complete excitation matrices involving a very large number of excited states have to be constructed and diagonalized. Formally, the numerical cost of diagonalizing the full TDDFT equations scales as $O(N^6)$, because of the tetradic nature of the RPA matrix [35]. This problem was overcome via a restricted excitation window (REW) approach, where calculations are performed within a smaller model subspace of single excitations from the relevant core orbitals that can be selected based on energies or orbital number [18]. This is a valid Ansatz because core excitation energies are well separated from pure valence–valence excitations. Similar implementations have also been reported and validated by other groups [14–17]. Further details of our approach can be found in Ref. [18], and for the details of TDDFT and the relevant working equations, we refer the interested reader to comprehensive reviews on the subject [30–38].

For the Gaussian basis set-based molecular calculations, the actinide atoms (U, Np and Pu) were represented with small core Stuttgart relativistic ECPs (replaces 60 core electrons) and corresponding basis sets containing g-functions [39, 40], while the ligand atoms (Cl and O) were represented with the aug-cc-pvtz basis [41, 42]. The actinide basis sets are of triple zeta quality. In general and as expected, the absolute positions of the core spectra are dependent on the choice of exchange-correlation functional as well as the basis set [19]. We have tested the B3LYP (20 % HF) [43–46], PBE0 (25 % HF) [28] and Becke Half-and-Half (50 % HF) [47] functionals, respectively. The PBE0 functional [58] gave the best agreement with experiment for all the actinide complexes considered in this study [25]. The spectra were shifted by ~58.4 eV to match experiment and uniformly Lorentzian broadened (1.0 eV). Unlike valence and charge-transfer (CT) excitations where range-separated exchange-correlation functionals offer significant improvement, they offer little or no improvement in the strongly localized core spectra. For the molecular calculations with the environment, the actinide complex in each case was placed in a set of point charges (determined using a Mulliken analysis), representing the host crystalline solid. The actinide hexahalides MCl_6^{2-} (M = U, Np, Pu) were performed with unrestricted DFT (triplet, quartet and quintet), and the calculated ground state $\langle S^2 \rangle$ values were 2.0116 (exact: 2.0), 3.7725 (exact: 3.75) and 6.0833 (exact: 6.0), respectively. The $UOCl_5^{1-}$ system was calculated with restricted DFT. The observed small spin contamination could be an indication of the presence of higher spin state character. In addition, the spin–

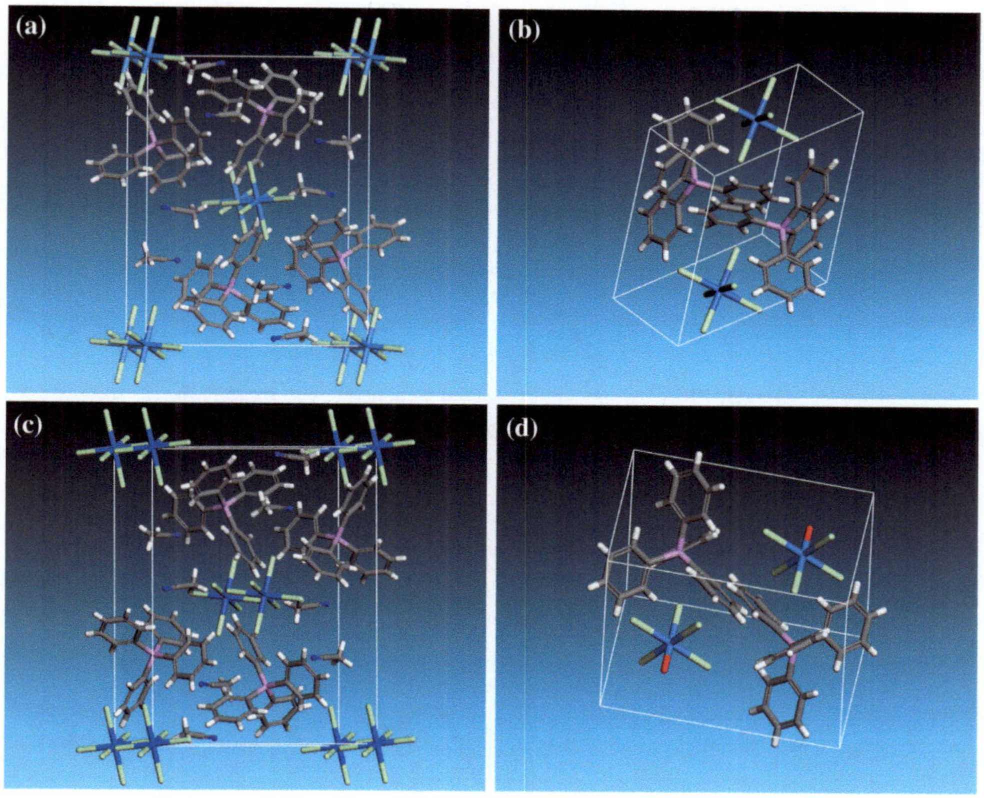

Fig. 1 Crystal structures of **a** $(Ph_4P)_2UCl_6{\cdot}4CH_3CN$, **b** $(Ph_4P)_2NpCl_6$, **c** $(Ph_4P)_2PuCl_6{\cdot}4CH_3CN$ and **d** $(Ph_4P)_2UOCl_5$ as reported in Refs. ([56], private communication with S. Kozimor)

orbit coupling for UCl_6^{2-} and $NpCl_6^{2-}$ is approximately 0.20–0.25 eV [48, 49] and will also result in a number of closely spaced excited states that may not be experimentally resolvable [25] as the natural width of Cl K-edge transitions is ~0.64 eV [50]. We cannot assess the importance of these contributions because our current implementation does not include spin-orbital and multiple excited-state electronic configuration effects [51]. However, the key features and the overall shape of the spectra are in qualitative agreement with experiment as we will discuss in the Results section.

To capture the role of dynamical motions of the atoms, planewave-based Car-Parrinello AIMD simulations were performed with the PBE exchange-correlation functional [52, 53]. Norm-conserving Troullier–Martins pseudopotentials [54] with relativistic corrections were used for the actinides, and Hamann pseudopotentials [55] were used for the other atoms in the system. Other calculation parameters include an energy cutoff of 100 Ry, density cutoff of 200 Ry, a simulation time step of 0.121 fs, and a fictitious electron mass of 600 au, respectively. The unit cell dimensions, symmetry and the atomic coordinates were taken from experimental data [56], [$UOCl_5^-$ via private communication with S. Kozimor]. Replacing all hydrogen atoms with deuterium allows for larger integration time steps. The simulation temperature was set at 300 K and controlled using a Nose–Hoover thermostat [57]. Approximately 10 ps of dynamics was collected for each of the structures. In our sampling, the first 3 ps was considered the equilibration phase and ignored.

Cl K-edge core absorption spectra calculations using REW-LR-TDDFT were performed on all the complexes using: (1) experimental molecular geometries without the environment, (2) experimental molecular geometries with the environment, and (3) 100 molecular geometry snapshots extracted from periodic AIMD runs, which were averaged to form the final spectra. Three hundred excited states from the Cl 1*s* states were calculated for each system. We note that optimized geometries do not change the overall shape and nature of the spectrum. All calculations were performed with NWChem 6.3 [28, 29].

3 Results

In this section, we present and discuss our calculated REW-LR-TDDFT Cl K-edge spectra of actinide hexahalides MCl_6^{2-} (M = U, Np, Pu). The structures of the different complexes are presented in Fig. 1.

3.1 UCl_6^{2-}

The ground state is an open-shell triplet ($5f^2$) where the HOMO/LUMO character of the alpha electrons is

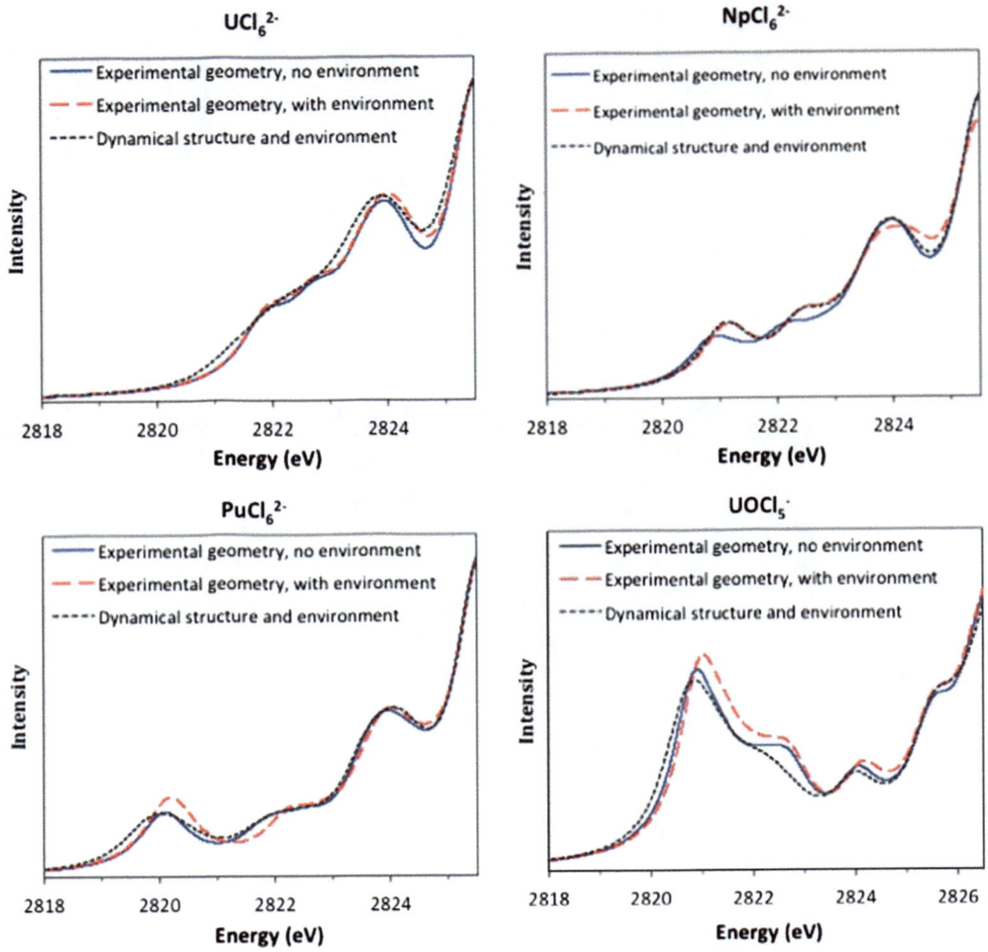

Fig. 2 Cl K-edge spectra of UCl_6^{2-}, $NpCl_6^{2-}$, $PuCl_6^{2-}$ and $UOCl_5^{1-}$ in the gas phase, in the presence of the host crystal and including dynamical effects using AIMD simulations

dominated by Cl 3*p* (HOMO−2), U 5*f* (HOMO−1), U 5*f* (HOMO), U 5*f* (LUMO), U 5*f* (LUMO+(1–4)), while orbitals containing U 6*d* were found at higher energy. For the beta electrons, we observe the Cl 3*p* (HOMO−2), Cl 3*p* (HOMO−1), Cl 3*p* (HOMO), U 5*f* + Cl 3*s* (LUMO) and U 5*f* and U 5*f* + Cl 3*s*/3*p* (LUMO+(1–4)), respectively. All the higher states are a mixture of U and Cl states.

The shape of the spectrum (Fig. 2) remains largely unaffected with and without the environment. The shoulder between 2,821.75 and 2,823 eV, which consists of two close peaks separated by ~0.75 eV and composed of excitations from the Cl 1*s* → U 5*f* states, is mostly smoothened out when the averaged spectrum is calculated using AIMD snapshots. The prominent peak at ~2,824 eV arises from excitations from the Cl 1*s* → U 6*d* + Cl 3*s*/3*p* states. On the whole, our calculated spectrum is in better agreement with experiment, in terms of the peak intensities and locations, compared with published gas-phase computational results [25].

3.2 $NpCl_6^{2-}$

The ground state is an open-shell quartet ($5f^3$) where the HOMO/LUMO character of the alpha electrons is dominated by Cl 3*p* (HOMO−3), Np 5*f* (HOMO−2), Np 5*f* + Cl 3*p* (HOMO−1), Np 5*f* + Cl 3*p* (HOMO), Np 5*f* (LUMO), Np 5*f* (LUMO+(1–3)). While the 5*f* orbitals in uranium were found to be atomic like, a much larger mixing of 5*f* with the Cl 3*p* is found for Np. For the beta electrons, we observe the Cl 3*p* (HOMO−3), Cl 3*p* (HOMO−2), Cl 3*p* (HOMO−1), Cl 3*p* (HOMO), and Np 5*f* (LUMO) and Np 5*f* and Np 5*f* + Cl 3*s*/3*p* (LUMO+(1–3)), respectively. The Np 6*d* orbitals are found above the Np 5*f* orbitals, and all the higher states are a mixture of Np and Cl orbitals.

The shoulder (Fig. 2) which is composed of two closely spaced features in UCl_6^{2-} is clearly separated by ~1.5 eV and red-shifted in $NpCl_6^{2-}$. The first feature ~2,821 eV arises from transitions from Cl 1*s* → Np 5*f* states, while the second peak ~2,822.5 eV is dominated by transitions

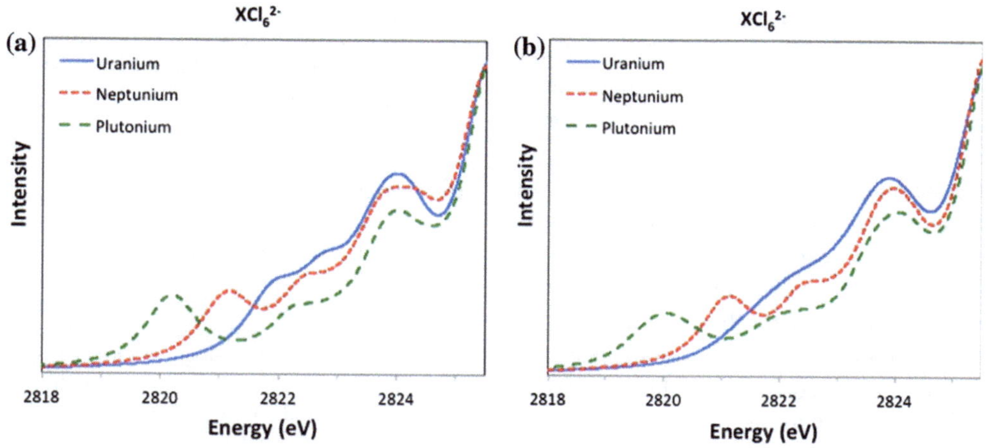

Fig. 3 Comparative Cl K-edge spectra of UCl_6^{2-}, $NpCl_6^{2-}$ and $PuCl_6^{2-}$ in the presence of the host environment. **a** Compares the spectra including the host environment. **b** Compares the spectra averaged over 100 configurations extracted from AIMD runs. The three spectra in each figure are normalized relative to each other such that the peaks ~2,826 have the same intensity. Note that the shoulder in UCl_6^{2-} between ~2,821.75 and ~2,823 eV, which consists of two peaks separated by ~0.75 eV is mostly smoothened out in the averaged spectra. These peaks are not resolved in the experimental data [25]

from Cl 1*s* → Np 5*f* + Cl 3*s*/3*p* states, respectively. The third peak ~2,824 eV is composed of excitations from Cl 1*s* → Np 6*d* + Cl 3*s*/3*p* states. Although there are differences with and without the environment and including dynamics, the overall shape and the number of features are preserved.

3.3 $PuCl_6^{2-}$

The ground state is an open-shell quintet ($5f^4$) where the HOMO/LUMO character of the alpha electrons is dominated by Cl 3*p* (HOMO−3), Pu 5*f* + Cl 3p (HOMO−2), Pu 5*f* + Cl 3*p* (HOMO−1), Pu 5*f* + Cl 3*p* (HOMO), Pu 5*f* (LUMO) and Pu 5*f* (LUMO+(1–2)). A further increase in mixing of the 5*f* with the Cl 3*p* is observed for Pu compared with both U and Np. For the beta electrons, we observe the Cl 3*p* (HOMO−3), Cl 3*p* (HOMO−2), Cl 3*p* (HOMO−1), Cl 3*p* (HOMO), Pu 5*f* + Cl 3*s*/3*p* (LUMO) and Pu 5*f* and Pu 5*f* + Cl 3*s*/3*p* (LUMO+(1–2)). The Pu 6*d* orbitals are found above the Pu 5*f* orbitals, and higher states are a mixture of Pu and Cl orbitals.

There are two clear early features (Fig. 2) separated by ~2.0 eV and further red-shifted compared with $NpCl_6^{2-}$. The first feature ~2,820 eV arises from transitions from Cl 1*s* → Pu 5*f* states, while the second peak ~2,822 eV is dominated by transitions from Cl 1*s* → Pu 5*f* + Cl 3*s*/3*p* states, respectively. The third peak ~2,824 eV is composed of excitations from Cl 1*s* → Pu 6*d* + Cl 3*s*/3*p* states. As in the $NpCl_6^{2-}$ spectra, there are differences in the spectra; however, the overall nature of the spectrum remains largely unaffected as in $PuCl_6^{2-}$ even with inclusion of dynamics.

The three actinide hexahalide spectra are compared in Fig. 3. There is a clear trend in the first and second features that is a direct consequence of the behavior of the 5*f* electrons in the actinide series. In addition to not being shielded by the filled 6*s* and 6*p* shells, the 5*f* electrons also shield each other poorly from the nucleus resulting in a lowering of their orbital energies with increasing atomic number. This is reflected in the increased spacing and red-shifting of the first and second features as one goes from UCl_6^{2-} → $PuCl_6^{2-}$. The second peak, as discussed above, is a result of the mixing of the actinide 5*f* orbitals with the Cl 3*s*/3*p* states, which increases as one moves from U to Pu. The prominent feature at ~2,824 eV is mostly unaffected in the three spectra, reflecting the much smaller shift in orbital energies compared with the 5*f* in the U, Np and Pu series.

3.4 $UOCl_5^{-}$

The ground state has a closed-shell configuration ($5f^0$) where HOMO/LUMO character is dominated by Cl 3*p* states (HOMO−2, HOMO−1, HOMO), while the unoccupied states are dominated by U 5*f* orbitals (LUMO, LUMO+1, LUMO+2 and LUMO+3). Higher states are a mixture of U, Cl and O orbitals.

The computed Cl K-edge spectra (Fig. 2) show two prominent features ~2,821 eV (Cl 1*s* → U 5*f*) and ~2,822.5 eV (Cl 1*s* → U 5*f* + Cl 3*s* + O 2*p*) that are distinct from the UCl_6^{2-} complex. The Cl 1*s* → U

$6d$ + Cl $3s/3p$ transitions are split into two peaks one at ~2,824 eV and a second (shoulder) at ~2,826 eV. These are similar in nature to the analogs in the actinide hexahalide complexes and also located in the same position in the spectrum. On the whole, the spectra with and without the host environment and including dynamical effects are very similar. The agreement with experiment is reasonable.

However, we do not observe the shoulder at ~2,820.5 eV that is clearly visible in the experimental data [25] by either introducing the crystal field via the host environment or dynamics. We analyzed this further by asking the question, what structural changes are needed to reproduce the experimental spectrum and the shoulder in particular?

Experimentally, the axial U–O and U–Cl bonds are 1.78 and 2.43 Å, and the equatorial U–Cl bonds are all 2.54 Å, respectively. If the axial U–Cl bond is elongated by ~0.25 Å and two of the equatorial U–Cl bonds (opposite to each other) are elongated by ~0.25 Å, all the experimental features are reasonably captured including the shoulder. However, these bond length elongations (~10 %) are unphysical and cannot be accounted for by a weak host crystal field, dynamics and different oxidation states (which only accounts for ~0.08 Å per charge of charge). It is possible that this feature may be related to spin-flip effects. We do not have a way to assess this at the moment as our TDDFT implementation does not include these effects.

4 Conclusions

In summary, we have applied our recently developed REW-LR-TDDFT approach to compute the Cl K-edge spectra of actinide hexahalides MCl_6^{2-} (M = U, Np, Pu) and the $UOCl_5^-$ complex in the gas phase as well as in the environment of the host crystal. We have reported the computed Cl K-edge spectra for $NpCl_6^{2-}$ and $PuCl_6^{2-}$ for the first time. In order to capture the influence of thermal effects and atomic motion, we have also calculated the averaged spectra over snapshots extracted from periodic planewave-based room temperature (300 K) AIMD simulations. We find that our UCl_6^{2-} results are in better agreement with experiment compared with recently published gas-phase computational results [25]. We also discuss the systematic trends in the spectra over the U, Np and Pu series as well as the spectrum of the $UOCl_5^-$ complex.

Acknowledgments This research was funded by the BES Heavy Element Chemistry program in the Division of Chemical Sciences, Geosciences, and Biosciences, Office of Basic Energy Sciences, US Department of Energy. All the calculations were performed using the Molecular Science Computing Capability at EMSL, a national scientific user facility sponsored by the US Department of Energy's Office of Biological and Environmental Research and located at Pacific Northwest National Laboratory (PNNL). PNNL is operated for the Department of Energy by the Battelle Memorial Institute under Contract DE-AC06-76RLO-1830. Discussions with and access to unpublished $UOCl_5$ structural data from Stosh Kozimor (LANL) are gratefully acknowledged.

References

1. Stöhr J (2003) NEXAFS spectroscopy. Springer, New York, pp 8–47
2. de Groot F, Kotani, A (2008) Core Level Spectroscopy of Solids. CRC Press, Boca Raton, FL:1–37, 457–462
3. Penfold T, Milne C, Chergui M (2013) Adv Chem Phys 153:1–41
4. Rehr JJ, Albers RC (2000) Rev Mod Phys 72:621–654
5. Rehr JJ, Ankudinov A (2005) Coord Chem Rev 249:131–140
6. Pham V, Tavernelli I, Milne C, van der Veen R, D'Angelo P, Bressler C, Chergui M (2010) Chem Phys 371:24–29
7. Atta-Fynn R, Johnson D, Bylaska E, Ilton E, Schenter G, de Jong WA (2012) Inorg Chem 51:3016–3024
8. Ågren H, Carravetta V, Vahtras O, Pettersson L (1994) Chem Phys Lett 222:75–81
9. Villaume S, Ekstrom U, Ottosson H, Norman P (2010) Phys Chem Chem Phys 12:5596–5604
10. Triguero L, Pettersson L, Ågren H (1998) Phys Rev B 58:8097–8110
11. Cavalleri M, Ogasawara H, Pettersson L, Nilsson A (2002) Chem Phys Lett 364:363–370
12. Iannuzzi M, Hutter J (2007) Phys Chem Chem Phys 9:1599–1610
13. Prendergast D, Galli G (2006) Phys Rev Lett 96:215502
14. Ray K (2007) DeBeer George S, Solomon E, Wieghardt K, Neese F. Chem Eur J 13:2783–2797
15. Stener M, Fronzoni G, de Simone M (2003) Chem Phys Lett 373:115–123
16. Besley N, Asmuruf F (2010) Phys Chem Chem Phys 12:12024–12039
17. Liang W, Fischer SA, Frisch MJ, Li X (2011) J Chem Theory Comput 7:3540–3547
18. Lopata K, Van Kuiken BE, Khalil M, Govind N (2012) J Chem Theory Comput 8:3284
19. Van Kuiken BE, Valiev M, Daifuku SL, Bannan C, Strader ML, Cho H, Huse N, Schoenlein RW, Govind N, Khalil M (2013) J Phys Chem A 117:4444–4454
20. Lopata K, Govind N (2011) J Chem Theory Comput 7:1344–1355
21. Zhang Y, Biggs JD, Healion D, Govind N, Mukamel S (2012) J Chem Phys 137:194306
22. Healion D, Zhang Y, Biggs JD, Govind N, Mukamel S (2012) J Phys Chem Lett 3:2326–2331
23. Ikeno H, de Groot FMF, Stavitski E, Tanaka I (2009) J Phys: Condens Matter 21:104208
24. Roemelt M, Maganas D, DeBeer S, Neese F (2013) J Chem Phys 138:204101
25. Minasian SG, Keith JH, Batista ER, Boland KS, Clark DL, Conradson SD, Kozimor SA, Martin RL, Schwarz DE, Shuh DK, Wagner GL, Wilkerson MP, Wolfsberg LE, Yang P (2012) J Am Chem Soc 134:5586–5597
26. Spencer LP, Yang P, Minasian SG, Jilek RE, Batista ER, Boland KS, Boncella JM, Conradson SD, Clark DL, Hayton TW, Kozimor SA, Martin RL, MacInnes MM, Olson AC, Scott BL, Shuh DK, Wilkerson MP (2012) J Am Chem Soc 135:2279–2290
27. Neidig ML, Clark DL, Martin RL (2013) Coord Chem Rev 257:394–406
28. Valiev M, Bylaska EJ, Govind N, Kowalski K, Straatsma TP, Van Dam HJJ, Wang D, Nieplocha J, Apra E, Windus TL, de Jong WA (2010) Comp Phys Comm 181:1477–1489

29. NWChem 6.3. http://www.nwchem-sw.org. Accessed 1 June 2013
30. Hirata S, Head-Gordon M (1999) Chem Phys Lett 302:375–382
31. Hirata S, Head-Gordon M (1999) Chem Phys Lett 314:291–299
32. Bauernschmitt R, Haser M, Treutler O, Ahlrichs R (1997) Chem Phys Lett 264:573–578
33. Bauernschmitt R, Ahlrichs R (1996) Chem Phys Lett 256:454–464
34. Stratmann R, Scuseria G, Frisch MJ (1998) Chem Phys 109:8218–8224
35. Tretiak S, Isborn C, Niklasson A, Challacombe MJ (2009) J Chem Phys 130:054111
36. Casida ME (1995) In: Chong PD (ed) Recent Advances in Density Functional Methods. World Scientific Publishing, River Edge, NJ, Vol. 1, Chap 5:155–192
37. Marques MA, Ullrich CA, Nogueira F, Rubio A, Burke K, Gross EK (eds) (2006) Time-dependent density functional theory (Lecture Notes in Physics), 1st edn. Springer, New York
38. Marques MA, Maitra NT, Nogueira FM, Gross EK, Rubio A (2012) Fundamentals of time-dependent density functional theory. Springer, Berlin
39. Cao X, Dolg M, Stoll H (2003) J Chem Phys 118:487–496
40. Cao X, Dolg M (2004) J Molec Struct (Theochem) 673:203–209
41. Dunning TH Jr (1989) J Chem Phys 90:1007–1023
42. Woon DE, Dunning TH Jr (1993) J Chem Phys 98:1358–1371
43. Becke AD (1993) J Chem Phys 98:5648–5652
44. Lee C, Yang W, Parr RG (1988) Phys Rev B 37:785–789
45. Vosko SW, Wilk L, Nusair M (1980) Can J Phys 58:1200–1211
46. Stephens PJ, Devlin FJ, Chabalowski CF, Frisch MJ (1994) J Phys Chem 98:11623–11627
47. Becke AD (1993) J Chem Phys 98:1372–1377
48. Wagner W, Edelstein N, Whittaker B (1977) Inorg Chem 16:1021–1026
49. Menzel ER, Gruber JB (1971) J Chem Phys 54:3857–3864
50. Krause MO, Oliver JH (1979) J Phys Chem Ref Data 8:329–338
51. Moore KT, van der Laan G (2009) Rev Mod Phys 81:235–298
52. Perdew JP, Burke K, Ernzerhof M (1996) Phys Rev Lett 77:3865–3868
53. Perdew JP, Burke K, Ernzerhof M (1997) Phys Rev Lett 78:1396
54. Troullier N, Martins JL (1990) Phys Rev B 43:1993–2006
55. Hamann DR (1989) Phys Rev B 40:2980–2987
56. Minasian SG, Boland KS, Feller RK, Gaunt AJ, Kozimor SA, May I, Reilly SD, Scott BL, Shuh DK (2012) Inorg Chem 51:5728–5736
57. Frenkel D, Smit B (2001) Understanding molecular simulation, 2nd edn: from algorithms to applications. Academic Press, San Diego
58. Adamo C, Barone V (1999) J Chem Phys 110:6158–6170

Theor Chem Acc (2014) 133:1446
DOI 10.1007/s00214-014-1446-0

REGULAR ARTICLE

Accurate atomization energies from combining coupled-cluster computations with interference-corrected explicitly correlated second-order perturbation theory

Konstantinos D. Vogiatzis · Robin Haunschild · Wim Klopper

Received: 2 September 2013 / Accepted: 7 January 2014 / Published online: 30 January 2014

Abstract Interference-corrected explicitly correlated second-order perturbation theory (INT-MP2-F12) is applied to accelerate the convergence to the complete-basis-set limit of coupled-cluster computations. Adding energy terms obtained from INT-MP2-F12 theory to the energies obtained from coupled-cluster singles-and-doubles (CCSD) computations yields a mean absolute deviation (MAD) from explicitly correlated CCSD results below 1 kJ/mol for a test set of 106 molecules. A composite scheme for the computation of atomization energies is assessed. This scheme is denoted as CCSD(T)+F12+INT and consists of the CCSD model with perturbative triples (CCSD(T)) supplemented with INT-MP2-F12 corrections, using a quadruple-zeta quality basis set (cc-pVQZ-F12). The composite scheme achieves chemical accuracy with respect to experimentally derived or computed reference values. Using Boys localized molecular orbitals, the MAD of the CCSD(T)+F12+INT/cc-pVQZ-F12 atomization energies from the reference values is below 1 kJ/mol for the G2/97 test set.

Keywords Atomization energy · Performance assessment · Test set · Coupled-cluster theory · Explicit correlation · Interference effects

1 Introduction

The *intrinsic error* of an ab initio method (i.e., the difference between theory and experiment) can only be revealed by approaching the complete-basis-set (CBS) limit of the respective method [1, 2]. This can be achieved either by extrapolations based on a series of computations in a systematic hierarchy of basis sets (e.g., Dunning's correlation-consistent basis sets [3]) or by the use of explicitly correlated wave functions [4–9].

Today, coupled-cluster theory with iterative single and double excitations followed by a non-iterative, perturbative treatment of triple excitations to the fifth order (CCSD(T)) [10] constitutes the "gold standard" of quantum chemistry. However, its *inherent high accuracy* is only obtained close to the CBS limit. Rather than computing this limit directly, for example, by using explicitly correlated coupled-cluster theory, one can combine small-basis-set CCSD(T) energies with results from low-level perturbation theory at the CBS limit, as follows:

$$\delta E_{\mathrm{CCSD(T)/CBS}} \approx \delta E_{\mathrm{MP2/CBS}} + (\delta E_{\mathrm{CCSD(T)}} - \delta E_{\mathrm{MP2}})_{\text{small basis}} \tag{1}$$

(the notation δE is used for correlation energies while total energies will be noted without the δ symbol). δE_{MP2} denotes the second-order Møller–Plesset (MP2) correlation energy, and typically, a double- or triple-zeta-quality basis can be regarded as "small basis." Expressions similar to Eq. (1) have been used on numerous occasions in the past. These are too numerous to be discussed here, but some of

Dedicated to Professor Thom Dunning and published as part of the special collection of articles celebrating his career upon his retirement.

K. D. Vogiatzis · W. Klopper (✉)
Institute of Nanotechnology, Karlsruhe Institute of Technology (KIT), Campus North, P.O. Box 3640, 76021 Karlsruhe, Germany
e-mail: klopper@kit.edu

K. D. Vogiatzis · R. Haunschild · W. Klopper
Institute of Physical Chemistry, Karlsruhe Institute of Technology (KIT), Campus South, P.O. Box 6980, 76049 Karlsruhe, Germany

the early works are, for example, mentioned in Refs. [11, 12]. Furthermore, combinations of different ab initio methods in conjunction with different basis sets, similar to Eq. (1), are key ingredients to many (modern) composite schemes such as the Gaussian-*n* model chemistries of Curtiss et al. [13–16], the correlation-consistent composite approach of Wilson et al. [17–19], the CBS methods of Petersson et al. [20–26], the focal-point analysis of Allen et al. [27, 28], the multicoefficient correlation methods of Truhlar et al. [29–31], the HEAT approach of Tajti, Harding et al. [32, 33], and the Weizmann-*n* theories of Martin et al. [34–36]. Noteworthy is also the study by Dunning and Peterson [37] in which various perturbation theories (e.g., third-order Møller–Plesset theory) are used to approximate the basis-set dependence of CCSD(T) theory.

The right-hand side of Eq. (1) can be interpreted as the CCSD(T) energy in a small basis plus a basis-set-truncation correction from MP2 theory. This correction improves significantly the accuracy of conventional CCSD(T) atomization energies. For example, in previous work [38], we have reported for 73 atomization energies a root-mean-square deviation (RMS) of 23 and 6.5 kJ/mol, without and with second-order basis-set-truncation corrections, respectively. Experimentally derived values from the Active Thermochemical Tables (ATcT) [39, 40] were used as reference. We found that the raw second-order corrections overestimated the atomization energies. Therefore, their contribution was scaled down using an *empirical* factor $f = 0.78$, yielding an RMS deviation of only 1.2 kJ/mol for the same set of molecules.

An alternative approach for scaling down the second-order corrections is the inclusion of interference factors [21, 41–43]. Interference factors f_{ij} are computed from first principles from the first-order wave function for individual orbital pairs i, j. They relate the basis-set-truncation error of the MP2 energy to the corresponding error of the CCSD method [44]. In our approach, the MP2 basis-set-truncation error is calculated as the difference between the explicitly correlated MP2 energy (MP2-F12) [45] and the conventional MP2 energy. Explicitly correlated theory incorporates the electronic-cusp behavior into the wave function and provides molecular electronic energies close to the CBS limit [5–9]. Thus, using interference factors, the CCSD(T)/CBS correlation energy is estimated as

$$\delta E_{\mathrm{CCSD(T)/CBS}} \approx \delta E_{\mathrm{CCSD(T)}} + \sum_{ij} f_{ij}\left(e_{ij}^{\mathrm{MP2-F12}} - e_{ij}^{\mathrm{MP2}}\right), \tag{2}$$

where $e_{ij}^{\mathrm{MP2\text{-}F12}}$ and e_{ij}^{MP2} are pair-correlation energies. All terms on the right-hand side of Eq. (2) are computed in the same basis set, which in the present work is either one of the cc-pVXZ basis sets of Dunning [3] or one of the cc-pVXZ-F12 basis sets of Peterson et al. [46].

We rewrite Eq. (2) as

$$\begin{aligned}\delta E_{\mathrm{CCSD(T)/CBS}} &\approx \delta E_{\mathrm{CCSD(T)+F12+INT}} \\ &= \delta E_{\mathrm{CCSD(T)}} + \delta E_{\mathrm{F12}} + \delta E_{\mathrm{INT}};\end{aligned} \tag{3}$$

$$\delta E_{\mathrm{F12}} = \sum_{ij}\left(e_{ij}^{\mathrm{MP2-F12}} - e_{ij}^{\mathrm{MP2}}\right); \tag{4}$$

$$\delta E_{\mathrm{INT}} = \sum_{ij}(f_{ij} - 1)\left(e_{ij}^{\mathrm{MP2-F12}} - e_{ij}^{\mathrm{MP2}}\right). \tag{5}$$

Recently, the applicability of this method, denoted as CCSD(T)+F12+INT, for the accurate description of non-covalent interactions was studied [47]. The CCSD(T)+F12+INT method was tested on the S22 benchmark database (revision S22B) [48, 49], and a mean absolute deviation (MAD) of 0.06 kcal/mol (0.25 kJ/mol) was reported.

In the present article, we assess the performance of the CCSD(T)+F12+INT method with respect to computations of atomization energies. Since atomization energies represent energy differences of systems containing different numbers of paired electrons, they pose a challenge for pair-electron theories and thus for interference-corrected MP2. They also require a high degree of flexibility in the description of the short-range interactions [50].

Details of the computational protocol are given in Sect. 2, followed by the calculation of atomization energies. The statistics for the 106-molecule [51] test set, the six-membered atomization energy (AE6) test set of Lynch and Truhlar [52], and the G2/97 [53, 54] test set are presented in Sect. 3 Finally, a summary is presented in Sect. 4.

2 Computational details

In this study, the performance of the CCSD(T)+F12+INT method has been assessed for three test sets. For the 106-molecule test set of Bakowies [51], the equilibrium geometries of all molecules were obtained from previous studies [4, 38] at the all-electron CCSD(T) level in the correlation-consistent core–valence triple-ζ basis set (cc-pCVTZ) of Woon and Dunning [3, 55]. For the hydrogens contained in these molecules, the cc-pVTZ basis was used. All of these 106 molecules are closed-shell species, and therefore, a restricted Hartree–Fock (RHF) determinant was used as reference for the correlation treatments. The geometries used for the AE6 test set had been obtained at the QCISD/MG3 level of theory by Truhlar et al. [56]. For the G2/97 test set, the B3LYP/6-31G(2*df*, *p*) geometries were taken from Ref. [57].

Unrestricted Hartree–Fock (UHF) references were used for the open-shell systems and RHF references for the closed-shell systems. In both cases, point-group symmetry was used. A "CABS-singles" correction was added to the Hartree–Fock energy, which is an energy correction obtained from single excitations into an auxiliary basis set (see, e.g., Ref. [58]).

All calculations were performed with the TURBOMOLE program package [59, 60]. Two families of basis sets were considered: the cc-pVXZ basis sets (X = D, T, Q, 5) of Dunning [3] and the cc-pVXZ-F12 basis sets (X = D, T, Q) of Peterson et al. [46]. Both families have been developed within a correlation-consistent methodology, the second specifically for explicitly correlated calculations. All calculations were performed in the frozen-core (fc) approximation. Core/core–valence corrections are added separately, computed in the cc-pCVQZ-F12 basis [61, 62]. This basis was also chosen for the Li, Be, and Na atoms for the corresponding systems in the G2/97 test set, because only the 1s orbital of Na was kept frozen and no orbitals at all for Li and Be.

As complementary auxiliary basis set (cabs), the "OptRI" sets of Peterson et al. were used [61–63]. The aug-cc-pwCV($X + 1$)Z cbas basis sets of Hättig [64] (aug-cc-pV($X + 1$)Z for the hydrogens, cc-pV($X + 1$)Z for Na) were used for the robust fitting of both the electron-repulsion and the F12-specific integrals. The auxiliary basis sets (cbas) of Ref. [64] were also used for the calculations with the cc-pCVQZ-F12 basis for the core/core-valence correction. The aug-cc-pV($X + 1$)Z jkbas basis sets of Weigend [65] were used for the two-electron contributions to the Fock matrix.

The interference-corrected MP2-F12 calculations (INT-MP2-F12) were carried out with spin-aligned [21] canonical or Boys localized orbitals (all terms of our computational protocol are orbital-invariant except the δE_{INT} term). For all explicitly correlated calculations (coupled-cluster and INT-MP2-F12), the orbital-invariant ansatz with full optimization of the F12 amplitudes was used, including "spin-flip" amplitudes [66–68] for open-shell systems, with the F+K approximation [45] for the commutator between kinetic energy and Slater-type geminal $f_{12} = \gamma^{-1}\{1 - \exp(-\gamma r_{12})\}$. Six Gaussian functions were used to approximate the Slater-type geminal. Throughout this article, the δE_{F12} and δE_{INT} terms were computed with ansatz 2 and approximation A of F12 theory [45]. The exponents γ were taken from the work of Peterson et al. [61–63]: $\gamma = 0.9\ a_0^{-1}$ for cc-pVDZ-F12, $\gamma = 1.0\ a_0^{-1}$ for cc-pVTZ-F12, and $\gamma = 1.1\ a_0^{-1}$ for cc-pVQZ-F12. The default exponent $\gamma = 1.4\ a_0^{-1}$ was used for Dunning's cc-pVXZ basis sets.

3 Results and discussion

3.1 106-molecule test set

The 106-molecule test set consists of the 105 molecules benchmarked by Bakowies [51] plus H_2. See Refs. [4, 38] for further details. In the following two subsections, we shall first look at the correlation-energy contribution to the atomization energy (at the CCSD and CCSD(T) levels) and then look at the total atomization energy (including the Hartree–Fock energy and other corrections).

3.1.1 Correlation-energy contribution to the atomization energy

Table 1 reports the statistical analysis of the raw CCSD correlation-energy contributions to the atomization energies with respect to the CCSD(F12)/def2-QZVPP reference values of Ref. [4]. As expected, the deviations from the reference values are significant due to the slow basis-set convergence of the conventional CCSD method. In the quintuple-zeta basis, the RMS is still ca. 10 kJ/mol. All deviations are positive, which means that the correlation-energy contributions to the atomization energies are *underestimated* at the conventional CCSD level.

Table 2 reports the statistical analysis of the CCSD+F12 method, in which the correlation energy is computed according to

$$\delta E_{\mathrm{CCSD+F12}} = \delta E_{\mathrm{CCSD}} + \delta E_{\mathrm{F12}}. \tag{6}$$

Addition of only the correction from MP2-F12 theory is improving the results, but the accuracy is not satisfactory. Nevertheless, for all basis sets, the MAD and the RMS values are significantly smaller than without δE_{F12} term. The RMS of the cc-pVDZ-F12, cc-pVTZ-F12, and cc-pVQZ-F12 basis sets is reduced by a factor of four, three, and two, respectively, compared to the results without δE_{F12} term. The negative values of the mean errors indicate

Table 1 Statistics of the deviations (in kJ/mol) of the computed correlation-energy contributions calculated with the conventional CCSD for the 106-molecule test set

Basis set	Mean error	MAD	RMS	Max error	Molecule	
cc-pVDZ	115.48	115.48	122.59	211.88	31	CH_4N_2O
cc-pVDZ-F12	95.71	95.71	101.53	178.30	72	C_4N_2
cc-pVTZ	49.89	49.89	52.71	100.91	104	N_2O_4
cc-pVTZ-F12	40.40	40.40	42.61	75.87	104	N_2O_4
cc-pVQZ	19.91	19.91	21.12	44.20	104	N_2O_4
cc-pVQZ-F12	17.17	17.17	18.16	33.51	104	N_2O_4
cc-pV5Z	9.71	9.71	10.38	22.74	104	N_2O_4

Table 2 Statistics of the deviations (in kJ/mol) of the computed correlation-energy contributions calculated with the CCSD+F12 method for the 106-molecule test set

Basis set	Mean error	MAD	RMS	Max error	Molecule	
cc-pVDZ	−45.54	45.54	49.60	127.58	104	N_2O_4
cc-pVDZ-F12	−23.51	23.51	24.87	52.99	104	N_2O_4
cc-pVTZ	−17.75	17.75	18.88	42.43	104	N_2O_4
cc-pVTZ-F12	−15.35	15.35	16.16	26.25	66	C_3H_8
cc-pVQZ	−10.73	10.73	11.31	22.28	104	N_2O_4
cc-pVQZ-F12	−9.66	9.66	10.18	16.77	72	C_4N_2
cc-pV5Z	−6.43	6.43	6.77	13.09	72	C_4N_2

Table 3 Statistics of the deviations (in kJ/mol) of the computed correlation-energy contributions calculated with the CCSD+F12+INT method for the 106-molecule test set

Basis set	Mean error	MAD	RMS	Max error	Molecule	
cc-pVDZ	−2.82	10.31	14.03	44.89	37	C_2F_4
cc-pVDZ-F12	10.31	10.57	11.96	24.32	70	C_4H_4
cc-pVTZ	3.09	4.13	4.17	9.87	98	H_3NO
cc-pVTZ-F12	2.66	2.83	3.17	7.00	70	C_4H_4
cc-pVQZ	−0.29	0.93	1.27	5.12	39	C_2HF_3
cc-pVQZ-F12	−0.27	0.67	0.87	3.01	103	N_2O_3
cc-pV5Z	−0.67	0.73	0.92	3.03	103	N_2O_3

that the addition of only the δE_{F12} term *overestimates* the electron-correlation contribution to the atomization energy.

Much higher accuracy is achieved by adding the δE_{INT} correction from interference-corrected MP2-F12 theory (Table 3). The statistics for all basis sets are greatly improved, in particular for the quadruple- and quintuple-zeta basis sets. Even the RMS of the triple-zeta basis sets is within the limits of "chemical accuracy," that is, below 1 kcal/mol (4.18 kJ/mol). Figure 1 shows the normal (Gaussian) distribution of the electron-correlation contributions to the atomization energies for the 106 molecules with respect to the CCSD(F12)/def2-QZVPP reference values. The Gaussians corresponding to the CCSD+F12+INT level are shown with dashed lines. Figure 1a includes the distributions obtained in the cc-pVTZ-F12 basis and Fig. 1b in the cc-pVQZ-F12 basis.

The molecules with the largest errors show large D1 diagnostics [69]. At the CCSD(T)/cc-pVQZ-F12 level, for example, the D1 values of dinitrogen trioxide (103.N_2O_3) and tetraoxide (104.N_2O_4) are 0.078 and 0.069, respectively.

In Table 4, we show the statistics of the CCSD(T)+F12+INT correlation-energy contributions to the atomization energies, computed according to Eq. (2). These contributions are compared with the corresponding results of Ref. [4], in which a (T) correction was added to the CCSD(F12)/def2-QZVPP energy. This (T) correction was obtained by extrapolating the corresponding energies computed in the cc-pCVQZ and cc-pCV5Z basis sets. In the small double- and triple-zeta basis sets, the statistics of the CCSD+F12+INT approach look somewhat better than those of the CCSD(T)+F12+INT approach due to the poor (T)-triples energies in these basis sets, but the performance of the approaches with and without triples is very similar in the larger quadruple- and quintuple-zeta basis sets.

We conclude this section by noting that the cc-pVQZ-F12 basis set yields a mean absolute deviation of about

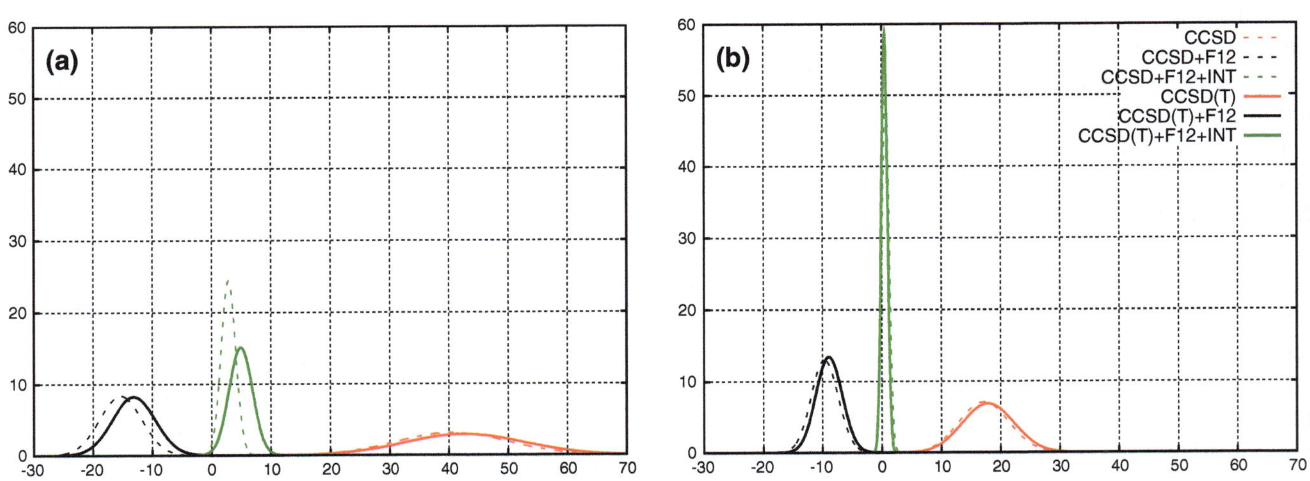

Fig. 1 Normal distributions of electron-correlation contributions to atomization energies (in kJ/mol) for CCSD(T) (*red*), CCSD(T)+F12 (*black*), and CCSD(T)+F12+INT (*green*). *Dashed lines* correspond to CCSD, *full lines* to CCSD(T). Results are shown for the basis sets **a** cc-pVTZ-F12 and **b** cc-pVQZ-F12

1 kJ/mol, that is, it yields CCSD and CCSD(T) electron-correlation contributions of chemical accuracy (<4.18 kJ/mol).

3.1.2 Total atomization energy

The *total* atomization energies of the 106 molecules will be discussed next. The results of the present work will be compared with those from Ref. [38]. For 73 of these molecules, accurate experimentally derived values are available from the ATcT database (cf. Ref. [38]).

In order to compare with experimentally derived values, a number of corrections to the total CCSD(T) energy should be taken into account. These include higher-order corrections for full triple excitations, calculated as the difference between CCSDT/cc-pVTZ and CCSD(T)/cc-pVTZ correlation energies (δE_T), and quadruple excitations, calculated as the difference between CCSDT(Q)/cc-pVDZ and CCSDT/cc-pVDZ correlation energies ($\delta E_{(Q)}$). In addition, scalar relativistic effects (ΔE_{MVD}), spin–orbit interactions for the atoms (ΔE_{SO}), and anharmonic zero-point vibrational energies ($\Delta E_{ZPVE} + \Delta E_{Anh}$) must be included. All of these corrections (denoted as "Other" in Table 5) were taken from Ref. [38] and are added to the present work's CCSD(T)+F12+INT energies.

The correction for core/core–valence correlation effects was *not* taken from Ref. [38] but rather recomputed in the course of the present work, in which the cc-pCVQZ-F12 basis was used,

$$\delta E_{CV} = \delta E_{ae-CCSD(T)/cc-pCVQZ-F12} - \delta E_{fc-CCSD(T)/cc-pCVQZ-F12}. \quad (7)$$

That is, δE_{CV} is obtained as the difference between the all-electron (ae) and frozen-core (fc) CCSD(T) energies in the cc-pCVQZ-F12 basis [61]. Table 6 shows δE_{CV} values for a few representative cases in comparison with the values of Ref. [38], showing that the cc-pCVQZ-F12 basis yields more accurate values (i.e., closer to the cc-pCV(Q5)Z extrapolated value) than the cc-pCVQZ basis used in Ref. [38].

Finally, the CABS-singles correction to the Hartree–Fock energy, as obtained in the INT-MP2-F12 calculations, is taken into account. All energy contributions of the present work's composite scheme are the following:

$$E_{\text{Composite Scheme}} = E_{\text{Hartree–Fock}} + \Delta E_{CABS} + \delta E_{CCSD(T)} + \delta E_{F12} + \delta E_{INT} + \delta E_{CV} + \delta E_T + \delta E_{(Q)} + \Delta E_{ZPVE} + \Delta E_{Anh} + \Delta E_{MVD} + \Delta E_{SO}. \quad (8)$$

Table 5 includes the individual components of the present work's composite scheme, obtained with the cc-pVQZ-F12 basis set. These components are the $E_{CCSD(T)}$ energy (which includes the Hartree–Fock energy), the second-order corrections INT and F12, as they have been described previously, and the CABS-singles correction. Additionally, the core/core–valence contributions to the atomization energies are also given in Table 5 ("CV" column). The remaining corrections of the composite scheme of Eq. (8) are collected in the column "Other,"

$$\Delta E_{Other} = \delta E_T + \delta E_{(Q)} + \Delta E_{ZPVE} + \Delta E_{Anh} + \Delta E_{MVD} + \Delta E_{SO}, \quad (9)$$

and the column "Total" contains the total atomization energies after summation of the individual terms. In the column "Ref," the atomization energies as computed in Ref. [38] are shown. The MAD of our composite scheme from the values of Ref. [38] is 1.3 kJ/mol, with an RMS value of 1.5 kJ/mol. The last two columns include the atomization energies from the ATcT database, where available, and the errors of our composite scheme with respect to the ATcT values. With respect to these values, the MAD is 1.75 kJ/mol and the RMS is 2.15 kJ/mol. The maximum deviation is found for cyclopropane (5.7 kJ/mol). Hence, our composite scheme does indeed achieve chemical accuracy. The two previous computational studies [4, 38] on the atomization energies of this test set reported MADs of 0.9 and 1.2 kJ/mol, respectively. In the first of these previous works, an empirical scaling factor $f_{int} = 0.78$ was applied. In the second, the main contribution to the correlation energy was obtained from explicitly correlated CCSD(F12) calculations.

Table 4 Statistics of the deviations (in kJ/mol) of the computed correlation-energy contributions calculated with the CCSD(T)+F12+INT method for the 106-molecule test set

Basis set	Mean error	MAD	RMS	Max error	Molecule
cc-pVDZ	15.86	17.00	18.70	37.02	98.H_3NO
cc-pVDZ-F12	16.93	16.93	18.03	33.74	70.C_4H_4
cc-pVTZ	7.73	7.85	8.46	16.02	98.H_3NO
cc-pVTZ-F12	4.90	4.90	5.22	11.85	104.N_2O_4
cc-pVQZ	1.32	1.50	1.81	4.56	104.N_2O_4
cc-pVQZ-F12	0.46	0.66	0.84	3.12	104.N_2O_4
cc-pV5Z	−0.20	0.49	0.64	−2.19	39.C_2HF_3

3.2 AE6 test set

Besides the 106-molecule test set, alternative and more compact representative test sets exist in the literature. The six-membered atomization energy (AE6) test set of Lynch and Truhlar [52] is an example of such a

Table 5 Atomization energy components (in kJ/mol) of the 106-molecule test set

Nr.[a]	Molecule		CCSD(T)	ΔE_{CABS}	INT[b]	F12[b]	CV	Other[c]	Total	Ref.[d]	ATcT	Error
1	CFN	Cyanogen fluoride	1,255.8	0.3	−7.3	20.6	8.0	−27.9	1,249.5	1,250.9	1,250.3 ± 1.7	−0.8
2	CFN	Isocyanogen fluoride	959.0	0.3	−6.6	19.7	6.2	−22.4	956.2	957.0	959.2 ± 2.7	−3.0
3	CF_2	Singlet difluoromethylene	1,069.2	0.3	−4.3	14.3	2.6	−21.9	1,060.2	1,060.0	1,059.1 ± 0.8	1.1
4	CF_2O	Carbonyl fluoride	1,735.7	0.4	−4.9	23.4	7.1	−43.6	1,718.1	1,717.2	1,718.4 ± 0.9	−0.3
5	CF_4	Tetrafluoromethane	1,980.0	0.5	−5.9	25.1	5.9	−56.3	1,949.3	1,949.5	1,947.9 ± 0.6	1.4
6	CHF	Singlet fluoromethylene	880.1	0.2	−4.5	12.1	2.3	−33.5	856.7	857.0		
7	CHFO	Formyl fluoride	1,667.4	0.3	−5.8	21.2	6.5	−58.4	1,631.2	1,631.0	1,631.4 ± 0.9	−0.2
8	CHF_3	Trifluoromethane	1,901.9	0.4	−4.4	22.9	5.5	−75.6	1,850.7	1,849.2	1,848.7 ± 0.9	2.0
9	CHN	Hydrogen cyanide	1,289.7	0.2	−7.5	18.6	7.4	−41.7	1,266.7	1,268.5	1,268.3 ± 0.2	−1.6
10	CHN	Hydrogen isocyanide	1,227.1	0.3	−7.6	18.8	6.6	−41.2	1,204.0	1,205.7	1,207.0 ± 0.6	−3.0
11	CHNO	Cyanic acid	1,686.2	0.3	−9.6	27.0	9.4	−57.9	1,655.4	1,657.1	1,657.2 ± 1.0	−1.8
12	CHNO	Isocyanic acid	1,788.4	0.3	−8.9	27.5	9.8	−57.3	1,759.8	1,760.4	1,761.0 ± 0.4	−1.2
13	CHNO	Formonitrile oxide	1,494.8	0.3	−9.5	26.9	10.2	−50.3	1,472.4	1,473.9	1,474.1 ± 1.2	−1.7
14	CHNO	Isofulminic acid	1,437.7	0.4	−9.4	26.2	7.7	−54.7	1,407.9	1,409.4	1,410.2 ± 1.0	−2.3
15	CH_2	Singlet methylene	750.0	0.1	−5.1	9.9	1.9	−42.6	714.2	715.4	7,14.9 ± 0.2	−0.7
16	CH_2F_2	Difluoromethane	1,813.2	0.3	−6.1	20.7	5.3	−91.8	1,741.6	1,741.5	1,741.7 ± 0.8	−0.1
17	CH_2N_2	Cyanamide	1,997.5	0.4	−11.7	32.6	11.1	−89.9	1,940.0	1,941.5		
18	CH_2N_2	3H-Diazirine	1,821.1	0.3	−11.6	31.5	8.3	−87.0	1,762.6	1,764.3		
19	CH_2N_2	Diazomethane	1,858.0	0.4	−11.0	31.4	10.6	−80.8	1,808.6	1,809.9		
20	CH_2O	Formaldehyde	1,548.4	0.2	−7.3	18.7	5.9	−70.9	1,495.0	1,496.6	1,495.8 ± 0.2	−0.8
21	CH_2O	Hydroxymethylene	1,332.0	0.3	−7.1	18.3	4.2	−70.8	1,276.9	1,277.9	1,277.8 ± 1.1	−0.9
22	CH_2O_2	Dioxirane	1,690.7	0.3	−8.9	25.8	5.9	−86.4	1,627.4	1,627.9	1,629.6 ± 1.7	−2.2
23	CH_2O_2	Formic acid	2,072.7	0.4	−9.3	27.4	7.8	−91.7	2,007.3	2,008.7	2,008.4 ± 0.3	−1.1
24	CH_2O_3	Performic acid	2,214.4	0.4	−10.0	33.8	8.2	−98.1	2,148.7	2,148.9		
25	CH_3F	Fluoromethane	1,752.4	0.3	−7.1	18.3	5.2	−106.7	1,662.4	1,663.4	1,665.1 ± 0.6	−2.7
26	CH_3N	Methanimine	1,814.6	0.3	−9.6	24.0	7.2	−104.6	1,731.9	1,733.6	1,733.5 ± 1.0	−1.6
27	CH_3NO	Formamide	2,343.9	0.4	−11.3	33.0	9.8	−120.3	2,255.5	2,256.6		
28	CH_3NO_2	Methyl nitrite	2,470.6	0.4	−12.4	38.2	8.3	−126.7	2,378.4	2,379.0		
29	CH_3NO_2	Nitromethane	2,479.1	0.5	−11.8	38.9	9.9	−131.1	2,385.5	2,385.3		
30	CH_4	Methane	1,743.5	0.2	−7.7	15.9	5.3	−118.6	1,638.6	1,640.9	1,642.2 ± 0.1	−3.6
31	CH_4N_2O	Urea	3,104.8	0.6	−15.8	46.9	13.2	−170.1	2,979.6	2,981.0		
32	CH_4O	Methanol	2,125.9	0.3	−9.5	24.3	6.5	−136.7	2,010.8	2,012.4	2,012.7 ± 0.2	−1.9
33	CH_5N	Methylamine	2,408.7	0.4	−12.0	29.6	7.9	−168.4	2,266.2	2,268.4	2,269.0 ± 0.5	−2.8
34	CO	Carbon monoxide	1,072.4	0.2	−4.9	12.7	4.5	−13.9	1,071.0	1,072.5	1,072.1 ± 0.1	−1.1
35	CO_2	Carbon dioxide	1,607.3	0.3	−7.7	22.1	8.2	−32.8	1,597.4	1,599.1	1,598.2 ± 0.1	−0.8
36	C_2F_2	Difluoroacetylene	1,585.4	0.3	−7.4	23.1	12.6	−40.4	1,573.6	1,574.8	1,577.0 ± 1.7	−3.4
37	C_2F_4	Tetrafluoroethylene	2,432.6	0.5	−8.3	32.2	12.3	−67.4	2,401.9	2,402.2	2,405.2 ± 1.0	−3.3
38	C_2HF	Fluoroacetylene	1,640.4	0.3	−7.0	20.6	11.4	−55.9	1,609.8	1,610.7	1,612.3 ± 1.0	−2.5
39	C_2HF_3	Trifluoroethylene	2,415.0	0.5	−5.2	29.9	11.7	−84.9	2,367.0	2,363.5		
40	C_2H_2	Acetylene	1,672.8	0.2	−7.8	18.4	10.3	−70.0	1,623.9	1,625.5	1,626.2 ± 0.2	−2.3
41	$C_2H_2F_2$	1,1-Difluoroethylene	2,423.5	0.4	−7.2	27.5	11.2	−102.0	2,353.4	2,352.7		
42	C_2H_2O	Ketene	2,200.8	0.3	−8.8	25.7	11.9	−84.2	2,145.7	2,146.7	2,147.3 ± 0.2	−1.6
43	C_2H_2O	Oxirene	1,878.1	0.3	−10.3	26.4	10.5	−77.2	1,827.8	1,828.8		
44	$C_2H_2O_2$	Glyoxal	2,620.4	0.3	−12.1	33.7	12.0	−99.5	2,554.8	2,557.1	2,555.3 ± 0.6	−0.5
45	C_2H_3F	Fluoroethylene	2,372.7	0.3	−8.0	25.3	10.5	−118.2	2,282.6	2,282.5	2,278.4 ± 1.7	4.2
46	C_2H_3FO	Acetyl fluoride	2,922.3	0.4	−9.2	33.4	11.9	−133.1	2,825.7	2,825.1		
47	C_2H_3N	Acetonitrile	2,544.9	0.3	−11.4	30.7	12.8	−119.8	2,457.5	2,459.2		
48	C_2H_3N	Methyl isocyanide	2,443.7	0.3	−11.3	30.5	11.6	−120.4	2,354.4	2,355.6		

Table 5 continued

Nr.[a]	Molecule		CCSD(T)	ΔE_{CABS}	INT[b]	F12[b]	CV	Other[c]	Total	Ref.[d]	ATcT	Error
49	C_2H_4	Ethylene	2,334.3	0.2	−10.0	22.9	10.0	−133.5	2,223.9	2,226.3	2,225.9 ± 0.2	−2.0
50	C_2H_4O	Acetaldehyde	2,803.4	0.3	−10.8	31.0	11.3	−147.4	2,687.8	2,688.9	2,688.9 ± 0.4	−1.1
51	C_2H_4O	Oxirane	2,693.9	0.3	−12.0	31.6	11.0	−153.3	2,571.5	2,573.0	2,573.9 ± 0.5	−2.4
52	$C_2H_4O_2$	Acetic acid	3,323.0	0.5	−12.2	39.6	13.2	−166.0	3,198.1	3,198.5	3,199.3 ± 1.5	−1.2
53	$C_2H_4O_2$	Methyl formate	3,253.3	0.4	−12.0	38.9	12.5	−166.8	3,126.3	3,126.5	3,125.2 ± 0.6	1.1
54	C_2H_5F	Fluoroethane	2,988.2	0.4	−10.6	30.4	10.4	−181.6	2,837.2	2,838.0	2,838.5 ± 1.9	−1.3
55	C_2H_5N	Aziridine	2,976.9	0.4	−14.1	37.2	12.6	−186.0	2,827.0	2,828.7		
56	C_2H_6	Ethane	2,955.0	0.3	−12.3	28.0	10.3	−198.2	2,783.1	2,786.0	2,787.2 ± 0.2	−4.1
57	C_2H_6O	Dimethyl ether	3,308.5	0.4	−13.7	35.7	11.1	−211.5	3,130.5	3,132.6	3,132.4 ± 0.5	−1.9
58	C_2H_6O	Ethanol	3,358.7	0.4	−12.8	36.3	11.6	−212.3	3,181.9	3,183.0	3,182.8 ± 0.3	−0.9
59	C_2N_2	Cyanogen	2,059.1	0.3	−12.5	32.6	15.1	−40.9	2,053.7	2,056.5	2,055.8 ± 0.5	−2.1
60	C_3H_3N	Acrylonitrile	3,147.2	0.3	−13.0	37.6	17.6	−133.2	3,056.5	3,057.7		
61	C_3H_4	Allene	2,909.4	0.3	−12.6	29.7	15.4	−146.0	2,796.2	2,798.9	2,800.9 ± 0.5	−4.7
62	C_3H_4	Cyclopropene	2,816.7	0.3	−11.7	31.0	15.3	−148.8	2,702.8	2,704.3	2,705.1 ± 1.0	−2.3
63	C_3H_4	Propyne	2,914.9	0.3	−11.4	30.4	15.7	−146.9	2,803.0	2,804.1	2,805.6 ± 0.5	−2.6
64	C_3H_6	Cyclopropane	3,533.6	0.3	−14.1	36.1	15.7	−217.7	3,353.9	3,356.1	3,359.7 ± 0.6	−5.8
65	C_3H_6	Propene	3,565.6	0.4	−13.2	35.0	15.2	−209.5	3,393.5	3,395.1	3,395.0 ± 0.4	−1.5
66	C_3H_8	Propane	4,175.5	0.4	−15.9	39.9	15.3	−272.0	3,943.2	3,945.9	3,944.6 ± 0.4	−1.4
67	C_3O_2	Carbon suboxide	2,734.0	0.3	−11.7	35.8	18.9	−57.9	2,719.4	2,721.3		
68	C_4H_4	Butatriene	3,491.4	0.3	−14.2	36.8	20.5	−156.9	3,377.9	3,380.1		
69	C_4H_4	Cyclobutadiene	3,386.6	0.4	−15.1	38.8	18.9	−159.8	3,269.8	3,272.3		
70	C_4H_4	Tetrahedran	3,272.1	0.4	−15.3	40.0	21.6	−160.3	3,158.5	3,160.4		
71	C_4H_4	Vinylacetylene	3,526.4	0.3	−13.4	37.3	20.6	−160.8	3,410.4	3,411.6		
72	C_4N_2	Dicyanoacetylene	3,265.3	0.3	−15.9	47.0	26.0	−68.3	3,254.4	3,256.1		
73	FH	Hydrogen fluoride	587.1	0.2	−2.9	6.7	1.0	−26.8	565.3	565.9	566.0 ± 0.0	−0.7
74	FHO	Hypofluorous acid	652.2	0.2	−4.9	12.8	1.2	−37.4	624.1	624.3	624.0 ± 0.4	0.1
75	FHO_2	Fluoroperoxide	860.4	0.3	−6.6	19.4	1.6	−47.2	827.9	827.4		
76	FH_2N	Monofluoroamine	1,057.1	0.3	−7.5	19.2	2.5	−73.3	998.3	999.0		
77	FH_3N_2	Fluorohydrazine	1,686.3	0.4	−11.2	32.7	5.5	−119.4	1,594.3	1,594.1		
78	FNO	Nitrosyl fluoride	880.8	0.3	−6.4	20.1	2.3	−17.3	879.8	878.7		
79	F_2	Difluorine	156.4	0.1	−2.5	6.0	0.1	−6.0	154.1	153.8	154.6 ± 0.2	−0.5
80	F_2N_2	Difluorodiazene (cis)	1,029.1	0.3	−8.2	27.3	3.5	−31.2	1,020.8	1,019.7		
81	F_2N_2	Difluorodiazene (trans)	1,023.3	0.3	−8.4	27.2	3.6	−31.1	1,014.9	1,013.5		
82	F_2O	Difluorine monoxide	379.6	0.2	−3.9	12.8	0.4	−13.7	375.4	373.5	373.3 ± 0.7	2.1
83	F_2FO_2	Perfluoroperoxide	616.2	0.3	−5.4	19.3	0.7	−15.7	615.4	612.3	609.7 ± 0.8	5.7
84	F_3N	Trifluoroamine	844.3	0.3	−4.6	21.5	1.2	−30.5	832.2	828.9		
85	HNO	Nitrosylhydride	844.1	0.3	−7.5	18.7	2.7	−35.3	823.0	824.3	823.6 ± 0.1	−0.6
86	HNO_2	Nitrous acid (cis)	1,281.2	0.4	−9.4	26.6	3.7	−52.0	1,250.5	1,251.7	1,251.5 ± 0.4	−1.0
87	HNO_2	Nitrous acid (trans)	1,282.6	0.4	−9.1	26.5	3.8	−51.6	1,252.6	1,253.0	1,253.3 ± 0.1	−0.7
88	HNO_2	Nitrous acid, H-NO_2	1,249.3	0.4	−8.9	27.2	4.9	−57.1	1,215.8	1,216.4		
89	HNO_3	Nitric acid	1,591.7	0.5	−9.4	34.7	6.1	−70.9	1,552.7	1,552.0	1,551.6 ± 0.2	1.1
90	HN_3	Hydrogen azide	1,353.4	0.5	−11.5	32.7	7.8	−53.5	1,329.4	1,330.8	1,329.7 ± 0.6	−0.3
91	H_2N_2	Diazene (cis)	1,195.9	0.4	−10.0	24.8	4.1	−71.5	1,143.7	1,145.4	1,143.5 ± 0.9	0.2
92	H_2N_2	Diazene (trans)	1,218.4	0.4	−10.2	24.8	4.2	−73.1	1,164.5	1,166.1	1,165.8 ± 0.7	−1.3
93	H_2N_2	Diazene (iso)	1,114.0	0.3	−9.7	25.2	4.7	−69.3	1,065.2	1,066.1	1,065.1 ± 0.9	0.1
94	H_2N_2O	Nitrosamide	1,586.8	0.4	−11.7	32.8	6.2	−84.2	1,530.3	1,531.4		
95	H_2O	Water	964.3	0.3	−5.7	12.7	1.9	−57.1	916.4	917.6	917.8 ± 0.1	−1.4
96	H_2O_2	Hydrogen peroxide	1,109.6	0.3	−8.2	19.4	2.4	−70.2	1,053.3	1,055.0	1,055.2 ± 0.1	−1.9

Table 5 continued

Nr.[a]	Molecule		CCSD(T)	ΔE_{CABS}	INT[b]	F12[b]	CV	Other[c]	Total	Ref.[d]	ATcT	Error
97	NH_3	Ammonia	1,231.7	0.3	−8.0	17.9	3.1	−90.0	1,155.0	1,156.9	1,157.3 ± 0.1	−2.3
98	H_3NO	Ammonia oxide	1,380.4	0.4	−9.7	25.9	3.9	−108.8	1,292.1	1,290.5		
99	H_3NO	Hydroxylamine	1,483.8	0.4	−9.8	25.4	3.9	−106.4	1,397.3	1,398.7	1,398.7 ± 0.5	−1.4
100	N_2H_4	Hydrazine	1,806.9	0.5	−12.9	31.2	5.7	−139.9	1,691.5	1,694.0	1,695.6 ± 0.2	−4.1
101	N_2	Dinitrogen	936.4	0.2	−7.5	18.9	4.3	−12.7	939.6	940.8	941.1 ± 0.1	−1.5
102	N_2O	Nitrous oxide	1,104.1	0.4	−9.2	26.9	6.4	−27.3	1,101.3	1,102.8	1,102.0 ± 0.1	−0.7
103	N_2O_3	Dinitrogen trioxide	1,598.0	0.5	−10.7	40.4	6.6	−41.3	1,593.5	1,592.4	1,591.1 ± 0.2	2.4
104	N_2O_4	Dinitrogen tetraoxide	1,923.9	0.6	−15.9	48.7	9.1	−60.5	1,905.9	1,908.0	1,908.5 ± 0.2	−2.6
105	O_3	Ozone	590.1	0.3	−6.5	18.5	1.4	−8.2	595.6	594.9	596.1 ± 0.1	−0.5
106	H_2	Dihydrogen	456.2	0.1	−2.4	4.0	0.0	−26.1	431.8	432.7	432.1 ± 0.0	−0.3

[a] Same number and same molecule as in Ref. [51] except for dihydrogen

[b] Second-order corrections from the INT-MP2-F12 method

[c] Corrections to the electronic energy taken from Ref. [38]. See text for further explanation

[d] Total energy from Ref. [38]

Table 6 Basis-set convergence of the core/core–valence contribution (in kJ/mol) as obtained at the CCSD(T) level

Nr.[a]	Molecule		cc-pCVQZ[b]	cc-pCV5Z[b]	cc-pCVQZ-F12	cc-pCV(Q5)Z[b]
1	CFN	Cyanogen fluoride	6.92	7.26	7.21	7.61
9	CHN	Hydrogen cyanide	6.59	6.95	6.82	7.32
10	CHN	Hydrogen isocyanide	5.78	6.06	5.98	6.35
15	CH_2	Singlet methylene	1.59	1.65	1.65	1.71
20	CH_2O	Formaldehyde	5.21	5.43	5.36	5.65
30	CH_4	Methane	4.99	5.19	5.06	5.40
34	CO	Carbon monoxide	3.76	3.96	3.90	4.17
35	CO_2	Carbon dioxide	7.00	7.33	7.29	7.68
40	C_2H_2	Acetylene	9.60	10.10	9.86	10.62
73	FH	Hydrogen fluoride	0.77	0.77	0.75	0.77
74	FHO	Hypofluorous acid	0.64	0.62	0.67	0.60
79	F_2	Difluorine	−0.29	−0.32	−0.28	−0.35
92	H_2N_2	Diazene (trans)	3.34	3.44	3.53	3.56
95	H_2O	Water	1.61	1.63	1.61	1.66
97	NH_3	Ammonia	2.71	2.80	2.78	2.88
101	N_2	Dinitrogen	3.38	3.55	3.57	3.74
102	N_2O	Nitrous oxide	5.03	5.21	5.37	5.41
105	O_3	Ozone	0.24	0.17	0.38	0.09

The ROHF reference was used for the atoms

[a] Same number and same molecule as in Ref. [51]

[b] Data taken from Ref. [38]

small set. It has been proposed to represent the 109 atomization energies of the Database/3 data set [70]. The AE6 test set consists of the atomization energies of six molecules: SiH_4, SiO, S_2, propyne (C_3H_4), glyoxal ($C_2O_2H_2$), and cyclobutane (C_4H_8). The set contains a few but diverse cases, and it seems interesting to assess the performance of the CCSD(T)+F12+INT method.

Reference values, obtained at the (frozen-core) CCSD(T)(F12)/cc-pVQZ-F12 level of theory, were taken from Ref. [71]. Note that the core/core–valence contributions, higher-order terms (i.e., full triples and perturbative quadruples), and the scalar relativistic and spin–orbit effects have been subtracted from the values of Ref. [71] in order to be able to compare directly the CCSD(T) values.

Figure 2 shows the errors of the CCSD(T), CCSD(T)+F12, and CCSD(T)+F12+INT approaches, using the cc-pVQZ-F12 basis set, in comparison with the CCSD(T)(F12)/cc-pVQZ-F12 values of Ref. [71]. Similar to the 106-molecule test set, the three methods behave similarly for the atomization energies of the six molecules of the AE6 set. Conventional CCSD(T) energies have errors that exceed in some cases 20 kJ/mol. Addition of the raw δE_{F12} term overestimates the atomization energies. The interference-corrected CCSD(T)+F12+INT values are very accurate, however, showing a MAD of 0.8 kJ/mol and the RMS error of 1.0 kJ/mol. The maximum deviation is at about 2 kJ/mol for the S_2 molecule, whereas the other two methods show maximum errors for the cyclobutane molecule.

3.3 G2/97

The last collection of molecules for which the performance of the CCSD(T)+F12+INT method has been assessed is the G2/97 test set [53, 54]. It is composed of 148 diverse molecules. It includes some difficult cases, like the carbonyl fluoride (COF_2) molecule, whose experimental reference value has a high uncertainty [72], 29 radicals, aromatic moieties, some relatively "large" molecules (e.g., benzene), and molecules that are composed of atoms not included in the previous two sets (e.g., Li, Al, Na). All members of the G2/97 test set are listed in Table 7.

Recently, new theoretical reference values have been published for the G2/97 test set [73]. These accurate values have been calculated using a computational protocol based on explicitly correlated CCSD(T) theory. Corrections for higher excitations and core/core–valence correlation effects are accounted for. The main argument for using these theoretical reference values (and not experimentally derived ones) is that the high uncertainty of some of the

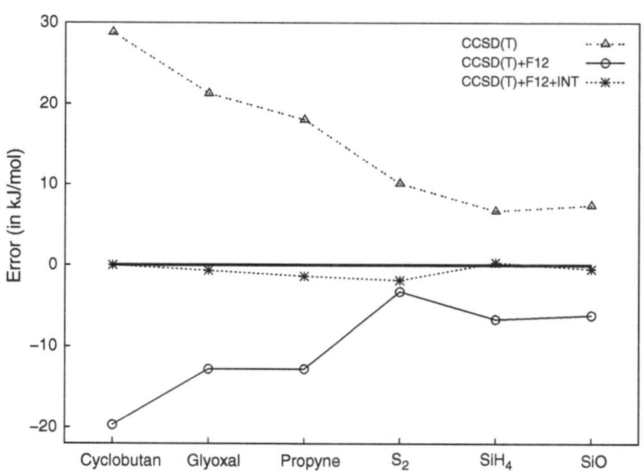

Fig. 2 Deviations from the computational reference values of the AE6 test set

Table 7 Atomization energies of the G2/97 test set (in kJ/mol) from the CCSD(T)+F12+INT method with canonical and localized orbitals (Boys)

Molecule		Canonical	Boys	Ref.[a]
$AlCl_3$	Aluminum trichloride	1,310.7	1,304.1	1,307.5
AlF_3	Aluminum trifluoride	1,805.9	1,801.6	1,802.9
BCl_3	Boron trichloride	1,354.2	1,348.2	1,350.9
BF_3	Boron trifluoride	1,964.4	1,960.2	1,961.8
BeH	Beryllium monohydride	213.8	211.7	212.0
CCl_4	Tetrachloromethane	1,314.1	1,309.9	1,310.2
CF_4	Tetrafluoromethane	1,999.5	1,996.3	1,997.1
CH	Methylidyne radical	350.5	350.5	350.3
CH_2Cl_2	Dichloromethane	1,549.6	1,546.3	1,547.1
CH_2F_2	Difluoromethane	1,827.7	1,825.2	1,825.7
CH_2O_2	Formic acid	2,091.0	2,088.7	2,089.5
CH_2O	Formaldehyde	1,559.5	1,558.3	1,558.9
CH_2	Singlet carbene	754.5	754.4	754.3
CH_2	Triplet carbene	793.7	793.7	793.6
CH_3Cl	Chloromethane	1,441.8	1,649.1	1,649.7
CH_3	Methyl radical	1,282.4	1,282.4	1,282.3
CH_3O	Hydroxymethyl radical	1,708.5	1,708.4	1,707.5
CH_3O	Methoxy radical	1,668.3	1,668.0	1,667.3
CH_3S	Methylthio radical	1,595.2	1,595.4	1,593.9
CH_4	Methane	1,751.8	1,752.0	1,752.0
CH_4O	Methanol	2,140.8	2,139.5	2,139.8
CH_4S	Thiomethanol	1,981.0	1,979.3	1,979.8
$CHCl_3$	Trichloromethane	1,441.8	1,435.6	1,436.4
CHF_3	Trifluoromethane	1,920.9	1,916.2	1,916.8
CHO	Formyl radical	1,161.5	1,161.4	1,160.9
CN	Cyano radical	743.6	743.3	742.9
CNH	Hydrogen cyanide	1,300.9	1,299.9	1,300.4
CNH_3O_2	Methyl nitrite	2,496.6	2,492.8	2,492.2
CNH_3O_2	Nitromethane	2,506.5	2,501.9	2,501.7
CNH_5	Methylamine	2,426.6	2,425.6	2,425.5
C_2Cl_4	Tetrachloroethylene	1,963.4	1,958.0	1,960.5
C_2F_4	Tetrafluoroethylene	2457.0	2452.9	2,453.7
C_2H	Ethynyl radical	1,097.9	1,097.6	1,097.8
C_2H_2	Acetylene	1,683.4	1,682.6	1,683.2
$C_2H_2O_2$	Glyoxal	2,641.9	2,639.3	2,640.4
C_2H_2O	Ketene	2,217.8	2,215.3	2,216.2
C_2H_3Cl	Vinyl chloride	2,265.8	2,262.8	2,263.7
C_2H_3	Vinyl radical	1,853.0	1,853.0	1,852.4
C_2H_3F	Vinyl fluoride	2,389.8	2,386.9	2387.5
C_2H_3O	Carbonyl methane	2,424.5	2,424.2	2,422.4
C_2H_3OCl	Acetyl chloride	2,788.3	2,782.3	2,783.4
C_2H_3OF	Acetyl fluoride	2,946.7	2,941.8	2,942.6
C_2H_4	Ethylene	2,347.2	2,346.7	2,347.1
$C_2H_4O_2$	Acetic acid	3,350.6	3,346.3	3,346.9
$C_2H_4O_2$	Methyl formate	3,280.3	3,275.8	3,276.4
C_2H_4O	Acetaldehyde	2,823.5	2,820.7	2,821.3
C_2H_4O	Oxirane	2713.4	2,712.1	2,712.3

Table 7 continued

Molecule		Canonical	Boys	Ref.[a]
C_2H_4S	Thiirane	2,608.7	2,606.9	2,607.2
C_2H_5Cl	Ethyl chloride	2,888.4	2,885.2	2,885.7
C_2H_5	Ethyl radical	2516.0	2,516.1	2,515.3
C_2H_5O	Ethoxy radical	2,908.7	2,907.3	2,906.3
C_2H_6	Ethane	2,970.7	2,970.5	2,970.5
C_2H_6O	Dimethyl ether	3,330.2	3,328.5	3,328.5
C_2H_6O	Ethanol	3,382.2	3,379.3	3,379.5
C_2H_6OS	Dimethyl sulfoxide	3,574.6	3,570.4	3,570.7
C_2H_6S	Dimethyl sulfide	3,206.2	3,204.1	3,204.3
C_2H_6S	Thioethanol	3,211.8	3,208.6	3,209.0
C_2N_2	Cyanogen	2,079.1	2,077.2	2,078.2
C_2NF_3	Trifluoroacetonitrile	2,675.8	2,668.8	2,669.9
C_2NH_3	Acetonitrile	2,564.3	2,562.2	2,562.7
C_2NH_5	Aziridine	3,000.1	2,998.6	2,998.4
C_2NH_5O	Acetamide	3,619.2	3,614.7	3,615.0
C_2NH_7	Dimethylamine	3,628.1	3,626.2	3,625.9
C_2NH_7	Ethylamine	3,662.1	3,660.2	3,660.1
C_3H_4	Allene	2,926.2	2,925.5	2,926.2
C_3H_4	Cyclopropene	2,836.1	2,834.4	2,834.9
C_3H_4	Propyne	2,933.7	2,931.6	2,932.2
C_3H_6	Cyclopropane	3555.6	3,554.3	3,554.4
C_3H_6	Propene	3,587.3	3,585.0	3,585.5
C_3H_6O	Acetone	4,080.7	4,077.3	4,077.8
C_3H_7Cl	1-Chloropropane	4,117.7	4,113.1	4,113.5
C_3H_7	Isopropyl radical	3754.8	3,754.7	3,753.5
C_3H_8O	Methoxyethane	4,572.5	4,568.8	4,568.7
C_3H_8O	Isopropyl alcohol	4,626.3	4,622.1	4,622.2
C_3H_8	Propane	4,199.5	4,197.8	4,197.9
C_3NH_3	Acrylonitrile	3,171.8	3,168.3	3,169.1
C_3NH_9	Trimethylamine	4,841.6	4,838.7	4,838.1
C_4H_{10}	Isobutane	5,435.0	5,431.5	5,431.5
C_4H_{10}	*n*-Butane	5,427.3	5,425.5	5,425.5
C_4H_4O	Furan	4,142.5	4,138.2	4,138.7
C_4H_4S	Thiophene	4,015.7	4,012.1	4,012.8
C_4H_6	1,3-Butadiene	4,215.5	4,213.4	4,214.2
C_4H_6	2-Butyne	4,178.8	4,176.7	4,177.3
C_4H_6	Bicyclo[1.1.0]butane	4,105.9	4,103.5	4,103.5
C_4H_6	Cyclobutene	4,169.6	4,167.4	4,167.8
C_4H_6	Methylenecyclopropane	4,134.0	4,131.3	4,131.6
C_4H_8	Cyclobutane	4,790.7	4,790.1	4,790.0
C_4H_8	Isobutene	4,829.7	4,826.8	4,827.1
C_4H_9	*tert*-Butyl radical	4,997.4	4,997.5	4,994.7
C_4NH_5	Pyrrole	4,465.1	4,461.3	4,461.4
C_5H_8	Spiropentane	5,350.0	5,348.8	5,348.7
C_5NH_5	Pyridine	5,156.3	5,152.1	5,152.3
C_6H_6	Benzene	5,696.6	5,693.5	5,693.9
Cl_2	Dichlorine	246.4	245.7	245.9
CO	Carbon monoxide	1,080.4	1,079.7	1,080.4

Table 7 continued

Molecule		Canonical	Boys	Ref.[a]
CO_2	Carbon dioxide	1,622.0	1,620.3	1,621.6
COF_2	Carbonyl fluoride	1,754.6	1,750.2	1,751.3
COS	Carbonyl sulfide	1,395.7	1,392.9	1,394.2
CS	Carbon monosulfide	712.4	711.3	712.5
CS_2	Carbon disulfide	1,162.0	1,159.7	1,161.4
FCl	Chlorine monofluoride	2,62.1	259.7	259.7
F_2	Difluorine	159.3	158.7	158.2
F_3Cl	Chlorine trifluoride	537.1	530.2	529.3
HCl	Hydrogen chloride	447.9	447.3	448.2
HF	Hydrogen fluoride	591.1	590.9	591.3
HOCl	Hypochlorous acid	693.1	691.1	691.2
HO	Hydroxyl radical	446.8	446.8	446.8
HS	Mercapto radical	365.2	365.1	365.2
H_2	Dihydrogen	457.9	457.9	457.7
H_2O_2	Hydrogen peroxide	1,121.1	1,120.7	1,120.6
H_2O	Water	971.5	971.4	971.8
H_2S	Hydrogen sulfide	765.9	765.5	766.2
LiF	Lithium fluoride	581.5	581.1	581.8
LiH	Lithium hydride	242.2	242.2	242.3
Li_2	Dilithium	100.7	100.7	101.2
Na_2	Disodium	71.1	71.1	71.6
NaCl	Sodium chloride	410.5	409.0	411.6
NF_3	Trifluoroamine	861.4	857.2	856.4
NH_2	Amino radical	760.4	760.4	760.2
NH_3	Ammonia	1,241.8	1,241.7	1,241.8
NH	Imidogen	346.1	346.0	345.8
NO_2	Nitrogen dioxide	942.8	942.6	940.8
NOCl	Nitrosyl chloride	794.9	791.2	791.4
NO	Nitric oxide	633.0	633.1	632.2
N_2	Dinitrogen	948.1	947.4	947.8
N_2H_4	Hydrazine	1,825.6	1,825.1	1,824.8
N_2O	Nitrous oxide	1,122.3	1,120.0	1,120.4
OCl	Monochlorine monoxide	268.2	268.1	266.4
OF_2	Difluorine monoxide	388.5	386.5	385.6
OS	Sulfur monoxide	523.2	523.2	522.4
O_2	Dioxygen	500.1	499.8	499.8
O_2S	Sulfur dioxide	1,081.2	1,078.7	1,079.8
O_3	Ozone	602.0	599.4	598.7
P_2	Diphosphorus	481.5	480.4	481.2
PF_3	Phosphorus trifluoride	1,528.7	1,524.1	1,524.4
PH_2	Phosphino radical	643.5	643.6	643.3
PH_3	Phosphane	1,009.6	1,009.2	1,009.3
S_2	Disulfur	428.6	428.0	428.6
$SiCH_6$	Methylsilane	2,626.6	2624.9	2,625.1
$SiCl_4$	Silicon tetrachloride	1,626.1	1,621.1	1,624.0
SiF_4	Silicon tetrafluoride	2,415.2	2,411.8	2,413.4
SiH_2	Singlet silylene	642.6	642.4	6,42.4

Table 7 continued

Molecule		Canonical	Boys	Ref.[a]
SiH_2	Triplet silylene	557.1	557.0	557.1
SiH_3	Silyl radical	954.0	954.0	953.8
SiH_4	Silane	1,357.6	1,357.4	1,357.6
SiO	Silicon monoxide	800.7	799.8	800.3
Si_2H_6	Disilane	2,239.8	2,238.7	2,239.1
Si_2	Disilicon	226.0	301.3	301.4
Mean Error		−1.72	0.28	
MAD		1.77	0.62	
RMS		2.32	0.88	
MAX		−7.8	3.4	

[a] From Ref. [73], higher excitations and core/core–valence energy contributions have been subtracted

experimental data makes the new reference data more accurate on average [73]. Therefore, the accuracy of our interference-corrected MP2-F12 model will be analyzed with respect to the values of Ref. [73]. Note that the reference values of Ref. [73], as presented in Table 7, do not contain electron-correlation contributions from higher excitations nor core/core–valence effects. They are plain CCSD(T)(F12)/cc-pVQZ-F12 values (cc-pCVQZ-F12 for Li, Be, and Na).

All results reported in this section have been obtained at the CCSD(T)+F12+INT/cc-pVQZ-F12 level of theory (cc-pCVQZ-F12 for Li, Be, and Na) using either canonical or Boys localized occupied Hartree–Fock orbitals.

Let us first consider the results obtained with canonical orbitals. The MAD obtained with these orbitals is 1.8 kJ/mol. This is within the limit of chemical accuracy, but larger than for the 106-molecule (1.3 kJ/mol) and the AE6 (0.8 kJ/mol) test sets. Table 7 shows that many of the molecules with large errors in the computed atomization energy are either linear or have other high molecular symmetry such as a threefold axis. Examples are F_3Cl (7.8 kJ/mol), $CHCl_3$ (5.4 kJ/mol), NF_3 (5.0 kJ/mol), and C_2NF_3 (trifluoroacetonitrile, 5.9 kJ/mol).

One drawback of the INT computations is that the results are not invariant to orbital rotations. This is particularly unpleasant for cases with degenerate canonical orbitals, which are not uniquely defined. Therefore, calculations of interference factors usually use localized occupied Hartree–Fock orbitals [21]. We have applied the Boys localization procedure to the G2/97 test set and present the results in Table 7. For open-shell cases, the α- and β-orbitals are localized separately and the two sets of orbitals are aligned in space as much as possible [21].

The Boys localization significantly reduces the errors of most of the cases that showed large deviations when canonical orbitals were employed. For example, the error of chlorine trifluoride is reduced from 7.8 to 0.9 kJ/mol and for pyridine from 4.0 to 0.2 kJ/mol. These corrections are also reflected in the statistics: The MAD and RMS values drop below 1 kJ/mol. In particular, the MAD from the Boys localized orbitals (0.6 kJ/mol) is about three times smaller than the one obtained with canonical orbitals. However, there are still a few cases that show errors larger than 2 kJ/mol: $AlCl_3$, BCl_3, $SiCl_4$, and the *tert*-butyl radical. It should also be noted that the orbital-invariant MP2-F12 calculations of the tetrachloroethylene (C_2Cl_4) and disodium (Na_2) molecules showed numerical problems with respect to the optimization of the F12 amplitudes (the **B** matrix was not positive definite). For these molecules, the atomization energies shown in Table 7 have been computed with the fixed-amplitudes ansatz.

4 Summary

The CCSD(T)+F12+INT method performs well in comparison with other explicitly correlated coupled-cluster calculations. The method was tested with respect to the atomization energies of the molecules of the 106-molecule, the AE6, and the G2/97 test sets. All mean absolute deviations were below 2 kJ/mol. In particular, for the G2/97 test set, the MAD is 1.8 kJ/mol, when *canonical* Hartree–Fock orbitals are used. This value is significantly reduced to 0.6 kJ/mol when a Boys localization of the occupied Hartree–Fock orbitals is performed. For the open-shell cases, the alignment of the α and β-orbitals is important to uniquely define the interference correction.

To summarize, a thermochemical composite scheme may be formulated starting from a conventional CCSD(T) calculation to which a basis-set-truncation correction is added, obtained from low-order perturbation theory. With respect to explicitly correlated CCSD(T)(F12) calculations, the present composite scheme saves computer time without much loss in accuracy.

Acknowledgments This work was supported by the DFG through the Center for Functional Nanostructures (CFN, Project No. C3.3). We are also grateful to the Bundesministerium für Bildung und Forschung (BMBF) through the Helmholtz Research Programme "Science and Technology of Nanosystems" and to the State of Baden-Württemberg for providing the necessary infrastructure. R.H. thanks the Carl-Zeiss-Stiftung for financial support.

References

1. Peterson KA, Dunning TH Jr (1995) J Phys Chem 99:3898–3901
2. Dunning TH Jr (2000) J Phys Chem A 104:9062–9080
3. Dunning TH Jr (1989) J Chem Phys 90:1007–1024
4. Klopper W, Bachorz RA, Hättig C, Tew DP (2010) Theor Chem Acc 126:289–304

5. Hättig C, Klopper W, Köhn A, Tew DP (2012) Chem Rev 112:4–74
6. Kong L, Bischoff FA, Valeev EF (2012) Chem Rev 112:75–107
7. Ten-no S (2012) Theor Chem Acc 131:1070
8. Ten-no S, Noga J (2012) WIREs Comput Mol Sci 2:114–125
9. Feller D, Peterson KA (2013) J Chem Phys 139:084110
10. Raghavachari K, Trucks GW, Pople JA, Head-Gordon M (1989) Chem Phys Lett 157:479–483
11. Helgaker T, Klopper W, Tew DP (2008) Mol Phys 106:2107–2143
12. Papajak E, Truhlar DG (2012) J Chem Phys 137:064110
13. Curtiss LA, Jones C, Trucks GW, Raghavachari K, Pople JA (1990) J Chem Phys 93:2537–2545
14. Curtiss LA, Raghavachari K, Trucks GW, Pople JA (1991) J Chem Phys 94:7221–7230
15. Curtiss LA, Raghavachari K, Redfern PC, Rassolov V, Pople JA (1998) J Chem Phys 109:7764–7776
16. Curtiss LA, Redfern PC, Raghavachari K (2007) J Chem Phys 126:084108
17. DeYonker NJ, Cundari TR, Wilson AK (2006) J Chem Phys 124:114104
18. DeYonker NJ, Wilson BR, Pierpont AW, Cundari TR, Wilson AK (2009) Mol Phys 107:1107–1121
19. Mahler A, Wilson AK (2013) J Chem Theory Comp 9:1402–1407
20. Petersson GA, Bennett A, Tensfeldt TG, Al-Laham MA, Shirley WA, Mantzaris J (1988) J Chem Phys 89:2193–2218
21. Petersson GA, Al-Laham MA (1991) J Chem Phys 94:6081–6090
22. Petersson GA, Tensfeldt TG, Montgomery JA Jr (1991) J Chem Phys 94:6091–6101
23. Montgomery JA Jr, Ochterski JW, Petersson GA (1994) J Chem Phys 101:5900–5909
24. Petersson GA, Malick DK, Wilson WG, Ochterski JW, Montgomery JA Jr, Frisch MJ (1998) J Chem Phys 109:10570–10579
25. Petersson GA, Frisch MJ (2000) J Phys Chem A 104:2183–2190
26. Wood GPF, Radom L, Petersson GA, Barnes EC, Frisch MJ, Montgomery JA Jr (2006) J Chem Phys 125:94106–94121
27. East ALL, Allen WD (1993) J Chem Phys 99:4638–4650
28. Császár AG, Allen WD, Schaefer HF III (1998) J Chem Phys 108:9751–9764
29. Fast PL, Sánchez ML, Truhlar DG (1999) Chem Phys Lett 306:407–410
30. Fast PL, Corchado JC, Sánchez ML, Truhlar DG (1999) J Phys Chem A 103:3139–3143
31. Fast PL, Corchado JC, Sánchez ML, Truhlar DG (1999) J Phys Chem A 103:5129–5136
32. Tajti A, Szalay PG, Császár AG, Kállay M, Gauss J, Valeev EF, Flowers BA, Vázquez J, Stanton JF (2004) J Chem Phys 121:11599–11613
33. Harding ME, Vázquez J, Ruscic B, Wilson AK, Gauss J, Stanton JF (2008) J Chem Phys 128:114111
34. Boese AD, Oren M, Atasoylu O, Martin JML, Kállay M, Gauss J (2004) J Chem Phys 120:4129–4141
35. Karton A, Rabinovich E, Martin JML, Ruscic B (2006) J Chem Phys 125:144108
36. Karton A, Martin JML (2012) J Chem Phys 136:124114
37. Dunning TH Jr, Peterson KA (2000) J Chem Phys 113:7799–7808
38. Klopper W, Ruscic B, Tew DP, Bischoff FA, Wolfsegger S (2009) Chem Phys 356:14–24
39. Ruscic B, Pinzon RE, Morton ML, von Laszevski G, Bittner SJ, Nijsure SG, Amin KA, Minkoff M, Wagner AF (2004) J Phys Chem A 108:9979–9997
40. Ruscic B, Pinzon RE, Morton ML, Srinivasan NK, Su MC, Sutherland JW, Michael JV (2006) J Phys Chem A 110:6592–6601
41. Nyden MR, Petersson GA (1981) J Chem Phys 75:1843–1862
42. Petersson GA, Nyden MR (1981) J Chem Phys 75:3423–3425
43. Vogiatzis KD, Barnes EC, Klopper W (2011) Chem Phys Lett 503:157–161
44. Ranasinghe DS, Petersson GA (2013) J Chem Phys 138:144104
45. Bachorz RA, Bischoff FA, Glöß A, Hättig C, Höfener S, Klopper W, Tew DP (2011) J Comput Chem 32:2492–2513
46. Peterson KA, Adler TB, Werner HJ (2008) J Chem Phys 128:84102–84114
47. Vogiatzis KD, Klopper W (2013) Mol Phys 111:2299–2305
48. Jurečka P, Šponer J, Černý J, Hobza P (2006) Phys Chem Chem Phys 8:1985–1993
49. Marshall MS, Burns LA, Sherrill CD (2011) J Chem Phys 135:194102
50. Helgaker T, Jørgensen P, Olsen J (2000) Molecular electronic structure theory. Wiley, Chichester
51. Bakowies D (2007) J Chem Phys 127:84105–84128
52. Lynch BJ, Truhlar DG (2003) J Phys Chem A 107:8996–8999
53. Curtiss LA, Raghavachari K, Redfern PC, Pople JA (1997) J Chem Phys 106:1063–1079
54. Curtiss LA, Redfern PC, Raghavachari K, Pople JA (1998) J Chem Phys 109:42–54
55. Woon DE, Dunning TH Jr (1995) J Chem Phys 103:4572–4586
56. Lynch BJ, Zhao Y, Truhlar DG (2005) J Phys Chem A 109:1643–1649
57. Haunschild R, Janesko BG, Scuseria GE (2009) J Chem Phys 131:154112
58. Köhn A, Tew DP (2010) J Chem Phys 132:024101
59. TURBOMOLE V6.5 2013, a development of University of Karlsruhe and Forschungszentrum Karlsruhe GmbH, 1989-2007, TURBOMOLE GmbH, since 2007; available from http://www.turbomole.com
60. Furche F, Ahlrichs R, Hättig C, Klopper W, Sierka M, Weigend F (2013) WIREs Comput Mol Sci doi:10.1002/wcms.1162
61. Hill JG, Mazumder S, Peterson KA (2010) J Chem Phys 132:054108
62. Hill JG, Peterson KA (2010) Phys Chem Chem Phys 12:10460–10468
63. Yousaf KE, Peterson KA (2008) J Chem Phys 129:184108–184115
64. Hättig C (2005) Phys Chem Chem Phys 7:59–66
65. Weigend F (2008) J Comput Chem 29:167–175
66. Ten-no S (2004) J Chem Phys 121:117–129
67. Bokhan D, Ten-no S, Noga J (2008) Phys Chem Chem Phys 10:3320–3326
68. Tew DP, Klopper W (108) Mol Phys 108:315–325
69. Janssen CL, Nielsen IM (1998) Chem Phys Lett 290:423–430
70. Lynch BJ, Truhlar DG (2003) J Phys Chem A 107:3898–3906
71. Haunschild R, Klopper W (2012) Theor Chem Acc 131:1112
72. Curtiss LA, Raghavachari K, Redfern PC, Pople JA (2000) J Chem Phys 112:7374–7383
73. Haunschild R, Klopper W (2012) J Chem Phys 136:164102

Theor Chem Acc (2014) 133:1440
DOI 10.1007/s00214-013-1440-y

REGULAR ARTICLE

On the mutual exclusion of variationality and size consistency

So Hirata · Ireneusz Grabowski

Received: 10 August 2013 / Accepted: 12 December 2013 / Published online: 23 January 2014

Abstract Why do variational electron-correlation methods such as truncated configuration-interaction methods tend to be non-size-consistent (non-size-extensive)? Why are size-consistent (size-extensive) methods such as Møller–Plesset perturbation and coupled-cluster methods non-variational? We conjecture that the variational and size-consistent properties are mutually exclusive in an ab initio electron-correlation method (which thus excludes the Hartree–Fock and density-functional methods). We analyze some key examples that support the truth of this conjecture.

Keywords Electron correlation · Size consistency · Size extensivity · Variational principle · Diagrams · Coupled-cluster theory · Møller–Plesset perturbation theory · Configuration-interaction theory

Dedicated to Professor Thom Dunning and published as part of the special collection of articles celebrating his career upon his retirement.

S. Hirata (✉)
Department of Chemistry, University of Illinois at Urbana-Champaign, 600 South Mathews Avenue, Urbana, IL 61801, USA
e-mail: sohirata@illinois.edu

S. Hirata
CREST, Japan Science and Technology Agency, 4-1-8 Honcho, Kawaguchi, Saitama 332-0012, Japan

I. Grabowski
Faculty of Physics, Astronomy and Informatics, Institute of Physics, Nicolaus Copernicus University, Grudziadzka 5, 87-100 Toruń, Poland

1 Conjecture

Let the ground-state wave function of a molecule, Ψ, be expanded as a linear combination of Slater determinants, $\{\Phi\}$, constructed with molecular orbitals (MO's) from the Hartree–Fock (HF) method [1–3],

$$\Psi = c_0\Phi_0 + \hat{C}_1\Phi_0 + \hat{C}_2\Phi_0 + \cdots + \hat{C}_n\Phi_0 \tag{1}$$

$$= c_0\Phi_0 + \sum_i^{\text{occ.}}\sum_a^{\text{virt.}} c_i^a\Phi_i^a + \frac{1}{(2!)^2}\sum_{i,j}^{\text{occ.}}\sum_{a,b}^{\text{virt.}} c_{ij}^{ab}\Phi_{ij}^{ab} + \cdots + \frac{1}{(n!)^2}\sum_{i_1,\ldots,i_n}^{\text{occ.}}\sum_{a_1,\ldots,a_n}^{\text{virt.}} c_{i_1\cdots i_n}^{a_1\cdots a_n}\Phi_{i_1\cdots i_n}^{a_1\cdots a_n}, \tag{2}$$

where $\hat{C}_k$ is the k-electron excitation operator, c's are the expansion coefficients, $\Phi_{i_1\cdots i_k}^{a_1\cdots a_k}$ is a k-electron excited determinant, "occ." and "virt." stand for occupied and virtual orbitals in the ground-state wave function (Φ_0) of the reference method, and n is the number of electrons. It is assumed that the number of virtual orbitals is greater than n.

In the context of this article, an ab initio electron-correlation method is considered *variational*, when the adjustable parameters in its wave function such as $\{c\}$ are determined so as to minimize the energy expectation value,

$$E = \langle\Psi|\hat{H}|\Psi\rangle, \tag{3}$$

with the constraint that Ψ remain normalized,

$$\langle\Psi|\Psi\rangle = 1. \tag{4}$$

It is equivalent to minimizing the following quotient with no constraint:

$$E = \frac{\langle\Psi|\hat{H}|\Psi\rangle}{\langle\Psi|\Psi\rangle}. \tag{5}$$

The energy thus obtained is an upper bound of the exact (full CI) value at a given basis set. In this article, a method is not variational when this upper-bound property is lost. Other authors [4, 5] have used the term *stationarity* to characterize a method whose energy is not an upper bound but is otherwise variational. We adopt this terminology. Variational methods include the HF, approximations in density-functional theory (DFT), truncated and full configuration-interaction (CI), multiconfigurational self-consistent field (MCSCF), complete active space self-consistent field (CASSCF), and multireference CI (MRCI) methods. Some of them also use orbitals as variational parameters, but we do not consider these degrees of freedom here because they only contribute marginally in capturing electron correlation.

An MO method is considered *size consistent* or, synonymously, *size extensive*, when the energy E obtained from it is asymptotically proportional to the volume V or the number of electrons n, i.e., $E = O(V) = O(n)$, in the limit $V \to \infty$ ($n \to \infty$). However, the more rigorous criterion of size consistency adopted in this work is based on the wave-vector counting [6, 7], which is equivalent to the well-known diagrammatic criterion [7, 8]. This criterion is a sufficient condition for (and thus stricter than) $E = O(V) = O(n)$; there can be a method that has an extensive energy, but is nonetheless deemed non-size consistent [9]. One of the authors discussed extensively these rigorous and other less rigorous size-consistency criteria and widespread misconceptions about them [7, 10]. Size-consistent methods include the HF, Møller–Plesset perturbation (MP), and coupled-cluster (CC) methods [1–3]. DFT is not *diagrammatically* size consistent, but its energy is extensive (see the "Appendix").

We conjecture that an ab initio electron-correlated MO method cannot be simultaneously variational and size consistent.

Here, the "method" excludes the HF method, which is simultaneously variational and size consistent, but does not describe any electron correlation. In fact, HF is variational merely with respect to its one-particle basis set; its variational space for describing electron correlation is null. Also excluded is DFT because it is not an ab initio MO theory and not subject to a diagrammatic size-consistency analysis [7, 10]. It is also more appropriately characterized as stationary than variational according to the aforementioned definitions [4] because a DFT energy is not an upper bound of the exact energy at a given basis set. We can paraphrase the conjecture as follows: *the only simultaneously variational and size-consistent MO methods are the ones that use single-determinant wave functions*. This version of the conjecture encompasses DFT.

We do not consider full CI, full CC, or infinite-order MP as a counterexample to this conjecture, even though they are simultaneously variational and size consistent by being exact at a given basis set. This is because these methods do not exist in the thermodynamic limit except as an abstract concept; they are not characterized by a finite number of equations with a finite number of terms at $n = \infty$. They are, therefore, outside the scope of the usual size-consistency analysis, and some additional considerations need to be made when discussing their size consistency, such as the effect of the number (not the values) of terms in their equations on size consistency (see below). Worse yet, full CI changes its mathematical definition depending on the number of electrons: it is equal to CI singles and doubles (CISD) for He ($n = 2$); CI singles, doubles, and triples for Li ($n = 3$); etc., and ultimately becomes ill-defined as $n \to \infty$, where it ceases to be a finite mathematical procedure. An infinite-order MP method does not even have a tangible definition. We will revisit full CI as it helps us understand how a variational method resists size consistency even near the exact limit.

We also confine ourselves to the methods that lend themselves to a second-quantized or diagrammatic description based on delocalized MO's in a crystal, to which the rigorous size-consistency test can be applied. The methods can be based on localized orbitals insofar as a basis transformation can be achieved without altering the nature of its approximations or obscuring the definition of its variational space (see the "Appendix"). Stochastic methods such as quantum Monte Carlo (QMC) [11–15] are not easily subjected to the analyses that have led to the conjecture, although they do not appear to be counterexamples, either; in fact, one of the authors with coauthors has proposed [16–18] a diagrammatically size-consistent variant of QMC, which is non-variational.

The variational property is useful (if not essential) in the studies of multiple potential energy surfaces that intersect and/or couple through vibronic interactions. It also simplifies the formalisms for an analytical evaluation of energy derivatives and molecular properties. Size consistency or extensivity is a must in applications to solids, liquids, and large molecules. It is, therefore, not surprising that theorists have pursued an electron-correlated method that is simultaneously variational and size consistent (e.g., [5, 19] recently). The objective of this article is to question the very existence of such a method within the framework of ab initio MO theory, where we expand a wave function as a linear combination of Slater determinants constructed with delocalized, symmetry-adapted MO's. In what follows, we analyze some key examples that seem to suggest that they may not exist.

2 Case study 1: CI

A truncated CI is not size consistent [1–3]. Although this is well known, it is instructive to analyze its origin, which has to do with the normalization of its wave function [7, 10]. A wave function of a crystal in the mth-order truncated CI approximation ($m < n$) is subjected to the standard normalization,

$$1 = |c_0|^2 + \sum_{i}^{\text{occ.}} \sum_{a}^{\text{virt.}} \sum_{\boldsymbol{k}_i} |c_{\boldsymbol{i}}^{\boldsymbol{a}}|^2 + \frac{1}{(2!)^2} \sum_{i,j}^{\text{occ.}} \sum_{a,b}^{\text{virt.}} \sum_{\boldsymbol{k}_i,\boldsymbol{k}_j,\boldsymbol{k}_a} \left|c_{\boldsymbol{ij}}^{\boldsymbol{ab}}\right|^2 + \cdots + \frac{1}{(m!)^2} \sum_{i_1,\ldots,i_m}^{\text{occ.}} \sum_{a_1,\ldots,a_m}^{\text{virt.}} \sum_{\boldsymbol{k}_{i_1},\ldots,\boldsymbol{k}_{a_{m-1}}} \left|c_{\boldsymbol{i}_1\ldots\boldsymbol{i}_m}^{\boldsymbol{a}_1\ldots\boldsymbol{a}_m}\right|^2, \quad (6)$$

where $\boldsymbol{k}_i$ is the wave vector of the $\boldsymbol{i}$th MO. Here, the bold letter $\boldsymbol{i}$ indicates that it is a composite index of energy band i and wave vector $\boldsymbol{k}_i$. In the second term of the right-hand side, the summation over $\boldsymbol{k}_i$ is taken, while that over $\boldsymbol{k}_a$ is not. This is because the latter is not independently variable owing to the periodic symmetry or, equivalently, the conservation of momentum, which reads

$$\boldsymbol{k}_a = \boldsymbol{k}_i + \boldsymbol{K}, \quad (7)$$

where $\boldsymbol{K}$ is an arbitrary integer multiple of a reciprocal lattice vector. Likewise, the conservation of momentum requires,

$$\boldsymbol{k}_a + \boldsymbol{k}_b = \boldsymbol{k}_i + \boldsymbol{k}_j + \boldsymbol{K}, \quad (8)$$

in the third term of the right-hand side, which, therefore, has the summation over $\boldsymbol{k}_i, \boldsymbol{k}_j$, and $\boldsymbol{k}_a$, but not over $\boldsymbol{k}_b$.

Each summation over a wave vector contributes to a factor of V (or n) to the volume (size) dependence of the term. This is so because the number of independent wave vectors in the reciprocal unit cell is proportional to the number of distinct states, which, in turn, bears a linear relationship with the volume of the crystal. Another way to understand this is to use the relationship,

$$\sum_{\boldsymbol{k}} = \frac{V}{(2\pi)^3} \int \mathrm{d}\boldsymbol{k}. \quad (9)$$

With this, we find that expansion coefficients for single-excited determinants, $c_{\boldsymbol{i}}^{\boldsymbol{a}}$, must have the asymptotic volume dependence of $V^{-1/2}$ so that the second term in the right-hand side of Eq. (6) is V^0 just like the left-hand side. Likewise, $c_{\boldsymbol{ij}}^{\boldsymbol{ab}}$ scales as $V^{-3/2}$ and, generally, the expansion coefficients for kth-order excited determinants ($k > 0$) is a $V^{1/2-k}$ quantity. This has been called "the normalization theorem" by one of the authors [10].

Diagrammatically, Eq. (6) can be expressed as

$1 = |c_0|^2 +$ 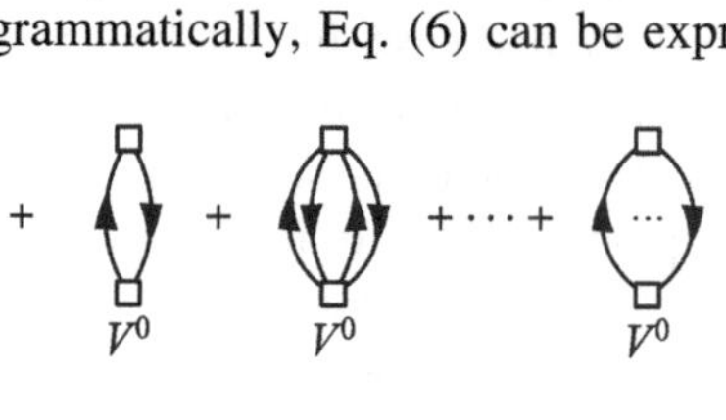 , (10)

where the $2k$-edge square vertex represents the k-electron excitation operator, $\hat{C}_k$ ($\hat{C}_k^\dagger$). Since each diagram must be closed (all orbital indexes are summed over) and has exactly two vertexes ($\hat{C}^\dagger$ and $\hat{C}$), it is also always connected. Each downgoing edge is associated with an occupied orbital $\boldsymbol{i}$ with a wave vector $\boldsymbol{k}_i$, while an upgoing edge with a virtual orbital $\boldsymbol{a}$ with a wave vector $\boldsymbol{k}_a$. It can be shown that there are ($n_{\text{edge}} - n_{\text{vertex}} + 1$) distinct (i.e., linearly independent) wave vectors in a closed, connected diagram, satisfying all momentum conservation conditions [6]. Here, n_{edge} is the number of edges, which is equal to the number of wave vectors, and n_{vertex} is the number of vertexes, which is equal to the number of momentum conservation conditions. Each of these conditions reduces the number of distinct wave vectors by one. In a connected diagram, one of the conditions is automatically satisfied if the remaining ($n_{\text{vertex}} - 1$) conditions are met. This explains the expression: ($n_{\text{edge}} - n_{\text{vertex}} + 1$).

Using this, we can arrive at the same, aforementioned conclusion that $2k$-edge vertex must be a $V^{1/2-k}$ quantity for Eq. (10) to be satisfied. For example, each vertex in the first diagram scales as $V^{-1/2}$. There is $(2 - 2 + 1) = 1$ distinct wave vector in this diagram, contributing to a factor of V^1 in its volume dependence. Together, this diagram is a V^0 quantity. The volume dependence shown right beneath each diagram in Eq. (10) is based on the assumption of the $V^{1/2-k}$ dependence of the k-electron excitation CI coefficients.

The exact electronic Hamiltonian is diagrammatically given by

$\hat{H} = E_{\text{HF}} +$ 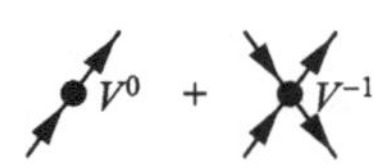 , (11)

where the first, second, and third terms in the right-hand side scale as V^1, V^0, and V^{-1}, respectively. The latter two scaling expressions are made explicit in Eq. (11). These are the fundamental properties of the Hamiltonian [7, 10, 20] and unchangeable by any approximation adopted in electronic structure theory. They originate from the fact that a Fock integral (the two-edge, filled-circle vertex) is thermodynamically intensive (V^0) and an antisymmetrized two-electron integral (the four-edge vertex) is inversely extensive (V^{-1}) [7, 10, 20].

With these, the energy of a truncated CI given by Eq. (3) is diagrammatically expressed as

$E = |c_0|^2 E_{\text{HF}} +$ 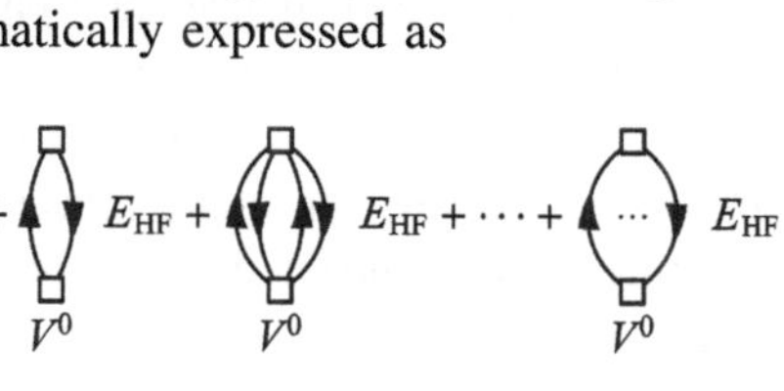 (12)

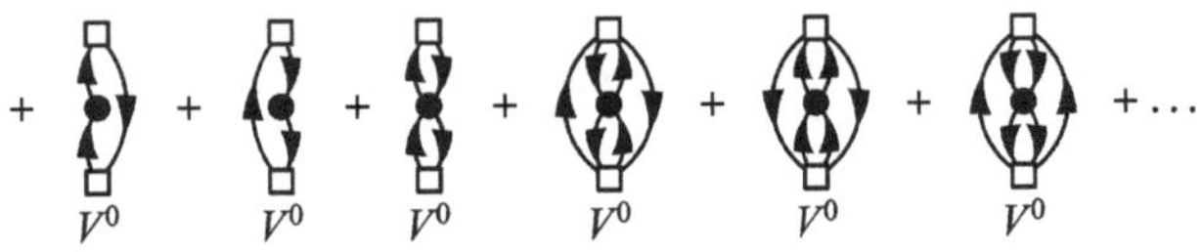

(13)

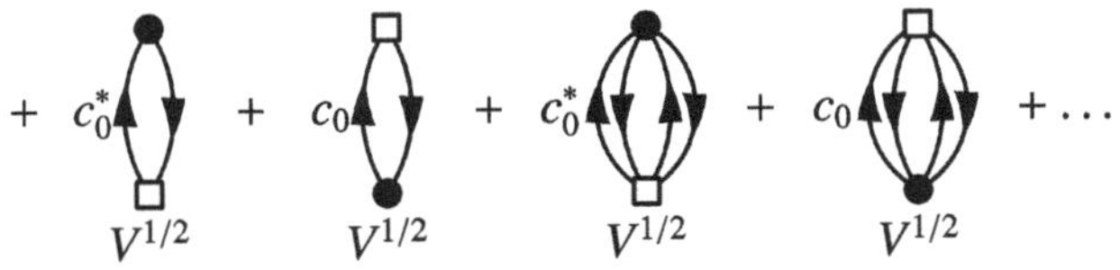

(14)

Each diagram is closed and either connected or disconnected. Every closed part is annotated with its volume dependence. This equation terminates after a finite number of terms because a diagram consists of exactly two excitation vertexes (one or both of which can be c_0) and up to one Hamiltonian vertex and there are a finite number of topologically distinct ways to connect and close them.

The terms in the first line, Eq. (12), are all disconnected, and their sum is E_{HF}. They, therefore, display the V^1 dependence, which is the correct size dependence of an extensive energy. The diagrams in the second line, Eq. (13), all scale as V^0. The first three of them are identified as the diagrams of an excitation energy in the CI singles (CIS) method [21, 22]. Hence, they also have the correct size dependence if they are interpreted as an excitation energy, which is an intensive quantity. The diagrams in the third line, Eq. (14), correspond to various contributions to the correlation energy in the ground state. They scale as non-physical $V^{1/2}$, as compared to the correct dependence of V^1. The $V^{1/2}$ dependence of the first term can be verified as follows: the Fock integral (filled circle) scales as V^0, while the one-electron excitation amplitude (open square) as $V^{-1/2}$. There is only one distinct wave vector in this diagram, contributing to a factor of V^1. The product of these three factors is $V^{1/2}$.

To summarize, minimizing the energy expression like Eq. (3) using $\{c\}$ as variational parameters is meaningful if and only if the wave function is normalized; otherwise, the energy would not be bounded from below and a minimum would not exist. This normalization requirement demands that the k-electron excitation amplitudes have the $V^{1/2-k}$ dependence. This dependence, in turn, makes the correlation energy have the $V^{1/2}$ dependence, which is in between extensive and intensive and hence non-physical. The method then becomes non-size consistent.

The same conclusion can be drawn from using Eq. (5) instead. This has been shown in many places (e.g., [1, 3, 7]), demonstrating the disconnected nature of the resulting amplitude equations. It will not be repeated here.

How does then full CI, at least formally or in an abstract sense, maintain size consistency while being variational? To answer this, let us consider the simplified example of a cluster of non-interacting n helium atoms, in which the effect of one-electron excitation is assumed negligible. For this system, full CI must take into account up to $2n$-electron excitations. Owing to the non-interacting and identical nature of the constituent atoms as well as the invariance of full CI wave function and energy with respect to a unitary transformation of orbitals, we can switch the basis orbitals to those localized on atoms, without changing the nature of the method. Then, the $\hat{C}_{2k}$ amplitudes in the localized basis are the product, $\prod_{l=1}^{k} \hat{C}_2^{[l]}$, where $\hat{C}_2^{[l]}$ is the exact two-electron excitation amplitudes of the lth atom. For example, the $\hat{C}_{2n}$ amplitude is factored as

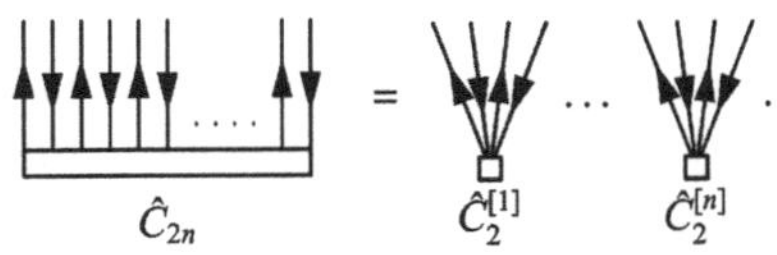

(15)

The size consistency of the full CI correlation energy is entirely due to the energy contribution from the highest rank, i.e., $2n$-electron excitation amplitude, where $2n$ is the number of electrons. This can be understood considering the following *reductio ad absurdum*: if a lower-ranked excitation operator could ensure size consistency, a truncated CI containing up to that operator would be size consistent. The energy diagram involving the $2n$-electron excitation vertex has the form:

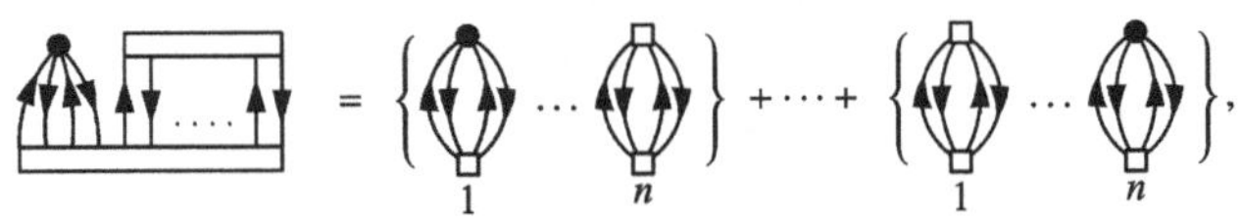

(16)

where $\{\ldots\}$ is one disconnected diagram and only the first and last (nth) diagrams are shown. The right-hand side is a sum of the n disconnected diagrams, the lth of which reports the correlation energy of the lth atom. Each ($\{\ldots\}$) of the n terms in the right-hand side scales as V^0, making the sum scale as nV^0 or V^1 (extensive) because V and n are proportional to each other. This proves the size consistency of full CI. Hence, the size consistency of full CI comes at the price of its equations having an infinite number of terms (in the right-hand side) or its diagrams having an infinite number of edges (in the left-hand side) in the thermodynamic limit ($n = \infty$). In this sense, full CI does not exist as a finite, well-defined procedure in the thermodynamic limit, and the statement about its size consistency is as meaningless as its practicability to an infinitely extended system. Full CI, therefore, does not qualify as a counterexample of the conjecture. The same applies to CASSCF, which performs an orbital-optimized full CI calculation within an active space.

3 Case study 2: CC

A CC method uses an approximate wave function obtained by acting an exponentiated excitation operator on the HF wave function [1–3, 23]:

$$\Psi_{\mathrm{CC}} = t_0 \exp(\hat{T}_1 + \hat{T}_2 + \cdots + \hat{T}_m)\Phi_0, \tag{17}$$

where $\hat{T}_k$ is the k-electron excitation operator. The most successful version of a CC method, the projection CC method [3, 23, 24], uses the intermediate normalization ($t_0 = 1$) [7, 10] and determines the excitation amplitudes non-variationally. It is size consistent [1, 3, 7, 10, 23].

What happens if we instead determine the excitation amplitudes variationally? That is, what happens if we minimize the energy expectation value of Eq. (3) in the above wave function, with the constraint that the latter is normalized as in Eq. (4)? This leads to variational CC [4, 5, 19, 25–30].

Substituting Eq. (17) into Eq. (4), we obtain a diagrammatic equation,

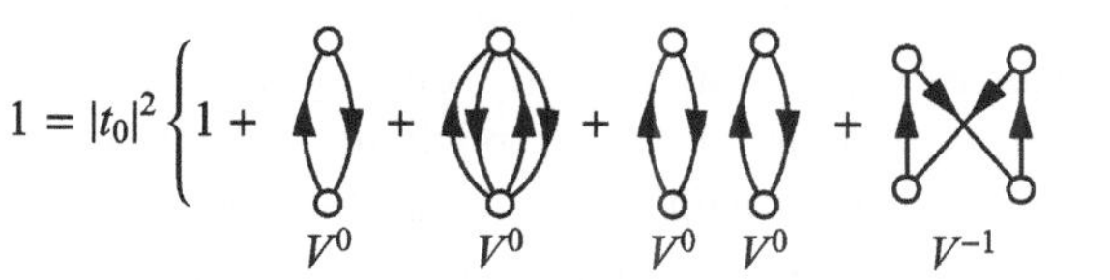

(18)

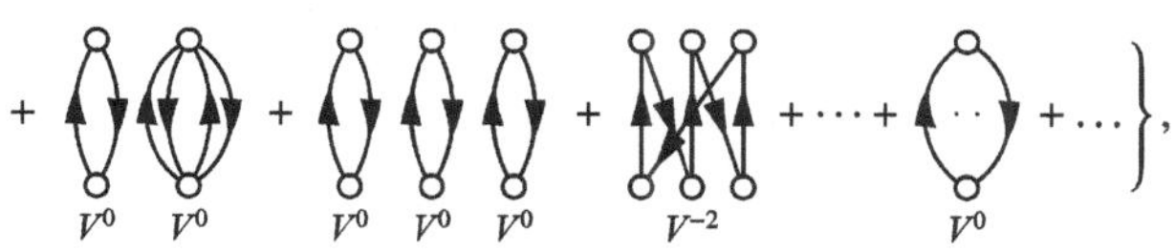

(19)

where the open-circle vertex with $2k$ edges designates the k-electron excitation operator, $\hat{T}_k$ ($\hat{T}_k^\dagger$). We encounter two immediate problems. One is the presence of an infinite number of terms in this equation, making it impossible to satisfy. That this equation does not terminate [4, 26] can be understood by the presence of the "accordion" diagrams such as the last diagram in the first line and the third in the second line; it is the one that is (or contains) a connected part with more than two vertexes. An infinite number of larger accordion diagrams can be generated by repeated $\hat{T}^\dagger\hat{T}$ insertions into a smaller one.

The other is the inconsistent volume dependence of the terms. For the first two diagrams to have the correct V^0 dependence (as does the left-hand side), the $2k$-edge open-circle vertex or $\hat{T}_k$ must scale as $V^{1/2-k}$, according to the identical logic leading to the same scaling for $\hat{C}_k$ [7, 10] (t_0 is a V^0 quantity). This, however, makes the last (accordion) diagram in the first line display the incorrect volume dependence of V^{-1}. The third (accordion) diagram in the second line also shows inconsistent V^{-2} dependence. If one instead chooses one of these accordion diagrams to have the correct V^0 dependence, the inconsistency remains (and, in fact, gets worse). In other words, accordion diagrams, with more than two intensive vertexes, violate the intensive diagram theorem [7, 10].

In this case, however, taking the thermodynamic limit ($V \to \infty$) seems to help resolve both problems. The accordion diagrams have incorrect volume dependence in ways that make them vanish in the $V \to \infty$ limit. Therefore, in that limit, Eq. (19) is simplified to

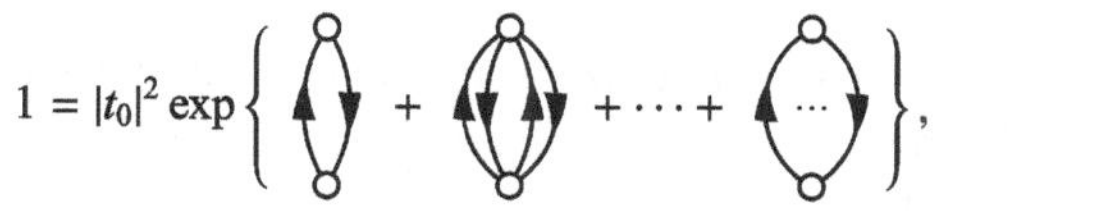

(20)

which has a finite number of terms in the argument of the exponential. Each term in the argument, furthermore, scales correctly as V^0, assuming that a k-electron excitation amplitude still scales as $V^{1/2-k}$.

However, this causes inconsistent volume dependence in the energy expression. Diagrammatically, it is

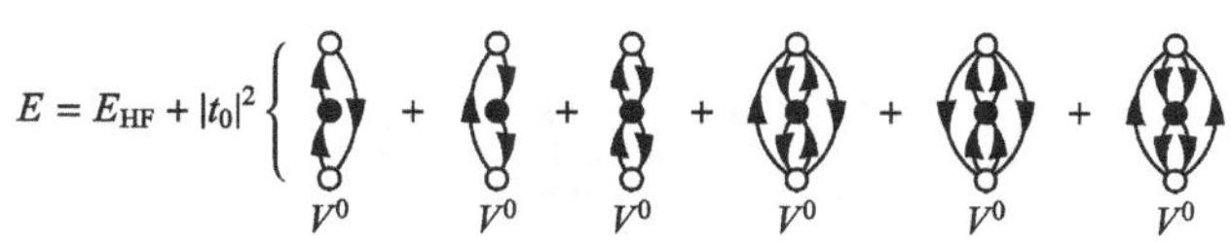

(21)

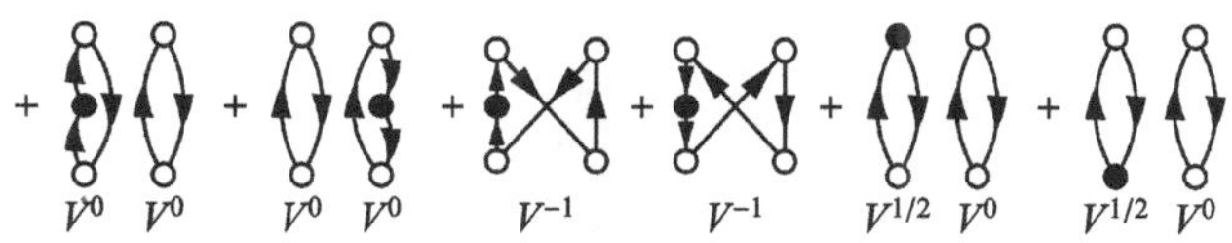

(22)

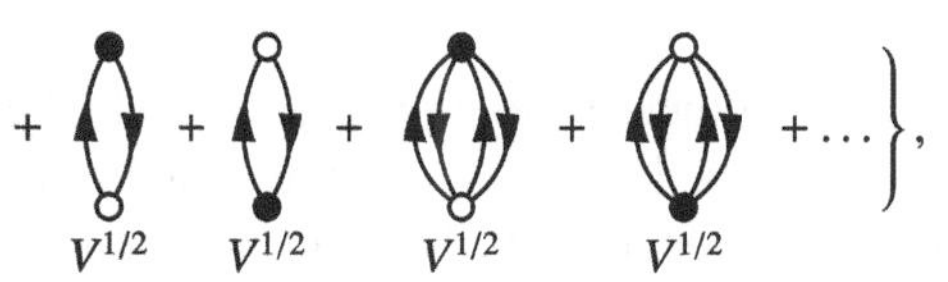

(23)

which again has an infinite number of terms [4]. We can circumvent this infinity by taking the thermodynamic limit. In that limit, all accordion diagrams such as the third and fourth diagrams in the second line (scaling as V^{-1}) vanish. The remaining terms, still infinitely many, are shown to be the product of a finite number of closed, connected energy diagrams and Eq. (20), which is unity, but has an infinite number of diagrams when expanded. In the thermodynamic limit, therefore, the above equation reduces to

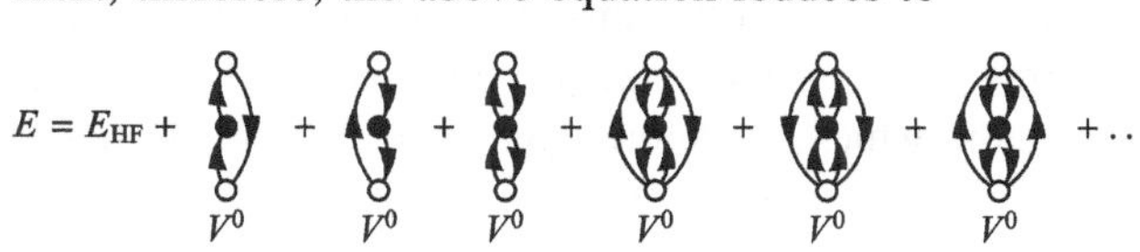

(24)

(25)

which consist of only a finite number of closed, connected diagrams. The diagrams in the first line scale as V^0. They may be considered as having the correct scaling if they are to be interpreted as intensive excitation energies. The diagrams in the second line describe the correlation energy in the ground state, but they scale incorrectly as $V^{1/2}$. Hence, variational CC is not size consistent. Note that the variational CC method of Nakatsuji and Hirao [31] has also been implicated [26] in the lack of size consistency.

As is well known [4, 26], it is possible to bypass the issue of normalization altogether by using

$$E = \frac{\langle \Psi_{\mathrm{CC}}|\hat{H}|\Psi_{\mathrm{CC}}\rangle}{\langle \Psi_{\mathrm{CC}}|\Psi_{\mathrm{CC}}\rangle} = \langle \Psi_{\mathrm{CC}}|\hat{H}|\Psi_{\mathrm{CC}}\rangle_{\mathrm{connected}}. \quad (26)$$

The rightmost means that only connected diagrams are retained, namely

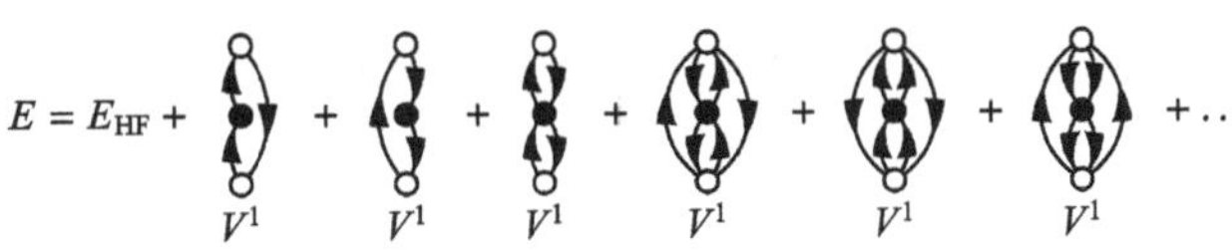

(27)

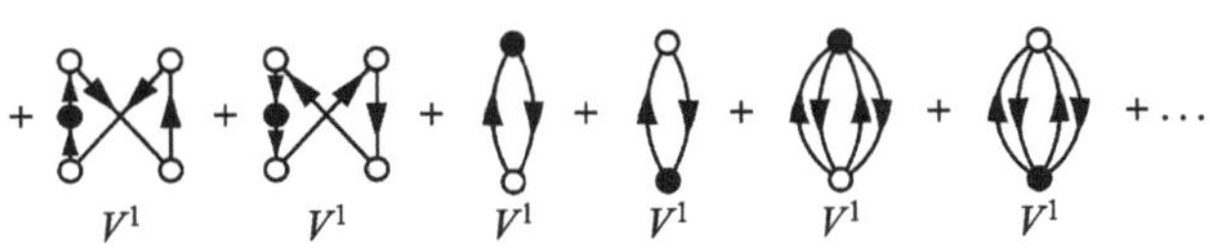

(28)

Since the excitation amplitudes no longer have to satisfy the normalization condition such as Eq. (4), they are allowed to have appropriate volume dependence that renders the above expression extensive. To do so, we require a k-electron excitation amplitude to have V^{1-k} dependence, according to the normalization theorem [7, 10]. This makes every diagram in the above equation have the consistent V^1 dependence, which can be verified diagram-by-diagram as before [7].

Note that the accordion diagrams such as the first two in the second line also have the correct V^1 dependence and cannot be eliminated by taking the thermodynamic limit unlike in the previous treatment. Since there are an infinite number of accordion diagrams, there are an infinite number of terms with correct size dependence in this energy expression, which, therefore, cannot be evaluated [4, 26]. In this sense, variational CC according to Eq. (26) does not exist as a finite, well-defined procedure.

The equations determining the excitation amplitudes as variational parameters are obtained by differentiating the above energy expression with an amplitude and equating it with zero. For instance, the differentiation with a one-electron excitation amplitude yields

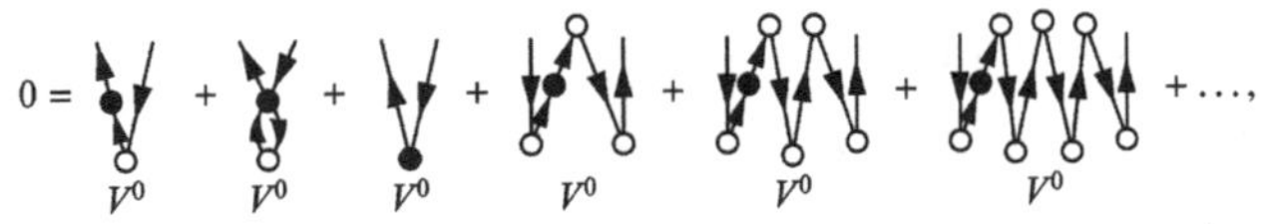

(29)

which consists of open, connected diagrams, all of which scale consistently as V^0. However, as has been recognized by others [4, 26] and expected from the preceding analysis, there are an infinite number of accordion diagrams in this equation, too, making this equation unsolvable unless it is truncated. A truncation would not impair size consistency, but it would result in the loss of variationality because we would no longer be satisfying the variational condition, from which the above equation is obtained.

Nearly all of these findings are already known [4, 26]. However, when put in the context of the conjecture, they seem to assume much greater significance; we can witness size consistency and variationality directly undermining each other in variational CC. Explicit normalization (as necessary for variationality) results in the loss of size consistency, while the connected expression leads to non-truncating equations or the loss of variationality when truncated.

4 Conclusion

The cause of non-size consistency is concisely explained by March et al. [6] as follows: "The reason in each case is that any finite-order wave function is only capable of describing a finite number of simultaneous pair excitations while a true many-body wave function must be capable of describing the simultaneous excitation of an infinite number of particles for $N \to \infty$." For a method to be variational, on the other hand, the weight of every basis function (a Slater determinant) in the wave function must be independently variable. These two requirements seem incompatible, which explains (if not proves) the mutual exclusion of variationality and size consistency in the methods examined above. Therefore, the difficulty of satisfying variationality in a size-consistent method tends to manifest itself as having to deal with an infinite number of terms or an infinite number of diagrammatic edges.

In truncated CI, the variationality limits the excitations considered to a finite rank, making the method non-size consistent. Full CI is formally size consistent and variational, but at the price of having to have an infinite number of variational parameters in the thermodynamic limit, which is impossible to determine except in an abstract sense. In projection CC, an exponential wave function allows excitations of an infinite number of electrons with a finite number of amplitudes in a size-consistent way. However, the weights of various excitations (Slater determinants) are coupled and not independently adjustable as necessary to make the energy variational. In variational CC, depending on the approach to the solution, one encounters either the difficulty of infinity or the loss of size consistency or variationality.

Acknowledgments We thank the anonymous reviewers and colleagues who read our manuscript and challenged the conjecture with many incisive questions, which became the basis of the Appendix. We are particularly indebted to Dr. David Ceperley for pointing out some of our misconceptions about QMC. The opinions expressed in this article are those of the authors alone and not necessarily shared by these colleagues. S.H. thanks Nicolaus Copernicus University for the financial support for his visit at its Institute of Physics in Toruń, where this work has been performed. S.H. is also supported by the US Department of Energy, Office of Science, Basic Energy Sciences under award number DE-FG02-11ER16211. I.G. is supported by Polish Committee for Scientific Research MNiSW under Grant No. N N204 560839. S.H. is a Camille Dreyfus Teacher-Scholar and a Scialog Fellow of the Research Corporation for Science Advancement.

Appendix: Questions and answers

In the following, we list questions, many of which are from the reviewers and colleagues who have read a version of this paper and challenged the conjecture. We offer our answers to these questions. The answers are increasingly speculative as we go down the list, and some may not have the rigor normally expected in a scientific paper. It is our hope that this serves as a basis of a fuller discussion and the ultimate proof or disproof of the conjecture.

Q: What is the difference between size consistency and size extensivity?

A: There is none, as regard to a method that computes the total energy in the ground state; they are synonymous and are based on one and the same diagrammatic criterion [7, 8]. One of the authors has extensively discussed the inadequacy of the supermolecule criterion often associated with the term "size consistency" in the past.

Q: Is DFT size consistent and variational?

A: No. DFT is not *diagrammatically* size consistent, but its energy is extensive, which can be shown as follows. An exchange-correlation energy per unit cell in a crystal is written as $\int f[\rho(\boldsymbol{r})]d\boldsymbol{r}$, where the domain of integration is one unit cell and f is the so-called exchange-correlation functional of electron density ρ. When one increases the volume of the crystal under the periodic boundary condition without changing ρ, the exchange-correlation energy per unit cell remains the same, which means that it is thermodynamically intensive. This, in turn, proves the extensivity of the total energy, assuming that the rest of the energy expressions are formally the same as in HF and are extensive.

DFT is not variational, but merely stationary, according to our adopted definitions of these terms [4]. This is because its energy can go below the full CI energy in the same basis set. An exception is the optimized effective potential (OEP) method [32, 33], which is diagrammatically size consistent and variational; OEP is sometimes viewed as the exact-exchange DFT [34–36]. It does not describe any electron correlation and is, therefore, not a counterexample of the conjecture.

Q: Is the Hylleraas functional size consistent and variational?

A: No. The Hylleraas functional [37, 38] (see also the Sinanoğlu equation [39]) is given by

$$E = \langle \Phi_0 | \hat{T}_2^\dagger \hat{H}_0 \hat{T}_2 | \Phi_0 \rangle + 2 \langle \Phi_0 | \hat{T}_2^\dagger \hat{H}_1 | \Phi_0 \rangle, \tag{30}$$

where $\hat{H}_0$ and $\hat{H}_1$ are the zeroth-order Hamiltonian and perturbation operator in the Møller–Plesset partitioning, $\hat{T}_2$ is a two-electron excitation operator, and Φ_0 is the HF reference wave function. The coefficients in $\hat{T}_2$ are determined variationally so as to minimize E, which can be easily shown to be the energy of second-order MP (MP2) [37, 38]. Being MP2, the Hylleraas functional is size consistent, but not variational. Extending the variational space spanned by $\hat{T}_2$ is equivalent to enlarging the basis set. In this sense and considering the fact that an MP2 energy can be lower than the full CI energy, we contend that the Hylleraas functional is only stationary [4].

The Hylleraas functional is often used as a basis of explicitly correlated MP2 (MP2-F12) [37, 38], in which the action of $\hat{T}_2$ yields,

$$\hat{T}_2 \Phi_0 = \frac{1}{(2!)^2} \sum_{i,j}^{\text{occ.}} \sum_{a,b}^{\text{virt.}} t_{ij}^{ab} \Phi_{ij}^{ab} + \frac{\hat{Q}_{12}}{(2!)^3} \sum_{i,j}^{\text{occ.}} \sum_{k,l}^{\text{occ.}} \sum_{a,b}^{\text{virt.}} F_{kl}^{ab} t_{ij}^{kl} \Phi_{ij}^{ab}, \tag{31}$$

with $F_{kl}^{ab} = \langle ab | F_{12} | kl \rangle$, where F_{12} is an explicit function of inter-electronic distance r_{12} such as $\exp(-\gamma r_{12})$. See Ref. [37] for the definition of $\hat{Q}_{12}$, which is unimportant here. The functional is made stationary with respect to $\{t_{ij}^{ab}\}$, $\{t_{ij}^{kl}\}$, and γ to approach the MP2 energy in the complete-basis-set limit, which is not an upper bound of the full CI energy. It is, therefore, still not a counterexample of the conjecture. Note that two-electron basis functions such as r_{12} or $\exp(-\gamma r_{12})$ can only couple two electrons at a time and span the same space the two-electron excitation operator does. Hence, they cannot fundamentally alter the non-variationality or size consistency of MP2.

In the same token, a method using a single-determinant wave function multiplied by a Jastrow factor (an exponential of a function of electron–electron and electron–

nucleus distances) should have the same characteristics as the projection or variational CC regarding variationality and size consistency. It cannot, therefore, simultaneously achieve size consistency and variationality, as we have seen in Sect. 3. It has been documented [26, 40] that nuclear calculations using a Jastrow factor do not have the upper-bound property and are, therefore, not variational according to our definition of the term. This is known as the "Emery difficulty" [26, 40].

Q: Can a size-consistency correction to a variational method restore size consistency?

A: There are size-consistency corrections to non-size-consistent methods such as CISD [41–46] and MRCI [45, 47]. These corrections, some related to Padé approximants [45], can "minimize" size-consistency errors and render the methods "approximately" size consistent, which means they are not size consistent. They may also be non-variational because the energies are not expectation values any more.

Q: Is local basis CI size consistent and variational?

A: A truncated CI method combined with an embedded-fragment scheme [48] can obviously yield an extensive total energy, giving a false impression that the method is simultaneously variational and size consistent. It is not variational because the energy thus obtained, which is not the energy expectation value in the global wave function, has no reason to be an upper bound of the exact energy of the whole molecule except in a special circumstance such as when there is no interaction between fragments. A counterexample for a special circumstance is, however, not a counterexample.

Another example of a variational local basis method is the generalized valence bond method. It is equivalent to MCSCF [1], which is not size consistent. Yet another example of a variational method that can employ a variety of bases including spatially local ones is the density matrix renormalization group method [49]. The underlying approximations in its wave functions are, however, often full CI [50] and CASSCF [51], neither of which is a counterexample, as explained in Sect. 2.

We have, however, been unable to generalize the definition of variationality in this conjecture to encompass all spaces including the determinant space and one-particle basis set. Classes of methods such as the Fock-space perturbation theory of Refs. [52, 53] are not easily subjected to this conjecture, though no claim seems to have been made that it is simultaneously size consistent and variational.

Q: If the conjecture is correct, how have variational methods been successfully applied to solids?

A: With a scalable algorithm running on a supercomputer, variational methods, which cannot be size consistent if the conjecture is correct, can still be applied to sufficiently large systems to address problems of condensed matter [11, 54, 55] including the issue of strong electron correlation in solids. The conjecture does not contradict this. Nonetheless, it is reasonable to expect a size-consistent method to be more accurate and efficient (these two properties are interchangeable) for solids than a similar non-size-consistent method; the development strategy emphasizing size consistency championed by Bartlett [56] has been successful in quantum chemistry despite the fact that it usually deals with relatively small molecules. In fact, the question of mutual exclusion and the conjecture emerged as a result of our desire to obtain size-consistent methods for strong correlation in solids. If a diagrammatically size-consistent method (that may or may not be variational) exists that can describe strong correlation in solids, we expect it to outperform the existing methods that are not size consistent. If a size-consistent method that works well for strong correlation is shown not to exist, it may even imply different size dependence of the strongly correlated portion of the energy.

References

1. Szabo A, Ostlund NS (1989) Quantum chemistry. McGraw-Hill, New York
2. Helgaker T, Jørgensen P, Olsen J (2000) Molecular electronic-structure theory. Wiley, Chichester
3. Shavitt I, Bartlett RJ (2009) Many-body methods in chemistry and physics. Cambridge University Press, Cambridge
4. Bartlett RJ, Noga J (1988) Chem Phys Lett 150:29
5. Robinson JB, Knowles PJ (2011) J Chem Phys 135:044113
6. March NH, Young WH, Sampanthar S (1967) The many-body problem in quantum mechanics. Cambridge University Press, Cambridge
7. Hirata S (2011) Theor Chem Acc 129:727
8. Goldstone J (1957) Proc R Soc Lond A 239:267
9. Hirata S, Keçeli M, Yagi K (2010) J Chem Phys 133:034109
10. Hirata S, Keçeli M, Ohnishi Y-Y, Sode O, Yagi K (2012) Annu Rev Phys Chem 63:131
11. Ceperley DM, Alder BJ (1980) Phys Rev Lett 45:566
12. Hammond BL, Lester WA Jr., Reynolds PJ (1994) Monte Carlo methods in ab initio quantum chemistry. World Scientific, Singapore
13. Luchow A, Anderson JB (2000) Annu Rev Phys Chem 51:501
14. Kalos MH, Whitlock PA (2008) Monte Carlo methods. Wiley-VCH, Weinheim
15. Kolorenč J, Mitas L (2011) Rep Prog Phys 74:026502
16. Willow SY, Kim KS, Hirata S (2012) J Chem Phys 137:204122
17. Willow SY, Kim KS, Hirata S (2013) J Chem Phys 138:164111
18. Willow SY, Hermes MR, Kim KS, Hirata S (2013) J Chem Theory Comput 9:4396
19. Cooper B, Knowles PJ (2010) J Chem Phys 133:234102
20. Hirata S, Ohnishi Y (2012) Phys Chem Chem Phys 14:7800
21. Foresman JB, Head-Gordon M, Pople JA, Frisch MJ (1992) J Phys Chem 96:135
22. Hirata S (2005) J Chem Phys 122:094105
23. Bartlett RJ, Musiał M (2007) Rev Mod Phys 79:291
24. Purvis III GD, Bartlett RJ (1982) J Chem Phys 76:1910
25. Arponen J (1983) Ann Phys 151:311
26. Pal S, Prasad MD, Mukherjee D (1983) Theor Chim Acta 62:523

27. Pal S, Prasad MD, Mukherjee D (1985) Theor Chim Acta 68:125
28. Haque A, Kaldor U (1986) Int J Quantum Chem 29:425
29. Kutzelnigg W (1998) Mol Phys 94:65
30. Xian Y (2005) Phys Rev B 72:224438
31. Nakatsuji H, Hirao K (1978) J Chem Phys 68:2053
32. Sharp RT, Horton GK (1953) Phys Rev 90:317
33. Talman JD, Shadwick WF (1976) Phys Rev A 14:36
34. Görling A (1999) Phys Rev Lett 83:5459
35. Ivanov S, Hirata S, Bartlett RJ (1999) Phys Rev Lett 83:5455
36. Hirata S, Ivanov S, Grabowski I, Bartlett RJ, Burke K, Talman JD (2001) J Chem Phys 115:1635
37. Shiozaki T, Valeev EF, Hirata S (2009) Annu Rep Comp Chem 5:131
38. Kong LG, Bischoff FA, Valeev EF (2012) Chem Rev 112:75
39. Sinanoğlu O (1962) J Chem Phys 36:3198
40. Emery V (1958) Nucl Phys 6:385
41. Brueckner KA (1955) Phys Rev 100:36
42. Langhoff SR, Davidson ER (1974) Int J Quantum Chem 8:61
43. Davidson ER, Silver DW (1977) Chem Phys Lett 52:403
44. Siegbahn EM (1978) Chem Phys Lett 55:386
45. Duch W, Diercksen GHF (1994) J Chem Phys 101:3018
46. Meissner L (1996) Chem Phys Lett 263:351
47. Meissner L, Grabowski I (1999) Chem Phys Lett 300:53
48. Gordon MS, Fedorov DG, Pruitt SR, Slipchenko LV (2012) Chem Rev 112:632
49. Chan GKL, Sharma S (2011) Annu Rev Phys Chem 62:465
50. Chan GKL, Head-Gordon M (2003) J Chem Phys 118:8551
51. Ghosh D, Hachmann J, Yanai T, Chan GKL (2008) J Chem Phys 128:144117
52. Kutzelnigg W (1981) Chem Phys Lett 83:156
53. Kutzelnigg W (1982) J Chem Phys 77:3081
54. Foulkes WMC, Mitas L, Needs RJ, Rajagopal G (2001) Rev Mod Phys 73:33
55. Booth GH, Grüneis A, Kresse G, Alavi A (2013) Nature 493:365
56. Bartlett RJ (1981) Annu Rev Phys Chem 32:359

Theor Chem Acc (2014) 133:1426
DOI 10.1007/s00214-013-1426-9

REGULAR ARTICLE

Accurate quadruple-ζ basis-set approximation for double-hybrid density functional theory with an order of magnitude reduction in computational cost

Bun Chan · Leo Radom

Received: 8 August 2013 / Accepted: 13 November 2013 / Published online: 10 December 2013

Abstract In the present study, we investigate a scheme for the approximation of quadruple-ζ (QZ) energies for double-hybrid density functional theory (DHDFT) procedures, using smaller double-ζ and triple-ζ basis sets. Such an approach would allow DHDFT/QZ energies to be estimated in cases where the explicit calculations might be too demanding computationally. We find this approach, denoted Q[D,T], to be very accurate for the MP2 same-spin (MP2SS) component and generally reasonable for the MP2 opposite-spin (MP2OS) component. The performance of the Q[D,T] approximation is quite insensitive to the type of basis sets used, as well as to the specific DHDFT procedure. Overall, we find that the approximation, when used in combination with the maug-cc-pVQ[D,T]Z basis sets, performs well for the calculation of relative energies. The use of explicit MP2OS/maug-cc-pVQZ energies together with the Q[D,T] MP2SS energies yields even better agreement with complete QZ energies, but at a somewhat greater computational cost. For a representative large system for DHDFT, namely C_{60}, we find that the Q[D,T] approximation leads to a reduction in CPU time by more than an order of magnitude when compared with the corresponding explicit QZ calculation, with little reduction in accuracy.

Keywords Computational chemistry · Double-hybrid DFT · MP2SS · MP2OS · Basis-set extrapolation · C_{60}

Dedicated to Professor Thom Dunning and published as part of the special collection of articles celebrating his career upon his retirement.

Electronic supplementary material The online version of this article (doi:10.1007/s00214-013-1426-9) contains supplementary material, which is available to authorized users.

B. Chan (✉) · L. Radom (✉)
School of Chemistry and ARC Centre of Excellence for Free Radical Chemistry and Biotechnology, University of Sydney, Sydney, NSW 2006, Australia
e-mail: chan_b@chem.usyd.edu.au

L. Radom
e-mail: radom@chem.usyd.edu.au

1 Introduction

The past few years has seen the rapid development of double-hybrid density functional theory (DHDFT) procedures [1–19] into a powerful tool for computational chemists, owing primarily to the generally superior accuracy and robustness of these methods when compared to conventional density functional theory (DFT) [12, 16, 17]. DHDFT procedures include wavefunction-type components, namely Hartree–Fock exchange (E_X^{HF}) and MP2 correlation (E_C^{MP2}), in addition to exchange-correlation components from DFT (E_X^{DFT}, E_C^{DFT}):

$$E_{DHDFT} = c_{DFTX}E_X^{DFT} + c_{HF}E_X^{HF} + c_{DFTC}E_C^{DFT} + c_{MP2}E_C^{MP2}$$

They are typically formulated by fitting the proportion of wavefunction-type and DFT components, i.e., the coefficients c, to certain training sets.

Most commonly, a quadruple-ζ (QZ) basis set is used in conjunction with the DHDFT method in the fitting procedure [1–4, 6, 7, 11–13, 17], with the rationale being that the incorporation of MP2 correlation necessitates the use of basis sets of such a size to ensure sufficient basis-set convergence. Goerigk and Grimme have shown that the performance of a number of DHDFT procedures that are optimized for a QZ basis set deteriorates significantly when used together with a triple-ζ basis set, but there are also some DHDFT methods that show a relatively mild basis-set dependence [12].

The inclusion of MP2 correlation in conjunction with the typical use of a QZ basis set in DHDFT methods has led to DHDFT procedures being significantly more demanding computationally than conventional DFT procedures. At present, these requirements have restricted the applicability of DHDFT methods to systems with tens of atoms. Indeed, we have recently performed a DHDFT/QZ calculation for C_{60} [20] and found it to be extremely demanding on the computational resources. In comparison, conventional DFT procedures are already applicable to systems with several hundred atoms.

We note that modern theoretical techniques such as the RI-MP2 approximation [21] can accelerate the computationally demanding MP2 component of DHDFT calculations. An alternative way of reducing the computational requirements for DHDFT methods is to reduce their reliance on large QZ basis sets. In this regard, a number of DHDFT procedures have been designed for use with smaller triple-ζ basis sets [5, 6, 8–10, 14–16, 18, 19], and some of them have been shown to perform comparably to DHDFT methods that have been optimized for QZ basis sets [6, 12, 16].

Nonetheless, most DHDFT procedures, including some of the latest ones with improved performance [17], have been designed for QZ basis sets. Although it is possible to reformulate these methods for triple-ζ basis sets so as to improve computational efficiency while preserving the general accuracy, such an approach would normally require thorough refitting for each procedure. It would be more desirable to be able to employ a simpler strategy that is adaptable to a wider range of DHDFT procedures to achieve the same goal, namely to obtain the optimal QZ results using smaller basis sets.

To this end, we note that Karton and Martin have employed explicitly correlated MP2 to enable the use of triple-ζ [22] or even double-ζ [23] basis sets for accurate DHDFT calculations. In the present study, we attempt to achieve such a goal using an alternative approach involving basis-set extrapolation. The application of basis-set extrapolation to DHDFT procedures has been investigated previously [6, 22, 24, 25], but its potential in reducing the demand on computational resources for a formally QZ calculation has not yet been exploited. We emphasize that the various approaches for lowering the computational cost are not exclusive but can be complementary, and a combination of these techniques may result in further improvements in the computational efficiency in future advances.

2 Computational details

Standard wavefunction-type and DFT calculations [26–29] were carried out with Gaussian 09 [30]. D2 and D3 dispersion corrections [31, 32] were obtained with the dftd3 program [31]. Geometries, zero-point vibrational energies and thermal corrections for 298 K enthalpies that are used in the present investigation were taken from previous studies [16, 33], while DHDFT single-point energies were obtained in the current investigation. The frozen-core approximation was employed in the evaluation of the MP2 correlation energy.

Basis sets that are external to the Gaussian 09 program were obtained from the EMSL basis-set library [34, 35]. We abbreviate the Dunning basis sets [36] cc-pV*n*Z as V*n*Z and aug-cc-pV*n*Z as AV*n*Z, where $n =$ T and Q. The use of aug′ in a basis set, for example, in aug′-cc-pVTZ, indicates the use of diffuse functions on non-hydrogen atoms, and we abbreviate aug′-cc-pV*n*Z as A′V*n*Z. The aug′-def2-*n*ZVPP basis-set description represents the def2-*n*ZVPP basis sets [37] augmented with diffuse functions taken from aug′-cc-pV*n*Z. They are abbreviated as A′*n*ZVPP. We have also employed the minimally augmented basis sets of Truhlar and co-workers, i.e., maug-cc-pV*n*Z [38], and these basis sets are abbreviated as maV*n*Z in the present study. They contain only the s and p components from the A′V*n*Z basis sets in their diffuse functions.

Two-point extrapolation schemes with the formula $E_L = E_{\mathrm{CBS}} + \mathrm{A}\ L^{-\alpha}$ were used in the evaluation of the CBS limit [39], where L is the cardinal number of the basis set, i.e., 2 for double-ζ and 3 for triple-ζ basis sets, etc., and α is an adjustable parameter. The values of α that were optimized in the present study were obtained by fitting to the G2-1 training set [40]. Following previous practice [16, 33, 41], our optimization procedures involved minimizing the average of the mean absolute deviation (MAD) from benchmark values and the standard deviation (SD) of these deviations. All relative energies are reported in kJ mol^{-1}.

3 Results and discussion

3.1 Approach for basis-set approximation

In the present study, we employ the two-point basis-set extrapolation formula $E_L = E_{\mathrm{CBS}} + \mathrm{A}\ L^{-\alpha}$ to estimate QZ basis-set values using smaller basis sets [39]. Alternative formulae have been examined for the extrapolation of DHDFT energies, and it has been found that they yield similar results [24]. In the specific extrapolation formula used in the present study, the optimal value for the exponent α depends on the theoretical procedure and the basis sets. For example, a value of 5 is typically used for the extrapolation of Hartree–Fock energies [42]. For the extrapolation of coupled-cluster correlation energies, a value of 3 is often used for α, for example, in the W2 procedure [42]. However, for the analogous W1 method in which smaller basis sets are employed, an optimal value of

3.22 has been determined by fitting to the W2 correlation energies [42]. For the extrapolation of MP2 correlation energies, the optimal α values for the various basis-set combinations have been determined to be 2.1 (AV[D,T]Z), 2.5 (AV[T,Q]Z), 2.7 (AV[Q,5]Z) and 2.8 (AV[5, 6]Z) [43]. In contrast, optimal α values of $\sim$2.9 have been obtained for the MP2 correlation energies in several DHDFT procedures, showing only a very mild dependence on the basis sets employed for the extrapolation [24].

In preliminary investigations, we have found that approximating the DFT component of the QZ energies using the corresponding DZ and TZ energies is rather poor. Thus, we do not recommend the use of the extrapolation approximation for the DFT component of DHDFT procedures. Our findings are in accord with the results of previous studies that showed a somewhat erratic basis-set convergence for some DFT procedures [44, 45]. We note that only the MP2 correlation energies were extrapolated to the CBS limit in relevant previous studies [6, 22].

In the present study, we use triple-ζ and QZ basis sets to obtain our benchmark CBS energies (CBS[T,Q]). We use an α value of 3 for the MP2 opposite-spin (MP2OS) and MP2 same-spin (MP2SS) components. For the CBS extrapolation with double-ζ and triple-ζ basis sets, we determine the optimal α values for the MP2OS and MP2SS components by fitting to the corresponding CBS[T,Q] values. In the ideal case where the resulting CBS[D,T] values are identical to the CBS[T,Q] values, one can reverse the formula $E_L = E_{\mathrm{CBS}} + A\ L^{-\alpha}$ and use the CBS[D,T] values, the triple-ζ energies and the α[T,Q] values to obtain the exact QZ energies. The quality of such an approximation depends, obviously, on the quality of CBS[D,T] as an approximation to CBS[T,Q].

In principle, it would be appropriate to first rigorously determine the optimal α values for the CBS[T,Q] extrapolation, rather than simply using the typical value of 3 for correlation energies (MP2). However, the objective of the present study is not to obtain the most accurate CBS values but to use smaller double-ζ and triple-ζ basis sets to approximate QZ values, and we find that the quality of the resulting approximate QZ values is not very sensitive to the choice of α[T,Q] values, as long as the corresponding α[D,T] values are optimized accordingly. For instance, for the approximation of the DSD-B-LYP/A′VQZ procedure, when an α[T,Q] value of 5 is used for the MP2 components and the associated α[D,T] values are refitted, the resulting method has an MAD that is virtually identical to that for the method derived from an α[T,Q] value of 3.

3.2 Approximation of CBS[T,Q] total energies by CBS[D,T]

We have explored the approximation of CBS[T,Q] energies by CBS[D,T] for two recent DHDFT procedures, namely DSD-B-LYP [11] and DSD-PBE-P86 [17], in conjunction with a number of basis sets. The optimal α[D,T] values for the G2-1 set are shown in Table 1. We can see that the optimized α[D,T] values vary slightly for the different basis sets examined. For the MP2OS component, the optimal α[D,T] values are all $\sim$2.3, while a range of $\sim$3.0–3.3 is found for MP2SS. On the other hand, the optimized α[D,T] values for DSD-PBE-P86 are remarkably similar to those for DSD-B-LYP, which is indicative of a very mild dependence of the optimal extrapolation exponents on the DHDFT procedure itself.

How good are the CBS[D,T] values as approximations to CBS[T,Q] (Table 2)? The performance of the CBS[D,T] approximation is in general quite good, with MADs that are no larger than 4.2 kJ mol^{-1} (for MP2OS/A′[D,T]ZVPP). The approximation is better for MP2SS than for MP2OS, with MADs of $\sim$1–2 kJ mol^{-1}, in general, for the former and $\sim$3–4 kJ mol^{-1} for the latter. This points toward a quite smooth and, as we shall see, rapid basis-set convergence for the MP2SS component. Such an observation is consistent with the theoretical analysis of Kutzelnigg and Morgan [46], which demonstrated for helium-type species a more rapid L^{-6} convergence for SS energies compared with an L^{-4} convergence for OS energies. The larger MADs for MP2OS are also consistent with the fact that MP2OS energies are typically larger than the

Table 1 Optimized exponents α[D,T] for the MP2OS and MP2SS components for the approximation of CBS[T,Q] DHDFT energies by CBS[D,T]

	V[D,T]Z	maV[D,T]Z	A′V[D,T]Z	A′[D,T]ZVPP
DSD-B-LYP				
MP2OS	2.27	2.22	2.34	2.29
MP2SS	3.06	3.08	3.34	3.17
DSD-PBE-P86				
MP2OS	2.26	2.21	2.33	2.29
MP2SS	3.05	3.07	3.34	3.18

Two-point extrapolation with the formula $E_L = E_{\mathrm{CBS}} + A\ L^{-\alpha}$, for [T,Q], $\alpha = 3$ for the MP2 components

Table 2 Mean absolute deviations (kJ mol^{-1}) for the MP2OS and MP2SS components of the CBS[D,T] DHDFT total energies from the CBS[T,Q] values for the G2-1 set of species

	V[D,T]Z	maV[D,T]Z	A′V[D,T]Z	A′[D,T]ZVPP
DSD-B-LYP				
MP2OS	3.6	3.2	2.8	4.2
MP2SS	1.6	0.9	0.9	1.5
DSD-PBE-P86				
MP2OS	3.6	3.1	2.8	4.2
MP2SS	1.6	0.9	0.9	1.5

corresponding MP2SS energies (by approximately a factor of three) [47]. Notably, we see that the MADs for the two DHDFT procedures are very close to one another for both components and all basis sets examined, and this supports the insensitivity of the approximation to the details of the DHDFT method. We note that, among the four types of basis sets examined, the CBS[D,T] approximations perform the best when used in conjunction with A′V[D,T]Z, although the use of other types of basis set does not yield results that are substantially worse for the MP2 components.

3.3 Performance of the approximate quadruple-ζ DHDFT Q[D,T] approach on the G2-1 set of relative energies

In the previous section, we have seen that the CBS[D,T] extrapolation provides a quite adequate approximation to CBS[T,Q] for MP2SS, but the performance is somewhat less good for the MP2OS component. Accordingly, we also expect that the quality of the approximate QZ component energies obtained with the DZ and TZ energies, i.e., Q[D,T] energies, to be quite good for MP2SS but less so for MP2OS when compared with explicit QZ energies. However, in chemical applications, the calculated total energies are almost always a means to obtain relative energies, which usually benefit from error compensation.

How well does the Q[D,T] approximation perform for relative energies? We have examined the performance of a range of basis-set approximation schemes for DSD-B-LYP for the thermochemical properties in the G2-1 set. We note that the general observations are very similar for DSD-PBE-P86 (electronic supplementary material). In addition to the use of the Q[D,T] approximation for one or more components, we have also investigated the replacement of QZ energies by TZ energies. The MADs from experimental benchmarks are shown in Table 3. Before we proceed to a discussion of the performance of the various approximation schemes, we first note that the general performance of DSD-B-LYP is rather insensitive to the type of explicit QZ basis set used (column A), with the exception of the expected very poor performance of the VQZ basis set for electron affinities (EAs) due to the lack of diffuse functions.

Table 3 Mean absolute deviations (kJ mol^{-1}) from experimental benchmark values for DSD-B-LYP with various explicit and approximate quadruple-ζ basis sets for the G2-1 set of thermochemical properties

	A	B	C	D	E	F	G	H
DFT	Q	Q	Q	Q	Q	Q	Q	T
MP2OS	Q	Q	Q[D,T]	Q[D,T]	Q	T	T	T
MP2SS	Q	Q[D,T]	Q	Q[D,T]	T	Q	T	T
V*n*Z								
G2-1	11.5	11.6	11.7	11.9	12.0	14.5	15.2	21.1
ΔH_f	6.3	6.3	6.5	6.6	6.5	9.9	10.4	14.7
IE	7.8	7.8	8.0	8.0	8.2	9.9	10.3	10.9
EA	29.3	30.0	29.8	30.5	30.7	33.6	35.1	52.8
PA	2.7	2.6	2.6	2.5	2.6	2.3	2.3	3.7
maV*n*Z								
G2-1	6.4	6.4	6.5	6.6	6.6	8.4	8.8	9.7
ΔH_f	5.6	5.6	5.6	5.7	5.6	7.6	8.0	10.4
IE	7.4	7.4	7.5	7.5	7.6	9.2	9.5	9.3
EA	7.3	7.5	7.7	7.9	7.9	10.2	10.9	10.8
PA	3.1	3.1	3.1	3.0	3.0	2.5	2.5	2.7
A′V*n*Z								
G2-1	6.3	6.3	6.3	6.3	6.5	8.2	8.5	9.9
ΔH_f	6.0	6.0	6.1	6.2	6.1	8.6	8.9	12.4
IE	7.3	7.2	7.3	7.3	7.5	8.8	9.0	8.8
EA	6.1	6.0	5.9	5.8	6.4	8.0	8.4	8.3
PA	3.6	3.7	3.8	3.8	3.6	3.2	3.2	3.8
A′*n*ZVPP								
G2-1	6.2	6.1	5.8	5.7	6.3	7.8	8.0	8.4
ΔH_f	5.7	5.7	5.2	5.3	5.7	7.3	7.5	8.5
IE	7.3	7.2	7.1	7.0	7.5	9.0	9.2	9.1
EA	6.2	6.0	5.4	5.1	6.5	8.2	8.7	8.7
PA	3.4	3.5	3.7	3.8	3.3	2.9	2.9	3.1

ΔH_f = heat of formation, IE = ionization energy, EA = electron affinity, PA = proton affinity, Q = explicit QZ basis set, Q[D,T] = approximate QZ basis set obtained from extrapolation using DZ and TZ basis sets, T = explicit TZ basis set

For the various QZ basis-set approximations, we can see that the use of Q[D,T] for the MP2SS component (column B) leads to MADs very close to those of the full QZ calculations in nearly all cases. The largest difference occurs for the EAs for the VQ[D,T]Z basis sets, but even in this case the MAD for the approximate method (30.0 kJ mol^{-1}) is still quite close to that for the full QZ calculations (29.3 kJ mol^{-1}). When the Q[D,T] approximation is applied to the MP2OS component (column C), the MADs are also quite close to those for full QZ for all four types of basis sets. The replacement of full QZ values with Q[D,T] energies for both MP2OS and MP2SS (column D) leads to MADs that are again quite close to those in columns A, B and C.

When the QZ basis set for the MP2SS component is replaced by a TZ basis set (column E), the resulting methods have MADs that are quite close to those for the corresponding full QZ procedures (column A). This is indicative of a rapid basis-set convergence for this component in terms of the calculation of relative energies, which is in accord with the close agreement between the corresponding extrapolated methods in column B and the full QZ procedures. On the other hand, when the QZ energies for the MP2OS component are replaced by TZ values (columns F and G), we find significant differences in the performance when compared with those in column A, regardless of the type of basis set employed. In addition, the use of TZ basis sets for all components (column H), i.e., conventional TZ calculations, leads to further significant deviations of the MADs from the full QZ values, and this points toward a significant overall basis-set effect. By comparing the results in columns C and F, it can be seen that the Q[D,T] approximation represents a cost-effective improvement over TZ calculations for MP2OS.

3.4 Further assessments on larger data sets

We now further evaluate the performance of the Q[D,T] approximation for the larger G2/97 test set [48, 49]. We focus on the approximation with the maVQ[D,T]Z basis sets for the MP2SS component (Table 3, column B), as well as the one that also uses the same approximation for MP2OS (column D). We deem these approaches capable of providing a generally reliable and cost-effective means for obtaining accurate energies. The results in Table 3 suggest that such an approximation choice is likely to be also adequate for the other three types of basis sets. We have examined four DHDFT procedures, namely DSD-B-LYP [11], DSD-PBE-P86 [17], DSD-PBE-PBE [17] and DSD-B-P86 [17]. To obtain the Q[D,T] energies, we use α[D,T] values obtained as the averages of those for DSD-B-LYP and DSD-PBE-P86 (Table 1) for all DHDFT procedures, i.e., 2.217 for MP2OS and 3.073 for MP2SS, as these values appear to be rather insensitive to the DHDFT method.

Table 4 Mean absolute deviations (kJ mol^{-1}) from experimental benchmark values for the various DHDFT procedures with explicit maVQZ and approximate maVQ[D,T]Z basis sets for the G2/97 set of thermochemical properties

	DSD-B-LYP	DSD-PBE-P86	DSD-PBE-PBE	DSD-B-P86
(A) Full QZ				
G2/97	7.9	7.3	8.3	7.5
ΔH_f	8.4	7.2	8.4	8.4
IE	8.2	8.6	9.0	7.7
EA	6.7	6.4	7.7	5.7
PA	3.6	2.8	3.6	2.6
(B) DFT(Q) + MP2OS(Q) + MP2SS(Q[D,T])				
G2/97	7.9	7.4	8.3	7.6
ΔH_f	8.5	7.2	8.4	8.5
IE	8.1	8.5	9.0	7.7
EA	6.8	6.5	7.7	5.7
PA	3.6	2.8	3.6	2.6
(C) DFT(Q) + MP2OS(Q[D,T]) + MP2SS(Q[D,T])				
G2/97	8.1	7.5	8.4	8.0
ΔH_f	8.7	7.5	8.5	9.3
IE	8.2	8.5	8.9	7.7
EA	7.1	6.8	8.1	6.0
PA	3.6	2.7	3.6	2.7

We can see that, for all the DHDFT procedures, the use of the Q[D,T] approximation only for the MP2SS component (block B) leads to MADs that are virtually identical to those for the corresponding full QZ calculations (block A) (Table 4). The application of the additional Q[D,T] approximation to the MP2OS component (block C) leads to somewhat larger deviations from the MADs for full QZ calculations, but the respective values in blocks A and C remain quite close to one another. These results are consistent with the observations in Tables 2 and 3 and support the use of the approximation of QZ energies by extrapolating DZ and TZ values for the MP2OS and (especially) MP2SS components for DHDFT procedures. Importantly, the use of generic α[D,T] values leads to very similar results for all the DHDFT procedures considered, including DSD-PBE-PBE and DSD-B-P86 for which the optimal α[D,T] values are not determined explicitly, which points toward the general applicability of such approximations.

We note that, while many DHDFT procedures are parameterized in conjunction with QZ basis sets and may therefore perform best when used with QZ basis sets (see [16] for an example of the impact of using a "mismatched" basis set on the performance of a DHDFT procedure), the extrapolation to the complete-basis-set (CBS) limit may in some cases lead to improved results [24, 50]. It is of interest to ask how the CBS extrapolation performs more generally when applied to a DHDFT procedure that is fitted with QZ basis sets. To this end, we have examined the performance of an approximation to DSD-PBE-P86/CBS in which the DFT component is obtained with the maVQZ basis set and the MP2 correlation energies are obtained at the CBS limit by extrapolating with maVTZ and maVQZ. For the G2/97 set, this method gives an overall MAD (11.1 kJ mol^{-1}) that is significantly larger than that for DSD-PBE-P86/maVQZ (7.3 kJ mol^{-1}). This difference is dominated by the substantially poorer performance for ΔH_f's for the CBS method (16.0 kJ mol^{-1}) than for the QZ procedure (7.2 kJ mol^{-1}). The performance for PAs is also slightly poorer with CBS (3.1 vs. 2.8 kJ mol^{-1} for QZ), but the MADs for IEs and EAs are lower with CBS (7.4 and 5.2 kJ mol^{-1} for IEs and EAs, respectively) than with QZ (8.6 and 6.4 kJ mol^{-1} for IEs and EAs, respectively). We note that these results are consistent with our previous observations, e.g., that the use of a QZ basis set with a method that has been fitted with a TZ basis set leads to poorer performance [16]. In general, we deem the use of a QZ basis set more appropriate for a DHDFT method parameterized for QZ, even though in some cases the use of CBS yields somewhat better performance.

We now investigate the application of the basis-set approximation schemes for the calculations of relatively large systems, where the benefit from a reduction in the demand on computational resources is starting to be more significant. In a recent study, the heat of formation of C_{60} was estimated at the DSD-PBE-B86/VQZ level by means of isodesmic-type reactions [20]. We will use a set of 13 reactions from that study to assess the performance of the basis-set approximations. We have examined the application of the approaches shown in Table 4, namely (1) full QZ, (2) QZ for DFT and MP2OS, and Q[D,T] for MP2SS, and (3) QZ for DFT, and Q[D,T] for MP2OS and MP2SS, to the DSD-PBE-B86 procedure with the V*n*Z (A–C) or maV*n*Z (D–F) basis sets. The resulting reaction energies are shown in Table 5.

For the reactions that do not involve C_{60} (reactions 1–9), we find that the use of Q[D,T] for MP2SS (columns B and E) provides very good approximations to the corresponding full QZ reaction energies in columns A and D, respectively. Among these cases, the largest deviation (LD) is just 0.3 kJ mol^{-1} for reactions 4 and 7 with the V*n*Z basis sets (column B). When the Q[D,T] approximation is also used for the MP2OS component (columns C and F), the deviations of the reaction energies from the values in columns A and D are slightly larger. For reactions 1–9, the LD in column C is that for reaction 4 (2.0 kJ mol^{-1}), and the corresponding LD in column F is −1.5 kJ mol^{-1} for reaction 9. In general, the use of the maVQ[D,T]Z approximation (columns E and F) leads to slightly better agreement with the full maVQZ reaction energies (column D) than the use of VQ[D,T]Z (columns B and C) for approximating the full VQZ energies in column A.

Table 5 DSD-PBE-P86 vibrationless reaction energies (kJ mol^{-1}) for a set of 13 isodesmic-type reactions calculated in conjunction with a variety of explicit and approximate quadruple-ζ basis sets

	A[a]	B	C	D	E	F
DFT	VQZ	VQZ	VQZ	maVQZ	maVQZ	maVQZ
MP2OS	VQZ	VQZ	VQ[D,T]Z	maVQZ	maVQZ	maVQ[D,T]Z
MP2SS	VQZ	VQ[D,T]Z	VQ[D,T]Z	maVQZ	maVQ[D,T]Z	maVQ[D,T]Z
1. $C_{10}H_8 + C_2H_4 \rightarrow 2C_6H_6$	−40.0	−40.0 (0.1)	−39.6 (0.4)	−40.3	−40.2 (0.1)	−39.9 (0.3)
2. $C_{14}H_{10} + 2C_2H_4 \rightarrow 3C_6H_6$	−70.6	−70.4 (0.2)	−69.6 (1.0)	−71.0	−70.9 (0.1)	−70.3 (0.7)
3. $C_{14}H_{10} + C_6H_6 \rightarrow 2C_{10}H_8$	9.5	9.5 (0.0)	9.7 (0.2)	9.5	9.5 (0.0)	9.6 (0.1)
4. $C_{18}H_{12} + 3C_2H_4 \rightarrow 4C_6H_6$	−105.6	−105.2 (0.3)	−103.6 (2.0)	−106.2	−106.0 (0.2)	−104.9 (1.3)
5. $C_{18}H_{12} + 2C_6H_6 \rightarrow 3C_{10}H_8$	14.6	14.6 (0.1)	15.3 (0.8)	14.6	14.6 (0.0)	14.9 (0.3)
6. $C_{18}H_{12} + C_{10}H_8 \rightarrow 2C_{14}H_{10}$	−4.4	−4.4 (0.0)	−4.1 (0.3)	−4.4	−4.4 (0.0)	−4.3 (0.1)
7. $C_{20}H_{10} + 5C_2H_4 \rightarrow 5C_6H_6$	−368.2	−367.9 (0.3)	−366.9 (1.2)	−368.9	−368.8 (0.2)	−368.3 (0.6)
8. $C_{20}H_{10} + 5C_6H_6 \rightarrow 5C_{10}H_8$	−168.0	−168.1 (−0.1)	−168.7 (−0.7)	−167.6	−167.8 (−0.2)	−168.6 (−1.0)
9. $C_{20}H_{10} + 5C_{10}H_8 \rightarrow 5C_{14}H_{10}$	−215.3	−215.6 (−0.2)	−217.2 (−1.9)	−215.1	−215.3 (−0.2)	−216.6 (−1.5)
10. $C_{60} + 10C_6H_6 \rightarrow 6C_{20}H_{10}$	−483.8		−491.8 (−7.9)			−488.0
11. $C_{60} + 10C_{10}H_8 \rightarrow 8C_{20}H_{10}$	−147.9		−154.3 (−6.5)			−150.7
12. $C_{60} + 10C_{14}H_{10} \rightarrow 10C_{20}H_{10}$	282.8		280.1 (−2.7)			282.5
13. $C_{60} + 10C_{18}H_{12} \rightarrow 12C_{20}H_{10}$	669.6		673.8 (4.2)			672.9

Deviations from the corresponding full QZ values are shown in parentheses

[a] [20]

For the reactions that involve C_{60} (reactions 10–13), somewhat larger deviations are found for the energies in column C when compared with the full VQZ values (column A). Presumably, like most approximate methods, a larger number of species involved in these reactions can sometimes lead to a greater accumulation of deviations. To this end, we have analyzed the corresponding deviations for the G2/97 set of relative energies (ΔH_f, IE, EA and PA). We find that the deviations are notably larger for ΔH_f (LD = 9.1 kJ mol^{-1}) than for IE (LD = 1.8 kJ mol^{-1}), EA (LDs = −2.8 kJ mol^{-1} for atoms and −2.0 kJ mol^{-1} for molecules) and PA (LD = −0.6 kJ mol^{-1}). However, it is noteworthy that, when the Q[D,T] approximation is applied only to the MP2SS component, the resulting relative energies deviate from the full QZ values by less than 0.8 kJ mol^{-1} for all the data in the G2/97 set.

3.5 Illustrative application to larger systems

Having thoroughly evaluated the performance of the Q[D,T] approach for the approximation of explicit QZ energies and identified the circumstances under which it works best, we now apply the approximate method to systems that are particularly demanding on computational resources when an explicit QZ basis set is used. Thus, we have calculated the IE and EA of C_{60} using the DSD-PBE-P86 procedure, with the DFT component obtained with the maVQZ basis set and the MP2OS and MP2SS components calculated using the corresponding Q[D,T] approximation. We note that, while the experimental ΔH_f for C_{60} is associated with a large uncertainty (2,560 ± 100 kJ mol^{-1}), the experimental IE (7.57 ± 0.01 eV, 730.4 ± 1.0 kJ mol^{-1}) and EA (2.666 ± 0.001 eV, 257.2 ± 0.1 kJ mol^{-1}) values for C_{60} are quite accurate [51]. Our calculated IE (736.5 kJ mol^{-1}) and EA (255.7 kJ mol^{-1}) at the approximate DSD-PBE-P86/maVQZ level, i.e., DSD-PBE-P86/maVQ[D,T]Z, are in reasonably good agreement with the experimental values.

In using the Q[D,T] approximation, what savings in computational resources have we achieved? For C_{60}, the explicit DSD-PBE-P86/VQZ calculation involving 3,300 basis functions and performed in the previous study [20] consumed 3,383 h of CPU time with an allocated memory of 42 GB and an allocated scratch disk space of 377 GB. For the reactions in Table 5 that involve C_{60}, the calculation of C_{60} itself dominates the computational time. For example, reaction 13 (C_{60} + 10 $C_{18}H_{12}$ → 12 $C_{20}H_{10}$) took a total CPU time of 3,650 h for the explicit DSD-PBE-P86/VQZ computation of all three species involved.

In the present study, we placed the same constraints as in the previous work on the memory and scratch disk for our calculations for C_{60}, $C_{60}^{\bullet+}$ and $C_{60}^{\bullet-}$, and we recorded the CPU time for our approximate QZ calculations (Table 6). For our approximate DSD-PBE-P86/VQ[D,T]Z calculation, the total CPU time consumed for the DFT/VQZ, DHDFT/VDZ and DHDFT/VTZ components is 254 h, which represents more than an order of magnitude of savings when compared with the full VQZ calculation. Likewise, the total CPU time consumed for the calculation of reaction 13 is 307 h, which is also an order of magnitude less than the explicit VQZ calculations. We note that this computational saving is achieved with little loss of accuracy: 673.8 versus 669.6 kJ mol^{-1} (Table 5). The maVQ[D,T]Z calculation for C_{60} took a total CPU time of 514 h, which is approximately twice that for the VQ[D,T]Z calculation, but it is still substantially less than for the explicit VQZ calculation, and it has the advantage of incorporating the effect of diffuse functions, which is expected to be essential for the calculation of the EA.

Table 6 CPU time (h) for the calculation of C_{60}, $C_{60}^{\bullet+}$ and $C_{60}^{\bullet-}$ with DSD-PBE-P86 (DHDFT) using explicit QZ basis sets, approximate QZ basis sets (i.e., Q[D,T]Z) and their components (DFT and DHDFT)

	C_{60}	$C_{60}^{\bullet+}$	$C_{60}^{\bullet-}$
DHDFT/VQZ	3,383		
DHDFT/VQ[D,T]Z	254		
DFT/VQZ	144		
DHDFT/VDZ	7		
DHDFT/VTZ	103		
DHDFT/maVQ[D,T]Z	514	1,529	1,065
DFT/maVQZ	272	575	548
DHDFT/maVDZ	25	96	59
DHDFT/maVTZ	218	859	459

The maVQ[D,T]Z calculations for $C_{60}^{\bullet+}$ and $C_{60}^{\bullet-}$ consumed a total of 1,529 and 1,065 h of CPU time, respectively. Thus, they are still less costly than the explicit VQZ calculation for C_{60}. As the explicit VQZ calculation for C_{60} is more than an order of magnitude more costly than the corresponding VQ[D,T]Z scheme, we can expect a full DSD-PBE-P86/maVQZ calculation for $C_{60}^{\bullet+}$ to be more than an order of magnitude more costly than the corresponding maVQ[D,T]Z calculation and hence to consume well over 15,000 h of CPU time! We note that, for the approximate VQZ calculation for C_{60} and the approximate maVQZ calculations for C_{60} and $C_{60}^{\bullet-}$, the longest calculation in each case is actually the DFT/QZ component rather than the DHDFT/TZ calculation. On the other hand, for the approximate maVQZ calculation for $C_{60}^{\bullet+}$, the DHDFT/maVTZ component took notably longer than the DFT/maVQZ component.

These examples clearly demonstrate the utility of the Q[D,T] approximation for the MP2 components for DHDFT procedures, and the considerable cost savings associated with such a scheme. We reiterate that further reduction in the demand for computational resources can be achieved with techniques such as the resolution-of-the-identity approximation for both the DFT and the MP2 components. In the cases where the Q[D,T] approximation is expected to be less good for the MP2OS component, e.g., ΔH_f via atomization energies for large molecules, it may be more desirable to employ explicit QZ energies for this component. Such a scheme would involve carrying out conventional DHDFT calculations with DZ and TZ basis sets, and an SOS-MP2-type DHDFT/QZ calculation [52].

The SOS-MP2 protocol, as implemented in some computational chemistry packages such as Q-Chem [53], has a fourth-order scaling as opposed to the fifth-order scaling for conventional MP2. To this end, we have briefly examined the use of SOS-RI-MP2/VDZ for the calculation of C_{60} in comparison with the corresponding RI-MP2/VDZ calculation. The Hartree–Fock component in these calculations consumed $\sim$2,081 s of CPU time. The correlation calculation for SOS-RI-MP2 took 1,685.3 s, while for RI-MP2 it took 2,635.6 s. Therefore, while the inclusion of an additional MP2OS/QZ component necessarily increases the cost, we expect the total computational resources required for such a protocol to be still notably less than that for full QZ calculations for large systems. We also reiterate that the Q[D,T] approximation does not exclude the use of additional cost-saving techniques for DHDFT calculations, such as the DB-RI-MP2 procedure of Head-Gordon and co-workers [54] or the MP2[K] procedure of Deng and Gill [55].

4 Concluding remarks

In the present study, we have investigated a scheme for the approximation of QZ energies for double-hybrid DFT (DHDFT) procedures, using smaller double-ζ and triple-ζ basis sets with a recast of the conventional extrapolation formula $E_L = E_{\text{CBS}} + A\ L^{-\alpha}$. The following key findings emerge from our investigation.

1. The approximation (denoted Q[D,T]) requires different optimal α exponents for the MP2OS and MP2SS components. The α values are not very sensitive to the type of basis set used and are also rather insensitive to the specific DHDFT procedure. We find that the performance of the QZ approximation for the MP2 components is not very sensitive to the specific type of basis sets used.
2. Overall, we find that a DHDFT scheme in which the DFT component is calculated with the maug-cc-pVQZ basis set, and the MP2OS/QZ and MP2SS/QZ components are approximated by maug-cc-pVQ[D,T]Z, generally performs well for the evaluation of relative energies. This scheme employs optimal α[D,T] values of 2.217 for MP2OS and 3.073 for MP2SS in conjunction with the standard α[T,Q] value of 3 for both MP2 components. These parameters are applicable not only to DSD-B-LYP and DSD-PBE-P86 for which the values are optimized, but also to DSD-PBE-PBE and DSD-B-P86 for which the values are independently applied without reoptimization.
3. While the approximation of MP2OS/maug-cc-pVQZ and MP2SS/maug-cc-pVQZ by the maug-cc-pVQ[D,T]Z approach, overall, is quite good for the calculation of relative energies, particularly for the MP2SS component, the performance is less good for some species. In these cases, the use of explicit MP2OS/maug-cc-pVQZ energies can help to overcome such a shortcoming, while still maintaining significant savings in computational requirements by the use, for example, of SOS-MP2.
4. For a prototypical large system for DHDFT, namely C_{60}, we find that the use of the Q[D,T] approximation for MP2OS and MP2SS leads to a reduction in CPU time by more than an order of magnitude when compared with the explicit DSD-PBE-P86/cc-pVQZ calculation, with little loss of accuracy. Such a saving in computer time has enabled the calculation of the more expensive $C_{60}^{\bullet+}$ and $C_{60}^{\bullet-}$ ions at the approximate DSD-PBE-P86/maug-cc-pVQ[D,T]Z level with resources notably less than those for the explicit DSD-PBE-P86/cc-pVQZ calculation for C_{60}.

Acknowledgments We gratefully acknowledge funding (to L.R.) from the Australian Research Council (ARC) and generous grants of computer time (to L.R.) from the National Computational Infrastructure (NCI) National Facility and Intersect Australia Ltd.

References

1. Grimme S (2006) Semiempirical hybrid density functional with perturbative second-order correlation. J Chem Phys 124:034108/1–034108/16
2. Grimme S, Schwabe T (2006) Towards chemical accuracy for the thermodynamics of large molecules: new hybrid density functionals including non-local correlation effects. Phys Chem Chem Phys 8:4398–4401
3. Tarnopolsky A, Karton A, Sertchook R, Vuzman D, Martin JML (2008) Double-hybrid functionals for thermochemical kinetics. J Phys Chem A 112:3–8
4. Karton A, Tarnopolsky A, Lamere JF, Schatz GC, Martin JML (2008) Highly accurate first-principles benchmark data sets for the parametrization and validation of density functional and other approximate methods. Derivation of a robust, generally applicable, double-hybrid functional for thermochemistry and thermochemical kinetics. J Phys Chem A 112:12868–12886
5. Benighaus T, Lochan RC, Chai JD, Head-Gordon M (2008) Semiempirical double-hybrid density functional with improved description of long-range correlation. J Phys Chem A 112:2702–2712

6. Chai JD, Head-Gordon M (2009) Long-range corrected double-hybrid density functionals. J Chem Phys 131:174105/1–174105/13
7. Graham DC, Menon AS, Goerigk L, Grimme S, Radom L (2009) Optimization and basis-set dependence of a restricted-open-shell form of B2-PLYP double-hybrid density functional theory. J Phys Chem A 113:9861–9873
8. Zhang Y, Xu X, Goddard WA III (2009) Doubly hybrid density functional for accurate descriptions of nonbond interactions, thermochemistry, and thermochemical kinetics. Proc Natl Acad Sci USA 106:4963–4968
9. Zhang IY, Luo Y, Xu X (2010) XYG3s: speedup of the XYG3 fifth-rung density functional with scaling-all-correlation method. J Chem Phys 132:194105/1–194105/11
10. Zhang IY, Luo Y, Xu X (2010) Basis set dependence of the doubly hybrid XYG3 functional. J Chem Phys 133:104105/1–104105/12
11. Kozuch S, Gruzman D, Martin JML (2010) DSD-BLYP: a general purpose double hybrid density functional including spin component scaling and dispersion correction. J Phys Chem C 114:20801–20808
12. Goerigk L, Grimme S (2011) Efficient and accurate double-hybrid-meta-GGA density functionals–evaluation with the extended GMTKN30 database for general main group thermochemistry, kinetics, and noncovalent interactions. J Chem Theory Comput 7:291–309
13. Sharkas K, Toulouse J, Savin A (2011) Double-hybrid density-functional theory made rigorous. J Chem Phys 134:064113/1–064113/9
14. Brémond E, Adamo C (2011) Seeking for parameter-free double-hybrid functionals: the PBE0-DH model. J Chem Phys 135:024106/1–024106/6
15. Zhang IY, Xu X, Jung Y, Goddard WA III (2011) A fast doubly hybrid density functional method close to chemical accuracy using a local opposite spin ansatz. Proc Natl Acad Sci USA 108:19896–19900
16. Chan B, Radom L (2011) Obtaining good performance with triple-ζ-type basis sets in double-hybrid density functional theory procedures. J Chem Theory Comput 7:2852–2863
17. Kozuch S, Martin JML (2011) DSD-PBEP86: in search of the best double-hybrid DFT with spin-component scaled MP2 and dispersion corrections. Phys Chem Chem Phys 13:20104–20107
18. Mohajeri A, Alipour M (2012) B2-PPW91: a promising double-hybrid density functional for the electric response properties. J Chem Phys 136:124111/1–124111/4
19. Zhang IY, Su NQ, Brémond ÉAG, Adamo C, Xu X (2012) Doubly hybrid density functional xDH-PBE0 from a parameter-free global hybrid model PBE0. J Chem Phys 136:174103/1–174103/8
20. Karton A, Chan B, Raghavachari K, Radom L (2013) Evaluation of the heats of formation of corannulene and C_{60} by means of high-level theoretical procedures. J Phys Chem A 117:1834–1842
21. Feyereisen M, Fitzgerald G, Komornicki A (1993) Use of approximate integrals in ab initio theory. An application in MP2 energy calculations. Chem Phys Lett 208:359–363
22. Karton A, Martin JML (2011) Basis set convergence of explicitly correlated double-hybrid density functional theory calculations. J Chem Phys 135:144119/1–144119/7
23. Goerigk L, Karton A, Martin JML, Radom L (2013) Accurate quantum chemical energies for tetrapeptide conformations: why MP2 data with an insufficient basis set should be handled with caution. Phys Chem Chem Phys 15:7028–7031
24. Chuang YY, Chen SM (2011) Infinite basis set extrapolation for double hybrid density functional theory 1: effect of applying various extrapolation functions. J Comput Chem 32:1671–1679
25. Chuang YY, Chen SM (2012) Infinite basis set extrapolation for double hybrid density functional theory 2: effect of adding diffuse basis functions. J Chin Chem Soc 59:1094–1103
26. Hehre WJ, Radom L, Schleyer PvR, Pople JA (1986) Ab initio molecular orbital theory. Wiley, New York
27. Koch W, Holthausen MC (2001) A chemist's guide to density functional theory, 2nd edn. Wiley, New York
28. Cramer CJ (2002) Essentials of computational chemistry: theories and models. Wiley, Chichester
29. Jensen F (2007) Introduction to computational chemistry, 2nd edn. Wiley, Chichester
30. Frisch MJ, Trucks GW, Schlegel HB, Scuseria GE, Robb MA, Cheeseman JR, Scalmani G, Barone V, Mennucci B, Petersson GA, Nakatsuji H, Caricato M, Li X, Hratchian HP, Izmaylov AF, Bloino J, Zheng G, Sonnenberg JL, Hada M, Ehara M, Toyota K, Fukuda R, Hasegawa J, Ishida M, Nakajima T, Honda Y, Kitao O, Nakai H, Vreven T, Montgomery JA Jr, Peralta JE, Ogliaro F, Bearpark M, Heyd JJ, Brothers E, Kudin KN, Staroverov VN, Keith T, Kobayashi R, Normand J, Raghavachari K, Rendell A, Burant JC, Iyengar SS, Tomasi J, Cossi M, Rega N, Millam JM, Klene M, Knox JE, Cross JB, Bakken V, Adamo C, Jaramillo J, Gomperts R, Stratmann RE, Yazyev O, Austin AJ, Cammi R, Pomelli C, Ochterski JW, Martin RL, Morokuma K, Zakrzewski VG, Voth GA, Salvador P, Dannenberg JJ, Dapprich S, Daniels AD, Farkas O, Foresman JB, Ortiz JV, Cioslowski J, Fox DJ (2010) Gaussian 09 revision c01. Gaussian Inc, Wallingford
31. Grimme S, Antony J, Ehrlich S, Krieg H (2010) A consistent and accurate ab initio parametrization of density functional dispersion correction (DFT-D) for the 94 elements H-Pu. J Chem Phys 132:154104/1–154104/19
32. Grimme S, Ehrlich S, Goerigk L (2011) Effect of the damping function in dispersion corrected density functional theory. J Comput Chem 32:1456–1465
33. Chan B, Deng J, Radom L (2011) G4(MP2)-6X: a cost-effective improvement to G4(MP2). J Chem Theory Comput 7:112–120
34. Feller D (1996) The role of databases in support of computational chemistry calculations. J Comput Chem 17:1571–1586
35. Schuchardt KL, Didier BT, Elsethagen T, Sun L, Gurumoorthi V, Chase J, Li J, Windus TL (2007) Basis set exchange: a community database for computational sciences. J Chem Inf Model 47:1045–1052
36. Dunning TH Jr (1999) Gaussian basis sets for use in correlated molecular calculations I. The atoms boron through neon and hydrogen. J Chem Phys 90:1007–1023
37. Weigend F, Ahlrichs R (2005) Balanced basis sets of split valence, triple zeta valence and quadruple zeta valence quality for H to Rn: design and assessment of accuracy. Phys Chem Chem Phys 7:3297–3305
38. Papajak E, Leverentz HR, Zheng J, Truhlar DG (2009) Efficient diffuse basis sets: cc-pVxZ+ and maug-cc-pVxZ. J Chem Theory Comput 5:1197–1202
39. Halkier A, Helgaker T, Jørgensen P, Klopper W, Koch H, Olsen J, Wilson AK (1998) Basis-set convergence in correlated calculations on Ne, N_2, and H_2O. Chem Phys Lett 286:243–252
40. Curtiss LA, Raghavachari K, Trucks GW, Pople JA (1991) Gaussian-2 theory for molecular energies of first- and second-row compounds. J Chem Phys 94:7221–7230
41. Chan B, Radom L (2012) W1X-1 and W1X-2: W1-quality accuracy with an order of magnitude reduction in computational cost. J Chem Theory Comput 8:4259–4269
42. Parthiban S, Martin JML (2001) Assessment of W1 and W2 theories for the computation of electron affinities, ionization potentials, heats of formation, and proton affinities. J Chem Phys 114:6014–6029
43. Hill JG, Peterson KA, Knizia G, Werner HJ (2009) J Chem Phys 131:194105/1–194105/13
44. Wang NX, Wilson AK (2004) The behavior of density functionals with respect to basis set I. The correlation consistent basis sets. J Chem Phys 121:7632–7646

45. Prascher BP, Wilson AK (2007) The behaviour of density functionals with respect to basis set V. Recontraction of correlation consistent basis sets. Mol Phys 105:2899–2917
46. Kutzelnigg W, Morgan JD III (1992) Rates of convergence of the partial-wave expansions of atomic correlation energies. J Chem Phys 96:4484–4508
47. Grimme S, Goerigk L, Fink RF (2012) Spin-component-scaled electron correlation methods. WIREs Comput Mol Sci 2:886–906
48. Curtiss LA, Raghavachari K, Redfern PC, Pople JA (1997) Assessment of Gaussian-2 and density functional theories for the computation of enthalpies of formation. J Chem Phys 106:1063–1079
49. Curtiss LA, Redfern PC, Raghavachari K, Pople JA (1998) Assessment of Gaussian-2 and density functional theories for the computation of ionization potentials and electron affinities. J Chem Phys 109:42–55
50. Chan B, Radom L (2012) Approaches for obtaining accurate rate constants for hydrogen abstraction by a chlorine atom. J Phys Chem A 116:3745–3752
51. NIST Chemistry WebBook (2011) National Institute of Standards and Technology Gaithersburg, MD 20899. http://webbook.nist.gov. Accessed Feb 2013
52. Jung Y, Lochan RC, Dutoi AD, Head-Gordon M (2004) Scaled opposite-spin second order Møller–Plesset correlation energy: an economical electronic structure method. J Chem Phys 121:9793–9802
53. Shao Y, Molnar LF, Jung Y, Kussmann J, Ochsenfeld C, Brown ST, Gilbert ATB, Slipchenko LV, Levchenko SV, O'Neill DP, DiStasio RA, Lochan RC, Wang T, Beran GJO, Besley NA, Herbert JM, Lin CY, Van Voorhis T, Chien SH, Sodt A, Steele RP, Rassolov VA, Maslen PE, Korambath PP, Adamson RD, Austin B, Baker J, Byrd EFC, Dachsel H, Doerksen RJ, Dreuw A, Dunietz BD, Dutoi AD, Furlani TR, Gwaltney SR, Heyden A, Hirata S, Hsu CP, Kedziora G, Khalliulin RZ, Klunzinger P, Lee AM, Lee MS, Liang W, Lotan I, Nair N, Peters B, Proynov EI, Pieniazek PA, Rhee YM, Ritchie J, Rosta E, Sherrill CD, Simmonett AC, Subotnik JE, Woodcock HL, Zhang W, Bell AT, Chakraborty AK, Chipman DM, Keil FJ, Warshel A, Hehre WJ, Schaefer HF, Kong J, Krylov AI, Gill PMW, Head-Gordon M (2006) Advances in methods and algorithms in a modern quantum chemistry program package. Phys Chem Chem Phys 8:3172–3191
54. Steele RP, DiStasio RA, Shao Y, Kong J, Head-Gordon M (2006) Dual-basis second-order Møller-Plesset perturbation theory: a reduced-cost reference for correlation calculations. J Chem Phys 125:074108/1–074108/11
55. Deng J, Gill PMW (2011) A new approach to dual-basis second-order Møller–Plesset calculations. J Chem Phys 134:081103/1–081103/4

Theor Chem Acc (2014) 133:1417
DOI 10.1007/s00214-013-1417-x

REGULAR ARTICLE

A perspective on the localizability of Hartree–Fock orbitals

Ida-Marie Høyvik · Kasper Kristensen · Thomas Kjærgaard · Poul Jørgensen

Received: 14 September 2013 / Accepted: 25 October 2013 / Published online: 19 November 2013

Abstract A common perception about molecular systems with a nonlocal electronic structure (as manifested by a nonlocal Hartree–Fock (HF) density matrix), such as conjugated π-systems, is that they can only be described in terms of nonlocal molecular orbitals. This view is mostly founded on chemical intuition, and further, this view is strengthened by traditional approaches for obtaining local occupied and virtual orbital spaces, such as the occupied Pipek–Mezey orbitals, and projected atomic orbitals. In this article, we discuss the limitations for localizability of HF orbitals in terms of restrictions posed by the delocalized character of the underlying density matrix for the molecular system and by the orthogonality constraint on the molecular orbitals. We show that the locality of the orbitals, in terms of nonvanishing charge distributions of orbitals centered far apart, is much more strongly affected by the orthogonality constraint than by the physical requirement that the occupied orbitals must represent the electron density. Thus, the freedom of carrying out unitary transformations among the orbitals provides the flexibility to obtain highly local occupied and virtual molecular orbitals, even for molecular systems with a nonlocal density matrix, provided that a proper localization function is used. As an additional consideration, we clear up the common misconception that projected atomic orbitals in general are more local than localized orthogonal virtual orbitals.

Dedicated to Professor Thom Dunning and published as a part of the special collection of articles celebrating his career upon his retirement.

I.-M. Høyvik (✉) · K. Kristensen · T. Kjærgaard · P. Jørgensen
Department of Chemistry, qLEAP Center of Theoretical Chemistry, Aarhus University, Langelandsgade 140, 8000 Aarhus C, Denmark
e-mail: idamh@chem.au.dk

Keywords Local orbitals · Local correlation · Projected atomic orbitals · Delocalization

1 Introduction

Since the seminal papers by Lennard-Jones and Pople [1] and Hall [2] on the significance of local occupied orbitals, localized molecular orbitals (MOs) have received an increasing amount of attention. The use of localized MOs ranges from understanding concepts related to molecular bonding to more practical considerations, where the local orbitals constitute an efficient basis in which electron correlation effects may be efficiently expressed. Due to the popularity of the application of localized MOs, a vast number of localization schemes have been proposed, see e.g., [3–16]. The most popular of these schemes are the ones of Edmiston and Ruedenberg [5], Boys [3–5], and Pipek and Mezey [8]. More recent developments in the area of orbital localization functions have focused on imposing strict constraints on the spatial extent of the orbitals, e.g., by minimizing powers of the orbitals' second central moments [14] and powers of the orbitals' fourth central moments [16]. Powers act as a penalty that ensures all orbitals are local, i.e., no single orbital in the set is delocalized. As a result, these localization functions have proven less system dependent than the standard ones. Until recently, only occupied orbitals could be localized since the traditional optimization algorithm—the Jacobi sweeps of iterations—is not capable of localizing the more complicated virtual space. However, with the introduction of the trust-region algorithm for orbital localization functions [17], both local occupied and local virtual orbitals may be generated, even for large molecular systems.

The advances made for localization function optimization—as achieved by the trust-region approach and the introduction of robust localization functions—open up for explorations of the localizability of orthonormal Hartree–Fock (HF) orbitals. Such explorations have previously not been considered, and the concept of MO localizability has therefore often been overwhelmed by chemical intuition, especially for "delocalized" systems. By delocalized systems, chemists refer to e.g., conjugated π-systems, where the electrons are considered to be delocalized throughout the π-system. This interpretation is strongly connected to MO diagrams and associated canonical orbitals. However, all quantum mechanical observables calculated at the HF level of theory are invariant to a unitary transformation among the occupied and among the virtual HF orbitals. It is precisely this freedom that is used when localizing HF orbitals. An orbital-invariant definition of the locality of a molecular system comes from considering the HF density matrix, i.e., the system is considered to be delocalized if the HF density matrix (and hence the physical electron density) is not local in nature. Thus, through the HF density matrix, which is constructed from occupied MO coefficients obtained from a HF calculation, the degree of delocalization is tied closely together with the MOs.

Since the HF optimization condition defines the separation of the orbital space into an occupied and a virtual part, where the occupied HF orbitals describe the HF density matrix, it represents a physical limitation to the localizability of the occupied as well as the virtual orbitals. However, the HF optimization condition is not the only requirement that is important for orbital locality. The locality of MOs is also limited by requiring that all orbitals are orthogonal to each other. The orthogonality requirement—a mathematical limitation—provides a convenient framework to work within, but it is a very strict requirement seen from a locality point of view. Orthogonalization of the orbitals ultimately leads to orbital tails, and nodal planes are required to fulfill the orthogonality conditions. Orbital tails result in nonvanishing charge distributions for orbitals which are centered far apart and consequently make sparsity, e.g., for local correlation methods more difficult to achieve. It is therefore of interest to quantify whether the HF optimization condition or the orthogonality requirement is the main source of the orbital tails.

In this paper, the physical and mathematical limitations on the localizability of orthonormal HF orbitals will be explored and compared by carrying out calculations where the two effects are separated. To investigate how the locality of the electronic structure of a molecule (as manifested by the locality of the HF density matrix) impacts the orbital locality, we will here study two molecules both of which are made up of carbon and hydrogen, but which are very different in terms of electronic structure. One molecule is a diamond structure ($C_{331}H_{216}$), and the other is a graphene sheet ($C_{106}H_{28}$), see Fig. 1. Diamond is an example of a molecule that is an electric insulator and has a localized electronic structure, while the graphene sheet is a prototypical example of a delocalized system. Graphene thus represents a molecular system for which the existence of local HF orbitals is not an obvious matter. Here, we investigate the impact of the nonlocal density matrix of graphene on orbital locality and compare it to the constraints accompanying the orthogonality requirement. We demonstrate that it is indeed possible to localize both occupied and virtual orbitals for graphene and that the orthogonality constraint is a bigger restriction on locality than the HF optimization condition. The common misconception that it is not possible to obtain localized HF orbitals for delocalized systems is probably also influenced by the fact that the very system-dependent Pipek–Mezey orbitals and projected atomic orbitals have traditionally been used to span the occupied and virtual orbital space, e.g., in local correlation approaches. We address this issue by demonstrating that the fourth moment localized orbitals are in general more local—and much less system dependent—than the Pipek–Mezey orbitals and projected atomic orbitals.

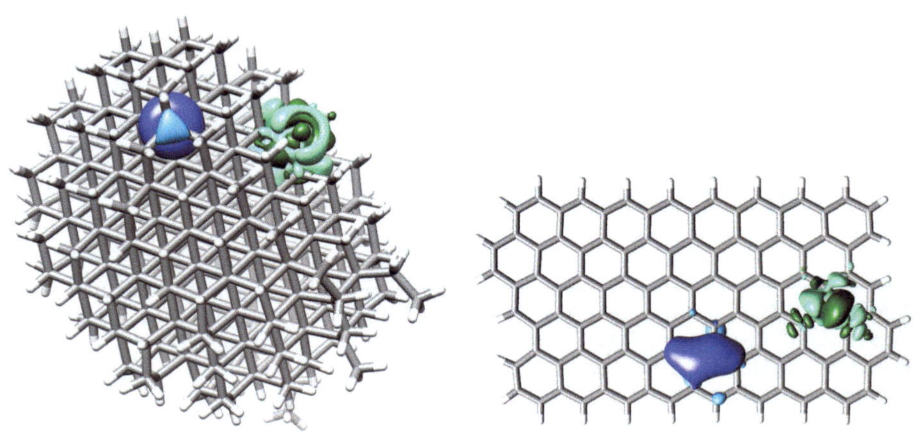

Fig. 1 A visualization of diamond (*left*) and graphene (*right*). The least local-occupied (*blue*) and least local virtual (*green*) LMOs of Sect. 3 are also shown. The plots are made using the UCSF Chimera program package [22]

The structure of the paper is as follows. In Sect. 2, the locality of diamond and graphene is analyzed in the atomic orbital (AO) basis. In Sect. 3. we compare the locality of diamond and graphene in the MO basis with emphasis on the physical and mathematical limitations for orbital locality. In Sect. 4, we compare the localized orbitals employed in Sect. 3 to Pipek–Mezey orbitals and projected atomic orbitals. All illustrations are carried out using Dunning's cc-pVDZ basis [18]. In Sect. 5, a summary and some concluding remarks are given.

2 Background

To emphasize the different physical nature of the diamond and graphene molecules, we look at the locality of the systems in the AO basis. In particular, we look at the HF density matrix and the Fock matrix. Assuming that the HF optimization condition is satisfied, the HF density matrix is defined by the elements

$$D_{\mu\nu} = \sum_i C_{\mu i} C_{\nu i} \tag{1}$$

where the index i runs over the occupied HF orbitals, μ and ν are AO indices, and $\mathbf{C}$ is the MO coefficient matrix. The Fock matrix may be expressed in terms of the density matrix as

$$F_{\mu\nu} = h_{\mu\nu} + \sum_{\rho\sigma} D_{\rho\sigma}(2g_{\mu\nu\rho\sigma} - g_{\mu\sigma\rho\nu}) \tag{2}$$

where $h_{\mu\nu}$ is the usual one-electron integral and $g_{\mu\nu\rho\sigma}$ denotes the two-electron integral in a Mulliken notation,

$$g_{\mu\nu\rho\sigma} = \int \chi_\mu(\mathbf{r}_1)\chi_\nu(\mathbf{r}_1)\frac{1}{r_{12}}\chi_\rho(\mathbf{r}_2)\chi_\sigma(\mathbf{r}_2)d\mathbf{r}_1 d\mathbf{r}_2 \tag{3}$$

To visualize the spatial locality of the density matrix (Eq. 1) and the Fock matrix (Eq. 2), we have in Fig. 2 (left) plotted the distance decay of the density matrix elements, and in Fig. 2 (right), the distance decay of the Fock matrix elements. The values plotted in Fig. 2 are the maximum absolute values (max$|D_{\mu\nu}|$ and max$|F_{\mu\nu}|$) as a function of distance between the atomic orbitals $|\mu\rangle$ and $|\nu\rangle$, $R_{\mu\nu}$.

The density matrix element decay plotted in Fig. 2 (left) reflects the different electronic structures of the two molecules. The density matrix elements for diamond decay uniformly, while the density matrix element decay for graphene is very slow, resulting in an increasingly large gap between the size of density matrix elements for diamond and graphene as the distance $R_{\mu\nu}$ increases. This shows why graphene is considered to be a molecule with a delocalized electronic structure. The difference in the density matrices for diamond and graphene is reflected in the distance decay of the Fock matrix elements, as seen in Fig. 2 (right). To identify the terms in the Fock matrix responsible for the slow distance decay, we have in Fig. 3 plotted the maximum absolute one-electron ($h_{\mu\nu}$), Coulomb ($2\sum_{\rho\sigma} D_{\rho\sigma}g_{\mu\nu\rho\sigma}$) and exchange ($\sum_{\rho\sigma} D_{\rho\sigma}g_{\mu\sigma\rho\nu}$) contributions to the AO Fock matrix (see Eq. 2) separately for diamond (left) and graphene (right). Clearly, the exchange contribution is dominating at large distances for both systems, and it is ultimately responsible for the slow Fock matrix decay for graphene. This happens because the Coulomb and one-electron terms are killed off by the charge distribution $\chi_\mu(\mathbf{r})\chi_\nu(\mathbf{r})$ that vanishes for large $R_{\mu\nu}$, whereas this charge distribution is not present in the exchange term where the decay is therefore controlled by the decay of the density matrix, which is slow for graphene.

Thus, the locality of the density matrix (and hence the Fock matrix) strongly depends on the nature of the electronic structure of the physical system. The density matrix is invariant with respect to unitary transformations among the occupied orbitals. However, in spite of that, the common perception among chemists is that it is not possible to describe a delocalized molecular system using highly localized molecular orbitals. Using diamond and graphene

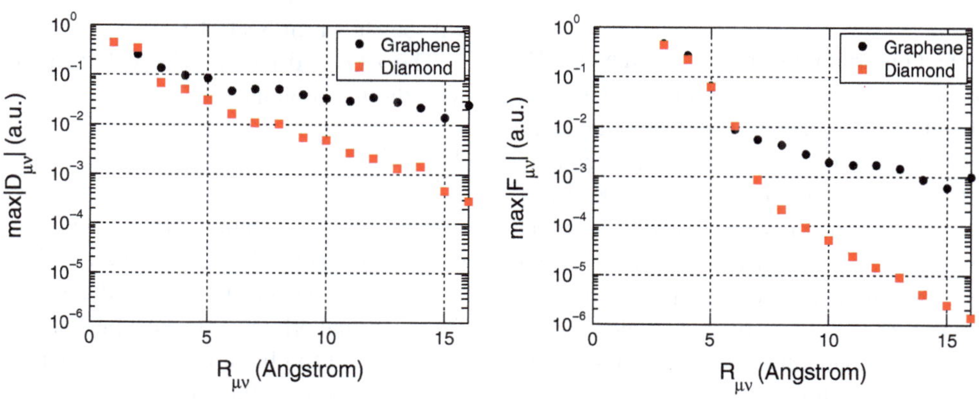

Fig. 2 The distance decay of density matrix elements (*left*) and the distance decay of the Fock matrix in the AO basis (*right*) for diamond and graphene

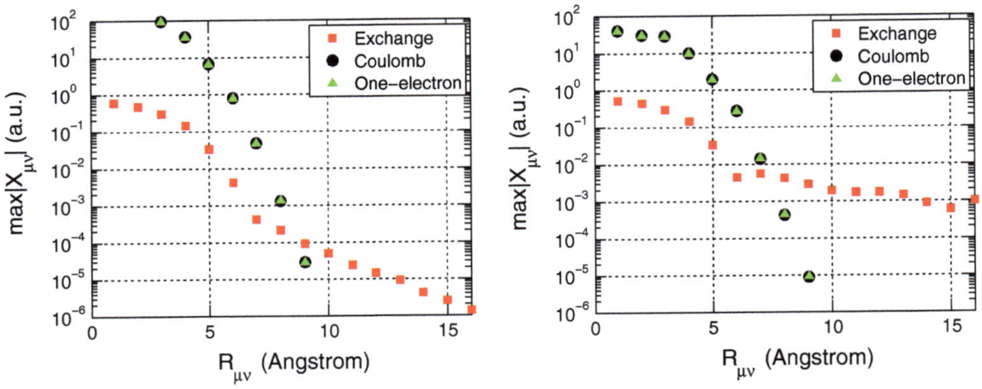

Fig. 3 The distance decay of one-electron, Coulomb and exchange term entering the Fock matrix, plotted for diamond (*left*) and graphene (*right*). Note that there is a factor two on the Coulomb term which is consistent with the definition of the AO Fock matrix in Eq. 2

as examples, we now explore the impact of the nature of the density matrix has on the localizability of the HF orbitals (physical limitation) and compare it to the impact of requiring the basis of MOs to be orthogonal (mathematical limitation).

3 The localizability of Hartree–Fock orbitals

The goal of this section is to elucidate how the physical and mathematical restrictions—relative to each other—affect the localizability of the HF orbitals. In Sect. 3.1, we introduce the orbital bases which will be used to separate the physical and mathematical effects from each other. Locality measures are given in Sect. 3.2, and in Sect. 3.3, we give numerical illustrations for examining the physical and mathematical restrictions. In Sect. 3.4, the distance decay of Fock matrix elements in the MO basis is examined, since this decay is interesting from a local correlation point of view.

3.1 Bases used to separate physical and mathematical effects on orbital locality

To separate the effect on orbital locality imposed by the HF optimization condition (which separates the orbital space into an occupied and a virtual orbital space) from the effect of the orthogonality requirement, bases which exhibits none, one or both conditions are needed. The AOs, $\{|\mu\rangle\}$, constitute a natural starting point, serving as a basis which is nonorthogonal and does not satisfy an occupied–virtual partitioning of the orbital space. To obtain an orthogonal set of orbitals—which does not satisfy an occupied–virtual partitioning—we may orthogonalize the set of AOs. In this work, we consider the Löwdin symmetric orthogonalization [19]

$$|\bar{\mu}\rangle = \sum_{\nu} |\nu\rangle [\mathbf{S}^{-1}]_{\nu\mu} \tag{4}$$

where $\mathbf{S}$ is the overlap matrix in the atomic orbital basis, $S_{\mu\nu} = \langle\mu|\nu\rangle$. Although the set $\{|\bar{\mu}\rangle\}$ gives the set of orthogonalized atomic orbitals (OAOs) which, in a least squares sense, most closely resembles the AOs [20], they do not necessarily represent the most spatially local set of orthogonal orbitals spanning the same space as the AOs. To improve the spatial locality, we explicitly localize the OAO set $\{|\bar{\mu}\rangle\}$ by an orthogonal transformation among the orbitals,

$$|\bar{\mu}^{\mathrm{L}}\rangle = \sum_{\nu} |\bar{\nu}\rangle \mathrm{U}_{\nu\mu} \tag{5}$$

where the orthogonal transformation matrix $\mathbf{U}$ is determined by minimizing the sum of the second power of the orbitals' fourth central moments [16]. We elaborate on this choice of localization function in Sect. 3.2. The resulting set of localized orthogonalized atomic orbitals (LOAOs), $\{|\bar{\mu}^{\mathrm{L}}\rangle\}$, hence represents a set of spatially localized orthogonal orbitals that does not satisfy an occupied—virtual partitioning of the orbital space. The improved locality of LOAOs compared to OAOs is significant, as will be shown in Sect. 3.3

The AOs and the LOAOs will be used for comparison with the HF orbitals

$$|p\rangle = \sum_{\mu} |\mu\rangle C_{\mu p} \tag{6}$$

where $p, q, \ldots$ denote orbital indices of unspecified occupation. Henceforth, the indices $i, j, \ldots$ are used to denote occupied orbitals and indices $a, b, \ldots$ are used to denote virtual orbitals. The occupied and virtual HF orbitals are localized to obtain the localized HF molecular orbitals (LMOs) using orthogonal transformations among the occupied orbitals and among the virtual orbitals

$$|i^{\mathrm{L}}\rangle = \sum_j |j\rangle \mathrm{U}'_{ji} \tag{7}$$

$$|a^{\mathrm{L}}\rangle = \sum_b |b\rangle \mathrm{U}''_{ba} \tag{8}$$

where the orthogonal transformation matrices $\mathbf{U}'$ and $\mathbf{U}''$—as was the case for $\mathbf{U}$ in Eq. 5—are determined by minimizing the sum of the second power of the orbitals' fourth central moments (see Sect. 3.2). The superscript "L" is used to differentiate localized HF orbitals from standard HF orbitals (e.g., canonical MOs). However, for the remainder of this paper, all HF orbitals considered are localized and we will for simplicity omit the superscript "L" for the localized HF orbitals.

Hence, we have three different bases—the AOs, LOAOs, and LMOs—which represents three different scenarios with respect to the HF optimization criterion and the orthogonality constraint. For an overview of the notation and properties of the three bases, see Table 1.

3.2 The orbital localization function and locality measures

The orthogonal transformations in Eqs. 5, 7 and 8 are determined by minimizing the second power of the fourth central moment [16]

$$\xi = \sum_{\mathrm{P}} \langle \varphi_P | (\hat{\mathbf{r}} - \langle \varphi_{\mathrm{P}} | \hat{\mathbf{r}} | \varphi_{\mathrm{P}} \rangle)^4 | \varphi_{\mathrm{P}} \rangle^2 \tag{9}$$

where $\{|\varphi_{\mathrm{P}}\rangle\}$ refers to the set of LOAOs, occupied LMOs, or virtual LMOs. The localization function is optimized using the trust-region implementation described in [17]. The reason for using the fourth central moment localization function is the spatial compactness of the orbitals obtained, and the fact that the locality of the orbitals generated using this function shows minimal basis set and system dependence. The minimal system dependence is especially important since it is of interest to explore the localizability of HF orbitals for systems with very different bonding structures.

To quantify the spatial locality of a normalized orbital $|\varphi_{\mathrm{P}}\rangle$ the second and fourth moment orbital spreads may be used. The second moment orbital spread

Table 1 An overview of the notation and properties of the three bases employed

Basis	Notation	Orthogonal?	Occupied-virtual?
AOs	$\{|\mu\rangle\}$	No	No
LOAOs	$\{|\bar{\mu}^{\mathrm{L}}\rangle\}$	Yes	No
LMOs	$\{|p\rangle\}$	Yes	Yes

$$\sigma_2^{\mathrm{P}} = \sqrt{\langle \varphi_{\mathrm{P}} | (\hat{\mathbf{r}} - \langle \varphi_{\mathrm{P}} | \hat{\mathbf{r}} | \varphi_{\mathrm{P}} \rangle)^2 | \varphi_{\mathrm{P}} \rangle} \tag{10}$$

describes the orbital bulk locality, while the fourth moment orbital spread [16]

$$\sigma_4^{\mathrm{P}} = \sqrt[4]{\langle \varphi_{\mathrm{P}} | (\hat{\mathbf{r}} - \langle \varphi_{\mathrm{P}} | \hat{\mathbf{r}} | \varphi_{\mathrm{P}} \rangle)^4 | \varphi_{\mathrm{P}} \rangle} \tag{11}$$

puts more emphasis on the orbital density in the tail region of the orbital. Comparing the size of σ_4^{P} relative to σ_2^{P}, thus, gives an indication of how rapidly the orbital tail is decaying. If σ_4^{P} and σ_2^{P} are of comparable sizes, the orbital density in the tail region is small and consequently the orbital density distribution has a thin tail. Contrary, if σ_4^{P} is large compared with σ_2^{P} the orbital density distribution has a thick tail.

Although the orbital spreads give detailed information on the locality of the orbitals, it is useful to supplement this information by considering how the charge distribution between two orbitals, $\varphi_{\mathrm{P}}(\mathbf{r})\varphi_{\mathrm{Q}}(\mathbf{r})$, decay as a function of the distance, R_{PQ}, between the orbital centers,

$$R_{\mathrm{PQ}} = |\langle \varphi_{\mathrm{P}} | \hat{\mathbf{r}} | \varphi_{\mathrm{P}} \rangle - \langle \varphi_{\mathrm{Q}} | \hat{\mathbf{r}} | \varphi_{\mathrm{Q}} \rangle| \tag{12}$$

Since the overlap between two orthonormal orbitals by construction is 1 or 0, the overlap of the LOAOs and LMOs cannot be used as a measure to characterize the charge distribution. We, therefore, use the numerical overlap, Ω_{PQ}, as a measure of the magnitude of the charge distributions,

$$\Omega_{\mathrm{PQ}} = \int |\varphi_{\mathrm{P}}(\mathbf{r})||\varphi_{\mathrm{Q}}(\mathbf{r})| d\mathbf{r} \tag{13}$$

The numerical overlap Ω_{PQ} will be plotted as a function of R_{PQ}, and since the orbital sets contain large numbers of orbitals, only the largest Ω_{PQ} will be plotted for a given R_{PQ} distance interval.

3.3 Numerical illustrations for separating physical and mathematical effects

To separate the physical effect imposed by the HF optimization condition from the orthogonality constraint, we consider the diamond and graphene molecules introduced in Sect. 2 and report the locality of the three different bases (AOs, LOAOs and LMOs) in terms of the maximum orbital spreads (Eqs. 10 and 11) and numerical overlaps between orbitals as a function of distance between the orbital centers (Eq. 13). The orbital spreads are tabulated in Table 2 and the numerical overlaps are plotted in Fig. 4. To get an idea on the significance of the orbital spreads, the LMOs of diamond and graphene are illustrated in Fig. 1. In Table 2, the locality of the Löwdin symmetric orthogonalized AO (without any explicit localization) is also given, illustrating that the gain in locality by localization is around 2 a.u. in terms of orbital spread. This locality improvement is

Table 2 The maximum σ_2 and σ_4 values for the AOs, LOAOs, and LMOs, where the LOAOs and LMOs have been localized by minimizing Eq. 9

	Diamond		Graphene	
	σ_2^{max} (a.u.)	σ_4^{max} (a.u.)	σ_2^{max} (a.u.)	σ_4^{max} (a.u.)
AOs	2.9	3.1	2.9	3.1
OAOs	4.3	5.9	3.9	5.1
LOAOs	2.4	3.4	2.3	3.3
LMOs, occupied	1.8	2.5	2.8	4.1
LMOs, virtual	2.8	4.0	3.1	3.9
PMO, occupied	1.9	2.7	4.1	7.8
PAO	3.1	4.5	6.0	10.3

We also give values for the projected atomic orbitals (PAOs) and occupied orbitals obtained using the Pipek–Mezey localization function (PMO) in Eq. 16

essential to compare the HF orbital locality to that of orthogonalized AOs. Table 2 also contains information about the Pipek–Mezey orbitals and projected atomic orbitals which will be discussed in Sect. 4.

We will now discuss the main points from the results in Table 2 and Fig. 4. The discussion is separated in separate discussions about bulk and tail regions of the orbitals.

3.3.1 Bulk locality

The orbital bulks (which are measured by σ_2) for the LOAOs and LMOs are, in general, more local than the bulk of the AOs. This is possible since the LOAOs and LMOs are combinations of AOs on different centers, and this provides the flexibility to generate orbitals with more spatially confined bulks. Looking at the numerical overlaps in Fig. 4, we see the effect of the improved bulk locality, since, for small R_{PQ}, the Ω values for the LOAOs and LMOs are in general below the Ω values for the AOs. This is particularly pronounced for the LOAOs and the occupied LMOs for diamond. Further, the σ_2^{max} of the diamond occupied LMOs is lower than the σ_2^{max} of the LOAOs, while the virtual orbitals of diamond exhibit slightly larger σ_2^{max} values than the LOAOs. Hence, it is possible to obtain better bulk locality for the occupied orbitals for diamond than for the LOAOs. Thus, for diamond, the occupied–virtual partitioning of the orbital space enables a set of occupied orbitals that are more local than the LOAOs, whereas the virtual orbitals necessarily will have a locality comparable to (or less than) that of the LOAOs. This is evident since the occupied and virtual orbitals collectively must span the same space as the LOAOs, and if the HF optimization conditions leave the occupied space described by easily localizable orbital components, the virtual space still has to account for the rest. For graphene, the delocalized nature of the system manifests itself in a set of occupied and virtual orbitals with similar locality and which is comparable to that of the LOAOs. The reason for a less local bulk for the graphene occupied orbitals compared to the diamond occupied orbitals is the difference in bond structure. For diamond, bonding primarily take place between two carbon atoms, whereas in graphene bonding is between multiple carbon atoms (see Fig. 1). Hence, the delocalized character of graphene only requires a small increase in bulk locality compared with diamond and consequently the occupied–virtual partitioning does not—for neither diamond nor graphene—poses a restriction on bulk locality which is so strong that it is not possible to obtain a locality similar to that of the LOAOs.

3.3.2 Tail locality compared to the AOs

When considering the tails of the orbitals, we must consider the sizes of σ_4^{max} relative to σ_2^{max} in Table 2. Since $\sigma_2^{max} \approx \sigma_4^{max}$ for the AOs while $\sigma_4^{max} > \sigma_2^{max}$ for the LOAOs and LMOs, the LOAOs and LMOs have more orbital density in the tail region. This is also evident from a

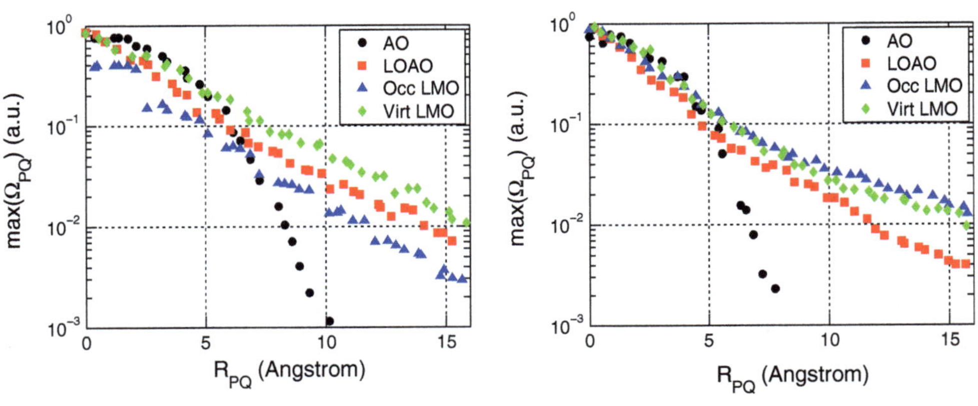

Fig. 4 The numerical overlaps $\Omega_{\mu\nu}$ (*black*), $\Omega_{\bar{\mu}^L\bar{\nu}^L}$ (*red*), Ω_{ij} (*green*), and Ω_{ab} (*blue*) plotted as a function of the distance between orbital centers for diamond (*left*) and graphene (*right*)

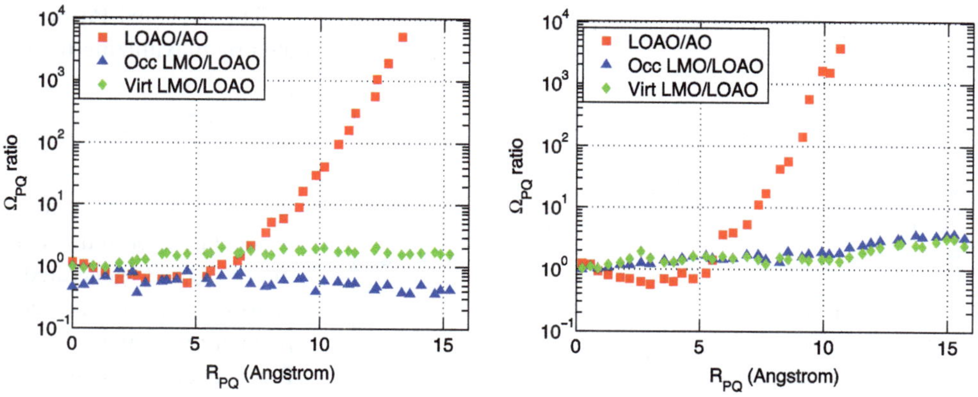

Fig. 5 The ratios $\frac{\Omega_{\bar{\mu}^L\bar{\nu}^L}}{\Omega_{\mu\nu}}$ (*red*), $\frac{\Omega_{ij}}{\Omega_{\bar{\mu}^L\bar{\nu}^L}}$ (*green*), and $\frac{\Omega_{ab}}{\Omega_{\bar{\mu}^L\bar{\nu}^L}}$ (*blue*) plotted as a function of distance for diamond (*left*) and graphene (*right*)

conceptual point of view, since the LOAOs and the LMOs are orbitals expanded in the basis of AOs, and therefore, the LOAOs and LMOs contain contributions from AOs on different atomic centers, with the magnitude of the contribution determined by the expansion coefficients. Since an AO is a single atom-centered function, it has—in this context—no tail. This is substantiated by the numerical overlap plots or the AOs in Fig. 4, where the $\Omega_{\mu\nu}$ values are seen to rapidly decay to zero (to be more precise, the very rapid tail decay of the AO tails is governed by the underlying Gaussian functions in the cc-pVDZ basis). All sets of LMOs presented are seen to have similar tail decay behavior.

To illustrate the effects on the orthogonalization and on the occupied–virtual splitting explicitly and separately, we have in Fig. 5 plotted the distance dependence of the size of $\Omega_{\bar{\mu}^L\bar{\nu}^L}$ relative to $\Omega_{\mu\nu}$ (red curve), the size of Ω_{ij} relative to $\Omega_{\bar{\mu}^L\bar{\nu}^L}$ (green curve), and the size of Ω_{ab} relative to $\Omega_{\bar{\mu}^L\bar{\nu}^L}$ (blue curve). Hence, the red curve explicitly expresses the effect of the orthogonalization on the distance decay of charge distributions, while the green and blue curves explicitly expresses the effect of requiring the HF optimization condition to be fulfilled. As is seen from Fig. 5 (right), the dominating curve beyond 6–7 Å for both diamond (left) and graphene (right) is the red curve. Hence, already for rather small separations, the orthogonalization effect is seen to be strong and the distance decay of orbital charge distributions for larger distances is thus governed by the orthogonalization requirement and not by the physical separation of the orbitals into an occupied and a virtual space.

3.4 The impact of orbital localizability on the MO Fock matrix

In Sect. 2, it was shown that the AO Fock matrix elements for graphene exhibit a slow decay compared with the AO Fock matrix elements for diamond, while in the previous section, we saw that this difference does not impede the construction of a set of localized HF orbitals for graphene. However, the localizability of the orbitals was seen to be strongly restricted by requiring them to be orthogonal. The purpose of this section is to see the effect of the localizability of MOs on the distance decay of Fock matrix elements in the MO basis. The motivation for this investigation is the importance of LMOs for use in local correlation methods, where the MO Fock matrix elements enters, e.g., in the coupled cluster amplitude equations. The locality of the Fock matrix elements therefore directly affects the locality of the amplitude equations. The distance decay of the maximum absolute MO Fock matrix elements for diamond and graphene is plotted in Fig. 6.

We first consider the occupied–occupied Fock matrix decay. The slower decay for the AO Fock matrix of graphene compared with diamond in Fig. 2 (right) is also manifested for the occupied–occupied Fock matrix elements in Fig. 6. The matrix elements in the AO and MO bases are qualitatively of the same order of magnitude, but the decays in the MO basis are more uniform, leading to a smaller difference between graphene and diamond in the MO basis than in the AO basis.

The virtual–virtual Fock matrix elements also decay systematically and are very similar to graphene and diamond in accordance with their similar $\sigma_2^{\max}$ and $\sigma_4^{\max}$ values in Table 2. Interestingly, the virtual–virtual Fock matrix elements are seen to be one or two orders of magnitude larger than the corresponding occupied–occupied Fock matrix elements for diamond. This—in conjunction with Fig. 2—indicates that it is primarily the underlying AO Fock matrix decays which dictates the occupied–occupied Fock matrix decay, while the virtual orbital locality is the primary source of the decay of the virtual–virtual Fock matrix elements. An implication of this is that the difference between the virtual–virtual and AO Fock matrix decays is much greater for diamond than for graphene. Thus, the loss of locality in the Fock matrix when going from the AO to the MO basis is more detrimental for systems with a localized electronic structure.

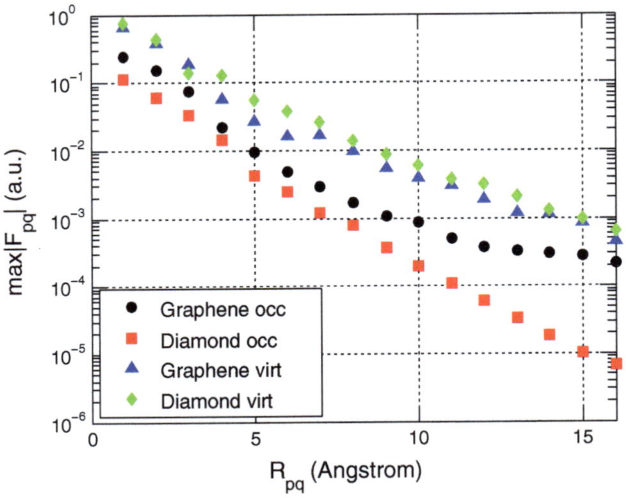

Fig. 6 The distance dependence of the maximum absolute occupied–occupied and virtual–virtual Fock matrix elements for diamond and graphene for the LMOs of Sect. 3

4 Comparisons to Pipek–Mezey orbitals and projected atomic orbitals

It is shown in the previous section that HF orbitals may be localized irrespective of the locality of the density matrix using the fourth moment orbital localization function [16]. In this section, we compare the LMOs of the previous section to orbitals traditionally used, e.g., in local correlation approaches. The occupied LMOs of the previous section are compared to occupied orbitals obtained using the Pipek–Mezey localization function [8], while the virtual LMOs are compared to the redundant and nonorthogonal projected atomic orbitals (PAOs).

The PAOs are generated by projecting the occupied space out of the AO basis

$$|\mu'\rangle = \left(1 - \sum_i |i\rangle\langle i|\right)|\mu\rangle \tag{14}$$

In contrast to the iteratively localized LMOs, the locality of the PAOs is a consequence of their construction from the AOs. It is a common misconception that PAOs constitute the most local representation of the virtual orbital space. Why this is not the case may be seen by rewriting Eq. 14 as

$$|\mu'\rangle = |\mu\rangle - \sum_{\nu\rho} |\nu\rangle D_{\nu\rho} S_{\rho\mu} \tag{15}$$

From Eq. 15, it is clear that a PAO ($|\mu'\rangle$) is an AO ($|\mu\rangle$) with a correction which depends on the density matrix elements through the AO overlap matrix ($\sum_{\nu\rho} |\nu\rangle D_{\nu\rho} S_{\rho\mu}$). Hence, $|\mu'\rangle$ will only be properly local if the density matrix is local in nature, whereas poor locality of the PAOs is inevitable if the system has a nonlocal density matrix.

The occupied Pipek–Mezey orbitals (PMOs) are generated by maximizing the sum of squared orbital populations [8]

$$\xi_{PM} = \sum_i \sum_A (Q_{ii}^A)^2 \tag{16}$$

where A runs over atomic centers and the index i runs over all occupied orbitals. Traditionally, the orbital population matrix $\mathbf{Q}^A$ is generated using the Mulliken population analysis, but it has been shown [21] that the Löwdin population analysis yields better results. Therefore, the Pipek–Mezey results presented here have been obtained using an implementation employing the Löwdin population analysis.

The locality results for the PAOs and PMOs are presented in terms of maximum orbital spreads and numerical overlaps as a function of orbital distance, as introduced in Sect. 3. The maximum orbital spreads are listed in Table 2, and the plots of the numerical overlaps are given in Fig. 7. The occupied and virtual LMOs of Sect. 3 are included for comparison.

Comparing first the PMOs to the occupied LMOs is seen that they exhibit very similar $\sigma_2^{\max}$ and $\sigma_4^{\max}$ values for diamond. For graphene, on the other hand, the PMOs yield significantly larger $\sigma_2^{\max}$ and $\sigma_4^{\max}$ values than the occupied LMOs. This trend is also evident from Fig. 7, where the numerical overlaps for diamond decay very similarly for the two sets of orbitals, while, for graphene, the decay is slow for the PMOs compared with the occupied LMOs. For graphene, the PMOs are thus significantly delocalized in space compared with the LMOs.

A comparison of the PAOs to the virtual LMOs using Table 2 and Fig. 7 leads to a similar conclusion. For diamond, the PAOs and virtual LMOs exhibit similar locality. For graphene, the PAOs are delocalized, while the virtual LMOs are localized in space. In particular, we note that for graphene, the $\sigma_4^{\max}$ value for the PAOs is almost three times larger than that for the virtual LMOs, indicating that the PAOs have very long and thick orbital tails. In Fig. 7 (right), this is manifested by the very slow decay of the numerical overlap for the PAOs. This behavior for the PAOs is not surprising, considering the very different decays of the density matrices for diamond and graphene in Fig. 2 and the direct dependence of the PAOs on the density matrix shown in Eq. 15.

In conclusion, the locality of the PMOs and PAOs are strongly system dependent. For non-conjugated systems like diamond, PMOs and PAOs are local, while they are delocalized for conjugated systems like graphene. On the other hand, using a proper localization function, such as the one in Eq. 9, it is indeed possible to generate localized (and much less system dependent) HF orbitals. The common misconception that it is not possible to generate localized

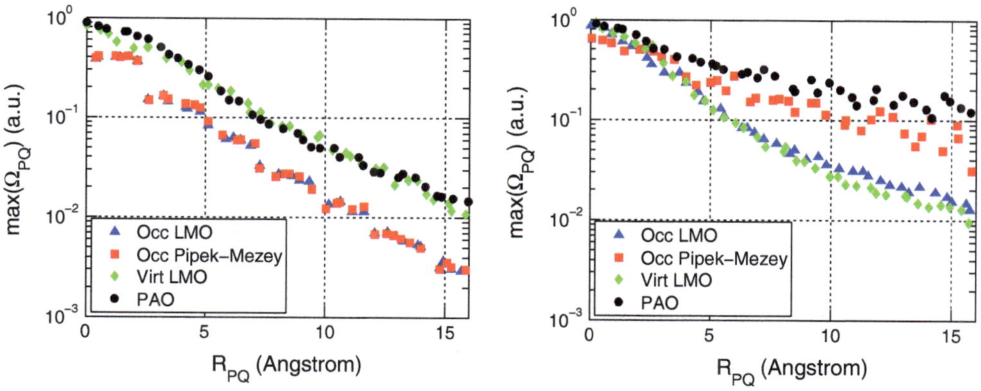

Fig. 7 The distance decay of numerical overlaps Ω_{PQ} for the occupied Pipek–Mezey orbitals and the projected atomic orbitals for diamond (*left*) and graphene (*right*). The distance decay of the numerical overlaps for the LMOs of Sect. 3 are included for comparison

HF orbitals for delocalized systems may partly be attributed to the fact that in local correlation methods, PMOs and PAOs have often been used to span the occupied and virtual orbital spaces.

5 Summary and conclusions

In this paper, the limits on the localizability of HF orbitals are explored in terms of two limiting effects: the physical limitation associated with the HF optimization condition (the occupied HF orbitals must represent the physical electron density) and the mathematical orthogonality constraint. The orthogonality constraint is investigated by comparing LOAOs to atomic orbitals, while the effect of the HF optimization condition is analyzed by comparing localized HF orbitals to LOAOs. Graphene and diamond structures are used as model systems since they constitute two different types of molecules. Graphene has a delocalized electronic structure, and this is reflected by the very slow distance decay of the HF density matrix elements. Diamond, on the other hand, has a more local electronic structure as manifested by the relatively rapid distance decay of the HF density matrix elements.

The results show that the physical constraint (as imposed by the HF optimization condition) leads to a difference in bulk locality of diamond and graphene occupied orbitals, but that the difference is of only 1 a.u. in terms of orbital spreads. The difference reflects the different bonding structures of graphene and diamond, but it is minor compared with the differences in locality of the density matrices for the two systems. The bulk of the virtual orbitals of both systems is of similar locality, and the tail decay behaviors for both occupied and virtual orbitals of graphene and diamond are similar. Thus, there are only minor differences between the localizability of the HF orbitals for graphene and diamond.

The loss in locality of the HF orbitals compared with the localized orthogonalized orbitals is small and is mainly due to the loss in flexibility, since orbital transformations must be carried out among the virtual and among the occupied HF orbitals in contrast to among the entire basis for the LOAOs. On the other hand, the tails of the LOAOs are seen to be quite extensive compared with the atomic orbitals. This substantiates the severity of the orthogonality constraint in terms of spatial locality of the orbitals, since orthogonalization alone yields a slow tail decay. Thus, outside the bulks of the orbitals—which is only slightly system dependent—the orthogonality constraint is the limiting factor for localizability.

Further, standard approaches to obtain local virtual and local occupied spaces—the projected atomic orbitals and Pipek–Mezey occupied orbitals—are shown to generate poorly localized orbitals for graphene but perform well for diamond. The misconception that projected atomic orbitals in general generate orbitals of good locality is pointed out and justified by showing the explicit dependence of projected atomic orbitals on the HF density matrix. Projected atomic orbitals of poor locality are thus inevitable for systems with a nonlocal HF density matrix. This has previously fueled the notion of the limited localizability for delocalized systems (such as graphene), but here it is shown that this effect is merely a result of their construction and is unrelated to the actual localizability of molecular orbitals for such systems.

Thus, contrary to common belief, we have demonstrated that the localizability of HF orbitals is not strongly affected by the locality of the density matrix. The invariance of the density matrix with respect to transformations of the occupied orbitals allows the density matrix to be described by highly localized molecular orbitals. In contrast, requiring orbitals to be orthogonal pose, a severe restriction on the form of the orbitals and is the main reason for slowly decaying orbital tails.

Acknowledgments The research leading to these results has received funding from the European Research Council under the European Union's Seventh Framework Programme (FP/2007-2013)/ ERC Grant Agreement no. 291371

References

1. Lennard-Jones JE, Pople JA (1950) Proc R Soc 202(1069):166
2. Hall G (1950) Proc R Soc 202:336
3. Boys SF (1960) Rev Mod Phys 32:296
4. Boys SF (1966) Quantum theory of atoms, molecules and solid state. Academic, New York, p 253
5. Edmiston C, Ruedenberg K (1963) Rev Mod Phys 35(3):457
6. Caldwell D, Reddington P, Eyring H (1979) Proc Natl Acad Sci 76:3042
7. Reed AE, Weinhold F (1985) J Chem Phys 83:1736
8. Pipek J, Mezey PG (1989) J Chem Phys 90:4916
9. von Niessen W (1972) J Chem Phys 56:4290
10. Subotnik JE, Dutoi AD, Head-Gordon M (2005) J Chem Phys 123:114108
11. Aquilante F, Pedersen T, de Meras AS, Koch H (2006) J Chem Phys 125:174101
12. Weber V, Hutter J (2008) J Chem Phys 128(6). doi:10.1063/1.2828507
13. Ziółkowski M, Jansík B, Jørgensen P, Olsen J (2009) J Chem Phys 131:124112
14. Jansík B, Høst S, Kristensen K, Jørgensen P (2011) J Chem Phys 134:194104
15. de Silva P, Giebultowski M, Korchowiec J (2012) Phys Chem Chem Phys 14(2):546
16. Høyvik IM, Jansík B, Jørgensen P (2012) J Chem Phys 137:224114
17. Høyvik IM, Jansík B, Jørgensen P (2012) J Chem Theory Comput 8:3137
18. Dunning TH (1989) J Chem Phys 90(2):1007. doi:10.1063/1.456153
19. Löwdin PO (1949) J Chem Phys 18(3):365
20. Carlson BC, Keller JM (1957) Phys Rev 105(1):102
21. Høyvik IM, Jansík B, Jørgensen P (2013) J Comput Chem 34(17):1456
22. Pettersen EF, Goddard TD, Huang CC, Couch GS, Greenblatt DM, Meng EC, Ferrin TE (2004) J Comput Chem 25(13):1605. doi:10.1002/jcc.20084.url:http://dx.doi.org/10.1002/jcc.20084

Theor Chem Acc (2014) 133:1449
DOI 10.1007/s00214-014-1449-x

REGULAR ARTICLE

Computing optical rotation via an *N*-body approach

Taylor J. Mach · T. Daniel Crawford

Received: 16 October 2013 / Accepted: 10 January 2014 / Published online: 25 January 2014

Abstract Properties of four chiral compounds—(*S*)-methyloxirane, (*S*)-methylthiirane, (*S*)-2-chloropropionitrile, and (*M*)-dimethylallene—centered in a solvation shell of six to seven water molecules have been computed using time-dependent density functional theory at several wavelengths using a many-body expansion. Interaction energies, total system dipole moments, and dynamic dipole polarizabilities converge rapidly and smoothly, exhibiting only minor oscillations with higher-body contributions. At three-body truncation of the expansion, errors in such properties as compared to the full cluster typically fall to less than 1 % (and much smaller in most cases). Specific optical rotations, however, are found to converge much more slowly and erratically, requiring five-body contributions to obtain errors less than 5 % in three of four test cases, and six-body terms for (*S*)-methylthiirane. The source of this behavior is found to be the wide variation of both magnitude and sign of the specific rotation with changes in the configuration of individual solute/solvent clusters. Thus, unlike simpler properties such as energies or dipole moments, where each fragment makes a small, same-sign contribution to the total property, specific rotations typically involve much larger contributions that partly cancel in the many-body expansion. Thus, the computational costs of molecular dynamics simulations of explicit solvation, for example, will be only partially alleviated by such expansions.

Keywords Many-body expansion · Optical rotation · Solvation · Response properties

Dedicated to Professor Thom Dunning and published as part of the special collection of articles celebrating his career upon his retirement.

Electronic supplementary material The online version of this article (doi:10.1007/s00214-014-1449-x) contains supplementary material, which is available to authorized users.

T. J. Mach · T. D. Crawford (✉)
Department of Chemistry, Virginia Tech, Blacksburg, VA 24061, USA
e-mail: crawdad@vt.edu

1 Introduction

Over the last two decades, the quality with which optical properties of chiral molecules [1–8] can be computed has increased dramatically, including optical rotation [3, 9–14], electronic [15–25], and vibrational [26–33] circular dichroism, and Raman optical activity [34–44]. In the 16 years, since Polavarapu [9] published the first *ab initio* calculations of optical rotation, for example, comparison with *vapor phase* experimental measurements [45–51] has revealed that such response properties often require high levels of electron correlation [13, 52–55], large one-electron basis sets (including diffuse functions), [10, 56–59] and, in some cases, vibrational corrections [52–55, 58, 60–66]. As a result of this effort, the number of remaining discrepancies between theory and experiment for gas-phase measurements of optical rotation has been significantly reduced [14, 45, 67].

However, the vast majority of experimental measurements of such properties are made in condensed phases, either in solvent or in the neat state, and attempts to simulate such environments accurately for chiroptical properties have seen only limited success. Continuum-based solvent models [68–71], for example, have been reported to reproduce experimental trends (as opposed to absolute values) only for those systems in which electrostatic solute–solvent interactions are dominant [54, 55, 64, 72, 73].

However, in cases where molecule-specific interactions are significant, implicit solvent models are not reliable. Microsolvation studies by Xu et al. [74, 75] on the paradigmatic chiral compound methyloxirane have revealed that both the magnitude and sign of the computed optical rotation can vary substantially depending on the number and configuration of solvent molecules in the region closest to the solute. In addition, combined molecular dynamics and density functional studies by Beratan et al. [76, 77] on the same compound in water and benzene have demonstrated that the solute can produce a surprisingly strong transfer of chirality to the solvent.

Clearly, a careful accounting of explicit solute–solvent interactions—at least in the cybotactic region—is vital to the accurate simulation of the exquisitely sensitive chiroptical response. However, the cost of such calculations is prohibitively expensive at present. Given the high-degree polynomial scaling of conventional quantum chemical methods—$\mathcal{O}(N^3)$ for Kohn–Sham density functional theory (DFT) [78] and $\mathcal{O}(N^5)$ for coupled cluster theory [79, 80]—inclusion of even a few water molecules, much less larger solvents such as benzene or cyclohexane, is far beyond the current state of the art. The cost of such computations is further exacerbated by the need to average properties of large numbers of snapshots of solute–solvent configurations along dynamical trajectories. While reduced- or even linear-scaling density functional and coupled cluster methods exist [81–86], their prefactors—and thus their algorithmic crossover points—lie too far out in terms of system size for them to provide any significant advantage for such computations.

One possible means of reducing the computational cost of explicit solvation models is to apply a many-body expansion to the (relatively) weak interactions of the solute and solvent components. This approach involves decomposition of the total energy, E, of the complete solute–solvent system into N-body contributions, *viz.*

$$E = \sum_{I}^{N} E_I + \sum_{I<J}^{N} V_{IJ} + \sum_{I<J<K}^{N} V_{IJK} + \cdots \quad (1)$$

where

$$V_{IJ} = E_{IJ} - E_I - E_J \quad (2)$$

and

$$V_{IJK} = E_{IJK} - V_{IJ} - V_{IK} - V_{JK} - E_I - E_J - E_K, \quad (3)$$

etc. In the above equations, E_I is the energy of fragment/monomer I, E_{IJ} is the energy of the dimer composed of fragments I and J, etc. Thus, the $V_{IJ\ldots}$ represent N-body corrections, and truncation of the expansion at a particular term may yield a cost-effective approximation to a computation on the complete system, provided that the concomitant loss of accuracy is sufficiently small. Such expansions have been widely used in quantum chemistry for decades, and we particularly note numerous applications of such an approach to water and other molecular clusters [87–92]. Often the scheme is enhanced by embedding the cluster in an electrostatic (or other simple) potential [90, 93], or making use of theory and/or basis-set hierarchies that treat higher-body terms using less expensive quantum chemical methods [91, 94]. Furthermore, while most applications of Eq. (1) have focused on (interaction) energies, extensions to first-order properties such as dipole moments [90] and energy gradients [91] have also been recently reported. For the properties investigated thus far, the many-body expansion is found to converge rapidly, with truncation at three-body terms yielding small errors—often a fraction of a percent—relative to the much more expensive computation performed on the aggregate system [90, 91].

The goal of the present work is to examine the convergence of the many-body expansion for higher-order solute properties such as frequency-dependent dipole polarizabilities and optical rotations. As test cases, we have chosen several small chiral compounds whose chiroptical response has been well studied, both in vapor and in condensed phases. However, our purpose here is not to reproduce experimental measurements, but instead to determine if truncation of Eq. (1) is still effective for properties that are much more sensitive to the quality of the wave function than are energies or first derivatives.

2 Computational details

The molecular dynamics program Gromacs [95] was used to place each of the four test solutes—(*S*)-methyloxirane, (*S*)-methylthiirane, (*S*)-2-chloropropionitrile, and (*M*)-dimethylallene—in a fixed configuration of water molecules. From these snapshots, only those solvent molecules lying completely within a sphere of radius of 5.5 Å from the solute's geometric center were retained, yielding clusters of six or seven water molecules surrounding the solute. (See the supporting information for coordinates of each cluster.) For each test case, interaction energies, dipole moments, dipole polarizabilities, and specific optical rotations were computed using the many-body expansion of Eq. (1).

The specific rotation of a chiral compound may be computed using the Rosenfeld optical activity tensor [1, 96],

$$\mathbf{G}'(\omega) = -\frac{2\omega}{\hbar}\sum_{j\neq 0}\frac{\mathrm{Im}(\langle\psi_0|\boldsymbol{\mu}|\psi_j\rangle\langle\psi_j|\mathbf{m}|\psi_0\rangle)}{\omega_{j0}^2 - \omega^2}, \quad (4)$$

where $\boldsymbol{\mu}$ and $\mathbf{m}$ are the electric and magnetic dipole operators, respectively, and ω is the frequency of plane-polarized light. The summation in Eq. (4) runs over the excited electronic (unperturbed) wave functions, ψ_j, each associated with an excitation frequency, ω_{j0}. The Rosenfeld tensor was computed at four wavelengths—355, 436, 589, and 633 nm—using time-dependent density functional theory (TD-DFT) with the B3LYP [97, 98] functional in Gaussian 09 [99]. Gauge-including atomic orbitals (GIAOs) [100, 101] were employed with the length representation of the electric dipole operator to ensure origin independence of the computed rotations. All computations were performed with the aug-cc-pVDZ basis set of Dunning and co-workers based on previous studies of its efficacy, especially for DFT [56, 59].

The specific rotation is related to the trace of the $\mathbf{G}'$ tensor [4],

$$[\alpha]_\omega = \frac{(72.0 \times 10^6)\hbar^2 N_A \omega}{c^2 m_e^2 M} \times \left[\frac{1}{3}\mathrm{Tr}(\mathbf{G}')\right] \tag{5}$$

where $\mathbf{G}'$ and ω are given in atomic units, N_A is Avogadro's number, c is the speed of light (m/s), m_e is the electron rest mass (kg), and M is the molecular mass (amu). In order to obtain the specific rotation of the solute within the solvent cluster, the many-body expansion of Eq. (1) was used to obtain the $\mathbf{G}'$ tensor for the complete solute–solvent system at a selected truncation level. The coordinate system of the full molecular cluster was held fixed for all fragment computations to ensure that the separate tensors could be combined. The final specific rotation was obtained from this composite $\mathbf{G}'$ tensor, with the mass M taken to be only that of the solute.

In order to automate use of the many-body expansion of Eq. (1), a local version of the PSI4 [102] program was modified to generate separate calculations automatically for all required fragment combinations based on a master input file containing the coordinates of the complete solvent–solute system. The individual fragment computations can then be carried out independently and in parallel fashion. The Python front end of PSI4 was modified to monitor the status of these automatically generated jobs, compiling the results as the various tasks complete. When all data required for a particular truncation level have been collected, the corrections are automatically computed by Python functions tied into the PSI4 driver. Furthermore, these functions are completely general, enabling easy extension to other properties of interest or higher levels of theory or approximation.

3 Results and discussion

3.1 (*S*)-2-Chloropropionitrile

The interaction energy of (*S*)-2-chloropropionitrile (Table 1) with seven water molecules in a configuration extracted from a molecular dynamics simulation is well represented by the truncated many-body expansion of Eq. (1). Limiting the expansion to only two-body terms yields an error of nearly 12 % as compared to a computation on the full cluster. However, extension to three-body contributions reduces the error to ca. 1 %, and four-body terms are accurate to 0.10 %, comparable to the convergence observed for water clusters by Tschumper [103].

Table 1 Absolute percent errors for (*S*)-2-chloropropionitrile in a cluster of water molecules relative to the full eight-body calculation

N-body truncation	Interaction energy	Dipole moment	Polarizability (633 nm)	Specific rotation (633 nm)
1	–	8.46	2.47	106.66
2	11.75	2.55	2.90	26.91
3	1.06	0.11	0.34	13.05
4	0.10	0.01	0.41	13.27
5	0.15	0.01	0.30	4.18
6	0.07	0.01	0.06	1.88
7	0.02	0.00	0.01	0.80

For dipole moments, the convergence of the many-body expansion is somewhat faster, with two-body truncation yielding percent errors of less than 3 % relative to the full cluster. Three-body truncation of the dipole moment is accurate to within 0.11 %, and higher-order terms are essentially negligible. These results for dipole moments are consistent with trends observed by Truhlar et al. [90] for dipole moments of chloride or hydrogen fluoride in aqueous solution, as well as for pure hydrogen fluoride clusters. For solvated chloride, for example, they observed errors of under 3 % at the two-body truncation (referred to as pairwise additive in Ref. [90]) without employing electrostatic embedding.

The convergence of the expansion for frequency-dependent polarizabilities (evaluated at 633 nm) starts even better than interaction energies and dipole moments, with a simple sum of monomer polarizabilities giving an error of only 2.5 %. However, extension of the sum to pairwise contributions actually increases the error slightly to ca. 3 %. While three-body components quickly reduce the error to under 0.5 %, subsequent corrections are somewhat larger than observed for interaction energies and dipole moments.

Experimental and computational studies of the chiroptical properties of (*S*)-2-chloropropionitrile have been reported before by Wiberg et al. [50] and by Kowalczyk et al. [57]. The latter reported modest basis-set dependence of both DFT (B3LYP) and coupled cluster methods. The specific rotation of the (*S*)-2-chloropropionitrile plus water cluster is found to be significantly larger and of opposite sign than that of the isolated solute molecule (+156.4 vs.

−12.3 deg dm^{-1} (g/mL)$^{-1}$ at 633 nm; see Ref. [57]). Although the cluster is not energetically optimum, the large perturbation introduced by the solvent cage is additional evidence of the importance of molecule-specific interactions on this property.

Unfortunately, convergence of the specific rotation of the solvated chloropropionitrile with many-body contribution (Table 2) is substantially slower than observed for the other properties. Employing solely monomers yields the incorrect sign of the rotation (e.g., −10.4 vs. +156.4 deg dm^{-1} (g/mL)$^{-1}$ at 633 nm). Incorporation of dimers in the expansion produces a qualitatively correct specific rotation, but with a quantitative error of 27–33 % for the wavelengths considered here. Extension to trimers cuts this error roughly in half, with the computed rotations still smaller in magnitude than those obtained for the full cluster. However, inclusion of four-body terms overshoots the correct value yielding the same percent error of approximately 13 %. Only when five-body terms are employed does the error finally decrease to under 5 % for the specific rotation and then to under 2 % at the six-body truncation.

3.2 (*S*)-Methyloxirane

The many-body expansion converges for the interaction energy of (*S*)-methyloxirane (Table 3) with a surrounding cluster of water molecules slightly faster than for (*S*)-2-chloropropionitrile, with the pairwise truncation of Eq. (1) giving an error of less than 7 %, as shown in Table 3. The errors are slightly oscillatory for (*S*)-methyloxirane, however, with that arising from the four-body expansion (0.14 %) slightly higher than from the three-body components (0.09 %). Nevertheless, the errors beyond trimers are negligible for the interaction energy.

The dipole moment of the (*S*)-methyloxirane plus water cluster requires dimers to reproduce that of the full cluster to within 2 %, and, unlike the interaction energy, the errors decay smoothly with the many-body expansion. In addition, just as for (*S*)-2-chloropropionitrile, the 633-nm dipole polarizability of (*S*)-methyloxirane converges more slowly than either the dipole moment or interaction energy expansions, and it exhibits an oscillatory pattern similar to that observed for the interaction energy.

(*S*)-methyloxirane has long served as a challenging test case for quantum chemical models of optical activity [52–54, 76, 77, 104]. Its vapor-phase specific rotation is small and bisignate (+7.49 ± 0.30 deg dm^{-1} (g/mL)$^{-1}$ at 355 nm and −8.39 ± 0.20 deg dm^{-1} (g/mL)$^{-1}$ at 633 nm; see Ref. 49) and thus exceedingly sensitive not only to the choice of theoretical method and basis set, but also to the inclusion of molecular vibrational corrections, which were found to be necessary to reproduce experimental results. In addition, the optical activity of (*S*)-methyloxirane exhibits strong solvent dependence, such that even the sign of its liquid-phase specific rotation varies with the choice of solvent [105].

Like its (*S*)-2-chloropropionitrile counterpart, the specific rotation of (*S*)-methyloxirane contained within a cluster of seven water molecules (270.7 deg dm^{-1} (g/mL)$^{-1}$ at 633 nm) is larger than that of the isolated molecule. Furthermore, the rotation is monosignate (positive) at all wavelengths down to 355 nm, unlike the gas-phase property. The specific rotation of (*S*)-methyloxirane converges much more slowly in the many-body expansion (Table 4) than the other properties considered here, though the convergence is slightly better relative to (*S*)-2-chloropropionitrile. Monomer, dimer, and trimer truncations approach the full value from below with the last yielding a fortuitously small error (as small as 0.08 % at 589 nm). However, like (*S*)-2-chloropropionitrile, the series is not converged at this point: the four-body truncation overshoots and the error increases to 5–8 %, depending on the wavelength. Expansion to six-body terms is necessary to converge the error relative to the full cluster down to around 1 %.

Table 2 Specific rotations [deg dm^{-1} (g/mL)$^{-1}$] and absolute percent errors (APE) relative to the full eight-body calculation for (*S*)-2-chloropropionitrile in water

N-body truncation	355 nm		436 nm		589 nm		633 nm	
	$[\alpha]_\omega$	APE	$(\alpha)_\omega$	APE	$(\alpha)_\omega$	APE	$(\alpha)_\omega$	APE
1	−52.4	109.23	−27.5	107.88	−12.4	106.81	−10.4	106.66
2	380.6	32.95	246.2	29.35	132.4	27.22	114.3	26.91
3	488.6	13.92	302.2	13.29	158.0	13.19	136.0	13.05
4	642.8	13.24	394.4	13.15	206.2	13.31	177.1	13.27
5	547.4	3.55	336.1	3.58	174.8	3.96	149.9	4.18
6	577.2	1.70	355.1	1.88	184.5	1.41	159.3	1.88
7	572.2	0.81	351.4	0.82	183.7	0.95	157.6	0.80
8	567.6	–	348.5	–	182.0	–	156.4	–

Table 3 Absolute percent errors for (*S*)-methyloxirane in a cluster of water molecules relative to the full eight-body calculation

N-body truncation	Interaction energy	Dipole moment	Polarizability (633 nm)	Specific rotation (633 nm)
1	–	12.34	2.06	69.88
2	6.57	1.80	2.94	19.35
3	0.09	0.22	0.24	0.36
4	0.14	0.09	0.60	7.52
5	0.07	0.04	0.52	3.81
6	0.05	0.01	0.18	1.12
7	0.01	0.00	0.00	0.00

3.3 (*M*)-Dimethylallene

The convergence of the many-body expansion for the interaction energy, dipole moment, and dipole polarizability of (*M*)-dimethylallene in a cluster of water molecules (Figs. 1, 2; Table 5) is very similar to that observed for (*S*)-2-chloropropionitrile and (*S*)-methyloxirane. Indeed, for all three properties, the series converges slightly faster, and the three-body truncation already provides errors of less than 1% as compared to the full cluster. The series oscillates only slightly for the interaction energy and polarizability with a slight increase in the error between three- and four-body contributions, but the errors remain small enough to be negligible.

(*M*)-Dimethylallene is unique among the chiral molecules considered here in that it exhibits a stereogenic axis rather than a stereogenic center. Its specific rotation was the subject of a detailed investigation in 2008 by Wiberg et al. [51] who reported a surprisingly large difference between its measured values in the neat liquid and vapor. At long wavelengths (633 and 589 nm), the specific rotation of the isolated molecule (-108.4 and -125.1 deg dm^{-1} (g/mL)$^{-1}$; see Ref. [51]) is approximately the same as that of the solute plus water snapshot extracted from the dynamical simulation. For the shorter wavelengths, however, the rotation is much larger for the cluster, suggesting that the lowest excited state shifts to lower energy due to the interaction with the nearby water molecules.

The specific rotation of (*M*)-dimethylallene, on the other hand, converges more slowly than either (*S*)-2-chloropropionitrile or (*S*)-methyloxirane, though the qualitative behavior is essentially the same (Table 6). At long wavelengths, the specific rotation of the cluster is approached from below (i.e., from more negative values), with increasing many-body truncation, and the four-body expansion overshoots as before. The behavior is somewhat different at shorter wavelengths, however, with the many-body expansion approaching the full value from above and exhibiting greater oscillation. This is a result of the fact that the Rosenfeld tensor diverges in the vicinity of a resonance, and the lowest excitation energy is clearly also sensitive to the truncation of the series. In the case of (*M*)-dimethylallene, errors as large as 20 % (corresponding to -21 deg dm^{-1} (g/mL)$^{-1}$ at 633 nm) remain even at the five-body truncation level.

3.4 (*S*)-Methylthiirane

The many-body expansion for the interaction energy, dipole moment, and dipole polarizability of (*S*)-methylthiirane in a cluster of six water molecules (Table 7) converges more smoothly than for (*S*)-methyloxirane and (*M*)-dimethylallene, with no oscillation apparent with increasing truncation level. The specific rotation of (*S*)-methylthiirane plus water, on the other hand, is much more difficult to model accurately. Vapor-phase specific rotations were measured in 2005 by Wilson et al. [49] who reported $+36.5 \pm 1.7$ deg dm^{-1} (g/mL)$^{-1}$ at 633 nm and $+64.7 \pm 2.3$ deg dm^{-1} (g/mL)$^{-1}$ at 355 nm for the (*R*) enantiomer. Crawford et al. [61] reported DFT and coupled cluster specific rotations as well as electronic

Table 4 Specific rotations (deg dm^{-1} (g/mL)$^{-1}$) and absolute percent errors (APE) relative to the full eight-body calculation for (*S*)-methyloxirane in water

N-body truncation	355 nm		436 nm		589 nm		633 nm	
	$(\alpha)_\omega$	APE	$(\alpha)_\omega$	APE	$(\alpha)_\omega$	APE	$(\alpha)_\omega$	APE
1	408.3	65.08	215.0	67.42	97.1	69.55	81.5	69.88
2	928.0	20.63	528.6	19.89	256.9	19.44	218.3	19.35
3	1,185.3	1.38	666.6	1.02	318.6	0.08	269.7	0.36
4	1,235.2	5.65	696.7	5.59	340.0	6.65	291.0	7.52
5	1,156.0	1.12	647.4	1.89	311.0	2.47	260.4	3.81
6	1,163.6	0.48	662.5	0.40	318.4	0.14	273.7	1.12
7	1,174.7	0.48	661.0	0.18	320.4	0.48	270.7	0.00
8	1,169.1	–	659.8	–	318.8	–	270.7	–

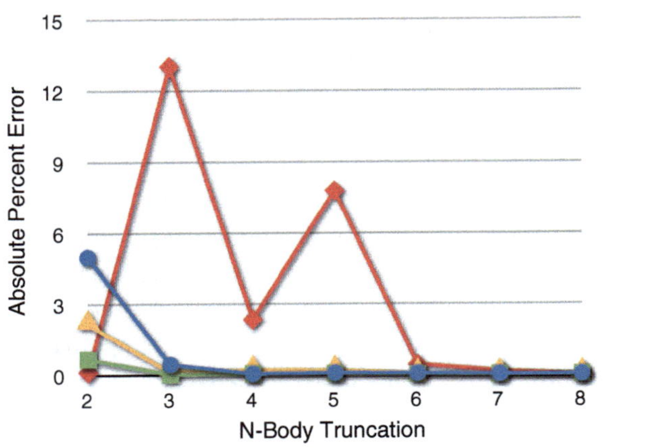

Fig. 1 Absolute percent error in the interaction energy, dipole moment, polarizability (355 nm), and optical rotation (355 nm) for (*M*)-dimethylallene solvated by seven water molecules as a function of *N*-body approximation level. The monomer approximation is excluded due to its large error

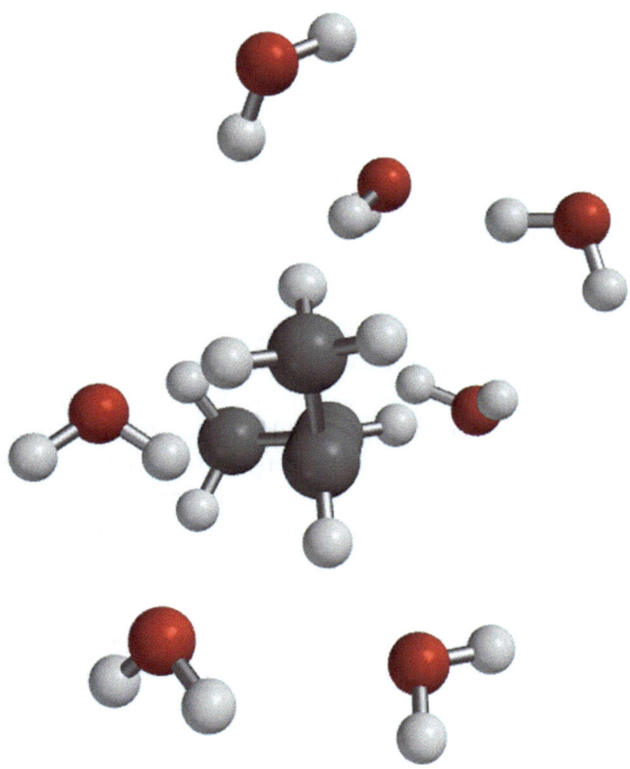

Fig. 2 (*M*)-Dimethylallene solvated by seven water molecules within a 5.5 Å radius of the solute's geometric center. Coordinates of the structure are found in the supporting information

circular dichroism spectra for (*S*)-methylthiirane in 2007 and found that vibrational corrections were necessary to produce reasonable comparison with experiment in the long wavelength regime, but the same corrections distorted the comparison for shorter wavelengths.

Table 5 Absolute percent errors for (*M*)-dimethylallene in a cluster of water molecules relative to the full eight-body calculation

N-body truncation	Interaction energy	Dipole moment	Polarizability (355 nm)	Specific rotation (355 nm)
1	–	11.27	2.68	75.66
2	4.97	0.66	2.13	0.12
3	0.43	0.01	0.11	12.99
4	0.03	0.04	0.19	2.33
5	0.06	0.04	0.17	7.78
6	0.04	0.01	0.03	0.42
7	0.01	0.00	0.02	0.15

The specific rotation of (*S*)-methylthiirane in the cluster of six water molecules modeled here is bisignate, with a strong positive Cotton pole as the wavelength of the polarized field becomes shorter. The specific rotation of the isolated molecule, [61] however, is monosignate according to both experiment and coupled cluster theory, though the B3LYP functional finds a bisignate dispersion due to its underestimation of the energy of lowest excited state. The convergence of the many-body expansion of the specific rotation of the (*S*)-methylthiirane plus water cluster is more erratic than that of the other species investigated here. At 633 nm, for example, the rotation shifts from -42.5 deg dm^{-1} (g/mL)$^{-1}$ using only monomers to $+83.9$ deg dm^{-1} (g/mL)$^{-1}$ using dimers, and then back to -23.1 deg dm^{-1} (g/mL)$^{-1}$ with trimers. At 355 nm, on the other hand, the variation is even more pronounced (though without concomitant sign changes), with a one-body truncation of $+74.0$ deg dm^{-1} (g/mL)$^{-1}$, two-body truncation of $+665.2$ deg dm^{-1} (g/mL)$^{-1}$, three-body $+1.1$ deg dm^{-1} (g/mL)$^{-1}$, and four-body $+138.4$ deg dm^{-1} (g/mL)$^{-1}$, before the series finally begins to approach the untruncated result. The relatively small total rotations at long wavelengths produce very large corresponding percent errors—more than 2,000 % for the pairwise expansion at 589 nm, for example.

4 Conclusions

When describing the interactions between solute and solvent, the venerable many-body expansion converges rapidly and smoothly for energies, dipole moments, and dipole polarizabilities—that is, properties that are strongly localized and/or easily partitioned into local contributions. On the other hand, for properties such as optical rotations, whose values (both magnitude and sign) depend strongly on more distant interactions as well as basis-set effects, Eq. (1) converges much more slowly. We have demonstrated that for four paradigmatic chiral compounds—(*S*)-methyloxirane, (*S*)-methylthiirane, (*S*)-2-chloropropionitrile, and

Table 6 Specific rotations [deg dm^{-1} (g/mL)$^{-1}$] and absolute percent errors (APE) relative to the full eight-body calculation for (*M*)-dimethylallene in water

N-body truncation	355 nm		436 nm		589 nm		633 nm	
	$(\alpha)_\omega$	APE	$(\alpha)_\omega$	APE	$(\alpha)_\omega$	APE	$(\alpha)_\omega$	APE
1	−190.5	75.66	−253.6	25.53	−163.9	23.65	−143.8	32.16
2	−783.4	0.12	−422.7	24.15	−187.6	41.56	−157.0	44.33
3	−884.1	12.99	−394.8	15.96	−156.6	18.17	−129.6	19.15
4	−800.7	2.33	−323.2	5.09	−123.7	6.67	−99.9	8.16
5	−843.4	7.78	−392.6	15.31	−155.5	17.28	−129.8	19.27
6	−785.7	0.42	−334.2	1.86	−133.4	0.67	−108.9	0.13
7	−781.4	0.15	−341.6	0.32	−131.9	0.52	−108.2	0.55
8	−782.5	–	−340.5	–	−132.5	–	−108.8	–

Table 7 Absolute percent errors for (*S*) in a cluster of water molecules relative to the full seven-body calculation

N-body truncation	Interaction energy	Dipole moment	Polarizability (589 nm)	Specific rotation (589 nm)
1	–	11.51	3.37	797.96
2	8.31	1.68	1.98	2,028.61
3	0.37	0.08	0.36	393.10
4	0.10	0.01	0.07	119.60
5	0.03	0.00	0.02	20.76
6	0.01	0.00	0.01	26.76

Table 8 Specific rotations (deg dm^{-1} (g/mL)$^{-1}$) and absolute percent errors (APE) relative to the full seven-body calculation for (*S*) in water

N-body truncation	355 nm		436 nm		589 nm		633 nm	
	$(\alpha)_\omega$	APE	$(\alpha)_\omega$	APE	$(\alpha)_\omega$	APE	$(\alpha)_\omega$	APE
1	74.0	41.23	−50.7	464.48	−47.3	797.96	−42.5	647.85
2	665.2	428.46	262.0	1,785.14	101.5	2,028.61	83.9	1,574.83
3	1.1	99.12	−35.7	356.77	−26.0	393.10	−23.1	306.56
4	138.4	9.93	7.5	45.71	−11.6	119.60	−10.9	91.30
5	114.4	9.10	14.2	1.83	−4.2	20.76	−4.8	15.05
6	125.5	0.30	12.4	11.11	−6.7	26.76	−6.7	17.24
7	125.9	–	13.9	–	−5.3	–	−5.7	–

(*M*)-dimethylallene—each contained in a small cluster of water molecules, the expansion is highly oscillatory and erratically convergent. Admittedly, the appearance of (relatively) strong hydrogen bonding interactions between the solute and solvent (and between solvent molecules themselves) makes the choice of water a worst-case scenario, and Eq. (1) may exhibit better behavior for more weakly interacting solvents such as cyclohexane. However, even in such cases, the transfer of chirality from the solute to the solvent (e.g., through the chiral hole analyzed by Beratan et al. [77]) yields widely disparate optical rotations among the various *N*-body fragments. As a result, the many-body expansion will still be exceedingly sensitive to truncation.

It may be possible to relieve at least some of the computational burden of explicit solvation models through, e.g., multi-theory QM/MM schemes [91, 93, 106] or distance-based criteria to eliminate minor contributions to the many-body expansion [107]. However, the non-locality of chiroptical properties on which such schemes rely is still poorly understood. This work suggests that the "chiral imprint" onto the solvent extends sufficiently far out from the solute as to require models of condensed phase dynamics that are much more robust—and potentially much more costly—than previously hoped. This is particularly discouraging considering the large number of molecular dynamics snapshots that must be averaged in the case of optical rotation to achieve a balanced representation of the opposing sign contributions from different molecular configurations [108].

Acknowledgments This work was supported by Grants from the US National Science Foundation: CHE-1058420 and a Multi-User Chemistry Research Instrumentation and Facility (CRIF:MU) award CHE-0741927. The authors are also grateful for helpful discussions with Prof. Gregory S. Tschumper (U. Mississippi).

References

1. Barron LD (2004) Molecular light scattering and optical activity, 2nd ed. Cambridge University Press, Camridge
2. Koslowski A, Sreerama N, Woody RW (2000) Theoretical approach to electronic optical activity. In: Berova N, Nakanishi K, Woody RW (eds) Circular dichroism: principles and applications, 2nd ed, Chapter 3. Wiley, New York, pp 55–95
3. Pecul M, Ruud K (2005) Adv Quantum Chem 50:185–212
4. Crawford TD (2006) Theor Chem Acc 115:227–245
5. Polavarapu PL (2007) Chem Rec 7:125–136
6. Crawford TD, Tam MC, Abrams ML (2007) J Phys Chem A 111:12057–12068
7. Autschbach J (2009) Chirality 21:E116–E152
8. Crawford TD (2012) High-accuracy quantum chemistry and chiroptical properties, vol 1. Wiley, New York
9. Polavarapu PL (1997) Mol Phys 91:551–554
10. Cheeseman JR, Frisch MJ, Devlin FJ, Stephens PJ (2000) J Phys Chem A 104:1039–1046
11. Grimme S (2001) Chem Phys Lett 339:380–388
12. Polavarapu PL (2002) Chirality 14:768–781
13. Ruud K, Stephens PJ, Devlin FJ, Taylor PR, Cheeseman JR, Frisch MJ (2003) Chem Phys Lett 373:606–614
14. Crawford TD, Stephens PJ (2008) J Phys Chem A 112:1339–1345
15. Grimme S, Peyerimhoff SD, Bartram S, Vögtle F, Breest A, Hormes J (1993) Chem Phys Lett 213:32–40
16. Grimme S (1996) Chem Phys Lett 259:128–137
17. Pulm F, Schramm J, Hormes J, Grimme S, Peyerimhoff SD (1997) Chem Phys 224:143–155
18. Grimme S, Waletzke M (1999) J Chem Phys 111:5645–5655
19. Carnell M, Grimme S, Peyerimhoff SD (1994) Chem Phys 179:385–394
20. Hansen AE, Bak KL (1999) Enantiomer 4:455
21. Pedersen TB, Koch H (2000) J Chem Phys 112:2139–2147
22. Autschbach J, Ziegler T, van Gisbergen SJA, Baerends EJ (2002) J Chem Phys 116:6930–6940
23. Autschbach J, Jorge FE, Ziegler T (2003) Inorg Chem 42:2867–2877
24. Pecul M, Ruud K, Helgaker T (2004) Chem Phys Lett 388:110–119
25. Grimme S (2004) Calculation of the electronic spectra of large molecules. In: Lipkowitz KB, Larter R, Cundari TR (eds) Reviews in computational chemistry, vol 20, Chapter 3, VCH Publishers, New York, pp 153–218
26. Stephens PJ, Jalkanene KJ, Devlin FJ, Chabalowski CF (1993) J Phys Chem 97:6107–6110
27. Bak KL, Jørgensen P, Helgaker T, Ruud K, Aa. Jensen HJ (1993) J Chem Phys 98:8873–8887
28. Bak KL, Jørgensen P, Helgaker T, Ruud K (1994) Faraday Discuss 99:121–129
29. Bak KL, Bludský O, Jørgensen P (1995) J Chem Phys 103:10548–10555
30. Cheeseman JR, Frisch MJ, Devlin FJ, Stephens PJ (1996) Chem Phys Lett 252:211–220
31. Devlin FJ, Stephens PJ, Cheeseman JR, Frisch MJ (1997) J. Phys. Chem. A 101:9912–9924
32. Stephens PJ, Devlin FJ (2000) Chirality 12:172–179
33. Devlin FJ, Stephens PJ, Scafato P, Superchi S, Rosini C (2002) Chirality 14:400–406
34. Polavarapu PL (1990) J Phys Chem 94:8106–8112
35. Nafie LA (1997) Ann Rev Phys Chem 48:357–386
36. Helgaker T, Ruud K, Bak KL, Jørgensen P, Olsen J (1994) Faraday Discuss 99:165–180
37. Barron LD, Hecht L, McColl IH, Blanch EW (2004) Mol Phys 102:731–744
38. Bouř P (2001) J Comp Chem 22:426–435
39. Pecul M, Rizzo A (2003) Mol Phys 101:2073–2081
40. Pecul M, Ruud K (2005) Int J Quantum Chem 104:816–829
41. Crawford TD, Ruud K (2011) ChemPhysChem 12:3442–3448
42. Hug W (2001) Chem Phys 264:53–69
43. Luber S, Reiher M (2008) Chem Phys 346:212–223
44. Ruud K, Thorvaldsen AJ (2009) Chirality 21:E54–E67
45. Müller T, Wiberg KB, Vaccaro PH (2000) J Phys Chem A 104:5959–5968
46. Müller T, Wiberg KB, Vaccaro PH, Cheeseman JR, Frisch MJ (2002) J Opt Soc Am B 19:125–141
47. Wiberg KB, Vaccaro PH, Cheeseman JR (2003) J Am Chem Soc 125:1888–1896
48. Wiberg KB, Wang YG, Vaccaro PH, Cheeseman JR, Trucks G, Frisch MJ (2004) J Phys Chem A 108:32–38
49. Wilson SM, Wiberg KB, Cheeseman JR, Frisch MJ, Vaccaro PH (2005) J Phys Chem A 109:11752–11764
50. Wiberg KB, Wang Y, Wilson SM, Vaccaro PH, Cheeseman JR (2005) J Phys Chem A 109:3448–3453
51. Wiberg KB, Wang Y, Wilson SM, Vaccaro PH, Jorgensen WL, Crawford TD, Abrams ML, Cheeseman JR, Luderer M (2008) J Phys Chem A 112:2415–2422
52. Tam MC, Russ NJ, Crawford TD (2004) J Chem Phys 121:3550–3557
53. Ruud K, Zanasi R (2005) Angew Chem Int Ed Engl 44:3594–3596
54. Kongsted J, Pedersen TB, Strange M, Osted A, Hansen AE, Mikkelsen KV, Pawlowski F, Jørgensen P, Hättig C (2005) Chem Phys Lett 401:385–392
55. Kongsted J, Pedersen TB, Jensen L, Hansen AE, Mikkelsen KV (2006) J Am Chem Soc 128:976–982
56. Stephens PJ, Devlin FJ, Cheeseman JR, Frisch MJ, Bortolini O, Besse P (2003) Chirality 15:S57–S64
57. Kowalczyk TD, Abrams ML, Crawford TD (2006) J Phys Chem A 110:7649–7654
58. Pedersen TB, Kongsted J, Crawford TD (2009) Chirality 21:E68–E75
59. Mach TJ, Crawford TD (2011) J Phys Chem A 115:10045–10051
60. Ruud K, Taylor PR, Åstrand P-O (2001) Chem Phys Lett 337:217–223
61. Crawford TD, Tam MC, Abrams ML (2007) Mol Phys 105:2607–2617
62. Mort BC, Autschbach J (2005) J Phys Chem A 109:8617–8623
63. Mort BC, Autschbach J (2006) J Phys Chem A 110:11381–11383
64. Kongsted J, Ruud K (2008) Chem Phys Lett 451:226–232
65. Crawford TD, Allen WD (2009) Mol Phys 107:1041–1057
66. Pedersen TB, Kongsted J, Crawford TD, Ruud K (2009) J Chem Phys 130:034310
67. Lahiri P, Wiberg KB, Vaccaro PH, Caricato M, Crawford TD (2013) Angew Chem Int Ed Engl. doi:10.1002/anie.201306339
68. Tomasi J, Persico M (1994) Chem Rev 94:2027–2094
69. Cramer CJ, Truhlar DG (1999) Chem Rev 99:2161–2200
70. Tomasi J, Cammi R, Mennucci B, Cappelli C, Corni S (2002) Phys Chem Chem Phys 4:5697–5712

71. Tomasi J, Mennucci B, Cammi R (2005) Chem Rev 105:2999–3093
72. Mennucci B, Tomasi J, Cammi R, Cheeseman JR, Frisch MJ, Devlin FJ, Gabriel S, Stephens PJ (2002) J Phys Chem A 106:6102–6113
73. Pecul M, Marchesan D, Ruud K, Coriani S (2005) J Chem Phys 122:024106
74. Su Z, Xu Y (2007) Angew Chem Int Ed Engl 46:6163–6166
75. Losada M, Nguyen P, Xu Y (2008) J Phys Chem A 112:5621–5627
76. Mukhopadhyay P, Zuber G, Goldsmith M, Wipf P, Beratan DN (2006) Chem Phys Chem 7:2483–2486
77. Mukhopadhyay P, Zuber G, Wipf P, Beratan DN (2007) Angew Chem Int Ed Engl 46:6450–6452
78. Parr RG, Yang W (1989) Density-functional theory of atoms and molecules. Oxford University, New York
79. Shavitt I, Bartlett RJ (2009) Many-body methods in chemistry and physics: MBPT and coupled-cluster theory. Cambridge University Press, Cambridge
80. Crawford TD, Schaefer HF (2000) An introduction to coupled cluster theory for computational chemists. In: Lipkowitz KB, Boyd DB (eds) Reviews in computational chemistry, vol 14, Chapter 2, VCH Publishers, New York, pp 33–136
81. Choi C, Ruedenberg K, Gordon M (2001) J Comp Chem 22:1484–1501
82. Gordon MS, Federov DG, Pruitt SR, Slipchenko LV (2011) Chem Rev 112:632–672
83. Schütz M, Werner H-J (2001) J Chem Phys 114:661–681
84. Russ NJ, Crawford TD (2004) Chem Phys Lett 400:104–111
85. Russ NJ, Crawford TD (2008) Phys Chem Chem Phys 10:3345–3352
86. McAlexander HR, Mach TJ, Crawford TD (2012) Phys Chem Chem Phys 14:7830–7836
87. Hankins D, Moskowitz JW (1970) J Chem Phys 53:4544–4554
88. Xantheas SS (1994) J Chem Phys 100:7523–7534
89. Christie RA, Jordan KD (2005) n-Body decomposition approach to the calculation of interaction energies of water clusters. In: Wales DJ (eds) Intermolecular forces and clusters II, vol 116. Springer, Berlin, pp 27–41
90. Leverentz HR, Maerzke KA, Keasler SJ, Siepmann JI, Truhlar DG (2012) Phys Chem Chem Phys 14:7669–7678
91. Bates DM, Smith JR, Janowski T, Tschumper GS (2011) J Chem Phys 135:044123
92. Richard RM, Lao KU, Herbert JM (2013) J Phys Chem Lett 4:2674–2680
93. Theel KL, Wen S, Beran GJO (2013) J Chem Phys 139:081103
94. Ferré N, Assfeld X (2003) J Mol Struct (Theochem) 632:83–90
95. Pronk S, Páll S, Schulz R, Larsson P, Bjelkmar P, Apostolov R, Shirts MR, Smith JC, Kasson PM, van der Spoel D, Hess B, Lindahl E (2013) Bioinformatics 29:845–854
96. Rosenfeld L (1928) Z Phys 52:161–174
97. Becke AD (1993) J Chem Phys 98:5648–5652
98. Lee C, Yang W, Parr RG (1988) Phys Rev B 37:785–789
99. Frisch MJ et al (2009) Gaussian 09 revision A.1. Gaussian Inc., Wallingford, CT
100. London F (1937) J Phys Radium 8:397–409
101. Ditchfield R (1974) Mol Phys 27:789–807
102. Turney JM et al (2012) WIREs Comput Mol Sci 2:556–565
103. Tschumper GS (2006) Chem Phys Lett 427:185–191
104. Lipparini F, Egidi F, Cappelli C, Barone V (2013) J Chem Theory Comp 9:1880–1884
105. Kumata Y, Furukawa J, Fueno T (1970) Bull Chem Soc Jpn 43:3920–3921
106. Bates DM, Smith JR, Tschumper GS (2011) J Chem Theory Comp 7:2753–2760
107. Dahlke EE, Truhlar DG (2007) J Chem Theory Comp 3:1342–1348
108. Kundrat MD, Autschbach J (2008) J Chem Theory Comp 4:1902–1914

Theor Chem Acc (2014) 133:1544
DOI 10.1007/s00214-014-1544-z

REGULAR ARTICLE

Quantitative estimation of uncertainties from wavefunction diagnostics

Matthew K. Sprague · Karl K. Irikura

Received: 26 September 2013 / Accepted: 12 July 2014 / Published online: 7 August 2014

Abstract Coupled-cluster calculations with large basis sets are used widely to make predictions of gas-phase thermochemistry. Wavefunction diagnostics are sometimes used to indicate whether or not there is problematic multireference character that may cause errors. Here, we investigate whether existing diagnostics, as well as diagnostics proposed by us, can be used to estimate these errors quantitatively. We calculate the atomization energy of 50 molecules, including known multireference molecules such as CN, C_2, O_3, *ortho*-benzyne, formaldehyde oxide, and hydrogen trioxy radical. In addition to the c_0^2, T_1, D_1, and %TAE[(T)] diagnostics, we assess the Hartree–Fock HOMO–LUMO gap, maximum occupation number defect, first vertical excitation energy, a direct estimate of multireference effects, and combinations of diagnostics as indicators of errant thermochemistry.

Keywords T_1 · D_1 · %TAE[(T)] · Atomization energy · Non-dynamical correlation · Thermochemistry

Dedicated to Professor Thom Dunning and published as part of the special collection of articles celebrating his career upon his retirement.

Electronic supplementary material The online version of this article (doi:10.1007/s00214-014-1544-z) contains supplementary material, which is available to authorized users.

M. K. Sprague · K. K. Irikura (✉)
Chemical Sciences Division, National Institute of Standards and Technology, Gaithersburg, MD 20899-8320, USA
e-mail: karl.irikura@nist.gov

M. K. Sprague
e-mail: matthew.sprague@nist.gov

1 Introduction

Theoretical predictions of ideal-gas molecular energetics are commonplace. The most popular theory in quantitative work is coupled-cluster theory with a perturbative correction, CCSD(T) [1, 2]. This is applied together with a large one-electron basis set, or a series of sets that can be extrapolated to the limit of a complete basis set (CBS). This practice has been summarized by Dunning [3], Peterson [4], Feller [5], and Dixon [6], to name a few. More recently, "explicitly correlated" treatments of electron correlation [7] have become easy to use. These new methods perform markedly better than conventional calculations for basis sets of similar size, reaching the CBS limit faster than traditional methods [8, 9]. Following the model of Dunning's correlation-consistent basis sets, new basis sets have been developed specifically for explicitly correlated calculations [8].

In some cases, CCSD(T) predictions are noticeably less accurate than usual. Examples include F_2, O_3, and transition metal compounds [10, 11]. Often, this is because Hartree–Fock (HF) theory produces such a poor description of the electronic structure that CCSD(T), which is built atop the HF wavefunction, is unable to compensate. To detect this problem, many wavefunction diagnostics have been proposed. Examples include the T_1 and D_1 diagnostics by Lee and coworkers [12, 13]; the B1, GB1, ROD, and M diagnostics by Truhlar and co-workers [14–16]; %TAE[(T)] by Martin and co-workers [17]; and the natural orbital occupation number diagnostic developed by Gordon and co-workers [18]. Recently, a diagnostic for hybrid density functional theory has been proposed by Fogueri et al. [19] based upon the dependence of atomization energies upon the mixing coefficient. The informatics analysis derived from density matrix renormalization group methods also can be used as a diagnostic [20]. Likewise, the wavefunction variance is promising, albeit expensive [21].

Wavefunction diagnostics are increasingly popular. For example, the paper presenting the T_1 diagnostic has accumulated about 1,000 citations [13]. This reflects increasing demand for information about the reliability of ab initio predictions of molecular properties. Despite the popularity of diagnostics, their interpretation is somewhat vague. The most common current interpretation is that triggering a diagnostic indicates that there *may* be an unusual problem with the associated prediction [11–13]. Of course, this statement is meaningful only if the *failure* to trigger the diagnostic provides a clear signal that the prediction is reliable. In other words, to adopt the terminology used in medical research, a diagnostic test is valuable to the extent that it produces no false negatives. This is a useful interpretation for many applications, where it can be misleading or costly if a bad value is taken to be reliable.

Wavefunction diagnostics often disagree; one diagnostic indicates a possible problem while another does not. This has been shown recently by Jiang et al. [22]. Additional examples of conflicting diagnostics can be found throughout the literature [19, 23, 24]. In some cases, pairs of diagnostics are believed to complement each other and to work best when used together [12, 25]. However, it is unclear *how* diagnostics should be combined.

In this paper, we seek a quantitative way of using wavefunction diagnostics to estimate uncertainties associated with computed energetics. We compute the atomization energies of molecules in which significant multireference character has been observed. The resulting errors, taken as the deviations from precise experimental values, are compared with various wavefunction diagnostics. Note that this is different from many previous studies of diagnostics, in which the focus is on detailed components of the molecular energy and not on agreement with experiments.

2 Computational strategy

2.1 Choice of wavefunction diagnostics

Eight diagnostics are examined in this study. Four are traditional or formally published diagnostics: $1 - c_0^2$ (where c_0 is the largest configuration expansion coefficient in a CASSCF wavefunction), the T_1 and D_1 diagnostics from single-reference coupled-cluster calculations [12, 13], and %TAE[(T)] (the fractional difference between atomization energies from corresponding CCSD(T) and CCSD calculations) [17]. For conciseness, henceforth we shall denote the last of these as %(T). The T_1, D_1, and %(T) diagnostics are particularly interesting because they are obtained with no additional effort or computational expense.

Non-dynamical correlation is often described as a consequence of near-degeneracy of orbitals [26]. This suggests that the HOMO–LUMO gap, labeled here as H–L gap, could be a useful diagnostic for estimating uncertainty on computed thermochemistry. Alternatively, the vertical excitation energy to the lowest excited state, here labeled E_{vert}, might be useful.

HF is a poor description of the wavefunction whenever a multi-configuration calculation yields natural orbital occupation numbers that are substantially different from the HF values. The size of this difference could be another diagnostic. Here, we use the natural orbital occupation numbers from a multireference configuration interaction (MRCI) calculation (details are in the following section). We define the maximum occupation number defect (MOND) as the largest difference between the MRCI and RHF occupation numbers. A small MOND indicates that HF gives a good description of the wavefunction. A large MOND indicates that the wavefunction has significant multireference character. This is similar to using natural orbital occupation numbers from other theoretical methods. For example, Bak et al. [27] used a CCSD(T) density matrix in their study of closed-shell thermochemistry.

The last diagnostic we consider, ΔE_{exc}, is one in which we attempt to isolate the effect of non-dynamical correlation upon the molecular energy. It is described in the following sub-section. It cannot be a *practical* diagnostic because it requires expensive multireference calculations.

Associated with each diagnostic is a threshold that divides calculations into two categories: reliable and unreliable. For established diagnostics, we use the following thresholds: $T_1 < 0.02$ [13], $D_1 < 0.06$ [11], $\%(T) < 10\ \%$ [11], and $1 - c_0^2 < 0.10$ [11]. Based on the linear correlation between $1 - c_0^2$ and MOND (see Online Resource 1, Figure SI-1), we choose a threshold of MOND < 0.12. There are no established thresholds for E_{vert} or the H–L gap; we choose $E_{vert} > 4.50$ eV and H–L gap > 0.58 hartree based upon the results presented in this paper. When combining two diagnostics, we first divide each value by its corresponding threshold. The resulting unitless, normalized values can then be combined mathematically.

2.2 ΔE_{exc} as an estimate of multireference effects

In the complete active space approach, with n active electrons distributed throughout m active orbitals, a common notation is CASSCF(n, m), where normally $m \leq n < 2m$. (The choice of active space can be challenging; good discussions have been published by Schmidt and Gordon [28] and by Veryazov et al. [29].) When $m < n < 2m-1$, states with higher spin have fewer active orbitals that are nominally vacant. Consequently, fewer low-spin coupled open-shell configurations are possible and one would expect

HF to be a better description than for low-spin states [30]. High-spin states are generally better described by HF theory because more of the electron correlation is of the exchange type, which is included in HF theory [31]. This principle has been applied effectively by Krylov in developing spin-flip methods [32]. If the ground state is troublesome, a simpler excited state can be used for the CCSD(T) calculations. For example, experimental singlet–triplet splittings can be used to correct computed ionization energies [33]. Similarly, Hernández-Lamoneda et al. [34] used this type of correction for low-spin states of $(O_2)_2$.

Consider a low-spin ground state with significant non-dynamical correlation and a high-spin excited state with negligible non-dynamical correlation. If a single-reference method cannot adequately account for non-dynamical effects, then it should do well for the excited state but place the ground state too high in energy. The excitation energy to the high-spin state (E_{exc}) will be smaller than as calculated using a comparable multireference method. We propose that this difference in excitation energies (ΔE_{exc}) can be used as a proxy for multireference effects on thermochemistry, as illustrated in Fig. 1. Expressed as an equation,

$$\Delta E_{\text{exc}} = E_{\text{exc}}(\text{multireference}) - E_{\text{exc}}(\text{single-reference}) \quad (1)$$

where E_{exc} is the vibrationless, vertical excitation energy. $\Delta E_{exc} > 0$ suggests that non-dynamical correlation predominates in the ground state (Fig. 1), and conversely $\Delta E_{exc} < 0$ suggests that dynamical correlation predominates in the ground state. It is the former situation that may be problematic for single-reference calculations.

We have chosen CCSD(T*)-F12a/VTZ-F12 as the single-reference method and internally contracted MRCISD-F12+Q/VTZ-F12 (multireference configuration interaction [35] with Davidson correction [36, 37]) as the

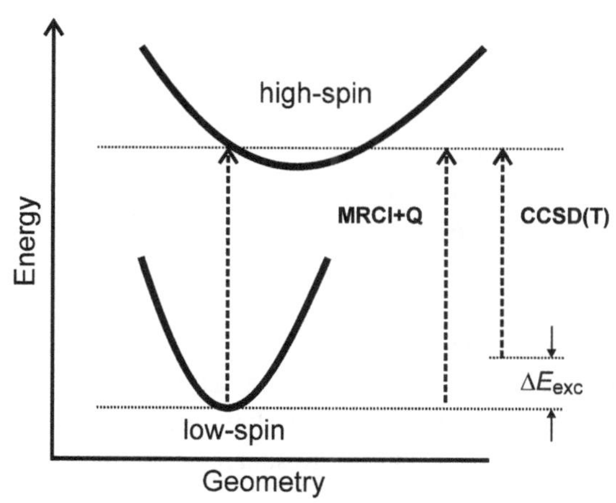

Fig. 1 ΔE_{exc} is the difference between multireference and single-reference values for the vertical excitation energy from the ground state to a higher-spin excited state. This is used as a proxy for multireference effects

multireference method. Although we feel that this is the best choice available to us, it is imperfect because these methods differ by more than the number of reference configurations. CCSD(T) includes single and double excitations, some triples, a perturbative estimate for other triples, some quadruples, and some higher-order excitations [1, 38]. Single-reference CISD+Q includes only singles, doubles, and an estimate for some quadruples [36]. Thus, CCSD(T) includes more dynamical correlation than CISD+Q. Compared with CISD+Q, multireference MRCISD+Q adds non-dynamical correlation and additional dynamical correlation, but may still be expected to include less dynamical correlation than CCSD(T).

There is no guarantee that all high-spin excited states will automatically have less multireference character than the ground state. We consider the upper state acceptable if it has a smaller MOND than the ground state.

2.3 Choice of calculations for energetics

Our intention here is to investigate the usefulness of diagnostics for predicting the accuracy of moderately high-level atomization energies, as compared with experimental benchmarks. Atomization reactions break all chemical bonds and the associated electron pairs and are challenging to predict accurately. Thus, atomization energies are a stringent test of theoretical methods. The energies of isodesmic or isogyric reactions, which benefit more from cancelation of errors [39], are easier to compute accurately. Nonetheless, atomization energies remain popular for ab initio thermochemistry, presumably because of their convenience. Our computational methods are described in the following section.

In the present study, we use deviations from experimental atomization energies to measure the practical value of wavefunction diagnostics. This requires experimental atomization energies that are sufficiently precise for the deviations to be meaningful. For experimental atomization energies, we required relative uncertainties of ≤ 0.1 %. The average fractional uncertainty for the molecules we chose was 0.03 %. In absolute terms, the average uncertainty was 0.2 kJ/mol and the maximum was 1.4 kJ/mol. The scarcity of high-precision experimental values was a severe constraint.

3 Computational details

Single-reference calculations were frozen-core RHF-UCCSD(T*)-F12a/VTZ-F12 [8, 40]. The RHF-UCCSD(T) method uses a restricted HF reference to perform an unrestricted coupled-cluster calculation [41, 42]. This was selected based upon observations by Knizia et al. [40] and by Bross et al. [43], which lead us to expect good performance partly through fortuitous cancelation of small errors. Scaling factors for (T*) were taken from MP2/MP2-F12 differences, following the software defaults, and are slightly different for each system. Because the F12 approach is so effective, these results are expected to be close to the complete basis limit [44]. The core-valence contribution to the atomization energy was estimated at the RHF-UCCSD(T*)-F12a/cc-pCVTZ-F12 level and basis set [40, 45] as the difference between frozen- and active-core energies.

Multireference calculations were frozen-core, internally contracted MRCI-F12+Q/VTZ-F12 [36, 37]. All MRCI calculations are from CASSCF references; active spaces are specified in Online Resource 1, Table SI-2.

Our calculations include no relativistic effects. One consequence is that the energies correspond to averages over the spin–orbit levels of the ground term. Corrections are needed for many atoms and molecules to compensate for this over-estimation of their energies. The spin–orbit corrections used here, from section II.C.2 of the NIST Computational Chemistry Comparison and Benchmark Database [10], are (in kJ/mol): C, 0.354; O, 0.933; F, 1.611; Cl, 3.513; OH, 0.833; NO, 0.717; CH, 0.167; CF, 0.461; FO, 1.173.

HOMO–LUMO gaps were computed using RHF/aug-cc-pVTZ geometries and energies. The lowest vertical excitation energy, E_{vert}, was computed using TD-B3LYP/aug-cc-pVTZ at the B3LYP/aug-cc-pVTZ geometry.

Calculations were performed on the 50 molecules listed in Table 1. The list includes several known "problem cases" with significant multireference character in the ground state [11, 17, 19, 22, 46]. Experimental atomization energies were taken from the evaluated data in the Active Thermochemical Tables (ATcT) [47].

All CCSD(T*)-F12a and MRCI-F12+Q calculations were performed using the MOLPRO 2012.1 software package [48].[1] TD-B3LYP calculations were performed using Gaussian 09 [49].

For notational convenience, the symbols ΣD_e and ΣD_0 are used to refer to equilibrium and zero-point atomization energies, respectively, of polyatomic molecules as well as diatomic molecules.

Diatomic molecules. The equilibrium geometry was computed at the RHF-UCCSD(T*)-F12a/VTZ-F12 level. Potential energy curves were computed by the same method, up to 1,000 cm^{-1} above the energy minimum using a distance increment of 0.5 pm. Spectroscopic constants (ω_e, $\omega_e x_e$, $\omega_e y_e$, B_e, α_e) were computed from a fitted sextic polynomial, as implemented in MOLPRO 2012.1 [48]. Vibrational zero-point energy (ZPE) was computed from the spectroscopic constants using a standard formula [50].

Polyatomic molecules. Geometries and harmonic vibrational frequencies were computed at the RHF-UCCSD(T*)-F12a/VTZ-F12 level except as noted otherwise. ZPE was estimated using the harmonic approximation (without empirical scaling), and applying an anharmonic correction estimated from vibrational configuration interaction (VCI) [51], up to quadruple excitations and three-mode couplings (3M-VCISDTQ), at the B3LYP/aug-cc-pVTZ level using the Vosko–Wilk–Nusair III (VWN3) correlation functional ("B3LYP3" in MOLPRO) [52]. For comparison with the VCI (see Online Resource 1, Table SI-3), anharmonic corrections were also estimated from vibrational self-consistent field (VSCF) calculations [53], vibrational MP2 (VMP2) [54], and second-order vibrational perturbation theory (VPT2) [55], all at B3LYP/aug-cc-pVTZ (spin-restricted for VSCF, VCI, VMP2, spin-unrestricted for VPT2). The atomization energy at the equilibrium geometry (ΣD_e) was computed at the RHF-UCCSD(T*)-F12a/VTZ-F12 level. VPT2 calculations were performed in Gaussian 09 [49]; all other calculations were performed in MOLPRO 2012.1 [48].

Difficulties and special cases. Unusual problems were encountered in some of the calculations, as described below:

1. We were unable to find a high-spin state with a lower MOND than the ground state for seven molecules: CO_2, FOOF, CH_2OH, CH_3OH, O_2, ClO, H_2. Consequently, we could not calculate ΔE_{exc}.
2. MRCI calculations failed for HN_3. Consequently, we could not calculate ΔE_{exc} or MOND.
3. %(T) cannot be computed for H_2 because it has only two electrons.
4. VSCF calculations failed for CH_2; ν_2 increased from 1,092 to 17,650 cm^{-1}. VPT2 did not suffer from the same problem, so we used VPT2 for the anharmonic correction to the ZPE.
5. For NO_3, *cis*-HO_3, and *trans*-HO_3, all of the anharmonic calculations yielded a ZPE greater than the harmonic calculation due to low frequencies changing drastically (see Online Resource 1 for details, pages 56–58). For this reason, we chose to use only the harmonic ZPE in calculating ΣD_0 for these three molecules.
6. To reduce computational expense on the VCI calculations for CH_2OH, CH_3OH, *ortho*-benzyne, and

[1] Certain commercial materials and equipment are identified in this paper in order to specify procedures completely. In no case does such identification imply recommendation or endorsement by the National Institute of Standards and Technology, nor does it imply that the material or equipment identified is necessarily the best available for the purpose.

Table 1 Computed and experimental atomization energies (ΣD_0, kJ/mol), fractional error in ΣD_0 compared to the literature (*f*), and diagnostic values (HOMO–LUMO gap in hartree, E_{vert} in eV, ΔE_{exc} in kJ/mol, others unitless) for 50 molecules

Molecule	ΣD_0 (calc)[a]	ΣD_0 (ATcT)[b]	$f \times 10^3$	T_1	D_1	%(T)	$1 - c_0^2$ (CAS)	MOND (MRCI)	H–L gap (RHF)	E_{vert} (TD-B3LYP)	ΔE_{exc}
Diatomics											
H_2	433.1	432.068[c]	2.5	0.006	0.009	[d]	0.012	0.035	0.649	9.93	[d]
CH	335.5	334.6 ± 0.1	2.8	0.009	0.016	1.1	0.046	0.087	0.432	0.47	0.6
HCl	429.1	427.644 ± 0.006	3.4	0.007	0.012	1.4	0.012	0.042	0.507	6.75	1.1
NH	328.4	327.9 ± 0.2	1.8	0.007	0.010	1.4	0.017	0.044	0.567	1.87	1.8
HF	566.9	565.97 ± 0.08	1.7	0.009	0.015	1.5	0.010	0.033	0.683	9.01	−7.6
OH	426.4	425.62 ± 0.03	1.7	0.008	0.014	1.6	0.013	0.038	0.569	4.20	0.2
CO	1,073.1	1,072.03 ± 0.06	1.0	0.018	0.037	3.2	0.057	0.064	0.621	5.88	−4.4
CF	546.2	545.5 ± 0.2	1.2	0.018	0.044	3.9	0.088	0.048	0.434	4.00	−9.0
N_2	941.8	941.15 ± 0.05	0.7	0.014	0.028	4.2	0.072	0.079	0.711	7.25	−2.1
CN	744.5	745.3 ± 0.2	−1.0	0.053	0.149	5.9	0.107	0.097	0.592	1.13	−1.0
NO	626.9	626.83 ± 0.07	0.1	0.021	0.051	6.3	0.064	0.088	0.482	5.67	1.7
Cl_2	239.5	239.242 ± 0.004	1.0	0.009	0.020	7.6	0.024	0.068	0.470	2.46	9.7
O_2	491.8	493.688[c]	−3.8	0.008	0.014	7.7	0.030	0.054	0.610	1.53	[e]
ClF	253.3	252.50 ± 0.09	3.3	0.012	0.029	8.4	0.028	0.067	0.535	2.70	2.5
ClO	264.7	265.35 ± 0.04	−2.3	0.046	0.156	9.8	0.034	0.062	0.514	3.62	[e]
FO	211.6	213.8 ± 0.2	−10.4	0.029	0.099	13.2	0.048	0.079	0.657	4.24	1.3
C_2	604.0	602.5 ± 0.1	2.5	0.039	0.088	13.3	0.291	0.373	0.343	−2.27[f]	−5.7
F_2	153.4	154.5 ± 0.1	−7.4	0.012	0.029	19.8	0.069	0.124	0.748	2.95	−5.4
Triatomics											
CH_2	753.1	752.5 ± 0.2	0.7	0.012	0.020	0.5	0.026	0.036	0.434	0.44	−3.1
NH_2	715.2	713.7 ± 0.2	2.0	0.008	0.012	1.4	0.026	0.042	0.514	2.27	0.1
H_2O	918.9	917.8 ± 0.2	1.2	0.009	0.017	1.5	0.022	0.038	0.541	6.53	−8.0
HCN	1,270.1	1,268.32 ± 0.06	1.4	0.015	0.030	3.0	0.076	0.083	0.529	5.94	−7.8
CO_2	1,600.3	1,598.18 ± 0.06	1.3	0.018	0.045	3.6	0.079	0.067	0.594	7.55	[e]
HOCl	658.4	656.36 ± 0.03	3.2	0.011	0.024	4.1	0.038	0.070	0.468	2.84	4.4
ClCN	1,171.5	1,166.5 ± 0.5	4.2	0.014	0.030	4.5	0.076	0.080	0.513	5.77	−261
HO_2	694.3	694.5 ± 0.2	−0.3	0.037	0.128	4.6	0.051	0.066	0.534	1.13	3.3
HNO	824.9	823.6 ± 0.1	1.7	0.016	0.043	5.0	0.078	0.105	0.447	−0.44[f]	1.4
HOF	624.9	624.5 ± 0.2	0.6	0.013	0.035	5.2	0.058	0.093	0.563	3.24	1.3
N_2O	1,103.3	1,102.0 ± 0.1	1.2	0.020	0.046	7.1	0.097	0.096	0.563	5.24	1.9
NO_2	927.0	927.35 ± 0.07	−0.4	0.025	0.065	8.7	0.095	0.113	0.559	2.94	1.0
OF_2	373.2	374.4 ± 0.3	−3.2	0.016	0.041	14.8	0.092	0.098	0.662	2.87	6.8
FOO	537.1	544.0 ± 0.3	−12.7	0.044	0.144	16.0	0.167	0.240	0.662	2.07	4.3
O_3	589.0	596.13 ± 0.04	−12.0	0.027	0.076	17.9	0.168	0.249	0.489	−1.43[f]	6.5
Tetratomics											
NH_3	1,159.5	1,157.24 ± 0.04	1.9	0.008	0.015	1.3	0.031	0.040	0.456	5.63	−14.6
H_2CO	1,498.4	1,495.6 ± 0.1	1.8	0.015	0.044	2.1	0.048	0.081	0.467	3.25	−5.4
HCCH	1,627.9	1,625.9 ± 0.2	1.2	0.014	0.031	2.1	0.089	0.087	0.447	5.11	−3.2
HOOH	1,056.0	1,055.21 ± 0.06	0.8	0.012	0.023	3.3	0.058	0.079	0.512	4.57	5.6
HN_3	1,332.8	1,329.7 ± 0.6	2.3	0.020	0.053	5.7	0.114	[g]	0.434	4.25	[e]
cis-HO_3	922.4[h]	922 ± 1	0.0	0.043	0.157	7.6	0.118	0.131	0.558	2.37	5.4
trans-HO_3	923.3[h]	924.4 ± 0.7	−1.3	0.043	0.148	8.2	0.140	0.173	0.543	2.42	5.1
ONNO	1,254.8	1,262.0 ± 0.2	−5.7	0.020	0.053	10.7	0.235	0.315	0.403	−0.38[f]	0.9
NO_3	1,129.2[h]	1,131.8 ± 0.2	−2.2	0.020	0.086	11.4	0.169	0.155	0.637	1.48	−37
FOOF	609.7	610.9 ± 0.4	−1.9	0.027	0.092	17.9	0.189	0.174	0.625	3.12	[e]

Table 1 continued

Molecule	ΣD_0 (calc)[a]	ΣD_0 (ATcT)[b]	$f \times 10^3$	T_1	D_1	%(T)	$1 - c_0^2$ (CAS)	MOND (MRCI)	H–L gap (RHF)	E_{vert} (TD-B3LYP)	ΔE_{exc}
Larger polyatomics											
CH_4	1,644.5	1,642.08 ± 0.08	1.4	0.008	0.013	0.7	0.039	0.042	0.577	9.35	1.3
C_2H_4	2,229.8	2,225.8 ± 0.2	1.8	0.012	0.034	1.3	0.043	0.097	0.414	4.17	−1.5
CH_3OH	2,016.3	2,012.2 ± 0.2	2.1	0.009	0.020	1.3	0.050	0.040	0.479	6.01	[e]
CH_2OH	1,619.4	1,616.1 ± 0.4	2.1	0.017	0.054	1.6	0.041	0.043	0.380	3.38	[e]
Benzene	5,470.7	5,463.8 ± 0.4	1.3	0.011	0.030	2.0	0.113	0.108	0.368	3.84	−119
o-Benzyne	4,672.2	4,663 ± 1	2.0	0.014	0.042	2.8	0.188	0.173	0.385	1.47	−1.6
H_2COO	1,526.9	1,524 ± 1	2.1	0.044	0.177	4.7	0.108	0.144	0.435	0.99	1.2

[a] CCSD(T*)-F12a/VTZ-F12, with core-valence and spin–orbit corrections, anharmonic ZPE

[b] Ref. [47]

[c] Uncertainty on ΣD_0 (ATcT) for H_2 and O_2 reported as <±0.0005 kJ/mol

[d] Since CCSD is equivalent to FCI for H_2, %(T) and ΔE_{exc} were not calculated

[e] Suitable high-spin states were not found for O_2, ClO, CO_2, HN_3, FOOF, CH_3OH, or CH_2OH

[f] TD-B3LYP calculation predicted states lower in energy than the ground state for C_2, HNO, O_3, and ONNO

[g] MRCI calculations failed for HN_3, therefore, MOND was not calculated

[h] Harmonic ZPE used to calculate ΣD_0 for *cis*-HO_3, *trans*-HO_3, and NO_3

benzene, we included only two-mode couplings (the "ZPVE" macro in MOLPRO's surface generation program).

7. To further reduce computational expense on the VCI calculations for *ortho*-benzyne and benzene, anharmonic corrections were computed using the aug-cc-pVDZ basis set rather than aug-cc-pVTZ.
8. TD-B3LYP predicts triplet electronic states lower than the singlet ground states of C_2, HNO, O_3, and ONNO; therefore, E_{vert} is negative for these molecules.

4 Results

4.1 Correlation of diagnostics with fractional error in atomization energy

Table 1 contains ΣD_0 (calculated and experimental), the fractional error of the calculated atomization energy compared to experiment (denoted as *f*, Eq. 2), and values of wavefunction diagnostics: T_1, D_1, %(T), $1 - c_0^2$, MOND, H–L gap, E_{vert}, and ΔE_{exc}.

$$f \equiv \frac{\left(\Sigma D_{0,\text{calc}} - \Sigma D_{0,\text{expt}}\right)}{\Sigma D_{0,\text{expt}}} = \frac{\Sigma D_{0,\text{calc}}}{\Sigma D_{0,\text{expt}}} - 1 \tag{2}$$

Our test set consists of 18 diatomic molecules, 15 triatomic, 10 tetratomic, and 7 larger polyatomics. The test set includes molecules traditionally considered "problem cases" with regard to multireference character, such as C_2, O_3, HO_3, NO dimer (ONNO), formaldehyde *O*-oxide (CH_2OO), and *ortho*-benzyne (C_6H_4) [18, 30, 56–62]. Details of the calculations for each molecule (electronic energies, ZPE, spin–orbit and core-valence corrections, excitation energies, and ground and excited state wavefunction diagnostics) can be found in Online Resource 1, Table SI-1.

Figure 2 shows the correlation of the unsigned fractional error in the calculated atomization energy, $|f|$, with %(T) (Fig. 2a) and T_1 (Fig. 2b). Correlation of all diagnostics to both the unsigned ($|f|$) and signed (f) errors can be found in Online Resource 1, Figure SI-2. Threshold values for the diagnostics are marked with a vertical red line.

As stated in Sect. 1, a diagnostic is most useful if it produces no false negatives. Let RMS_{neg} be the RMS value of *f* for all molecules with negative diagnostics. We define a false negative as a calculation with negative diagnostic and $|f| > 2 \times RMS_{neg}$. Table 2 contains the number of false negatives and the fraction of negatives that are false for each diagnostic for our test set of 50 molecules. All of the diagnostics contain only one or two false negatives (2–8 % of the negatives are false).

An even more useful diagnostic would not only be free of false negatives, but would also only be triggered when there is a problem with the calculated atomization energy; in other words, it would be free of false positives. Table 2 contains the number of false positives and the fraction of positives that are false for each diagnostic for our test set of 50 molecules. We define a false positive as a calculation with a positive diagnostic and $|f| < 2 \times RMS_{neg}$. Here, %(T) performs best, with only four false positives (44 % of the positives are false). MOND is second best (7 false positives,

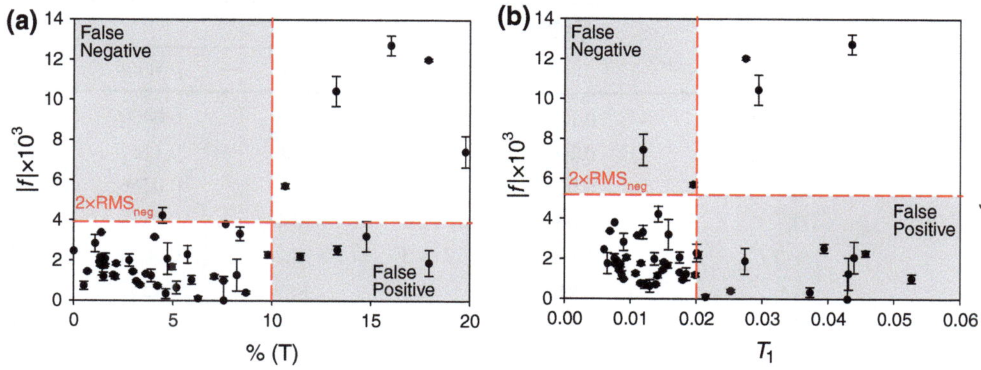

Fig. 2 Relationship of unsigned fractional error on atomization energy, |*f*|, with diagnostics: %(T) (Fig. 2a) and T_1 (Fig. 2b). Thresholds for diagnostics are marked with a *vertical dashed line* [11, 13]. The *horizontal dashed line* marks twice the RMS value of *f* for molecules with negative diagnostics (i.e., below threshold). Experimental uncertainties from ATcT [47] are shown by the *error bars*

Table 2 RMS values of *f* for negative and positive diagnostics, contrast ratio, number of false negatives and positives, and fraction of negative and positive diagnostics that are false

Diagnostic	Thresh	$10^3 \times \mathrm{RMS}_{neg}$	$10^3 \times \mathrm{RMS}_{pos}$	Contrast ratio	False Neg	False Pos	%Neg false	%Pos false
MOND	0.12	2.6	6.2	2.4	1	7	3	64
$1 - c_0^2$	0.10	2.9	5.4	1.9	2	11	5	85
T_1	0.02	2.6	5.5	2.1	2	12	6	80
D_1	0.06	2.5	5.9	2.3	2	10	5	77
%(T)	10 %	1.9	7.7	4.0	1	4	2	44
ΔE_{exc}	0 kJ/mol	2.5	4.3	1.7	1	26	5	87
E_{vert}	4.5 eV	1.9	4.3	2.3	1	29	6	85
H–L gap	0.58 au	5.4	2.9	0.5	1	36	8	97
A[%(T), T_1]	1.0	2.0	5.8	2.9	1	11	3	69
G[%(T), T_1]	1.0	2.0	6.0	3.0	1	10	3	67
Q[%(T), T_1]	1.0	2.0	5.8	2.9	1	11	3	69
H[%(T), T_1]	1.0	2.3	6.3	2.8	1	8	3	67
max[%(T), T_1]	1.0	2.1	5.5	2.7	1	13	3	72
min[%(T), T_1]	1.0	2.7	7.7	2.8	2	3	5	50

or 64 % of the positives are false). For the remaining diagnostics, 75 % or more of the positives are false.

Figure 2 includes error bars reflecting the uncertainties in the experimental atomization energies reported in the ATcT [47]. In most cases, the error bars do not cross the $2 \times \mathrm{RMS}_{neg}$ boundary, suggesting that better experiments for these molecules will not produce additional false positives or false negatives.

4.2 Assessment of diagnostic performance

For a diagnostic to convey information, it must divide the molecules into two groups (negative diagnostic, positive diagnostic) with a statistically significant difference in the distribution of fractional errors. This fundamental requirement is met if the two groups differ in mean, variance, or both.

We test whether the mean fractional errors of the two groups are unequal using Welch's *t* test [63]. This is similar to the well-known Student's *t* test, but does not assume equal variances for the groups. We test whether the variances of the two groups are unequal using the Brown–Forsythe test [64, 65]. This is similar to the better-known *F* test, but is more robust to non-normality. For both statistical tests, the null hypothesis is that the two groups have the same distribution. Both tests produce a *p* value, which is the probability of obtaining a test statistic equal to or greater than that observed if the two groups actually do have the same distribution. Table 3 contains the statistical results for each diagnostic. At the 95 % confidence level, all of the diagnostics except ΔE_{exc} successfully partition the data into groups with different error distributions. Thus, all the diagnostics except ΔE_{exc} merit further analysis.

Table 3 Test variables (t, W), degrees of freedom (DOF), and p values for tests of statistical significance for inequality of means (Welch's t test) and variances (Brown–Forsythe)

Diagnostic	Thresh	Welch's t test			Brown–Forsythe test[a]	
		t	DOF	p value	W	p value
MOND	0.12	2.3	11.1	0.0409	15.1	0.0003
$1 - c_0^2$	0.10	1.7	14.6	0.1167	5.0	0.0299
T_1	0.02	2.4	16.7	0.0266	6.2	0.0165
D_1	0.06	2.6	13.7	0.0218	7.9	0.0071
%(T)	10 %	4.3	8.2	0.0024	31.5	0.0000
ΔE_{exc}	0 kJ/mol	1.4	46.9	0.1557	2.5	0.1229
E_{vert}	4.5 eV	3.1	40.1	0.0037	8.7	0.0049
H–L gap	0.58 au	2.7	14.9	0.0159	7.1	0.0102
A[%(T), T_1]	1.0	4.0	16.0	0.0011	23.1	0.0000
G[%(T), T_1]	1.0	3.9	14.8	0.0015	24.9	0.0000
Q[%(T), T_1]	1.0	4.0	16.0	0.0011	23.1	0.0000
H[%(T), T_1]	1.0	3.4	12.0	0.0053	15.7	0.0002
max[%(T), T_1]	1.0	3.8	18.4	0.0012	20.4	0.0000
min[%(T), T_1]	1.0	2.6	5.2	0.0473	8.6	0.0051

[a] For the Brown–Forsythe test, degrees of freedom were 1 and 48

The estimate of uncertainty on computed thermochemistry is most meaningful for the "best performing" diagnostic. Three metrics of diagnostic performance that we consider are as follows:

1. The fraction of all negatives that are false, and the fraction of all positives that are false, as described in the previous section.
2. RMS_{neg}. Smaller values of RMS_{neg} allow us to assign a smaller uncertainty to a computed value when the diagnostic is not triggered.
3. The ratio of RMS for molecules that trigger/do not trigger the diagnostic, denoted here as the contrast ratio. A large contrast ratio indicates a meaningful difference in uncertainty when the diagnostic is triggered, whereas a contrast ratio of 1 indicates that the diagnostic does not convey any additional information about uncertainty.

From Table 2, we note that %(T) performs best according to these metrics (smallest fraction of false negatives, smallest fraction of false positives, smallest RMS_{neg}, largest contrast ratio). The other diagnostics, including T_1 and D_1 (the other two diagnostics obtained "for free" from a CCSD(T) calculation), perform worse than %(T) in all of these categories. For the %(T) diagnostic, $RMS_{neg} = 0.19$ %.

Note that although we chose a H–L gap threshold of 0.58 hartree (which minimizes the p value for Welch's t test for out test set), the contrast ratio is always <1, regardless of threshold. Because of this poor performance, we do not consider the H–L gap to be a useful diagnostic.

4.3 Combinations of diagnostics

As stated in Sect. 1, there have been recommendations to consider multiple diagnostics when determining the reliability of a computed property. Here, we consider combinations of T_1, D_1, %(T), MOND, and $1 - c_0^2$. Because the diagnostics use different numerical scales, we first normalize the values to the diagnostic threshold. The reduced diagnostic x is defined as

$$x = \frac{X}{X_0} \tag{3}$$

where X is the value of the diagnostic and X_0 is the diagnostic threshold from Table 2.

We examined the following simple combinations for pairs of diagnostics: arithmetic mean ($A[x_1, x_2]$), geometric mean ($G[x_1, x_2]$), harmonic mean ($H[x_1, x_2]$), quadratic mean ($Q[x_1, x_2]$), maximum of the reduced diagnostics ($max[x_1, x_2]$), and minimum of the reduced diagnostics ($min[x_1, x_2]$). The threshold for each combined diagnostic was 1.

$$A = \frac{x_1 + x_2}{2} \tag{4}$$

$$G = \sqrt{x_1 x_2} \tag{5}$$

$$H = \frac{2}{x_1^{-1} + x_2^{-1}} \tag{6}$$

$$Q = \sqrt{\frac{x_1^2 + x_2^2}{2}} \tag{7}$$

Table 2 compares the performance of %(T) to the six combinations of %(T) with T_1. (Values of the combined diagnostics for each molecule, and plots of $|f|$ and f versus the six combinations for all pairs of the individual diagnostics can be found in Online Resource 1 Table SI-4 and Figure SI-3, and Online Resource 2 Spreadsheet SI-1). None of these combinations of diagnostics perform better than %(T), based on the metrics in the previous section.

4.4 Estimation of atomization energy uncertainty based on %(T)

%(T) was the best performing diagnostic according to the metrics in the previous sections; therefore, we use it for our atomization energy uncertainty analysis. We estimate the uncertainty (twice the relative standard uncertainty) to be $2 \times \mathrm{RMS}_{\mathrm{neg}}$ for molecules that do not trigger the %(T) diagnostic (this interval includes 98 % of the molecules with negative diagnostics). Similarly, we use $2 \times \mathrm{RMS}_{\mathrm{pos}}$ as the uncertainty for molecules that do trigger the diagnostic (this interval includes all molecules with a positive diagnostic). This translates to uncertainties in atomization energies of ±0.4 % for %(T) < 10 %, and ±1.5 % for %(T) > 10 %.

5 Discussion

5.1 Limitations of our study

One of the most severe limitations that our study faces is the lack of precise experimental atomization energies. The ATcT [47] provided us with atomization energies known to 0.1 %; however, we were limited to relatively small molecules containing only H, C, N, O, F, and Cl. Extension of our work to other classes of molecules, such as transition metals, would require more precise experimental studies.

Our main result in this work is an estimation of uncertainties on computed atomization energies based on wavefunction diagnostics. To this end, the uncertainties that we estimate (±0.4 % for %(T) < 10 % and ±1.5 % for %(T) > 10 %) are strictly applicable only to the method that we used: RHF-UCCSD(T*)-F12a/VTZ-F12 geometries and energies, with core-valence, spin–orbit, and anharmonic ZPE corrections. We did not include scalar relativistic effects, diagonal Born–Oppenheimer corrections (DBOC), or higher-order terms such as $T_4 + T_5$. These effects can be significant with respect to the atomization energies obtained. Feller et al. [66] and Karton et al. [67] computed these additional contributions for a number of molecules, including 43 of the molecules in this study. The root-mean-square magnitude of these contributions to atomization energy for molecules in our test set was 0.7 % (absolute value of 3.6 kJ/mol). These contributions correlate linearly with f (see Online Resource 2, Spreadsheet SI-2). It has been shown that %(T) correlates well with the $T_4 + T_5$ contribution [17], which may explain the strong performance of %(T) as a diagnostic for our computed atomization energies.

It should be reiterated that the purpose of our study was to use wavefunction diagnostics to assess atomization energy uncertainties, and not to perform a benchmark study on the performance of cutting-edge theory. Despite neglecting scalar relativistic effects, DBOC, and higher-order effects, our computed atomization energies were within 0.4 % (2 × RMS) of the experimental data for molecules that do not trigger the %(T) diagnostic. Furthermore, 44 of the 50 computed atomization energies were within 4 kJ/mol of the experimental value. It is likely that this agreement is obtained from fortuitous cancelation of errors in the CCSD(T*)-F12a method, as suggested by Knizia et al. [40] and by Bross et al. [43].

5.2 Multireference character versus multireference effects

In this work, we have attempted to quantify the effects of multireference character by ΔE_{exc}. One interpretation of wavefunction diagnostics is that they are a measure of multireference character [12, 13, 23]. Given this interpretation, we can ask whether multireference *character* (i.e., the diagnostic value) correlates with multireference *effects* (i.e., ΔE_{exc}). We do not observe any meaningful correlation between ΔE_{exc} and the other diagnostics (see Online Resource 1, Figure SI-1). A possible reason for this is that multireference character is not necessarily a predictor of multireference effects. A second possible reason is that ΔE_{exc} is not a good measure of multireference effects, since CCSD(T) and MRCISD+Q differ in the amounts of dynamical and non-dynamical correlation that they account for (see Computational Strategy section).

5.3 Correlation of diagnostics with raw errors in atomization energy

Many benchmark studies of thermochemical methods use the raw error (with units of kJ/mol) in energetic properties as the performance metric rather than fractional error (which is unitless). Some examples include studies by Jiang et al. [22], Martin and co-workers [17], and Pople, Curtiss, and co-workers [68, 69]. If we use raw errors in our analysis of diagnostics, then $1 - c_0^2$ appears to perform best, as it has the smallest $\mathrm{RMS}_{\mathrm{neg}}$ (1.9 kJ/mol) and the largest contrast ratio (2.7), although it has more false negatives than MOND, %(T), ΔE_{exc}, or E_{vert}. Combinations of the $1 - c_0^2$ and MOND diagnostics yield slightly better performance, although these diagnostics would be impractical since they require expensive MRCI calculations. Combining the $1 - c_0^2$ and %(T) diagnostics offers no improvement over $1 - c_0^2$ alone. All of the individual diagnostics divide the molecules into two groups with means, variances, or both that are statistically significantly different. Tables of diagnostic performance and statistical significance using raw energies in the analysis can be found in Online Resource 1, Tables SI-5 and SI-6.

Although one might conclude that $1 - c_0^2$ should be the diagnostic of choice based on the analysis of raw errors, this conclusion may not be general. Raw energetic properties scale with the size of the system. For example, Feller et al. [66] have shown that for n-alkanes, the following properties scale with molecule size: DBOC, T − (T) correction, higher-order corrections, and uncertainty of the atomization energy in the CCSD(T)/CBS limit. Similarly, our data show correlation between the raw error and the atomization energy (plotted in Online Resource 1, Figure SI-5). However, none of the diagnostics studied in this work are size-extensive. Therefore, as system size increases, the absolute error in atomization energy will increase without a corresponding increase in the diagnostic, and any relationship between the two will be lost. In contrast, the fractional error in atomization energy has the same scaling behavior as the diagnostics. For this reason, we base our conclusions on the analysis using fractional errors.

6 Conclusions

We have investigated how wavefunction diagnostics can be used to make quantitative estimates of the uncertainty on computed atomization energies. For our test set of 50 molecules and eight diagnostics ($1 - c_0^2$, MOND, T_1, D_1, %(T), ΔE_{exc}, RHF HOMO–LUMO gap, and E_{vert}), we find that %(T) performs best according to our metrics: lowest fraction of negatives/positives which are false, lowest rms of f for molecules below the diagnostic threshold, and largest ratio between the rms of f for molecules above and below the diagnostic threshold. Six simple combinations (arithmetic mean, geometric mean, harmonic mean, quadratic mean, max, min) of pairs of normalized diagnostics ($1 - c_0^2$, MOND, T_1, D_1, %(T)) did not perform better than %(T) alone.

On the basis of our %(T) data, we estimate the uncertainty on RHF-UCCSD(T*)-F12a/VTZ-F12 atomization energies, including core-valence, spin–orbit, and anharmonic ZPE corrections, to be ±0.4 % when %(T) < 10 %, and ±1.5 % when %(T) > 10 % (2 × RMS in each region). The uncertainties on atomization energies from other computational models may be different.

Acknowledgments This research was performed while MKS held a National Research Council Research Associateship Award at the National Institute of Standards and Technology. We thank the anonymous reviewers for copious suggestions and patience, and Drs. Russell D. Johnson III, Thomas C. Allison, and Yamil Simon for helpful discussions and comments on the manuscript.

References

1. Raghavachari K, Trucks GW, Pople JA, Head-Gordon M (1989) A 5th-order perturbation comparison of electron correlation theories. Chem Phys Lett 157(6):479–483. doi:10.1016/s0009-2614(89)87395-6
2. Watts JD, Bartlett RJ (1993) Triple excitations in coupled-cluster theory: energies and analytical derivatives. Int J Quantum Chem 48(Supplement 27):51–66. doi:10.1002/qua.560480809
3. Dunning TH Jr (2000) A road map for the calculation of molecular binding energies. J Phys Chem A 104(40):9062–9080. doi:10.1021/jp001507z
4. Peterson KA, Feller D, Dixon DA (2012) Chemical accuracy in ab initio thermochemistry and spectroscopy: current strategies and future challenges. Theor Chem Acc 131(1):1–20. doi:10.1007/s00214-011-1079-5
5. Feller D, Peterson KA, Dixon DA (2012) Further benchmarks of a composite, convergent, statistically calibrated coupled-cluster-based approach for thermochemical and spectroscopic studies. Mol Phys 110(19–20):2381–2399. doi:10.1080/00268976.2012.684897
6. Dixon DA, Feller D, Peterson KA (2012) Chapter one—a practical guide to reliable first principles computational thermochemistry predictions across the periodic table. In: Wheeler RA (ed) Annual reports in computational chemistry, vol 8. Elsevier, pp 1–28. doi:10.1016/B978-0-444-59440-2.00001-6
7. Hättig C, Klopper W, Köhn A, Tew DP (2011) Explicitly correlated electrons in molecules. Chem Rev 112(1):4–74. doi:10.1021/cr200168z
8. Peterson KA, Adler TB, Werner H-J (2008) Systematically convergent basis sets for explicitly correlated wavefunctions: the atoms H, He, B–Ne, and Al–Ar. J Chem Phys 128(8):084102. doi:10.1063/1.2831537
9. Hill JG, Peterson KA, Knizia G, Werner HJ (2009) Extrapolating MP2 and CCSD explicitly correlated correlation energies to the complete basis set limit with first and second row correlation consistent basis sets. J Chem Phys 131(19):194105. doi:10.1063/1.3265857
10. Johnson RD III (2013) NIST Computational Chemistry Comparison and Benchmark Database, NIST Standard Reference Database Number 101, Release 16a, August 2013. National Institute of Standards and Technology. http://cccbdb.nist.gov
11. Jiang WY, DeYonker NJ, Wilson AK (2012) Multireference character for 3d transition-metal-containing molecules. J Chem Theory Comput 8(2):460–468. doi:10.1021/ct2006852
12. Lee TJ (2003) Comparison of the T_1 and D_1 diagnostics for electronic structure theory: a new definition for the open-shell D_1 diagnostic. Chem Phys Lett 372(3–4):362–367. doi:10.1016/S0009-2614(03)00435-4
13. Lee TJ, Taylor PR (1989) A diagnostic for determining the quality of single-reference electron correlation methods. Int J Quantum Chem 36:199–207. doi:10.1002/qua.560360824
14. Schultz NE, Gherman BF, Cramer CJ, Truhlar DG (2006) Pd_nCO ($n = 1,2$): accurate ab initio bond energies, geometries, and dipole moments and the applicability of density functional theory for fuel cell modeling. J Phys Chem B 110(47):24030–24046. doi:10.1021/jp064467t
15. Tishchenko O, Zheng JJ, Truhlar DG (2008) Multireference model chemistries for thermochemical kinetics. J Chem Theory Comput 4(8):1208–1219. doi:10.1021/ct800077r
16. Zhao Y, Schultz NE, Truhlar DG (2006) Design of density functionals by combining the method of constraint satisfaction with parametrization for thermochemistry, thermochemical kinetics, and noncovalent interactions. J Chem Theory Comput 2(2):364–382. doi:10.1021/ct0502763

17. Karton A, Rabinovich E, Martin JML, Ruscic B (2006) W4 theory for computational thermochemistry: in pursuit of confident sub-kJ/mol predictions. J Chem Phys 125(14):144108. doi:10.1063/1.2348881
18. Gordon MS, Schmidt MW, Chaban GM, Glaesemann KR, Stevens WJ, Gonzalez C (1999) A natural orbital diagnostic for multiconfigurational character in correlated wave functions. J Chem Phys 110(9):4199–4207. doi:10.1063/1.478301
19. Fogueri U, Kozuch S, Karton A, Martin JL (2012) A simple DFT-based diagnostic for nondynamical correlation. Theor Chem Acc 132(1):1–9. doi:10.1007/s00214-012-1291-y
20. Boguslawski K, Tecmer P, Legeza Ö, Reiher M (2012) Entanglement measures for single- and multireference correlation effects. J Phys Chem Lett 3(21):3129–3135. doi:10.1021/jz301319v
21. Hanrath M (2008) Wavefunction quality and error estimation of single- and multi-reference coupled-cluster and CI methods: the H(4) model system. Chem Phys Lett 466(4–6):240–246. doi:10.1016/j.cplett.2008.10.046
22. Jiang W, DeYonker NJ, Determan JJ, Wilson AK (2011) Toward accurate theoretical thermochemistry of first row transition metal complexes. J Phys Chem A 116(2):870–885. doi:10.1021/jp205710e
23. Jiang W, Jeffrey CC, Wilson AK (2012) Empirical correction of nondynamical correlation energy for density functionals. J Phys Chem A 116(40):9969–9978. doi:10.1021/jp305341a
24. Zhao Y, Tishchenko O, Gour JR, Li W, Lutz JJ, Piecuch P, Truhlar DG (2009) Thermochemical kinetics for multireference systems: addition reactions of ozone. J Phys Chem A 113(19):5786–5799. doi:10.1021/jp811054n
25. Nielsen IMB, Janssen CL (1999) Double-substitution-based diagnostics for coupled-cluster and Møller–Plesset perturbation theory. Chem Phys Lett 310(5–6):568–576. doi:10.1016/S0009-2614(99)00770-8
26. Cramer CJ (2004) Essentials of computational chemistry: theory and models, 2nd edn. Wiley, New York
27. Bak KL, Jorgensen P, Olsen J, Helgaker T, Klopper W (2000) Accuracy of atomization energies and reaction enthalpies in standard and extrapolated electronic wave function/basis set calculations. J Chem Phys 112(21):9229–9242. doi:10.1063/1.481544
28. Schmidt MW, Gordon MS (1998) The construction and interpretation of MCSCF wavefunctions. Annu Rev Phys Chem 49:233–266. doi:10.1146/annurev.physchem.49.1.233
29. Veryazov V, Malmqvist PA, Roos BO (2011) How to select active space for multiconfigurational quantum chemistry? Int J Quantum Chem 111(13):3329–3338. doi:10.1002/qua.23068
30. Nguyen MT, Nguyen TL, Ngan VT, Nguyen HMT (2007) Heats of formation of the Criegee formaldehyde oxide and dioxirane. Chem Phys Lett 448(4–6):183–188. doi:10.1016/j.cplett.2007.10.033
31. Szabo A, Ostlund NS (1996) Modern quantum chemistry: introduction to advanced electronic structure theory. Dover Publications, New York
32. Krylov AI (2001) Size-consistent wave functions for bond-breaking: the equation-of-motion spin-flip model. Chem Phys Lett 338(4–6):375–384. doi:10.1016/s0009-2614(01)00287-1
33. Irikura KK, Hudgens JW, Johnson RD III (1995) Spectroscopy of the fluoromethylene radicals HCF and DCF by 2 + 1 resonance enhanced multiphoton ionization spectroscopy and by ab initio calculation. J Chem Phys 103:1303–1308. doi:10.1063/1.469807
34. Hernandez-Lamoneda R, Bartolomei M, Hernandez MI, Campos-Martinez J, Dayou F (2005) Intermolecular potential of the O_2-O_2 dimer. An ab initio study and comparison with experiment. J Phys Chem A 109(50):11587–11595. doi:10.1021/jp053728g
35. Shiozaki T, Knizia G, Werner HJ (2011) Explicitly correlated multireference configuration interaction: MRCI-F12. J Chem Phys 134(3):034113. doi:10.1063/1.3528720
36. Langhoff SR, Davidson ER (1974) Configuration interaction calculations on nitrogen molecule. Int J Quantum Chem 8(1):61–72. doi:10.1002/qua.560080106
37. Davidson ER, Silver DW (1977) Size consistency in the dilute helium gas electronic structure. Chem Phys Lett 52(3):403–406. doi:10.1016/0009-2614(77)80475-2
38. Ivanov VV, Lyakhy DI, Adamowicz L (2005) New indices for describing the multi-configurational nature of the coupled cluster wave function. Mol Phys 103(15–16):2131–2139. doi:10.1080/00268970500083283
39. Hehre WJ, Radom L, von Schleyer R P, Pople JA (1986) Ab initio molecular orbital theory. Wiley, New York
40. Knizia G, Adler TB, Werner H-J (2009) Simplified CCSD(T)-F12 methods: theory and benchmarks. J Chem Phys 130(5):054104–054120. doi:10.1063/1.3054300
41. Knowles PJ, Hampel C, Werner H-J (1993) Coupled cluster theory for high spin, open shell reference wave functions. J Chem Phys 99(7):5219–5227. doi:10.1063/1.465990
42. Knowles PJ, Hampel C, Werner H-J (2000) Erratum: "Coupled cluster theory for high spin, open shell reference wave functions" [J. Chem. Phys. **99**, 5219 (1993)]. J Chem Phys 112(6):3106–3107. doi:10.1063/1.480886
43. Bross DH, Hill JG, Werner H-J, Peterson KA (2013) Explicitly correlated composite thermochemistry of transition metal species. J Chem Phys 139(9):094302. doi:10.1063/1.4818725
44. Tew DP, Klopper W, Neiss C, Hättig C (2007) Quintuple-zeta quality coupled-cluster correlation energies with triple-zeta basis sets. PCCP 9(16):1921–1930. doi:10.1039/b617230j
45. Hill JG, Mazumder S, Peterson KA (2010) Correlation consistent basis sets for molecular core-valence effects with explicitly correlated wave functions: the atoms B–Ne and Al–Ar. J Chem Phys 132(5):054108. doi:10.1063/1.3308483
46. Varandas AJC (2008) Extrapolation to the complete-basis-set limit and the implications of avoided crossings: the X $^1\Sigma^+{}_g$, B $^1\Delta_g$, and B' $^1\Sigma^+{}_g$ states of C_2. J Chem Phys 129(23):234103. doi:10.1063/1.3036115
47. Ruscic B (2014) Active Thermochemical Tables (ATcT) values based on version 1.112 of the Thermochemical Network. http://atct.anl.gov/index.html
48. Werner HJ, Knowles PJ, Knizia G, Manby FR, Schütz M, Celani P, Korona T, Lindh R, Mitrushenkov A, Rauhut G, Shamasundar KR, Adler TB, Amos RD, Bernhardsson A, Berning A, Cooper DL, Deegan MJO, Dobbyn AJ, Eckert F, Goll E, Hampel C, Hesselmann A, Hetzer G, Hrenar T, Jansen G, Köppl C, Liu Y, Lloyd AW, Mata RA, May AJ, McNicholas SJ, Meyer W, Mura ME, Nicklaß A, O'Neill DP, Palmieri P, Peng D, Pflüger K, Pitzer R, Reiher M, Shiozaki T, Stoll H, Stone AJ, Tarroni R, Thorsteinsson T, Wang M (2012) MOLPRO, version 2012.1, a package of ab initio programs
49. Frisch MJ, Trucks GW, Schlegel HB, Scuseria GE, Robb MA, Cheeseman JR, Scalmani G, Barone V, Mennucci B, Petersson GA, Nakatsuji H, Caricato M, Li X, Hratchian HP, Izmaylov AF, Bloino J, Zheng G, Sonnenberg JL, Hada M, Ehara M, Toyota K, Fukuda R, Hasegawa J, Ishida M, Nakajima T, Honda Y, Kitao O, Nakai H, Vreven T, Jr. JAM, Peralta JE, Ogliaro F, Bearpark M, Heyd JJ, Brothers E, Kudin KN, Staroverov VN, Kobayashi R, Normand J, Raghavachari K, Rendell A, Burant JC, Iyengar SS, Tomasi J, Cossi M, Rega N, Millam JM, Klene M, Knox JE, Cross JB, Bakken V, Adamo C, Jaramillo J, Gomperts R, Stratmann RE, Yazyev O, Austin AJ, Cammi R, Pomelli C, Ochterski JW, Martin RL, Morokuma K, Zakrzewski VG, Voth GA, Salvador P, Dannenberg JJ, Dapprich S, Daniels AD, Farkas Ö, Foresman JB, Ortiz JV, Cioslowski J, Fox DJ (2009) Gaussian 09, Revision C.01. Gaussian, Inc.

50. Irikura KK (2007) Experimental vibrational zero-point energies: diatomic molecules. J Phys Chem Ref Data 36(2):389–397. doi:10.1063/1.2436891
51. Neff M, Rauhut G (2009) Toward large scale vibrational configuration interaction calculations. J Chem Phys 131(12):124129. doi:10.1063/1.3243862
52. Vosko SH, Wilk L, Nusair M (1980) Accurate spin-dependent electron liquid correlation energies for local spin density calculations: a critical analysis. Can J Phys 58(8):1200–1211. doi:10.1139/p80-159
53. Roy TK, Gerber RB (2013) Vibrational self-consistent field calculations for spectroscopy of biological molecules: new algorithmic developments and applications. PCCP 15(24):9468–9492. doi:10.1039/C3CP50739D
54. Yagi K, Hirata S, Hirao K (2007) Efficient configuration selection scheme for vibrational second-order perturbation theory. J Chem Phys 127(3):034111. doi:10.1063/1.2748774
55. Barone V (2005) Anharmonic vibrational properties by a fully automated second-order perturbative approach. J Chem Phys 122(1):014108. doi:10.1063/1.1824881
56. Booth GH, Cleland D, Thom AJW, Alavi A (2011) Breaking the carbon dimer: the challenges of multiple bond dissociation with full configuration interaction quantum Monte Carlo methods. J Chem Phys 135(8):084104. doi:10.1063/1.3624383
57. Varandas AJC (2012) Ab initio treatment of bond-breaking reactions: accurate course of HO_3 dissociation and revisit to isomerization. J Chem Theory Comput 8(2):428–441. doi:10.1021/ct200773b
58. Ivanic J, Schmidt MW, Luke B (2012) High-level theoretical study of the NO dimer and tetramer: has the tetramer been observed? J Chem Phys 137(21):214316. doi:10.1063/1.4769226
59. Qi B, Su KH, Wang YB, Wen ZY, Tang XY (1998) Gaussian-2 calculations of the thermochemistry of Criegee intermediates in gas phase reactions. Acta Phys Chim Sin 14(11):1033–1039. doi:10.3866/PKU.WHXB19981114
60. Cremer D, Gauss J, Kraka E, Stanton JF, Bartlett RJ (1993) A CCSD(T) investigation of carbonyl oxide and dioxirane—equilibrium geometries, dipole-moments, infrared-spectra, heats of formation and isomerization energies. Chem Phys Lett 209(5–6):547–556. doi:10.1016/0009-2614(93)80131-8
61. Wenthold PG (2010) Thermochemical Properties of the Benzynes. Aust J Chem 63(7):1091–1098. doi:10.1071/ch10126
62. Karton A, Kettner M, Wild DA (2013) Sneaking up on the Criegee intermediate from below: predicted photoelectron spectrum of the CH_2OO^- anion and W3-F12 electron affinity of CH_2OO. Chem Phys Lett 585:15–20. doi:10.1016/j.cplett.2013.08.075
63. Welch BL (1947) The generalization of 'Student's' problem when several different population variances are involved. Biometrika 34(1–2):28–35. doi:10.1093/biomet/34.1-2.28
64. NIST/SEMATECH e-Handbook of Statistical Methods, Sect. 1.3.5.10—Levene Test for Equality of Variances. http://www.itl.nist.gov/div898/handbook/eda/section3/eda35a.htm. Accessed 7 Sept 2014
65. Brown MB, Forsythe AB (1974) Robust tests for the equality of variances. J Am Stat Assoc 69(346):364–367. doi:10.1080/01621459.1974.10482955
66. Feller D, Peterson KA, Dixon DA (2008) A survey of factors contributing to accurate theoretical predictions of atomization energies and molecular structures. J Chem Phys 129(20):32. doi:10.1063/1.3008061
67. Karton A, Daon S, Martin JML (2011) W4-11: a high-confidence benchmark dataset for computational thermochemistry derived from first-principles W4 data. Chem Phys Lett 510(4–6):165–178. doi:10.1016/j.cplett.2011.05.007
68. Pople JA, Head-Gordon M, Fox DJ, Raghavachari K, Curtiss LA (1989) Gaussian-1 theory—a general procedure for prediction of molecular-energies. J Chem Phys 90(10):5622–5629. doi:10.1063/1.456415
69. Curtiss LA, Redfern PC, Raghavachari K (2007) Gaussian-4 theory. J Chem Phys 126(8):084108. doi:10.1063/1.2436888

Theor Chem Acc (2014) 133:1438
DOI 10.1007/s00214-013-1438-5

REGULAR ARTICLE

What makes differences between intra- and inter-molecular charge transfer excitations in conjugated long-chained polyene? EOM-CCSD and LC-BOP study

Jong-Won Song · Kimihiko Hirao

Received: 27 September 2013 / Accepted: 5 December 2013 / Published online: 20 December 2013

Abstract We performed intra- and inter-molecular charge transfer (CT) excitation calculations of H_2N–$(CH{=}CH)_n$–NO_2 (**a**) and its equidistant H_2N–$H\cdots H$–NO_2 (**b**) using EOM-CCSD ($n = 1$–9), time-dependent (TD) long-range corrected (LC) density functional theory (DFT) ($n = 1$–10). It was shown that LC-BOP and LCgau-BOP outperform all the tested DFT functionals on inter- and intra-CT excitation energy and oscillator strength, regardless of CT interaction distance (R). Decomposition of TD-DFT optical excitation energies of (**a**) and (**b**) into HOMO–LUMO gap and excitonic binding energy disclosed that HOMO–LUMO gap reduction resulting from delocalization of HOMO and LUMO through bridged polyene conjugation is mainly responsible for the decreasing of intra-molecular CT excitation energy with chain number, while inter-molecular CT increases linearly with $-1/R$, which is wholly due to the decrease in excitonic energy between HOMO and LUMO. We found that success of exchange correlation functional on long-distanced intra-molecular CT calculations depends on correct descriptions of (1) Koopmans' energy of donor and acceptor and (2) excitonic energy between donor and acceptor, and (3) correct far-nucleus asymptotic behavior, $-1/R$. We found that LC scheme can satisfy (3), but needs an appropriate choice of long-range parameter able to satisfy (1) and (2). On the other hand, the pure, conventional hybrid, and screened hybrid functionals show near-zero intra- and inter-molecular excitonic energy regardless of R, which means optical band gap coincide with HOMO–LUMO gap. Therefore, we conclude that 100 % long-range Hartree–Fock exchange inclusion is indispensable for correct descriptions of intra-molecular CT excitations as well as inter-molecular CT.

Keywords Range separation · Long-range corrected DFT · Charge transfer excitation · TD-DFT · HOMO–LUMO gap · Excitonic energy · LC-DFT

1 Introduction

Charge transfer (CT) excitation is a core mechanism in photochemical process, such as photosynthesis, nonlinear optics, and photovoltaics. Modern researches on photochemistry demand theoretical chemists' well-predictive theoretical tools for inter-molecular and intra-molecular CT (inter-CT and intra-CT, respectively) excitation energies on both finite systems and condensed state materials [1, 2]. When we try to invent well-predictive theoretical methods, we should keep in mind that comprehending theoretical reasons behind the experimental phenomena precedes suggesting appropriate theoretical methods.

We have been well-acquainted with failures of conventional density functional theory (DFT) [3] in predicting meaningful inter-CT excitations, which result from incompetence of DFT exchange correlation functional to provide correct far-nucleus asymptotic behavior like

Dedicated to Professor Thom Dunning and published as part of the special collection of articles celebrating his career upon his retirement.

Electronic supplementary material The online version of this article (doi:10.1007/s00214-013-1438-5) contains supplementary material, which is available to authorized users.

J.-W. Song · K. Hirao (✉)
Computational Chemistry Unit, RIKEN Advanced Institute for Computational Science, 7-1-26, Minatojima-minami-machi, Chuo-ku, Kobe, Hyogo 650-0047, Japan
e-mail: hirao@riken.jp

$$\lim_{R\to\infty} v_{xc}^{\sigma}(\mathbf{R}) = -\frac{1}{R}, \quad (1)$$

where v_{xc}^{σ} is exchange correlation potential, $\mathbf{R}$ is the distance vector between donor (D) and acceptor (A), and $R = |\mathbf{R}|$ [4, 5]. Recently, we suggested the long-range corrected (LC) [6, 7] DFT whose exchange potential operator is separated into DFT and Hartree–Fock (HF) exchange using an error function as

$$\frac{1}{r_{12}} = \underbrace{\frac{1-\mathrm{erf}(\mu r_{12})}{r_{12}}}_{\mathrm{DFT}} + \underbrace{\frac{\mathrm{erf}(\mu r_{12})}{r_{12}}}_{\mathrm{HF}} \quad (2)$$

LC-DFT can provide correct asymptotic behavior of Eq. (1) and, therefore, can reproduce inter-CT excitation energy and its oscillator strength successfully [8].

However, when it comes to "intra-molecular" CT excitation, several questions remain about applicability of DFT functionals to intra-CT excitations. One of them is why intra-CT excitation energy decreases as R increases while inter-CT energy increases having an asymptotic behavior of Eq. (1) (see Fig. 2) [9–12]. Moreover, there has been no rigorous study to answer about whether 100 % long-range HF exchange of LC-DFT is indispensable for describing correct intra-CT excitations. Actually, some reported that Coulomb-attenuating method (CAM) B3LYP [13] which does not have a 100 % HF exchange in long range can reproduce intra-CT excitations [14–16]. In addition, there are some reports that smaller-range separation parameter (μ) than our determined value (0.47) is favored for intra-CT excitation of polyacene, polythiophene, polyparaphenylene, and so on [11, 12, 17–19]. However, to the best of our knowledge, no rigorous theoretical study was performed to find out whether 100 % long-range HF exchange is indispensable for describing correct intra-CT excitations of long-chained conjugated molecules or not and why intra-CT energy decreases with R, contrary to inter-CT excitation energy.

In this sense, polyene which has double bond conjugation is thought to be an appropriate system to study these points [20, 21]. However, it has not been studied in comparison with high-accuracy electron correlation method, such as EOM-CCSD method [22–27], even though it is the simplest conjugation structure in chemistry. Of course, EOM-CCSD method is not an absolute reference for excitation energy and oscillator strength calculations of polyene [28], and EOM-CCSD is reported not to be perfect $-1/R$ asymptotic behavior of charge transfer excitation energies due to its not fully size-extensive problem. However, EOM-CCSD is a valuable accessible method to examine CT excitations of TD-DFT. In addition, numerical deviations from $-1/R$ asymptotic behavior are expected to be small enough to be neglected [29].

In this article, we will perform high-accuracy EOM-CCSD calculations of intra-CT excitations of polyene having $-NH_2$ and $-NO_2$ as electron–donor and acceptor with increasing chain number. Then, we will check whether 100 % long-range HF exchange is essential in intra-CT energy by comparing the results of a variety of LC-DFT, CAM-B3LYP, and conventional hybrid functionals with EOM-CCSD ones. Then, we will decompose excitation energy obtained using time-dependent (TD) DFT into Koopmans' orbital energies and excitonic binding energy [10] to find out what makes intra-CT excitation energy differ from inter-CT excitation energy and what factors are essential for exchange correlation functional to describe intra-CT excitations correctly.

2 Theory

From Refs. [12, 17], we can approximately get optical CT excitation gap between HOMO and LUMO as

$$\omega_{\mathrm{CT}}^{\mathrm{TD}} = \varepsilon_{\mathrm{LUMO}} - \varepsilon_{\mathrm{HOMO}} - (\mathrm{HH}|O(r_{12})|\mathrm{LL}) + (\mathrm{HL}|O(r_{12})|\mathrm{HL}). \quad (3)$$

Here, $O(r_{12})$ is HF exchange operator. For example, $O(r_{12})$ of B3LYP is c_x/r_{12} ($c_x = 0.2$), that of LC-DFT is $\mathrm{erf}(\mu r_{12})/r_{12}$, that of CAM-B3LYP is $[\alpha + \beta\mathrm{erf}(\mu r_{12})]/r_{12}$, and that of pure functional is 0. Normally, optical CT excitation gap obtained using TD-DFT is smaller than $I^{\mathrm{D}} - A^{\mathrm{A}}$ fundamental gap (I^{D} is the vertical ionization potential of the D and A^{A} is the vertical electron affinity of the A). The difference is known to result from binding energy between the interacting excited electron and hole pair which is called exciton. Here, we will call this excitonic binding energy. Therefore, excitonic binding energy between D and A is

$$E^{\mathrm{exciton}} = \omega_{\mathrm{CT}}^{\mathrm{TD}} - (I^{\mathrm{D}} - A^{\mathrm{A}}), \quad (4)$$

where E^{exciton} is excitonic energy between two orbitals (e.g., HOMO and LUMO) [10, 30].

In LC-DFT calculations, it is reported that negative Kohn–Sham (KS) orbital energy of HOMO is close to I^{D} and that of LUMO is close to A^{A}. In our previous study, we showed that HOMO–LUMO gaps of tested molecules predicted by LC-BOP and LCgau-BOP are very close to their corresponding fundamental gaps calculated by CCSD(T) [31]. We also showed KS HOMO of LC functionals appropriately incorporates both relaxation and correlation effects. Hence, we can assume that LC-DFT can provide applicable $I^{\mathrm{D}} - A^{\mathrm{A}}$ fundamental gaps with KS HOMO–LUMO gaps. Therefore, in the exact DFT exchange correlation functional,

$$(I^{\mathrm{D}} - A^{\mathrm{A}}) = \varepsilon_{\mathrm{LUMO}} - \varepsilon_{\mathrm{HOMO}}. \quad (5)$$

and

$$E^{\text{exciton}} = \omega_{\text{CT}}^{\text{TD}} - (\varepsilon_{\text{LUMO}} - \varepsilon_{\text{HOMO}}). \tag{6}$$

In CT excitation case,

$$E^{\text{exciton}} = -(\text{HH}|O(r_{12})|\text{LL}) \tag{7}$$

because (HL|$O(r_{12})$|HL) will disappear quickly with increase in R, in that overlap between HOMO and LUMO is too small. However, excitonic energies of pure and screened hybrid DFT functionals become zero, in that (HH|$O(r_{12})$|LL) also disappear quickly. In case of LC-DFT, (HH|$O(r_{12})$|LL) becomes $-1/R$. Therefore, inter-CT excitations, excitonic energy of LC functionals [31–33] will be like

$$E^{\text{exciton}} \approx -1/R. \tag{8}$$

3 Computational details

To get appropriate geometries of amino (NH_2–) and nitro (NO_2–) end-caped polyene for intra-CT excitations, we optimized geometries of H_2N–$(CH{=}CH)_n$–NO_2 ($n = 1$–10) using LCgau-BOP/cc-pVDZ [34]. LCgau functional is one of the LC functionals, in which the short-range HF exchange integral is included by incorporating a Gaussian attenuation with two parameters (a and k). In G2 set [35] and polyene geometry test, LCgau-BOP was found to provide the lowest error [36, 37] and good bond length alterations [38], respectively. With LCgau-BOP-optimized geometries, we performed EOM-CCSD calculations of H_2N–$(CH{=}CH)_n$–NO_2 using cc-pVDZ ($n = 1$–9) and cc-pVTZ ($n = 1$–5) basis sets to obtain standard intra-CT excitations. We also note that there is a recent report saying that CAM-B3LYP reproduces BLA values of polyenes without end-caps most close to CCSD(T)/6-31G* [39].

To make geometries for inter-CT excitation between –NH_2 and –NO_2 with distances identical to those of intra-CT excitation, we first eliminated bridged polyene with –NH_2 and –NO_2 end-caps remained and added –H with the bond distance of 1.0 Å to –NH_2 and –NO_2. Then, we performed EOM-CCSD calculations to obtain inter-CT excitations between NH_3 and HNO_2.

Then, we performed TD-DFT calculations ($n = 1$–10) of H_2N–$(CH{=}CH)_n$–NO_2 and its equidistant H_2N–H···H–NO_2 using cc-pVTZ basis set. Exchange correlation functionals assessed for CT excitations are LC-BOP [40–42], LC-BOP12 [37], LCgau-BOP [34], LCgau-B97 [19], LC(0.33)-BOP [8], LC-ωPBE [43], ωB97X [44], CAM-B3LYP [13], B3LYP [45], HSE [46], Gau-PBEh [47], and PBE [48] functional. Note that optimized range separation parameters (μ) used for LC-BOP, LC-BOP12, LCgau-BOP, LC-ωPBE, ωB97X, LC(0.33)-BOP and CAM-B3LYP, and LCgau-B97 are $\mu = 0.47$, $\mu = 0.42$, $\mu = 0.42$, $\mu = 0.4$, $\mu = 0.3$, $\mu = 0.33$, and $\mu = 0.20$, respectively.

The LCgau scheme and OP correlation functional [49] calculations were performed on the development version of Gaussian09 program, and other calculations were executed on the official version of Gaussian09 program [50].

4 Results and discussions

4.1 Oscillator strength

Figure 1 shows the oscillator strengths obtained using HF, CIS, EOM-CCSD, and various TD-DFT calculations. As seen in the figure, calculated oscillator strengths can be classified into three groups: the first group is underestimated ones of PBE, B3LYP, HSE, and Gau-PBEh, the second is slightly overestimated ones of HF and CIS, and the third is appropriately estimated ones of LC functionals, such as LC-BOP, LC-BOP12, LCgau-BOP, LCgau-B97, and CAM-B3LYP. This shows that LC scheme is essential for obtaining correct oscillation strengths. Of the well estimated ones, CAM-B3LYP shows the smallest oscillator strengths. LC-BOP12 and LCgau-BOP with $\mu = 0.42$ show the closest oscillator strengths to EOM-CCSD/cc-pVTZ ones.

We should note that the strongest transitions are CT excitations between HOMO and LUMO in $n = 1$–10, but, from some larger chain number, excitations showing the strongest oscillator strength can be transferred to other CT or local transition band as expected in Ref. [21]. For example, in the EOM-CCSD excitations from HOMO to LUMO + 1 which has inner double bond character,

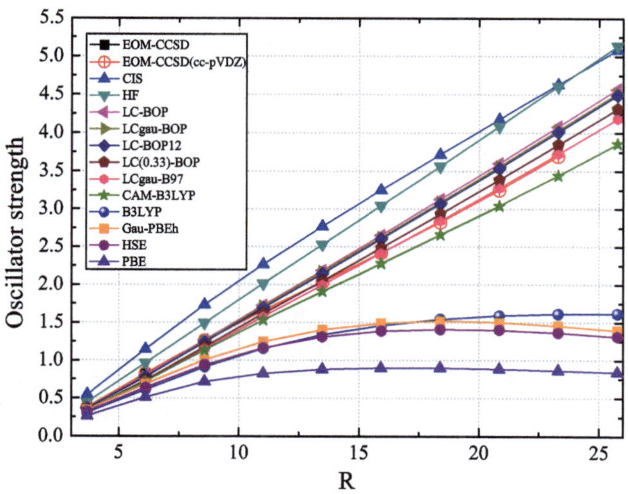

Fig. 1 Oscillator strengths of CT excitations calculated using TD-HF, CIS, EOM-CCSD, and various TD-DFTs over the distance (R; Å) between –NH_2 and –NO_2 of polyene ($n = 1$–10)

Table 1 CT excitation energies (eV) between NH_3 and HNO_2 calculated using TD-HF, CIS, EOM-CCSD, and various TD-DFTs

n	Distance, R	EOM-CCSD	CIS	HF	LC-BOP	LCgau-BOP	LC-ωPBE	ωB97X	LCgau-B97	CAM-B3LYP	B3LYP	Gau-PBEh	HSE	PBE
1	3.67721	7.78	9.54	9.54	7.26	7.00	6.77	6.22	5.05	5.20	3.46	4.78	3.72	2.04
2	6.11327	8.91	10.47	10.47	8.20	7.95	7.73	7.17	5.86	5.69	3.36	4.43	3.54	1.71
3	8.56319	9.45	10.99	10.99	8.72	8.48	8.26	7.70	6.38	6.00	3.40	4.34	3.49	1.63
4	11.01565	9.77	11.31	11.31	9.04	8.80	8.58	8.02	6.70	6.20	3.44	4.31	3.46	1.60
5	13.47082	9.99	11.53	11.53	9.25	9.01	8.79	8.23	6.92	6.33	3.47	4.30	3.45	1.59
6	15.92723	10.14	11.68	11.68	9.41	9.16	8.94	8.38	7.07	6.43	3.50	4.29	3.44	1.58
7	18.38685	10.25	11.79	11.79	9.52	9.28	9.05	8.49	7.18	6.50	3.52	4.28	3.44	1.58
8	20.84318	10.34	11.88	11.88	9.60	9.36	9.14	8.58	7.27	6.55	3.53	4.28	3.43	1.58
9	23.30343	10.41	11.95	11.95	9.67	9.43	9.21	8.65	7.34	6.60	3.54	4.28	3.43	1.57
10	25.76305	10.46	12.00	12.00	9.73	9.49	9.27	8.70	7.39	6.63	3.55	4.28	3.43	1.57

oscillator strength of polyene with $n = 3$ is 0.0477, but that with $n = 9$ is 0.6097.

4.2 Excitation energy

Tables 1 and 2 show inter-CT excitation energies increase as R increases, but intra-CT energies decrease. Figure 2 shows that these behaviors of both intra- and inter-CT energies are linearly dependent on $1/R$. It is noticeable that deviations of intra-CT energies of various DFT functionals from EOM-CCSD are smaller than those of inter-CT energies, in that intra-CT energies are smaller than inter-CT energies.

Interestingly, HF, CIS, and LC functionals with various μ values show almost identical slopes of inter-CT energies versus $1/R$. On the other hand, the DFT functionals without 100 % long-range HF exchange show slower slopes to $-1/R$ than LC functionals. For example, CAM-B3LYP shows a slower slope than LC functionals. Screened hybrid functionals, such as Gau-PBEh and HSE, and pure functional, PBE, show near-zero slope versus $1/R$. In contrast, all of the calculated intra-CT energies equally decrease linearly with $-1/R$ [9], even though the slopes become stable from $n = 3$–5 (Fig. 2b).

Based on the slope of CT energies versus $1/R$, calculated intra-CT energy results of all the methods can also be classified into three groups, which are similar with those of oscillator strengths: the first ones are DFT functionals with LC scheme, such as PBE, B3LYP, HSE, and Gau-PBEh, which underestimate intra-CT energies and have the steepest slope to $-1/R$, the second one consists of HF and CIS, which show slightly overestimated intra-CT energy and steeper slope to $-1/R$ than LC functionals, and the third ones are LC functionals, which show the slope similar with the EOM-CCSD ones.

Actually, LC functionals also slightly underestimate intra-CT energies, compared with EOM-CCSD/cc-pVDZ. However, intra-CT energies of LC-BOP with $\mu = 0.47$ are closer to EOM-CCSD/cc-pVTZ rather than EOM-CCSD/cc-pVDZ, in that intra-CT energy of EOM-CCSD/cc-pVTZ is lower than EOM-CCSD/cc-pVDZ by about 0.1 eV, even though only CT energies of polyenes of $n = 1$–5 are available for comparison. Therefore, of all the tested functionals, LC-BOP provides both intra- and inter-CT excitation energies closest to EOM-CCSD ones regardless of R. LC-BOP12 and LCgau-BOP show similar tendency for inter- and intra-CT excitation energies with the EOM-CCSD ones. On the other hand, LC functionals with small range separation parameter, such as LCgau-B97 and LC(0.33)-BOP ($\mu = 0.2$ and $\mu = 0.33$, respectively), and CAM-B3LYP with incomplete long-range HF exchange provide underestimated intra- and inter-CT excitation energies. The pure, conventional hybrid and screened hybrid functionals, such as PBE, B3LYP, HSE, and Gau-PBEh, show quite underestimations.

4.3 Decomposition of CT excitation energies to Koopmans and excionic energies

Let us see first inter-CT excitation energies. HOMO–LUMO energy gaps ($\varepsilon_{LUMO} - \varepsilon_{HOMO}$) calculated using all the tested DFT functionals and HF method are nearly constant, independently of change of R (Fig. 3a). However, the inter-CT excitonic energies increase as a $-1/R$ shape and equally converge to zero as R goes to infinity (Fig. 4a). In detail, the inter-CT excitonic energy change with R of LC functionals is the identical and has the same slope versus $-1/R$.[1] In other words, inter-CT excitonic energies of LC functionals are independent of μ parameter. On the other hand, the slopes of inter-CT excitonic energies of B3LYP and CAM-B3LYP versus $-1/R$ are $c^{B3LYP} = 0.2$ and $c^{CAM\text{-}B3LYP} = 0.65$, respectively, which were obtained

[1] See Supplementary Material.

Table 2 CT excitation energies (eV) of polyenes (n = 1–10) calculated using TD-HF, CIS, EOM-CCSD and various TD-DFTs

n	Distance, R	EOM-CCSD (cc-pVTZ)	EOM-CCSD	CIS	HF	LC-BOP	LCgau-BOP	LC-BOP12	LC-ωPBE	ωB97X	LC (0.33)-BOP	LCgau-B97	CAM-B3LYP	B3LYP	Gau-PBEh	HSE	PBE
1	3.67721	5.25	5.43	6.14	5.89	5.28	5.25	5.21	5.22	5.15	5.04	5.05	5.04	4.85	5.09	4.94	4.62
2	6.11327	4.67	4.80	5.30	5.06	4.59	4.65	4.57	4.58	4.51	4.43	4.37	4.36	4.11	4.30	4.16	3.84
3	8.56319	4.25	4.36	4.72	4.50	4.19	4.15	4.11	4.13	4.05	3.99	3.90	3.87	3.54	3.72	3.59	3.26
4	11.01565	3.97	4.07	4.34	4.13	3.89	3.85	3.83	3.83	3.74	3.69	3.58	3.53	3.13	3.29	3.16	2.82
5	13.47082	3.79	3.88	4.08	3.88	3.69	3.64	3.63	3.63	3.53	3.49	3.34	3.29	2.82	2.96	2.83	2.47
6	15.92723		3.76	3.89	3.70	3.55	3.50	3.49	3.48	3.37	3.34	3.17	3.11	2.57	2.69	2.56	2.19
7	18.38685		3.67	3.76	3.57	3.45	3.39	3.39	3.38	3.26	3.23	3.05	2.97	2.37	2.48	2.34	1.96
8	20.84318		3.61	3.66	3.47	3.37	3.31	3.31	3.30	3.18	3.16	2.96	2.88	2.21	2.30	2.16	1.78
9	23.30343		3.56	3.58	3.39	3.31	3.25	3.25	3.24	3.11	3.09	2.88	2.80	2.08	2.16	2.01	1.62
10	25.76305			3.52	3.33	3.26	3.20	3.20	3.19	3.06	3.04	2.83	2.75	1.98	2.04	1.88	1.49

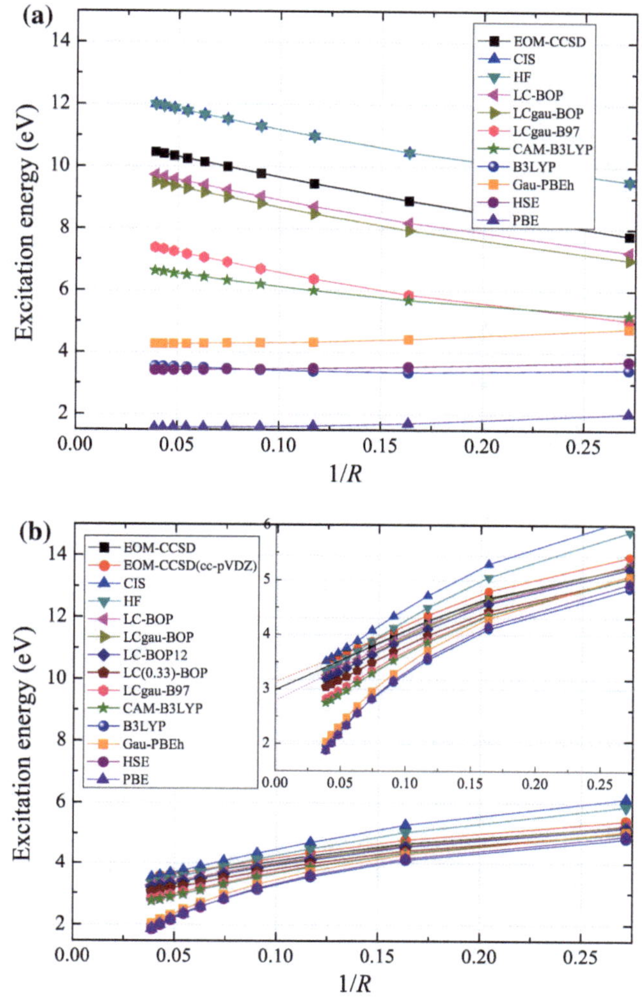

Fig. 2 Dependence of CT excitation energies (eV) calculated using TD-HF, CIS, EOM-CCSD, and various TD-DFTs over $1/R$ (Å^{-1}) **a** between $-NH_2$ and $-NO_2$ of polyene (n = 1–10) and **b** between NH_3 and HNO_2

by estimating slopes of E^{exciton} of Eq. (8) versus $-1/R$ and assuming $c^{\text{LC-DFT}}$ to be 1 [30]. Note that the slopes are exactly identical with the parameter values for HF exchange inclusion of hybrid functionals (Those of B3LYP and CAM-B3LYP are $c_{\text{HF}} = 0.2$ and $\alpha + \beta$=0.65, respectively). The c values of pure and screened hybrid functionals are almost zero. These results indicate that 100 % long-range HF exchange inclusion is indispensable for achieving exact $-1/R$ asymptotic behavior of inter-CT energy.

Meanwhile, we should note that, in inter-CT excitations, the errors from HOMO–LUMO gaps are shown to be larger than those from excitonic energy. Therefore, the performances of DFT exchange correlation functionals on inter-CT excitation energy seem to mainly depend on accurate assumptions of Koopmans' orbital energies of D and A.

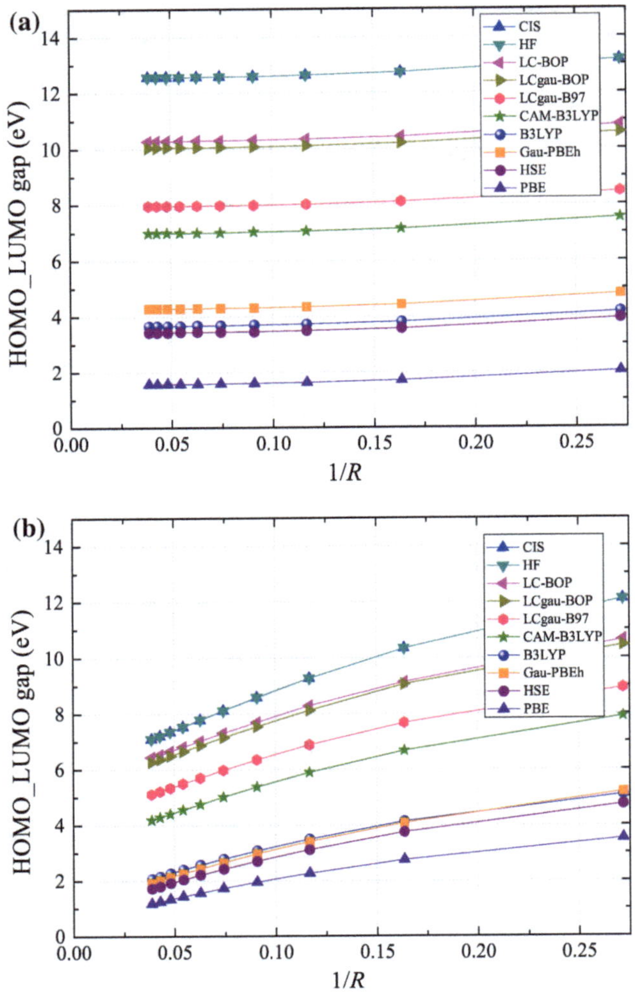

Fig. 3 Dependence of HOMO–LUMO gaps (eV) of HF and various DFTs over $1/R$ (Å^{-1}) **a** between $-NH_2$ and $-NO_2$ of polyaene (n = 1–10) and **b** between NH_3 and HNO_2

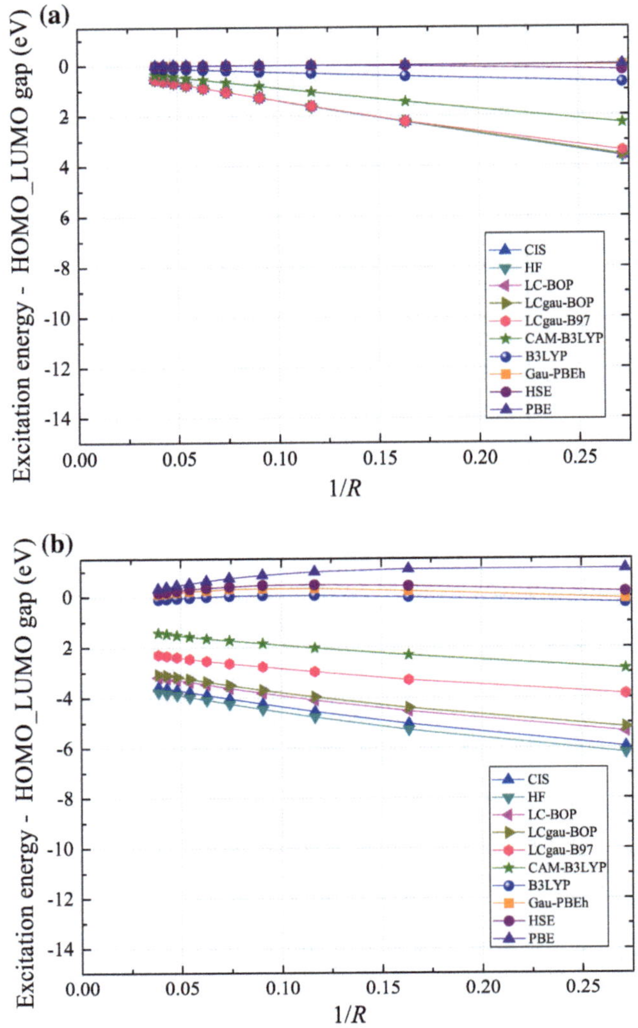

Fig. 4 Differences between CT excitation energies and HOMO–LUMO gaps (eV) over $1/R$ (Å^{-1}) **a** between $-NH_2$ and $-NO_2$ of polyaene (n = 1–10) and **b** between NH_3 and HNO_2

On the other hand, unlike the case of inter-CT excitations, HOMO–LUMO gaps of intra-CT excitations decrease linearly with $-1/R$ (Fig. 3b). This shows that, D and A are connected with conjugated polyene chain in intra-CT excitation cases, and therefore, Koopmans' orbital energies of D and A are deeply affected by conjugation polyene chain length, R. From the point that the excitonic energies normally increase or remain constant with increase in R, we can deduce that decreasing of HOMO–LUMO gaps with R is mainly responsible for decreasing of intra-CT excitation energy with R.

Figure 5 shows why HOMO and LUMO gaps decrease with R in intra-CT excitations. The shapes of HOMO and LUMO of LC-BOP are similar with those of HF, and those of B3LYP are similar with those of PBE. That is, the HOMO and LUMO shapes of PBE and B3LYP are delocalized more than those of HF and LC-BOP. Since orbital delocalization decreases its orbital energy, we can easily rationalize that HOMO and LUMO gaps of PBE and B3LYP are smaller than those of HF and LC-BOP. Therefore, HOMO and LUMO delocalization through bridged double bond results in the decrease in intra-CT excitation energy. Note that the HOMO–LUMO gaps of $H_2N–CH=CH–NO_2$ (n = 1) are very similar with those of inter-CT excitations. This shows that correct estimations of HOMO–LUMO gaps affected by combinations of D and A are indispensable for accurate calculations of intra-CT excitation energy, let alone inter-CT.

As the case of inter-CT excitation calculations, pure, screened hybrid, and conventional hybrid functionals show constantly near-zero excitonic energies, irrespective of R [9] (Fig. 4b). This shows that pure, screened hybrid, and conventional hybrid functionals reproduce optical CT excitation energies which are quite similar with HOMO–LUMO gap [51]. This actually means that no excitonic energy is included in optical CT excitation energies obtained from these functionals irrespective of R. On the

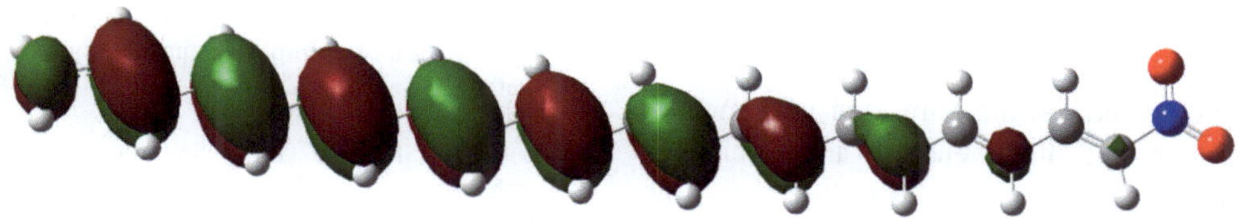

(a)

(b)

(c)

(d)

Fig. 5 Shapes of HOMO of **a** HF, **b** PBE, **c** LC-BOP, and **d** B3LYP

other hand, LC functionals are shown to provide excitonic energy which increases with R. However, unlike inter-CT excitation cases, intra-CT excitonic energies obtained from LC functionals depend on μ parameter. In detail, LC functionals with large μ value show largely negative multitude of excitonic energy.

To explain μ dependency of intra-CT excitonic energy obtained from LC functionals, we can consider (HH|$O(r_{12})$|LL) term in Eq. (7). In inter-CT excitations, HOMO and LUMO are localized to D and A, which makes (HH|$O(r_{12})$|LL) dependent only on the distance, R, between D and A. Consequently, (HH|$O(r_{12})$|LL) will become Coulombic interaction having distance of R, between two point charges centered on D and A. However, in intra-CT excitations, HOMO and LUMO are delocalized broadly through conjugation chain as seen in Fig. 5. Therefore, (HH|$O(r_{12})$|LL) cannot become simply Coulombic interaction between two point charges, but rather (HH|$O(r_{12})$|LL) depends on μ value since $O(r_{12})$ includes μ parameter. Overlapping between HOMO and LUMO of intra-CT excitation is larger than that of inter-CT excitation, which consequently will increase (HH|$O(r_{12})$|LL).

Finally, overall intra-CT excitonic energies obtained from LC functionals are non-negligibly large, in contrast with inter-CT excitation. Therefore, correct estimations of excitonic energy, let alone HOMO–LUMO gap, are important in accurate calculations of intra-CT excitation energy. In addition, in intra-CT excitations, both of excitonic energy and HOMO–LUMO gap are deeply related to μ value.

5 Conclusions

In this study, we compared TD-DFT results of intra-molecular charge transfer excitations between polyene bridged NH_2 and NO_2 and inter-molecular charge transfer excitations between equidistant NH_3 and HNO_2 with EOM-CCSD results and tried to find out why intra-molecular charge transfer energy differs from inter-molecular charge transfer. We found that the differences between inter- and inter-molecular charge transfer excitations result mainly from delocalization of HOMO and LUMO through polyene conjugation chain, which reduces HOMO–LUMO gap and increases the multitude of negative excitonic energy of intra-molecular charge transfer energy. As a result, intra-molecular charge transfer excitation energy decreases. On the other hand, since energy gaps between HOMO and LUMO which are not connected with polyene conjugation bridges do not change with R increases, inter-molecular charge transfer excitation energy is affected only by excitonic energy when R increases.

LC-BOP functional showed the intra- and inter-molecular charge transfer excitation energy and oscillator strength of tested systems closest to EOM-CCSD ones. The successful factors of LC-BOP are inclusion of long-range 100 % HF exchange and appropriate μ value of 0.47 from the point that CAM-B3LYP and LC functionals with small μ value fail in reproducing correct intra- and inter-molecular charge transfer excitation energies.

In inter-molecular charge transfer excitations, the LC scheme with μ larger than 0.2 can provide a correct asymptotic behavior of $-1/R$ between donor and acceptor which provides correct excitonic energy. However, to describe correct intra- and inter-molecular charge transfer excitation energies, a proper choice of μ value for LC scheme is compulsory, in that a proper μ value enables LC scheme to provide not only appropriate HOMO–LUMO gaps for both inter- and intra-molecular charge transfer excitations, but also correct excitonic energy for intra-molecular charge transfer excitations. On the other hand, the pure, conventional hybrid, and screened hybrid functionals show near-zero excitonic energy irrespective of R, which makes no difference between HOMO–LUMO gap and optical charge transfer excitation energy.

From this study, we are keenly aware that correct charge transfer excitation calculations of LC functionals on various chemical systems are deeply dependent on μ value, which is one of the problems the LC scheme is facing [52]. Eagerly hoping to propose an improved LC scheme which is free from a choice of range separation parameter, we will continue to study intra- and inter-molecular charge transfer excitations of various conjugation systems and with various combinations of donor and acceptor end-caps using LC-DFT functionals as a future work. In addition, it was reported that differences between EOM-CCSD and EOM-CC3 [53] in polyenes without end-caps are not so small [54]. Therefore, we also hope to assess TD-DFT functionals with higher electron correlation methods, such as EOM-CC3, for long-chained $-NH_2$ and $-NO_2$ end-capped polyenes in the near future.

Acknowledgments We would like to pay honors to Prof. Dunning for his great achievements. This research was supported by the Japan Society for the Promotion of Science (JSPS) KAKENHI Grant Number 23225001. The numerical calculations were conducted on the RIKEN Cluster of Clusters (RICC).

References

1. Jono R, Fujisawa J, Segawa H, Yamashita K (2011) Theoretical study of the surface complex between TiO_2 and TCNQ showing interfacial charge-transfer transitions. J Phys Chem Lett 2(10): 1167–1170. doi:10.1021/Jz200390g
2. Jono R, Yamashita K (2012) Two different lifetimes of charge separated states: a porphyrin–quinone system in artificial photosynthesis. J Phys Chem C 116(1):1445–1449. doi:10.1021/Jp2075034
3. Kohn W, Sham LJ (1965) Self-consistent equations including exchange and correlation effects. Phys Rev 140(4A):A1133–A1138
4. Tozer DJ, Handy NC (1998) Improving virtual Kohn–Sham orbitals and eigenvalues: application to excitation energies and static polarizabilities. J Chem Phys 109(23):10180–10189. doi:10.1063/1.477711
5. Dreuw A, Weisman JL, Head-Gordon M (2003) Long-range charge-transfer excited states in time-dependent density functional theory require non-local exchange. J Chem Phys 119(6):2943. doi:10.1063/1.1590951
6. Savin A (1996) In: Seminario JM (ed) Recent developments and applications of modern density functional theory. Elsevier, Amsterdam, p 327
7. Iikura H, Tsuneda T, Yanai T, Hirao K (2001) A long-range correction scheme for generalized-gradient-approximation exchange functionals. J Chem Phys 115(8):3540. doi:10.1063/1.1383587
8. Tawada Y, Tsuneda T, Yanagisawa S, Yanai T, Hirao K (2004) A long-range-corrected time-dependent density functional theory. J Chem Phys 120(18):8425. doi:10.1063/1.1688752
9. Tretiak S, Igumenshchev K, Chernyak V (2005) Exciton sizes of conducting polymers predicted by time-dependent density functional theory. Phys Rev B 71(3). doi:10.1103/Physrevb.71.033201
10. Wong BM (2009) Optoelectronic properties of carbon nanorings: excitonic effects from time-dependent density functional theory. J Phys Chem C 113(52):21921–21927. doi:10.1021/Jp9074674
11. Wong BM, Hsieh TH (2010) Optoelectronic and excitonic properties of oligoacenes: substantial improvements from range-separated time-dependent density functional theory. J Chem Theory Comput 6(12):3704–3712. doi:10.1021/Ct100529s
12. Salzner U, Aydin A (2011) Improved prediction of properties of pi-conjugated oligomers with range-separated hybrid density functionals. J Chem Theory Comput 7(8):2568–2583. doi:10.1021/Ct2003447
13. Yanai T, Tew DP, Handy NC (2004) A new hybrid exchange-correlation functional using the Coulomb-attenuating method (CAM-B3LYP). Chem Phys Lett 393(1–3):51–57. doi:10.1016/j.cplett.2004.06.011
14. Peach MJG, Benfield P, Helgaker T, Tozer DJ (2008) Excitation energies in density functional theory: an evaluation and a diagnostic test. J Chem Phys 128(4):044118. doi:10.1063/1.2831900
15. Plotner J, Tozer DJ, Dreuw A (2010) Dependence of excited state potential energy surfaces on the spatial overlap of the Kohn–Sham orbitals and the amount of nonlocal Hartree–Fock exchange in time-dependent density functional theory. J Chem Theory Comput 6(8):2315–2324. doi:10.1021/Ct1001973
16. Dev P, Agrawal S, English NJ (2012) Determining the appropriate exchange-correlation functional for time-dependent density functional theory studies of charge-transfer excitations in organic dyes. J Chem Phys 136(22). doi:10.1063/1.4725540
17. Kuritz N, Stein T, Baer R, Kronik L (2011) Charge-Transfer-Like $\pi \rightarrow \pi^*$ excitations in time-dependent density functional theory: a conundrum and its solution. J Chem Theory Comput 7(8):2408–2415. doi:10.1021/Ct2002804
18. Rohrdanz MA, Herbert JM (2008) Simultaneous benchmarking of ground- and excited-state properties with long-range-corrected density functional theory. J Chem Phys 129(3):034107. doi:10.1063/1.2954017
19. Song J-W, Peng DL, Hirao K (2011) A semiempirical long-range corrected exchange correlation functional including a short-range Gaussian attenuation (LCgau-B97). J Comput Chem 32(15): 3269–3275. doi:10.1002/Jcc.21912

20. Peach MJG, Tellgrent EI, Salek P, Helgaker T, Tozer DJ (2007) Structural and electronic properties of polyacetylene and polyyne from hybrid and coulomb-attenuated density functionals. J Phys Chem A 111(46):11930–11935. doi:10.1021/Jp0754839
21. Anne FB, Purpan FD, Jacquemin D (2013) Charge-transfer in quasilinear push–pull polyene chains. Chem Phys Lett 581:52–56. doi:10.1016/j.cplett.2013.07.021
22. Koch H, Jorgensen P (1990) Coupled cluster response functions. J Chem Phys 93(5):3333–3344. doi:10.1063/1.458814
23. Stanton JF, Bartlett RJ (1993) The equation of motion coupled-cluster method—a systematic biorthogonal approach to molecular-excitation energies, transition-probabilities, and excited-state properties. J Chem Phys 98(9):7029–7039. doi:10.1063/1.464746
24. Koch H, Kobayashi R, Demeras AS, Jorgensen P (1994) Calculation of size-intensive transition moments from the coupled cluster singles and doubles linear-response function. J Chem Phys 100(6):4393–4400. doi:10.1063/1.466321
25. Kallay M, Gauss J (2004) Calculation of excited-state properties using general coupled-cluster and configuration-interaction models. J Chem Phys 121(19):9257–9269. doi:10.1063/1.1805494
26. Krylov AN (2008) A study of the dynamics of contaminants in the own external atmosphere of orbital stations. Russ J Phys Chem B 2(5):827–833. doi:10.1134/S1990793108050266
27. Bartlett RJ (2012) Coupled-cluster theory and its equation-of-motion extensions. Wiley Interdiscip Rev Comput Mol Sci 2(1):126–138. doi:10.1002/wcms.76
28. Schreiber M, Silva-Junior MR, Sauer SPA, Thiel W (2008) Benchmarks for electronically excited states: CASPT2, CC2, CCSD, and CC3. J Chem Phys 128(13):134110. doi:10.1063/1.2889385
29. Musial M, Bartlett RJ (2011) Charge-transfer separability and size-extensivity in the equation-of-motion coupled cluster method: EOM-CCx. J Chem Phys 134(3):034106. doi:10.1063/1.3511783
30. Kronik L, Stein T, Refaely-Abramson S, Baer R (2012) Excitation gaps of finite-sized systems from optimally tuned range-separated hybrid functionals. J Chem Theory Comput 8(5):1515–1531. doi:10.1021/Ct2009363
31. Kar R, Song J-W, Hirao K (2013) Long-range corrected functionals satisfy Koopmans' theorem: calculation of correlation and relaxation energies. J Comput Chem 34(11):958–964. doi:10.1002/Jcc.23222
32. Livshits E, Baer R (2007) A well-tempered density functional theory of electrons in molecules. Phys Chem Chem Phys 9(23):2932. doi:10.1039/b617919c
33. Tsuneda T, Song J-W, Suzuki S, Hirao K (2010) On Koopmans' theorem in density functional theory. J Chem Phys 133(17):174101. doi:10.1063/1.3491272
34. Song J-W, Tokura S, Sato T, Watson MA, Hirao K (2007) An improved long-range corrected hybrid exchange-correlation functional including a short-range Gaussian attenuation (LCgau-BOP). J Chem Phys 127(15):154109. doi:10.1063/1.2790017
35. Curtiss LA, Raghavachari K, Redfern PC, Pople JA (1997) Assessment of Gaussian-2 and density functional theories for the computation of enthalpies of formation. J Chem Phys 106(3):1063–1079
36. Song J-W, Watson MA, Hirao K (2009) An improved long-range corrected hybrid functional with vanishing Hartree–Fock exchange at zero interelectronic distance (LC2gau-BOP). J Chem Phys 131(14):144108. doi:10.1063/1.3243819
37. Song J-W, Hirao K (2013) Long-range corrected density functional theory with optimized one-parameter progressive correlation functional (LC-BOP12 and LCgau-BOP12). Chem Phys Lett 563:15–19. doi:10.1016/j.cplett.2013.01.064
38. Song J-W, Watson MA, Sekino H, Hirao K (2008) Nonlinear optical property calculations of polyynes with long-range corrected hybrid exchange-correlation functionals. J Chem Phys 129(2):024117. doi:10.1063/1.2936830
39. Jacquemin D, Adamo C (2011) Bond length alternation of conjugated oligomers: wave function and DFT benchmarks. J Chem Theory Comput 7(2):369–376. doi:10.1021/Ct1006532
40. Becke AD (1988) Density-functional exchange-energy approximation with correct asymptotic behavior. Phys Rev A 38(6):3098–3100
41. Song J-W, Hirosawa T, Tsuneda T, Hirao K (2007) Long-range corrected density functional calculations of chemical reactions: redetermination of parameter. J Chem Phys 126(15):154105. doi:10.1063/1.2721532
42. Tsuneda T, Hirao K (1997) A new spin-polarized Colle–Salvetti-type correlation energy functional. Chem Phys Lett 268(5–6):510–520
43. Vydrov OA, Scuseria GE (2006) Assessment of a long-range corrected hybrid functional. J Chem Phys 125(23):234109. doi:10.1063/1.2409292
44. Chai J-D, Head-Gordon M (2008) Systematic optimization of long-range corrected hybrid density functionals. J Chem Phys 128(8):084106. doi:10.1063/1.2834918
45. Becke AD (1993) Density-functional thermochemistry. 3. The role of exact exchange. J Chem Phys 98(7):5648–5652
46. Heyd J, Scuseria GE, Ernzerhof M (2003) Hybrid functionals based on a screened Coulomb potential. J Chem Phys 118(18):8207. doi:10.1063/1.1564060
47. Song J-W, Yamashita K, Hirao K (2012) Gaussian attenuation hybrid scheme applied to the Ernzerhof–Perdew exchange hole model (Gau-PBEh). J Chem Phys 137(24):244105. doi:10.1063/1.4772401
48. Perdew JP, Burke K, Ernzerhof M (1996) Generalized gradient approximation made simple. Phys Rev Lett 77(18):3865–3868
49. Tsuneda T, Suzumura T, Hirao K (1999) A new one-parameter progressive Colle–Salvetti-type correlation functional. J Chem Phys 110(22):10664–10678
50. Frisch MJ, Trucks GW, Schlegel HB, Scuseria GE, Robb MA, Cheeseman JR, Scalmani G, Barone V, Mennucci B, Petersson GA, Nakatsuji H, Caricato M, Li X, Hratchian HP, Izmaylov AF, Bloino J, Zheng G, Sonnenberg JL, Hada M, Ehara M, Toyota K, Fukuda R, Hasegawa J, Ishida M, Nakajima T, Honda Y, Kitao O, Nakai H, Vreven T, Montgomery JA Jr, Peralta JE, Ogliaro F, Bearpark M, Heyd JJ, Brothers E, Kudin KN, Staroverov VN, Kobayashi R, Normand J, Raghavachari K, Rendell A, Burant JC, Iyengar SS, Tomasi J, Cossi M, Rega N, Millam NJ, Klene M, Knox JE, Cross JB, Bakken V, Adamo C, Jaramillo J, Gomperts R, Stratmann RE, Yazyev O, Austin AJ, Cammi R, Pomelli C, Ochterski JW, Martin RL, Morokuma K, Zakrzewski VG, Voth GA, Salvador P, Dannenberg JJ, Dapprich S, Daniels AD, Farkas O, Foresman JB, Ortiz JV, Cioslowski J, Fox DJ (2009) Gaussian09, Revision A.1. Gaussian, Inc, Wallingford, CT
51. Izmaylov AF, Scuseria GE (2008) Why are time-dependent density functional theory excitations in solids equal to band structure energy gaps for semilocal functionals, and how does nonlocal Hartree–Fock-type exchange introduce excitonic effects? J Chem Phys 129(3):034101. doi:10.1063/1.2953701
52. Baer R, Livshits E, Salzner U (2010) Tuned range-separated hybrids in density functional theory. Annu Rev Phys Chem 61:85–109. doi:10.1146/annurev.physchem.012809.103321
53. Koch H, Christiansen O, Jorgensen P, deMeras AMS, Helgaker T (1997) The CC3 model: an iterative coupled cluster approach including connected triples. J Chem Phys 106(5):1808–1818. doi:10.1063/1.473322
54. Schreiber M, Silva MR, Sauer SPA, Thiel W (2008) Benchmarks for electronically excited states: CASPT2, CC2, CCSD, and CC3. J Chem Phys 128(13):131410. doi:10.1063/1.2889385

Theor Chem Acc (2014) 133:1450
DOI 10.1007/s00214-014-1450-4

REGULAR ARTICLE

Unimolecular and hydrolysis channels for the detachment of water from microsolvated alkaline earth dication (Mg^{2+}, Ca^{2+}, Sr^{2+}, Ba^{2+}) clusters

Evangelos Miliordos · Sotiris S. Xantheas

Received: 29 October 2013 / Accepted: 11 January 2014 / Published online: 7 February 2014

Abstract We examine theoretically the three channels that are associated with the detachment of a single water molecule from the aqueous clusters of the alkaline earth dications, $[M(H_2O)_n]^{2+}$, M = Mg, Ca, Sr, Ba, $n \leq 6$. These are the unimolecular water loss ($M^{2+}(H_2O)_{n-1}$ + H_2O) and the two hydrolysis channels resulting the loss of hydronium ($[MOH(H_2O)_{n-2}]^+ + H_3O^+$) and Zundel ($[MOH(H_2O)_{n-3}]^+ + H_3O^+(H_2O)$) cations. Minimum energy paths (MEPs) corresponding to those three channels were constructed at the Møller–Plesset second order perturbation (MP2) level of theory with basis sets of double- and triple-ζ quality. We furthermore investigated the water and hydronium loss channels from the mono-hydroxide water clusters with up to four water molecules, $[MOH(H_2O)_n]^+$, $1 \leq n \leq 4$. Our results indicate the preference of the hydronium loss and possibly the Zundel-cation loss channels for the smallest size clusters, whereas the unimolecular water loss channel is preferred for the larger ones as well as the mono-hydroxide clusters. Although the charge separation (hydronium and Zundel-cation loss) channels produce more stable products when compared to the ones for the unimolecular water loss, they also require the surmounting of high-energy barriers, a fact that makes the experimental observation of fragments related to these hydrolysis channels difficult.

Keywords Alkaline earth dication aqueous clusters · Unimolecular dissociation · Potential energy curve · Electronic structure · Hydrolysis channel

Dedicated to Professor Thom Dunning and published as part of the special collection of articles celebrating his career upon his retirement.

Electronic supplementary material The online version of this article (doi:10.1007/s00214-014-1450-4) contains supplementary material, which is available to authorized users.

E. Miliordos · S. S. Xantheas (✉)
Physical Sciences Division, Pacific Northwest National Laboratory, 902 Battelle Boulevard,
P.O. Box 999, MS K1-83, Richland, WA 99352, USA
e-mail: sotiris.xantheas@pnnl.gov

1 Introduction

The structure and stability of small aqueous clusters of alkaline earth metal dications (Mg^{2+}, Ca^{2+}, Sr^{2+}, and Ba^{2+}) has been the subject of numerous experimental [1–18] and theoretical [4, 8, 11, 14–36] studies aimed at providing archetypal models for their aqueous solvation. It is generally accepted that the first solvation shell of these ions in water is saturated with six water molecules [6, 10, 11, 13, 14]. The speciation of these metal cations in an aqueous environment depends both upon their electronic properties as well as the characteristics of the aqueous environment [37]. The second ionization potential (IP) of the alkaline earth metals (save Mg) [38] is below the first ($^2B_1$) IP of water (12.6206 ± 0.0020 eV) [39], and this determines the position of the lower asymptote that correlates with the formation of the aqueous complex [36]. Depending on the pH of an aqueous solution, the speciation ranges [37] from $M^{2+}(H_2O)_n$ to $[M_x(OH)_y]^{(2x-y)+}$, M = Mg, Ca, Sr or Ba. The following two channels:

$$M^{2+}(H_2O)_n \rightarrow \left[M(H_2O)_{n-1}\right]^{2+} + H_2O \tag{1}$$

$$M^{2+}(H_2O)_n \rightarrow \left[MOH(H_2O)_{n-2}\right]^{+} + H_3O^+ \tag{2}$$

describe the unimolecular dissociation and hydrolysis mechanisms, respectively. The first one corresponds to the process of pure water loss originating from the exchange of ligands between the first and second solvation shells, whereas the second one is usually referred to as the charge separation reaction in the literature. In the latter case, the escaping water molecule initially creates a hydrogen bond with one of the other water molecules attached to the metal in the first solvation shell and subsequently detaches a H^+ ion from one of the first shell ligands producing H_3O^+. In the following, we will refer to this second channel as the one corresponding to the loss of hydronium. The remaining singly charged metal hydroxide cluster and hydronium fragments experience Coulombic repulsion and can break apart [11, 14, 16, 17, 29]. It has been demonstrated that the net charge on the metal remains practically unchanged revealing a bonding picture of the type [24, 29] $M^{2+}(H_2O)_m(OH^-)$. It should be mentioned that the charge separation pathway has been observed experimentally only for the lighter metal clusters [1, 6, 7, 12]. For instance, the $[Ca(H_2O)]^{2+}$ complex was elusive during collision-induced experiments until recently [16], due to the charge separation $[Ca(H_2O)]^{2+} + H_2O \rightarrow CaOH^+ + H_3O^+$ reaction [1, 12]. Similarly, the largest cluster size for which this reaction has been reported to occur [7] is four water molecules for Mg and two for the other three metals.

Experimentally, doubly charged aqueous clusters of the alkaline earth metal ions were produced by electrospray ionization, and special effort has been devoted to the measurement of their hydration energies. To this end, Armentrout and co-workers have recently employed the collision-induced dissociation (CID) combined with the guided ion beam tandem mass spectrometry techniques to study the aqueous clusters of Mg, Ca, and Sr [11, 14, 16, 17]. CID was also used by Barran et al. [6] to examine the Mg clusters, while Williams and co-workers applied the blackbody infrared dissociation (BIRD) methodology to study the aqueous clusters of all four metals [5, 9]. Finally, Kerbale and co-workers used high-pressure mass spectrometry (HPMS) to also study clusters of all four metal dications [2, 3].

From the theoretical point of view, most of the previous calculations have been reported using the density functional theory (DFT) with a variety of functionals and basis sets [8, 11, 14, 16, 17, 27–30, 33–35]. Ab initio results have been reported at the (restricted) Hartree–Fock (RHF) and Møller–Plesset second order perturbation (MP2) levels of theory [4, 8, 11, 16, 17, 19–21, 23–25, 31, 33, 35]. The previous theoretical results reported in the literature have mainly focused on obtaining the optimal geometries of the $[M(H_2O)_n]^{2+}$ clusters, including their various isomers, and their energies with emphasis in estimating their hydration energies. The majority of the previous studies have dealt with the clusters of the first two metal atoms, Ca and Mg [4, 8, 11, 16, 17, 19–21, 23, 25, 27–35], while considerably less attention has been paid to aqueous clusters of Sr [14, 19–21, 23, 29] and Ba [19, 23, 29] dications. The structures and hydration energies of the $[M(H_2O)_n(OH)]^+$ ions have also been reported in the literature for M = Mg, Ca, and Ba [18, 22, 26]. Finally, we would like to mention that only a handful of studies have previously theoretically examined the charge separation process described by reaction (2). Specifically, Beyer et al. [29] in 1999 studied both of the chemical reactions (1) and (2) for all four metals, while Peschke et al. [28] in the same year reported analogous results for the cases of Mg and Ca. In both articles, transition states and energy barriers were reported for clusters with just two water molecules. More than 10 years later, Armentrout and co-workers examined the stationary points (equilibrium structures and transition states) of the $[Mg(H_2O)_{3,4}]^{2+}$, $[Ca(H_2O)_{2,3}]^{2+}$, and $[Sr(H_2O)_2]^{2+}$ systems [14, 16, 17]. However, no minimum energy paths (MEPs) have yet been reported in the literature for those and larger clusters.

In the present study, we report the MEPs for the reactions (1) and (2), for all four alkaline earth metals with $n = 2$, 3 water molecules at the MP2 level of theory. For Mg, we additionally included the MEPs for $n = 4$. In the cases of $n = 3$, 4, we also constructed the MEPs relative to the following reaction, which involves the Zundel cation (Zundel-cation loss channel):

$$M^{2+}(H_2O)_n \rightarrow \left[MOH(H_2O)_{n-3}\right]^+ + H_3O^+(H_2O). \quad (3)$$

Reactions (2) and (3) can be both considered as charge separation processes. For $n > 4$, we report only the relative energetics of the reactants and products of reactions (1), (2), and (3). Furthermore, the full MEPs for the removal of one additional water molecule from the product $[M(H_2O)_m(OH)]^+$ cations for $m = 2$ are reported. We show that the water loss reaction prevails for this case, as the charge separation reaction leading to $M(OH)_2 + H_3O^+$ is not energetically favorable.

The article is organized as follows: In Sect. 2, we outline the computational methods used in the present study. In Sect. 3, we describe the MEPs for the $[M(H_2O)_n]^{2+}$ species, where M = Mg, Ca, Sr, Ba and $n = 2$, 3, or 4 (only for Mg). In Sect. 4, we present the results of the energetics of reactions (1–3) for all four metals up to six water molecules. Finally, in Sect. 5, we examine the water loss process of the $[M(H_2O)_{n-1}(OH)]^+$ species. Section 6 summarizes our findings.

2 Computational details

All calculations reported in this study were carried out at the Møller–Plesset second order perturbation level of theory (MP2). The 1*s* orbitals of oxygen and the 1*s*2*s*2*p* orbitals of Mg and Ca were kept frozen. We used two different basis sets of double- and triple-ζ quality for H, O, Mg, and Ca. Specifically, for O and H, we used Dunning's aug-cc-pVDZ and aug-cc-pVTZ basis sets [40, 41], for Mg, the cc-pV(D + *d*)Z and cc-pV(T + *d*)Z [42], and for Ca, the cc-pVDZ and cc-pVTZ basis sets [43]. For Sr and Ba, we used the small-core relativistic pseudopotentials ECP28MDF and ECP46MDF, which replace the 1*s*2*s*2*p*3*s*3*p*3*d* and 1*s*2*s*2*p*3*s*3*p*3*d*4*s*4*p*4*d* electrons, respectively [44]. These pseudopotentials are combined with the 8*s*8*p*5*d*4*f* and 9*s*9*p*6*d*4*f* Gaussian functions to construct the molecular orbitals of the remaining electrons [44]. From now on, we use the acronyms ADZ and ATZ to denote the various basis sets as follows: ADZ = cc-pV(D + *d*)Z/(Mg), cc-pVDZ/(Ca), ECP28MDF-f/(Sr), ECP46MDF-f/(Ba), aug-cc-pVDZ/(O,H) and ATZ = cc-pV(T + *d*)Z/(Mg), cc-pVTZ/(Ca), ECP28MDF/(Sr), ECP46MDF/(Ba), aug-cc-pVTZ/(O,H).

All calculations, including the geometry optimizations, were performed with no symmetry constraints to exclude any artificially converged highly symmetric structures. To further verify that the located structures are real minima, we also calculated the harmonic vibrational frequencies at the equilibrium structures at the MP2/ADZ level. These structures compare very well with the ones already reported in the literature (when available). Zero-point energy (ZPE) corrections were estimated using these harmonic frequencies, i.e., no anharmonicities are included. For the frequencies, we used the atomic mass of the most abundant isotope, namely 1.00783 (^{1}H), 15.9949 (^{16}O), 23.98505 (^{24}Mg), 39.9626 (^{40}Ca), 87.9056 (^{88}Sr), and 137.905 (^{138}Ba). The MEPs as a function of the *R*(M–O) distance are obtained at the MP2/ADZ level of theory by optimizing the rest of the internal coordinates for each value of *R*. The MP2/ATZ//MP2/ADZ energetics of the various stationary points of the MEPs are used in the discussion of their features. All calculations were performed with the MOLPRO [45] and Gaussian09 [46] electronic structure codes.

3 Minimum energy paths of $[M(H_2O)_n]^{2+}$

In the following, we report the MEPs for the systems with the following stoichiometry: $[Mg(H_2O)_{2,3,4}]^{2+}$ (Sect. 3.1), $[Ca(H_2O)_{2,3}]^{2+}$ (Sect. 3.2), and $[Sr(H_2O)_{2,3}]^{2+}$, $[Ba(H_2O)_{2,3}]^{2+}$ (Sect. 3.3). All optimized structures, their energies, and harmonic vibrational frequencies are given in the Supporting Information.

3.1 $[Mg(H_2O)_{2,3,4}]^{2+}$

Figure 1 shows the MEPs at the MP2/ADZ level of theory for both the water loss and the charge separation channels of the $[Mg(H_2O)_2]^{2+}$ system. The MEP was constructed by pulling one water molecule (bearing the O_1 atom) away from the ion and optimizing the rest of the geometrical parameters for every Mg–O_1 distance. Along the unimolecular water loss channel, this ligand can be detached either with or without making a hydrogen bond with the second water molecule. These two possibilities are shown in the MEP of Fig. 1: The dotted line with the solid circles for *R*(Mg–O) >3 Å traces the (higher energy) path along which no hydrogen bond is formed. On the other hand, there is a lower energy path for 3 Å < *R*(Mg–O_1) < 4 Å due to the formation of a hydrogen bond followed by an increase in the energy to the $[Mg(H_2O)]^{2+} + H_2O$ asymptote (dashed line with open circles in Fig. 1). This hydrogen-bonded minimum lies 31.4 kcal/mol higher in energy than the $Mg^{2+}(H_2O)_2$ global minimum; and it is stabilized with respect to that global minimum by a small barrier of 2.3 kcal/mol. By forcing the water molecule to be further detached from the ion, the charge separation channel [reaction (2)] opens up, facilitated by the loss of a H^+ ion from the first solvation shell. This MEP is shown with a solid line and filled square symbols in Fig. 1 (for *R*(Mg–O) >4 Å). Observe that the charge separation MEP is always lower in energy than the water loss MEPs. Moving from the H-bonded structure and after surmounting a barrier of 8.5 kcal/mol at *R*(Mg–O) ~5.5 Å, the charge separation MEP falls off as 1/*R* to the $MgOH^+ + H_3O^+$ asymptote due to Coulombic repulsion. Due to the 1/*R* dependence, even for a distance of *R*(Mg–O) ~15 Å, the energy is 20 kcal/mol higher than the asymptotic limit, which lies 4.7 kcal/mol lower than the energy of the $Mg^{2+}(H_2O)_2$ minimum.

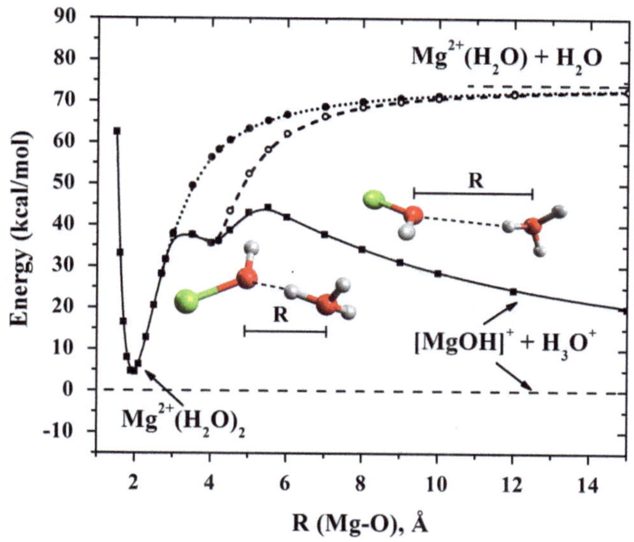

Fig. 1 Minimum energy paths of the $[Mg(H_2O)_2]^{2+}$ system

Our MP2 relative energies compare favorably with the DFT values of Peschke et al. [28] and Beyer et al. [29]. In particular, relative to the equilibrium structure, our

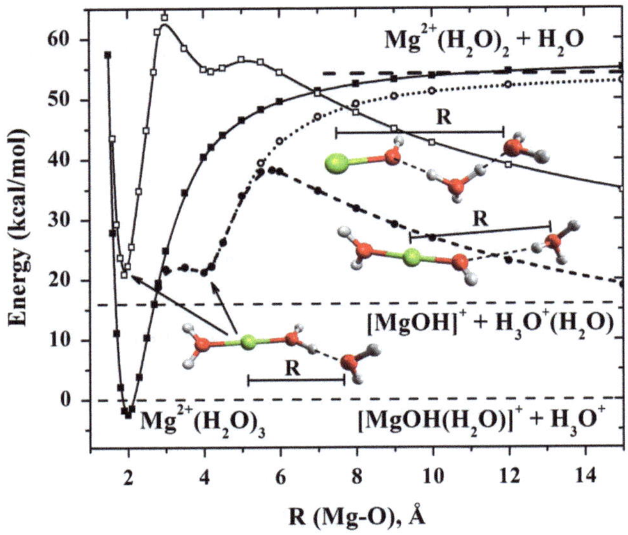

Fig. 2 Minimum energy paths of the $[Mg(H_2O)_3]^{2+}$ system. The *dashed line* with the *filled circles* dissociates to $[MgOH(H_2O)]^+$ + H_3O^+, whereas the *solid line* with the *open squares* to $[MgOH]^+$ + $H_3O^+(H_2O)$

H-bonded minimum is at 31.4 kcal/mol, whereas B3LYP gives a value of 31.5 kcal/mol [29]. The transition state of the charge separation reaction at ~5.5 Å is presently at 39.9 kcal/mol, while B3LYP with two different basis sets gives 36.3 [28] and 39.4 kcal/mol [29], respectively. It should be mentioned that those DFT barriers included zero-point energy corrections (unlike ours, since we performed ZPE corrections only for the minima). To summarize, the lowest energy path for the detachment of a water molecule from the $Mg^{2+}(H_2O)_2$ minimum corresponds first to the formation of a hydrogen bond, $Mg^{2+}(H_2O)...H_2O$ and then to the dissociation to $MgOH^+$ + H_3O^+, i.e., it starts off as a unimolecular water loss and ends up to the charge separated products.

We turn now to the $[Mg(H_2O)_3]^{2+}$ case and the corresponding MEPs shown in Fig. 2. In that case, we have three different adiabatic channels: (a) the water loss, reaction (1), (b) the hydronium loss, reaction (2), and (c) the Zundel-cation loss, reaction (3) channels. In the former two cases, the distance R of the x axis is measured from the Mg center to the oxygen atom of the departing water molecule, as in Fig. 1. In the latter case, we, however, measure the distance R from the Mg center to the oxygen atom of the Zundel cation further away from Mg. We would like to remind once again that all other geometrical parameters are optimized along the MEPs for every R(Mg–O) value without the use of any symmetry constraints.

We first describe the MEPs of Fig. 2 for the $[Mg(H_2O)_2]^{2+}$ + H_2O water loss channel starting from large R distances and moving down to the global minimum at R(Mg–O) ~2 Å. As before, the water can approach the Mg^{2+} positively charged center by interacting or not with the rest of the water molecules. The solid line (filled squares) starting from the $Mg^{2+}(H_2O)_3$ equilibrium structure and ending in the $Mg^{2+}(H_2O)_2$ + H_2O asymptote represents the path where there are no hydrogen bonds of the departing water with any of the other ligands. On the other hand, the formation of a H-bond makes the $[Mg(H_2O)_2]^{2+}...H_2O$ attractive interaction more efficient, and thus, the respective MEP (dotted line, open circles) decreases faster going to the H-bonded shallow minimum at ~4 Å. Pushing the water molecule of the second solvation shell further toward the metal center leads to the global $Mg^{2+}(H_2O)_3$ minimum via a negligible barrier.

For the lowest $[Mg(H_2O)(OH)]^+$ + H_3O^+ adiabatic path (dashed line, filled circles in Fig. 2), there is a Coulombic repulsion responsible for the $1/R$ shape of the MEP for R(Mg–O) >5.8 Å. At this distance, the positively charged hydronium experiences the negative local charge of the oxygen atom of the OH group. Recall that the $Mg^{2+}(H_2O)(OH^-)$ bonding picture has been previously suggested in the literature [24, 29]. At this point, one of the hydrogen atoms of H_3O^+ migrates to the oxygen of the OH group and the MEP turns over decreasing in energy. By the time the H atom has moved close to OH, we have the formation of a H-bond, and therefore, this path crosses the MEP coming from the $[Mg(H_2O)_2]^{2+}$ + H_2O channel (dotted line, open circles).

We finally examine the $MgOH^+$ + $H_3O^+(H_2O)$ channel which lies ~15.8 kcal/mol higher than the $[Mg(H_2O)(OH)]^+$ + H_3O^+ asymptote. For reasons of clarity in Fig. 2, for this case, we show the MEP in terms of the distance R between the Mg center and the oxygen atom of the more distant water molecule. For long distances, we again observe the $1/R$ repulsive behavior (solid line, open squares). As the Zundel-cation approaches, the negative charge of OH is not screened anymore from the Mg^{2+} center, and one of its hydrogen atoms moves to the negatively charged OH group, as in the previous case. As a result, the cluster with two hydrogen bonds, schematically drawn as $Mg^{2+}(H_2O)...(H_2O)...(H_2O)$, is formed (this is the minimum located at R = 4.2 Å lying 54.6 kcal/mol above the lowest asymptote). By pushing further in the most distant water molecule, i.e., decreasing R, that molecule eventually is attached to Mg producing the $[Mg(H_2O)_2]^{2+}...H_2O$ minimum (last ligand in the second solvation shell of the Mg center). Notice that now since R corresponds to the distance between Mg and the O of the first shell, the minimum is at R = 1.9 Å. This interaction is reported here for the first time in the literature.

In summary, removing a water molecule from the $[Mg(H_2O)_3]^{2+}$ minimum structure leads to the $[MgOH(H_2O)]^+$ + H_3O^+ fragments. This process is facilitated via the formation of a hydrogen-bonded intermediate, and its subsequent decomposition over a barrier lying 23.7 and 40.7 kcal/mol above the $Mg^{2+}(H_2O)_3$ minimum, respectively. The corresponding numbers at the DFT [B3LYP/6-311 + G(2d,2p)//B3LYP/6-311 + G(d,p)] level, corrected for ZPE, are 22.5 and 38.7 kcal/mol [17], in good agreement with our present results. The dissociation energies for reactions (1–3) are 2.3, 18.5, 56.9 kcal/mol, respectively. Estimates of 0.2 and 55.2 kcal/mol for the first and third channels have been previously reported by Carl et al. [16].

We next consider the case with four water molecules. The corresponding MEPs are shown in Fig. 3, and they describe the same channels as the ones in Fig. 2 for the $n = 3$ case; in this respect, the MEPs of the two figures bear similar features. One difference is that the energy gaps between the first dissociation channel (hydronium loss) and the next two (Zundel cation and water loss) have now decreased appreciably from 15 and 55 kcal/mol for the $n = 3$ case, now being 5 and 35 kcal/mol, respectively. As a result, the MEPs for the $n = 4$ case are packed closer together than for $n = 3$. Along the channel that removes a water molecule from the $Mg^{2+}(H_2O)_4$ minimum, the hydrogen-bonded intermediate $[Mg(H_2O)_3]^{2+}...H_2O$ at $R = 4$ Å and energy 7.7 kcal/mol (cf. Fig. 3) is initially formed. Compared to Fig. 2, the MEP leading to hydronium loss (dashed line with filled circles in Fig. 3) lies above the one leading to water loss (dotted line with open circles) for distances between 4 and 8 Å; this will certainly play an important role in the overall dynamics and kinetics of processes (1) and (2). Taking into account the Zundel-cation loss channel, which also passes through the above H-bonded structure, will further add to the complexity. The relative energetics of the several stationary points in Fig. 3 are within ~4 kcal/mol with the DFT results of Carl et al. [17].

A few interesting observations regarding the MEPs of $[Mg(H_2O)_n]^{2+}$, $n = 2, 3, 4$ (Figs. 1, 2, 3) are in order. First, the transition from the $Mg^{2+}(H_2O)_n$ to the $[Mg(H_2O)_{n-1}]^{2+}...H_2O$ structure becomes easier with increasing n, since the energy difference decreases from 31.3, to 23.7, to 18.0 kcal/mol. In contrast, the barrier from $[Mg(H_2O)_{n-1}]^{2+}...H_2O$ to $[Mg(H_2O)_{n-2}(OH)]^{2+}...H_3O^+$ increases with n, from 8.7 ($n = 1$) to 17.0 ($n = 2$) to 28.7 ($n = 3$) kcal/mol. The MEPs of Figs. 1, 2, and 3 demonstrate the in situ formation of $[Mg(H_2O)_n]^{2+}$ from $[Mg(H_2O)_{n-1}]^{2+} + H_2O$. However, although the binding energy of the water molecule decreases with n mostly due to the increasing steric repulsion, the binding energy of the $Mg^{2+}(H_2O)_n$ global minimum with respect to the lowest energy fragments (hydronium loss asymptote) is increasing, from −4.5 ($n = 1$) to 2.5 ($n = 2$) to 10.3 ($n = 3$) kcal/mol, i.e., the hydrated metal cluster is becoming more stable with respect to the lowest dissociation asymptote.

3.2 $[Ca(H_2O)_{2,3}]^{2+}$

The MEPs of the $[Ca(H_2O)_2]^{2+}$ and $[Ca(H_2O)_3]^{2+}$ systems are shown in Figs. 4 and 5, respectively. We adopt the same definition for the distance R as for the Mg case (vide supra). Similar to Mg, the MEP connecting the $Ca^{2+}(H_2O)_n$ with the $Ca^{2+}(H_2O)_{n-1} + H_2O$ is smooth when no hydrogen-bonded intermediates are formed (see the solid lines with the filled symbols in Figs. 4, 5). The formation of the hydrogen-bonded intermediate results in an energy lowering and alters the MEPs for $R > 3$ Å (dashed line

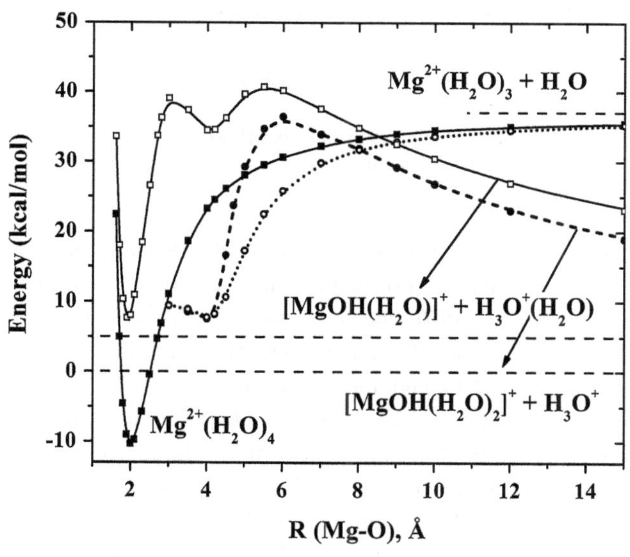

Fig. 3 Minimum energy paths of the $[Mg(H_2O)_4]^{2+}$ system. The *dashed line* with the *filled circles* dissociates to $[MgOH(H_2O)_2]^+ + H_3O^+$, whereas the *solid line* with the *open squares* to $[MgOH(H_2O)]^+ + H_3O^+(H_2O)$

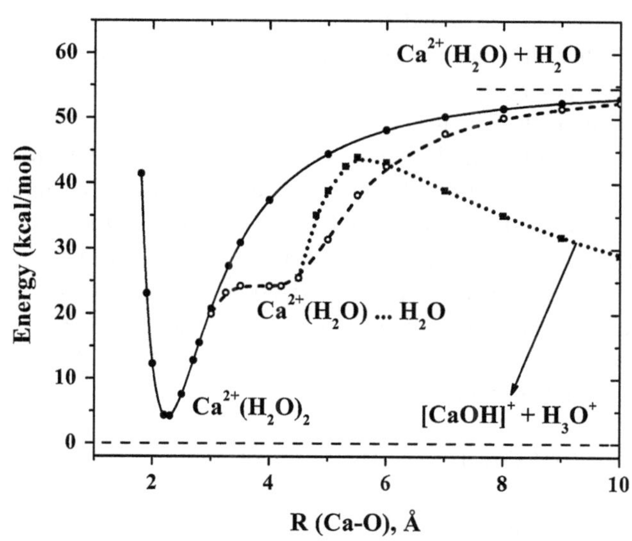

Fig. 4 Minimum energy paths of the $[Ca(H_2O)_2]^{2+}$ system. The *dotted line* with the *filled squares* dissociates to $[CaOH]^+ + H_3O^+$

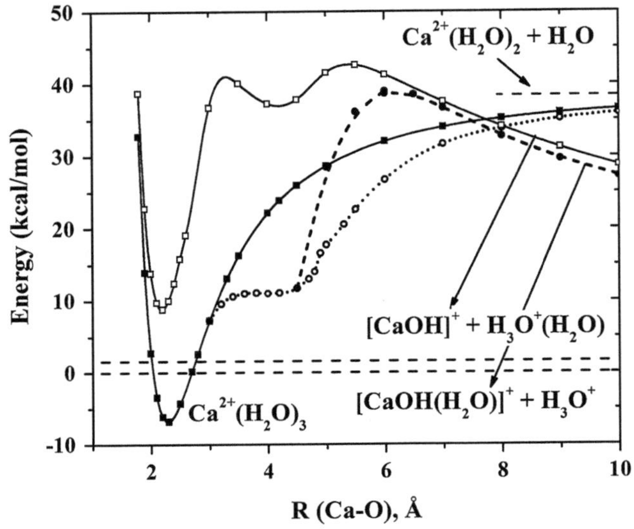

Fig. 5 Minimum energy paths of the $[Ca(H_2O)_3]^{2+}$ system. The *dashed line* with the *filled circles* dissociates to $[CaOH(H_2O)]^+ + H_3O^+$, whereas the *solid line* with the *open squares* to $[CaOH]^+ + H_3O^+(H_2O)$

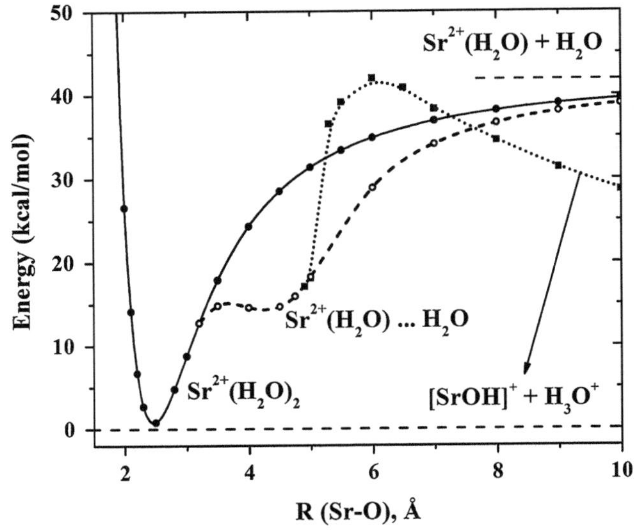

Fig. 6 Minimum energy paths of the $[Sr(H_2O)_2]^{2+}$ system. The *dotted line* with the *filled squares* dissociates to $[SrOH]^+ + H_3O^+$

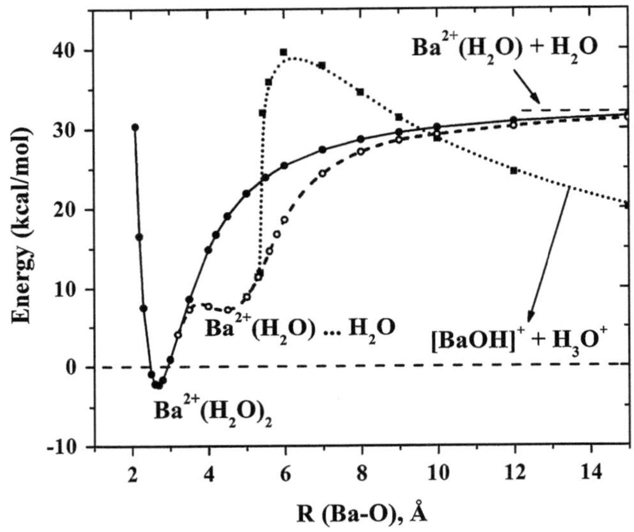

Fig. 7 Minimum energy paths of the $[Ba(H_2O)_2]^{2+}$ system. The *dotted line* with the *filled squares* dissociates to $[BaOH]^+ + H_3O^+$

with open circles in Fig. 4 and dotted line with open circles in Fig. 5).

The $Ca^{2+}(H_2O)...H_2O$ hydrogen-bonded complex can further dissociate via two different paths: an endothermic one leading to $Ca^{2+}(H_2O) + H_2O$ and an exothermic one to $[CaOH]^+ + H_3O^+$ overcoming a barrier of 19.7 kcal/mol. This energy barrier is at least twice as much as the one for the Mg case (8.7 kcal/mol). The difference between the energy levels of $Ca^{2+}(H_2O)_2$ and $[CaOH]^+ + H_3O^+$ is 5.0 kcal/mol, very similar to that for Mg.

As regards the $Ca^{2+}(H_2O)_2...H_2O$ hydrogen-bonded intermediate, it can follow three different routes to dissociation. The first one to $Ca^{2+}(H_2O)_2 + H_2O$ requires an energy of 25.5 kcal/mol, while the other two leading to the hydronium or Zundel-cation loss are exothermic by roughly the same amount, 11.0 and 9.5 kcal/mol, albeit with rather large energy barriers, 27.6 and 33.8 kcal/mol, respectively. The former barrier height is larger than that of Mg (17.0 kcal/mol), whereas the latter one is smaller than the one for Mg (42.7 kcal/mol). The Zundel-cation channel loss proceeds via the $[Ca(H_2O)]^{2+}...H_2O...H_2O$ intermediate (solid line with open squares in Fig. 5), which lies 43.5 kcal/mol above the $Ca^{2+}(H_2O)_3$ global minimum. Our relative energetics are in agreement with the ZPE-corrected DFT values of Peschke et al. [28], Beyer et al. [29], and Carl and Armentrout [16].

3.3 $[Sr(H_2O)_{2,3}]^{2+}$ and $[Ba(H_2O)_{2,3}]^{2+}$

For the Sr and Ba dications, we first examine the cases with two water molecules, the MEPs of which are shown in Figs. 6 and 7. The stationary points of these MEPs have been previously reported by Beyer et al. [29] and Carl et al. [14] at the DFT level of theory with the B3LYP functional. The energy differences between $Sr^{2+}(H_2O)_2$ and $Sr^{2+}(H_2O)...H_2O$ (H-bonded intermediate) and between $Sr^{2+}(H_2O)...H_2O$ and the transition state to $Sr^+(OH) + H_3O^+$ are 13.8 (vs. 13.5 [29], 13.6 [14]) kcal/mol and 27.4 (vs. 23.2 [29], 24.1 [14]) kcal/mol. The corresponding values for Ba are 9.6 (vs. 9.5 [29]) and 32.3 (vs. 29.2 [29]) kcal/mol. The H-bonded intermediate was found to be stabilized monotonically from Mg to Ba with respect to the global $M^{2+}(H_2O)_n$ minimum, while the barrier to the dissociation to hydronium increases from Mg to Ba. However, the final dissociation energy from the equilibrium aqueous

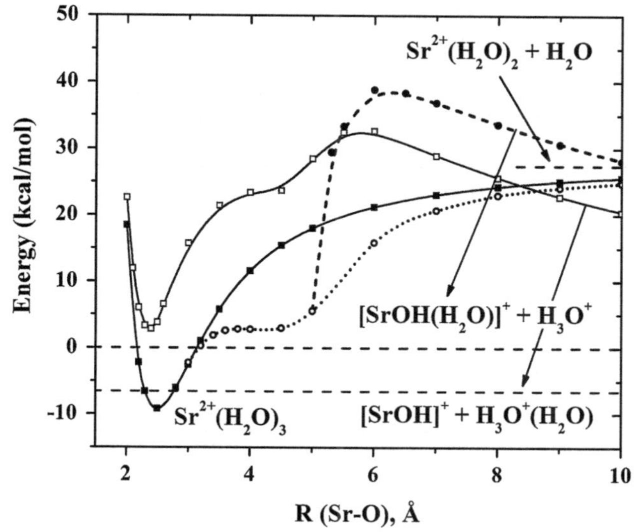

Fig. 8 Minimum energy paths of the $[Sr(H_2O)_3]^{2+}$ system. The *dashed line* with the *filled circles* dissociates to $[SrOH(H_2O)]^+ + H_3O^+$, while the *solid line* with the *open squares* to $[SrOH]^+ + H_3O^+(H_2O)$

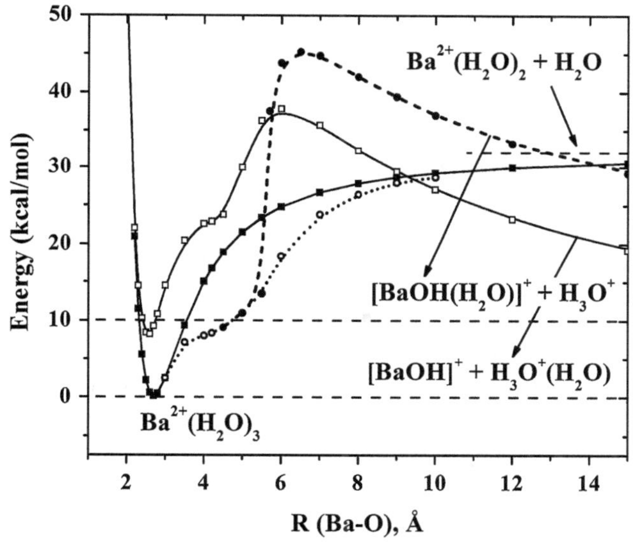

Fig. 9 Minimum energy paths of the $[Ba(H_2O)_3]^{2+}$ system. The *dashed line* with the *filled circles* dissociates to $[BaOH(H_2O)]^+ + H_3O^+$, while the *solid line* with the *open squares* to $[BaOH]^+ + H_3O^+(H_2O)$

complex to the hydronium products remains small within the range of ±5 kcal/mol.

Similar trends are observed for the tri-coordinated aqueous metal clusters. The respective MEPs are depicted in Figs. 8 and 9, and to the best of our knowledge, there is no previous reference regarding their stationary points in the literature. Briefly, the $M^{2+}(H_2O)_3$ minimum lies 12.0/8.4 kcal/mol (M = Sr/Ba) below the $M^{2+}(H_2O)_2...H_2O$ intermediate (albeit not a real minimum for Ba), which in turn is 26.1/24.4 kcal/mol below the water loss asymptote. The transition from $M^{2+}(H_2O)_2...H_2O$ to the $MOH(H_2O)]^+ + H_3O^+$ asymptote proceeds over a barrier of 36.0/42.6 kcal/mol, while the channel leading to $MOH^+ + H_3O^+(H_2O)$ proceeds via a smaller energy barrier (29.8/29.6 kcal/mol) and is associated with a larger exothermicity (9.6/8.2 kcal/mol).

When going from Mg^{2+} to Ba^{2+}, the ionic radii increase and consequently the charge density decreases along this direction, therefore weakening the interaction between $M^{2+}(H_2O)_2$ and H_2O. Consequently, the binding energy is expected to become smaller going from Mg to Ba in complete agreement with Figs. 2, 5, 8, and 9. In addition, the H-bonded $M^{2+}(H_2O)_2...H_2O$ intermediate energetically approaches the equilibrium $M^{2+}(H_2O)_3$ structure at the same time decreasing the already small barrier between those two stationary points. Indeed, according to Fig. 9, this transition for Ba is barrierless, and the $Ba^{2+}(H_2O)_2...H_2O$ "intermediate" is not a local minimum. The case is exactly the same for the "doubly" H-bonded intermediate, $M^{2+}(H_2O)...H_2O...H_2O$, which is 57.1 kcal/mol higher than the $M^{2+}(H_2O)_3$ minimum for Mg and decreases to 46.1, 32.6, and 22.9 kcal/mol for Ca, Sr, and Ba, respectively. The barrier from the former to the latter structure is also decreasing and it actually vanishes for Sr and Ba. A final remark has to do with the relative order of the hydronium and Zundel-cation fragments: For Mg, the hydronium loss channel produces fragments that are lower in energy by 16.3 kcal/mol. On the other hand, Ba prefers the Zundel-cation loss process by 10.0 kcal/mol; Ca and Sr are somewhere in-between.

4 Energetics of the channels corresponding to the water, hydronium, and Zundel-cation loss

In this section, we report the energetics of reactions (1), (2), and (3) as a function of both the metal atom M and the number n of water molecules in the cluster. The absolute energies and geometric structures of all molecular species involved in this section are reported in the Supporting Information. Setting as zero of the energy scale, the energy of the $M^{2+}(H_2O)_n$ minima, the energies of the products of reactions (1), (2), and (3) are listed in Table 1. Besides our own values, we also include available theoretical and experimental ones from the literature for reaction (1). The available data for reaction (2) are limited in the literature to the smallest clusters and have already been discussed earlier. We are not aware of any data related to reaction (3) previously reported in the literature.

In Table 1, we list our MP2 results using two different basis sets of double- and triple-ζ quality (MP2/ADZ and MP2/ATZ, see Sect. 2). In general, the two basis sets give values differing no more than a couple of kcal/mol in the

Table 1 Energy differences (kcal/mol) between the $M^{2+}(H_2O)_n$ minima (set as the zero of the energy scale) and the asymptotes corresponding to the unimolecular water loss $M^{2+}(H_2O)_{n-1} + H_2O$, the hydronium loss $[MOH(H_2O)_{n-2}]^+ + H_3O^+$, and the Zundel-cation loss $[MOH(H_2O)_{n-3}]^+ + H_3O^+(H_2O)$ channels, M = Mg, Ca, Sr, Ba and n = 1–6

n	MP2/ ADZ	MP2/ATZ[a]	Theory[b]	Expt.[c]	MP2/ADZ	MP2/ATZ[a]	MP2/ADZ	MP2/ATZ[a]
	$[Mg(H_2O)_{n-1}]^{2+} + H_2O$				$[Mg(H_2O)_{n-2}(OH)]^+ + H_3O^+$		$[Mg(H_2O)_{n-3}(OH)]^+ + H_3O^+(H_2O)$	
1	77.9	78.7 (76.8)	78.3					
2	69.6	69.8 (67.6)	69.5		−4.7	−5.3 (−7.3)		
3	56.9	57.3 (54.9)	56.9	53.3 (3.0)	2.3	2.0 (−0.4)	18.5	17.9 (14.1)
4	46.8	46.8 (44.4)	46.2	42.4 (2.5)	10.0	9.7 (7.8)	15.3	14.7 (10.5)
5	32.8	32.4 (29.4)	31.6	27.7 (2.1)	11.1	10.4 (8.0)	9.0	8.0 (3.6)
6	30.7	30.3 (27.8)	24.9	23.3 (1.8)	19.1	18.3 (15.6)	8.0	6.6 (2.3)
	$[Ca(H_2O)_{n-1}]^{2+} + H_2O$				$[Ca(H_2O)_{n-2}(OH)]^+ + H_3O^+$		$[Ca(H_2O)_{n-3}(OH)]^+ + H_3O^+(H_2O)$	
1	58.1	57.1 (55.3)			53.9			
2	50.7	49.9 (47.9)	47.1	49.6 (4.2)	−5.1	−8.2 (−8.8)		
3	46.6	45.2 (43.0)	43.0	40.6 (2.1)	6.2	4.1 (3.3)	7.8	2.9 (0.6)
4	41.8	40.3 (38.6)	37.3	33.7 (2.1)	15.6	13.8 (13.3)	14.4	10.3 (8.4)
5	33.7	32.6 (30.3)	30.6	26.8 (1.8)	19.8	18.2 (17.3)	15.5	12.3 (10.0)
6	30.6	30.3 (28.5)	25.3	23.5 (2.1)	24.5	25.1 (24.4)	16.6	14.4 (12.2)
	$[Sr(H_2O)_{n-1}]^{2+} + H_2O$				$[Sr(H_2O)_{n-2}(OH)]^+ + H_3O^+$		$[Sr(H_2O)_{n-3}(OH)]^+ + H_3O^+(H_2O)$	
1	46.5	48.2 (46.6)	48.1	48.2 (1.4)				
2	40.6	41.7 (39.9)	43.1	41.0 (1.2)	−1.8	−5.6 (−6.4)		
3	38.1	39.1 (37.3)	39.0	34.4 (1.2)	8.7	5.7 (5.0)	2.6	−0.5 (−2.5)
4	33.9	34.3 (32.7)	35.2	29.7 (1.2)	16.4	13.9 (13.7)	8.9	5.8 (4.1)
5	28.9	28.7 (26.5)	29.3	24.4 (0.9)	21.2	18.4 (17.7)	11.6	8.5 (6.6)
6	26.8	26.5 (24.7)	26.8	22.4 (0.7)	26.1	23.8 (23.5)	14.3	10.8 (8.9)
	$[Ba(H_2O)_{n-1}]^{2+} + H_2O$				$[Ba(H_2O)_{n-2}(OH)]^+ + H_3O^+$		$[Ba(H_2O)_{n-3}(OH)]^+ + H_3O^+(H_2O)$	
1	38.9	41.0 (39.5)	41.3					
2	33.8	35.3 (33.6)	37.2		1.0	−5.0 (−6.1)		
3	32.8	33.9 (32.3)	33.7		10.1	5.5 (5.1)	0.0	−5.2 (−7.4)
4	29.1	29.8 (28.4)	30.5	25.6 (0.9)	16.0	12.9 (13.2)	5.5	1.2 (−0.1)
5	25.3	25.3 (23.5)	26.2	21.1 (0.6)	19.8	17.1 (17.6)	7.6	4.1 (3.2)
6	23.0	23.3 (21.8)	23.9	17.9 (0.6)	23.3	21.6 (22.6)	9.0	6.3 (5.8)

[a] ZPE-corrected values are given in parentheses

[b] *Ab initio* values from the literature. Values for Mg and Ca are from Ref. 33 (MP2/6-311 ++G(3d,3p)//MP2/6-311 ++G(3,3)), while values for Sr and Ba are from Ref. 23 (MP2/6-31 + G*//RHF/6-31 + G*)

[c] Experimental hydration enthalpies from the literature. Mg values from Ref. 17 (CID), Ca values from Ref. 16 (CID), Sr values from Ref. 14 (CID), and Ba values from Ref. 5 (BIRD). Uncertainties are reported in parentheses

water loss channel for all four metals; our results are in agreement with previous MP2 calculations with different basis sets. The value of 24.9 kcal/mol for the Mg case with $n = 6$ given by Ref. [33] differs appreciably from our 30.7 and 30.3 kcal/mol values. Our numbers agree, however, better with the 28.5 kcal/mol value of Ref. [23] at the MP2/6-311 + G* level of theory. Our ZPE uncorrected energy differences are always larger than the experimental hydration energies (see Table 1), as expected, but there is better agreement when ZPE is included. Especially for the smaller clusters, our ZPE-corrected values are within the experimental error. For larger clusters, the difference between our values and experiment is of the order of 2–3 kcal/mol and it is due to the fact that our calculations are still missing some portion of electron correlation.

The two channels described by reactions (2) and (3) involve the mono-hydroxides of the four metals under consideration and the hydronium or the hydrated hydronium (Zundel) cations. An interesting observation is the lowering of the energy difference between $M^{2+}(H_2O)_n$ and the

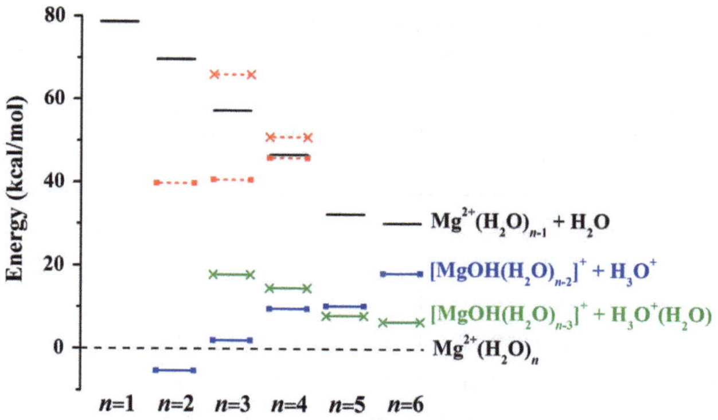

Fig. 10 Relative energies of the $[Mg(H_2O)_n]^{2+}$ (*dashed line* at zero energy), $[Mg(H_2O)_{n-1}]^{2+} + H_2O$ (plain *solid lines*), $[MgOH(H_2O)_{n-2}]^{+} + H_3O^{+}$ (*solid lines* with *filled squares*), and $[MgOH(H_2O)_{n-3}]^{+} + H_3O^{+}(H_2O)$ (*solid lines* with cross symbols) species. The *dashed lines* with the *filled square* symbols correspond to the barriers of the hydronium loss channel and the *dashed lines* with the cross symbols to the barriers of the Zundel-cation loss channel

asymptotes of the two channels going from the ADZ to the ATZ basis sets. This holds true for all four metals, and it can be up to 5.2 kcal/mol for the $Ba^{2+}(H_2O)_n \rightarrow [Ba(OH)]^{+} + H_3O^{+}(H_2O)$ energy difference. A similar conclusion was previously noted by Rao et al. [33] for similar systems: "An increase in the quality of the basis set from double-ζ to triple-ζ has a significant effect on the sequential binding energies, irrespective of the geometries used."

Figures 10, 11, 12 and 13 display pictorially the MP2/ATZ results of Table 1. The x axis traces the number n of the water molecules while the y axis the relative energy (kcal/mol). In those figures, the dashed lines correspond to the energy of the $M^{2+}(H_2O)_n$ minima (set as the zero of the energy scale), the plain solid lines to the products of the water loss reaction (1), the solid lines with the closed squares indicate the hydronium loss reaction (2), the dashed lines with the filled squares indicate the barrier height going from the reactants to the products of reaction (2), and finally, the solid and dashed lines with the "×" symbol pertain to the products and the transition state of the Zundel-cation loss reaction (3), respectively. The barrier heights are taken from Figs. 1, 2, 3, 4, 5, 6, 7, 8, and 9 at the MP2/ADZ level of theory. The various asymptotes are also listed on the right hand side of Figs. 10, 11, 12, and 13. In general, the trends with n shown in Figs. 10, 11, 12, and 13 are similar for all metals, except for the channel described by reaction (3) for Mg (shown with the solid lines with the "×" symbol) that crosses the one described by reaction (2) between $n = 4$ and 5. We will discuss the origin of this difference in the subsequent Sect. 5.

It is clear that the water loss channel is endothermic for every possible case of M and value of n. Note that this channel is, as expected, less endothermic for large n values since the insertion of a water molecule to the larger ion–water complexes causes additional steric repulsion.

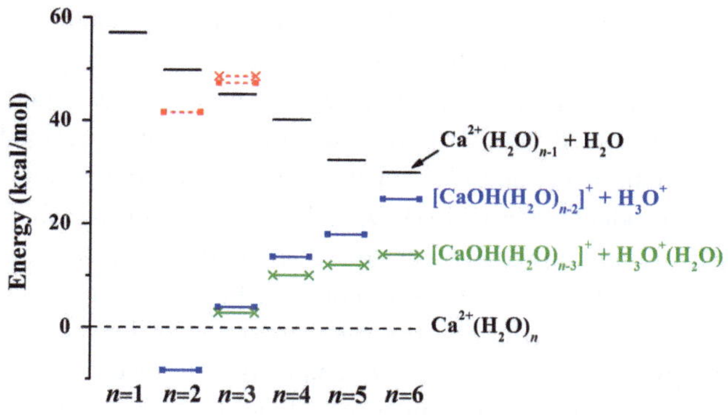

Fig. 11 Similar to Fig. 10, but for Ca

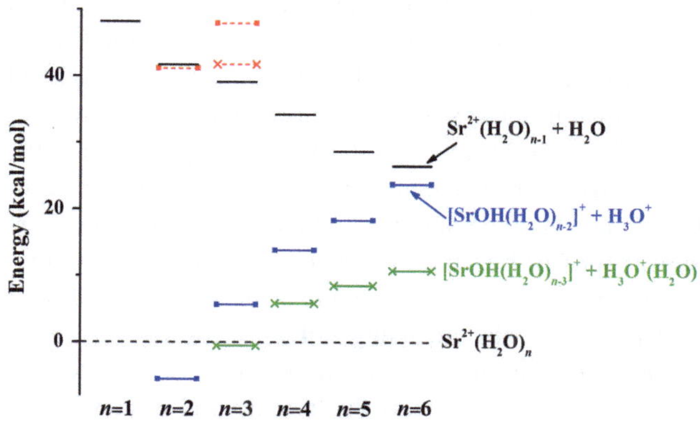

Fig. 12 Similar to Fig. 10, but for Sr

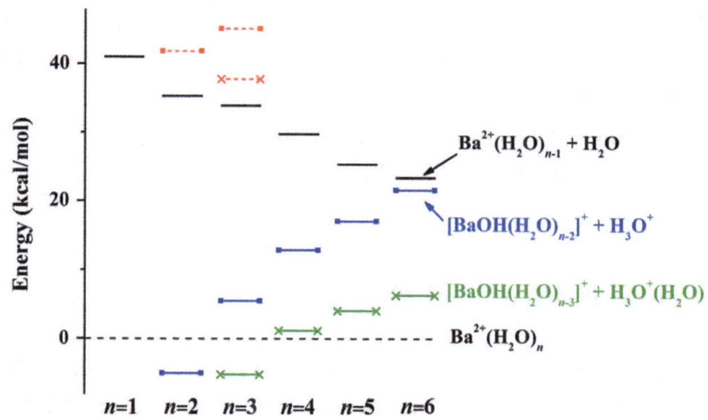

Fig. 13 Similar to Fig. 10, but for Ba

Additionally, for the same value of n, the water loss channel follows a monotonic decrease in the binding energy of the complex going from Mg to Ba. The reason is that all four metals bear the same charge but have a different ionic radius. For instance, the charge density on Ba^{2+} is smaller than that of Mg^{2+} due to its larger size. Therefore, the "affinity" between Ba^{2+} and water is smaller.

The two other channels are competitive for Mg and Ca, whereas for Sr and Ba, the Zundel loss channel produces more stable products (for $n \geq 3$). However, the hydronium and Zundel-cation loss fragments are more difficult to observe experimentally. The reason that the water loss channel is almost exclusively observed experimentally can be mainly attributed to the large energy barriers that need to be surmounted in order to produce the fragments described by reactions (2) and (3). Specifically, for Mg, the ground-state products $[Mg(OH)(H_2O)_{n-2}]^+ + H_3O^+$ for $n = 2$, 3, and 4 are lower by at least 40 kcal/mol from the $Mg^{2+}(H_2O)_{n-1} + H_2O$ asymptote (see Fig. 10). For $n = 4$, however, the barrier height is comparable to the dehydration energy, and therefore, the two channels described by reactions (1) and (2) are competitive. Judging from the trends observed for $n = 1$–3, we expect the barrier heights to be larger than the dissociation energy to neutral water for larger n. Consequently, the highest number for reaction (2) to be observed is $n = 4$, in complete agreement with the experimental observations [7]. Following the same premise and noting that the energy barriers follow an increasing trend as n goes from 2 to 3 (Figs. 11, 12, 13), we can speculate that the cluster size for the reactions (2) and (3) to occur are $n = 2$ for Ca, Sr, and Ba. The only case where the Zundel cation can appear for Ca, Ba, or Sr is for $n = 3$. More accurate predictions necessitate the calculation of the transition states for larger n values, also including ZPE corrections.

Observe that the energy difference between the two charge separation channels (hydronium and Zundel-cation loss reactions) is bigger for larger clusters and for heavier metal cations. For the same reasons discussed earlier, the binding energy of a single water molecule to a $[MOH(H_2O)_n]^+$ cluster is decreasing with increasing n and atomic number of the metal. Hence, a water molecule prefers (when n or the atomic number increase) to stay more attached to H_3O^+ with a binding energy of 33.7 kcal/mol (at MP2/ATZ) rather than to $[MOH(H_2O)_n]^+$.

5 Mono-hydroxide $[M(H_2O)_n(OH)]^+$ clusters

In this section, we examine the case of the mono-hydroxide water complexes of the titled metal dications, $[MOH(H_2O)_n]^+$. We specifically report the energy needed to detract a water molecule from these species (see Table 2). Additionally, for the case of $n = 2$, we constructed the corresponding MEPs (see Fig. 14). It is shown that for $n = 2$ the lowest energy path is the plain water loss channel and that the products of the hydronium loss process, $M(OH)_2 + H_3O^+$, are much higher in energy.

In Table 2, we list the dissociation energies of the following reaction:

$$[MOH(H_2O)_n]^+ \rightarrow [MOH(H_2O)_{n-1}]^+ + H_2O. \quad (4)$$

Similar to the case of the $M^{2+}(H_2O)_n$ clusters, the dissociation energy is a decreasing function of both n and the ionic radii. Namely, it is larger for Mg, which has the smallest ionic radius, and smaller for Ba. It is also larger for $n = 1$ and smaller for $n = 4$. Interestingly, the range of the binding energies, i.e., the difference between the $n = 1$ and $n = 4$ cases, is larger for Mg (27.6 kcal/mol), decreasing to 9.6 (Ca), 6.8 (Sr), and 4.5 (Ba) kcal/mol. In general, we observe smaller differences between the results with the ADZ and ATZ basis sets than for the case of the $M^{2+}(H_2O)_n$ clusters. Our ATZ results for Ba are in agreement with the CCSD(T) results of Ref. [18], while

Table 2 Water detachment energies (kcal/mol) corresponding to the reaction $[MOH(H_2O)_n]^+ \rightarrow [MOH(H_2O)_{n-1}]^+ + H_2O$, M = Mg, Ca, Sr, Ba and $n = 1$–4

n	MP2/ADZ	MP2/ATZ[a]	Theory[b]	MP2/ADZ	MP2/ATZ[a]	Theory[b]
		Mg			Ca	
1	50.0	50.0 (48.0)	59.1	35.3	33.0 (30.9)	36.7
2	39.1	39.1 (36.3)	46.2	32.5	30.6 (28.6)	34.3
3	31.7	31.7 (29.1)	38.4	29.5	28.2 (26.3)	31.7
4	22.6	22.4 (20.2)	27.8	25.9	23.4 (21.4)	
		Sr			Ba	
1	27.6	27.9 (26.0)		23.7	23.4 (21.1)	23.8
2	26.1	26.0 (24.0)		23.2	22.4 (20.3)	21.4
3	24.1	24.2 (22.4)		21.5	21.1 (19.1)	20.3
4	21.8	21.1 (18.9)		19.5	18.9 (16.8)	

[a] ZPE-corrected values are given in parentheses

[b] *Ab initio* values from the literature: Ref. 22 for Mg (MP4SDTQ/6-31G*//SCF/6-31G*), Ref. 26 for Ca (MP2//SCF, 9s7p3d/(Ca), 6-31G*/(O,H)), and Ref. 18 for Ba (CCSD(T)//DFT/*m*PW1PW91)

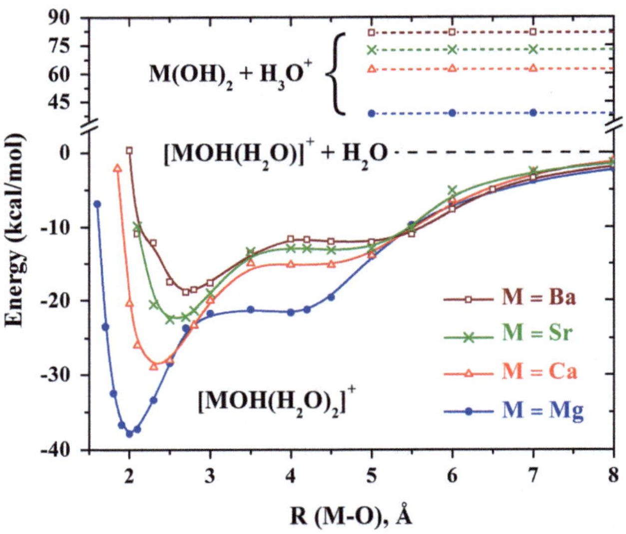

Fig. 14 Minimum energy paths for the $[MOH(H_2O)_2]^+$ clusters, where M = Mg, Ca, Sr, and Ba. The *dashed lines* on the *top* part of the Figure mark the various $M(OH)_2 + H_3O^+$ asymptotes

they differ more than 5 kcal/mol for Mg from the MP4SDTQ results of Ref. [22]. The MP2 results from Ref. [26] for Ca are in good agreement with ours.

It is noteworthy that a water molecule binds stronger to $M^{2+}(H_2O)_{n+1}$ compared to $[MOH(H_2O)_n]^+$ (see Tables 1, 2) due to the stronger Coulombic interaction with a doubly charged metal core. The suggested [24, 29] electronic structure $M^{2+}(H_2O)_n(OH^-)$ indicates the same charge–dipole interaction between the additional water and the metal center for the two systems. However, the negatively charged OH^- group can cause the opposite effect by reducing the binding energy.

The results of Table 2 further explain the qualitative difference seen in the trend of the energetics of channel (3) for Mg (indicated by the solid lines with the "×" symbol in Fig. 10) when compared to the rest of the metals (Figs. 11, 12, 13). The energy levels of reaction channel (3) cross the ones for channel (2) between $n = 4$ and 5 for Mg, whereas for all other metals, there is no such crossing. The reason is that the $[MOH(H_2O)_n]^+$ water detachment energies for the Mg $n = 1$ and 2 clusters (cf. Table 2) are larger than the hydronium–water interaction energy (33.7 kcal/mol at MP2/ATZ), whereas for all other metals, they are smaller than this value, and therefore, the energy levels for channel (3) always lie below the ones for channel (2) for Ca, Sr, and Ba.

In Fig. 14, we show the MEPs corresponding to the removal of one water molecule from $[MOH(H_2O)_2]^+$. These MEPs are constructed by fixing the distance between the metal and the oxygen of the departing water molecule at different lengths and optimizing all other geometrical parameters. After a plateau of width larger than 1 Å for distances $R > 3$ Å, the MEPs converge smoothly to the $[MOH(H_2O)]^+ + H_2O$ asymptote. The aforementioned plateau is caused by the hydrogen bond between the water that is attached to the metal and the one that is departing. The energy of the fragments $M(OH)_2 + H_3O^+$ is shown with dashed lines for the various metals and ranges from ~38 kcal/mol (Mg) to ~78 kcal/mol (Ba) above the $[MOH(H_2O)]^+ + H_2O$ lowest energy asymptote.

The hydration energy of 33.7 kcal/mol (MP2/ATZ) for H_3O^+ is generally larger (except for Mg and $n = 2$) than the binding energy of a water to $[MOH(H_2O)_{n-1}]^+$ indicating that the $M(H_2O)_{n-3}(OH)_2 + H_3O^+(H_2O)$ fragments will be closer in energy to the $[MOH(H_2O)_{n-1}]^+ + H_2O$ ones. Therefore, the possibility exists for $[MOH(H_2O)_n]^+$ ($n \geq 3$) to dissociate to $M(H_2O)_{n-3}(OH)_2 + H_3O^+(H_2O)$, especially for Ba and large n. The Supporting Information includes the geometries and energies of the $M(OH)_2$ molecules as well as the energy required to detach a hydroxide from $[M(H_2O)_n(OH)]^+$, i.e., the dissociation energies for the $[M(H_2O)_n(OH)]^+ \rightarrow [M(H_2O)_n]^{2+} + OH^-$ reaction.

6 Synopsis

We examined the MEPs along which a water molecule can be detached from the pure and mono-hydroxide water complexes of the alkaline earth metals Mg^{2+}, Ca^{2+}, Sr^{2+}, and Ba^{2+}. Three different channels were investigated: one leading to the unimolecular water loss and two leading to the loss of a hydronium and a Zundel (singly hydrated hydronium) cation. For the cases with two, three, and four (only for Mg) water molecules, we constructed MEPs at the MP2/ADZ level of theory, while for the larger systems, we investigated only the reactants and products of the above three processes at the MP2/ATZ level. In the case of the mono-hydroxide compounds, we report the MEPs for the systems with two water molecules and the energetics with up to four water molecules.

Despite the voluminous work reported in the literature for the systems studied presently, this is the first systematic work encompassing all four metals with up to six water molecules and three dissociation channels. Our results are consistent with several experimental observations, such as the fact that only the small clusters dissociate to fragments involving the hydronium cation. Additionally, we propose the possibility of the production of the Zundel cation for the heavier metals with three water molecules. The large energy barriers of the charge separation channels for the largest complexes prevent them from being observed, although they are more exothermic than the water loss ones. Finally, mono-hydroxide water clusters prefer the pure water loss channel, because it is energetically more favorable.

Acknowledgments We acknowledge useful discussions with Drs. Nikolai Petrik and Gregory Kimmel of PNNL. This work was supported by the US Department of Energy, Office of Basic Energy Sciences, Division of Chemical Sciences, Geosciences and Biosciences. Pacific Northwest National Laboratory (PNNL) is a multi-program national laboratory operated for DOE by Battelle. This research used resources of the National Energy Research Scientific Computing Center, which is supported by the Office of Science of the US Department of Energy under Contract No. DE-AC02-05CH11231.

References

1. Spears KG, Fehsenfeld FC (1972) J Chem Phys 56:5698
2. Blades AT, Jayaweera P, Ikonomou MG, Kerbale P (1990) J Chem Phys 92:5900
3. Peschke M, Blades AT, Kerbale P (1998) J Phys Chem A 102:9978
4. Pye CC, Rudolph WW (1998) J Phys Chem A 102:9933
5. Rodriguez-Cruz SE, Jockush RA, Williams ER (1999) J Am Chem Soc 121:8898
6. Barran PE, Walker NR, Stace AJ (2000) J Chem Phys 112:6173
7. Shvartsburg AA, Michael Sin KW (2001) J Am Chem Soc 123:10071

8. Markham GD, Glusker JP, Bock CW (2002) J Phys Chem B 106:5118
9. Wong RL, Paech K, Williams ER (2004) Int J Mass Spectrom 232:59
10. Bush MF, Saykally RJ, Williams ER (2007) Chem Phys Chem 8:2245
11. Carl DR, Moison RM, Armentrout PB (2007) Int J Mass Spectrom 265:308
12. Carl DR, Moison RM, Armentrout PB (2009) J Am Soc Mass Spectrom 20:2312
13. Bush MF, O'Brien JT, Prell JS, Wu C-C, Saykally RJ, Williams ER (2009) J Am Chem Soc 131:13270
14. Carl DR, Chatterjce BK, Armentrout PB (2010) J Chem Phys 132:044303
15. Cabavillas-Vidosa I, Rossa M, Pino GA, Ferrero JC, Cobos CJ (2012) Phys Chem Chem Phys 14:4276
16. Carl DR, Armentrout PB (2012) J Phys Chem A 116:3802
17. Carl DR, Armentrout PB (2013) Chem Phys Chem 14:681
18. Cabavillas-Vidosa I, Rossa M, Pino GA, Ferrero JC, Cobos CJ (2013) J Phys Chem A 117:4997
19. Kaupp M, Schleyer PR (1992) J Phys Chem 96:7316
20. Klobukowski M (1992) Can J Chem 70:589
21. Bauschlicher CW Jr, Sodupe M, Partridge H (1992) J Chem Phys 96:4453
22. Watanabe H, Iwata S, Hashimoto K, Misaizu F, Fuke K (1995) J Am Chem Soc 117:755
23. Glendening ED, Feller D (1996) J Chem Phys 100:4790
24. Kaufman Katz A, Glusker JP, Beebe SA, Bock CW (1996) J Am Chem Soc 118:5752
25. Markham GD, Glusker JP, Bock CL, Trachtman M, Bock CW (1996) J Chem Phys 100:3488
26. Watanabe H, Iwata S (1997) J Phys Chem A 101:487
27. Pavlov M, Siegbahn EM, Sandström M (1998) J Phys Chem A 102:219
28. Peschke M, Blades AT, Kebarle P (1999) Int J Mass Spectrom 185/186/187:685
29. Beyer M, Williams ER, Bondybey VE (1999) J Am Chem Soc 121:1565
30. Dudev T, Lim C (1999) J Phys Chem A 103:8093
31. Merrill GN, Webb SP, Bivin DB (2003) J Phys Chem A 107:386
32. Dang LX, Schenter GK, Glezakou VA, Fulton JL (2006) J Phys Chem B110:23644
33. Sirinivasa Rao J, Dinadayalane TC, Leszczynski J, Narahavi Sastry G (2008) J Phys Chem A 112:12944
34. Lei XL, Pan BC (2010) J Phys Chem A 114:7595
35. Gonzalez JD, Florez E, Romero J, Reyes A, Restrepo A (2013) J Mol Model 19:1763
36. Miliordos E, Xantheas SS (2013) Phys Chem Chem Phys. doi:10.1039/c3cp53636j
37. Baes CFJ, Mesmer RE (1986) The hydrolysis of cations. Krieger Publishing Company, Malabar, FA
38. Moore CE (1971) Atomic energy levels as derived from the analysis of optical spectra. COM-72-50283, US Department of Commerce, National Technical Information Service, National Bureau of Standards, Washington, DC
39. Lias SG, Bartmess JE, Liebman JF, Holmes JL, Levin RD, Mallard WG (1988) J Phys Chem Ref Data 17(Suppl 1):1
40. Dunning TH Jr (1989) J Chem Phys 90:1007
41. Kendall RA, Dunning TH Jr, Harrison RJ (1992) J Chem Phys 96:6796
42. Woon DE, Peterson KA, Dunning TH Jr, in preparation, and as implemented in MOLPRO (ref [45])
43. Koput J, Peterson KA (2002) J Phys Chem A 106:9595
44. Lim IS, Stoll H, Schwerdtfeger P (2006) J Chem Phys 124:034107
45. MOLPRO, version 2010.1, a package of ab initio programs, Werner HJ, Knowles PJ, Knizia G, Manby FR, Schütz M, Celani P, Korona T, Lindh R, Mitrushenkov A, Rauhut G, Shamasundar KR, Adler TB, Amos RD, Bernhardsson A, Berning A, Cooper DL, Deegan MJO, Dobbyn AJ, Eckert F, Goll E, Hampel C, Hesselmann A, Hetzer G, Hrenar T, Jansen G, Köppl C, Liu Y, Lloyd AW, Mata RA, May AJ, McNicholas SJ, Meyer W, Mura ME, Nicklass A, O'Neill DP, Palmieri P, Pflüger K, Pitzer R, Reiher M, Shiozaki T, Stoll H, Stone AJ, Tarroni R, Thorsteinsson T, Wang M, Wolf A. http://www.molpro.net
46. Gaussian 09, Revision D.01, Frisch MJ, Trucks GW, Schlegel, HB, Scuseria GE, Robb MA, Cheeseman JR, Scalmani G, Barone V, Mennucci B, Petersson GA, Nakatsuji H, Caricato M, Li X, Hratchian HP, Izmaylov AF, Bloino J, Zheng G, Sonnenberg JL, Hada M, Ehara M, Toyota K, Fukuda R, Hasegawa J, Ishida M, Nakajima T, Honda Y, Kitao O, Nakai H, Vreven T, Montgomery JAJr, Peralta JE, Ogliaro F, Bearpark M, Heyd JJ, Brothers E, Kudin KN, Staroverov VN, Kobayashi R, Normand J, Raghavachari K, Rendell A, Burant JC, Iyengar SS, Tomasi J, Cossi M, Rega N, Millam NJ, Klene M, Knox JE, Cross JB, Bakken V, Adamo C, Jaramillo J, Gomperts R, Stratmann RE, Yazyev O, Austin AJ, Cammi R, Pomelli C, Ochterski JW, Martin RL, Morokuma K, Zakrzewski VG, Voth GA, Salvador P, Dannenberg JJ, Dapprich S, Daniels AD, Farkas Ã, Foresman JB, Ortiz JV, Cioslowski J, Fox DJ (2009) Gaussian, Inc., Wallingford, CT

Theor Chem Acc (2014) 133:1431
DOI 10.1007/s00214-013-1431-z

REGULAR ARTICLE

Stacking of the mutagenic DNA base analog 5-bromouracil

Leo F. Holroyd · Tanja van Mourik

Received: 10 October 2013 / Accepted: 25 November 2013 / Published online: 14 December 2013

Abstract The potential energy surface of the stacked 5-bromouracil/uracil (BrU/U) dimer has been investigated in the gas phase and in solution (water and 1,4-dioxane), modeled by a continuum solvent using the polarizable continuum model. Minima and transition states were optimized using DFT (the M06-2X density functional and the 6-31+G(d) basis set). Six stacked gas-phase BrU/U minima were located: four in the face-to-back orientation and two face-to-face. The global minimum in the gas phase is a face-to-face structure with a twist angle of 60° and a zero-point energy-corrected interaction energy of −10.7 kcal/mol. The BrU/U potential energy surface is geometrically and energetically similar to that of U/U (Hunter and Van Mourik in J Comput Chem 33:2161, 2012). Energy calculations were also performed on experimental geometries of stacked dimers (47 containing BrU stacking with either adenine, cytosine, guanine or thymine and 51 containing thymine also stacking with one of those four bases) taken from DNA structures in the Protein Data Bank. Single-point interaction energies were computed at different levels of theory including MP2, CCSD(T) and DFT using the mPW2PLYP-D double-hybrid functional augmented with an empirical dispersion term, using basis sets ranging from aug-cc-pVDZ to aug-cc-pVQZ. No strong evidence was found for the suggestion that the mutagenicity of BrU is due to enhanced stacking of BrU compared to the corresponding stacked dimers involving thymine.

Keywords 5-Bromouracil · DFT · Density functional theory · Stacking · Dispersion · Mutagenicity

Dedicated to Professor Thom Dunning and published as part of the special collection of articles celebrating his career upon his retirement.

Electronic supplementary material The online version of this article (doi:10.1007/s00214-013-1431-z) contains supplementary material, which is available to authorized users.

L. F. Holroyd · T. van Mourik (✉)
EaStCHEM School of Chemistry, University of St Andrews, North Haugh, St Andrews KY16 9ST, UK
e-mail: tanja.vanmourik@st-andrews.ac.uk

1 Introduction

The structure of DNA is crucial for its replicatory function. One of the two major forces determining the structure of the helix—the better understood one—is the specific Watson–Crick base pairing (*inter*strand hydrogen bonding) of adenine (A) with thymine (T) and of guanine (G) with cytosine (C). However, it is by now well established that the intrastrand π-stacking of adjacent bases is also crucial and may be quantitatively even more significant than base pairing as a thermodynamic driving force for the stabilization of DNA [1–4].

A comprehensive understanding of stacking in vivo is yet to be achieved, since it turns out to be a rather subtle blend of electrostatic, dispersive, solvent-driven and other effects, and is highly geometry-dependent. However, recent progress on both the theoretical and experimental sides has been impressive—see, e.g., Šponer et al. [4, 5] for reviews of insights from quantum chemical studies, and Kool [1] for an earlier review of experimental knowledge.

To investigate the balance of different energetic contributions to stacking, one very fruitful experimental approach has been to substitute bases from beyond the natural set of nucleic acid components [uracil (U), thymine (T), adenine (A), cytosine (C) and guanine (G)]. For example, Kool et al. [6] studied the effects of adding various natural and non-natural dangling nucleosides into DNA 7-mers (dangling nucleosides lie at the end of a DNA strand but have no hydrogen bonding partner on the other strand, i.e., they take part in stacking but not base pairing). They found that stacking strength correlates quite well with surface area, and to a certain extent with polarizability, but does not appear to require a large dipole moment (in fact it decreases with polarity for constant surface area).

Despite recent progress, the challenge of correctly describing stacking interactions in computational chemistry remains non-trivial and is a key criterion in the development and assessment of new correlated electronic structure methods (especially density functionals).

The London dispersion force, a second-order correlation effect, is undoubtedly a major stabilizing component of stacking [1, 4, 7–10] (and of aromatic interactions in general) [11], but it represents a particular difficulty for DFT. Most standard GGA and hybrid GGA functionals, such as B3LYP, do not capture dispersion at all and cannot be used for π-stacked systems [5, 12, 13]. MP2 does capture dispersion, but tends to overbind these kinds of complexes as it approaches the complete basis set limit, unless the coupled cluster correction (the difference between MP2 and, typically, CCSD(T) calculated in a small basis set) is included [5, 14].

Along with some other methods, MP2 is also highly susceptible to basis set superposition error (BSSE) when small basis sets are used, causing an artificial attraction which is often highly geometry-dependent and can therefore produce false energy minima or hide true ones [15–17]. The counterpoise procedure (CP) of Boys and Bernardi [18] is therefore recommended.

In this study of the stacking of 5-bromouracil, uracil and thymine, we use the highly parameterized non-local density functional M06-2X [19] and the dispersion-corrected double-hybrid functional RI-mPW2PLYP-D [20, 21]. M06-2X contains a large number of parameters obtained through the use of training sets that contain dispersion-dominated complexes. It is therefore expected to describe dispersion effects well, and recent studies have supported this [22–27]. Hunter and Van Mourik used this functional to explore the energy landscapes of the stacked U/U (uracil) and T/T (thymine) dimers, and found an extremely good agreement between the CP-corrected M06-2X/6-31+G(d) interaction energies and the available CCSD(T)/CBS data [28]. In preliminary calculations for the present study, we calculated the interaction energies of five of Hunter and Van Mourik's M06-2X/6-31+G(d)-optimized U/U and T/T dimers using RI-mPW2PLYP-D/aug-cc-pVTZ, and noted that they were all reproduced to within 0.5 kcal/mol.

Our group is interested in the thymine analogue 5-bromouracil (BrU) due to its well-established mutagenicity [29–32]. The mutagenic behavior of BrU is believed to be a result of its ability to incorporate into DNA in place of thymine, and then mispair with guanine, rather than adenine. The conventional model of the BrU-G mispair contains BrU in its enolized form (containing a hydroxyl group at O4), paired with guanine in pseudo-Watson–Crick fashion via three hydrogen bonds (Scheme 1, left). The assumption is that the bromine substituent somehow stabilizes the enol tautomer, and the enol-containing mispair is protected against enzymatic repair due to its stereochemical match to the correct G-C pair, hence the greater tendency of BrU to form stable mutagenic mispairs.

Arguing against the enol model, however, various experimental groups have noted that the observed BrU-G mispairs in synthetic polynucleotides contain BrU hydrogen bonding with G in wobble pair (non-Watson–Crick) fashion, implying the canonical keto tautomer of BrU (Scheme 1, right). They have suggested that the stabilization of the mispair is caused by enhanced stacking compared to the corresponding T-G and U-G mispairs, and that this, rather than tautomerization, lies behind the mutagenicity of BrU [30–32].

The present paper describes a comprehensive search of the energy landscape of the stacked dimer formed by BrU and U, stacking in both face-to-back and face-to-face orientations (see below for the definition of these terms). This work complements the earlier study by Hunter and Van Mourik, in which all minima and most transition states

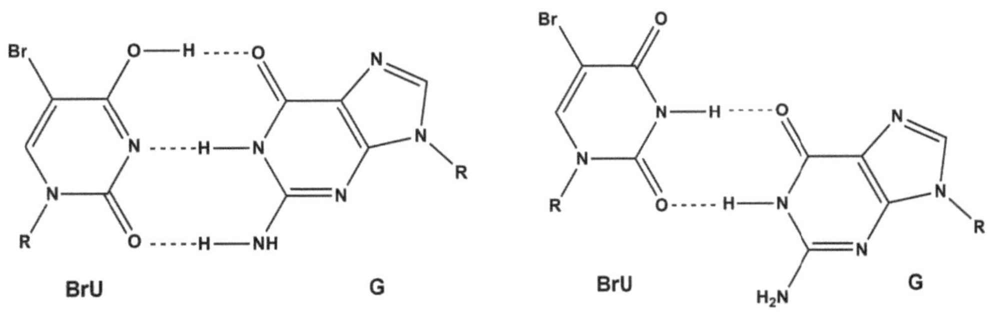

Scheme 1 The enolized (*left*) and wobble pair (*right*) structures of the 5-bromouracil-guanine base mispair

(TSs) for the stacked U/U and T/T dimers were located [28]. In the current paper, we will also present calculations comparing the relative stacking strengths of BrU and T in fixed (non-optimized) geometries taken from experimental DNA structures, which contain only face-to-back stacking. Considered together, these studies allow us to explore the effect of a bromine substituent on the stacking behavior of pyrimidine-type nucleic acid bases.

2 Methodology

2.1 BrU/U dimers

5-Bromouracil/uracil stacked structures were created by placing the two base molecules in a parallel configuration. Two geometries were considered: face-to-back (one molecule exactly on top of the other) and face-to-face (with U rotated by 180° around the C2–C5 axis so that its N1 and N3 atoms are above the N3 and N1 atoms, respectively, of BrU; see Scheme 2 for the atom labeling). Dummy atoms (X) were placed in the geometric centers of each ring. The vertical separation between the bases was chosen as 3.4 Å, as this is a typical value in B-form DNA [33]. Rotational energy profiles were created by rotating one base around the geometric center of its ring, while keeping the other fixed, by varying the O2(BrU)–X(BrU)–X(U)–O2(U) torsion angle (τ_{twist}) between 0 and 355° with a step size of 5°. All other geometrical parameters were held constant.

Interaction energies were calculated at each step using the M06-2X [19] functional and the 6-31+G(d) basis set. The counterpoise (CP) method [18] was used to correct for BSSE. The CP-corrected interaction energies follow from:

$$\Delta E^{CP} = E^{CP}_{BrU/U}(BrU/U) - E_{BrU}(BrU) - E_U(U) \quad (1)$$

The subscripts refer to the molecular system, while the round brackets indicate whether the calculation was done at the optimized geometry of the stack (BrU/U) or at the

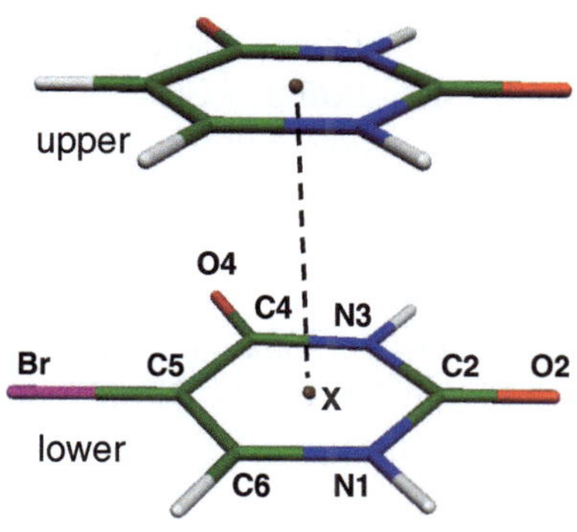

Scheme 2 Atom labeling in the stacked BrU/U dimer

monomer-optimized geometries (BrU and U). The CP-corrected stacked dimer energies are obtained as follows:

$$E^{CP}_{BrU/U} = E^{noCP}_{BrU/U}(BrU/U) - BSSE \quad (2)$$

$$BSSE = E^{\{BrU/U\}}_{BrU}(BrU/U) + E^{\{BrU/U\}}_{U}(BrU/U) - E^{\{BrU\}}_{BrU}(BrU/U) - E^{\{U\}}_{U}(BrU/U) \quad (3)$$

The superscripts in curly brackets indicate whether the calculation is done in the basis set of the dimer, {BrU/U}, or in the monomer basis sets, {BrU} and {U}. The monomer geometries in the scans were taken to be M06-2X/6-31+G(d)-optimized geometries (optimized in the gas phase for the gas-phase scans or in the corresponding solvent in the solution-phase scans described below). With this choice of monomer geometries, $E^{\{B\}}_B(BrU/U)$ and $E_B(B)$ (B = U or BrU) are identical in Eqs. 1–3. The calculations were performed using the "Scan" and "Counterpoise" keywords in Gaussian 09 [34].

The calculations were performed in the gas phase and also in solution. The solution was modeled by the polarizable continuum model (PCM), using the integral equation formalism variant (IEFPCM), via the "SCRF" keyword in Gaussian 09. The cavity containing the solute was constructed by placing spheres around each solute atom (including all hydrogens) using universal force field (UFF) radii. Note that this procedure does not yield the free energy of stacking in solution, but rather a modified electronic energy that differs from the gas phase term due to the electrostatic interaction of the solute with the reaction field of the solvent, the screening of the electrostatic interaction of the bases, and other effects such as that associated with creating the solute cavity [4]. Two different solvents were used: water (dielectric constant $\varepsilon = 78.3553$) and 1,4-dioxane ($\varepsilon = 2.2099$), the latter of which was chosen because it is a typical solvent of low dielectric constant used in the crystallization of bio-organic compounds.

The minima obtained in the potential energy scans were subsequently fully geometry-optimized at the same level of theory, M06-2X/6-31+G(d). TSs were located by using the synchronous transit-guided quasi-newton (STQN) method [35], using the QSTn ($n = 2$ or 3) keywords, in some cases followed by conventional transition state searches using the "Opt = TS" keyword. The geometry optimizations and transition state searches were carried out using Gaussian's "Tight" convergence criteria. All calculations employed Gaussian's "Ultrafine" integration grid (containing 99 radial shells and 590 angular points per shell). Harmonic vibrational frequencies were computed at the same levels of theory to verify the nature of the stationary points (minima or TSs) and to compute zero-point energy (ZPE) corrections to the interaction energies.

In the gas-phase calculations, the counterpoise correction was applied throughout the geometry optimizations and transition state searches. In the PCM calculations, the counterpoise correction was applied only to the (non-counterpoise) optimized structures, in separate calculations without the "SCRF" keyword, as it is not possible to implement counterpoise during a PCM calculation.

In addition to the optimized twist angle, we also computed the vertical distance between two stacked bases (analogous to the rise of a helical base step), the horizontal displacement from the position where one base is exactly on top of the other (analagous to the slide) and the angle between the two planes (analogous to the tilt) of the optimized stacked dimers. The fixed values of these parameters during the rotational scans had been as follows: vertical separation = 3.4 Å, horizontal separation = 0 Å and angle between planes = 0°. The optimized geometrical parameters were calculated with an in-house Fortran 90 program, using a method that has also been previously described in Ref. 28. The vertical distance was calculated as the scalar product of the vector connecting the midpoints of the base rings ($\overrightarrow{m_1m_2}$; calculated as the average of the atomic coordinates of the C2, C4 and C6 atoms) and the plane vector of one of the bases (where a plane is defined by the C2, C4 and C6 atoms). This corresponds to the component of $\overrightarrow{m_1m_2}$ that is perpendicular to the base plane. As the two bases are not completely parallel in the optimized structures, this scalar product depends on which base's plane vector is used. The vertical distance presented herein is therefore computed as the average of the two values obtained when using the first or the second base's plane vector. The horizontal distance is calculated as the component of the vector connecting the two midpoints ($\overrightarrow{m_1m_2}$) that is in the plane of one base. Like for the vertical separation, the horizontal separation is calculated as the average of the two values obtained when using the first or the second base's plane vector. The angle between the planes of the two bases is simply the angle between their plane vectors.

The gas-phase minima obtained with M06-2X/6-31+G(d) were subsequently re-optimized with the dispersion-corrected double-hybrid functional RI-mPW2PLYP-D [20, 21] using ORCA 2.8 [36]. This functional, developed by Grimme, consists of a modified Perdew–Wang exchange term with a 55 % Hartree–Fock contribution, a LYP correlation term with a 25 % perturbative (RI-MP2) contribution employing resolution of the identity (RI) [37] and a D2-type dispersion correction [38]. The optimizations were performed using the aug-cc-pVDZ basis set. Additionally, single-point energies were calculated using the aug-cc-pVTZ basis set, both for the re-optimized structures and for the original structures obtained with M06-2X.

2.2 Experimental BrU and T stacked structures

To obtain further information on the stacking strength of BrU relative to the naturally occurring DNA bases, we have also performed energy calculations on experimental geometries of stacked dimers taken from DNA structures in the Protein Data Bank (PDB) [39]. To this end, we located 28 PDB entries for experimental DNA molecules containing dBrU nucleotides in the chain. From these structures, 47 different stacked dimers containing BrU stacking with either A, C, G or T were created. These dimers were created by isolating two adjacent nucleotides within a strand (i.e., deleting all other nucleotides, as well as any waters), then removing all the sugar-phosphate backbone atoms, and saturating the bases by adding hydrogen atoms in positions corresponding to the canonical tautomers. The monomers were not methylated at any position: the N1 atoms were simply capped with hydrogen. The hydrogen positions were optimized as described below.

Since we are concerned with DNA (as opposed to RNA), the most relevant comparison to be made is between the stacking strengths of BrU and T—in the context of BrU's mutagenicity, it is necessary to explain why the BrU-G mispair is more stable than the T-G mispair in DNA. We therefore also isolated 51 dimers containing T stacking with A, C, G or T, from 17 separate DNA polynucleotides in the PDB (four of which were also among the BrU-containing entries). All the PDB entries used to generate structures in this study are listed in the Supplementary Material, Table S1. The resulting total of 98 stacked dimers can be sorted into fifteen types according to sequence: A/BrU, BrU/A, C/BrU, BrU/C, G/BrU, BrU/G, T/BrU, BrU/T, A/T, T/A, C/T, T/C, G/T, T/G and T/T. The base listed first lies toward the 5′ end of the strand, while the base listed second lies toward the 3′ end. Note that in DNA, any given base pair of type X/Y (using this notation) is geometrically distinct from Y/X.

To perform the calculations, firstly, the H atomic positions were geometry-optimized, while keeping all other atoms frozen, using the "ModRedundant" keyword in Gaussian 09. These partial optimizations were performed on the stacked base dimers (rather than the isolated monomers) at the M06-2X/6-31+G(d) level of theory. The absence of full optimization of the dimers results in the use of unrelaxed experimental geometries, meaning that any experimental errors will also be present in the structures used in the calculations.

Single-point interaction energies of the 98 stacked dimers were then calculated at the RI-mPW2PLYP-D/aug-cc-pVTZ level of theory. The counterpoise-corrected interaction (stacking) energies E_{int} were defined as:

$$E_{\text{int}} = E_{\text{AB}} - E_{\text{A}} - E_{\text{B}} \quad (4)$$

where E_{AB}, E_A and E_B are the energies of the complete dimer, the first base (toward the 5′ end of the helix) and the second base (toward the 3′ end), respectively. The energies of each isolated monomer were calculated in the presence of ghost orbitals associated with the other monomer. This "vertical" counterpoise method takes into account BSSE but not deformation energy, which we have neglected because the monomers are in non-optimized geometries taken directly from experimental data. Due to memory issues, the counterpoise correction for one of the A/T dimers (PDB 1A35) could not be performed with RI-mPW2PLYP.

Single-point interaction energies were also calculated at the following levels of theory: df-MP2/aug-cc-pVDZ, df-MP2/aug-cc-pVQZ and df-LCCSD(T0)/aug-cc-pVDZ. These calculations, which employed the density fitting (df) and (for coupled cluster only) local (L) orbital approximations [40], were performed using MOLPRO 2010.1 [41]. The T0 extension in df-LCCSD(T0) indicates the inclusion of perturbative noniterative local triple excitations [42, 43]. Orbital localization was carried out according to the Pipek-Mezey procedure [44]. The vertical counterpoise correction was applied through the use of ghost orbitals, as above. By focal-point analysis [45], assuming that the coupled cluster correction to df-MP2 is already converged at the aug-cc-pVDZ basis set level, the df-LCCSD(T0)/aug-cc-pVQZ energies of each dimer and monomer were estimated as:

$$E^{\text{aug-cc-pVQZ}}_{\text{df-LCCSD(T0)}} \approx E^{\text{aug-cc-pVQZ}}_{\text{df-MP2}} + E^{\text{aug-cc-pVDZ}}_{\text{df-LCCSD(T0)}} - E^{\text{aug-cc-pVDZ}}_{\text{df-MP2}} \quad (5)$$

where the quantity $[E^{\text{aug-cc-pVDZ}}_{\text{df-LCCSD(T0)}} - E^{\text{aug-cc-pVDZ}}_{\text{df-MP2}}]$ is the coupled cluster correction, ΔCC. It is established that this correction converges with respect to basis set faster than the coupled cluster energy itself does [5, 46]. The MP2 method systematically overbinds aromatic stacked clusters, but ΔCC (which is nearly always repulsive) is known to remedy this [5, 14].

3 Results

3.1 BrU/U dimers

Figure 1 shows the M06-2X/6-31+G(d) gas-phase interaction energy of the face-to-back BrU/U stacked dimer as a function of the τ_{twist} angle. The BSSE is fairly large, varying between 3.05 kcal/mol (when $\tau_{\text{twist}} = 320°$) and 3.67 kcal/mol (when $\tau_{\text{twist}} = 95°$). Energy minima are located at $\tau_{\text{twist}} = 85°$, 170°, 205° and 290°. Energy maxima are located at $\tau_{\text{twist}} = 5°$, 120°, 185° and 240°. The interaction energies around the highest maximum, at 5°, are positive (repulsive), presumably due to the use of a rigid scan. Full relaxation of the monomer geometries and intermonomer geometry would lead to negative interaction energies. Note that all fully optimized minima and TSs (including the one near 5° (with $\tau_{\text{twist}} = 9°$))have negative interaction energies (Table 1).

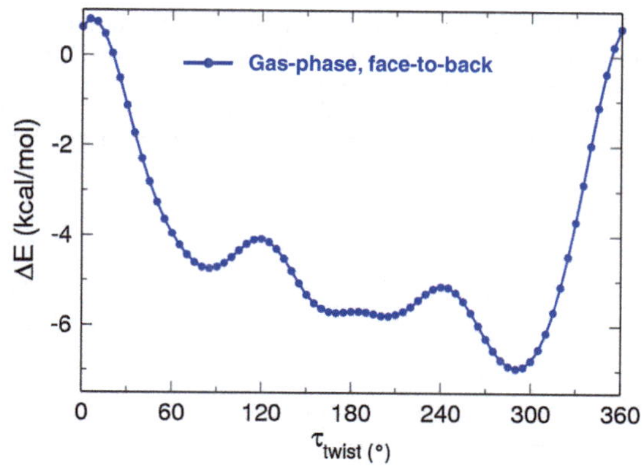

Fig. 1 Interaction energies as a function of the twist angle for the gas-phase face-to-back BrU/U stacked dimer calculated at the M06-2X/6-31+G(d) level of theory

Figure 2 shows the interaction energies of the face-to-back BrU/U stacked dimer in PCM solvent (water and 1,4-dioxane) as a function of the τ_{twist} angle. The BSSE must be calculated without taking into account the influence of the solvent. In both solvents, there are four minima, located at $\tau_{\text{twist}} = 90°$, 155°, 210° and 295° in water and 90°, 165°, 205° and 290° in 1,4-dioxane. The maxima are at 0°, 125°, 180° and 245° in water and 5°, 120°, 185° and 245° in 1,4-dioxane.

The water curve is much flatter than the gas and 1,4-dioxane curves, with only 2.3 kcal/mol separating the highest maximum (at 0°) from the deepest minimum (at 295°). This reduction in the twist dependence of stacking in a highly polar solvent has been observed before and is due to solvent screening, which contributes a free energy term that is stabilizing for geometries with unfavorable electrostatic interactions and vice versa, and thus largely compensates for the electrostatic contribution to stacking [4].

Full geometry optimization of the four gas-phase minima yielded three distinct structures, since optimization from $\tau_{\text{twist}} = 165°$ and $\tau_{\text{twist}} = 205°$ yielded identical structures. All three optimized minima were confirmed as true minima by the absence of negative frequencies of the calculated harmonic vibrational modes. Table 1 shows the interaction energies and geometrical parameters of all face-to-back minima (and TSs) in the gas phase. All interaction energies are negative, including those of the TSs. The most stable structure has a twist angle of 287°, and its CP-corrected, ZPE-corrected M06-2X/6-31+G(d) interaction energy is −9.8 kcal/mol. However, it is almost isoenergetic with the minimum at $\tau_{\text{twist}} = 184°$, which has an

Table 1 Interaction energies (in kcal/mol) and structural parameters (distances in Å, angles in degrees) of the face-to-back BrU/U minima and transition states in the gas phase calculated at the M06-2X/6-31+G(d) level of theory

τ_{twist} (initial)	τ_{twist} (final)	Angle between planes	Vertical separation	Horizontal separation	Im. Freq.[a]	ΔE^{CP}	ΔE_0^{CP}
5	9	18	2.94	3.72	1	−5.7	−5.3
85	68	20	3.12	0.91	0	−7.9	−7.2
120	189	4	3.15	1.92	1	−8.3	−7.7
115[b]	183	2	3.00	2.37	0	−8.9	−8.2
205	184	9	3.09	1.51	0	−10.1	−9.5
240	226	6	3.22	0.69	1	−7.5	−7.0
290	287	12	3.06	0.89	0	−10.6	−9.8

[a] Number of imaginary frequencies

[b] This minimum was obtained by "following" the imaginary frequency of the transition state listed above it

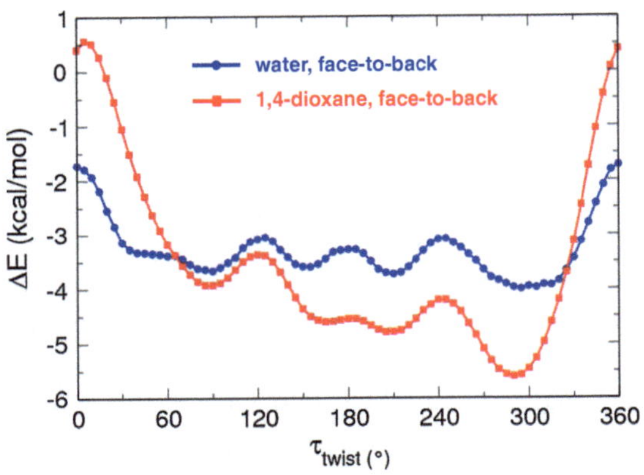

Fig. 2 Interaction energies as a function of the twist angle for the face-to-back BrU/U stacked dimer in PCM solvent (water and 1,4-dioxane) calculated at the M06-2X/6-31+G(d) level of theory

interaction energy of −9.5 kcal/mol. The third minimum is less stable ($\Delta E^{CP} = -7.2$ kcal/mol). None of these structures are perfectly planar or parallel, as can be seen from the nonzero values of the horizontal separation of the two bases and the angles between their planes. All three minima also exhibit a decrease in vertical separation, to ~3.1 Å, compared with the fixed value of 3.4 Å in the rotational scans.

Transition states were successfully located by launching QSTn searches from three of the four maxima in the gas-phase energy profile. However, the TS obtained from $\tau_{twist} = 120°$ (optimized twist: 189°) has an interaction energy of −7.7 kcal/mol, more negative than the minimum obtained with $\tau_{twist} = 68°$ (−7.2 kcal/mol), suggesting that it does not connect that minimum with any other. By "following the imaginary frequency"—performing geometry optimizations on structures obtained from the TS by adding or subtracting the magnitudes of the displacement of each atom in the imaginary vibrational mode—two minimum-energy structures were obtained: one was identical to the minimum with $\tau_{twist} = 184°$, while the other was similar to that one ($\tau_{twist} = 183°$) but with a larger horizontal separation (2.37 Å as opposed to 1.51 Å) and a stabilization energy of −8.2 kcal/mol. This structure thus represents a fourth minimum on the BrU/U gas-phase face-to-back stacking energy surface. It is presumably linked to the minimum with $\tau_{twist} = 184°$ via the TS with $\tau_{twist} = 189°$, which has an intermediate horizontal separation of 1.92 Å. The optimized minima and TSs are depicted in Fig. 3.

Four energy minima and four TSs were obtained in the case where water was used as a PCM solvent. Their geometrical parameters and interaction energies are listed in Table 2. The most stable minimum has an optimized twist of 293° and an interaction energy of −5.2 kcal/mol, i.e., it is geometrically similar to the gas-phase minimum but less tightly bound. Note that the TS obtained from $\tau_{twist} = 245°$ has an optimized twist angle of 313°, which does not lie between those of the minima it apparently connects (242° and 293°).

The situation with the optimized face-to-back structures in 1,4-dioxane, as shown in Table 3, is more complicated. Only three distinct energy minima were found, because optimization from $\tau_{twist} = 205°$ yielded the same structure as starting from $\tau_{twist} = 290°$, with an optimized twist of 288°. Optimization from the minimum with $\tau_{twist} = 90°$ originally yielded a structure with one imaginary frequency, but "following" this frequency with one set of imaginary atomic displacements resulted in optimization to a true minimum (the one shown in the table). (Using the other set of displacements resulted in another, similar structure with one imaginary frequency.)

Four distinct TSs in 1,4-dioxane were found. However, one of these, located by starting from $\tau_{twist} = 245°$, was not stacked but L-shaped (angle between planes = 98°), with an apparent N–H···Br interaction. "Following" the imaginary frequency of this TS yielded, for one set of displacements, the already-discovered energy minimum structure with $\tau_{twist} = 288°$, and for the other set of displacements, a structure which was planar and appeared to contain a bifurcated hydrogen bond involving bromine.

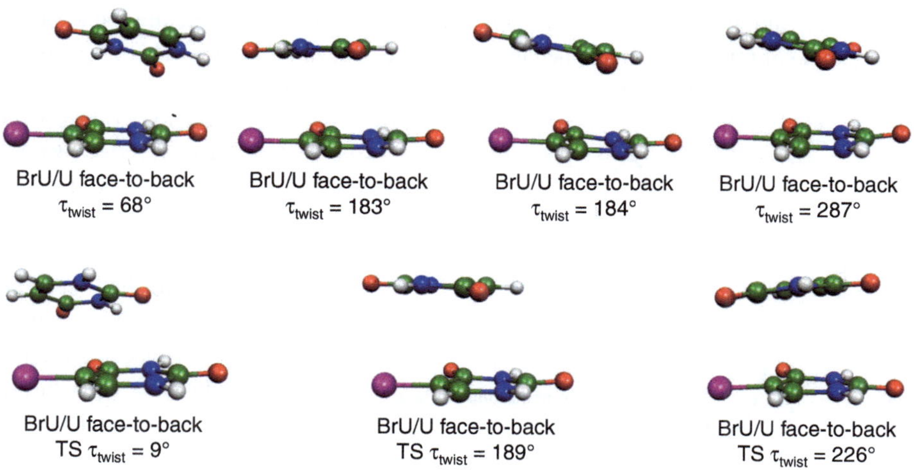

Fig. 3 Structures of the stacked face-to-back BrU/U minima (*top*) and transition states (*bottom*) optimized in the gas phase at the M06-2X/6-31+G(d) level of theory

Table 2 Interaction energies (in kcal/mol) and structural parameters (distances in Å, angles in degrees) of the face-to-back BrU/U minima and transition states in water PCM solvent calculated at the M06-2X/6-31+G(d) level of theory

τ_{twist} (initial)	τ_{twist} (final)	Angle between planes	Vertical separation	Horizontal separation	Im. Freq.	ΔE^{CP}	ΔE_0^{CP}
0	331	10	3.10	2.24	1	−4.3	−4.0
90	67	6	3.15	1.41	0	−4.4	−4.0
125	122	1	3.10	1.18	1	−4.3	−4.0
155	130	2	3.16	1.19	0	−5.1	−4.7
180	182	1	3.24	0.23	1	−3.5	−3.3
210	242	3	3.16	1.13	0	−4.6	−4.1
245	313	5	3.17	1.87	1	−3.8	−3.6
295	293	6	3.09	1.12	0	−5.5	−5.2

Table 3 Interaction energies (in kcal/mol) and structural parameters (distances in Å, angles in degrees) of the face-to-back BrU/U minima and transition states in 1,4-dioxane PCM solvent calculated at the M06-2X/6-31+G(d) level of theory

τ_{twist} (initial)	τ_{twist} (final)	Angle between planes	Vertical separation	Horizontal separation	Im. Freq.	ΔE^{CP}	ΔE_0^{CP}
5	2	3	3.14	1.47	1	−2.3	−2.1
90	74	13	3.14	0.52	0	−5.7	−5.3
120	95	6	3.13	0.67	1	−5.5	−5.2
165	183	6	3.08	1.58	0	−7.7	−7.3
185	221	2	3.17	1.07	1	−6.0	−5.7
245	355	98	3.36	4.50	1	−5.7	−5.3
290	288	9	3.04	1.03	0	−7.9	−7.6

Figure 4 shows the gas-phase interaction energy of the face-to-face BrU/U stacked dimer as a function of the τ_{twist} angle. The interaction energies are negative throughout. The BSSE is again fairly large, varying between 2.92 kcal/mol (when $\tau_{twist} = 45°$) and 3.58 kcal/mol (when $\tau_{twist} = 270°$). Energy minima are located at $\tau_{twist} = 55°$, 210° and 290°. Energy maxima are located at $\tau_{twist} = 130°$, 235° and 355°.

Figure 5 shows the interaction energies of the face-to-face BrU/U stacked dimer in PCM solvents as a function of the τ_{twist} angle. All energies are negative throughout. In water, there are four minima, located at $\tau_{twist} = 80°$, 210°, 270° and 330°, and four maxima, at 5°, 120°, 245° and 290°. In 1,4-dioxane, there are also four minima, at 50°, 215°, 290° and 325°, and four maxima, at 130°, 240°, 295° and 355°. The water curve is again much flatter than the gas and 1,4-dioxane curves.

For the face-to-face gas-phase structures, only two distinct energy minima could be located, because the optimizations starting from $\tau_{twist} = 210°$ and $\tau_{twist} = 290°$ yielded the same structure, with a twist angle of 296° and an interaction energy of −8.5 kcal/mol. The other minimum, with an optimized twist angle of 60°, has the most negative CP-corrected, ZPE-corrected M06-2X/6-31+G(d) interaction energy of any

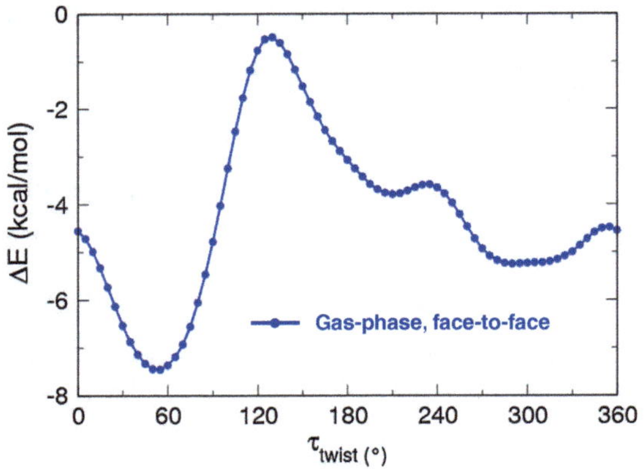

Fig. 4 Interaction energies as a function of the twist angle for the gas-phase face-to-face BrU/U stacked dimer calculated at the M06-2X/6-31+G(d) level of theory

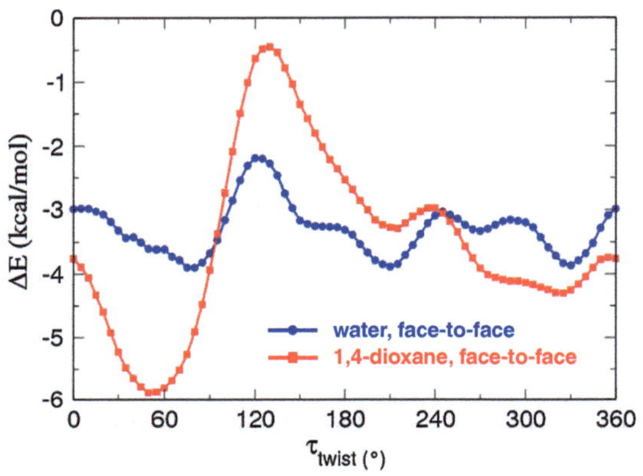

Fig. 5 Interaction energies as a function of the twist angle for the face-to-face BrU/U stacked dimer in PCM solvent (water and 1,4-dioxane) calculated at the M06-2X/6-31+G(d) level of theory

structure found in this study (−10.7 kcal/mol), and therefore appears to be the global minimum for the BrU/U stacked dimer. Because only two minima were located, we were only able to perform two QST3 transition state searches (starting from τ_{twist} = 130° and 355°), both of which were successful. However, a standard transition state search starting from τ_{twist} = 235° yielded an L-shaped TS (angle between planes = 71°). "Following" the imaginary frequency yielded, for one set of atomic displacements, a planar hydrogen (H)-bonded energy minimum structure, and for the other set of displacements, another L-shaped structure (angle between planes = 65°), this one an energy minimum. The optimized minima and TSs are shown in Fig. 6.

As with the face-to-back structures, the value of the vertical separation decreased from 3.4 to ~3.1 Å during all optimizations and transition state searches of face-to-face structures, i.e., the two bases moved closer together. All the energies and geometries of the gas-phase face-to-face structures are listed in Table 4.

For the face-to-face structures in water, four energy minima and four TSs were obtained, as listed in Table 5. As in the face-to-back case, the interaction energies of the optimized face-to-face structures are smaller in the two solvents than in the gas phase.

For the face-to-face structures in 1,4-dioxane, three distinct energy minima were located from the first round of geometry optimizations, as the optimization starting from τ_{twist} = 330° yielded the same minimum as that starting from 290°, with an optimized twist angle of 298°. Three TSs were located. However, two of these (obtained by starting from τ_{twist} = 130° and 240°) have slightly greater stabilization energies (−4.02 and −4.12 kcal/mol respectively) than the minimum obtained from τ_{twist} = 215° (−3.99 kcal/mol), though they would both be expected to connect this minimum with another and therefore lie

Fig. 6 Structures of the stacked face-to-face BrU/U minima (*top*) and transition states (*bottom*) optimized in the gas phase at the M06-2X/6-31+G(d) level of theory

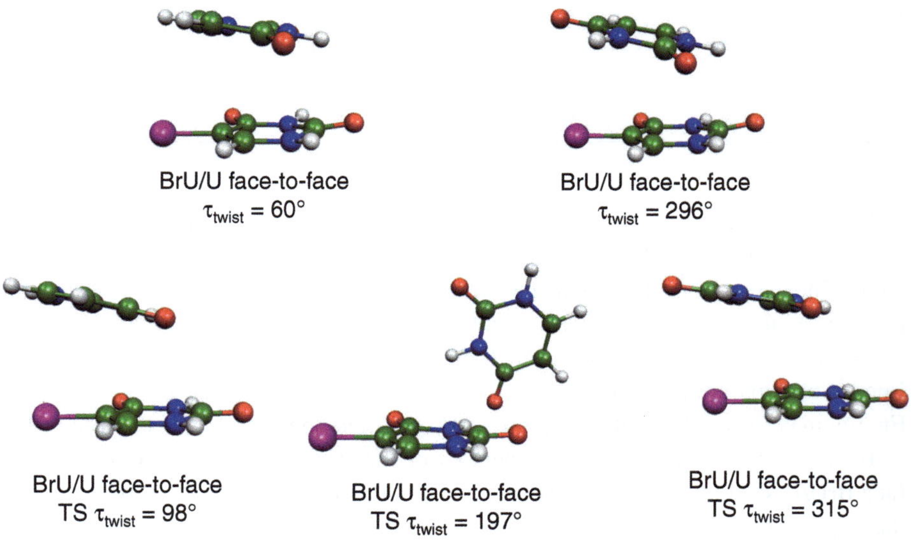

Table 4 Interaction energies (in kcal/mol) and structural parameters (distances in Å, angles in degrees) of the face-to-face BrU/U minima and transition states in the gas phase calculated at the M06-2X/6-31+G(d) level of theory

τ_{twist} (initial)	τ_{twist} (final)	Angle between planes	Vertical separation	Horizontal separation	Im. Freq.	ΔE^{CP}	ΔE_0^{CP}
55	60	12	3.04	0.87	0	−11.5	−10.7
130	98	14	3.14	2.36	1	−6.5	−6.1
235	197	71	3.15	3.73	1	−6.2	−5.7
290	296	18	3.11	1.23	0	−9.1	−8.5
355	315	10	3.10	2.23	1	−7.8	−7.3

Table 5 Interaction energies (in kcal/mol) and structural parameters (distances in Å, angles in degrees) of the face-to-face BrU/U minima and transition states in water PCM solvent calculated at the M06-2X/6-31+G(d) level of theory

τ_{twist} (initial)	τ_{twist} (final)	Angle between planes	Vertical separation	Horizontal separation	Im. Freq.	ΔE^{CP}	ΔE_0^{CP}
5	43	10	3.10	1.09	1	−4.8	−4.6
80	71	6	3.08	1.14	0	−5.6	−5.2
120	119	1	3.16	1.31	1	−4.3	−4.0
210	181	6	3.19	1.28	0	−4.5	−4.0
245	250	5	3.21	1.17	1	−3.9	−3.6
270	254	3	3.14	1.30	0	−4.8	−4.2
295	336	3	3.17	1.13	1	−4.2	−4.0
330	17	2	3.10	1.10	0	−5.1	−4.6

energetically above it. Furthermore, the optimized twist angle of the TS obtained from $\tau_{twist} = 240°$ is 196°, which does not lie between those of the optimized minima it would be expected to connect (209° and 298°).

For the TS with optimized $\tau_{twist} = 165°$, "following" the imaginary frequency yielded two new stacked energy minima with optimized twist angles of 136° and 178°, the former having a large horizontal separation (2.96 Å). Both of these minima are more stable than the respective TS. However, for the TS with optimized $\tau_{twist} = 196°$, the same procedure yielded the minimum with $\tau_{twist} = 178°$ and the already-discovered minimum with $\tau_{twist} = 209°$. The latter structure has a very slightly greater stabilization energy than the TS (by 0.01 kcal/mol) before the zero-point and counterpoise corrections are applied, but these corrections actually reverse the order of stability, such that the TS with $\tau_{twist} = 196°$ ($\Delta E_0^{CP} = -4.1$ kcal/mol) is 0.1 kcal/mol more stable than the minimum with $\tau_{twist} = 209°$ ($\Delta E_0^{CP} = -4.0$ kcal/mol), even though the attempt to optimize the TS along the appropriate reaction coordinate led to that minimum. Clearly, the use of PCM renders this area of the potential energy surface extremely flat, leading to complications in the identification of turning points, which is exacerbated by the necessity to apply both the energetic corrections (zero-point and counterpoise) to the optimized energies a posteriori. In addition, the counterpoise correction is not entirely accurate as it cannot be applied in combination with the PCM method. All optimized structural parameters and energies for face-to-face stacking in 1,4-dioxane are listed in Table 6.

We also note that the TS that was located by starting from a twist angle of 295° had a rather different optimized twist angle (214°).

Table 7 shows the RI-mPW2PLYP-D/aug-cc-pVTZ interaction energies of the gas-phase structures, using the geometries obtained from M06-2X/6-31+G(d) as well as those re-optimized with RI-mPW2PLYP-D/aug-cc-pVDZ. Re-optimization of the face-to-face minimum with $\tau_{twist} = 296°$ resulted in its rearrangement into a planar H-bonded structure. The other five minima remained stacked and only underwent slight geometrical changes. The interaction energies for those five minima always agree to within 1.0 kcal/mol with the corresponding M06-2X/6-31+G(d) energies. The ordering of the two lowest-energy face-to-back structures is changed by RI-mPW2PLYP-D: the minimum with $\tau_{twist} = 184°$ is now the most stable. However, the difference is negligible, being less than 0.2 kcal/mol even for the re-optimized structures.

Note that the RI-mPW2PLYP-D calculations included corrections for BSSE but not for deformation energies or zero-point vibrations (since the energies were not calculated at the same level of theory as the optimizations). With M06-2X, the ZPE corrections to the interaction energies range from 0.6 to 0.8 kcal/mol in the face-to-back structures and from 0.7 to 0.9 kcal/mol in the face-to-face structures (these are not the ZPE corrections to the total energies but to the energy of interaction, i.e., $E_{ZPE}^{B1/B2}$ (B1/B2) $- E_{ZPE}^{B1}$ (B1) $- E_{ZPE}^{B2}$ (B2)).

Table 6 Interaction energies (in kcal/mol) and structural parameters (distances in Å, angles in degrees) of the face-to-face BrU/U minima and transition states in 1,4-dioxane PCM solvent calculated at the M06-2X/6-31+G(d) level of theory

τ_{twist} (initial)	τ_{twist} (final)	Angle between planes	Vertical separation	Horizontal separation	Im. Freq.	ΔE^{CP}	ΔE_0^{CP}
50	63	10	3.03	1.04	0	−8.6	−8.0
130	165	14	3.28	1.72	1	−4.1	−4.0
130[a]	178	10	3.19	1.24	0	−4.5	−4.2
130[a]	136	21	3.04	2.96	0	−6.4	−6.0
215	209	9	3.19	0.30	0	−4.1	−4.0
240	196	12	3.22	0.34	1	−4.2	−4.1
295	214	18	3.16	2.61	1	−2.5	−2.2
325	298	11	3.07	1.48	0	−6.7	−6.3
355	341	2	3.15	1.27	1	−5.1	−4.9

[a] These minima were obtained by "following" the imaginary frequency of the transition state listed above them

Table 7 RI-mPW2PLYP-D/aug-cc-pVTZ interaction energies (in kcal/mol) and structural parameters (distances in Å, angles in degrees) of the six gas-phase BrU/U minima identified above

ΔE^{CP} (RI-mPW2PLYP-D)		Twist	Angle between planes	Vertical separation	Horizontal separation
M06-2X geom.[a]	Re-optimized[b]				
Face-to-back minima					
−9.9 (−10.1)	−10.0	184 (184)	10 (9)	3.10 (3.09)	1.51 (1.51)
−9.7 (−10.6)	−9.8	287 (287)	12 (12)	3.09 (3.06)	0.87 (0.89)
−7.9 (−7.9)	−8.1	67 (68)	22 (20)	3.24 (3.12)	0.68 (0.91)
−8.5 (−8.9)	−8.6	183 (183)	2 (2)	3.05 (3.00)	2.34 (2.37)
Face-to-face minima					
−9.1 (−9.1)	−22.0[c]	n/a (296)	n/a (18)	n/a (3.11)	n/a (1.23)
−11.1 (−11.5)	−11.2	59 (60)	12 (12)	3.08 (3.04)	0.76 (0.87)

In parentheses are the original counterpoise-corrected, nonzero-point-corrected energies and optimized structural parameters from M06-2X/6-31+G(d) geometry optimizations

[a] M06-2X/6-31+G(d)-optimized geometries

[b] RI-mPW2PLYP-D/aug-cc-pVDZ-optimized geometries

[c] Planar H-bonded

3.2 Experimental BrU and T stacked structures

Table 8 shows the mean RI-mPW2PLYP-D/aug-cc-pVTZ interaction energies of each of the 15 types of experimental stacked dimer investigated, whereas Table 9 shows the df-MP2/aug-cc-pVDZ, df-MP2/aug-cc-pVQZ, df-LCCSD(T0)/aug-cc-pVDZ and estimated df-LCCSD(T0)/aug-cc-pVQZ interaction energies. The number of data points contributing to each mean value for df-MP2 and df-LCCSD(T0) (i.e., the number of representative structures of each type of dimer) is as follows: 8 × A/BrU, 8 × BrU/A, 4 × C/BrU, 4 × BrU/C, 8 × G/BrU, 7 × BrU/G, 3 × T/BrU, 5 × BrU/T, 9 × A/T, 5 × T/A, 5 × C/T, 8 × T/C, 7 × G/T, 7 × T/G and 10 × T/T. As mentioned in the Methodology section, the counterpoise correction for one of the A/T dimers could not be performed with RI-mPW2PLYP, so the mean stacking energy for dimers of this type is calculated from eight rather than nine data points for this method.

Table 8 Mean RI-mPW2PLYP-D/aug-cc-pVTZ interaction energies ΔE^{mean} (kcal/mol) of different types of experimental stacked dimer

Type X/Y	ΔE^{mean}	Type Y/X	ΔE^{mean}	All X+Y[a]	ΔE^{mean}
BrU/A	−6.4	A/BrU	−6.6	BrU+A	−6.5
T/A	−5.4	A/T	−6.3	T+A	−5.9
BrU/C	−4.2	C/BrU	−7.6	BrU+C	−5.9
T/C	−3.9	C/T	−5.6	T+C	−4.6
BrU/G	−5.6	G/BrU	−4.9	BrU+G	−5.2
T/G	−4.6	G/T	−4.7	T+G	−4.7
BrU/T	−3.0	T/BrU	−4.3	BrU+T	−3.5
T/T	−3.2			T+T	−3.2

X = T or BrU; Y = A, C, G or T

[a] All combinations of X and Y

It can be seen that the average RI-mPW2PLYP-D stacking interaction is stronger in each type of BrU-containing dimer than in the corresponding type of

Table 9 Mean df-MP2/aug-cc-pVDZ, df-MP2/aug-cc-pVQZ, df-LCCSD(T0)/aug-cc-pVDZ and estimated df-LCCSD(T0)/aug-cc-pVQZ interaction energies (kcal/mol) for each type of experimental dimer

Type	ΔE^{mean}(df-MP2)		ΔE^{mean}(df-LCCSD(T0))	
	Aug-cc-pVDZ	Aug-cc-pVQZ	Aug-cc-pVDZ	Aug-cc-pVQZ
Type X/Y				
BrU/A	−7.1	−7.5	−6.2	−6.7
T/A	−5.9	−6.5	−4.9	−5.4
BrU/C	−5.0	−5.4	−4.1	−4.5
T/C	−4.7	−5.3	−3.7	−4.3
BrU/G	−6.3	−6.8	−5.5	−6.0
T/G	−5.4	−5.8	−5.0	−5.3
BrU/T	−4.0	−4.8	−3.1	−3.9
T/T	−4.0	−4.7	−3.2	−3.8
Type Y/X				
A/BrU	−8.2	−9.1	−7.0	−7.9
A/T	−7.7	−8.5	−7.4	−8.1
C/BrU	−8.2	−8.9	−7.2	−7.9
C/T	−6.1	−6.7	−5.4	−6.0
G/BrU	−6.7	−7.4	−5.5	−6.3
G/T	−5.9	−6.7	−4.8	−5.6
T/BrU	−5.5	−6.1	−4.7	−5.4
All X+Y[a]				
BrU+A	−7.6	−8.3	−6.6	−7.3
T+A	−7.1	−7.7	−6.5	−7.2
BrU+C	−6.6	−7.2	−5.6	−6.2
T+C	−5.2	−5.8	−4.4	−5.0
BrU+G	−6.5	−7.1	−5.5	−6.1
T+G	−5.7	−6.2	−4.9	−5.4
BrU+T	−4.6	−5.3	−3.7	−4.5
T+T	−4.0	−4.7	−3.2	−3.8

X = T or BrU; Y = A, C, G or T

[a] All combinations of X and Y

T-containing dimer for all except one case (BrU/T vs. T/T). But the differences are mostly marginal: only in two cases does BrU stack more strongly than T by greater than 1 kcal/mol on average (C/BrU vs. C/T with a difference of 2.0 kcal/mol and T/BrU vs. T/T with a difference of 1.2 kcal/mol). The table also shows the mean stacking energies for all combinations of T and BrU with A, C, G and T, i.e., collapsing the distinction between dimers of type X/Y and Y/X. When the data are grouped in this fashion, mean stacking energies are always higher for BrU than for T, but only in the case of cytosine does the difference exceed 1 kcal/mol.

Focusing on the estimated df-LCCSD(T0)/aug-cc-pVQZ energies (Table 9), the average stacking is stronger in BrU-containing dimers than in the corresponding T-containing dimers in all except one case (A/BrU vs. A/T). But again the differences are small: the biggest differences are C/BrU versus C/T (with a difference of 1.8 kcal/mol), T/BrU versus T/T (1.6 kcal/mol) and BrU/A versus T/A (1.2 kcal/mol). When X/Y and Y/X dimers are grouped together (T/A with A/T, BrU/A with A/BrU etc.), the mean stacking energies are always higher for BrU, but the largest difference (for cytosine) is only 1.1 kcal/mol.

4 Discussion and conclusions

We have investigated the potential energy surface of the stacked BrU/U dimer in the gas phase and in PCM solvent. When considered in combination with data that were previously obtained on the U/U dimer, these results allow us to examine the energetic effect of a bromine substituent on the stacking behavior of uracil. We have supplemented these data with a comparison of the interaction energies of experimental structures of stacked dimers in DNA containing BrU and/or thymine. This information may be useful in elucidating the cause of BrU's tendency to mispair with guanine in DNA, resulting in point mutations.

If BrU stacks significantly more strongly than U and T, this could account for the greater stability of the BrU-G mispair against enzymatic repair (which is implied by the mutagenicity of BrU), regardless of the mispair's hydrogen-bonding pattern. The conventional view is that BrU-G mispairing is driven by enolization of the brominated base, forming a tautomer that mimics cytosine in its hydrogen bonding. However, the greater stacking strength of BrU—if proved real—would provide indirect evidence against this view, by making it unnecessary to invoke the formation of a non-canonical enol tautomer of BrU (which is disfavored, at least in the gas phase).

The notion that brominated pyrimidines engage in stronger π–π stacking interactions than their non-brominated counterparts makes intuitive physical sense: bromination tends to increase molecular polarizability, and, as mentioned in the Introduction, Kool et al. [6] demonstrated a qualitative correlation between polarizability and stacking strength. Further supporting the importance of polarizability, it was shown in the 1960s and 1970s that halogenation at the 5-position causes a marked increase in the association tendency of uracil with adenine (in order of increasing van der Waals radius of the halogen) [47], that bromination of uracil stabilizes polymers of adenine and uracil [48], and that bromination of cytosine stabilizes polymers of inosine and cytosine [49]. On the theoretical side, Hobza et al. [50] calculated that the parallel-displaced dimers of C_6Br_6 and C_6Cl_6 (which adopt a stacking-like geometry) are considerably more strongly bound than those of C_6H_6 or C_6F_6, and by DFT-SAPT analysis they attributed this extra stabilization to a greater dispersion energy related to the increased polarizability.

We have located six stacked energy minima for BrU/U in the gas phase: four in the face-to-back orientation and two face-to-face. We have also located a number of TSs on the BrU/U stacking energy surface. The global minimum on the gas-phase M06-2X/6-31+G(d) stacking energy surface of BrU/U is a face-to-face structure with a τ_{twist} angle of 60° and a counterpoise-corrected, zero-point-corrected interaction energy of −10.7 kcal/mol. This dimer is geometrically similar to the U/U global minimum previously located by Hunter and Van Mourik [28], which has a τ_{twist} angle of 48°, but the BrU/U stacking energy of −10.7 kcal/mol is modestly greater than the U/U energy, which is −8.9 kcal/mol at the same level of theory. In the BrU/U global minimum, the Br atom sits atop the H6 atom of U, and the C=O groups of the two bases are staggered such that no oxygen atom sits atop another. This arrangement presumably minimizes steric and electronic repulsion while still allowing a large favorable dispersive interaction between the π-systems. Due to the replacement of the H5 atom in uracil with bromine in BrU, the symmetry of the U/U rotational profile is lost, and all the gas-phase minima are energetically distinct. Generally speaking, however, there are no large differences between the interaction energies of the two systems.

Intrastrand stacking interactions in DNA are face-to-back, and typical twist angles range from 25° to 40° [33]. (Note that face-to-face stacking occurs in cross-strand stacking interactions). The BrU/U gas-phase minimum that is most representative of intrastrand stacking in vivo is therefore the face-to-back structure with a twist angle of 68°. This dimer has an M06-2X/6-31+G(d) stabilization energy of −7.2 kcal/mol. The equivalent U/U structure from the previous study [28] has a twist angle of 72° and a stabilization energy of −7.5 kcal/mol. This suggests that BrU may actually stack less strongly than U in physiologically relevant conformations, though it should be remembered that stacked dimer geometries in vivo do not generally correspond to gas-phase minima.

Turning to the experimentally derived dimers, the fairly consistent result from all levels of theory is that on average BrU stacks more strongly than T against each type of base, but only by a small amount (typically <1 kcal/mol). Therefore, our results do not provide any strong evidence to support the view, proposed by some experimental groups as an alternative to the conventional "rare tautomer" model of BrU-G mispairing [30–32], that the mutagenicity of BrU is caused by an enhanced stacking ability compared to thymine.

Acknowledgments The authors gratefully acknowledge EaStCHEM for computer time on the EaStCHEM Research Computing Facility and NSCCS (the EPSRC UK National Service for Computational Chemistry Software) for computer time on the Columbus computing facility.

References

1. Kool ET (2001) Annu Rev Biophys Biomol Struct 30:1–22
2. Yakovchuk P, Protozanova E, Frank-Kamenetskii MD (2006) Nucleic Acids Res 34:564–574
3. Bommarito S, Peyret N, SantaLucia J Jr (2000) Nucleic Acids Res 28:1929–1934
4. Šponer J, Šponer JE, Mládek A, Jurečka P, Banáš P, Otyepka M (2013) Biopolymers 99:978–988
5. Šponer J, Riley KE, Hobza P (2008) Phys Chem Chem Phys 10:2595–2610
6. Guckian KM, Schweitzer BA, Ren RX-F, Sheils CJ, Tahmassebi DC, Kool ET (2000) J Am Chem Soc 122:2213–2222
7. Friedman RA, Honig B (1995) Biophys J 69:1528–1535
8. Czyżnikowska Ż, Zaleśny R, Ziółkowski M, Gora RW, Cysewski P (2007) Chem Phys Lett 450:132–137
9. Morgado CA, Jurečka P, Svozil D, Hobza P, Šponer J (2009) J Chem Theory Comput 5:1524–1544
10. Smith QA, Gordon MS (2011) J Phys Chem A 115:11269–11276
11. Lima CFRAC, Rocha MAA, Gomes LR, Low JN, Silva AMS, Santos LMNBF (2012) Chem Eur J 18:8934–8943
12. van Mourik T, Gdanitz RJ (2002) J Chem Phys 116:9620–9623
13. van Mourik T (2004) Chem Phys 304:317–319
14. Kozuch S, Gruzman D, Martin JML (2010) J Phys Chem C 114:20801–20808
15. Holroyd LF, van Mourik T (2007) Chem Phys Lett 442:42–46
16. van Mourik T, Karamertzanis PG, Price SL (2006) J Phys Chem A 110:8–12
17. Shields AE, van Mourik T (2007) J Phys Chem A 111:13272–13277
18. Boys SF, Bernardi F (1970) Mol Phys 19:553–566
19. Zhao Y, Truhlar DG (2008) Theor Chem Acc 120:215–241
20. Schwabe T, Grimme S (2006) Phys Chem Chem Phys 8:4398–4401
21. Schwabe T, Grimme S (2007) Phys Chem Chem Phys 9:3397–3406
22. Korth M, Grimme S (2009) J Chem Theory Comput 5:993–1003
23. Hongo K, Cuong NT, Maezono R (2013) J Chem Theory Comput 9:1081–1086
24. Morgado CA, Jurečka P, Svozil D, Hobza P, Šponer J (2010) Phys Chem Chem Phys 12:3522–3534
25. Gu J, Wang J, Leszczynski J (2011) Chem Phys Lett 512:108–112
26. Zhao Y, Truhlar DG (2008) Acc Chem Res 41:157–167
27. Hohenstein EG, Chill ST, Sherrill CD (2008) J Chem Theory Comput 4:1996–2000
28. Hunter RS, van Mourik T (2012) J Comput Chem 33:2161–2172
29. Davidson RL, Broeker P, Ashman CR (1988) Proc Natl Acad Sci USA 85:4406–4410
30. Brown T, Kneale G, Hunter WN, Kennard O (1986) Nucleic Acids Res 14:1801–1809
31. Driggers PH, Beattie KL (1988) Biochemistry 27:1729–1735
32. Sternglanz H, Bugg CE (1975) Biochim Biophys Acta 378:1–11
33. Hassan MAE, Calladine CR (1997) Phil Trans R Soc Lond A 355:43–100
34. Frisch MJ, Trucks GW, Schlegel HB, Scuseria GE, Robb MA, Cheeseman JR, Scalmani G, Barone V, Mennucci B, Petersson GA, Nakatsuji H, Caricato M, Li X, Hratchian HP, Izmaylov AF, Bloino J, Zheng G, Sonnenberg JL, Hada M, Ehara M, Toyota K,

Fukuda R, Hasegawa J, Ishida M, Nakajima T, Honda Y, Kitao O, Nakai H, Vreven T, Montgomery Jr JA, Peralta JE, Ogliaro F, Bearpark M, Heyd JJ, Brothers E, Kudin KN, Staroverov VN, Kobayashi R, Normand J, Raghavachari K, Rendell A, Burant JC, Iyengar SS, Tomasi J, Cossi M, Rega N, Millam JM, Klene M, Knox JE, Cross JB, Bakken V, Adamo C, Jaramillo J, Gomperts R, Stratmann RE, Yazyev O, Austin AJ, Cammi R, Pomelli C, Ochterski JW, Martin RL, Morokuma K, Zakrzewski VG, Voth GA, Salvador P, Dannenberg JJ, Dapprich S, Daniels AD, Farkas Ö, Foresman JB, Ortiz JV, Cioslowski J, Fox DJ (2009) Gaussian 09, Revision A02. Gaussian Inc. Wallingford CT
35. Peng C, Schlegel HB (1993) Israel J Chem 33:449–454
36. Neese F (2012) The ORCA program system. WIREs Comput Mol Sci 2:73–78
37. Kendall RA, Früchtl HA (1997) Theor Chem Acc 97:158–163
38. Grimme S (2004) J Comput Chem 25:1463–1473
39. RCSB Protein Data Bank, www.pdb.org
40. Werner H-J, Manby FR, Knowles PJ (2003) J Chem Phys 118:8149–8160
41. Werner H-J, Knowles PJ, Knizia G, Manby FR, Schütz M, Celani P, Korona T, Lindh R, Mitrushenkov A, Rauhut G, Shamasundar KR, Adler TB, Amos RD, Bernhardsson A, Berning A, Cooper DL, Deegan MJO, Dobbyn AJ, Eckert F, Goll E, Hampel C, Hesselmann A, Hetzer G, Hrenar T, Jansen G, Köppl C, Liu Y, Lloyd AW, Mata RA, May AJ, McNicholas SJ, Meyer W, Mura ME, Nicklass A, O'Neill DP, Palmieri P, Peng D, Pflüger K, Pitzer R, Reiher M, Shiozaki T, Stoll H, Stone AJ, Tarroni R, Thorsteinsson T, Wang M, MOLPRO version 2010.1, a package of ab initio programs. see http://www.molpro.net
42. Schütz M (2000) J Chem Phys 113:9986–10001
43. Schütz M, Werner H-J (2000) Chem Phys Lett 318:370–378
44. Pipek J, Mezey PG (1989) J Chem Phys 90:4916–4926
45. Marshall MS, Burns LA, Sherrill CD (2011) J Chem Phys 135:194102–194110
46. Pitoňák M, Janowski T, Neogrády P, Pulay P, Hobza P (2009) J Chem Theory Comput 5:1761–1766
47. Nakano NI, Igarashi SJ (1970) Biochemistry 9:577–583
48. Riley M, Paul A (1971) Biochemistry 10:3819–3825
49. Howard FB, Frazier J, Miles HT (1969) J Biol Chem 244:1291–1302
50. Trnka J, Sedlak R, Kolář M, Hobza P (2013) J Phys Chem A 117:4331–4337

Theor Chem Acc (2014) 133:1541
DOI 10.1007/s00214-014-1541-2

REGULAR ARTICLE

Quantum Monte Carlo investigation of the H-shift and O_2-loss channels of *cis*-2-butene-1-peroxy radical

Zhiping Wang · Dmitry Yu Zubarev · William A. Lester Jr.

Received: 2 October 2013 / Accepted: 11 July 2014 / Published online: 12 August 2014

Abstract Owing to importance in combustion processes, O_2-loss and 1, 6-H-shift in *cis*-2-butene-1-peroxy radical have been investigated. Energies for these processes and the barrier height of the latter are computed using the diffusion Monte Carlo (DMC) method. The DMC energy for the 1, 6-H-shift was determined to be 4.56 ± 0.19 kcal/mol with barrier height of 26.79 ± 0.20 kcal/mol. The energy for O_2-loss was found to be 14.93 ± 0.24 kcal/mol. Quantitative differences between the findings of the present study and previous CBS-QB3 results indicate a discrepancy between high-level methods for the resonance-stabilized radicals. Further study is needed to identify the origin of these differences.

Dedicated to Professor Thom Dunning and published as part of the special collection of articles celebrating his career upon his retirement.

Electronic supplementary material The online version of this article (doi:10.1007/s00214-014-1541-2) contains supplementary material, which is available to authorized users.

Z. Wang · W. A. Lester Jr. (✉)
Chemical Sciences Division, Lawrence Berkeley National Laboratory, Berkeley, CA 94720, USA
e-mail: walester@lbl.gov

Present Address:
Z. Wang
Molecular Foundry, Lawrence Berkeley National Laboratory, Berkeley, CA 94720, USA

D. Y. Zubarev · W. A. Lester Jr.
Department of Chemistry, Kenneth S. Pitzer Center for Theoretical Chemistry, University of California, Berkeley, CA 94720-1460, USA

Present Address:
D. Y. Zubarev
Department of Chemistry and Chemical Biology, Harvard University, Cambridge, MA 02138, USA

Keywords Cis-2-butene-1-peroxy radical · Fixed-node diffusion Monte Carlo · Restricted open-shell trial wavefunction

1 Introduction

The increasing use of non-traditional petroleum-based fuels presents new challenges for combustion modeling. Understanding the energetics of elementary reactions that lead to or delay autoignition of biodiesel is critical to modeling the related combustion phenomena [1–3]. During low-temperature combustion processes, alkenes tend to lose a hydrogen atom from allylic sites to form allylic radicals that subsequently undergo reaction with O_2 to produce allylic peroxy radicals. As temperature increases, peroxy radicals become thermally unstable and undergo various transformations that can contribute to autoignition making them critical intermediates in spontaneous combustion [4, 5]. The addition of O_2 to various allylic radicals and subsequent reactions have been studied both experimentally and theoretically [6, 7].

A recent study of 2-butene-1-peroxy radical reaction channels [8] suggested that the 2-butene-1-peroxy radical, like other allylic peroxy radicals, appears to lose O_2 rather than undergo unimolecular reactions. The 1, 6-H-shift is the only reaction with the potential to compete to any extent with O_2-loss and contribute to autoignition (Reactions 1, 2).

(1)

Table 1 Reaction energies (E_{rxn}) and barrier height (E_0) with different methods

Method	E_{rxn}(O_2-loss), kcal/mol	E_{rxn}(H-shift), kcal/mol	E_0(H-shift), kcal/mol
RO-B3LYP/VTZ-ECP	19.39	6.64	26.81
FN DMC	19.32 ± 0.24	5.48 ± 0.19	30.18 ± 0.20
FN DMC + ZPE	14.93 ± 0.24	4.56 ± 0.19	26.79 ± 0.20
CBS-QB3, Ref. [8]	17.70	0.79	20.90

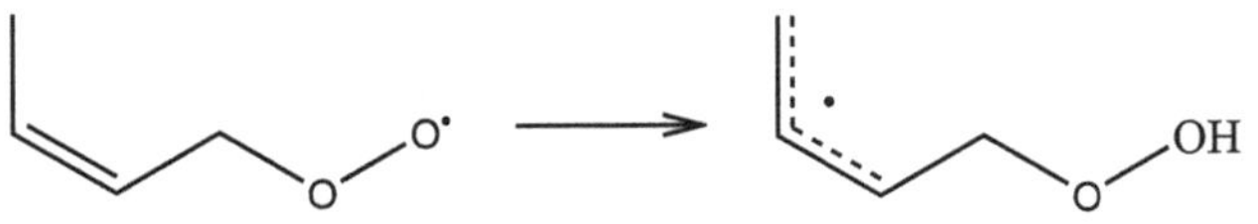

(2)

From calculations using the composite CBS-QB3 method [8], the barrier height for 1,6-H-shift is ~3 kcal/mol higher than the endothermic energy for barrierless O_2-loss. That such a small energy difference can be reliably achieved was demonstrated in a recent study of the OH bond dissociation energy in phenol where the difference between highly correlated theoretical levels was of order 4 kcal/mol [9].

To clarify the fate of the 2-butene-1-peroxy radical, we used fixed-node diffusion Monte Carlo (FN DMC) to compute the barrier height and energy of the 1,6-H-shift and energy of O_2-loss. In the following sections, we describe the computational methodology used in this study and discuss the findings.

2 Computational details

Coordinates of reactants and products of the two pertinent channels as well as the transition state of 1, 6-H-shift were taken from Ref. [8]. Following the methodology of Ref. [8], zero-point energy (ZPE) corrections for the FN DMC results were determined at B3LYP/6-311G(d, p) level and the ZPE values were scaled by a factor of 0.99.

The FN DMC [10–12] trial wave functions were a product of antisymmetric and symmetric components. The former was chosen as a single determinant constructed from Kohn–Sham orbitals. The latter was a 29-parameter Schmit-Moskowitz-Boys-Handy (SMBH) correlation function [13, 14]. This expansion includes electron–electron and electron–nucleus terms and increases computational efficiency by reducing the variance of the local energy. The SMBH correlation function parameters were energy optimized using the linear optimization VMC algorithm of Toulouse and Umrigar [15, 16]. The GAMESS [17] ab initio package was used to generate the restricted open-shell (RO) B3LYP [18, 19] orbitals using the triple-zeta energy-consistent pseudopotentials and basis sets of Burkatzki et al. [20].

An FN DMC algorithm with small time-step errors was employed [21]. To maintain a high acceptance ratio, the electrons were moved one at a time. A simple branching algorithm [22] was used to duplicate walkers with large weights and to eliminate those with small weights. All QMC calculations were performed using the Zori code [23]. A FN DMC simulation ensemble of ~200,000 walkers was chosen to ensure reduction of the population bias. Calculations were performed at 0.04, 0.02, 0.01, and 0.005 Hartree^{-1} time steps and were run until the stochastic error bars were below 0.0003 Hartree^{-1}. Weighted quadratic least-squares fits were used to extrapolate energy values to zero time step.

A set of benchmark FN DMC calculations provide a context for assessing the accuracy of the present results [24]. In that work, FN DMC computations performed for extended G1 molecular test set relied on single-determinant trial wave functions that were constructed using Stevens–Basch–Krauss (SBK) pseudopotentials and a valence basis set of quality similar to 6-311++G(2d,2p). The study reported the mean absolute deviation (MAD) of 3 kcal/mol for atomization energies and acknowledged the type of pseudopotential as one of the error sources. It also estimated the error associated with the use of SBK pseudopotential in FNDMC as 2 kcal/mol.

The present study relies on BFD pseudopotential and basis set which were optimized specifically for FN DMC calculations in order to reduce the errors associated with pseudopotentials developed for wave function-based quantum-chemical methods and DFT [20]. The performance of the BFD pseudopotential and basis sets was assessed with respect to V5Z/CCSD(T) calculations of binding energies for a set of 26 diatomic molecules. The MAD of 0.5 kcal/mol was reported.

A recent evaluation of the atomization energies for the G2 set [25] found MAD of 3 kcal/mol in the case of single-determinant trial wave functions formed from canonical orbitals and using BFD pseudopotential and basis sets. The errors associated with the subset of carbon based molecules, however, did not exceed 2 kcal/mol.

3 Results and discussion

Results of the FN DMC calculations are reported in Table 1 along with the findings of Ref. [8]. Results of RO-B3LYP calculations that were obtained in trial wave function preparation are also included. The FN DMC energy of O_2-loss is 14.93 ± 0.24 kcal/mol and that of H-shift is 4.56 ± 0.19 kcal/mol with a barrier 26.79 ± 0.20 kcal/mol, all ZPE-corrected.

First, we note that FN DMC values before ZPE correction are in reasonable agreement with RO-B3LYP calculations for the trial wave function (Table 1). Reaction energies lie within 1.2 kcal/mol, and the barrier heights are within 3.4 kcal/mol. This means that the RO-B3LYP and FN DMC approaches recover comparable amounts of dynamic correlation for the system of interest.

Second, there is an unexpected quantitative difference between ZPE-corrected FN DMC and CBS-QB3 values. The energies of O_2-loss and H-shift disagree by 2.77 and 3.77 kcal/mol, respectively, and H-shift barriers disagree by 6.89 kcal/mol. This is surprising in view of a recent study of phenoxy-radical formation, where comparable levels of ab initio computation predicted dissociation energies within 0.4 kcal/mol of each other [9]. Here, the aforementioned discrepancies mean that FN DMC barrier of the H-shift is higher than the endothermic energy of O_2-loss by 11.86 kcal/mol, which significantly exceeds the value of 3.2 kcal/mol from the CBS-QB3 calculation. The FN DMC predicts a 1, 6-H-shift energy of 4.56 kcal/mol which is 3.77 kcal/mol larger than the CBS-QB3 prediction of 0.79 kcal/mol.

Considering the competition between the 1, 6-H-shift isomerization of *cis*-2-butene-1-peroxy radical and the O_2-loss channel, the absolute values of the differences between FN DMC and CBS-QB3 predictions are large enough to change the qualitative picture. The barrier height of the H-shift reaction is increased by 5.89 kcal/mol at the FN DMC level of theory relative to CBS-QB3. A decrease in the barrier height by the same amount yields 15.01 ± 0.20 kcal/mol which is statistically indistinguishable from the O_2-loss reaction energy of 14.93 ± 0.24 kcal/mol (see Table 1).

To resolve the discrepancy between FN DMC and CBS-QB3 results, it would be insightful to investigate the use of alternative trial functions for DMC including restricted and unrestricted open-shell functions. In addition, it may be informative to optimize orbitals using the variational Monte Carlo method. The present methodology is typical for the majority of FN DMC studies.

4 Conclusions

We have used FN DMC to study the fate of an important intermediate in spontaneous combustion, *cis*-2-butene-1-peroxy radical, with specific consideration of two critical channels: 1, 6-H-shift isomerization and O_2-loss. The DMC energy for the 1, 6-H-shift was determined to be 4.56 ± 0.19 kcal/mol with barrier height of 26.79 ± 0.20 kcal/mol. The energy for O_2-loss was found to be 14.93 ± 0.24 kcal/mol. The differences between the findings of the present study and previous CBS-QB3 results indicate a discrepancy between two high-level methods for the resonance-stabilized radicals. This finding suggests that an accurate assessment of how competitive the 1, 6-H-shift is can benefit from further study.

Acknowledgments W.A.L. was supported by the Director, Office of Energy Research, Office of Basic Energy Sciences, Chemical Sciences, Geosciences and Biosciences Division of the US Department of Energy, under Contract No. DE-AC03-76F00098. This research used computational resources of the National Energy Research Scientific Computing Center, which is supported by the Office of Science of the US Department of Energy under Contract No. DE-AC02-05CH11231.

References

1. Firestone D (ed) (2006) Physical and chemical characteristics of oils, fats, and waxes, 2nd edn. American Oil Chemists Society, Urbana
2. Buda F, Bounaceur R, Warth V, Glaude PA, Fournet R, Battin-Leclerc F (2005) Combust Flame 142:170–186
3. Hayes CJ, Merle JK, Hadad CM (2009) Adv Phys Org Chem 43:79–134
4. Pilling MJ (ed) (1997) Low-temperature combustion and autoignition. Elsevier, Amsterdam
5. Battin-Leclerc F, Herbinet O, Glaude P, Fournet R, Zhou Z, Deng L, Guo H, Xie M, Qi F (2010) Angew Chem Int Ed 49:3169–3172
6. Zádor J, Taatjes CA, Fernandes RX (2011) Prog Energy Combust Sci 37:371–421
7. Lee J, Bozzelli JW (2005) Proc Combust Inst 30:1015–1022
8. Dibble TS, Sha Y, Thornton WF, Zhang F (2012) J Phys Chem A 116:7603–7614
9. Wang J, Domin D, Austin B, Zubarev DY, McClean J, Frenklach M, Cui T, Lester WA Jr (2010) J Phys Chem A 114:9832–9835
10. Anderson JB (1975) J Chem Phys 63:1499–1503
11. Hammond BL, Lester WA Jr, Reynolds PJ (1994) Monte Carlo methods in ab initio quantum chemistry: quantum Monte Carlo for molecules. World Scientific, Singapore
12. Foulkes WMC, Mitas L, Needs RJ, Rajagopal G (2001) Rev Mod Phys 73:33–83
13. Boys SF, Handy NC (1969) Proc R Soc London Ser A 310:63–78
14. Schmidt KE, Moskowitz JW (1990) J Chem Phys 93:4172–4178
15. Umrigar CJ, Toulouse J, Filippi C, Sorella S, Henning RG (2007) Phys Rev Lett 98:110201
16. Toulouse J, Umrigar CJ (2007) J Chem Phys 126:084102
17. Schmidt MW, Baldridge KK, Boatz JA, Elbert ST, Gordon MS, Jensen JH, Koseki S, Matsunaga N, Nguyen KA, Su SJ, Windus TL, Dupuis M, Montgomery JJA (1993) J Comput Chem 14:1347–1363
18. Becke AD (1993) J Chem Phys 98:1372–1377
19. Becke AD (1993) J Chem Phys 98:5648–5652
20. Burkatzki M, Filippi C, Dolg M (2007) J Chem Phys 126:234105
21. Umrigar CJ, Nightingale MP, Runge KJ (1993) J Chem Phys 99:2865–2890
22. Reynolds PJ, Ceperley DM, Alder B, Lester WA Jr (1982) J Chem Phys 77:5593–5603
23. Aspuru-Guzik A, Salomon-Ferrer R, Austin B, Perusquia-Flores R, Griffin M, Oliva RA, Skinner D, Domin D, Lester WA Jr (2005) J Comput Chem 26:856–862
24. Grossman JC (2002) J Chem Phys 117:1434–1440
25. Petruzielo FR, Toulouse J, Umrigar CJ (2012) J Chem Phys 136:124116

MIX
Papier aus verantwortungsvollen Quellen
Paper from responsible sources
FSC® C105338

If you have any concerns about our products,
you can contact us on
ProductSafety@springernature.com

In case Publisher is established outside the EU,
the EU authorized representative is:
Springer Nature Customer Service Center GmbH
Europaplatz 3, 69115 Heidelberg, Germany

Printed by Libri Plureos GmbH
in Hamburg, Germany